Global Geostrategic and Politico-Military Perspectives Through Millennia Past

Volume Two

Global Geostrategic and Politico-Military Perspectives Through Millennia Past

Volume Two

Major C.B. Verma (Retd.)

PENTAGON PRESS

GLOBAL GEOSTRATEGIC AND POLITICO-MILITARY PERSPECTIVES THROUGH MILLENNIA PAST

Major C.B. Verma (Retd.)

First Published in 2018

ISBN 978-93-86618-20-7 (Set)

Published by

PENTAGON PRESS
206, Peacock Lane, Shahpur Jat
New Delhi-110049
Phones: 011-64706243, 26491568
Telefax: 011-26490600
email: rajan@pentagonpress.in
website: www.pentagonpress.in

Printed at Aegean Offset Printers, Greater Noida, U.P.

In the Lotus Feat of His Divine Holiness
Shri Sai Baba of Shirdi (India)

Contents

VOLUME ONE

VOLUME TWO

Foreword

Study of the history of human civilisation, with all its explorative, investigative, and deductive challenges, cannot be but a rather complex exercise. It is a challenge that has enthused the interest of great scholars, explorers, scientists and philosophers down the ages for them to devote their lifetime to unearth, and more importantly, infer and record the story of mankind. The reward for that excursion comes by the way of the joys of discovery and spreading of knowledge among the naturally inquisitive humans. Indeed, the saga of evolution of the human society, politics, faiths and philosophies, with all its ups and downs, helps the contemporary generation in understanding the past and derive lessons from it to its future advantages. Obviously, therefore, while compilation of the story of civilisation is an intellectually as well as physically back-breaking endeavour, study of that compilation must be a very invigorating and satisfying experience for the readers.

Only a scholar of Major C B Verma's intellect, commitment and perseverance could plunge into a project so extensive, and complete it with the publication of his seminal work under the title of '**Global Geostrategic and Politico-Military Perspectives Through Millennia Past**' in two volumes of knowledge reservoir. It has taken the noted scholar, soldier, entrepreneur and socio-political observer over six decades to organise his time to do so. Admittedly, the facts discussed in these volumes are known and recorded, but the salience of this unique publication lies, firstly, in its analyses of events, and secondly, in presenting to the reader a storehouse of composite, multi-discipline and duly-compiled information in just two volumes. It is hoped that the hard-to-find features of this book would find the best utilisation from the academic as well as the knowledgeable world.

Volume One of the book starts with a projection of the ancient era, around 4500 BCE,and proceeds to cover the period up to the mediaeval age. It goes on to describe the phenomenon of 'birth of empires' wherein the Egyptian, Mesopotamian, Chinese, Indus Valley, Sematic and Aryan, Greek, Roman and European empires and civilisations, as the case be, are delved into. Notably, since the mankind's military history is closely intertwined with civilisational developments, at this stage the author has offered a rare insight into the military aspects that have shaped the world. The Volume, composed in six Chapters, ends with an exposition on the phenomenon of rise of the Western Empires and the advent of democracy.

Volume Two begins with the advent of what may be termed as the Modern period. It describes and analyses the courses and effects of the Industrial Revolution of the 17th and 18th Century before delving into the two landmark events of the 20th Century, namely the Great War or World War I (1914-18) and World War II (1939-45), which have shaped the world as it is seen today.

Appropriately, the geo-political, socio-cultural and politico-economic developments of that period is also discussed in continuation, thus building a holistic picture for the readers to look at. At this stage, a well-informed exposition on the role of modern facets of intelligence and politico-strategic application of Special Forces follow. Obviously, the author has highlighted these two modern-day attributes in the context of their salience as tools of geo-political power-play.

The later parts of Volume Two of the book are devoted to the examination of contemporary – described as 'super age' – developments in science and technology, and its assimilative effects on the society, particularly upon the Indian environment. Thus the assimilation of scientific inventions, computers and automated information processing, information technology, space sciences, etc. into the human societal construct has been delved into. Having described these landmark developments shaping up the modern era, the book shifts into an analytical role in interpretation and prognosis. In the true context of the modern times being conditioned by the overall effects of such multi-directional developments, this line of articulation is appropriate and timely. Accordingly, this crucial role is performed by touching upon the key nuances of nuclear, biological and chemical warfare, the scourge of terrorism and the 'great game' of international power-politics. Finally, the discussion is wound up with some thoughts over India's political, military and leadership attributes and challenges, with few hints for undertaking of reforms in line with the calls of the day.

A study so extensive and a work crafted over a long period of six decades requires many explanations and updates. The author has met that requirement through his 'Conclusions' and followed it up with a 'Post-script'. That indeed is an appropriate step to keep the readership in focus.

If it is a rewarding experience, the study of the history of human civilisation is also a serious business that requires quest and perseverance. One is sanguine that this painstaking contribution from Major C B Verma, and the missionary zeal with which the noted publisher Shri Rajan Arya has given shape to it, would raise readership interests, and its volumes would find a place of pride among the works of similar salience.

Jai Hind,

Lt Gen Gautam Banerjee, PVSM, AVSM, YSM (Retd)
Member Advisory Board and Distinguished Fellow
The Vivekananda International Foundation
New Delhi-110003

November 2017

Acknowledgement

I will always be gratefully indebted to Late Lt Gen. S K Sinha, PVSM, AVSM (Retd), former Governor of Jammu & Kashmir and Assam, for his most valuable guidance and inspiration, and for all the time he spared for me at his home, whether in Patna or New Delhi. He was indeed a 'living legend'. In fact, he had very kindly offered to write a 'Foreword' for this book and had taken the pains of meticulously going through the manuscript. Unfortunately, he was unable to do so because of his sudden demise.

I am also sincerely grateful to Lt Gen Gautam Banerjee, PVSM, AVSM, YSM (Retd) for his lifelong brotherly friendship, and for being a close family friend for many decades. I must say with heartfelt gratitude that this book has finally seen the light of the day due to his singular considerations and efforts. I do not have words to express my gratitude to him.

I remain indebted to Late Professor S Prasad, former Professor and Head of Department, Mechanical Engineering, College of Military Engineering, Pune, and Founder-Director of Kamla Nehru Institute of Technology, Sultanpur (UP). He had very kindly shared his insights about the fundamental aspects of modern technology with me.

Similarly, I must acknowledge the affection and invaluable guidance on many aspects that I was lucky to receive from Swami Madhusudan, Professor of Yoga at the Lal Bahadur Shastri National Academy of Administration, Mussoorie, (Dehradun), and the founder of the "Adhyatyam Yoga Mission" in Haridwar.

The United Services Institution (USI) Library, New Delhi, deserves a special mention for its valuable collections, which I could access being a life member for more than 50 years.

Last but not the least, I most sincerely wish to express my special thanks to Mr. Rajan Arya, Founder & CEO of Pentagon Press, New Delhi, for bearing with me and having my manuscript diligently edited.

Major C B Verma (Retd)

CHAPTER 7

Industrial Revolution, Its Multiple Influences, Two World Wars, Emerging Numerous Geopolitical, Socio-Cultural-cum-Socio-Political and Socio-Economic Developments, Multiple-Aspects of Intelligence Services and Concepts of the Special Forces

In course of our continuing discussions, and also in different contexts, we have since noted that from its very beginning, man's living history has been in the making with unbroken continuity. And at the same time, we have also noticed that the various phenomena of this continuous making of man's living history in this world have been highly complex, multi-dimensional and interconnected, and at times directly, indirectly or even remotely, mutually influencing as well.

The 'engine' behind all these historic occurrences of the relentless and gradually intensifying human activities since triggering many kinds of developments and events, were man's always active and, the 'multiply-creative and analytically thinking human-mind', as exclusively gifted to him by Mother Nature. And in all these processes soon as simultaneously parallel phenomena, the multiply advancing human-civilisations themselves took command of these human activities and developments. And in these multiple contexts, historically therefore, since the very advent of and the continuing advancements of human civilisations in the numerous geographical areas of the world, multiply consequent many developments and events occurring from time to time and their fallouts, as since times kept on rolling-by, have also been continuously providing many newer inputs to on-going making of the man's living history. In other words, we may also say that the multiply captivating saga of all the above mentioned inter-related as well as mutually-influencing human-activities, developments and events, continuing advancements of various human-civilisations and then, their far-reaching and also thereby multiply consequential fallouts in totality, constitute man's living history since continuously in the making as well.

Historically, therefore, we would soon be discussing that when times came during the later medieval-era period of the unprecedented **'Industrial-Revolution'**, the life as well as living history of man momentously occurred and thereby, totally changed the history of this world. **But regardless of all these mind-boggling developments and events, one dimension remains rather universally unaltered, and that is they all do remain always interrelated and mutually influencing as well.**

And specifically in these precise contexts, **Alvin Toffler** has mentioned that, "More important, we will search out the hidden connections among events that on the surface seem unrelated. It does little good to forecast the future of semiconductors or energy, or the future of the family (even one's own family), if the forecast springs from the premise that everything will remain unchanged. For nothing will remain unchanged. The future is fluid, not frozen. It is constructed by our shifting and changing daily decisions, and each event influences all others.... (*Now, on the multiple-aspects of super-industrialism he says that*) It is clearly impossible to know at this stage which combination of technologies will prove most useful for what tasks, but the array of tools and fuels available to us will surely be staggering, with more and more exotic possibilities becoming commercially plausible as oil prices climb." (Toffler, 1981, pp. 145 & 151)

From the historic records in this regard and, as since rendered in the scholarly books on history by the various learned authors we do historically find **that the broadly precise time-frame for the advent and subsequently, the momentous fructification of 'Industrial-Revolution', has been the second half of the seventeenth century (i.e. 1750/60 to 1800 A.D. or so)**. But it is also true that since the relentlessly occurring the numerous kinds of historically momentous events and developments and thereby, also triggering earlier unimagined and far-reaching fallouts for the mankind in this world do also in addition, always happen to be whether closely, indirectly or even remotely inter-related to the similar earlier phenomena of the past. Historically, these links may remain somewhat hidden and hence, may not be readily apparent but some kinds of the consequent links to the past are always there, and even though perhaps, these may remain buried in the perpetually accumulating dusts of time. But at the same time, in the precise historic contexts of the 'Industrial-Revolution' it is also true that historically gradual but the fast sprouting and hence, the rapidly expanding manifestation of this unprecedented revolution of altogether different kind, did momentously begin to fructify broadly during the last fifty years of the seventeenth century. In addition also, it remains equally true that the above mentioned aspects of historic links as well as, interconnectivity and even influencing relationships among the various events and developments, occurring over the ongoing passages of time and, along with the advent of human-civilisations do remain and historically survive as well.

We may now very relevantly recall and as also discussed in the earlier chapters in their different contexts that, as the gradual and accelerated sprouting-cum-growth of the various human-civilisations began to manifest in different far-off areas of the world, since the remote past of the history itself; **we have since noted that to begin with during the ancient-era itself; the copper, the wheel, bronze and the iron—were invented and fruitfully discovered during the 6000 B.C., 4000 B.C., 3000 B.C. and 1500-1600 B.C. respectively. And then subsequently during the historic medieval-era, the gunpowder and artillery were discovered during 1260 A.D. and 1415 A.D. respectively**. And historically therefore, each of these momentous discoveries and inventions, both singularly as well as collectively, contributed in their very far-reaching manners in the continuing multi-dimensional advancements of the mankind and thereby, the various human-civilisations. All these momentously historic developments and events have since much less to do with the freaks of history **but much more to do, with the 'relentlessly proactive and multiply-creative-**

cum-analytically thinking' human-mind. And therefore contextually, some of above prominently listed inventions and discoveries along with also the many more kinds of subsequent inventions as well as innovations by the multiply-creative ingenuities of the human-minds, again in their own multiply contributing manners; have also been responsible and thereby, pushing the otherwise since multi-dimensionally advancing human-civilisations towards historically the much later and even perhaps predestined—the medieval-era phenomena of the triggering and also multiply fast fructification of the 'Industrial-Revolution'.

No wonder that during our discussions earlier and also in different contexts, we have since already noted that the multi-dimensionally continuous making of the man's living-history always remain very vast and complex as well. And in these continuing phenomena of the relentless discoveries, inventions and innovations; since also momentously doing the requisite ground-work and thereby, preparing the suitable canvas for further continuing advancements of the human-civilisations, in different geographical areas of the world: their many later periods' far-reaching consequences and fallouts, generally do not at all, appear to being visible in the beginning itself. Hence, that precisely is the reason, as to why the man himself is seldom able to fully see through and grasp the multiply momentous consequences, fallouts and influences of his many history-making discoveries, inventions and innovations in the world.

Now broadly in the above multiple contexts, and while coming back to the historic aspects of the commencement of 'Industrial-Revolution' we find that **Professor S.R. Sharma** has written that, "**Modern industrialism has given a new trend to human civilisation had its birth in England in the eighteenth century. That movement is usually referred to as the Industrial Revolution**. Though of late some writers have criticised the use of the term 'Revolution', as too misleading, no more suggestive or comprehensively adequate expression has been found. Equally misleading is to suggest that the Industrial Revolution began in a particular year or even decade. But considering that the several important things which gave it its peculiar character occurred all together in a crowded fifty years or so, it would not be wrong to assign the genesis of this great movement to the latter half of the eighteenth century. That was also the period of other momentous happenings such as the Seven Years' War, the War of American Independence, and the French Revolution." (Sharma, 1948, pp. 385 &386)

Historically in the medieval-era Europe's continental contexts precisely in this regard, it was not that England was techno-scientifically, or even as per the analytically objective considerations concerning the multiple-aspects of the path-breaking human attributes for fructifying some innovative ingenuities, towards any new kinds of breakthrough since triggered by the creative dimensions of human-genius, **the English people were commanding a decisive edge over other Europeans like the French, Germans, the Dutch, Belgians, or the Italians, and so on**. Hence it has never been so, that the British were humanly any 'superior' to the other Europeans. It is another matter that altogether in a somewhat different perspectives earlier and discussed in the previous sixth chapter that, while competitively pursuing her overseas highly ambitious imperialistic adventures in different far-off continents of the world, many a times she geopolitically proved to be strategically more shrewd, calculative and cool in the inevitable and always challenging geostrategic and politico-military domain, over her other no less imperially ambitious European rivals. But in the multiple-aspects of Industrial-Revolution, what made the difference during the period of the second half of eighteenth century, was some unique geographical attributes of the British Isles. **These British Islands are territorially separated from the Europe's continental landmass, by since the well-known English Channel. The geographical feature came out to be**

again decisively advantageous, in case of the momentous and relatively much more unhampered fructification of Industrial-Revolution in the Great Britain. In a context, one may even call it a 'freak of the history', or 'freak of Nature'; but that is a different matter altogether.

In the multiply strategic perspectives and, as we look into the geographical map of Europe we notice the overriding facts that, Britain being an islands-nation on the western edge of main European continental mass this unique feature since down the ages, has kept her largely insulated from the direct military threats from her other European rivals. And on top of it, the British has always had very strong and dedicated traditions for the focused pursuits of Naval Powers and capabilities. **We may also say that the British has since been able to geostrategically and politico-militarily develop a fruitfully strong naval-culture and traditions. Inherent compulsions in this regard, are obvious and hence understandable. These being the precise reasons as to why and how the British Royal Navy have always then, been so powerful with vast reach and, historically maintained an edge over her other no less seafaring and imperially ambitious European rivals during those historic times of the overseas imperialistic pursuits.** And therefore as per the strategic tenets of matters-military during the historic medieval-age; the British were able to inflict some severe blows to their other powerful adversaries and at the same time to their great advantage, without much fear or worry for any similar retaliation. Here too, the historic extent of strategic vision and shrewdness lies in the British Nation's meaningful awareness of all these since situational advantages and thereby geostrategic and politico-militarily, making full use of these attributes. **And precisely at this point of our discussions in continuity; in a very relevant context, we may also say that since the reasonably well-groomed and well-placed the British attributes concerning the multiple-aspects of 'essential strategic-culture and thoughts', have always had decisive say and also, being multiply contributory in this regard.** In these specific perspectives interestingly, the man's living-history also shows us that during those highly chequered historic times when the aspects of Industrial-Revolution were fast fructifying in Europe including the U.K. in some lead and also, while the whole continent remained politico-militarily in a ferment, the British were able to take full advantage of her relative peace at home and thereby, more effectively develop and imbibe the multiple-fruits of the then fast manifesting Industrial Revolution in Europe. Historically these in turn, were able to substantially change the socio-cultural, socio-psychological and, the socio-political face of the British Nation, and consequently thereby also totally alter the subsequent paths of making of the British living-history, both in the European as well as global contexts.

Precisely in the above contexts, **Professor S.R. Sharma** says that, "It was this rare stimulus which quickened the pace of English industry to such an extent, towards the close of the eighteenth century that it almost looked like a *revolution*. **In the words of Professor Hammond, 'That revolution was marked by the dissolution of the old village, by the transformation of the textile industries, by changes of a different kind in the pottery industries, and by a great concentration of capital and power in the industries connected with iron, steel and coal'.** In short, that revolution converted England from being the 'granary of the North' (as the Romans found her) into the workshop of the world... **It was under these circumstances that a series of mechanical inventions appeared: Hargreave's 'spinning jenny' in 1764, Arkwright's 'water frame' in 1769, Crompton's 'mule' in 1779, Cartwright's 'power looms' in 1785, Whitney's 'gin' in 1792, etc. And more than anything else, the application of steam power to all departments of industry, including manufacture as well as transport—rendered possible by the genius of Watt (1769) and Stephenson (1814)—ushered in the era of large-scale production and distribution with all their inevitable**

consequences.... It has made the modern world what it is.... Suffice it here to observe that we owe all our comforts, conflicts, dangers, and outlooks to what was happening in the Western World during the past two hundreds or so." (Sharma, 1948, pp. 386 & 387)

These are the facts of history, concerning the multiple-aspects of historically the unprecedented 'Industrial-Revolution' then fructifying so fast indeed with the far-reaching fallouts in numerous ways for the mankind and hence, entirely changing the man's ways of life and also, the multiple-challenges henceforth confronting him in the world including those of the wars. **And moreover, the momentously unprecedented fructification of the multiple-aspects of Industrial-Revolution was ironically never so smooth and all that peaceful as well as also, perhaps for furtherance of the maintenance of any social tranquility for the mankind, since rather paradoxically the common masses almost everywhere sooner or later in the world, got perforce sucked into, in many ways in indeed the overwhelming whirlpool of this Industrial Revolution and its multiply the highly consequential soon to manifest after-effects.** In fact, on one side the human societies found many old jobs towards employment getting withered-away and on another side, they found new kinds of very hard-pressed and hard-bound rather some regimented types of jobs coming to the man but in lesser numbers, so that even many of the then contemporary experts were unable to fathom-out the inevitably resultant economic complexities in this regard. In a way, almost anybody who found himself or herself being affected by this, was seen to almost overwhelmingly being swept away. **Human societies the world over, found their entire socio-cultural and socio-economic fabrics of existence and living suddenly getting totally altered and thereby, even their socio-psychological landscapes were becoming somewhat turmoil-ridden, and hence the mankind with a shock found them for itself, substantially difficult to fully reconcile with these sweeping kinds of the multi-dimensional upheavals.**

In these precise contexts as well as with great historicity, we find that since the civilisational beginning itself, almost all such continuing breakthroughs have been multiply equipping the man and ironically at the same time, in his both constructively enriching and at the same time, no less in his destructive spheres too. In fact, the destructive spheres do not at all mean only the aspects of war or any kind of the 'clashes of arms', but at many times these perhaps no less consequential destructive aspects, may also have unwarlike dimensions but with some devastating socio-cultural and socio-psychological fallouts for human societies in various regions of the world. **Historically, like the phenomenal discovery of 'wheel' by the man during the remote ancient-era itself, the late medieval-era gradual but fast, advent of the highly momentous 'Industrial-Revolution' too have indeed been multi-dimensionally, great turning-points in the multiple-paths for the continuing advancements of human civilisations in the world and thereby, on-going making of the man's living-history**. But these two as well as, many other inventions, discoveries and innovations down the ages are all in the historically continuing chain, are interrelated and mutually-influencing as well. And the same historic aspects are equally true in continuity during the post-Industrial-Revolution, i.e. present modern high-tech age and then, same would also be similarly true in the future as well.

Jawaharlal Nehru has written that, "The three great industries grew up in England—textile, iron and coal—and factories sprung up in the coal areas and other suitable places. The face of England changed. Instead of the green and pleasant countryside, there grew up in many places these new factories with their long chimneys belching forth smoke and darkening the neighbourhood. They were not beautiful to look at, these factories surrounded by mountains of coal and heaps of refuse. Nor were the new manufacturing towns, growing around the factories things of beauty. They were put up anyhow, the only object of the owners being to get on with the making of money. They were ugly and large and dirty, and the starving workers had to put up with these as well as with the terribly unwelcome conditions in the factories.... (*similar to*) the squeezing out of the small farmers by the big landowners and growth of unemployment, resulting in riots and lawlessness in England. The new industries made matters worse to begin with. Agriculture suffered and unemployment increased. Indeed, as each new invention came it resulted in replacing manual labour by mechanical devices. This often led to workers being discharged and caused great resentment among them. Many of them came to hate the new machines, and they even tried to break them. The **machine-wreckers** these people were called. Machine-wreckers have quite a long history in Europe, going back to the sixteenth century, when a simple machine loom was invented in Germany. In the old book written by an Italian priest in 1679 it is stated about the loom that the Town Council of Danzig 'being afraid that the invention might throw a large number of workmen on the streets, had the machine destroyed, and the inventor secretly strangled or drowned'.... It is possible that if this machine would have come into use when it was first invented, other inventions would have followed, and the machine age would have come sooner than it did. But the mere fact that it was not used shows that the conditions were not then ripe for it. When the conditions were ripe, then the machinery established itself in spite of numerous riots in England. It was natural for the workers to feel resentment at the machine. Gradually they came to learn that the fault did not lie with the machine, but with the way it was used for the profit of a few persons.... The collapse of the cottage industries was not sudden, but it was rapid enough. By the end of the century—that is, by about 1800—the big factories were much in evidence. About thirty years later steam railways began in England with Stevenson's famous engine named the *Socket*. And so the machine went on advancing all over the country and in almost all departments of industry and life... It is interesting to note that all the inventors, many of whom I have not mentioned, came from the class of manual workers. It is from this class that many of the early industrial leaders came. But the result of their inventions and the factory system that followed was to make the gulf between the employer and the worker wider still." (Nehru, 1934, pp. 406 & 407)

As the numerous centuries have passed we may notice that to begin with some initial inventions-cum-innovations resulted in increasing the efficiency and productive effectiveness of the **man's muscle-power** itself, and as these phenomena further progressed gradually the multiply more and more productive uses of the **animal-power** came-in, in the service of man and thereby in many ways, resulting in improving his comforts and the qualities of life itself. Historically, the consequent fallouts of the above mentioned two ancient-era civilisational developments have been such that the man himself and his mass-based employability were not affected and, had rather increased. So the multiple-impacts to man's socio-cultural and socio-economic harmony got further improved. But historically it appears that to begin with, during the late medieval-era advent of the 'Industrial-Revolution' this perhaps didn't turn out to be so, and consequently man found himself getting somewhat adversely affected and also, his life getting unsettled in many ways. In these historic perspectives perhaps the key-factor appears to be the role and mode of power needed to facilitate the man in doing his job, chores and improving his productivity as well. Earlier in this civilisational arena the power-base since available to the man for his initial need-based employments, were the human muscle-power and then subsequently, in tandem with the innovative uses of animal-power as well.

But specifically during the much later medieval-era times of the 'Industrial-Revolution', socio-culturally as well as socio-economically, the entire scenario for the man in this regard, got altered with its unexpectedly numerous fallouts and naturally affecting differently the various human-societies the world over and thereby, the man found himself totally unprepared to be able to harmoniously take command of the fast developing situation. In fact the central-reason for all these unsettling consequences historically, happened to be the **advent of mechanical-power**; which in turn causing a large-scale replacement of the earlier noted sources of energy in this regard. **Hence in these specific contexts, we may also say that the principal instrument to since trigger all these, the then unprecedented and also totally unexpected multiply good as well as bad developments to begin with; was the productive employment of mechanical-power. And this particular aspect historically, sets apart the 'Industrial-Revolution' from the other similarly earlier events and developments**. The gradually developing possibilities for the innovative uses of mechanical-power in the agriculture too, had its consequential fallouts. Age-old cottage-industries in Europe to begin with and subsequently in other continents too, began to be fast getting replaced by the newly emerging factory-system for the all kinds of production to meet various human-needs in life. Hence, traditional artisans, weavers and craftsmen got marginalized and reduced to mere wage-earners in the somewhat dark corners of factories and astonishingly, man found himself rather being swallowed by these 'monstrous' factories and thereby, getting reduced to a kind of inconsequential cog in the various noisy machines since in operation in the factories. If we critically delve even deeper into these multiple-aspects, we may find in the beginning phases of these historic phenomena the yet somewhat hidden but fast sprouting seeds of the subsequent and gradually more and more automation, in the numerously multi-dimensional areas of industrial activities.

During the beginning stages of 'Industrial-Revolutions', these highly consequential as well as far-reaching socio-economic and industrial fallouts were no small matters and understandably then, the mankind didn't find itself even preliminarily prepared to somewhat fruitfully face these new civilisational developments and events, as these were too overwhelming and hence, the man found himself perforce sucked into their multiply gigantic kinds of whirlpools. In the new kinds of the fast emerging industrial-arena man lacked awareness of its multiple dimensions, the requisite training and skills to fit-in large numbers both directly and indirectly into this factory-system since coupled with and driven by the sole motives to keep-on making profits and earn more and more money, on the part of owners of the factories. In fact in many ways, the owners themselves got conquered by their increasing greed, resulting in their even some degree of callousness and dehumanization. And naturally all these unprecedented developments led to socially, the highly disturbing-cum-unsettling times in the human societies across the world.

We now live in the high-tech era of the 21st century world, and may not fully realise those historic times and the various kinds of upheavals then taking place, consequent to multiply far-reaching fallouts of the 18th century Industrial Revolution. But even then we may find the 'seeds' of the present-era rather no less disturbing situations in altogether different manners, in today's superbly automated and super industrialism-based socio-economic and socio-political scenario in the so-called advanced nations. In the bygone times of Industrial Revolution itself, as since times keep-on unendingly rolling-by, man is continuously found to be achieving more and more mindboggling scientific inventions, discoveries and, technological breakthroughs, in turn leading to many kinds of civilisational advancements but at the same time in spite of his consistent efforts, he has not been able to even marginally eradicate the roots causes for the above mentioned many kinds of socio-cultural, socio-economic and socio-political upheavals since continuing to erupt from time to time, in the various corners of the world. The earlier talked about multiple-aspects of the same age-old greed, still come into play in the economic-cum-industrial scenario across the world and hence, continue to overwhelmingly dominate the human-societies everywhere. May we say that, this too is another kind of a vice-like grip, which has still been menacingly holding the man as helpless prey? Well whatever, but facts do continue to remain the facts regardless. In these larger and highly engaging perspectives; it may not at all be out of context to look into certain universally timeless and precise thoughts of **Mahatma Gandhi** in this regard. Both geopolitically as well as socio-economically in the larger human contexts, **Mahatma Gandhi has said that this world since as Created by God, has enough always to meet everyone's need but not his greed. In addition, so far as the mankind's economic-cum-industrial arena is concerned; Gandhiji has given the multiply-superb and proactive concepts of the 'Trusteeship'**, to effectively and also rather permanently take care of the above discussed socio-economic disturbances and socio-political upheavals. These concepts are indeed capable of neutralizing the evils of greed-based and unchecked aspects of capitalism and at the same time, not giving any ground to surface the menacing communism as political thoughts for taking hold of the reins of governance anywhere. But have the civilisational advancements reached that high to really be able to reform the man enough and, also enable him to be adequately civilised; to be able to sincerely grasp the values, depths and dimensions of Gandhiji's these universally timeless concepts and thoughts? Well ironically the realities are that, these happen to be still the tall-orders for the man himself.

Now so far as the multiple-aspects of the above discussed **Industrial Revolution** are concerned, we have so far been concentrating on Europe only; but across the Atlantic the USA which had since already achieved her political independence in 1783/84 and due to the various earlier historic reasons and also the on-going migrations of the substantial European population and thereby, continuing multiple-links with the Europe, she too was in no way lagging behind in substantially receiving the fruits of Industrial Revolution. **Vast geographical areas and colossal natural resources of USA, was other decisive inputs for her and all these in combination; enabled her to substantially race ahead fast and in these paramountcy aspects, the rest is part of the recent and fairly well-known living-history of nearly past three centuries in the world.**

In the above USA's contexts precisely, **Richard Overy** (pages 228 and 229) has written that, "The centuries between 1800 and 1900 witnessed a 'market revolution' in the USA. Small farms and workshops gave way to a national economy where manufacturers produced for a distant market place. Better transportation, vast resources and favourable government policies were all crucial in creating, by 1900, the world's most productive economy... In 1800 the USA had been internationally recognized as a nation for only 17 years. Most of its estimated population of 5,298,000 dwelt in small communities or farms: there were only 73 towns and cities with a population of over 2500, of which just one—New York—had over 50,000 inhabitants. By 1900, the nation had been transformed. The estimated population of the USA had grown to 76,094,000 (an increase of 1337 per cent). There were now 827 settlements with a population between 2500 and 50,000, and 75 cities of more than 50,000 people (including three of more than 1,000,000). The statistics of economic growth during the century show equally large increases: the net tonnage entering US ports grew by 3400 per cent; total imports by 922 per cent; and exports by 2011 per cent... In 1800, not only was the USA primarily an agricultural nation, but the most farm crops were grown for local consumption. Farm families produced either for themselves, or for their neighbours. There was little chance, because of the high transportation, of moving towards a more national market-oriented type of economy. According to one estimate of 1816, shipping a ton of goods from Europe to America cost approximately $9; the same amount would enable it to be transported only 14.5km (9 miles) by land... Better transportation made possible a **'market revolution'**. After 1800, there were many improvements in communications. The building of the Cumberland road (1811-18) between Cumberland, Maryland and Vandalia on the Ohio river symbolized the drive to build turnpike roads that, by the time the road construction boom collapsed in 1821, had seen 6400km (4000 miles) of such roads constructed. Return voyages by the Clermond between New York and Albany in 1807 and the Washington between Louisville and New Orleans in 1817 inaugurated an era of steamship navigation on rivers and inland lakes. The successful completion of the Eric Canal in 1825 also prompted a major boom in canal-building in subsequent years.... The most crucial of all was the arrival of railroad. By 1840, the USA had 5355 (3328 miles) of railroad track (compared to 2896km/18 miles for the whole of Europe. A further 46,660km (29,000 miles), were built during the next 20 years. The last four decades of the 19th century saw a great expansion in railroad construction and completion of five new transcontinental railroads. By 1900, the USA had 311,094km (193,346 miles) of rail track carrying about 142 billion ton-miles of freight (compared to 39 billion in 1866). The railroad which had played so significant a part in creating a national market both recognized it and regularized it with the introduction of standard time-zones (1883) and a common track gauge."

In these momentously multiple-contexts, the above quoted paragraphs do indeed historically provide us a significant data-based panoramic view of the then prevailing as well as, the fast developing and expanding socio-economic, industrial and infra-structural scenario in the USA, by the end of 19th century. **During those historic times and based on the multi-dimensionally rather fast fruition of the 'Industrial-Revolution' additional and allied kinds of revolutions too, got fructified in both the USA and Europe like—the new management-systems for conducting business and running the industries, Banking-system, both internal-cum-overseas market-revolution, fast transforming as well as multiply expanding infra-structural facilities and assets and so on.** These revolutionary and far-reaching changes, multiple improvements as well as developments in the transport and communication facilities and then the fast and interdisciplinary progress in the multi-dimensional arena of the more and more scientific-cum-technological research, inventions, discoveries, and continuous innovations had all begun to multiply advance simultaneously as rather parallel phenomena, being always interrelated and mutually influencing as well. All these now modern-era phenomena indeed, were giving both new additional dimensions as well as directions, to since the already continuing multi-dimensional advancements of human-civilisations; and hence the world had already begun changing totally rather fast and consequently, the same phenomena now is perhaps, unendingly continuing with more and more additionally manifesting and unprecedented complexities in this regard, in the present new-age of the super high-tech and super-industrialism.

We may also say that these momentous phenomena are indeed becoming so accelerated as well as acquiring such overwhelming complexities; that it is becoming rather a challenge to decipher as to who is driving whom—the man or these momentous phenomena themselves, in the multiple-paths of the consequent developments-cum-progresses—towards the unprecedented and the so called human-civilisational advancements in this world. In some perhaps understandably intellectual haste, these issues may be tempted to rather get brushed aside as matters of academic conjectures only, but their highly consequential as well as far-reaching fallouts for the mankind the world over and also being so visible, are indeed too weighty to be brushed aside so lightly in this regard. **In the presently fast-moving new-age man finds himself getting overwhelmingly conquered and also so relentlessly being driven—onto his socio-cultural, geopolitical and politico-militarily multi-dimensional 'canvas'; that now the unprecedentedly**

dangerous and fast manifesting situations around him, are refusing to be under his grip and command.

We have also talked about the inherent 'greed' in man and therefore interestingly indeed, as we delve further in these subject-matters, we do notice that this 'greed' also has its own inherent capacities to trigger always some newer kinds of the situation-based: the highly selfish-cum-callous 'greedy ideas' in human-minds and ironically, regardless of the consequences for the larger human-societies. This has been so since the beginning of human evolution on this earth and, would perhaps inherently remain so always for the man. Hence precisely in the contexts of the owners of factories consequent to the Industrial-Revolution, profit-motives do mostly remain unchecked and therefore, socio-psychologically rather overwhelming in this regard. These were historically the precise reasons as to why the **highly greedy concepts of the 'laissez-faire' got so 'innovatively' sprouted in the typically 'fertile' minds of the then owners as well as managements of the factory-system** to begin with in England and then, in other European countries too. In addition later, these rather humanly unsettling concepts got fructified further with somewhat socio-economically devastating manner in the USA as well. These pioneers of industrialism then had developed their quite unethically greed-based desires, to run their factories and conduct their business as much as possible, without any interference from the government of the day. It would rather remain in fact, these greedy thoughts always, irreconcilable in many ways.

Contextually hinting at the powerful and hidden 'inhuman-aspects' in the above mentioned concepts of the so-called *'laissez-faire'*, **Jawaharlal Nehru** has mentioned that, "They developed a new philosophy which called *'laissez-faire'*—that is, they wanted to do just what they liked in their business without any interference from government. By having started factories to make things before other countries had done so they had got a lead, and all they wanted was a free field to make money. *Laissez-faire* became almost a semi-divine theory which was supposed to give an opportunity to everybody if he could take advantage of it. Each man and woman was to fight the rest of the world to go ahead, and if many fell in the struggle, what did it matter... (*No regards for*) the progress of co-operation between man and man, which had been the basis of civilisation. But the ***laissez-faire*** and the new capitalism brought the law of the jungle. **'Pig philosophy', Carlyle called it. Who laid down this new law of life and business? Not the workers. The poor fellows had little to say in the matter. It was the successful manufacturers at the top who wanted no interference with their success in the name of foolish sentiment**. So in the name of liberty and rights to property they objected even to the compulsory sanitation of private houses and interference with the adulteration of goods." (Nehru, 1934, 409 & 410)

The germination and the then rather fast fruition of '**laissez-faire**' too by itself; has also been a kind of momentous fallout of the man's above discussed, always the proactive mental phenomena on the part of man himself; as **one of the since numerous fallouts of the Industrial-Revolution. Some sort of being the proactive concepts and thoughts regarding 'capitalism', have always been present in the human-societies since ages but the multiply momentous consequences of the fast fructification of 'Industrial-Revolution' during the second half of the eighteenth century onward, gave altogether different meaning and dimensions to the multiple-aspects of 'capitalism'.** The Industrial-Revolution triggered the socio-economic and industrial concepts of the factory-system, which the man had not known so far, as the man earlier since ages, has been familiar with living in the largely self-contained village-oriented communities and hence; occupied along with the farming, in also various other kinds of productive activities in his age-old and broadly home-based numerous cottage-industries. But the much later advent of Industrial-

Revolution and thereby, dawn of the big many machines with their complex design and operating systems and coupled with big factory-complexes had almost totally altered the entire scenario in this regard, resulting in man himself getting swept-away to unfamiliar situation and ways of life, for which he found himself largely unprepared. This also as a result made the poor and uncared in the societies to suffer even more, while getting sucked into the far-reaching and largely very unsettling developments around them, being beyond their understanding.

Needless to say that setting-up of these factories always needed huge amount of money and hence, the same had to be mobilized by various channels of borrowings, and other avenues. These in turn, resulted in the beginning of the systems of banking and gradually other newer kinds of subsequently surfacing financial institutions and systems of management. But these factories always needed large financial and other kinds of investments and at the same time, they had long gestation periods as well, and this meant huge sums of money perforce getting locked for long periods of time, before they could generate profitable returns. In addition, to essentially facilitate all these little later profitable developments in multiple ways; expansive and the far-reaching many kinds of infrastructural facilities and assets, like transport, communications, warehouses, ports and so on had to be built-up suitably, again with additional huge investments and thereby, made operational for use in this regard. Initially some persons, who happened to be enterprising enough, were able to set-up factories for the competitively advantageous production of goods in relatively larger quantities for supplies to the markets. And as a result, those persons who had shown themselves to be more enterprising soon became substantially rich enough to set-up more factories and expand their industrial 'empire' and hence interestingly, these very people to begin with came to be known as the 'capitalists' in their respective societies.

We may recall that in the multiple-contexts of the man's vast and highly complex history that one thing leads to another and hence, many things lead to many other different things as well as newer civilisational developments and events, along with the always continuing multi-dimensional advancements of the human-civilisations. While elaborating these in our specific contexts at this precise point we do notice that the man's some inherent aspects of his **entrepreneurship** and **ingenuities**, imaginative harnessing of funds and other essential resources, for setting-up the factories resulted in the birth of the concepts of **'capitalism'** as well as that of the **'industrialism'**. And these in turn subsequently soon, resulted in the fast paced births of the **capitalists** as well as the **industrialists** in the world. In these multiple perspectives, the man's living-history shows us that initially to begin with; England again found herself in more advantageous positions in comparison to her other European rivals in this regard, as per the then prevailing overall geostrategic and politico-military situations. **England being already having established herself as the leading seafaring imperial-power across the world, with her strong Royal Navy and, fast expanding-cum-consolidating overseas Empires, since triggered by the multiple-aspects of the British imperialism; was already having substantial edge over her other European rivals, in availing the fruits of fast fructifying the 'Industrial-Revolution'.**

Industries always need assured and large viable sources for the supplies of power, raw-materials, transport and shipping facilities, huge capital and money for the investments and then at the same time; assured outlets as markets to profitably sell the finished products, to make more and more money. Well in these historic perspectives; as we continue to critically delve deeper into the overall geopolitical as well as socio-cultural-cum-economic scenario, then prevailing during the second half of eighteenth and the whole of nineteenth century world, we do clearly notice that the Britain was already in decisive leads over her other European imperial-rivals, both in the

European continent and other overseas continents of Asia, Africa and Australia as well. She was in a sort of commanding position to maintain her decisive geostrategic and politico-military supremacy almost across the world; though always being in stiff competition with her other European imperial-rivals, especially the French; and at the same time, in spite of she already having lost her American colonial-empire. **Her vast colonial empires in Asia, Africa and Australia and all the more so, her 'Jewel in the Crown'—India; came in handy to the British imperialists, capitalists and industrialists; as the most valuable as well as very lucrative sources of raw-materials and in addition, as the ready and vast captive markets for the sale of their finished industrial products, such as the textiles, iron and steel, railway-systems and equipment, dyes and so on, with colossal profits**. The British did historically exploit these extensively large and fast emerging situations to the hilt and made huge money and thereby; with all sorts of exploitations of the imperial-colonies, Britain became more and more rich, at the cost of numerous colonial natives. So while the colonies across the world stagnated and got impoverished, the Britain got multiply rich and, remarkably strong and prosperous as well.

With respect to the multiple-contexts of our these discussions and with the focused reference to the multiple-aspects of capitalism, industrialism as well as the ruthlessly exploitative imperialism; and thereby at this present precise point, it may appear to be especially relevant to quote the famous sayings of the three since globally well-known British thinkers and scholars themselves in this regard. These famous quotes are indeed timeless, with their own historicity, intellectual depth and contextual dimensions. They do also contain their substantially analytical penetration into these indeed, important historic issues, which we have since been discussing at present. Interestingly in some larger perspectives here, we may also say that the truths: in any respects as well as pertaining to the any particular historic developments and events; have their own multiple kinds of ways and means to always get glaringly surfaced principally, to objectively ensure the multiple enrichment of man's living-history since continuously being in the making too.

Adam Smith, since known the world over as father of the classical concepts of Economics, in his book *Wealth of Nations* has said that, "The government of an exclusive company of merchants (*i.e. principally meaning the then East India Company here*) is perhaps the worst of all governments for any country whatever... It is the interest of the East India Company considered as sovereign that the European goods which are carried to their Indian dominions should be sold there as cheaply as possible, and the Indian goods which are brought from there should be sold there as dear as possible. But the reverse of this is their interests as merchants. As sovereigns their interest is exactly the same with that of the country which they govern. As merchants their interest is directly opposite to that interest."

And then another famous and great British thinker-cum-statesman since belonging to the nineteenth century and named **Benjamin Disraeli** in precisely the similar contexts has very reflectively opined and also given analytically a very thoughtful comparative views on the rich and poor; always interactively coming face to face with each other in various manners, in socio-culturally quite different societies the world over. Hence he says that, "Two nations; between whom there is no intercourse and no sympathy; who are ignorant of each other's habits, thoughts and feelings, as if they were dwellers in different zones, or inhabitants of different planets; who are formed by a different breeding, are fed by different food, are ordered by different manners, are not governed by the same laws...the Rich and the Poor."

Another well-known British personality named Sir Charles Metcalfe, since then engaged in the service of the British East India company in Indian subcontinent had been able to study

and also develop some critically analytical insight into the then prevailing socio-cultural as well as socio-economic lives in India. And hence during the 1830 itself, in these momentous contexts and with indeed a historic insight he has said that, "The village communities are little republics having nearly everything they want within themselves; and almost independent of foreign relations. They seem to last where nothing else lasts. This union of the village communities, each forming a separate little State in itself...is in a high degree conducive to their happiness, and to the enjoyment of a great portion of freedom and independence." (*Source for all the above quotes:* Nehru, 1934, pp. 483, 467 & 489).

In these contexts, historically it is also true that initially these indeed substantially far-reaching and multiply influencing phenomena; did remain confined to Europe and largely to begin with, within the Western Europe only; but soon these began making deep inroads as expected, into the neighbouring larger territorial segments of the Eastern Europe as well.

But in the larger perspectives of multi-dimensional subject-matters of the 'Industrial-Revolution', as we begin to critically delve even further deeper into these historically highly phenomenal developments, we do relevantly notice two additional and, further highly consequential eventful developments in this regard. The first one being that, almost the whole of Africa and Asia, with perhaps the sole exception of Japan; were under the locally situation-based various kinds of the colonial-rules of the leading as well as ruthlessly rival and seafaring European Imperial-Powers. Hence, these colonies and their imperially subjugated native-populations across the world were in no position to take requisite initiatives and begin participating in this onward civilisational-march, as then taking place in Europe in the multiple-contexts of availing since the fast emerging fruits of the highly phenomenal 'Industrial-Revolution' of the late eighteenth and almost the whole of nineteenth centuries. But in spite of these the then historic restrictions and imperially forced limitations since, imposed upon the subjugated colonial-natives of Asia and Africa these continents too interestingly, could not be kept entirely insulated for long even directly or indirectly, from these historically highly consequential phenomena of the 'Industrial-Revolution. History has its own ways of making and influencing the lives of mankind in this world, regardless of man's from to time, own since situation-based: putting the various kinds of imposed differentiations among themselves. In fact, that remains under the supreme-command of since the unending and also fully incomprehensible—the evolutionary phenomena of the mother-Nature herself.

During in fact the early modern-era multi-dimensionally historic and highly eventful Industrial Revolution, as now being discussed in its so many contexts and also since as stated above, almost the whole of Asia and Africa and so on, were already under the colonial-rule of the various ruthlessly rival European Imperial-Powers. In these multiple-contexts and at this precise point; we do also notice that the already earlier medieval-era historic fructification of the multiple-aspects of otherwise the age-old 'imperialism', as relentlessly being triggered by those rival and highly ambitious European Imperial-Powers; those since specifically European kinds of the 'imperialism' were already at its zenith in the far-off different continents of the world, in the multiple-forms of their overseas vast colonial-empires. These colonial-empires—as established and consolidated in those far-off continents, the concerned European Imperial Powers principally led by the British, always needed to consolidate with any possible further expansion and at the same time, also ensure their better and better imperial governance in their best imperialistically advantageous interests; largely irrespective of the much care of both the short and long-term interests of the native populations living in those colonies, and without their any meaningful say in the governance of their lives and countries. As the man's living-history has been showing from the advent of multiple-aspects of since the age-old 'imperialism' these remain the inevitable multiple-consequences wherever any kind of the imperialism has taken roots and spread its tentacles, for the numerous kinds of rather ironical and ruthless exploitation of man by the man himself.

Whatever all these may be but, the 'imperialism' in its numerous incarnations and connotations still paradoxically keep-on lingering the world over and also, differently in different situations. For example, contextually as we have discussed little earlier during the historic phenomena of multiple fruition and consolidation of multiple-fruits of the **Industrial-Revolution** in the USA and, as initially since having brought from the Western Europe and thereby, another phenomenal development of the **'Market-Revolution'** soon dawned over there with its own multi-dimensional fallouts. We do also critically observe that historically indeed in this phenomenal development of Market-Revolution, initially hidden but the inherent 'seeds' of another unprecedentedly new kind of the multiple-aspects of **'Market-cum-commercial-Imperialism'**, were also destined to soon begin sprouting and manifesting in numerously multiple-ways across the world, with their own further newer kinds of consequences for the mankind. Hence daringly almost the whole of nineteenth century in the above discussed aspects of imperial-colonies across the world interestingly, the two parallel kinds of imperialism; i.e. historically the already entrenched feudally-colonial kind as well as, this new kind of the 'Market-cum-commercial imperialism' began proceeding ahead in tandem, as since triggered by the concerned European Imperial-Powers themselves in their colonies.

During now the bygone era of imperialism and the overseas colonial Empires across the world; the concerned European Imperial-Powers, for the more effective governance as well as more efficient exploitation of resources and wealth of their colonies, they themselves perforce had to sooner or later bring-in the multiple fruits of since the highly phenomenal 'Industrial-Revolution, into their colonies in the far-off continents of both Asia and Africa; and thereby in these historic processes naturally, any forced or otherwise any materializing barriers in this regard, had to perforce get demolished. In other words interestingly, this is another one of the numerous since situation-triggered ways of the continuous making of man's living-history in this world. So far as the continent of Australia since as located in the south-eastern meeting waters of the Indian and Pacific Oceans across the Indian Peninsula was concerned, the British found hardly any worthwhile civilised native population over there and hence, only a vast and very richly natural-resources-gifted landmass to be very beneficially colonised and, which they indeed did so to the hilt. Geopolitically to begin with, Australia soon became initially the very advantageous vast colony mainly for the white penal-settlements of the British-Empire. With time white as well as also other kinds of population kept-on migrating in this relatively vast and virgin continent and thereby, a new, big and sovereign democratic nation of Australia gradually got born. Historically during the recent past, these were the new and indeed phenomenally civilisational developments and events in the world.

Man's vast, highly complex and multi-dimensional living-history since continuously always in the making too; very emphatically as well as perceptively shows us that, birth of the sovereign, multicultural and democratic Nation of Australia in the recent past and in the present modern-era itself; is **a direct fruition of the momentous 'Industrial-Revolution'** when interestingly in fact, the multi-faceted concepts and practices of geopolitically-cum-strategically indeed very shrewd multiple-aspects of the British-imperialism across the transcontinental British-Empires, were historically at its zenith in spite of their loss of the American colonial-empire across the Atlantic. During those times, the British were relentlessly in the pursuits of maritime explorations and discoveries of new lands and territories for their geostrategic and politic-military colonisation.

The far-reaching multiple-fruits of Industrial Revolution further helped them in their highly adventurous and focused endeavour, by providing them new kinds of tools and thereby, enhanced their wherewithal in this regard. Australian learned professor and historian, **John Hirst**, *Sense and Nonsense in Australian History*, has informatively narrated the saga of Australia, with deep insight and objective historicity.

In the above mentioned book, **Professor John Hirst** has indeed talked very perceptively about numerous dimensions regarding so many things concerning the historic making of the new Australian Nation, in a very engaging manner and has written that, "By our standards it is an outrage, but I have this historian's quirk of knowing that the people who flogged criminals would think our practice of locking criminals up in all male institutions for thirty years or more cruel and unnatural. So I hesitate to condemn the past, but I recognise that I run the danger of not seeing horror when it is truly there.... History shows that there is no such thing as absolute power—F. Eggleston, *State Socialism in Victoria*, p. 315.... Mild, just but exact in discipline; he was father to his people who were attached to him from affection and obedient from confidence.—Tribute to James Cook on contemporary monument, copy in Maritime Museum, Greenwich.... Surely of all the 'rights of man', this right of ignorant man to be guided by the wiser, to be gently and or forcibly held in the true course by is the indisputable.—**Thomas Carlyle**, *Chartism*, chapter 6...Thomas Carlyle, the historian-philosopher of the nineteenth century, became my favourite guide. He lived through the industrial revolution and hated its tendency to reduce all human relations to a cash nexus. But he did not believe that the liberal and democratic programs of his day would meet humankind's deepest needs. (*But at the same time it is also equally or even more momentously true that so far; the humankind is yet perceive, invent, fruitfully conceptualise and, imbibe anything superior to the concepts and tenets of liberal democracy.*).... Multiculturalism, with contradiction at its heart, has a great capacity for the absurd. Its exponents are now developing an absurdist history of Australia. This is how it runs: Once upon a time there was a small inward looking, intolerant, racist Anglo-Celtic nation. It began to take migrants from Europe and then Asia. The migration program was an outstanding success. The nation turned into a diverse, open, tolerant society. The new migrants created the tolerance. Where is the absurdity in this? That a nation which hated 'wogs' invited hundreds of thousands of Italians and Greeks to come to its shores? No nations, like individuals, may under great urgent necessity do what is distasteful to them. The rationale for the migration policy was to boost Australia's population and economy as a defence against a renewed threat from the north. Better the dark skinned Greek than the Japanese. No, the absurdity lies in the claim that the migrants created the tolerance.... But if the strangers are numerous and carry with them a very different value system, could they then not have a transforming effect on the host society? The migrants were certainly numerous and over a short time constituted a reasonably high proportion of the old population, but they were not of one sort, they came from a number of different countries and were notoriously intolerant of each other. If Greeks and Turks, Serbs and Croats, Arabs and Jews live in peace in Australia, it is not because they brought tolerance with them. Let us abandon absurdity and explore a commonsense hypothesis: that the migration scheme has been a success because the roots of the tolerance lay in the society which invited the migrants to join it.... Unlike the United States and Canada, the English, Scots and Irish did not form separate enclaves. There was a remarkably even intermingling. This in itself suggests a high degree of toleration which was advanced further by the determination to keep sectarian rivalry out of community organisations. In the nineteenth century, hospitals, charities of all sorts, trade unions, friendly societies, sporting clubs, the mechanics institutes and the schools of arts which ran the libraries and the public halls—all these had Catholics and Protestants on their boards and among their members and clients and worked to keep them together. This was not easy. There was always low-level tension and sometimes spectacular brawls. Separate community organisations for Protestants and Catholics would have avoided the dissention, but threatened much worse (*yes indeed*)." (Hirst, 2005, pp. 5, 6, 11, 13 & 14)

How historic and precise the above quoted lines are, in providing a panoramic picture of Australia's chequered past and also, phenomena of the making as a modern Australian Nation-State in the world. Here we do also notice that there have been many kinds of undercurrents, which were indeed consisting of the historic socio-cultural, socio-psychological, as well as, socio-political threads having since brought by the diverse the then migrating populations along with them, who began coming to the Australian continent for getting settled. **These historic phenomena, mainly fructifying during the reasonably calm nineteenth century and also since coinciding with remarkably fast flowering of the multiple-fruits of Industrial-Revolution, were very much multi-dimensional in nature and also, interrelated as well as mutually influencing too.** In these multiple perspectives, as we continue to contextually delve deeper, we do notice perhaps two historic aspects being very much proactive, since the antiquity of man's living-history itself. That is to say that so far as any historicity of the any kind of civilisational making of a nation or an empire is concerned, there has to be principally a gradual process of the multiple events and developments, which multi-dimensionally go into its making—as in case of the Australian Nation-State; and/or, there have to be momentous and at times, even overwhelming historic past as well as multiple kinds of legacies perhaps going back to the very ancient of times, which may be forced to wait-cum-struggle for centuries, for the opportune moment and at the same time, appropriate situations to develop and begin triggering the overriding phenomena towards making or remaking of any particular Nation-State—as in case of the reincarnation of Israel in the Middle-East along the eastern coastlines of the Mediterranean, soon after the Second World War. **And therefore by just carving-out some territories by means of brute military might and, without any worthwhile historic-cum-geopolitical rationale in this regard may at the most create some kind of an 'artificial' nation and which in any case, can seldom be enduring on the totally beyond any manipulation: the test-bed of the time itself.** Hence, these two are also few of the interlinked-cum-mutually-influencing avenues of ongoing multiple-paths for the multi-dimensional and continuous making of history in this world.

Now again coming back to the multiple-aspects of the momentously historic Industrial-Revolution, we find that India in spite of being a very glorious and indeed an ancient civilisation as well as also, maintaining an unbroken civilisational continuity since thousands of years, over the man's larger and global civilisational-canvas itself, she did perforce miss this revolution and hence, got almost totally bypassed during its far-reaching and new kinds of history-making beginning in this regard. During those historic times of relatively the recent past India was an imperial-colony of the British, who were then ruling the oceanic-waves and hence, presiding over their indeed multi-coloured as well as multiply flexible transcontinental British-Empire. But in these contexts the imperial-British had themselves to perforce bring-in the multiple-fruits of Industrial-Revolution and thereby in these again momentously far-reaching processes, India too tremendously benefited in many ways. As a result, the multiple kinds of the fallouts too in this regard, were bestowed upon the historically rich and diverse civilisation of India and thereby enriching the same even more in many ways and, India indeed needs to be grateful for all these, in her own historic perspectives. **And at the same time it is also true that consequent to these phenomena of now the bygone times, the multi-dimensional foundations too, were being**

momentously laid for the subsequent making of the sovereign and democratic Indian Nation-State; enabling her little later to march-on fast towards some rapid economic-growth and much needed industrialization as well. In other words, we do also find and since already noted earlier a number of times in different historic contexts that the roots of the captivating present-era multi-dimensional developments and events indeed, lie buried into their own historic as well as highly chequered past.

Hence very perceptively and also quite appropriately, it has been said that on the expanding canvas of history one thing leads to another and also, many things lead to many different things. Therefore in the above larger contexts, **Professor S.R. Sharma of India** has mentioned that, "**England has been to the Modern World what Athens was to the Ancient**. Ideas, movements, and happenings in that Island sooner or later reflected themselves in rest of the world. England achieved parliamentary Democracy and all other countries have been striving ever since to emulate her example. **England started the Industrial Revolution and the whole world is still being transformed to her pattern**. England grew Imperialist and turned to Federalism for finding liberty in union, and nations are still trying to walk in her footsteps. Just as Rome and Christianity, gave unity to Christian Europe during the Middle Ages; so England and Science imparted unity to the Western Civilisation in later times. Hence, (*understandably it may be*) the very claim of England on our attention in dealing with the Expansion of Europe." (Sharma, 1948, pp. 389)

In the above lines we do notice that there is a panoramic glance of some phenomenal civilisational developments as well as multiple advancements being certainly indelible and also encompassing many centuries, but at the same time, these happen to be principally pertaining to the European civilisational canvas only. Hence, while broadening the purview of our presently on-going discussions precisely in this regard, we do also contextually notice that **India and even China—in spite of being two remarkably ancient as well as rich human-civilisations**, could not remain in a position to be counted upon and somewhat proactively get noticed, in the above mentioned and much later multi-dimensional 'great-games' of almost totally new kinds of the civilisational developments and events. If any particular Nation-State, in spite of its multiply rich-heritage and legacies, got overlooked or by-passed in any manner; then any possible lacunae perforce resides in that concerned Nation-State only, and certainly not in the broadly systemic as well as historically overwhelming phenomena of these very civilisational-advancements during its any particular timeline and, in any particular geographical areas of the world.

Paradoxically in this regard, in spite of the presence of almost all the essential civilisational ingredients, India has ironically been consistently found wanting geopolitically in a rather quite embarrassing manner in the above stressed subject-matters of the multiple-aspects of 'essential strategic-culture and thoughts', in the global comity of nations. And quite strangely as well as ironically again, India now as a sovereign and fast developing democratic Nation-State in the world, is perhaps regressively refusing to see reasons in this regard, and put their multi-dimensional acts purposefully together with the requisite strategic prudence and vision, so mandatory for her today. India as a modern Nation-State and now being master of her own destiny inescapably needs to meaningfully understand another universally crucial and timeless truth that **there are no permanent friends or enemies in this world, but always one's own paramountcy national-interests**. Contextually therefore here lies inescapably always the supreme importance of the cultivating, developing and purposefully imbibing and also updating in continuity—the multiple-aspects of 'essential strategic-culture and thoughts', for any concerned Nation-State.

Now coming back to the multiple-aspects of "Industrial-Revolution' **the highly consequential developments of the armaments and different kinds of weapon-platforms as well as, multiple advancements of the 'matters-military' too**; were proceeding at their own fast pace. In fact, almost all the fruits of Industrial-Revolution and thereby, multiple kinds of the momentous advancements in the numerous areas of sciences and technologies; were showing rather captivating capacities for their dual uses as well. Hence in these specifically larger perspectives and precisely during our presently on-going discussions; we may not overlook, that since the age-old rivalries in Europe, the all sorts of jealousies and thereby historically, a long and multiple kinds of legacies of the 'clashes of arms' among the various European imperial-powers in Europe just had not ever got cooled-down in any manner whatsoever. In fact, as the numerous fruits of the Industrial-Revolution as well as, the multiple-aspects of sciences and technologies began momentously fructifying, especially during the nineteenth century, those many kinds of the age-old rivalries and enmities instead of getting gradually pacified, ironically began rather getting alarmingly intensified. And therein geopolitically, **the multiple kinds of seeds of the First World War** were getting sown for their rather accelerated sprouting into the extremely rivalry-ridden and, traditional geostrategic and politico-military arena of the European continent. And consequently the traditionally and ironically in fact, rather irreconcilably-rival European Powers since, destined to fight this War, during the beginning decades of the twentieth century itself almost the whole of world was perforce going to be sucked into this War inevitably, and thereby very cruelly becoming instrumental in bestowing upon perhaps the entire humanity, historically an unprecedented kind of devastation.

Hence during the nineteenth century, as the multiple kinds of fruits of Industrial-Revolution got multiplied and being made available to the otherwise traditionally very rival European Nations their economic-cum-industrial powers and further as well as, the newer kinds of their 'geostrategic and politico-military' wherewithal and reach too, began getting increased manifolds. These the then unprecedentedly fast and newer kinds of developments in turn, did act as far-reaching causative-factors for momentously intensifying a number of historically highly consequential phenomena simultaneously, not only in Europe but also in the other far-off continents of Asia, Africa and Australia as well. So far as the USA was concerned she already being sovereign as well as safely located far away across the Atlantic, she remained unthreatened and hence busy in peacefully reaping the substantially beneficial economic-cum-industrial as well as, the matters-military oriented harvests. These aspects in fact gave the USA, tremendous opportunities to enhance her industrial-base and capacities for the highly profitable foreign-trade manifolds and therefore in the process, laying the multiple-foundations for her subsequent incarnation as an unchallenged Superpower in the world. So far as Europe was concerned, the overall scenario over there was more complex and fast becoming substantially volatile too as there were many kinds of the already intensifying undercurrents among traditionally the rivalry-ridden and leading European Imperial-Powers, but they too were not lagging behind in grabbing the multiple kinds of fruits since, already manifesting due to the numerous fallouts of Industrial-Revolution. And in the process, European nations too, as since led by the Britain, were fast industrializing themselves but their economic gains and the availability of markets for their highly profitable raw-materials, trade and commerce then, were largely dependent on their far-off and vast imperial-colonies in Asia and Africa. The Australian situations for the English were in fact quite different as this huge continental landmass during those times, had already become a kind of unchallenged and already highly beneficial and captive territories of the British Empire and thereby; destined to play subsequently very significant strategic roles, during the both World Wars, in the forthcoming twentieth century.

The whole of Europe led by Britain, was fast industrializing and multiply arming too, and also the prominent European Imperial-Powers remained busy in colonizing more and more territories in the both Asian and African continents, for expanding their empires and also making them more secured from any kinds of threats as well. In these indeed multiply expanding as well as, intensifying great-games then in full swing in Europe, the traditional lack of trusts among those rival European Imperial-Powers, was also alarmingly getting magnified in many ways. And consequently in these multi-dimensional processes, then historically getting so uncontrollably accelerated in Europe that paradoxically rather the relevantly strong but quite rivalry-ridden and traditionally otherwise being so proactive always, the concerned European geopolitical Power-Centres instead of being in any kind of control got themselves overwhelmed by numerous kinds of the highly consequential developments and events, then since fructifying so fast and also threateningly manifesting in the European continent. Further, we do notice that the momentously multi-dimensional recipe, for the soon destined to happen the First World-War (1914-18), was indeed getting completed fast.

Field Marshal Montgomery has very perceptively written that, "In the nineteenth century whole societies organized themselves for war in a more complete way than ever before, and armies were equipped with weapons far more powerful. The new warfare developed in Europe and America, but it penetrated to all parts of the world. In fact the years between 1815 and 1848 were a period of relative peace in Europe. However, during that time general developments were taking place which had a direct bearing on the emergence of modern war. The ideas of imperialism and nationalism were hardening; revolutionary changes in armies and equipment were born with the growth of population and new industrial techniques; new methods of communication increased the pace of life in general; and military theorists and politicians arose who rationalized and exploited these factors. Some trial was made of the new warfare in the fighting of the decade or so after 1848. Finally in the American Civil War (1861-5) and the French-Prussian War (1870-1), the new techniques anticipated the twentieth century... The most important background development was the rise in population—between 1830 and 1870 Europe's population increased by 30 percent. Mass population meant mass production of armaments and mass armies. Vastly increased numbers created new problems in transport and supply as well as tactics. The increasing influence of public opinion was another new factor which had to be taken into account by commanders and politicians. Pressure of population gave an impulse to imperialism... **In the years after 1815 Britain was the strongest power having emerged victorious from the Napoleonic wars besides being the leader of the Industrial Revolution, and it suited her that the world should be at peace. The British Navy, unchallengeable since Trafalgar, undertook to keep a Pax Britannica**.... Fighting was not, however, altogether avoided between 1815 and 1848; nationalism, allied with romanticism and liberalism, found outlets in numerous revolts. Most of these were suppressed by the great autocratic powers.... The arms race before 1914 ensured the continued development in military and naval technology. By 1900, rifles, pistols, carbines, and machine guns were as they were to be during the 1914/18 war. The principal development in the rifle were the magazine system of loading invented by **James Lee**, and smaller calibre bullets which, being lighter, travelled faster and on a flatter trajectory. Smokeless powder was adopted by the French in 1884—a development which was literally to transform the face of battlefields. There were more developments in explosives, with the production of dynamite in 1867 by **Alfred Nobel** and cordite in 1890. In 1883, **Hiram S. Maxim** patented his machine-gun, with which the recoil was used to load, fire, and eject continuously so long as the trigger was held back, the cartridges were stored in a flexible belt, and the barrel was cooled in a water jacket. In 1891 the British army adopted a special light model which weighed only 40 lb and fired 650 rounds a minute. **With slight modification Maxim's gun became the Victor's machine-gun, which was used in both world wars; more than any other weapon it was responsible for the character of trench warfare. In artillery, muzzle-loading gave way to breech-loading because long barrels gave velocity and range**. The British field artillery adopted 12 pounder guns. By 1890 most European armies were equipped with the guns which they were to use, with only minor modifications, in the 1914/18 war." (Montgomery, 1983, pp. 237 & 262) (*The above quoted paragraphs are indeed able to provide us highly informative developments in the field of armaments taking place in the European continent during the eventful nineteenth century. In addition, another important point historically is needed to be borne in our minds that since USA too was then, had already begun interacting very closely with Europe in many respects and hence, similar kinds of the parallel developments were also taking place over there as well.* ***And as since historically expected these far-reaching and armaments-oriented developments and inventions-cum-innovations were also subsequently going to almost totally alter perhaps, the entire gamut of warfare in the twentieth century and even beyond in the continuous course of time, as new developments kept-on manifesting in this regard.*** *Hence in the larger perspectives consequently we find that from then on, the subsequent phenomena of since the further multi-dimensional making of the man's living-history too, got eventfully changed. We would thereby, be viewing them rather in a substantially engaging manner as we proceed further-on here and in the next eighth chapter, both in the multiple-areas of matters-military as well as also in vast domain of the numerous kinds of human-endeavours, along with so many socio-cultural, socio-political, scientific and technological and, geopolitical initiatives-cum-activities*).

Jawaharlal Nehru, 1934 (p. 461) has written that, "It was the century (*i.e. the nineteenth century*) of marvellous mechanical progress. The Industrial Revolution brought in its train the Mechanical Revolution, and machines became more and more important in man's life. They did a great deal that man had done before, and eased his drudgery, and lessened his dependence on the elements, and produced wealth for him. **Science helped greatly, and travel and transport became swifter and ever swifter.** The railway came and displaced the stagecoach; steamship took the place of the sailing-ship, and then came the great ocean liner, powerful and stately going from continent to continent with speed and regularity. Towards the end of the century came the automobile and motor cars (*since pioneered in Germany by the engineers* ***Gottlieb Daimler*** *and* ***Karl Benz****—Richard Overy in his book on, 'The Times History of the World'*, page 220) spread out all over the world. And lastly came, the aeroplane. At the same time man began to control and utilize, a new wonder—electricity—and the telegraph and telephone appeared. All this made a vast difference to the world. As the means of communication developed and people travelled faster and faster, the world seemed to shrink and became much smaller. We are used to all this today, and we seldom think about it. But all these improvements and changes are newcomers to this world of ours; they have all come within the last 100 years (*i.e. presently around 185 years*)... It was also the century of Europe, or rather of Western Europe, and especially of England. The Industrial and Mechanical Revolutions had begun and progressed there, and they gave a great lead to Western Europe. England was predominant in sea power and industry, but gradually the other countries of Western Europe caught up with it. The United States of America also forged ahead with this new mechanical civilization, and the railroads carried them, westwards to the Pacific and made the huge country one nation. They were too busy with their own problems and their expansion to trouble themselves much about Europe and rest of the world. But they were strong enough to resent and prevent any interference from Europe." ***(Hence the Man's living-history will always ask from Nehru as to, why and how did he forget the above mentioned and also numerous similar other historic facts when his time came later, with his 'tryst with destiny' after 1947 as India's first Prime Minister***

***and, began presiding over the destiny of this Sovereign Democratic Republic during the beginning of its formative years?*).**

In the above multiple-contexts, as we may care to read the above quoted two somewhat long paragraphs together, since contextually picked-up from the books of two different authors and then we may clearly and in a rather focused manner visualise as to how the multiple-kinds of the fruits of Industrial Revolution, have been able to almost simultaneously trigger the many far-reaching as well as highly consequential developments, in the fields of armaments and also other areas of the matters-military and in addition, many different areas of the human-endeavour as well. **As we continue to delve deeper into the historically highly turmoil-ridden legacies of the past many centuries in Europe we may be able to in fact, see through as to how numerous kinds of causative-factors began sprouting so fast in Europe, and then maturing paradoxically rather to trigger the overwhelming manifestations of an uncontrollable 'action-stage' over which the largely preventable 'blood-bath' of the First World-War (1914-18) then, tragically became destined to be played by those rival countries; after getting regrouped into two different euphoria-ridden rival-groups, facing each other and then having drawn complex battle-lines between themselves**.

And historically the irony of ironies was that, the then politicians of the leading Imperial-Powers in Europe; who were callously calling the shots, **even began propagating that this war was to be fought to end all future wars**! Man perhaps is the strangest creature in the world and has strange ways of thinking, behaving and acting, as since we find those leading political leaders in Europe then, thinking, talking and responding to each other paradoxically in bizarre manners, prelude to the First World-War. It is not that historically, such ironical phenomena surfaced onto the highly complex and continuously expanding humankind's civilisational-canvas, for the first time as prelude to this first global-war. In fact contextually, if we may very discerningly begin to critically delve deeper into the man's vast living-history, we may then notice quite a number of such typically weird civilisational negations since the ancient of times, on the part of man himself, in this regard. That is in a way, another factual dimension of the manifestation of man's history

Now, in the above noted highly consequential aspects concerning the First World-War, **Field Marshal Montgomery** with his brilliant professionalism as well as intellectual prowess, has written with great historic precision that, "**On 24th June 1914 the Archduke Franz Ferdinand was assassinated in Bosnia, Austria, rightly supposing that pro-Serbian feeling in Bosnia accounted for the incident, delivered an ultimatum to Serbia on 25th July, and the following day declared war on her. Russia, the self-styled patron of the Slav peoples, mobilized against Austria on 30th July. Germany was an ally of Austria; France and Britain were allies of Russia, and so by early August Germany and Austria (the 'Central Powers') were aligned in war against France, Belgium, Britain and Russia (the 'Entente Powers'). Turkey emerged from secret into open alliance with Central Powers in September. Other countries entered the war later... The 1914/18 war was caused by a murder in the Balkans**. None of the statesmen or people of Europe positively desired war. Nobody consciously engineered it. A succession of diplomatic interchanges took place, in which the true object of national political leaders—Berchtold (Austria), Sazonov (Russia), Bethmann Hollweg (Germany), Viviani (Italy), and Asquith (Britain)—was to preserve the security of their countries. But they all miscalculated. Gestures intended to bluff and deter provoked violent reactions which were not anticipated: politicians were playing with diplomacy in an atmosphere which was in fact highly inflammable... First rivalries existed between the various powers. Britain was actually conscious of the challenge by Germany to her commercial and industrial power, and had been actively engaged in building up naval and merchant fleets. In France there was resentment at the events of 1870-1 and at Germany's possession of Alsace and Lorraine. Germany and Russia were rivals for influence in the Balkans. The Ottoman Empire in its ramshackle state had nothing considerable to gain from being hostile to anybody; but the Germans had been wooing the Turks—who had a score to pay off against Britain and Russia, the two powers who had patronized or bullied them all the way through the nineteenth century. While these antagonisms were by no means enough to make war inevitable, they did mean there was a highly charged atmosphere in which war might break out... **The situation was made still more dangerous by the fact that none of those responsible, and indeed few men in any nation, had any understanding of what war would mean. Nothing on a bigger scale than the recent small Balkan wars was imagined, and therefore the idea of war was generally taken lightly and irresponsibly. May be war would have been avoided; may be not; but, as it seems to me, nobody *tried* to prevent it**. The fatal step was the ordering of mobilization. From that point events got out of the control of individuals and became dictated by the war plans of the general staff. These plans because of the military doctrine of the time, was all offensive. Military escalation followed diplomatic insanity, and, as A.J.P. Taylor writes, the great armies which had been accumulated to provide security carried the nations to war by their own weight." (Montgomery, 1983, pp. 273 & 274)

Contextually in momentous historic perspectives in this regard, what could be more ironical than the above quoted highly perceptive narration given by one of the most famous and indeed all-time great military-commanders since so vividly having painted panoramic views of the eventful and very fast developments preceding the actual beginning of the 1914-18 World-War, with unprecedented tragic consequences then and, coupled with extreme human sufferings as well. In the above narrated deeply analytical perceptions, the author has provided the then quite paradoxical actions and initiatives at the political, military and diplomatic levels, in almost all the leading powers in Europe and, since not at all known for any kinds of enduring trusts and friendship among themselves earlier; especially pertaining to the dangerously eventful mid-months of the year 1914 itself. During whole of the nineteenth century and even earlier, specifically in the far-reaching and important areas of the matters-military, many new kinds of the professional thought-processes and developments had already begun in the European continent. In these areas, the Germans were then displaying some new kinds of militarily visionary thoughts and initiatives; which were soon adopted by the almost all countries of Europe to begin with, except perhaps in Great Britain. Around 1870 or so, the concepts of universal service with relatively a long tenure of enlistment in the military; were first promoted in Germany. Considerable population growth as well as new kinds of mass-output based industrial production systems in Europe then; enabled the political and military leaders over there, to raise and also professionally train quite larger armies to look after their pivotal geostrategic and politico-military imperatives vis-à-vis their arch-rivals. But politico-militarily there was another critical dimension to this concept as: those quite professional and large armies were not at their best in fighting various kinds of imperialistic wars for their Governments, in their imperial-colonies since situated in the distant and different overseas continents of the world. And therefore obviously, the Britain since having vast and quite diversified trans-continental British-Empire and thereby, due to her own overriding geopolitical and strategic compulsions; preferred to maintain a somewhat smaller but professionally fine-tuned permanent army as a military hub in this regard, around which any need based and large-scale very fast mobilization could be affected and thereby; even rather ironically inevitable arms-race then taking place in the European continent itself, could effectively be taken care-of perhaps in the best possible British national and also, their highly profitable imperial-interests as well. But as per the all kinds of geopolitical as well as their overriding military compulsions, the British were always strategically prudent and visionary enough to raise and effectively maintain their most powerful and indeed big blue-water Royal-Navy along with also very large merchant shipping-fleets.

In a way quite before the First World-War (1914-18) and hence during the nineteenth century itself we may in other words also say that many kinds of in fact, the **Revolution in Military-Affairs (i.e. RMA)** had already begun in the whole of Europe. And these militarily path-breaking aspects of the RMA were not limited to the fields of armaments and various kinds of the developments of weapon-platforms alone, but in many rather no less important areas of the military dimensions. **These included new kinds and colours of the military-uniforms, systems of training for both the men and officers of different levels, systems of command and control and then, pertaining to the improved systems of military staff-work at the various levels of military headquarters of the Army, Navy, Air Force and also in the Joint-Command headquarters. Interestingly for the first time, the British brought-in khaki colour uniforms in the Indian subcontinent itself and subsequently the khaki in different shades and textures, were gradually adopted by the numerous countries for their militaries in almost all over the world.** So far as the pioneering aspects of Staff-Colleges as training institutions were then concerned, the German and British took the lead in establishing their own distinct military-culture based institutions in this regard; which soon became inspiring role-models for the other countries as well. And later on, before and during even all the more big and highly destructive Second World-War; the systems and effectiveness in a way of the German General Staff, became a very fine as well as internationally admired example in this regard. To begin with, in these revolutionarily far-reaching matters-military oriented dimensions; the Nicholas General Staff Academy (Russian) since then located in St. Petersburg and, the British Staff College at Camberley soon became great rivals to the famous Prussian War Academy of the Germans. Similarly in other arms and services of the then Armed Forces in Europe the leading and intensely rival European nations had already set-up their own distinct kinds of the specialised military training institutions, for conducting the need-based and properly designed courses for the various kinds of training, for the military personnel and officers since, belonging to the different branches of their Armed Forces. In other words therefore, we may also say that the entire multi-dimensional aspects of these matters since pertaining to the continuous ongoing systems of training for various European Armed Forces, in turn became quite formal and also substantially professional as well.

And at the same time in these militarily paramountcy areas quite revolutionarily the age-old culture of some sort of the static stagnation both in thought-processes and actions on the ground, became things of the past and got almost totally replaced by the new and modern and, professionally dynamic ideas-cum-action-plans. And needless to say that in these rather vital areas concerning the matters-military the multiple kinds of the fruits and breakthroughs, as momentously consequent fallouts of the unprecedented Industrial-Revolution already having fructified and also multiply advancing and magnifying as well, over again the continuously expanding civilisational-canvas of the humankind, were indeed no less crucially important contributory-cum-triggering factors regarding multiple-areas and dimensions of the matters-military. In the multiple fields of sciences and technologies, the relentless research-cum-development oriented activities began progressing simultaneously both in economic-cum-industrial as well as, the matters-military areas too, with indeed numerously paths-breaking consequences. **Interestingly in somewhat larger perspectives in this regard we may also say in other words that, if on one side of the Industrial-Revolution, its fruits and fallouts and multi-dimensional scientific research and development activities were able to bring-in many kinds of the reliefs, comforts and improvements in the overall**

qualities of human-life then at the same time ironically perhaps, on the other side, numerous kinds of the war-oriented military-lethality too, were menacingly increasing in rather parallel manners; some had highly dangerous potentials to bring-in unprecedented range of deaths and destructions in the world. From points of views of indeed the purely humanistic considerations, we may also say that historically these were rather quite queer and also alarmingly sad situational developments for the mankind, and these thoughts with their own quite weighty perspectives; are definitely very relevant and also understandable—when we so clearly find the mankind itself ironically falling into the traps since, created by its own thoughts, actions and even rather initially well-intentioned, multi-dimensional as well as indeed, the so many developmental endeavours and breakthroughs, onto his own continuously expanding civilisational-canvas in the world. But these again being the mankind's civilisational-aspects, have their own age-old legacies and ironically continuous historicity of the culture of many kinds of conflicts and also the warfare as well. And hence these too again at the same time, do also constitute some of the vastly numerous ways of the multi-dimensionally continuous making of the man's living-history since, the very advent of human-civilisations in the world. **Specifically in these larger perspectives and, we may also recall that earlier too in some different and also in many allied contexts; we have since already noted that continuously multiplying pages of the man's living-history, remain always soaked in the mankind's relentless hard-work, tiresome labour, sweat, tears and the human-blood as well. And hence that's precisely the way, man's multi-dimensional living-history is continuously being made on the 'infinite-canvas' of time, since the beginning of history itself.**

While keeping in mind the contents, subject-matters and the multiple-dimensions of the above discussions and along with then, the extremely tragic range and quantum of the deaths and other allied tragic aspects regarding the consequences of First world-War (1914-18) we notice that **Arthur Birnie** (*The Art of War*, 1942, pp. 202, 203 and 204) has written that, "*General Characteristics of the War*—Among the many which gave this war its unprecedented character, three call for special mention:

(a) The number of combatants: Since 1870 all the great Continental nations had adopted universal military service, and they entered on the struggle with huge armies:

Germany	5,300,000
France	4,000,000
Austria	2.000,000
Russia	4,590,000
Italy	1,200,000

Great Britain and the United States started the war with small professional forces, but adopted compulsory service after the outbreak of hostilities. At the armistice their mobilized military strength was:

British Empire	2,002,000
United States	1,881,000 *and then....*

To maintain these armies at full field strength imposed a tremendous strain on the man-power of the belligerents. On the average, each of them called a fifth of its population. Germany mobilized 13,250,000 men, France 8,000,000, and Britain 9,500,000. The casualties were correspondingly high. In killed, wounded and prisoners German lost 7,000,000 (2,000,000 killed), France 5,000,000 (1,500,000 killed), the British Empire 3,000,000 (1,000,000 killed).

(b) Equally unprecedented was the consumption of munitions. Not only was extended use of existing war material, especially heavy artillery and machine guns, but new weapons were invented—aeroplanes, tanks, Lewis guns, trench mortars flame throwers, hand grenades, gas shells, etc. During the war Britain's annual expenditure on munitions, rose from 700,000 pounds to 64,000,000 pounds. France increased the number of her munitions workers from 25,000 to 1,500,000. The United States caught unprepared, required more than twelve months to equip its army for the field. **This new development greatly increased the importance of the economic factor in warfare. It became almost impossible for any but industrialized nations (*and even now it remains all the more true*) to wage war effectively**.

(c) From the combined effect of these two factors followed a third consequence—the deterioration of strategy. **The immense size of the combatant armies speedily put up an end to the war of movement**. Frontiers were completely covered and enemies had no flanks to turn. A decision could only be achieved by frontal attacks, and here the machine gun increased the strength of the defensive a thousand-fold and made breakthroughs almost impossible. **Three-fourths of the fighting consisted of the indecisive trench-warfare. None of the general staffs found a way out of this impasse, though the British, by their use of tanks, indicated a hopeful direction in which a solution might be sought**."

The above quoted historic lines do give us some very thought-provoking and indeed, quite disturbing factual-data concerning the First World-War, since also known as the 'Great War'. Earlier historically, mankind had not known such kinds of extensive deaths and destruction in any kinds of the manmade or natural calamities, or even any war. Concerning the manmade calamities, we may even with substantial authenticity say that, any kind of a major-cum-extensive war ranks perhaps far ahead of any others in this regard. And since ages, all kinds of the conflicts and wars have been extremely tragic and ironical but at the same time, rather on-going and distinct segments of the continuing civilisational legacies of the mankind, since the very advent of the human-civilisations. It is indeed, an extremely unbecoming civilisational paradox and it may even be termed also as a great civilisational negation on the part of mankind. These precisely are the reasons as to why in some contexts it has even been said that, 'man's history is the history of warfare'. In the larger perspectives in this regard, it perhaps at the human-level appears even to be a kind of great civilisational roadblock for the man; but that indeed remains historically regardless, a part of since the multi-dimensional and also continuing civilisational legacies-cum-multiple-progresses of the mankind, its culture of thought-processes and deeds/misdeeds. Hence while looking into these aspects from the points of views of the socio-culturally larger human-level refinements and at the same time, from the higher perspectives of the spiritual-level elevations of the mankind, we may find in the above thoughts, some contradictions too but that remains precisely the manners by which rather ironically, man's living-history is continuously being shaped over the infinite 'canvas of time' itself. In other words at this precise point, we may even be able to perhaps notice some kinds of the continuing struggles, between the already earlier talked-about, the 'angel' and the 'evil' since always present inside the man himself. These aspects may understandably be taken as matters of some conjecture, but that's too perhaps remains one of the ways to look into these historic and also since continuing civilisational paradoxes concerning the mankind itself.

And hence perhaps no wonder, that we do find in fact while looking into since the continuously multiplying pages of the man's living-history; the First World-War (1914-18), has historically been the first such big and so far then, the most extensively violent example in this regard. In these precise contexts, so far as the above quoted book of **Arthur Birnie** is concerned they do give some factual and very important information concerning the huge losses of military-lives and also some of the highly consequential military expenditures. With precise historic perceptions, Arthur Birnie also mentions about almost the total covering of frontlines by the networks of trenches and tactically-sighted the machinegun emplacements from both sides, did rather perplexingly create a situation of the stagnant stalemate, with perhaps no easy sight for achieving any kinds of the breakthroughs towards any decisive breakouts into homeland territories of the opposing forces, to secure a decision for conclusively ending this rather unprecedented kind of the extremely violent 'warlike cobweb'; since so menacingly manifesting on the European horizon as a kind of rather 'crippling-trap', for rather the entire humanity. But at the same time it also remains equally true in fact in much larger perspectives that the man's living-history has its inherently powerful nature to keep-on moving ahead regardless with the ongoing passages of time, by rather crushingly overtaking any kinds of obstacles on the way of its continuous making itself.

And precisely at this historic point, again the **fruits of 'Industrial-Revolution'** and the multi-dimensional advancements of the sciences as well as, industrial-technologies then, did provide requisite answers with suitable developments of the military-technologies and various kinds of the weapon-platforms and thus thereby, contextually giving births to the multiple-aspects of **'tank-warfare'** and, new dimensions to the **'warfare of military-maneuvers'**, at both the strategic and tactical levels. These far-reaching aspects then did provide new kinds of tools and avenues to the concerned military commanders and their staff-officers at different levels to visualise and craft new strategies and tactics towards making their operational-plans and thereby, for achieving favourable decisions in the battlefields. And hence as already emphasised in many different contexts earlier, these too are also additional examples of the multiple-paths of continuous making of history since, always remaining interrelated, mutually influencing and, multi-dimensional as well. In fact, from the points of views of studying purely the quite engaging and multiple historic-aspects of the First World-War (1914-18) as well as the specifically allied subject-matters concerning the multiple aspects of 'matters-military' in this regard, well-known military-history books covering these subjects-matters in contextually much greater historic details, since authored by the well-known experts for posterity are available for requisite studies and references and hence otherwise too, these aspects do naturally fall outside the scope of this book.

Learned military-historians have since studied and written about the important matters-military details, concerning the First World-War in rather two different segments: i.e. the 'Western Front' and the 'Eastern Front' and thereby contextually, provided us with distinct details regarding as to how wars on these two different fronts were fought; what were the different kinds of compulsions, challenges, quantum of troops and other kinds of military wherewithal on both sides; and also, how these were employed for action in the field and then in addition; the personality attributes of the top-level military commanders on both sides and hence, opposing each other's on the above mentioned two principal fronts. Well, indeed with comprehensive historic objectivity in retrospect and, not only from the angles of matters-military but in addition also, from larger socio-political as well as politico-military angles too we may even now say that, this war in many ways, was avoidable and in fact, not only concerning the colossal militarily 'wasted-efforts'; but also substantially as the inevitable consequences, turned out to be sheer wastage of the all kinds of very valuable resources as well as at the same time, tragic destructions of also indeed many very costly economic-cum-social assets. Historically since ages, any war has perforce been always a costly, violent and highly destructive business, and hence no wonder that, from the many kinds of socio-culturally prevailing viewpoints across the board, war also ironically remains indeed a great civilisational negation for the mankind. But it is also true that war at times, becomes inescapably unavoidable too and hence fighting the war becomes necessary for protecting the same indeed many invaluable human civilisational-aspects as well as multiply, the so much enrichment of possibly the entire humanity at large, from the menacing forces of evil.

And even at the risks of being somewhat out of context at this point and therefore, specifically in these aspects of our larger perspectives; while broadly viewing them, we do find the **legendary Indian Epic 'Bhagavad Gita' and the 'Mahabharata'** since belonging to the remote human civilisational antiquity; do provide a classic example in this regard and hence that's precisely the reason, as to why since many thousands of years; this immortal epic indeed, remains so enchantingly attractive and engaging for the mankind. **No wonder that therein, we do find superb and indeed immortal examples of battles between the good and the evil, between the virtuous and between the unscrupulous wrong-doers—in the *'Dharmachhatre Kuruchhethre'* i.e. in the humanly righteous domain of doing one's duties always and hence even so in the battlefields, when calls to doing the one's unwavering duties even at times, towards the inevitable clashes of arms and thereby, scripting the unique saga of fighting bravely and with selfless intentions for sacrifices; do therefore remain immortal, timeless and, inspiring as well.** But even while looking into historically in a perspective and, as per the timeless and rather a kind of the judgmental-framework—since contextually derived from the legendary-cum-Epical 'Mahabharata'; now the much later and modern-era's quite recently past multiple-aspects of the geopolitical-cum-matters-military dimensions as well as, causes and effects oriented history of the First World-War; can we as per the above mentioned parameters: say that these do remain even somewhat justifiable in this regard? But these too remain some kind of the historic conjecture and at the same time; it is also true that man's living-history and being a part therein—the living military-history too, do have numerous examples of the many kinds of such historic conjectures. But when we would come little later, to the multiple-aspects of Second World-War (1939-45), the story becomes different and thereby, all the more far-reaching as well as highly consequential saga of this war, do come out with substantial success to say; as per the above mentioned 'judgmental-framework' since derived from the 'Mahabharata' and hence, we would contextually be coming to these highly momentous historic aspects concerning the Second world-War little later.

Now again coming back to multiply historic-aspects of the relatively recent occurrence of First World-War (1914-18), and while looking into the various war-plans as well as thought-processes of the top-level political and military leaders on both sides; i.e. **the 'Central Powers' consisting of Germany, Austria and then, Turkey and opposing them the 'Entente Powers': consisting of France, Belgium, Britain and Russia** we do notice that one particular and substantially comprehensive 'War-Plan', turned out to be historically quite famous in this regard. This historic 'Operational-Plan' was conceived, and very comprehensively formulated with indeed a great historic vision during the last decade of eighteenth century itself in Germany and since, remains well-known as the famous **'Schlieffen Plan'**. The architect of the historic war-plan happened to be the legendary Chief of the German General Staff of those times, since known as the **Count Von Schlieffen**. He was himself a great student as well as scholar of the global and living military-history, spanning since from the very ancient of times and hence; while formulating his own plan, he could so very imaginatively take some ideas from Hannibal's plan of cavalry charge to smash the Roman cavalry, during the 216 B.C. in the battle field of Cannae. And the rest is history with its own at times rather strange ironies; but in spite of all these, man's living-history keeps manifesting and moving-on regardless.

Concerning the above noted famous war-plan, **Arthur Birnie** has written that, "In 1891 **Count Von Schlieffen** became Chief of the German General Staff. He was a scientific soldier, who had made a close study of Cannae, and hoped to imitate, on a large scale of a European campaign, the manoeuvre by which **Hannibal** won his great victory. The object of the Schlieffen Plan was to get the enemy into a position in which he could be shattered by a crushing blow, as the Roman legions were smashed by **Hasdrubal**'s cavalry. For years Germany had been preparing for a war on two fronts.... (*The pivotal principle involved had been*) **Get your enemy where you want him, hold him, and deal him a mortal blow**. It was the Schlieffen Plan which the German Staff used in 1914—with modifications. The Chief of Staff was the **younger Moltke, nephew of the victor of Sedan**. 'Keep the right wing strong' Schlieffen is said to have muttered on his deathbed. His successor disregarded this maxim. Not only did he reduce the number of army corps on the right, but he strengthened the left wing armies and assigned them a partially offensive role. These changes ruined Schlieffen's great strategic conception. (*When perhaps some sort of mediocrity overtakes and sets-aside the remarkable conceptual-vision of a genius in any particular given situation, all sorts of unexpected problems do take-over many kinds of rather the fast-moving emerging situations and hence inevitably thereby, any probable success mostly becomes at the most, a remotely distant possibility.*)." (Birnie, 1942, pp. 204 & 205)

So in these precise contexts concerning the First World-War; we find that many learned and well-known military-historians as well as the military commanders too, through their writings and autobiographies/biographies, have brought out rather many kinds historic 'ifs and buts and conjectures' regarding the conduct of operations and in additions also, systems-cum-evolutionary processes towards eventful developments of even crucial weapon-platforms. But at the same time, it is also momentously true that, everything and anything concerning the highly complex and multi-dimensional-cum-multi-directional and continuous phenomena of the making of man's vast living-history, cannot perforce remain under the control of mankind itself. As discussed a number of times in different contexts, it is true that if man is careful and conscious enough to be able to learn requisite lessons from history and thereby so, with his proactive and multiple strategic-vision, he puts himself in a some sort of and even at times far-reaching influencing position, to momentously influence the further making of the history up to at times, even a considerable extent including also his living military-history, which in any case, remains an inseparably integrative segment of that much larger and all-encompassing man's living-history. But at the same time, it remains also equally true as well as absolutely binding on the mankind itself that since the very beginning as well as the perpetually ongoing and highly complex-cum-multi-dimensional phenomena of the human-evolution itself, there have always been numerously far-reaching and also momentously sweeping events-cum-developments in this regard, which have perforce remained much beyond the control or any influencing reach of the man himself. These in any case, do always remain undeniable historic realities, but at the same time and as per the continuing dictates of man's living-history itself; it always equally remains rather mandatory on him to desirably maintain the requisite kinds of strategic-consciousness, to enable him to fruitfully keep learning from his history, in order to be in some kinds of the advantageous position towards shaping his future with the on-going passages of time in this world. Man's living-history in fact, never fears or favours or, even disfavours anyone in any manner whatsoever.

While keeping in our minds the multiple-dimensions concerning the above discussed some much broader perspectives during our presently on-going discussions regarding the 1914-18 First

World-War; historically we do notice that the USA entered into this war on the side of 'Entente-Powers', to fight against the Germans and Austrians. USA was in fact then in an unmatched position to employ rather her momentously far-reaching human-cum-material resources in this war in such a decisive manner that, this war could soon be brought to an end in favour of the 'Entente-Powers'. That's historically again due to some different reasons that, this war could be brought ironically to some kinds of a 'botched-up' end. And wherein thereby in fact as a great paradox of the history itself, in such a kind of the 'end' of this war tragically; the irrepressible seeds of the forthcoming and much more destructive and humanly devastating Second World-War, got sown onto the expanding civilisational-canvas of the mankind itself. **Hence, the seeds of history's future always remain buried in its historic past itself.**

Now specifically concerning the historically eventful processes of the USA's entry into the First World-War the famous and legendary US General of the Army (*i.e. Five-Star rank*) Douglas MacArthur in his autobiography named *Reminiscences*, remains a very valuable as well as also universally well received book in this regard. This book is not just an autobiography in fact, as since one of the very few most famous and all-time great military personalities as well as indeed outstanding military-commanders during both peace and war who could so decisively and momentously with his sheer multi-dimensional geostrategic and politico-military brilliance, were able to shape the making of man's living-history, with human-touch, sensitivity and, great humane feelings in their hearts. **In fact it won't be even any exaggeration to say that immediately after the Second World-War and since, in the defeated and surrendered Japan; General MacArthur as the Military-Governor of Japan, was so peacefully and, with his brilliant kinds of the highly effective socio-political as well as almost totally convincing persuasive powers was successfully able to smoothly and acceptably, turn the age-old and extremely feudalism-cum-militarism oriented the Japanese Nation-State into a truly vibrant-cum-liberal parliamentary democracy: wherein simultaneously somewhat similar to the already well tried and tested British model—i.e. a ceremonial Monarchy too was preserved, as the constitutional 'Head of State', within just few years, in an unique and historically enduring manner. This historically, was certainly no small feat in the then war-ravaged Japan which was unprecedentedly devastated by the extensive destructive powers of two atomic-bombs; since dropped onto her by none other than the U.S. military itself, which was fighting against the then Imperial Japanese Forces; under the command of General MacArthur himself, as the Supreme Commander of the multi-national Allied Forces' Pacific-Command in the Far-East, during the Second World-War**. Hence in these larger contexts, General MacArthur's autobiography is indeed a very vivid record of the great historic saga in many ways since, pertaining not only to the Second World-War but substantially much more; and not only concerning his Nation the USA, but also no less equally for the entire world.

In his above referred autobiography **General MacArthur** has mentioned that, "During this time (i.e. in the USA), the General Staff, in considering the pivotal question of the effective reliability of the National Guard for possible combat service abroad voted predominantly against the use of the Guard. I was the exception, believing that the Guard's strength could be increased by voluntary enlistments and its expanded units trained to combat effectiveness. I steadfastly shared my father's long-held belief in the citizen-soldier. **When war with Germany came (*i.e. the First World-War: 1914-18*), as I was certain that it would, National Guard Divisions should be able to fight proudly with the Regular Army... Congress declared war against Germany on April 6, 1917. It was the inevitable result of more than two years of indiscriminate sinking of American ships by German submarines. The loss of American lives in the torpedoing of the *Lusitania* had aroused the**

righteous wrath of the American people. President Wilson had done everything in his power to keep America out of the war, but Germany's threat to the freedom of the world forced Congress to throw America's gigantic industrial might and manpower into the conflict... The problem of the scope and nature of American troops became acute when a staff study was circulated to the effect that the ultimate strength to go was 500,000 and all Regulars. When the paper came to me, I was tired from overwork and indiscreetly endorsed it saying that I completely disagreed with its conclusions, but would not attempt to detail my reasons, as I felt no one would give them the slightest attention.... When I entered the Secretary's office (*since I was called by him, soon afterwards*) my heart sank. He was squatted deep down in his chair, puffing his pipe and reading that policy paper.... Finally he looked up and said in that slow, precise way of his, 'Major, I have just read your endorsement,' and stopped. I braced myself and cleared my throat to express my regret and apology, but before words came, he suddenly rose saying, 'I agree with you in this matter. Get your cap. We are going to the White House to place the whole question before the President for his decision.' I felt like walking in the air (*Interesting, isn't?*).... For over an hour, the Secretary and I took turns in advancing our arguments for use of unlimited force and employment of the National Guard to its full capacity. With great courtesy the President said, 'I am in general accord with your ideas. Baker put them into effect. And thank you Major, for your frankness.'... Later on, Secretary Baker sent for me. 'What is the best way', he said, 'to give maximum effect to the President's decision?' I suggested that we take units from the different states so that a division would stretch over the whole country like a rainbow:—from that time on it was known as the **Rainbow Division**. (*And like that, soon the USA was able raise a large number of such 'Rainbow Divisions'; based on the above brilliant idea, for their decisively successful employment during the historic First World-War*)." (MacArthur, 1965, pp. 51-53)

These lines give us glimpses of some of the innovative ways and means as to how to think *'out of the box'*, with indeed focused intentions to generate rather some smooth and workable ideas, towards shaping-up solutions to meet nationally and even internationally overriding challenges. The point here is that, in any kinds of the similarly given situation and, also facing rather some pressing or even perhaps some threatening challenges; if the concerned nation has geopolitically, some tangible geostrategic depth and inherent reserves of the population-based youthful manpower and other necessary resources; for quickly designing the desired solution towards effectively meeting those challenges then, it becomes inescapably imperative on that nation's top-level political as well as military leaderships to be strategically imaginative and fruitfully siege them without any delay, to save their Nation-State from possibly any national and also international embarrassments. Hence it is needless to say here that, specifically at this precise point of our on-going discussions in this regard, the above discussed thoughts pertaining to historically one of the very critical and also, geostrategic-cum-politico-militarily quite challenging situation of the recent past and in addition, pertaining to one of the most important nations, do have certain substantially pertinent messages to now the sovereign, democratic and also fast developing the Indian Nation-State as well. But at the same time and in these larger perspectives; the moot point here since rather inescapably surfacing—remains that does the present independent, modern and democratic India and also her politico-bureaucratic combine 'centers of power', are even now pragmatically conscious and sincerely serious enough to cultivate, evolve, develop and imbibe the multiple-aspects of **'essential strategic-culture and thoughts'** in quite necessarily quick-paced manner, for the Indian-Nation's multiply overall good as well as, the bright future? We would substantially be discussing these precisely India-specific important points and along with, their some overriding geopolitical dimensions in their various contexts further.

Now while coming back again to the multiple-aspects of the highly chequered historic saga of First World-War, we do also notice that one of the legendary British military-commanders, has himself provided some interesting but at the same time, very educative insight into the then existing lacunae in the British military. In his autobiographical book titled ***The Memoirs of Field Marshal Montgomery***, first published in the year 1958 and then brought as a Paperback Edition in 1960, by the Collins Fontana Books, London; and therein, in the (pages 35 and 36) the legendary British Field Marshal has very perceptively written that, "When the war (*i.e. the First World-War: 1914-18*) broke out I was a platoon commander. When it ended I was Chief of Staff (GSO I) of a Division and rising thirty-one, well able to think clearly, although my mind was still untrained. To an ambitious young officer with an enquiring mind, things seemed wrong... There was little contact between the generals and the soldiers. I went through the whole war on the Western Front, except during the period I was in England after being wounded; I never once saw the British Commander-in-Chief, neither French nor Haig, and only twice did I see an Army Commander... The higher staffs were out of touch with the regimental officers and with the troops. The former lived in comfort, which became greater as the distance of their headquarters behind the lines increased. There was no harm in this provided there was touch and sympathy between the staff and the troops. This was often lacking. At most large headquarters in back areas the doctrine seemed to me to be that the troops existed for the benefits of the staff. My war experience led to believe that the staff must be the servants of the troops, and a good staff officer must serve his commander and the troops but himself be anonymous... The frightful casualties appalled me. The so-called 'good fighting generals' of the war appeared to me to those who had a complete disregard for human life. There were of course exceptions and I suppose one such Plumer; I had only once seen him and had never spoken to him. There is the story of Sir Douglas Haig's Chief of Staff who was to return to England after the heavy fighting during the winter of 1917-18 on the Passchendaele front. Before leaving he said he would like to visit the Passchendaele Ridge and see the country. When he saw the mud and the ghastly conditions under which the soldiers had fought and died, he was horrified and said, 'Do you mean to tell me that the soldiers had to fight under such conditions?' And when he was told that it was so, he said, 'Why was I never told about this before?'... The fact that the Chief of Staff of the British Armies in Europe had no idea of the conditions under which the troops had to live, fight, and die, will be sufficient to explain the uncertainties that were passing through my mind when the war ended."

Concerning almost all the liberally democratic nations in the modern world, such lacunae and various kinds of the shortcomings concerning the multi-dimensionally complex-aspects of the **'matters-military'** both during the times of peace and war do in fact, mostly remain rather malignantly lingering. Man's living-history since ages and, over the always expanding civilisational-canvas, do show that in any kinds of dictatorial or totalitarian regime in a nation or societies, the systems of governance and thereby, the prevailing situations do perforce remain totally different. But at the same time, it is also true that in any such extremely oppressive regimes; there never remains any meaningfully humanistic concern-cum-care at any level of the political as well as military command-hierarchies, for both military and civil lives. But in case of any vibrant and healthy liberal democracy and indeed quite reassuringly, for the larger humanistic-cum-civilisational good; the situations do remain almost totally different. And any thinking citizen with some requisite awareness anywhere in today's modern-world and, rather irrespective of the any circumstantial compulsions would always remain in agreement that, this is precisely the way it should remain and be maintained as well. But when these otherwise quite important socio-political aspects are critically viewed in their wider contexts, we do also quite analytically notice that, these do pertain in fact, to mostly the peacetime situations and hence largely, concerning to some partial dimensions in this regard. **But in any case during wartime, many kinds of the compulsive conditions perforce get surfaced in such manners that, in the very fast moving and extremely violent scenario; many unexpected and also overwhelming situations get substantially out of control of both the political leaders and the top-level military-commanders since functioning both in the warzones as well as the headquarters and hence at times casualties too, for both the civil and military lives, become dauntingly problematic and so very difficult to be prevented**.

Hence contextually, the above discussed perceptions need to be critically viewed in their intended perspectives in this regard. But indeed the highly complex as well as multiple-aspects of the matters-military, do always have numerously larger dimensions too and hence; no liberally democratic nation, who has historically been able to purposefully display some kinds of the critically well-stationed and dynamic traditions of the multiple-aspects of **'essential strategic-culture and thoughts'** can ever afford to ignore them, in any given situation. In these multiple-contexts in fact, the universally larger point which may be seen to be surfacing now, is that in any unhindered and also prolonged periods of peace and well-being some liberally democratic Nation-States do tend to forget or rather ignore the always paramountcy multiple-aspects of the matters-military. And this is precisely the point where, the almost any possible kinds of the dangerous vulnerability with respect to the any concerned democratic nation tend to suddenly increase manifolds. And therefore in any applicable context; since rather the well-known proverbial saying that, **'prevention is the better form of cure'** can always be applied with substantial validity. And so in the precise contexts of our presently on-going discussions and, **with respect to the multiply larger dimensions concerning the matters-military; the necessary visionary-prudence, toward ensuring the some kinds of those 'prevention being the better form of cure' keeps getting automatically cultivated; by the multiple-aspects of 'essential strategic-culture and thoughts', if they have been purposefully vibrant and well-stationed in the thought-processes of the body-politic as well as the mechanics of governance of the any concerned democratic Nation-State.** And here again in these larger perspectives, there are many and precisely in fact, mandatory messages-cum-lessons for the modern, sovereign and fast developing democratic Indian Nation-State as well. We would be discussing these important aspects since being India-centric in greater details, in the end of the next eighth chapter.

In the above multiple-contexts we have since little earlier talked about in certain much larger dimensions, concerning the highly complex matters-military and hence at this precise point we may prefer to briefly look into specifically those very larger dimensions as well. In these precise respects, we may recall that **General George Washington, in fact the legendary founder of the sovereign Nation-State of the United States of America is learnt to have said that, 'If you want peace, then always be prepared for war'**. Specifically pertaining to the Armed Forces generally and quite understandably as well, it is said that, **'the more you sweat in peace, less you bleed in war'**. Now so far as always the quite pertinent perceptions concerning the question of, sweating of the people in military-uniform in course of their almost continuous training programme and other kinds of the soldiering-related activities are concerned, these are only few of the dimensions in this regard, and therefore equally important are the various kinds of the well-crafted training-programme for officers since stationed for service in the Armed Forces at various levels, **so that the concerned nation's military-forces do remain professionally always, a well-oiled and well-motivated fighting-machine**. These are again not in fact, any complete thought-processes in this regard, as to here—there are a number of larger very important and at the same time, no less relevantly paramountcy dimensions since specifically pertaining to the any concerned nation's very vital—the multiple-aspects of matters-military. **And these precisely remain concerned with any particular liberally democratic nation's political-class, their highly important and in fact substantially meaningful awareness towards again the same multiple-aspects of 'essential strategic-culture and thoughts' and in addition to that concerned nation's systems of decisions-making as well as, the governance as well.** And largely here lies the 'rub' in this regard and much more so specifically in case of the modern, sovereign and liberally democratic Indian Nation-State. So far as India and the Indian peoples even today are concerned, it is indeed very much a queer situation. Specifically in the larger contexts concerning the multiple-aspects of matters-military at this precise point, we may need to refer to the above quoted paragraphs from the well-known autobiography of the British Field Marshal Montgomery; wherein he has very perceptively referred to some of the militarily far-reaching and the then prevailing shortcomings, in rather some military-command-oriented culture of the very senior-level officers of the British Army, especially during the war years of the bygone First World-War era. Well, at the same time it remains also equally true that typical of the British peoples' multiply traditional visionary-prudence ethos, concerning the geopolitical as well as, the geostrategic and politico-military oriented compulsive dimensions, they couldn't have taken any prolonged time in their own kinds of rather the mandatory course-corrections. And needless to say that subsequently with respect to the British-Nation and, her both the political as well as military leaderships, the same aspects turned out to be indeed substantially visible during the Second World-War years.

Now so far as the multiply challenging developmental processes of both military armaments and the various kinds of weapon-platforms are concerned and that too, during the present fast-moving modern-era, they are perforce multi-disciplinary, highly complex and very tiresome jobs, having many kinds of the technical-cum-military trials and tribulations. While keeping specifically these multiple-perspectives in our minds, if we embark on a critically explorative journey through the humankind's continuously expanding civilisational-canvas since its antiquity; we would be able to historically quite informative-cum-panoramic glimpses of these highly chequered phenomena down the ages. And hence from the points of view of the larger civilisational considerations; here comes always the multiply triggering-roles of, already the earlier noted and discussed—the man's **'perpetually active and analytically thinking-cum-creative mind'** since, exclusively gifted by the mother-Nature to only the humankind. Well, these aspects since the very advent of man's living-history itself, are not true in case of only the multiple-aspects of matters-military alone; but also equally valid for all the domains of human-endeavours as well as the consequent civilisational advancements, in their multi-dimensional manners and also in addition, their own different kinds of rather the mandatory trials and tribulations. Hence these are also some of the numerous ways that, the human-civilisations have multiply been advancing relentlessly over the infinite 'canvas of time' since the very beginning itself. And while keeping these historically dynamic aspects in our minds; we do also very relevantly find that the man's living-history since continuously in the making, do contain substantially methodical living-records of the man's relentless multi-dimensional hard-work and the various kinds of developmental-efforts, towards his civilisations-related advancements. Now in these multiple-contexts precisely, while coming back to the historically highly turmoil-ridden times during the First World-War, we do notice the tremendously consequential developments of the **aeroplane** and the **tank**, as entirely the new kinds of weapon-platforms. Well, fruition of the multiply developmental processes of the aeroplane, were much more complex, challenging and time-consuming as well. And hence its action-oriented fruition could finally be seen during the later Second World-War era only. **But interestingly may be that, now the some**

very primitive-looking kinds of tanks did make their appearance in the battle fields of Western Front during the First World-war but during those extremely human-blood soaked fronts of the unending looking trench-lines, it was indeed a very far-reaching breakthrough to solve the then understandably suffocating stalemate, with an effective system to counter the deadly threats of the machine-guns, since deployed and tactically emplaced on both sides.

In the above specific contexts, concerning the historic and militarily paths-breaking developmental processes of the tanks, the highly informative thoughts of the legendary **American General George Patton**; have been very interesting **recorded in the Patton's biography, *Patton: Ordeal and Triumph.*** Incidentally, Patton too as then a young officer had actively participated in the First World-War. The author Farago Ladislas has written that, "By the time the United States was ready to it (*i.e. the First World War*), the war was no longer what it had set out to be. The precision clockwork of mass armies had run down and become mired in a solid line of trenches and barbed wire. The bankruptcy of a generation of military thought and planning became apparent. The British faced up to the crucial problem of this bloody stalemate. **They developed a seemingly utopian weapon that, they hoped, would be an antidote, as Captain B.H. Liddell Hart called it, to the defensive combination of machine guns and barbed wire... A colonel named Earnest D. Swinton hit the idea of a 'self-propelled vehicle, capable of crossing trenches and so armoured as to be impervious to machine-gun bullets'. Winston Churchill, then as the head of the Admiralty, became interested in it. He formed what he called the 'Landships Committee' to undertake experiments. Then W.A. Tritton and Lieutenant Walter G. Wilson were commissioned to design the first machine. To conceal this development from the Germans, a misleading pseudonym was sought for the new weapon. It came to be called 'tank' by its code name (*and interestingly, that code name itself became its universally accepted name subsequently.*)**... When first used, tanks represented a major surprise in the war. However, whatever was gained by the element of surprise was dissipated by the fact that, from the outset, the Allied High Command made very poor tactical use of the tank. This phase of its spasmodic evolution never ceased to appeal Patton. He later described it succinctly in an article in the *Cavalry Journal*. He recalled how the tank was simultaneously developed by both the British and the French, but along entirely different mechanical and tactical lines... 'The French' he wrote, 'following the lead of Ulysses, thought of their *chars d'assaut* as armoured carriers destined to transport groups of infantry, unscathed, across No Man's Land, through the wire and over the trenches and then disgorge them in the enemy's rear. The British, on the other hand, followed the Macedonian idea and constructed not carriers but mechanical fighters whose duty it was to shoot down resistance, smash wire and bridge trenches so as to render the infantry assault less possible.'... He went on, recalling in a sardonic mood the initial setback of the new weapon. 'Unfortunately for the French plan, that mutual esteem and confidence existing between allies prevented either nation from informing the other of its invention so that, when the French had some hundreds of machines almost ready for a surprise attack, the British spilled the beans by jumping off on the Somme on 15th September, 1916, with a handful of tanks.'... As a matter of fact, British General Headquarters attempted to change the outcome of the Battle of the Somme with only thirty-two tanks, all that were available for the operation. The Battle of Flanders in the autumn of 1917 offered no chance for success as far as the tank was concerned, chiefly because of the unfavourable terrain and the sparse use of the new weapon... By then it seemed that the tank would justify the scepticism of its opponents. However, it redeemed itself in the Battle of Cambrai on 20th November, 1917. Without preliminary bombardment, the British launched the battle with a surprise attack, employing 380 tanks. They penetrated the German lines without much trouble and were on the point of breaking through into the open country, but the exhausted troops were unable to take advantage of the situation. Although the Battle of Cambrai ended inconclusively, it did demonstrate that the tank could become an extremely useful weapon.... In the autumn of 1917, when the Army (*i.e. the U.S. Army*) decided at long last 'to adopt the new weapon'; the British and French already had thousands of them. But the United States had only two tanks it could call its own (*interesting, isn't it*?). They were custom built by the Ordinance Department in Washington from French blueprints. But the Ordinance had made so many modifications to the original design that the prototypes which resulted from this tinkering proved useless. They were lumbering and gawkish in appearance, and naïve in construction. One of them was propelled by a steam engine, of all things. (*Look at this, how strange are the ways the people behave while sitting in the corridors of power the world over.*).... Pershing (*i.e. the U.S. Army Chief of Staff then*) had asked the French and British to supply the tanks he needed, but they told him they had none to spare. 'As there was little or no prospect of obtaining tanks abroad,' Pershing noted, 'their manufacture at home was recommended and full details of requirements were accordingly cabled to Washington.' The War Department ordered thousands of tanks in the United States, but it took time to get the programme rolling. (On Armistice Day there were 23,405 tanks on order. But only twenty-six had actually been completed when the war ended.... *It is interesting indeed.*)". (Farago, pp. 49-52) (*Well whatever, the above quoted historic episode in short, concerning the highly chequered developmental processes of tanks, during the no less extremely violent and turmoil-ridden times of the First World-War itself was also in a way, no less magnificent account of the human ingenuity as well as indeed, man's the then techno-mechanical genius; which in turn was very soon going to bring-in revolutionary changes in the entire gamut of the land-warfare. In fact, in the numerously multiple-areas of the relentless human-endeavour from the very ancient of times itself; the roads for the all sorts of paths-breaking inventions, innovations and developments and also their road-maps have always been full of the pot-holes and road-blocks.*).

With respect to the above and in the larger perspectives regarding inventions and innovations it would perhaps be a matter of some great interest to look into the times and life of **ALFRED NOBEL**, who with his rare genius during the nineteenth century, was so much revolutionarily instrumental in triggering one of the most important inventions in the world, i.e.—**dynamite**, which he soon patented in his name and so in fact, succeeded in building a vast and highly prosperous industrial empire in Sweden. **Dynamite soon became a much sought-after material in mining, for other industrial purposes and also no less, in the multiple-fields of the matters-military even much before the early twentieth century First World-War**. But it was indeed, no easy affair for Alfred Nobel in achieving all this and in these respects, **Thomas Hellberg and Magnus Jansson**, in their well-researched and highly illustrative 'Coffee-Table Book', on 'Alfred Nobel', have indeed been able to give a pictorially panoramic glimpses of his extremely challenging and the risk-prone processes of trials and tribulations—towards fructifying his unprecedentedly revolutionary inventions and along with, his own life and times, in a very informative and also rather in some touchy manners.

The authors mentions that, "When Alfred rejoined his family in St. Petersburg (*Russia*) in 1852, Russia was about to enter the Crimean War. Immanuel's (*i.e. his father*) company, Fonderies and Ateliers Mecaniques Nobel and Fils, now employed more than a thousand people. The Russian army needed to be modernized and greatly expanded. With big government contracts the Nobel Company thrived as never before. The good years in Russia made it possible for Immanuel to repay his old debts in Sweden, and indeed, he was even able to help many of his friends on the other side of the Baltic Sea. During this time Alfred had no time to develop applications for nitroglycerine. He worked successively in all the departments of the family company and quickly learned how a larger company should be run... Immanuel borrowed huge sums of money to enable the company to expand to keep up with the flow of orders. Among other things, the Nobel Firm built Russia's first propeller-driven warship.... Most of the family fortune was immediately ploughed back into their enormously successful business... The Crimean War (1853-56) was fought in and around the Black Sea and therefore did not have much effect on life in St. Petersburg. Perhaps this was why Immanuel failed to see how the course of events was changing. When the war ended in 1856 the Nobels were suddenly confronted with disaster. Government contracts were given to other foreign firms—awarded mostly through bribes—and the otherwise well-run Nobel factory was abruptly faced with a liquidity problem.... In desperation, Immanuel once again sent young Alfred abroad, this time to try to arrange credits. Alfred travelled to London and Paris but was unsuccessful and soon returned home... In 1858 Fonderies and Ateliers Mecaniques Nobel and Fils went into bankruptcy... Immanuel returned to Sweden with Andrietta and his youngest son. Emil, Robert, Ludvig and Alfred remained in St. Petersburg to try to preserve the goodwill the Nobel had earned over the years. Their new company first operated under the name Nobel and Sons but was later changed to the Brothers Nobel.... Immanuel struggled with the same problem for a long time and in fact managed to detonate a solution consisting of 10% nitroglycerine and 90% gunpowder. He was enraged. He accused Alfred of virtually stealing his discovery... Alfred was deeply hurt by his father's letter. He felt his pride as an inventor and self-esteem had both been challenged... He sent a reply to Stockholm which adopted a rather superior tone and began with; 'My Dar Father', Alfred calmly refuted his father's accusations and described his own experiment in detail.... When he had read Alfred's letter, Immanuel immediately gave in. He realized his son was right, but his father's attack affected Alfred deeply. It was a long time before the relationship between father and son was restored to normal... In the same year Alfred decided to leave Russia. Despite his grudge against his father, he crossed the Baltic and went straight to his parental home in Stockholm.... On the 3rd of September 1864 Stockholm was shaken by a violent explosion. 140 kilograms of nitroglycerine exploded as the result of an unsuccessful experiment led by the young, chemist and inventor Alfred Nobel. The event left an indelible mark in the minds of Stockholmers, so that for decades afterwards time was reckoned 'before or after the Nobel explosion'... Despite the fact that people were killed and injured by the explosion, within only six weeks after the accident Alfred Nobel was able to gain sufficient capital to build the world's first nitroglycerine factory. The factory was to become initial building block in a vast industrial empire, and helped form the basis of one of the most important inventions in the world—dynamite... Alfred Nobel travelled endlessly throughout the world, alone and isolated, working tirelessly on new projects, many of which represented technological breakthroughs, while others were ahead of their time." (Hellberg and Jansson, pp. 28, 30 & back-cover page)

The above quoted paragraphs indeed, give us **glimpses of Alfred's scientific-cum-entrepreneurship-oriented genius and also, extent of the dimensions of his creative energy.** Needless to say that Alfred Nobel, in addition to being the founder of the **'Nobel-Prizes Foundation'** as since enunciated in his 'Will', he also happens to be during the evening of his life, the founder of the world-famous **'Bofors Arms Industry'**. In any case, if we may look into the detailed life-sketch of almost all the inventors and scientific explorers, since working tirelessly to unveil some of the contextually important elements into perhaps the endless mysteries of the mother-Nature, for larger good of the mankind, we would find almost the all of them full of such trying times and even full of a series of trials and tribulations that, all these in a first glance itself, would rather be looking more than enough as to 'break' any ordinary mortal. Hence, even for those rarely gifted men and women their extremely chequered journey to a sort of the universal 'Hall of Fame and glory', have perhaps seldom been any easy matter. It is also therefore, not possible to prepare any yardstick to perhaps measure their unwavering dedication to their respective causes, their will power and mental as well as physical tenacity in this regard. They are born for their focused aims and causes, live for them and, keep-on working for their fulfillments till in fact they breathe their last.

Now coming back to few of the historic facets of the First World-War, besides some legendary commanders and strategic-thinkers on both sides, one unique and very differently gifted British personality and also his certainly not so common geostrategic and politico-military exploits presently coming to mind, happens to be **Col, T.E. Lawrence**, since famously known as the **'Lawrence of Arabia'** as well. In fact from the German side as its ally the Turks, happened to be somewhat even amusingly rather hyperactive both in the southern Europe and then also, across the Mediterranean in the North Africa too. Turkey's such somewhat menacing adventures in any case, could not be ignored by the British for the sake of the protection of definitely so much strategic Suez-Canal, their oil-wells in the Middle-East and in addition also, for the security of their vast empires in Africa, Asia as well as Australia, and through which the then Britannia 'ruled the waves'. The sea-lanes passing through the Suez-Canal, being quite valuable for the international shipping and hence all the more so for the British and since through this only, they were speedily able to reach their 'Jewel in the Crown', i.e. INDIA too. Though during 1916, the British had already stationed a sizeable army of nearly 250,000 troops in Egypt, to meet these important strategic needs and also meet any kinds of the military emergencies, if ever caused by Turkey or any other possible hostile forces in the Middle-East. But during the highly challenging years of 1916-17, the British were already facing the Germans in Europe but in spite of this, they couldn't afford to take lightly the Turkish menace in North Africa. **They summoned then Captain T.E. Lawrence for his certain uncommon kinds of the highly consequential assignments in Egypt, since he was fairly known as an extra-ordinary expert on Arab affairs.** And the saga of what all Lawrence was able to do over there and also in a very short span of time, in the higher strategic interests of his British Nation's plans and ambitions in the Middle-East are now stored in the highly informative books, since already written by the various learned military-historians the world over for our need-based studies and research. The British then in North Africa, changed the top military field-commander and, sent **General Allenby** to assume the historic active command over there. Allenby was a professional with some remarkable theatre-level tactical innovations and also, an expert on the military-maneuver consisting of the very large body of troops. And he was able to put his military-acumen to the best use since, opportunities were available; and along with these, the brilliant exploits affected by Lawrence created historically a synergy in this regard, inevitably causing in the end—i.e. the forcing of Turks to surrender before the British and hence, Turkey was put out of action from this war.

Before Lawrence arrived in the Middle-East, the British in combination with the Sheriff of Mecca triggered an Arab revolt in Hejaz against the Turks in order to deviate, the Turks' attention away from the British Forces. During those times the Turks were controlling and operating the Hejaz Railway too and hence, the Turks retaliated with superior military force to crush the revolt and succeeded up to a considerable extent but the mindless atrocities since committed by the Turks against the Arabs, left deep marks on their psyche and making them fanatically opposed to the Turks. Even the religion was unable to bridge the enormity of enmity and the gap between them. When Lawrence arrived on the scene, he could see through the highly consequential emerging situation and hence with his unique expertise, was able to smell the

historic opportunity for ready exploitation according to his highly innovative plans. Lawrence being an expert on the guerilla-warfare and hence, was soon able to organise the Arab rebels into an independent and sizeable irregular-force and trained them in hit and run tactics against the Turks. This resulted in great harassment of the Turkish forces and also, some serious dislocation of their Hejaz railway. **In addition, Lawrence with his uncommon skill and with his active association with Faisal, son of Hussein, indulged in a kind of successful propaganda, resulting in the revolt expanding far and wide in both Arabia and Egypt and, even into the more and more Arab population.** Lawrence and Faisal then, with a great encircling movement, reached even Aqaba and easily conquered it; since it didn't happen to be strongly fortified. They further began hitting and raiding the Hejaz railway, seriously affecting the Turkish war-efforts and also, threatening the Turkish rear. Meanwhile encouraged by Lawrence's exploits, British forces under the command of General Allenby, planned to advance all the way across Gaza and into the Palestine, with great deception plans and, with extensive bombardments of the Turkish lines both from the land as well as, from the sea throwing the Turkish forces off-balance and also, seriously damaging their 'will to fight'. And with the capture of Jerusalem by the British, the Turks finally applied for the armistice and their surrender, and hence, leading to the making of a famous film the '**Lawrence of Arabia**'.

In retrospect, we find and as analytically recorded by some learned military-historians that historically the conceptual-framework as well as the multiple-aspects of operational-planning by either side, concerning the many historic campaigns since fought over on both the Western and Eastern fronts in Europe during the First World-War, now look quite bizarre and hence no wonder that, these in fact turned out to be rather ironically quite indecisive and at the same time also horribly and definitely avoidable loss of valuable human-lives and in addition, the tremendous wastage of the very valuable military and economic resources. In the above historic respects, **the campaigns of Verdun and Somme** since both fought simultaneously during the year 1916 in France, now turn out to be rather the classical cases of examples in this regard. In fact these in many ways, turned out to be the extremely costly and tragic consequences of the historically suffocating and perhaps very annoying aspects of the military-stalemate. It is a different matter that, when finally USA did enter the war with her massive industrial as well as military backup turning this war truly global that, the overall situations soon began turning towards some sort of the quick end to this certainly paradoxical war, with massive human tragedies.

Concerning the **historic campaigns of Verdun and at Somme in France** respectively, the General Editor **Christer Jorgensen** very reflectively as well as with great precision, mentioned that, "Verdun epitomises the attritional struggle that became characteristic of the World War I on the Western Front. Intended to destroy the French reserves, the battle drew the massive numbers of German troops as well.... Victory in the Great War would therefore be a matter of exhausting the other side, of making the cost continuing so high that peace became essential, at whatever price. The key here, again, was reserves. So long as the opposition had sufficient manpower available to feed into the combat zone, the war could go on. The German high command therefore came up with a plan to destroy the French reserves by drawing them into a 'meatgrinder'. The German plan was to attack something that the French had to defend, and to destroy their army with artillery and infantry attacks. The chosen objective was the fortified city of Verdun.... (*Now concerning again the historic battle at Somme in France*); it is mentioned that, 'The battle of the Somme is today synonymous with military folly and pointless bloodshed, but in truth there were good reasons to attempt the massive assault. The military balance at the time favoured the defender, but there seemed to be no other way to break the deadlock.... At the outset of the First World War, military thinkers expected a war of manoeuvre in which cavalry would play its traditional role as a striking arm. At first, something like this actually did happen, and in some areas of the Eastern Front the war retained a nineteenth century character, complete with the odd sabre melee between opposing cavalry brigade.... In the West however, it quickly became obvious that the defender had a huge advantage over the attacker. Things had been heading this way for some time. During the American Civil War and the Franco-Prussian War, the long-range accurate firepower of infantry rifles made assault by infantry or cavalry a hazardous business, machine guns tipped the balance further, though they were heavy and unable to redeploy quickly. The war on the Western Front took on many of the characteristics of a siege, with heavy entrenched forces on both sides fighting from behind barbed-wire obstacles... Between the big offensive, the war became one of raid and counter-raid, usually at night, with artillery pondering away at the opposing trenches. Infantry holding the forward positions were subjected to horrible conditions. Huddling in their muddy dugouts and enduring bombardment whilst being unable to reply. Being under fire and unable to fight back is one of the most morally draining experiences men can suffer and morale was, understandably a problem." (Jorgensen, 2011, pp. 191, 192, 199 &200)

The above described scenario indeed, paints a somewhat not so common kind of the any wartime picture. If all these historically even now remain so depressing a reading; one may very well be able to imagine, as to what all the then troops at the front on either side might have had to undergo in that war. **War has always been a very messy, extremely violent and multiply-tragic affair.** In the multiple-contexts of the matters-military if we may care to read the above quoted paragraphs' 'in-between-the-lines' with their specifically relevant historicity at this precise point, with somewhat analytical criticality then we may perhaps notice that, certain very pertinent lessons, most likely derived from the conduct of the already even earlier happened since eighteenth century American Civil War and also the Franco-Prussian War, had most likely not gelled fruitfully in the minds of the military-thinkers, who eventually mattered during the early twentieth century First World-War. These perceptions at the most may be matters of some conjecture only perhaps but many a times even some conjectures too, do in fact make some substantial contributions towards facilitating certain kinds of even differently contextual discussions on any particular subject-matter since taken in hand, concerning any specific study-cum-exploratory project. Whatever may be, but it also remains equally true that the advent of firearms had already taken place during the earlier eighteenth century itself and also, these had since been tried and extensively used with some deadly effect during the earlier wars too. Hence in spite of all these, remaining somewhat uncritically fixated with the bygone nineteenth century thought-processes in this regard, do create perhaps, some queer situations. **In fact, the living military-history and the valuable lessons therein, also indicate that it is professionally not enough to just win the war but also as far as possible, the same is needed to be achieved with minimum possible loss of human-lives as well as the minimal collateral damages to both the military and civil assets too.** Hence ages back during the very ancient-era itself and since already discussed in the earlier chapters, **Sun Tzu of China had even said very appropriately that the best victory in a war is that which is achieved by without actually fighting it.** This classical perception may look like perhaps, just some kinds of an idealistic situation but it is also equally true that even such so-called idealistic situation too, does at times materialise in the practical world and, with its own historic authenticity as well. The most recent and indeed, globally far-reaching as well as highly consequential example—as since happened during the recently passed twentieth century itself—is the historic winning of **once the**

terribly menacing Cold-War, by the USA and her other Western Allies over the then USSR and her communist allies in Europe. We may say that now the historic process of US victory in the above mentioned Cold-War without fighting the same too as hot-war, may be taken as another way of the making of man's living-history.

Any war howsoever it may be short or prolonged; as it begins some time due to certain kinds of the emerging situations in any case, also somehow has to end as well. In other words, if any otherwise on-going war does not end then quite paradoxically rather; how can a new war ever start? Again quite pathetically, this too is no less a queer situation. Now so far as the First World-War is concerned if its beginning was rather quite strange then, its end too happened to be no less queer and even substantially botched-up as well. Historically if the start-up of this first ever global war based on some very strange and flimsy grounds then at the same time, its end too happened to be no less bizarre. **The height was that and as generally perceived then, that this war was started and fought by the traditionally rival and also irreconcilably jealous—the then European Imperial-Powers in fact, to end all future wars. Sometimes historic facts are indeed, stranger than fiction**! Instead of preventing all future wars in fact quite strangely, in the botched-up end of the First World-War, the powerful seeds of the soon to be fought Second World-War, were perforce got sown into perhaps, the then inevitably consequent destiny of the modern-era's mankind itself. It turned out to be the next Second World-War was going to be much more tragic, violent and unimaginably many times more devastating as well.

Concerning the indeed, botched-up end of the First World-War, **Field Marshal Montgomery** has very precisely written that, "The powers found it no easier to make a satisfactory peace than it had been to win the war. As a last manoeuvre against France and Britain the Germans sent their request for an armistice to President Wilson of the United States. But France and Britain would not be done down. Peace was *dictated* from Versailles on 28th June. Alsace and Lorraine were returned to France; the British Empire and France acquired considerable territories under the guise of 'mandates'; a new Poland was created. The former Austro-Hungarian and Ottoman empires were abolished by the recognition of new national states. Germany lost comparatively little territory in Europe, but was disarmed and ordered to pay reparations. Russia after the Bolshevik Revolution was not accepted in the comity of civilised nations, and the territorial settlement of Brest-Litovsk was left in fact to stand. A League of Nations was set up, but Germany was not allowed to join for some years, and the United States chose not to. The lasting and dangerous effects of the 1914/18 war were a humiliated and resentful Germany and an outlawed and mistrustful Russia. The economic stability of the pre-1914 world was never quite restored. The conditions of the peace of Versailles were to prove profoundly unsatisfactory.... When Foch heard of the signing of the **Peace Treaty of Versailles** he observed: 'This is not peace; it is an armistice for 20 years.' He was right: a peace which had been dictated to humiliate and avenge could not last. Hitler made it clear that peace was merely a period of preparation for total war, and the Nazi state became a war machine tuned for action. **The principles held by the new German leaders—total war and *blitzkrieg* tactics—were in utter contrast to the defensive attitudes of the western democracies, where it was believed that war would be avoided**. I had many discussions with Sir Winston Churchill when it was all over; he was of the opinion that the tragedy could have been prevented: 'the malice of the wicked was reinforced by the weakness of the virtuous'. 'Thrice armed is he who hath his quarrel just'; but Hitler knew well that he is four times armed who gets his blow in first. 'The blitzkrieg burst on Poland in September 1939, and then on the west in May 1940'... The war which engulfed the world was totally different from the conflict of 1914/18, trench systems, barbed wire, siege warfare, all disappeared; I who fought in both can bear witness to this. During the war, crimes were committed by the Germans and Japanese which have no parallel in scale and wickedness in history. Moreover, the mighty weapon of air power brought destruction and misery to non-combatants on the home fronts by the hideous process of bombarding upon cities and industrial centres from the air—begun by the Germans, but well and truly repaid by the Allies as the war progressed. The 1939/45 war was, I suppose, the greatest tragedy in the history of mankind (*yes, indeed.*). It has been estimated that the total number of people killed from all causes were nearly 40 million—of whom at least 17 to 18 million were civilians. The amount of human suffering was beyond all belief." (Montgomery, 1983, pp. 307 & 309)

We may recall that in different contexts and also, at different points of our on-going discussions, we have already noted that the man's living-history since perpetually in the making regardless, spares, forgives or even forgets no one, under any circumstances. In fact this universally timeless truth irrespective, of the any historic-age in the history's timeline remains overwhelmingly true in the man's living-history, both in the international arena as well as, in any particular nation's intra-national spheres too. If the **historic 'Treaty of Versailles'** of the twentieth century itself for example, since full of flaws could not spare the international communities sooner or later, of its multi-dimensionally extreme consequences; then in the same manner, any serious socio-economic-cum-political, geopolitical or even geostrategic and politico-military kinds of blunder; if ever committed willingly or rather naively, by the top-level political leadership within any particular Nation-State too, can never spare the peoples of that particular concerned nation, from its inevitable setback and, the any possible sufferings, without any fear or favour in this regard. Many noted military as well as civil personalities the world over have very reflectively as well as differently, commented upon the briefly above discussed **'Treat of Versailles'**.

Jawaharlal Nehru too (1934, p. 794) has written that, "In the course of the treaty discussions at Versailles, it was posed on behalf of the Japanese Government that a clause recognizing racial equality be introduced into the treaty. This was not accepted. Japan was, however, consoled by the gift of Kiauchau in China. The big three were generous at the expense of a weak and humble ally like China. Because of this China did not sign the treaty... Such was the 'Treaty of Versailles', which put an end to 'the war to end war'. Philip Snowden, who later became Viscount Snowden, and a Cabinet Minister in England, made the following comment on the Treaty:

> 'The Treaty should satisfy brigands, imperialists, and militarists. It is death-blow to the hopes of those who expected the end of the war to bring peace. It is not a peace treaty, but a declaration of another war. It is the betrayal of democracy and of the fallen in the war. The Treaty exposes the true aims of the allies,'
>
> Indeed, the Allies, in their hatred and pride and greed, overreached themselves. They began to repent in after years when the consequences of their own folly threatened to overwhelm them. But it was too late then."

Yes indeed, it was too late by then. In fact by the time perhaps, the 'so called' victors of the First World War were able to gather some understanding of their rather colossal follies and also, some wartime dusts were able to settle-down on their mental-horizon; ironically on the larger-horizon of the then prevailing international scenario, the extremely menacing and dark clouds of then soon to occur the Second World-War had begun gathering and also being restless to smash even some semblance of the existing truce during those times, with their impending cloud-burst. **Man's living-history also has some strange habits to repeat itself, if the mankind somehow**

refuses to meaningfully learn perhaps, the self-manifesting important lessons of the history itself. But it all mostly remains very complex. Historically all the more concerning wars and the warfare in fact, the matters are seldom linear and simplistic for the peoples, on various sides of the 'play-fields' in this regard. In retrospect, it may even be argued that if more care and some sort of vision-oriented caution, would have visibly been exercised by the concerned architects of the 'Treaty of Versailles' the seeds of the forthcoming Second World-War, would have been perhaps prevented from sprouting or at the most certainly, delayed substantially so much that, some other kinds of the additional preventing initiatives too, could possibly be taken. Well historically, these are matters of some opinions only, and especially in these matters, opinions too differ in many ways. There are some schools of thoughts, which with their historic reasons; may not find that 'Treaty of Versailles', being not so much sharp-edged and they too have their own well-argued perceptions as well. Professor S.R. Sharma has mentioned that, "The work of Versailles was mixture of realism and idealism. The former was represented by **Clemenceau** and **Lloyd George** (*i.e. of France and Britain respectively*), and later by **President Wilson of America**. The redrawing of the map of the world and reparations were due to the former, and the **constitution of a League of Nations** was the achievement of the later." (Sharma, 1948, p. 417). But ironically in both the above aspects in this regard; greed, arrogance and even perhaps, rather utter lack of some geopolitical accommodativeness on the part of concerned victors were found wanting. Moreover naturally, both sides were seen with some glaring fault-lines and hence; no such historic 'designs' did stand the test of time and in addition, even the USA herself had shown no inclination to sign the charter of the League of Nations.

If we may care to somewhat delve deeper critically, into the making of the even rather some regressively fixated psyche regarding the socio-cultural as well as, the geostrategic and politico-military aspects concerning Germany in Europe, Japan in the Far-East and also their some Allies, we would be noticing that indeed, the overwhelmingly powerful and dictatorial 'blind-spots' of the historically lingering **militarism**, were always preventing them to see the reasons in the multi-dimensionally complex geopolitical domain of since mostly unpredictable and always changing—the games of international power-politics. The contextually finer points in this regard, are that **the strategically prudent and even rather broadly desirable aspects of remaining militarily stronger as being the instruments of deterrence and also, to prevent war is one thing but then, getting regressively fixated with some kinds of the illogical militarism in fact, remains very dangerous as well as, totally different**. In fact historically, the trouble with such kinds of the lingering militarism remains that it cannot be neutralised by the diplomatic dialogues and even with the concerned apex-level international discussions. Moreover, these have powerful tendencies as per some kinds of the emerging situations to get manifested fast, even with irrational violence and destructions wherever and also in whatever manner, these tendencies may sense any opportunities for their senseless manifestations. In fact they have to be perhaps, destroyed with some decisive completeness if necessary, by means of the perforce rather total war whether globally or within some need-based limited geographical areas. That has been some kind of the historical paradox too, and down the ages with their different textures as well as historicity; the man's living-history has numerous examples in this regard, to substantiate these perceptions. Hence contextually, while taking the maximum possible objective views, and also as per the above discussions historically again perhaps, it remains a matter of conjecture at the most that in any case, the Second World-War against the then Axis-Powers could have at all been prevented. Contextually, the differing opinions may be debated by different scholars and authors the world over, but at times the man's history also indicates that certain powerfully lingering fault-lines too, do remain quite strong to ever be logically wished away. Historically, the manifesting dimensions of the First World-War (1914-18) and subsequently then, those of the Second World-War (1939-45), and the making of their histories too, contain differing threads and hence, are substantially different as well.

The 'causes' for these historically momentous developments and events, and also manifesting in different dimensions have in fact been predating the above mentioned two historic World-Wars, and then in continuity and as too we would be seeing, some of them have been occurring even after the two wars on the living-history's continuing timelines and in addition then, seen to be spilling-over in the present twenty-first century as well. But as since discussed earlier too in different contexts, and also as per the multi-dimensional systems of the continuous making of man's living-history itself, they all do remain always interrelated and mutually-influencing as well. In order to substantiate our above mentioned inferences and that too contextually, we may quote, "**When war broke out in 1914,' wrote Basil Matthews in the *Review of Reviews*, May 1920, 'five empires of the despotic military type remained on the earth's surface. They were the German, the Austrian, the Turkish, the Russian, and the Japanese**. Today four out the five are smashed in irretrievable ruin. Japan alone remains. The old European order has gone—the new Asiatic Power, rich now beyond the dream of avarice, with its man-power unimpaired and its ambitions vaster than those of Alexander, leaps upon the stage fully equipped. On the face of it, then, the first and dominant facts of the world situation are in favour of the Orient.' But since the Orient today....has been the creation of the occident, we have to trace here the entire trend of World History in both the hemispheres... Of five empires referred to above—Germany, Austria, Russia, Turkey, and Japan—the most formidable were Germany in Europe, and Japan in Asia. Though Austria was the oldest imperial power in Europe, her power had been successively curtailed since her loss of Silesia. Russia had steadily grown at her expense. Italy successfully revolted against her in 1861: Austria retired from Germany and formed the Austro-Hungarian Dual Monarchy in 1867. She never recovered from the blow of Sadowa. Ever since then she was always tied to the apron-strings of Prussia. Her *Dual Alliance* with that country, affected in 1879, was to culminate in her sacrifices sustained during 1914-18, and finally in the Nazi *coup* of 1938. The Russian Empire crashed in 1917 after having sustained a series of internal and external shocks. The 'sick man of Europe', despite the crutches supplied to him, from time to time by England and France, had been too frequently amputated to survive for long. He could live only in his republican *avatar* of the Ata Turk, Kemal Pasha, in the post-war world." (Sharma, 1948, pp. 411 & 412)

As we may care to read deeper and in between the above quoted lines and also, in conjunction with the earlier discussed multiple-aspects of since the First World-War, two things may appear to be quite clear and that being, (a) in spite of the so many highly checkered historic developments and events, then taking place in fast successions during the latter half of the nineteenth century and also along with, during the first half of twentieth century too and then, (b) some kind of the so-called historic 'Treaty of Versailles' ironically, to rather end perhaps all future wars any substantially meaningful and enduring peace was still alluring the then Europe and thereby to the entire mankind in the world. The eighteenth century decades were in fact, the prolonged times of the all sorts of rather militarism-oriented—socio-economic-cum-political turmoil in the European continent. In these historic contexts then, many new kinds of regressively-fixated thought-processes; were being thrown-up in the European societies. We may recall that the multiply historic aspects of the age-old feudalism, vested-interests-oriented monarchies, militarism and coupled with, the then some powerful dimensions concerning the aspects of chivalry and so on, have been the traditional features of the extremely-medieval societies in Europe. These features then happened to be entrenched over there and at the same time also remained powerfully pronounced during the centuries prior to the dawn of democratic ideas and hence naturally, much before the advent of **'Industrial-Revolution'**. Seeds of the then initial kinds of the democracy and democratic ideals, and the social awareness, etc., were in fact, first sown-in, in England with the ideas of **'Magna-Carta'** and then it was just a matter of time, before the similar kinds of ideas gradually began spreading all over the European continent like some kinds of the sociological bush-fires.

If we may critically view some of the historic developments and events since listed in the above quoted paragraphs from the book of Professor S.R. Sharma; then we may notice them as part of the above described larger socio-economic-cum-political phenomena of those times, then taking place in Europe. In addition, those above quoted historic developments and events did also contain quite understandably, the obvious kinds of certain geostrategic and politico-military dimensions as well. In the wombs of the substantially complex—the above mentioned socio-economic-cum-political phenomena of many revolutionary and proactive ideas in the then European continent in were hidden in the sprouting seeds of one after another, the two historic World-Wars. So as discussed earlier as well, all these always happened to be interrelated as well as mutually-influencing too, since manifesting as per perhaps their predestined timelines, onto the man's continuously expanding civilisational-canvas in the world.

In retrospect now and also, contextually within a broader framework of their historicity as we critically tend to have **a panoramic-view over mainly the nineteenth century** Europe, we may find them to be, on the one hand, if very far-reaching as well as momentously history-making too, then on the other rather, no less frightening for the whole world as well. In other words, we may also say that in these precise contexts of the nineteenth century Europe, in fact the man's living-history since continuously in the making, had also been quite menacingly acquiring a kind of the 'sharp-cum-double-edged sword' for the entire mankind. What indeed, an extremely paradoxical situation then it was and at the same time, the whole humanity found itself rather overwhelmingly entrapped with apparently no escape-route available whatsoever. At times, man's living-history since continuously in the making over his 'expanding civilisational-canvas' in fact, manifests its power to be able to make the mankind almost totally helpless as well. But at the same time, after the certain momentous events and developments in the living-history does open some wide windows, for the man to be able to learn certain valuable lessons as well.

In the above specific contexts, **Jawaharlal Nehru, in his earlier referred book on 'Glimpses of World History'** has indeed opined historically almost similar perceptions, in his scholarly more precise and beautiful language (pages 580 and 581). Nehru has mentioned that, "We must now go back to Europe and have another look at the intricate and ever-changing picture of this continent during the nineteenth century.... (*And*) some of its leading characteristics—industrialism, and capitalism and imperialism and socialism, and nationalism and internationalism—also of democracy and science, and the tremendous revolution in methods of transport, and popular education and its product, the modern newspaper. All these things, and many more, made up the civilisation of Europe then—the *bourgeois* civilisation in which the new middle classes controlled the industrial machine under the capitalist system. This civilisation of *bourgeois* Europe went from success to success; it climbed height after height; and toward the end of the century it had impressed itself and the world with its might, when disaster came.... In Asia we have also seen in some detail this civilisation in action. Urged on by its growing industrialism, Europe stretched its arms to distant lands and tried to grab them and control them and generally to interfere with them to its own advantage. By Europe here I mean especially Western Europe, which had taken the lead in industrialism, and of all these western countries, England was for long the unquestioned leader, far ahead of the others, and profiting greatly by this lead."

So the **historically unprecedented 'Industrial-Revolution',** its momentous fallouts as well as the multiple-fruits then, onto the man's always expanding civilisational-canvas bestowed upon the traditionally rival-cum-ruthlessly competing Western European Imperial-Powers—such military wherewithal and earlier unknown kinds of the colonialism-oriented geopolitical ambitions and greedy ideas that no society or the nation situated in any remote corner of the world, could remain untouched and unaffected in some way or another for long. It was indeed, such an overwhelming sway of the fast changing times that almost all of them, either had to bend in a kind of submission or were broken—regardless of the consequences. **Historically besides being an era of the multi-dimensional progress concerning the human-civilisations on one hand then, and on the other, these were also the very disturbing times principally in the very vast as well as remarkably rich in the various kinds of natural resources—Asian and African continents; it was mostly one-sided affairs of rather callously getting exploited too. But interestingly in the larger perspectives, it is also said that there is 'no free-lunch' in this world. In a way it was paradoxical too that, if the traditionally rival and very greedy European Imperial-Powers were always looking onto their numerous colonies in Asia and Africa with their greedy eyes, situations for them back-home in Europe was far from smooth and comforting, since the existing many dimensions over there like—competitive imperialism, blind militarism, radically emerging nationalism and revolutionary ideas challenging the age-old feudalism and the entrenched vested-interests, traditional clashes of arms, the relatively new industrialism and so on—additionally one new dimension of rather the ruthless mercantilism with international economic-cum-industrial connotations too were getting added. In the process, the entire prevailing geostrategic and politico-military scenario was getting murkier, not only in Europe but as a consequence, for almost the whole world**. Europe was in a highly disturbing 'state of flux', with numerous volatile points. We may recall that in the different contexts, we have also discussed that the man's living-history since continuously in the making has its own perhaps inherent designs to unsparingly extract rather heavy prices from the man for his own omissions and commissions and hence, somewhat reflectively in other words and contextually we may say that this aspect along with its all historicity by itself, is a kind of its perpetually living-lesson of the history in this regard. Geopolitically, if the rulers in the home countries of the European Imperial-Powers were occupied in furthering their geostrategic and politico-military interests almost around the world then the merchants in Europe too, then experiencing the highly profitable rather 'commercial kinds of blood' in their teeth and accordingly found that, their trade was fast multiplying during the momentous nineteenth century, when the imperially-fixated colonialism was at its zenith in the colonies around the world.

Naturally as the things had to be, the international trade was moving principally through the sea-routes and therein, those rival European-Powers' Imperial-Navies had their own commanding roles to perform and facilitate trade and, so on in this regard. **During those highly chequered times, Britain's Royal-Navy was comparatively the strongest and thus, it enabled the British to decisively call the shots in the seas and almost monopolize the then international trade, commerce and banking-systems globally, through various sea-routes**. But all these aspects principally belonged to the politico-military-cum-commerce oriented one larger dimension alone and hence, this was not the end to the then multiply complex prevailing scenario in the European continent. Fruits of the already momentously on-going Industrial-Revolution, had generally begun to sprout fast in the Western European nations only and also principally led then, by Great Britain, whereas the nations in Eastern Europe including the vast landmass of Tsarist Russia, were still stuck in the older grooves of the medieval patterns of autocratic feudalism and the repetitive-rut of social lifestyles; quite unlike of the West-European countries, where fast as well as revolutionary changes were being experienced. But historically, as we may care to critically delve deeper into these momentous aspects over there, one aspect concerning the various kinds of entrenched vested-interests, happened to be largely common in the whole of Europe. **The Kings and Monarchs since belonging to the different European nationalities during those times had one selfish-streak in common among themselves, and i.e. to make relentless efforts for maintaining the status-quo and protect their thrones and the so-called 'divine rights' as erroneously perceived by them with rather their regressive mindsets, to rule their countries forever**.

What to talk of the Eastern Europe, if parliaments had already taken births in the leading West-European nations but there too these were in the powerful hands of rich and the feudal barons and hence in either case, the masses were left to toil, get exploited and keep-on suffering, and thereby everywhere the then emerging popular nationalism was forcibly suppressed. But any such suppression socio-politically, can only be temporary phenomena and therefore, as something like mass-based eruptions, had to forcibly come and consequently sooner or later thereby, fructify revolutions as well. Learned historians and thinkers have identified the year 1848 as a coincidence, the year of revolutions in Europe, and wherein again the **various kinds of the vested-interests as well as the imperialism-oriented Congress of Vienna earlier in 1815** too, contributed in no small measures and then, also triggering the 1830 risings at numerous locations in Europe forcing the European Kings and Monarchs to resort to the ruthless force to suppress them, in Austria, Belgium, France, Italy and so on. **So it all historically resulted in a kind of the ding-dong battles/struggles between the masses and their extremely callous rulers.** Interestingly, all these happened during mostly the latter half of the nineteenth century, i.e. years before the start of First World-War. But sooner or later, changes had to come in Europe; bringing-in perhaps all possible kinds of the far-reaching transformations in Europe, and subsequently in the multi-dimensional as well as multi-layered manners, and then also later in other parts of the world as well. In fact, momentous advent of the fruits of sciences and technologies and, coupled with the action-oriented ideas concerning democracy—as since quoted above from Nehru's book—brought-in a kind of the new era of since the popular education and modern newspapers, which together in-turn in as much areas as possible across the world, were indeed able to trigger the above mentioned transformations and also many kinds of the social-revolutions. An era of the multiply increased dissipation of information and also, the unprecedented rise in awareness-levels of the peoples concerned then, had fast begun to dawn on the humanity's enlarging civilisational-horizon. Today in the present twenty-first century it is no wonder since we are so familiar with the power and reach of the media in all sorts of manners, in the peoples' daily-lives across the world like relentless routines of 'information-bombardments', but during those highly turbulent times of the nineteenth century, all these though quite primitive then, were certainly revolutionary and indeed very far-reaching developments for the mankind.

The **1815 occurrence of the Congress of Vienna** had many dimensions and, the numerous sinisterly hidden motives too. In a way, this historic conference of the ruling potentates in Europe then was also shaping the ways since somewhat soon to occur, **the two World-Wars during twentieth century,** with previously unknown consequences for the entire mankind. In the above pages, we have since briefly discussed these historically momentous aspects and also subsequently with soon to manifest their far-reaching fallouts which in-turn, were indeed going to enormously alter the multiple-paths of making of man's history onto the multi-dimensional, and also perpetually expanding, 'civilisational-canvas' in the world.

About this Congress at Vienna, **Professor S.R. Sharma** has written that, "(*Concerning*) the immediate reactions of Austria and Prussia to the revolutionary outbreak in France.... The Challenge of Leopold II (brother and successor to Joseph II) to the revolutionaries was reinforced by alliances, at first with Prussia, then Russia, England, and all the rest of Europe. The ultimate result was the defeat of Napoleon and the humiliation of France in the Vienna settlement. **This settlement was as fateful in consequences as that of Utrecht a century earlier (1713) and Versailles a century later (1918). The high-priest of the Vienna Congress was the Austrian statesman Metternich. Few men have exercised such powerful influence over the destinies of a continent like this 'Napoleon of diplomacy'. The mere fact that Metternich presided over the deliberations of this most momentous gathering, where almost all the potentates of Europe were personally present, is sufficient indication of his importance**.... The first outward manifestation of the spirit of the Congress was the formation of the Quadruple Alliance between Austria, Prussia, Russia, and England. Its ostensible purpose was the defence of the Settlement; but in reality it sought to be the bulwark of reactionary 'Legitimism'. When England saw this sinister tendency, which was a negation of all her liberal principles, she withdrew from the 'Concert of Europe' and allowed her allies to cling to their Holy Alliance under the aegis of Czar Alexander I who had been privately characterised at the Congress of Vienna as 'half fool, half Bonaparte.'.... **It has been well observed by Professor Morse Stephens, that 'The doctrines of the French Revolution did more than the victories of Napoleon to destroy the political system of the eighteenth century**.' In the so called Holy Alliance eighteenth century dynasticism was on its last legs. **The subsequent history of Europe during the nineteenth century marks the triumph of Nationalism, Democracy, Liberalism, in country after country**.... When Paris hath a cold, it is said, the whole of Europe sneezes. But we might as well say that whenever there is to be a political earthquake in Europe it is first indicated by the French seismograph.... **There were national and democratic risings all over Europe in 1830 and 1848**." (Sharma, 1948, pp. 371 & 372)

In the larger contexts of the present twenty-first century and also along with the future-times as well, we do notice so many earlier indeed unthinkable transformations in the European continent and then in addition, even formation of the European-Union. But these are in fact very recent and precisely, some kinds of the gradual developments during the post-Second World-War years over there and not the least much more so, after the historic demise of the cold-war fixations between the Two Super-Powers as well as, disintegration of the USSR and hence, after the liquidation of Warsaw-Pact in Europe. Europe has now totally changed, with the withering-away of the multiple-aspects of age-old feudalism, imperialism and the uncontrolled militarism in the then European societies and their Nation-States, and has indeed come a long way, since the end of Second World-War in the year 1945. The historically such on-going phenomena is always very complex matters, consisting of a number of the momentous developments and events, bringing in numerous far-reaching changes in the subsequent multi-dimensional making the man's living-history in this world. But in fact they all, always remaining interrelated and mutually influencing as well. In these precise contexts, we may also say that historically **the highly momentous and very far-reaching advent of the 'Magna-Carta' in the year 1215 in England, was indeed the initial triggering point for all these subsequently gradual developments and events in Europe, in rather unpreventable manners.**

While keeping in mind the above broader perspectives, we do also notice that during the nineteenth century in Europe, the different countries had their own and different kinds of a sort of 'roadmaps' for bringing-in their socio-political revolutions and thereby, leading to their distinctly different socio-cultural changes as well as systemic transformations. In fact, in the different West-European nations, different degrees and patterns of the democratic thoughts and the conceptual-framework for governance, gradually began to fructify in their polity; but for the Eastern Europe, it was altogether a different scenario since missing the initial advent of 'Industrial-Revolution'. The nations of Eastern Europe during those momentous times then, happened to be largely stuck in the earlier medieval-era socio-cultural-cum-political environment. It was a queer situation since in these precise contexts the East-European nations were also missing the initial advent of democratic-thoughts as well and, howsoever otherwise these might have been quite turmoil-ridden to begin with, in the leading West-European countries. Hence understandably, these socio-political kinds of static situations in a way in the Eastern Europe had altogether different fallouts over there during the forthcoming twentieth century and, being quite dissimilar to those of the Western European nations. But even in the Western Europe then, **Germany had been altogether a different case in point** since the multiple-aspects of the age-old extremely feudalistic militarism and rather also perhaps rigidly fixated imperialism-oriented nationalism, happened to be much stronger and more pronounced. And hence consequently in these special kinds of prevailing mindsets in Germany since the year 1871 or so, we may be able to also discern the momentous seeds of the two World-Wars later ready to sprout, one after another during twentieth century, with totally unprecedented and extremely devastating consequences of destructions and deaths for the entire mankind.

In the above discussed momentously prevailing scenario in Europe, all along the nineteenth century certain noted personalities in different countries like Germany, Austria and so on indeed had rather exceptionally far-reaching and historic roles in the then European continent. In a way in the larger perspectives, **in any particular historic situation and timelines, personalities have always mattered and that too perhaps differently, in quite decisive manners**. Hence **in these specific contexts of since the multiply continuous processes of the on-going making of man's living-history, two aspects have always been having momentously decisive-roles to play.** In the above precise context in this regard, **one aspect has been the any given historic situation and, the second one remains the corresponding historic-personalities concerned**. History also quite informatively shows that at times, **any particularly outstanding personality and his momentously historic roles during his actively decisive-life, can turn out to be a double-edged sword with all sorts of far-reaching both positive and substantially rather sometimes, very negative fallouts, as his downfall as well as, causing extensive harms to his own and, the other affected nations—or in certain cases, even for the entire mankind**.

In the above discussed contexts and concerning the nineteenth century Germany in this regard, one historic personality who then happened to be instrumental in ensuring the unification of the German nation, has been **the legendary Bismarck**. A **British historian, Mrs. Louise Creighton** has written that, "**The rise of Count Bismarck (A.D. 1815-1898)**,—The Congress of Vienna (1815), which settled the affairs of Europe after the fall of Napoleon I, had bound together the different princes of Germany in a confederation. But the confederation did not mean very much, and the states were really quite independent of one another. **More and more Prussia became the leading power in Germany, and at last there arose a man in Prussia who was to succeed in uniting Germany in one great empire. This was Otto von Bismarck, who belonged to a noble Prussian family**. He was a big, strong man, who ate and drank a great deal, and could bear any amount of fatigue. He had a tremendous will, and when he wanted anything he knew how to get it. In his youth he was wild and loved to amuse himself, and enjoyed a free, open-air life. But he soon showed what a strong able man he was. He was sent as ambassador to Petersburg and then to Paris, so that he learnt much about the affairs of other nations, and their relations with one another, and began to see clearly what must be done to make Prussia a great power. From Paris he was called back to Berlin by his King, William I, who made him chief minister of Prussia. **His first care was to reform the army, for he had made up his mind to make war on Austria as soon as he could find an excuse, and to unite all Germany under the guidance of Prussia**....he said that the questions of the time would have to be decided by blood and iron, and he came afterwards to be called the man of blood and iron. The Prussian Parliament would not give him the money that he needed for the army, but he did not care about the rights of Parliament and managed to get his money without it. He paid no heed to the complaints of the members, who hated for the way in which he bullied them. (*Interesting, but historically so true.*)." (Creighton, 1917, pp. 190-192)

Contextually looking back into the earlier paragraph, where we have also talked about certain kinds of such attributes since perhaps, inherent into the man's living-history's some otherwise outstanding personalities with rather some quite uncommon leadership qualities and whose historic deeds/misdeeds were able to perforce bring-in great and extensively negative fallouts. Such personalities are even capable of creating such overwhelming situations during their rather exceptionally historic times that for the time being, they succeed in occupying seats of unquestioned dictatorship and thereby in a way, with their regressively-fixated mindsets and then in course of time do perforce drag theirs as well as, all other affected nations—in a kind of the history's some smashing kinds of 'whirlpool' as well. During the last twentieth century, and just after a couple of decades after in fact, the botched-up end of First World-War (1914-18) precisely, one such historic personality came on the scene in Germany itself, in the name of **Adolf Hitler** and, the rest is history. While touching the so-called end phases of the First World-War little earlier, we have since hinted upon the momentous seeds of the 1939–45 Second World-War and also there-in, the historically highly unsettling roles of Hitler too.

Professor S.R. Sharma has elucidated that, "Despite Bismarck's great triumphs over Austria and France. Germany was far from being a 'satisfied nation'. She had been the last in the race for colonies, and such places as she got 'in the sun' (her African colonies) were too scorching for her surplus population. Elsewhere she found herself anticipated by Anglo-Saxon cousins. The Industrial Revolution created for Prussia all the insistent demands—for raw materials and markets—that England had felt earlier: but her scope for extension was circumscribed. Hence her struggle for existence became desperate. Hence her philosophers like Nietzsche began to inculcate the doctrine of 'real politik'; and the patriotic aspirations of a united Germany turned from love of country to love of more country. Her new 'Kultur' tried to find expression in diplomacy and war.... France was not likely to reconcile herself in her loss of Alsace-Lorraine, the humiliation of Sedan, the German occupation of Paris, and the terms of the Treaty of Frankfort (1871). **Bismarck knew that France would continue to be Prussia's deadliest enemy**. So he began to weave a sinister web of diplomacy, every line of which was calculated to keep France isolated and weak.... **Bismarck had already in the year of Wilhelm's accession (1888) increased the German army by 800,000 'in shining armour'. The new Kaiser therefore set himself to the task of creating a great German navy; for without it his ambition of the Teutonic domination of the world (in commerce and colonisation) could not be achieved**. So Heligoland was purchased from England herself in 1890, to form a splendid naval base for Germany; the excellent Kiel Canal was constructed; and strong naval stations were also built at Borkum, Cuxhaven, and Wilhelmshaven. A series of Naval Bills were passed to carry out the Kaiser's naval programme. **In 1900, at the Paris exhibition, the Germans openly proclaimed to the world in gold letters '*Our future lies on the water*'....** The 'peaceful penetration' of the world by German missionaries and merchants in the mean while proceeded apace. For instance, while there were not more than 16,000 Germans in all their colonies at the accession of Wilhelm II, in Brazil alone there were not less than 350,000 Germans on the eve of the Great War. **At home science was put in commission to do its best—or worst.... All their knowledge, their thoroughness, their powers of organisation—for in this also they have been unsurpassed—were turned to the production of zeppelins, submarines, Krupp guns, mines, torpedoes, prison-gases, and other devices. Railways were constructed with broad sidings for troops and cannons; and a bargain was struck with Turkey for the extension of the German railway system to Baghdad—for penetration into the Orient**." (Sharma, 1948, pp. 412-414)

In the above quoted paragraphs, as we may begin to rather discerningly read in between the lines, we may clearly notice the rise of extremely fixated, almost totally unchecked multi-dimensional rise of the militarism-soaked ruthless German global-imperialism, during the nineteenth century onward itself with a kind of the almost totally changed emphasis and even having potentials for the international repercussions then, so far not yet known or experienced by the world at large. These far-reaching historic aspects had in their wombs, soon to overwhelmingly manifest but rather yet hidden then, the great as well as unprecedentedly violent turning-points for the further making of man's multi-dimensional living-history, along with the advent of twentieth century onward in the modern-era world.

With the end of Second World-War, world got changed so much that even history itself too in fact, very emphatically forced the mankind to rather sincerely see the reason and also meaningfully understand that—any further war of such global dimensions and devastating reach, won't be acceptable at all. In these precise perspectives, we may also say that consequent to the multiply relentless advancements of the sciences and technologies and, the Industrial-Revolution as well; **now the newly added nuclear-dimensions into the man's already various kinds of massively enlarging arsenal, did decisively bring-in, the effectively reassuring aspects of the deterrence ironically rather, towards even thinking of any kinds of the global-scale total war** Hence from then on, the man's 'expanding civilisational-canvas' too, had got changed in its textures and momentous dimensions. In fact proactively below the surface and yet somewhat visible as well; if one may care to discerningly peep into these presently on-going substantially momentous aspects that, now since a kind of the quiet end of the twentieth century's last decades', cold-war between the two erstwhile bitterly rival superpowers; at the entire multi-dimensional and multi-layered national-cum-international levels—geostrategic and politico-military thought processes and also the strategic-doctrines formulations—among the leading as well as most of the other fast developing democratic nations in the world too, got totally changed perhaps for the better.

Hence, the earlier widely accepted **Clausewizian thoughts** that war is an additional-cum-different kind of instrument for furtherance of the state-policy by other means, has become a thing of the past. Well in these larger perspectives, as we critically delve deeper into these far-reaching aspects of substantial importance for the mankind, it would be rather naïve to even think for a moment that after the end of Second World-War (1939-45), man has become more peace-loving and more 'civilised' as well in any manner. In fact, the truth is far from it. The plain reality is that, on the chessboard of international arena of in fact, now the multiply globalised great-games regarding the geostrategic and politico-militarily, the relentless pursuits by the new-era oriented competing nations, concerning their multi-dimensional power-politics; the momentous advent of nuclear weapons and their superbly high-tech oriented delivery-systems, even normally of the unthinkable-cum-unknowable dimensions; have decisively acted as the game-changers. All these have resulted in the substantial change in the 'nuts and bolts' of entire mechanics of the statecraft, geopolitics and the art and crafts of the conduct of diplomacy as well. What we find is that militarily, the more emphasis is on the effectively displayed deterrence and simultaneously then at the politico-diplomatic levels, the prudent strategic stress is given to promote the national-cum-international economic interests of any concerned nations. **Hence the nations now, strive to forge alliances principally to geopolitically nurse their both short as well as long terms economic and other multi-dimensional security interests and well-being, rather than lining-up for fighting any kind of the hot-war.** So in the age-old and overriding domain of the multiple-aspects of **'essential strategic-culture and thoughts'**; what is seen that the older positions of the goalposts have not only got changed but consequently, the new kinds of challenges too have emerged, such as the highly complex aspects of the 'gunboat-diplomacy' with numerous kinds of the new and additional dimensions, for the concerned nations.

Hence in the presently fast-moving as well as super high-tech oriented modern-era any nation, who has traditionally been having a flexibly well-placed—**the multiple-aspects of the proactive 'essential strategic-culture and thoughts'**, with their own inherent kinds of the prudently effective institutional-dynamics, will always be in substantially advantageous positions than her competitors/rivals, in the presently changed scenario for now the multiply complex great-games of international power-politics. In these precise contexts here, any particular nation's military wherewithal as well as, her traditionally dynamic-cum-disciplined culture of the matters-military would only act just as a purposeful deterrence, rather than perhaps ever coming into any kind of the direct action against any one; since the times and situations for any such thing, have now substantially got changed, may be for good of the mankind. **Hence, it is also equally true that the mankind in any case or, on its own otherwise in its modern-era incarnations, has not become any more peace-loving.** In the above discussed precise contexts and also in our overall larger perspectives as well, we may recall that, earlier we have since already emphasised rather a number of times the always overwhelming importance and values of the cultivating, evolving, developing, and proactively imbibing, the timeless aspects of the 'essential strategic-culture and thoughts' in one's entire gamut of the national well-being as well as, in the processes of one's national-governance, with reassuring and integral security-aspects, and peace.

And interestingly at the same time, we do also find other dimensions to this overall scenario, in the present-era comity of nations. And those being that since the older kinds of the unchecked feudalism, militarism, and the highly fixated-aspects of the politically hegemonic imperialism too, in the earlier commonly followed conventional sense and also, broadly up to the end of nineteenth century or so; have become things of the past. **But that does not mean that the multiple-aspects of the old and rather malignantly-lingering imperialism have vanished altogether since these have only altered their older textures and fixations and, resurfaced in its more perplexing as well as also no less overwhelming, kinds of the new incarnations**. As we continue to peep deeper in this regard, of the substantial importance we begin to clearly notice that now, the totally new kinds of the economic as well as internationally the trading-oriented and also no less overwhelming—the multiple-aspects of imperialism—have already begun to surface on the mankind's global horizon. And hence, some kinds of newly fixated great-games of the new kinds of exploitations—of the disadvantaged nations, by the comparatively more advantaged nations in this regard, are now being played, for satisfying the all sorts of justifiable or otherwise too, their economic greed and profits. In respect of the multiple-aspects of imperialism therefore, if earlier the motivating factors were the politico-military subjugation of the conquered nations for building/expanding of the erstwhile colonial-empires by the victor-nations for their different kinds of ambitions and motives, but now in the changed and also the new kinds of fast-moving scenario, the overriding motivations are principally, and even at times may be for the unethical economic gains and, corner rather maximum possible exploitative profits. And wherein at the same time, the new and also continuously advancing as well as revolutionarily improving the multi-dimensional scientific-cum-technological marvels, since onto the man's 'expanding civilisational-canvas' in this regard, are providing highly complex and earlier the unknown kinds of inputs as well.

In precisely the above multiple-contexts, **Mary Kaldor** in her book, *New and Old Wars: Organized Violence in a Global Era* (pages 13, 15, 28, 69 and 70), has written that, "As Clausewitz was fond of pointing out, war is a social activity. It involves the mobilization and organization of individual men, almost never women, for the purpose of inflicting physical violence; it entails the regulations of certain types of social relationships and has its own particular logic. Clausewitz, who was arguably the greatest exponent of modern war, insisted that war could not be reduced to either art or to science. Sometimes, he likened war to business competition and often used economic analogies; to illustrate his points.... Every society has its own characteristic form of war. What we tend to perceive as war, what policy-makers and military leaders define as war, is in fact, a specific phenomenon which took shape in Europe somewhere between the fifteenth and eighteenth centuries, although it has passed through several different phases since then. It was a phenomenon that was intimately bound up with the evolution of the modern state. It went through several phases,....from the relatively limited wars of the seventeenth and eighteenth centuries associated with the growing power of the absolutist state, to the more revolutionary wars of the nineteenth century such as the Napoleonic Wars or the American Civil War, both of which were linked to the establishment of nation-states, to the total wars of the early twentieth century, which were wars of alliances and, later, blocs. Each of these phases was characterized by a different mode of warfare, involving different types of military forces, different strategies and techniques, different relations and means of warfare. But despite these differences, war was recognizably the same phenomenon: a construction of centralized, 'rationalized' hierarchically ordered, territorialized modern state. As the centralized, territorialized modern state gives way to new types of polity emerging out of the global processes, so war, as we presently conceive it, is becoming an anachronism.... **The logical end point of the technological trajectory of modern war is, of course, weapons of mass destruction, particularly nuclear weapons. A nuclear war would be one in which force is applied in the extreme in a matter of minutes. But what rational purpose would ever justify their use**? In the post-war period, many strategic thinkers have grappled with this problem. Do not nuclear weapons nullify the premise of modern warfare—state interest? Finally in the post-war period, alliances have been rigidified so that the distinction between what is internal and what is external is also eroded. Already in the Second World War, it became apparent that individual nation-states could not fight wars unilaterally. This lesson was applied in the construction of the post-war alliances. Integrated command systems established a military division of labour in which only the superpowers had the independent capacity to wage full-scale wars. Essentially, European countries, in the post-war period, abandoned one of the essential attributes of sovereignty—the monopoly of legitimate organised violence—and, at least in Western Europe, what was effectively a transnational civil society was extended to a group of nations. There is widespread discussion about the social science finding that democracies do not go to war with each other. Interestingly enough, what is not discussed is the integration of military forces on a transnational basis which provides a practical constraint against war, Claus Offe makes a similar point about the 1989 revolutions in Eastern Europe; the reason they were so peaceful, he argues, was because the integration of military forces in the Warsaw Pact and this also explains the Romanian exception.... The political goals of the new wars are about the claim to power on the basis of seemingly traditional identities—nation, tribe, religion. Yet the upsurge in the politics of particularistic identities cannot be understood in traditional terms. It has to be explained in the context of a growing cultural dissonance between those who participate in transnational networks which communicate through e-mail, faxes, telephone and air travel, and those who are excluded from global processes and are tied to localities even though their lives may be profoundly shaped by those same processes." (Kaldor, 1999, 13, 15, 28, 69 & 70)

Now again coming back to the highly momentous historic-aspects of the nineteenth century, we may recall that in these specific contexts, so far we have briefly discussed them earlier, concerning the European continent only. But the man's living-history since also shows that the recently bygone **nineteenth century was in fact, no less momentous as well as history-making for other continents too in the world**. We may contextually prefer to critically delve into them as well briefly in a way to broadly crystallize the mosaic of the 'conceptual-framework' of the momentous reach and also, rather some kinds of the global dimensions of now the bygone nineteenth century. We may recall that earlier in different contexts, and also in broader terms and with their integral historicity as well, we have discussed that the various kinds of the numerously different ongoing developments and events taking place since almost simultaneously or in rather fast-paced successions and, in some far-off geographical locations/regions of the world may apparently appear to be isolated or unconnected but they all in some forms or other, are mostly inter-related and mutually influencing too. And at the same time now, we may broadly perceive that since the momentously unprecedented advent of the 'Industrial-Revolution', these interrelationship-cum-mutually-influencing aspects, among the numerously ongoing occurrence of the various history-making developments and events onto the expanding 'civilisational-canvas' of the mankind, have gone more and more complex and intensified. In a way it appears that these too, are some of the vastly numerous ways of since the multi-dimensionally ongoing making of the man's living-history in the world.

In the above multiple-contexts, as we may now begin to look into the various aspects of now the bygone nineteenth century times in **the USA**, we may notice that the **four-year historic American Civil-War (1861-65)** too, took place during that highly historic 19th century and, which geopolitically as well as economically, integrated the many American states and thus, forming the modern, large and highly vibrant, dynamic and sovereign Nation-State, i.e. United States of America. We have since already discussed earlier in some details, these the then history making aspects of North-American continent. In fact, even during the civil-war itself and also more so after it, when things got cooled down and thereby, bringing-in gradually very fast and enduring peace and also the highly desirable socio-economic-cum-political tranquility over there, the migrations from the various ethnic as well as socio-cultural entities of Europe got phenomenally intensified and thus, increasing the US populations manifolds within a matter of just few years. During those times of the nineteenth century in fact, as the newly emerged sovereign Nation-State of the USA has had a huge continental size landmass and thus, there happened to be lot of vacant territories and space since gifted with multiply vast natural resources, and so in the beginning years migrations were largely accepted and not resented upon. Europe was having surplus populations with many economic constraints and then, during the 1840s famine principally in the Ireland and also surfacing of refugees too in Europe, many Europeans rather en-masse, began to set-sail towards the USA, for the much greener pastures.

But in any case historically, there had to be a limit to all these and hence, the already settled Americans began to strongly resent such large inputs of the then multi-ethnic European populations since migrating into the USA and thus, forcing the US government to frame much stricter policies, through the legislative embargo on any more unchecked new immigrations. USA is also known as the sovereign Nation-State of the multi-ethnic-cum-multi-national immigrants. In these respects interestingly, while quoting the poet Walt Whitman, **Richard Overy** in his already earlier referred book has written that, "The **poet Walt Whitman**, writing in the 1850s, described America as 'not merely a nation but a teeming of nations'. Immigrants have made a massive contribution to American finance, society, politics and culture. American society today is recognizably and irreversibly multi-cultural. Yet anxieties persist among some Americans towards immigration both legal and illegal (the so-called 'wetbacks') from Mexico that has transformed much of the South-West and Far West into predominantly Spanish-speaking areas. (*Well in any case, these are perhaps inevitable fallouts in this regard; and hence, any nation has its unquestionable sovereign right to put some control-cum-checks and balances, since as deemed necessary by that concerned nation's government*)." (Overy, 2008, p. 235)

Interestingly in the above historic contexts of now the bygone nineteenth century immigration then in many ways, had been a major history-making phenomenon in the whole of North American continent. Hence, another continental-size modern and sovereign Nation-State of **Canada**, since located in the further **north of USA** and so in the Northern America itself has also got sizeable population inputs to begin with from Europe, during the nineteenth century. Along with the USA earlier, Canada too had been a British imperial-colony and thereby, forming a part of once the transcontinental British empire. As part of the new world as well as, once part of the erstwhile global British empire in the North American continent and also being the northern neighbour of the USA, it has its own history of the few centuries only and hence at the same time, somewhat akin to that of the USA itself; but there are some substantial historic differences too in this regard. Quite unlike the USA but, somewhat similar to once the **'Jewel in the Crown', i.e. INDIA**; historically with their typical and proactively well-placed—the multiple-aspects of **'essential strategic-culture and thoughts'** and thereby, with their geopolitical shrewdness and also as per their imperial-needs then, the British did set-up the **'Hudson Bay Company' in Canada**. This Company was then engaging principally in the 'fur-trade', to substantially increase and expand the British influence and profits and hence, began taking control of the maximum possible territories over there, as per their imperial-policies, since well-crafted with long-term strategic-cum-geopolitical shrewdness and also, along with their globally applicable visionary flexibility, for furthering the multiple-aims and aspirations of their once unmatched British empire.

But in spite of all the above imperialism-driven as well as the transcontinental geostrategic and politico-military wherewithal of the imperial-British, acquisition/subjugation of the vast tracts of alien lands in different and far-off continents, have had to rather inevitably bring-in different kinds of the problematic fallouts as well, for once the unmatched and indeed, the global British empire. That in fact remains part of the very nature of man's living-history since continuously in the making and as it is generally said, there is nothing like 'free-lunch' in the world. That may remain historically a different story and much less problematic situation if anywhere any such lands might have been virgin or, quite sparsely populated by rather some very primitive kinds of aboriginals; but in any case, substantial extents of the multiple-problems did surface in Canada too, and had to be tackled. And then the climatic conditions and many difficult economic problems in the beginning also, came in the way of comparatively smoother flow of immigrants from Europe.

So historically peculiar to Canada, the aboriginal population in the western parts over there, kept-on dominating during the nineteenth century and so some fierce wars had to be fought by the British, to take control of those territories from the aboriginals. The substantially decisive war in this regard, took place in the year 1812, which happened to be principally the war of British and hence not surprisingly; the French-speaking immigrants from Europe, since settled in areas of Quebec, Nova Scotia, Newfoundland etc. kept themselves away from this war. We may recall that in somewhat allied and also in much broader contexts, we have since discussed these aspects concerning the historic making of the sovereign, democratic and modern Nation-State of Canada, more relevantly in the last chapter itself.

The war in 1812 then, prompted the British to encourage more and more the loyalist immigrants and also the refusees, to reinforce the larger interests of British Empire in the Northern parts of North America. In fact contextually, the long-term strategic-vision as well as, the well-conceived flexibilities in this regard always serves much better, the larger interests of any concerned imperial-power. **So historically in the fitness of things then over there, the British true to their fairly matured multiple-aspects of the 'essential strategic-culture and thoughts', granted Canada the geopolitically effective status in fact, of the first self-governing colony in North America, way back in the year 1867 itself**. But there were yet many years to historically roll-by, before Canada could become democratically united. Meanwhile, size of the aboriginal population too kept-on reducing fast quite a lot, due to armed clashes and epidemic, etc., in favour of the European settlers. Now it may sound strange but during the momentously unsettling years of the nineteenth century; even their southern neighbour the US Americans too, began to threaten some adjacent territories in Canada. But the British troops stationed there, along with their Canadian volunteers since already settled there and interestingly, even some of the aboriginal warriors—countered the threats and succeeded in driving back the Americans.

It is a different story that now, the overall situations during the post-Second World-War era from the latter half of twentieth century onward, stands substantially changed and internal peace prevails in the whole of North American continent, having high economic growth and technological-cum-industrial developments and also along with in fact, purposeful open-borders between the USA and Canada. But if during the mid-nineteenth century, the British, Irish and the French immigrants could bring-in major transformations in the different provinces of Canada, Asian and African immigrants were never encouraged by the concerned Canadian authorities. It is a different matter that in spite of all these; fairly large African, Japanese, Chinese and also the Indian Sikhs came to settle into largely the state of British Columbia in Canada and, has been substantially contributing into the Canadian economy. But credit also indeed goes to the Canadian authorities that, Canada has now officially declared herself—a multi-cultural and multi-ethnic society and Nation-State in the world. And after the Second World-War, more immigrants from the present Caribbean islands, Africa, Asia, middle-East and even from the Latin America, came to settle in Canada to truly add more substance in fact, to the above declared official status of the Canadian Nation-State. **And in addition, even many socio-political problems among the English and French speaking European settlers too got sorted-out by resorting to somewhat the repeated processes of the referendums, to keep Canada politically united. These are no small feats in the man's modern living-history and interestingly, the last referendum was held as late as in the year 1995 itself.** Canada has now well-tried and tested and also, the strategically formulated socio-economic understandings with the USA and then a mutually trusted security-arrangements as well, to effectively deter perhaps even any remotely possible threat to begin with, towards the North American continent.

Now while continuing to discuss the multi-dimensional historicity of indeed, the bygone momentous nineteenth century; as we begin to contextually peep into the **South American continent**, we clearly notice that almost all the **Latin American Nation-States** being in a state of unsettled but, each one also being in somewhat different kinds of turmoil. Historically the pivotal point to begin with in this regard, has been that unlike principally the erstwhile British imperial colonies in North America; the then Latin American colonies since, belonging to the Spanish and Portuguese imperial-powers had different kinds of socio-cultural-cum-geopolitical cultures, lacking flexibility as well as, even little liberal approaches towards maintaining colonial relationship in governance. These aspects within rigid colonial framework of the exploitative manners of governance in fact, historically prevented in gradually building-up any meaningful socio-psychological as well as cultural-levels' purposefully enduring relationships, between the colonial subjects and their imperial-masters. And in our broader contexts now in hand, as we critically delve little deeper into these historic aspects in Latin America, we do notice the inevitability of the then impending turmoil-ridden fallouts over there, indeed during the nineteenth century itself. No wonder that as different kinds of the socio-politically far-reaching upheavals and revolutions began gathering momentum, back home in the home countries of those imperial-powers in Europe during the nineteenth century and even earlier as since already discussed contextually in this chapter before, the Portuguese and Spanish imperialists became unable to continue to hold onto their distant imperial-colonies, in South America. By the eighteenth centuries onward itself; in Brazil, Peru, Mexico, Argentina and other Latin American States, serious and even threatening cries for political freedom, substantial autonomy in governance, self-determination in the economic matters became louder and thereby, they also got disconnected from their European imperial masters, being mostly rigid and lacking in geopolitical-vision.

In fact writings on the wall then were getting clear and louder during the eighteenth century onward and more so during the nineteenth century. Further, consequent to the advent and fruits of 'Industrial Revolution' the path-breaking print-media surfaced, and hence, the newspapers too came into being bringing-in more information and awareness onto the peoples' mental horizon across the world, as new civilisational tools in their hands. So during the early decades of the nineteenth century itself and as also being inevitable in any case, the Spanish and Portuguese colonies in Latin America began breaking away from their European imperial-masters; who were themselves weakened due to the fallouts of Napoleonic wars in Europe earlier and also, consequent to their own internally unsettled conditions. In addition, these disturbances in the erstwhile imperial colonies in Latin America; were also rather indirectly aided and abated by the French and Americans too, in no small measures. During those turmoil-ridden early historic years of the nineteenth century, the Portuguese Monarch along with the members of his royal family fled his capital in Lisbon to escape harms and, onto all the way to the Brazilian capital city of Rio de Janeiro. Since this city was also the hub of his colonial empire in Latin America; interestingly he himself in fact, began presiding over a kind of the liquidation of his own colonies and thereby, transforming Brazil into an independent Nation-State. Hence, after return of the Monarch back to Lisbon again, to reoccupy the Portuguese throne over there his son prince Pedro I, while abandoning his claim to the Portuguese throne, and thus assuming the role of a Monarch stationed in Rio de Janeiro, himself declared the independence of Brazil and thereby, heralding the demise of the Portuguese colonies there.

But the Spanish imperialists, since their colonies in Latin America happened to be more expansive were initially in no mood to relent so readily and also, peacefully as well. Failing to crush the revolts against Spain there, violent and somewhat prolonged armed-clashes resulted between the Spanish imperial forces and the revolutionary forces in a number of places like Venezuela, New Granada (now known as Columbia), Peru, and so on. But the die was already cast and new living-history of South America, then fast in the making, was already unfolding. So the harassed and frustrated Spain was in fact, almost forced to accept defeat in the battlefields and hence, began granting political independence to her Latin American colonies. But political freedom so revolutionarily obtained by means of even fighting wars, were one thing for these newly independent Latin American Nation-States, but socio-politically as well as socio-psychologically, to be able to peacefully govern themselves by being even some sort of democracies in fact then, indeed did remain another 'cup of tea' in this regard. And for these Latin American nations initially, these were tall orders in spite of the all sorts of initiatives and efforts, by the then emerging political and military leaders over there. Hence, these unsettled conditions resulted in serious socio-political turmoil and thereby ironically, perforce throwing-up all sorts of military dictatorships in those nations and no wonder that, many of them turned out to be thoroughly corrupt and irresponsible in their mostly shortsighted acts and also their poor governance.

But as nearly half of nineteenth century rolled by, some semblance of stabilities came in some parts of South America, through the establishments of the 'oligarchic republics', which were brought through some kinds of the political understanding and alliances among the regional elites and the so called landed-cum-moneyed aristocrats. And things began to take some positive turns after their political independence and consequent to rapid market-oriented economic growth in the European and North American industrial nations, huge demands for the Latin American exports of the different kinds of raw-materials and mineral-ores, were created abroad. So the situation was no more so bleak and, all these attracted substantial foreign investments into their economies and also resulted in the mass migrations mainly of the South European ethnic-stock for settling there and, for participating in all kinds of economic activities.

In the above multiple contexts, **Richard Overy** he has written that, "From the mid-19th century there was greater stability, marked by the arrangements, 'oligarchic republics', essentially based on alliances of regional elites. (Brazil remained an Empire until the monarchy fell in 1889) The consolidation of nation-sates at this point was driven by export-led growth. Politics was no longer the zero-sum game of the immediate post-independence period. Rapid growth in the world-economy meant an increasing demand for Latin American exports and inflow of foreign investment, notably from Britain. Foreign capital helped rejuvenate the mining sector, and financed railway construction, the modernization of ports and domestic market consolidation. **Associated with greater political stability and economic opportunity, mass migration from Sothern Europe transformed agriculture in temperate zones, such as Brazil, the Argentine pampas, Uruguay, and central Chile and Columbia, Cuba, Spanish until 1898, also received large migrants. Commodity booms—Peruvian guano, Chilean nitrates, Brazilian coffee and rubber, River Plate cereals and meat—underwrote the new political order**." (Overy, 2008, pp. 238 & 239)

So far as the continent of **Australia**, the islands of **New Zealand** and the **Indian subcontinent** in this regard, are concerned, we may recall that we have since contextually discussed them even in somewhat broader perspectives, already in the previous chapter itself. It is a different matter that they all are now politically independent and sovereign Nation-States in their own well-accepted manners. But during the bygone nineteenth century and even earlier; all these happened to be valuable British colonies and hence, were very much part of the transcontinental British Empire. Now they are all vibrant democracies, except the north-western part of the Indian subcontinent, being the State of Islamic Republic of Pakistan; which came into being by its turmoil-ridden and some kind of rather a 'botched-up' birth and, consequent to India's partition in the year 1947, when the imperial British were hurriedly preparing to leave the subcontinent. During those times and after the devastating end of Second World-War, once indeed the unmatched global British Empire were fast getting rolled-back towards its final destination into the dustbin of man's living-history—for more study, research and to write-about. Like the USA and Canada in North America and also, somewhat the Latin American States; populations of now the sovereign Nation-States of Australia and New Zealand too, do mainly constitute of the mass-scale immigrants from Europe and other places. We have briefly discussed their these aspects in the last chapter and we have since also noted that, Australia began as the distant penal-settlement colonies for mainly the British white-convicts after its discovery through sea-routes and mapping of its eastern coastlines by the **legendary English explorer, James Cook** during second half of the eighteenth century itself.

Further, the aboriginals in Australia and also the **Maori tribals in New Zealand**, gradually and mainly by force, were evicted from their lands and driven away, for making attractively substantial spaces for the white-settlers from Europe. Those were ironically, the common norms during the imperial heyday of the European imperialism in the overseas colonial-empires. And historically these processes were in full swing during whole of the momentous nineteenth century and even for some decades beyond. Interestingly, on the socio-cultural-cum-economic front both Australia and the New Zealand, being as part of since earlier the global British Empire; gradually turned out to have remarkably rather, many similarities in their own specific manners and also geographically, the islands of New Zealand happens to be located somewhat in the south-east of the Australian continent in the Pacific Ocean. Historically then, approximately by end of the momentous nineteenth century if Australia turned out to be 'working men's paradise, principally for the white population in the British Empire' then also interestingly, New Zealand incarnated itself into a kind of the peaceful 'rural Britain of the south'. Both got developed into large-scale and very profitable sheep-farms for the production of wool and in addition, became exporters of grain, sugar, dairy products, frozen beef and also, many minerals as well, for Britain and the British Empire.

And at the same time during those bygone years, the British with their typical geostrategic and politico-military vision in this regard, were fast expanding and consolidating their imperial-rule in the Indian subcontinent with its numerous facets. In the last chapter itself, we have already touched-upon in some detail and from various historic angles, these subject-matters, since historically pertaining to the Indian subcontinent and so much interestingly, the historic **first Indian so-called war of independence/the 'Sepoy-Mutiny' too, happened during the nineteenth century itself, i.e. in the year 1857**. The multiple-aspects of British imperial-rule, having lasted for nearly three centuries, are a multi-dimensional saga of indeed unique historicity and so colourful in nature and thereby, if all these changed India for ever then, the British nation too, got unprecedentedly influenced and affected by his once 'Jewel in the Crown'. Volumes have been written about it with both sweet as well as, perhaps no less sour tastes in this regard, by various learned historians and authors both in India and Britain, for posterity.

In the above multiple-contexts regarding the Indian subcontinent and historically, in specific contexts of the Two World-Wars too **Jawaharlal Nehru** has written that, "One of the consistent aims of British policy in India was to create vested interests, which being of their own making, would rely upon them and become their supporters in India. In this way the feudal princes were strengthened and big *zamindars* and *taluqdar* class created, and even social conservatism was encouraged in the name of religious non-interference. All these vested interests were themselves interested in the exploitation of the country, and indeed could exist only because of this exploitation. The biggest vested interest created was that of British capital.... **A statement made by an Englishman, Lord Salisbury, who was Secretary of State for India, has often been quoted and, it is illuminating. He said in 1875 (*i.e. even after the contextually above mentioned 1857 war in India.*):**

> **'As India must be bled, the lancet should to the parts where the blood is congested, or at least is sufficient, not to those which are already feeble from want of it.'**

The British occupation of India and the policy they pursued here produced many results, some of which were not welcome to the British. But even individuals can seldom control all the results of their actions, much less can nations. Often enough among the results of certain activities are new forces which oppose those very activities, and fight them, and overcome them. **Imperialism produces nationalism; capitalism produces large aggregations of working men in factories, who unite and combat the capitalist owners. Government repression meant to stifle a movement and suppress people actually results often in strengthening and steeling them, and thus preparing them for final victory.**" (*How true it remains universally.*). (Nehru, 1934, p. 720)

The above quoted paragraphs historically reveal some indeed very strongly imperialism-obsessed dimensions of the British colonial-rule, as then being fast consolidated in the vast Indian Subcontinent as well, during the nineteenth century. Historically those were the times of seafaring European colonialists and wherein, the British were indeed globally the lead-players since, aided and abated by their naval-cum-army might and thereby; resulting in a sort of the golden-age of intercontinental British Empire. But come twentieth century and, numerous far-reaching events, especially in Europe began unfolding fast and thus just after the first decade of this century, 1914-18 First World-War came—bringing-in not only many changes in the then global scenario but, tragic devastations as well. We have since already discussed these multiple-aspects concerning this Great War and along with its numerous fallouts little earlier in this chapter itself. Further on, times were never the same again for the British Empire, and hence all the more so, for their imperial-rule over 'Jewel in the Crown' i.e. India. Added to all these, the freedom movements in colonial India for getting rid of the British imperial rule too, were fast gathering momentum especially when the Britain had already begun experiencing serious economic difficulties in the Continent itself after the 1914-18 war.

And precisely during those turbulent times for the British Raj in India in the year 1919 itself; the infamous and extremely demonic historic episode of the **'Jallianwala Massacre'** took place and thus, rather irreparably alienating the Indian natives from their imperial rulers. This was not only extremely inhuman and tragic but also at the same time, grossly unnecessary for the British imperialists then stationed in the Indian subcontinent as well. We may even say that this more severely damaged the socio-psychological-cum-political fabric of the relationships between the British and the Indians and thereby, seriously cracked some kind of the 'glass-house of their mutual trust' as well. **To be fair to the British, we may also say that this was certainly not any part of their imperial psyche under any normal circumstances but, to the great misfortune of the peaceful protesters gathered in Jallianwala Bag then and also, to the great shame as well as embarrassment to the British rulers, it was all an extremely foolish action on the part of a 'pig-headed' British army officer namely, Brigadier General Reginald Dyer, which seriously shook the foundation of the British imperial-rule in India**. And perhaps in spite of their some best of efforts, the British were never able to wash the stains of this sinful act on their part whether in India or back in their homelands, from their coat of arms.

Well, now it is nearing almost a hundred years since this humanly devastating tragedy took place on April 13, 1919, in the city of Amritsar in Indian State of Punjab, and hence much water has already flown down the Chenab and Jhelum rivers and accordingly now, as per the twenty-first century demands concerning the geopolitical strategic prudence, the sovereign and fast developing Indian nation-State is rather legitimately expectedly to display—otherwise a fast emerging Power's requisite geostrategic and politico-military sagacity, to be able to now take this extremely tragic episode of the bygone era in her stride—in some kind of her matured historic perspectives. Here again comes the crucially important interplay of the already contextually discussed rather more than ones, the same **multiple-aspects of the 'essential strategic-culture and thought'**. Well, crucially in such geopolitically relevant and also geostrategically important respects perhaps, a million dollar question does always arise that, has even now the sovereign and democratic Indian Nation-State been able to prove herself in any internationally well-accepted as well as also domestically, in any reassuring manner? **Needless to say that, in order to be able to do the same in some kinds of any worthwhile manner it is always mandatory to first be able to substantially cultivate, evolve, multiply develop and, imbibe the above mentioned multiple-aspects of the 'essential strategic-culture and thoughts', and then also simultaneously be able to proactively maintain them in their requisite stations at all levels as well as, in almost all the areas of India's national endeavour. Hence, precisely these aspects in turn only would be able to positively-cum-proactively—to throw-up the healthy and confident 'political-will' as well as, trigger the desired momentum in the leadership-dynamics at all levels and also in all areas of the national endeavour in the entire gamut of national governance in this regard**. At this precise point, we may also recall that we have since already emphatically discussed and stressed upon these paramountcy aspects, in different contexts earlier as well.

The fundamental point perhaps in any context is that, there is no use for any nation to regressively keep-on 'living' largely in past without any long-term tangible strategic purposes as well as, without any kind of the somewhat worthwhile furtherance of one's national interests. In fact one's multiple-aspects of the nationalism too, always need to be guided as well as fine-tuned with the no less essential aspects of requisite strategic-prudence rather than, be naively drenched largely in some sort of rather blinding sentimentalism only. Such shortsighted fixations in this regard, lead to at times: rather extremely damaging regressions and thereby; the first casualty is mostly the crucially important multiple-aspects of the 'essential strategic-culture and thoughts' in one's national being itself. **Hence, these far-reaching aspects are certainly no matters of taking to streets or even naively beating one's chest. Any dynamically fast developing, matured, self-assuring-cum-powerful sovereign democratic Nation-State always needs to be displaying internationally effective strategic-prudence and also, have proactive capacities to positively move along with the continuous march of man's living-history—since always in the making onto the man's 'expanding civilisational-canvas'**. These aspects in fact, are again some essential segments of the already earlier emphasised multiple-aspects of the 'essential strategic-culture and thoughts'. Therefore, all these need to be proactively as well as fruitfully taken in their intended perspectives.

During the February 2013 official visit of the **British Prime Minister David Cameron** to India, well-known Indian Journalist and writer, **M.J. Akbar**, in his short article has written that, "The 20th century was born (*for India indeed.*), at Jallianwalla Bagh. In a curious way, India should thank the butchers of Jallianwalla for ripping apart the last mask of British colonization.... When will Indians and Gorkhas apologise for killing fellow Indians at Jallianwalla? They continued to squeeze the trigger on unarmed, helpless civilians amid screams and shock until ammunition ran out.

When will brown bureaucrats of the Indian Civil Service, who found clever explanations for colonial exploitation apologise? British rule was never a solely British enterprise. It could not have survived a day without an obedient Indian comprador class, mostly purchased by nothing more glamorous than a salary. When will the *zamindars* and nawabs, who squeezed a famished peasantry to death and feasted in garden-palaces on the rewards, apologise?**The British used a million Mir Jafars, who queued up to serve, during their 150 years of true power. They had come a long way to rule, not turn the other cheek. A transfer of wealth to the 'mother country' was standard procedure in the era of European colonization, and not uniquely British. It must also be stressed that British rule, for all its faults, was much more humane than that of France in Algeria, Belgium in Congo, or the Dutch in Indonesia.... India's problem with history is a consistent unwillingness to do some serious research in a mirror**. The British did not establish their rule, step by careful step, merely because they were strong, they succeeded because Indians had become weak. How about a collective Indian apology on behalf of our recent forefathers?" (*Well, interesting but indeed, so precise and historically true as well.*). (Akbar, 24 February 2013)

Further in the similar perspectives, and in again in *The Times of India* another equally well-known commentator, strategic analyst and author, **Swapan Das Gupta**, with his substantial analytical skills as well as, a great sense of history; has written in his short article that, "There was a time when history was an engagement involving the dead, the living and the unborn. Today, thanks to the epidemic of prefixes (post-modernism, post-colonial, neo-liberal, et al), the story of human experience has been reduced to conversations involving tiny groups of 'professional' historians. The wider citizenry that should, ideally, have informed perceptions of their heritage and inheritance have been disdainfully left out of the process.... The results have been horrible. An India that was born relatively unconcerned with history has become even less so. An enlightened yet critical view of how our ancestors coped with (*or did not or even could not.*) coped with challenges and uncertainties have been replaced by either idyllic or prejudicial fantasies. **By far the most damaging contribution has been that of 'scientific' history which, thanks to its impersonal nature and inherent dryness, has virtually killed popular interest in the past. For the *aam aadmi (i.e. common man in India)*, history has become a Bollywood hand-me-down.... This perversion has had two consequences. For some, not least the political class, the rendering of the past has become aspect of contemporary politics—tales to be moulded and presented as facets of a contested nationhood. To the completely uninitiated, history has become an extension of mythology—a process that conveniently bypasses chronology and empirical rigour**. (*What a paradoxical irony indeed even in the sovereign democratic Nation-State, i.e. India—especially in super-age of the 21st century in the presently on-going modern-era. We would further be contextually discussing these indeed very far-reaching and weighty subjects in the last sub-chapter of the next eighth chapter*). By definition, any appreciation of the past involves a great deal of tentativeness. Yet, if mass reaction is any guide, everything from Shivaji to Gandhiji bound in unflinching certitudes.... Viewing history as series of certitudes forecloses awkward conclusions. **Like the present, there is no single reality that defines the past**, a point to consider the next time we make it a contemporary battlefield." (*The substantially crucial issues at this precise point too, do enduringly remain that—the above quoted indeed very far-reaching and critically perceptive comments need to be taken constructively as well as with no less strategically fruitful positivity and therefore; the essential prerequisites in this regard, do always exist again—in the already discussed a number of times, in different contexts—the same* ***multiple-aspects of 'essential strategic-culture and thoughts'***.). (Gupta, 24 February 2013)

Further on as we critically continue to somewhat panoramically survey the recently bygone nineteenth century and hence also, during the heyday of transcontinental imperialism of rather the ruthlessly rival and seafaring European Imperial-Powers for promoting and nurturing their differently competing colonial-agendas in various parts of the world; they resorted to even arbitrarily partition among themselves, the vast continent of **Africa** and, their this imperial programme began in the year 1880 itself. There was a kind of the greed-driven scramble for more territorial, mineral and other natural-resources assets over there—among those rival European Imperial-Powers. We have briefly discussed these aspects too, in the last chapter. For the then highly greedy and ambitious European imperialists to begin with, unlike the Indian and China subcontinents and so on the African continent in fact, happened to be relatively a virgin, vast and primitive landmass, again appearing to be rather readily available for exploitation by those European imperialists. But historically here too, to their utter dismay, the going didn't happen to be all that smooth and trouble-free for them in this regard.

Richard Overy in his book has mentioned that, "Little of Africa was directly ruled by Europeans in 1880; the French had been subjugating Algeria since 1830; there were small French and British colonies in West Africa; there were moribund Portuguese settlements in Angola and Mozambique. Only in the south, where the British Cape colonists were in competition with the Afrikaners of the Transvaal and Orange Free State, did political control extend far inland. Yet within two decades the continent had been seized and partitioned. Of the 40 political units to which it was reduced by 1913, 36 were completely under European control. France was the largest beneficiary, ruling nearly one-third of Africa's 30.3 million sq. km (11.7 million sq. miles).... Partition caused increasing bloodshed. Abyssinia routed the Italians at Adowa in 1896; some 120,000 Sudanese died in Britain's reconquest of the Mahdist state; and **Rhodes** settlers fought bitterly with Ndebele and Shona as they moved north. Conflicts climaxed with South African War (1899-1902) in which Britain with difficulty won the Transvaal gold fields and absorbed the Afrikaner republics. Elsewhere black Africans, although bitterly opposing the European powers, could never offer concerted resistance and for the most part were easily overcome." (Overy, 2008, pp. 253 & 255) (*We may also contextually recall that, we have since discussed these historic aspects then mostly pertaining to the reasonably vast and strategic territories of the South African region, in greater details in the last chapter itself*).

In the above ongoing multiply phenomenal perspectives concerning the recently bygone nineteenth century; the multi-dimensional and since continuously in the making—man's living-history also shows that this momentous century was no less path-breaking as well as, the entirely new kinds of onward history-making for the Nation-State of **JAPAN** too, and as located in the far-eastern region of the Asian continent. We may also recall that in the earlier chapters and more so in the last sixth chapter we have since already discussed in the numerously allied contexts, many historic aspects concerning Japan. And therein, we have also very relevantly noted that Japan has just not historically been like her neighbours, mainly China and all the more not like India in South-Asia, although both of them have been much bigger in the geographical sizes and also at the same time, both of these two Asian Giants since commanding in some continuity, far more ancient and indeed once the multiply glorious civilisations. Contextually in these historic perspectives, we also notice that unlike many other Asian, African and in addition, principally the Latin American societies-cum-Nation-States and to her credit, Japan could never be so overwhelmingly subjected to any kind of the exploitative colonialism, as earlier these had rather been both geopolitically-cum-militarily, sponsored and multiply promoted by the then rival and seafaring Imperial European Powers, for their own greed and ambitions. Hence, as we delve deeper into these historic aspects concerning the Japanese Nation-State, we do in addition also notice that unlike others, the historically momentous same nineteenth century indeed, did bring-in altogether different kinds of challenges-cum-opportunities too, for Japan and the dynamically unique in many ways, her highly industrious peoples as well. And therefore precisely in these very contexts, Japan's since relatively age-old and consistently well-stationed and also proactively vibrant—the multiple-aspects of 'essential strategic-culture and thoughts' too; have had many multi-dimensionally decisive and also highly successful roles to play in this regard. In other words we may also say that, the aspects of inherently dynamic socio-cultural-cum-psychological attributes of the Japanese peoples and along with, their above mentioned multiple-aspects of 'essential strategic-culture and thoughts', were indeed able to create a highly valuable synergy towards fructifying their unique achievements as well as, socio-economic-cum-industrial developments, against all possible odds. But historically, the singularly more decisive factors have been that, their above mentioned highly valuable attributes did also ensure that—Japan mostly remained sharply focused as per her geostrategic and politico-military imperatives and also, ensured clarity about their highest national-level interests and ambitions. And hence, we may say that, these multiple-aspects do also constitute her traditionally well-oiled pivotal-fulcrum of the 'making of modern and indeed multiply-dynamic Japan'.

Well, for some centuries during the medieval-era, Japan did remain under the prolonged and perhaps, perforce self-imposed isolation and consequently during those times' living-history of Japan also turned out to be highly chequered and somewhat turmoil-ridden. **Hence we may recall and, since also discussed in the last chapter, that during such times in the past, the highly historic Shogunate-era in the medieval Japan, did go a long way in shaping-up her peoples' multi-dimensional sense of confidence and thereby, chartered new paths towards the future making of Japan's indeed unique living-history in the world**. And as few centuries passed and, the historically momentous nineteenth century dawned on the man's expanding civilisational-canvas, Japan came out of her a kind of self-imposed isolation, with altogether new and also outward looking strategic-dynamism as well as, her unique and multiply-futuristic vision too. Historically these aspects since pertaining to Japan may be called unique as she didn't belong to the already earlier contextually discussed fraternity of the European Nation-States; and unlike almost all other Non-European nations, she was still able to successfully forge the multiple-aspects of her living-history quite differently and hence, reassuringly became capable of competing in the then on-going great games; with other European and also the newly American seafaring imperial powers on their terms, and irrespective of the obstacles and challenges on the way.

And precisely in these historic perspectives and then during the momentous nineteenth century, the mounting and also overwhelmingly imperialism-oriented pressures from those Western-Powers, indeed compulsively triggered the multi-faceted and fast-paced modernisation as well as industrialization of the Japanese Nation-State, which subsequently to their indeed unexpected dismay, was going to surprise and even shock those very Western-Powers. And so we may say in a way that the 'Japanese chicken had finally come home to roost'. But these emerging and fast-paced developmental-aspects had some disturbing fallouts for the Japanese Nation too, as they did bring-in some unsettling internal stresses and challenges for the power-centres in Japan. Behind Japan's rapid modernisation, principally the USA had her agenda to secure access to Japanese ports, for own commercial-cum-geopolitical gains. This compelled Japan to enter into a disadvantaged treaty with USA in the year 1858 and, soon also had to bring-back the direct imperial-rule in the year 1868, through Meiji Emperor over there. But by then Japan had already acquired many new industrial and agricultural productive facets and thereby, multiplying her strategic wherewithal and economic strengths in this regard.

As per her unique geostrategic and politico-military ingenuity and, coupled with her economic-cum-industrial genius, Japan had made those very Western-Powers, as her role-models for her own modernisation as well as the fast-paced processes for the multi-dimensional developments and, she had already reaped rich dividends and thereby she was definitely not any kind of the 'push-over' Nation-State for those alien imperialists. But interestingly along the way, Japan soon got her own multiply imperialistic-designs manifested and hence, also threatening other vulnerable nations in the neighbourhood. **History has always been showing that whenever any Nation-State is able to garner multiply economic, industrial and also, geostrategic and politico-militarily far-reaching wherewithal as her greatly commanding national-assets, she has seldom been able to effectively subjugate her consequently always and, overwhelmingly manifesting the urges to nurse the multiple-aspects of her own brand of imperialism whenever and wherever feasible and thereby, building her own colonial-empires**. What an irony all these remain always in many ways, for the man's living-history itself! In the process, Japan rapidly proceeded ahead towards more and more westernization, at the cost of her own traditional value-systems and culture, bringing-in many kinds of the internally widespread resentments too. Hence, Japan had to perforce and also prudently rather, correctively cater for all these socio-cultural stresses as well.

Precisely in the above multiple-contexts concerning Japan then, during the momentous nineteenth century, **Richard Overy** has very informatively mentioned that, "It was crucial to Japan's rapid modernization to bring these strengths into play. The abolition of feudalism, and the replacement in 1873 of feudal dues with cash payments based on the value of land, enabled wealth to be redirected to the purposes of central government. To encourage the transfer of technology and to stimulate investment, the government used this land-tax to build model factories in strategic and import-saving industries, such as steel and textiles, and became directly involved in the development of transport and communications. There was also indirect encouragement in the form of subsidies and tax privileges.... The new leaders introduced complementary administrative reforms: feudal domains were replaced by modern bureaucracy (1871); the feudal army was replaced by a conscript one (1873); a bicameral legislature provided the basis for political unity and stability (1889); a national education system was instituted (1872); and legal codes based on those of France and Germany were introduced (1882).... By 1918 (*i.e. at the end of First world-War*), when she had achieved major penetration of markets in China, the USA and elsewhere; she had become a major importer of raw materials and exporter of finished goods. Such developments meant considerable changes for Japanese society: by 1918 the population had risen to 55 million (from 35 million in 1873) and nearly one-third lived in towns of 10,000 or more.... National strengths brought expansion overseas, motives being both strategic and economic. Japan began by claiming neighbouring islands such as the Ryukyus and Korriles. The 'unequal treaties' were revoked in 1894. Then concerns that China was too weak to keep Russia out of Korea led to war (1894-5) in which Japan destroyed the Chinese forces and acquired Formosa (Taiwan). The Russian threat, however, was only resolved by Japan's overwhelming defeat of Russia on land and sea in 1904-5, which led to Japanese control over the Liaotung peninsula, extensive rights in southern Manchuria and the acquisition of southern Sakhalin. There followed the annexation of Korea (1910), the acquisition of former German territories in China (1914), and the 'Twenty-one Demands' (1915), which made sweeping though not wholly successful claims, over China. In 1919 the Paris Peace Conference confirmed most of the gains and granted Japan mandate to German colonies in the Pacific.... The end of the war (*i.e. 1914-18 First World War*) brought economic disruption, major disturbances in town and countryside, and the assassination in 1921 of the prime minister by a fanatic. These events presaged turbulent years ahead." (Overy, 2008, pp. 256 & 257)

Specifically in our on-going contexts at present, if we may care to delve little deeper and read in between the lines of above quoted paragraphs, then we do historically notice many of the then highly disturbing but also, far-reaching making of the further living-history concerning the Japanese nation. This history-making course—into which Japan was perhaps forced into by her own multi-dimensional genius and dynamism as well as, her unchecked and also overwhelmingly imperialism-oriented ambitions; was inevitably going to cause great harms and indeed crippling devastations in the near future, when soon in fact destined to occur the Second World-War (1939-45), menacingly came globally onto the horizon of entire humankind. This is in a way historically, another great irony of facts, towards the continuous making of the man's multiple-faceted living-history onto his expanding civilisational-canvas in the world. **Here in these historic contexts, we may also recall that earlier since contextually a number of times, we have already noted that history never favours or disfavours anyone anywhere and also, forgives or forgets none whatsoever.** But in these larger and highly consequent historic contexts while objectively speaking, we may also say that, like other Western highly ambitious Imperial-Powers, Japan was then no less industrialized as well as, economically advanced and therefore, she too had landed herself in no less compulsive situations; such as keeping her lines of communications and transportations secured and under control, for equally facilitating the unhampered movements of the import of raw-materials to homelands-based: internationally competitive factories and in return, exports of the various finished goods to her overseas markets and also at the same time, keeping those markets no less secured in the interests of her economy as well. So as per detached views in this regard we may also say that all these indeed were rather compulsive 'catch-22' situations for the concerned nations in these great-games and hence, these in fact do remain the then global economic dimensions of the unprecedentedly devastating Second World-War; which we would be briefly discussing now shortly.

And now we may prefer to take note that by the time this great-war (1914-18) came to its rather 'botched-up' end or even sometimes during little earlier than that, the no less consequential and also historic—the **'Russian-Revolution'** came menacingly as well as violently onto the then Tsarist Russia's skies. Meanwhile since the very beginning of the nineteenth century **China** too, though under highly exploitative subjugation of the leading Western Imperial-Powers and also Japan as well, in fact was passing through its own kinds of the extremely turbulent times. Among these unscrupulously greedy imperial-powers, each one had its own ruthlessly competing and self-serving agendas, with almost total disregard towards the then so much exploited and suffering Chinese masses. It was then a kind of the almost 'free for all' situations for scrambles for those alien imperialists and accordingly, we have since earlier in the last chapter already discussed these historic-aspects pertaining to the then China.

So far as the **Tsarist Russia** was concerned, its own indigenous imperialist and also no less callous-cum-corrupt Tsars, who were the traditional ruling royal-family since the medieval times, and hence during those times Russia remained mainly backward and stagnant agricultural-economy. And even otherwise too, as geographically trapped in the harsh Russian climatic conditions and, inhabited by the poor and suffering masses in addition, were also exploitatively lorded over by the ruthless landlords and feudal barons. And during those momentous times in Western Europe, **Karl Marx** came onto the German scene, with his new kinds and almost the unprecedentedly revolutionary and also globally applicable socio-economic-cum-political thoughts; as perhaps some new ideas for the entire humanity. **Ironically, it was a different matter that in Germany itself, it could never take any roots into the indigenous soil, but gradually along with**

the westerly winds the seeds of these Marxian thoughts and the revolutionary 'philosophy' had already begun taking roots into perhaps the readily available fertile socio-political soils of the Tsarist Russia, for eventual sprouting and then soon to be powerfully flowered with the great world-wide multiple-consequences, during almost the whole of twentieth century itself. This gave violent birth to once the mighty USSR and thereby, creation of the vast Euro-Asian Soviet-Empire rather ruthlessly by means of the geostrategically-cum-politico-militarily bringing into the geopolitical sway of its dictatorial supremacy, over almost the whole of neighbouring 'States or regions', since located both in Europe and the central Asia as well. In now the overall historic perspectives these far-reaching developments were rather understandably then, manifested as the highly complex as well as, geopolitically the multi-faceted and also indeed multiply-threatening concerns for the whole of capitalist-cum-democratic Western countries. So the first and also globally overwhelming—a kind of nightmare for the entire humanity and consequential fallouts cropped-up as the **advent of Cold-War**.

Then perhaps inevitably as the final results subsequently—leading to the political disintegration of Soviet-Empire and also withering-away of the once mighty super-power the USSR, with eventual culmination of the cold-war, resulted in favour of Western-Powers and also for rather the ideas of free, liberal and multi-party democratic thoughts-cum-ways of life. **But the multiply-complex processes for all these far-reaching historic developments, mainly during the whole of the second-half of twentieth century; had been rather principally driven by the ruthlessly attritional and globally unprecedented arms-race between once the NATO and Warsaw-Pact geopolitical-cum-military-Alliances, led respectively by USA and the USSR when the world-peace as well as, the very survival of humanity on this planet-Earth, were indeed hanging by a 'slender-thread'**. During those threateningly momentous times history had really become uncontrollably fast-paced and indeed worrisome for all concerned, since proactively engaged in **these highly complex great-games of the international power-politics**. And even more so, the above mentioned two irreconcilably rival superpowers; were not at all in any commanding position to even slightly exercise some semblance of control, or rather confidently influence the turn of those historic events and developments in any manner. In fact during those times, all these had indeed turned into an overwhelming 'Balance of Terror'. All these once mighty and very far-reaching developments, now form momentous segments of the very recent living-history of the mankind.

And historically therefore, and as per also the above and then in multiple-contexts of the 'Russian-Revolution' as well as, the First World-War too, **Professor S.R. Sharma** has mentioned that, "**Russia had to pay a very heavy price for her participation in the Great War**. She had, it will be remembered, taken up the sword on behalf of Serbia in 1914. **In spite of her earlier victories against Germany and Austria, the War entailed such sufferings and strain at home that, her domestic malcontents created a revolution**. The history of the Bourbons now repeated itself with the Romanoff's, Nicolas II and Alexandra playing the role of Louis XVI and Marie Antoinette. **The 'morning star' of this revolution was Carl Marx, as that of French Revolution had been Rousseau; its Danton was Lenin, its Jacobins the Bolsheviki**. To cut a long story short, on 25th October 1917 the Socialist Soviet Republic was proclaimed by the Communists under the leadership of **Nicolai Lenin**. Petrograd has (*had then once*) become Leningrad.... **With the death of Lenin in 1924 Russian communism entered a new phase. A terrible duel ensued between Trotsky the Jew and Stalin the Georgian for the Dictatorship over the Proletariat (workers, soldiers and peasants)**. After five years' struggle the Jewish journalist was ousted by Stalin ('the man of steel') in 1929. Trotsky went into exile from Russia, and stood for a World Revolution; Stalin initially stood for the preliminary consolidation of the Revolution within Russia. While the idealist revolutionary was roaming abroad, the practical revolutionary was transforming Russia (through a series of Five Year Plans), so as 'to catch up and surpass the capitalist countries' in industrial progress." (Sharma, 1948, p. 419)

So that had been the then historic processes of indeed the extremely ruthless making of once the mighty USSR—courtesy, the unchallenged, unchecked and rigid, communism-oriented dictatorship of Josef Stalin. But in our specific and broader perspectives, as we may begin to critically delve deeper into the above quoted lines, we historically first notice that both Stalin and Trotsky were indeed equally far from being any 'men of peace' since, both believed in only the ruthlessly violent means to achieve their regressively obsessed political ambitions with even some kinds of rather international connotations and that too with total disregard for the any consequential sufferings caused to the humanity. In fact it could never be otherwise in any manner since, both were die-hard products of the **'Marxian school of thoughts'**. Their individual kinds of the proactive perceptions regarding **the socio-economic-cum-political theories of Karl Marx** might have been at some variance and also, in their methodologies for putting them into practice towards their programme-based focused-systems, at the both national-cum-international levels in fact, ruthlessly sweeping implementations of their rather regressively fixated politico-economic ideologies. But, their conceptually broader aims were almost the same in this regard. Interestingly, it also remains the universally accepted and applicable fact that, the two swords can never be accommodated in a single scabbard and hence, either of them had to finally go out of Russia in exile to finally save his own life. **Subsequently, the stormy winds of Marxian communist revolution travelled gradually further south of the USSR into China and where in the highly chequered historic processes, altogether different kinds of prolonged struggles and thereby also, the highly complex and a series of the clashes of arms and wars ensued, and where different western Imperial-Powers too, muddled for their own imperialistic greed by fishing into the then troubled Chinese waters.** We may recall that in the earlier chapters since in various contexts, we have already discussed these historic-aspects of China wherein the communism took its victorious roots much later in 1049, even after the end of Second World-War. We would be briefly coming to them shortly later here and, even in the next eighth chapter.

In these precise contexts, while discussing some historic aspects of USSR it is rather contextually essential to also briefly touch upon the Soviet Russia's once most dreaded **'The extensive network of Gulag'**—now already dismantled but their daunting memories and the human-tortures-cum-blood stained narrative details do survive into the pages of man's living-history, principally through the Russian literary masterpieces. These precisely pertain to indeed the nauseating history of the **concentration-cum-hard-labour camps,** since very secretly located and clandestinely operated in different manners, all over Russian as well as in the neighbouring subjugated territories of the Soviet communist-regime during the twentieth century. In fact, such hard-labour camps were not new to only the Soviet regime, but even during the times of Tsarist Russia, such camps did exist for ruthlessly punitive purposes. But concepts of their functions as well as operative-systems took altogether new incarnations during the Soviet regime, where more and more such camps were set-up all over the places and, even by utilizing, the consequentially the then vacant sites and buildings of the various Russian Churches, and their Christian Monasteries since under the die-hard communism, any religion or religious pursuits had hardly any place to survive and flourish. **And historically moreover, during the more cruelly dictatorial regime of Stalin; these systems concerning the Soviet 'Gulag', gathered terribly more momentum and hence, multiply expanded**

as well. And they named them as 'Constructive Labour Camps', ironically for the political-reforms and ideological-education of the all kinds of defaulters as well as, the law-breakers in the Soviet 'Fatherland'. Anyway to objectively say the least now, these remain perhaps historically the unsurpassed and horrible symbols of the very recently bygone, the great civilisational-negations of the mankind. Now in these multiple-contexts, so far as the various aspects of the earlier noted Stalin's multi-dimensional and, the highly ambitious programme of achieving the 'accelerated industrial progress as well as, to even surpass the western capitalist countries' were concerned; there appeared to be nothing morally, humanly or otherwise on their face-values—anything wrong with Stalin's thought-processes in this regard but it was indeed not at all so, when critically reviewed from the numerous relevant angles. The irrefutable facts remain that, it was all rather demonically achieved with almost all kinds of ruthlessness and, at the cost of great human sufferings and also extremely cruel and blood-stained deaths of the millions of Russian lives in the hands of Stalin's officials and totally callous 'goons', at almost every level of the Soviet regime in USSR. Hence for the days to come all these do remain the extremely painful and totally inhuman and as well as, no less prolonged-saga in the Russian living-history of perhaps all times.

Specifically in the above contexts, famous **Russian author and also Nobel Laureate, Aleksandr I. Solzhenitsyn** has indeed given visibly so much painful narrations of Gulag, in almost all its, normally unimaginable dimensions. And therein the celebrated Russian author has very perceptively written that. "According to the estimates of émigré Professor of Statistics Kurganov, this 'comparatively easy' internal repression cost us, from the beginning of the October Revolution up to 1959, a total of....sixty six million—66,000,000—lives. We of course, cannot vouch for this figure, but we have none other that is official. And just as soon as the official figure is issued the specialists can make necessary critical comparisons.... In February, 1918, the Chairman of the Council of People's Commissars, Comrade Lenin, demanded that the number of places of imprisonment be increased and that repression of criminals be intensified, and in May, already going over to concrete guidance, he gave instructions that the sentence for bribery must be *not less than* ten years of prison and ten years of forced labour *in addition*, i.e., a total of twenty years. This scale might seem pessimistic at first: would forced labour really still be necessary after twenty years? But we know that forced labor turned out to be a long-lived measure and that even after fifty years, it would still be extremely popular... For many months after the October Revolution the prison personnel everywhere remained Tsarist, and the only new officials named were Commissars of prisons. The brazen jailors went so far as to create their own *trade union* ('The Union of Prison Employees') and established an *elective basis* for prison administration. (The only time, *it was* in all Russian history.) The prisoners were not to be left behind either—they, too, had their own internal self-government. (Circular of the People's Commissariat of Justice, April 24, 1918: prisoners, wherever possible, were to be brought into self-verification and self-supervision).... In August 1918, several days before the attempt on his life by Fanya Kaplan, Vladimir Ilyich Lenin wrote in a telegram to Yevgeniya Bosh and to Penza Provincial Executive Committee (they were unable to cope with a peasant revolt): 'Lock up all the doubtful ones (not guilty mind you, but doubtful—A.S.] in a *concentration camp* outside the city.' (And in addition 'carry out merciless mass terror'—this was before the decree)... Only on September 5, 1918, ten days after this telegram, was signed the Decree of the Red Terror published, which was signed by Petrovsky, Kursky, and Bonch-Bruyevich. In addition to the instructions on mass executions, it stated in particular: 'Secure the Soviet Republic against its class enemies by isolating them in *concentration camps*... So that is where this term—*concentration camps*—was discovered and immediately seized upon and confirmed—one of the principal terms of the twentieth century, and it was to have a big international future.

And this is *when* it was born—in August and September 1918. The word itself had already been used during World War I, but in relation to POW's and undesirable foreigners. But here in 1918 it was for the first time applied to the citizens of one's own country.... Consequently, if a prisoner had a year sentence and escaped and was caught, then his sentence was automatically extended to 1968. For a second escape from a concentration camp, execution was prescribed (and of course, was punctiliously carried out)." (Aleksandr, 1974-75, pp. 10, 11, 17 & 180)

Aleksandr Solzhenitsyn's voluminous books on highly cruel systems of Gulag, then prevailing in the very proactive and punctilious manners all over the USSR do give us, very engaging and detailed accounts, and which indeed make a substantially sad reading for anybody with normal human sensitivity. During the same twentieth century, but later than Lenin and Stalin a no less cruel dictator but also anti-communist came to power in Germany in the persona of **Adolf Hitler**. He quite craftily borrowed some leafs from the rigid doctrines of both **Lenin** and **Stalin,** for secretly setting-up his **historically different kinds of the 'concentration camps'**, for exclusive purposes of the unprecedentedly ruthless persecution of the Jews and also, to very clandestinely annihilate the Jewish men, women and children for the sole intentions to get rid of them. And broadly parallel to the times of Lenin and then, subsequently Stalin in Russia another revolutionary dictator, but rather much more 'confused and eccentric' too, came onto the Italian scene, in the name of **Benito Mussolini**, and we would later briefly be discussing about him and his Italy then. At this point, if we may believe that, a human-being cannot be turned into a demonically savage beast, but the above named persons and also similarly many more since the ancient times, rather irrefutably prove otherwise. We may recall that in the beginning chapters and may be in some different contexts, we have since mentioned that both the '**Angel**' and '**Demon**', simultaneously reside inside a man and hence perhaps circumstantially, it depends that which one of them come into an unchallenged and strongly commanding position inside the man and thereby rather decisively overwhelm him for charting-out the further paths of his life This aspect is in fact, universally applicable and contextually, for the both ordinary mortals as well as, the different kinds of historic and also the history-making personalities. That's precisely the way man has been evolved in course of time on this planet-Earth, and at the same time, Earth itself being really a tiny segment within the perpetually on-going and also, incomprehensively complex phenomena of the evolution of mother-Nature in the infinite Universe. Well, these are different stories altogether.

Now coming back to the twentieth century Soviet Union, when communism-oriented demonic dictatorship of the so-called 'proletariat' held its sway all over, the quite tragic life-story of **Svetlana Alliluyeva, Stalin's daughter** too, gives a somewhat 'keyhole' views of the then prevailing 'state of affairs' and also broadly of the environment, over there. In this precise context, **Martin Ebon's book, titled *Svetlana: The Story of Stalin's daughter*** remains of special interest and relevance. In this book the author has quoted from Svetlana's first press conference after her flight to freedom and on her arrival in the USA, thus—"When I left Moscow last December in order to convey the ashes of my late husband, Mr. Brijesh Singh, in his home in India, I fully expected to return to Russia within one month's time. However, during my stay in India I decided that I could not return to Moscow. It was my own decision, based on my own feelings and experiences, without anyone's advice, or help, or instruction.... The strongest struggle was going on in my heart all the time because I would have to leave my children and not see them for quite a long time. I did everything to force myself to return home. But it was all in vain..." (Ebon, 1974-75) (*In fact, the above referred book on Svetlana and her thought-processes, convey many aspects of not only her personal life but also, many facets of the ruthlessly dictatorial Soviet-Regime then so powerfully entrenched in the*

USSR and, also regarding her peoples as well. In the overall larger perspectives of human-lives anywhere in the world; the universally fundamental point remains always valid that, irrespective of the any kind of political ideology since being followed by the concerned Regime it is never even rationally workable in any manner, onto the man's larger and expanding civilisational-canvas, that any Regime can ever successfully treat the human-beings as some kinds of robot-like living creature, which can both politically and socio-culturally be programmed and manipulated, as per the dictates of the Regime itself. And these aspects remain precisely in fact the basic genesis perhaps, for any diehard Communist Regime's socio-political-cum-economic as well as, the multiply geostrategic and politico-military undoing. And the then USSR too, finally turned-out to be no exception in this regard).

As per perhaps the above perceptions as well as in the overall larger perspectives in this regard, **Martin Ebon** has written that, "(*and after her arrival in India*) Now and then representatives of the Soviet embassy, silent and ill at ease, came to see her. They talked secretly. She was more withdrawn than usual after they left. On January 16, Prime Minister Indira Gandhi visited the village. She was accompanied by an entourage. There was little chance for a quiet personal talk, but Svetlana gave Mrs. Gandhi to understand that she hoped to remain in India, although the Prime Minister did not encourage her in any way.... She wrote to an Indian friend, 'I do not want to ask now either the government of India or anybody else in India for asylum in India. Since I still have my valid Soviet passport, I cannot feel secure in India nor in any other country on which the USSR, can bring pressure.... To Brijesh Singh (*i.e. her husband*), and to Svetlana through his guidance, failure of the Marxist God impelled a return to traditions—whether Hindu or Christian—that permit the individual to place his own insights and needs into a pluralistic framework, leaving aside the limitations of dogma.... **Summing up her views on the excesses of the Stalin regime and those who were active in it, Svetlana said: 'I feel somewhat responsible for those horrible things, killing people unjustly: I feel that responsibility for this was and is the Party's, the regime, and the ideology as a whole'**...." (Ebon, 1974-75, pp. 14, 15, 27, 164 & 168)

In these precise perspectives now in our hands, we find that **Italy** though once being seat of the bygone-era's great and ancient Roman civilisation, was even before the coming of twentieth century, had got already stuck in the extremely dangerous and quite unsettling socio-economic-cum-political turmoil in the Western Europe. Her situation was then much more complex than those of other Western European imperialistic nations but also, no less problematic and revolutionary too, than that of the Russian Nation-State. Hence, it was all indeed a very queer situation then in Italy and historically at the same time highly problematic scenario in Italy, was already getting manifested itself even before the beginning of First World War. In fact even a couple of years before the commencement of World-War I, her both rather socio-economic and politico-militarily ill-planned and, somewhat also mismanaged war with Turkey (1911-12) ironically pushed her into a severe socio-economic-cum-political troubles and thus, trapping her into a serious crisis of the national governance too. Though militarily, Italy came out victorious in this some kind of a two-year localized war with Turkey and, even with forced annexation of Tripoli as a wartime booty and as part of the Italian colonies in Africa, which did temporarily bring some happiness to the diehard Italian imperialists and other bourgeois classes in Italy but, certainly failed to bring-in economic succor of any kind. And Italy kept on drifting fast from one kind of a crisis to even another, being still more serious and thereby seemed to be slipping into perhaps, towards some violent revolution. And in such fluid and dangerous situation to add even further socio-economic-cum-political wounds to the Italian Nation-State; dark-clouds of the already impending First World-War to begin with, began menacingly bursting all over the European continent; and obviously then Italy too in anyway, couldn't have remained untouched. But with some typical naivety the then power-centres in Italy wanted their country to remain aloof first, and tried to pose as being non-participant and neutral and then quite amazingly soon wanted to garner bigger monetary as well as, some territorial gains as colonies without seriously realising that in the bargain, the Italian Nation-State was ultimately going to rather exploitatively be trapped in the wartime great-games of either side of the clashing alliances of the more powerful European nations since at the outset itself, she had already refused to join her German Allies. Hence England and France, being the more resourceful nations offered Italy bigger 'carrots' by means of ready cash and in addition, promises for new colonies too after the war so that, Italy couldn't refuse from joining on their side to fight the war against Germany and her Allies. It is rightly said always that, **everything is a fair-game in both love and war** in this world. But to the great misfortune of Italy; soon **consequent to the Bolshevik revolution in Russia** and its multiply disturbing fallouts in Europe and that too, even before the war ended did upset the entire game-plan causing, severe disappointments and setbacks for the Italian nation.

All these brought-in fresh and much more severe economic crisis in Italy, making her perhaps even to look like a 'failed State' in Europe. All these caused a series of complicated labour-problems and also economically crippling strikes by the factory-workers; since engaged in different kinds of industries and forcing the factory-owners to declare lockouts. And in addition then, substantially the large numbers of Italian military-men since returning home from the war got demobilized, with no tangible jobs available for them and thus adding to the already huge numbers of unemployed in the nation. During those dangerously momentous times in Italy soon after the World-War I, at the political-level Socialist Party was having some sway, with middle-classes having leading roles, and there-in heated discussions ensued under the inspirations of both the French as well as Russian Bolshevik revolutions, as radically possible role-models for their guidance. **But none had any rationally workable ideas to even somewhat successfully steer Italy, from this highly menacing and problematic morass**. Well, in the substantially broader contexts in this regard and of course also as perhaps a historic conjecture, we may say that like India, Italy too having a legacy of the very rich and glorious civilisational past since belonging to the bygone ancient-era, had been lacking and may be probably to a lesser degree—the multiple-aspects of **'essential strategic-culture and thoughts'** and hence these too, had been pivotally responsible for the then Italian morass in Europe. But unlike India, Italy being in the Western Europe had not missed the multiply advantageous fallouts of indeed **historically unprecedented—the momentous 'Industrial-Revolution'** but perhaps the somewhat traditional lack of the other above mentioned aspects of the no-less paramountcy importance, was most likely preventing the then Italian leadership from pulling-out the nation from the pit of problems.

During those highly turbulent times in Italy there were historically two sets of the parallel and simultaneously advancing-cum-enlarging problems, then confronting Italy: one being the severely economic and the other happened to be almost the total lack of the multiply capable as well as, substantially visionary leadership to take command of the situation and successfully lead the nation. And so the recipe for further continuation of highly troubled-times for the Italian peoples and their Nation-State was unfortunately put actively in place. In the relevantly allied and much larger contexts, we may say that traditionally socio-cultural-cum-politico-economic multiple-aspects of the peoples concerned in any nation, do remain always very important, but the multiply visionary and indeed constructive as well as the far-reaching leadership aspects in any extraordinarily given situation, are always decisive and of the paramountcy importance. Now so far as contextually the historic aspects of Italy were concerned, during those dangerously unsettling and turbulent times, **Benito Mussolini** came onto the then indeed highly turmoil-ridden Italian-horizon, with his all sorts of highly fixated and largely discouraging and confused ideas. This subsequently proved him to be far from being capable of tangibly solving the nation's multiply-daunting problems. In any case interestingly, **Mussolini was no George Washington, Mahatma Gandhi or Ho Chin Min in any manner whatsoever**.

In precisely the above multiple-contexts concerning the Italian nation then, **Jawaharlal Nehru** has very informatively mentioned that, "...**certain volunteer groups that had been formed in 1919, out of the demobilized soldiers, by Benito Mussolini, *Fasci di combattimenti*, 'fighting groups', they were called, and their chief function was to attack, whenever an opportunity arose, socialists and radicals and their institutions. Thus they would destroy the printing press of a socialist newspaper, or attack a municipality or co-operative association under socialist or radical control. The big industrialists and upper bourgeoisie generally, began to patronize and finance these 'fighting groups', in their fight against labour and socialism. Even the government was indulgent towards them, as it wanted to break the power of the Socialist Party**.... Who was this Benito Mussolini who had organized these fighting groups or *Fasci di combattimenti*, or fascists, as we might call them for short? He was a young man then (he was just fifty then—*i.e. in 1933, when Nehru wrote this*—having been born in 1883) who had a varied and exciting career. His father was a blacksmith who was a socialist, and Benito therefore grew up with a socialist background. In his youth he became a fiery agitator, and was expelled from several Swiss cantons for the revolutionary propaganda. **He attacked the moderate socialist leaders violently for their moderation. He openly approved of the use of bombs and other methods of terrorism against the State**.... After the war, Mussolini stopped calling himself a socialist. He was at a loose end, disliked by his old party and having no influence with the working classes. He began to denounce pacifism and socialism and, at the same time, even the *bourgeois* State. He denounced every kind of State and, calling himself an 'individualist', praised anarchy. This was what he wrote. What he did was to found *Fascismo* or fascism, in March 1919, and enroll the out-of-work soldiers in his fighting squads. Violence was the creed of these groups and, as the government seldom interfered, they grew in daring and aggression.... At last, in October 1922, the fascist bands, directed by regular army generals, marched on Rome. The Prime Minister, who had so far tolerated fascist activities, now tolerated martial law. But it was too late, and the King himself was now on Mussolini's side. He (the King) vetoed the martial law decree, accepted his Prime Minister's resignation, and invited Mussolini to be the next Prime Minister and form his ministry. The fascist army reached Rome on October 30, 1922, and on the same day Mussolini arrived by train from Milan to become the Prime Minister.... **Although fascism had no principles or ideals, it had a definite technique of violence and terrorism, and it had a certain outlook on the past which helps us a little to understand it. Its symbol was an old imperial Roman symbol which used to be carried in front of the Roman Emperors and magistrates**. This was bundle of rods (*fasces* they were called, hence *Fascismo*) with an axe at the centre. **The fascist organization is also based on the old Roman model even the names used being the old ones. The fascist salute, the *fascista*, is the old Roman salutation with the raised and outstretched arm**. Thus the fascists looked back to imperial Rome for inspiration; they had the imperialist outlook. **Their motto was 'No discussion'—only obedience**, a motto suited to an army perhaps (*even in a soldierly professional, vibrant and institutionalised army, there are professionally explorative and knowledgeable discussions, before the final-orders are finalized and issued.*), but certainly not to a democracy. Their leader, Mussolini was ill, Duce the dictator. As their uniform they adopted a black shirt, and they were thus known as the 'black shirts'.... **Mussolini became the all-powerful dictator of Italy**. He was not only the Prime Minister, but at the same time he was the Minister for Foreign Affairs, the Interior, the Colonies, War, Marine, Air, and Labour. He was practically the whole Cabinet. The poor King retired into the background and was seldom heard of. Parliament was gradually pushed aside and became a pale shadow of itself. **The Fascist Grand Council dominated the stage, and Mussolini dominated the Fascist Grand council.** (Nehru, 1934, pp. 946, 948-952 & 958) (***So by all these, the recipe was indeed complete for another saga of a further period of historic turmoil in Italy.***). Let us look a little more deeply into Fascism and try to find out what it is. It glories in violence and hates pacifism. Mussolini writing in the *Encyclopedia Italiana* says:

> "Fascism does not believe in the necessity or utility of perpetual peace. Therefore it repudiates pacifism, which conceals a refusal to struggle and an essential cowardice—in face of sacrifice. War and war only, raises human energies to the maximum of tension and seals with its nobility the peoples who have the courage to accept it. All other trials are substitutes they do not place the individual before the choice of life and death."

In the above quoted lines historically, Nehru has been able to very informatively give us three main points, in his indeed well-known precise and scholarly language: (a) what were the overall socio-political crisis-ridden conditions, before Mussolini could grab unquestioned dictatorial powers in Italy during the 1920s; (b) what precisely means by Fascism and, how Mussolini himself perceived and defined it; (c) how and through what all means as well as manipulations, Mussolini finally succeeded in becoming Italy's Prime Minister to eventually become the dictator, with his typically unscrupulous craftiness and double-talks and so on. Consequent to his misdeeds, what all happened to the Italian Nation-State finally, till the end of year in 1945 would be looked at later while we delve into the topical aspects of the **historically momentous Second World-War (1939-45)** in this chapter itself. Briefly, in retrospect we clearly notice that by all counts, Mussolini was indeed 'tailor-made' to join Adolf Hitler of Germany; to jointly fight against the Allied Powers, when callings for the already impending Second World-War came. We may recall that in various contexts, we have interestingly observed that for almost all kinds of the historic developments and events, their seeds do remain mostly buried in the past, for their perhaps subsequently time-bound and further history-making incubation, onto the man's 'expanding civilisational-canvas' in the world.

Interestingly concerning Mussolini as a man and also on his personality traits, a well-known **Fellow of the Royal Historical Society and a famous orator during the Second World-War Era, Mr. E.L. Hasluck** has written that, "Benito Mussolini has been called 'the Sawdust Caesar'. At the time this epithet was first applied to him, there was very little sawdust in the Italian Dictator's make-up—and not much of the Caesar. **When one considers the corruption of the Old Italian Parliament, the incompetence of the pre-Fascist Governments and their inability to keep order—including an inability to suppress the blatantly militant activities of Fascism—it is necessary to admit that Mussolini's early work for Italy was by no means all destructive. Mussolini, for all his theatrical bombast, did more than any other man for half a century to weld the Italian people into a consciously united nation, and his development of colonial enterprise, his road building, his extension of agriculture and his suppression of Mafia brigandage in the south entitle him to some credit marks. It was when he began to follow the will-o'-wisp of imperial expansion that the *Duce* started down the slippery slope to ruin.**" (Hasluck, p. 113)

Now before coming to the multiply contextual aspects of the 1939-45 Second World-War, we may briefly look into another and also rather unprecedentedly far-reaching historic event and which too remains an altogether different kind of the momentously contributory causative-factors, towards the triggering processes of that most devastating the 1939-45 war itself. And obviously in these numerous contexts, that historic event happens to be precisely the **'Great Depression of 1929-31/32'**, thereby suddenly bringing-in worldwide politico-economically the previously unknown and so overwhelming devastations into the multi-dimensional-cum-multi-layered systems of the international trade and commerce since during by that time, already being proactively in-place—towards the multiple-functioning of the world-economy. The above mentioned in fact, the sudden historic event of depression is of global economic dimensions and different experts worldwide, have expressed somewhat varying kinds of their perceptively analytical thoughts in this regard. Well, those are principally the subject-matters as well as the knowledge-domain meant for the learned experts to discerningly debate and deliberate upon, for the illuminating benefits of mankind the world-over. The multi-dimensional complexities of the above mentioned 'Great Depression' during in fact, the then pre-Second World-War era and that too, within almost just a decade after since the earlier already talked-about indeed, the botched-up end of the First World-War (1914-18) can perhaps clearly be gauged by one historically astonishing reality that, when it suddenly manifested itself in the year 1929, even no leading experts on matters concerning the multiple-aspects of global-economics and systems of financial transactions in the front-line nations in Europe and America too, or even the official bigwigs being in-charge of the highest level matters-finance in those very nations; had any sort of inkling that this globally far-reaching economic disasters, were about to devastatingly visit the world—**politico-economically almost overturning the globally multiple economic-systems.**

Hence, when it did suddenly come in 1929, those leading European and American nations were yet to recover fully from indeed the shattering shocks of the recently bygone First World-War, and were yet to even substantially repair the collateral-cum-other kinds of damages inflicted upon them by this 1914-18 war, which now in some historic contexts, may also be called as a 'futile war'. Broadly the plain reality in this regard, was that on the sudden appearance of this unprecedented global economic-disaster; the above mentioned experts and also the concerned bigwigs in the corridors of powers in those affected leading nations, could have no perceptive clues as to how to come to the grips over this gigantic menace and thereby also be able to even partially steer their respective nations out of this so crippling an economic-morass, of indeed such global-reach and implications. In fact, it was something like a nightmare to begin with and hence, it really needed some outstanding ideas and indeed rare stroke of genius to shatter the cobwebs of this disaster for pulling out the affected nations and also, gradually but quickly, put the earlier global systems of economy back on their rails. What in fact the whole world initially saw through the extremely astonished eyes of both the leading economic and political experts as well as the ordinary peoples all over was **that on October 1929, the New York Exchange suddenly crashed and thereby, sending all sorts of indeed shattering shock waves into the world-economy.** And these in turn, unleashed totally mind-boggling and also uncontrollable chain-reactions of the extremely adverse events. In fact, the rich, industrialized and powerful nations in both Europe and America happened to be the hardest hit and moreover, especially those leading nations were subjected to much more damages and violence during the then recently bygone First World-War, just a decade ago. They were yet to fully recover and substantially repair their already shattered economy and consequently to begin with, this colossal depression came as too much of a burden on them to endure. There were numerously shattering fallouts, and in Europe and America alone—many millions of people were thrown out of their jobs and the various means of livelihood; and so the streets in their cities got almost choked with unemployed and, even hundreds of thousands of the all sorts of bankrupt peoples. Peoples all over, just didn't know what to do, where to go and how to even partially cope-up with the crises staring into their panicky eyes. **Even industrially underdeveloped and principally the agricultural nations like India too were not left untouched then, as the prices of their various farm-produces, had begun falling alarmingly. If in the industrially more developed western nations, their various kinds of the factory-produced finished goods were piling-up in their markets and lying unsold then in principally the agricultural-economies, their farm-products began accumulating and lying unsold**. And thereby in both the worlds numerously high cases of desperate suicides too, started getting reported. So almost throughout the world, people began losing all hopes, became depressed and desperate as their lives became untenable and full of the all sorts of sufferings.

One well-known British economist, **Professor Keynes** advocated new ideas that later came to be broadly known as his **'macro-economic theory'** and this concept came in handy in the USA too. Governments of these nations were inspired and also needed to perforce create jobs by starting the many necessary infra-structural as well as different developmental projects and even print more paper-currencies in a controlled and regulated manner, to pay their people for the jobs done by them so that, the much needed purchasing-powers do begin to somewhat sizably trickle into their hands and thereby, trigger their various kinds of the consumer-sector markets—to gather some mobility and start gradually and consistently rolling again as well. These since well-conceived initiatives did also result in a couple of other valuable fallouts and those being that prices soon stopped from falling and the further sliding-down of the concerned stock-exchanges too were checked. In spite of all these possible steps then being taken the 'wounds were too deep' that, when any further fall of the New York Stock-Exchange index was arrested by July 1933, the share-prices were able to command just about fifteen percent only of their 1929 market-prices. But by then all was not so bleak and the tides had indeed begun to turn, gradually and somewhat steadily for the better in this regard. But differently more affected and also at the same time, somewhat inward looking and more impatient nations: especially Germany and Japan took perhaps such drastic steps that soon; new kinds of the quite serious geostrategic and politico-military problems, began surfacing onto the already turmoil-ridden global-horizon of those highly chequered times. In retrospect now, in the larger perspectives since pertaining to these multiply momentous historic scenario we may somewhat discerningly opine that, all these were turning into a queer and again fast emerging and very fluid situations, and also containing rather some very quickly incubating seeds of the highly dangerous fallouts, i.e. something like probably a much devastating war for the already suffering mankind the world-over in near future.

This devastating global-conflict soon turned to be the 1939-45 Second World-War. In our contextually relevant perspectives, we would very soon begin to briefly discuss some specific aspects of this historic global war with its certain new as well as, the totally unexpected kinds of twists and turns in the multiple-paths of on-going and the further making of man's living-history, from the second half of the twentieth century onward. As the further years have been rolling-by; these multiply new paths of the onward makings of man's living-history now, soon turned-out to be **so momentous and far-reaching that altogether a new 'Super-Age' have since dawned in the presently ongoing modern-era and thereby, multi-dimensionally bringing-in highly complex and indeed the mind-boggling transformations all round**; and these in turn triggering multiply far-reaching qualitative changes in both the lives as well as, thought-processes of the mankind.

Precisely in these multiple-contexts concerning the above mentioned 'Great Depression' **Richard Overy** has written that, "The crash of 1929 was a symptom as much as a cause of the worldwide slump that followed. The world economy had been weakened by the First World War and the massive debts it generated. It proved impossible to re-introduce the gold standard system fully, or to revive an effective multilateral trading system. Over-production led to falling prices and declining profits. Despite the boom in America (*little earlier to this great depression only.*), business confidence elsewhere had been low and investment sluggish. When the US stock market crashed, it was against the background of an already declining and fragile world economy.... The effects unfolded slowly. **Protective tariffs were set up worldwide to save domestic industry. Even laissez-faire Britain adopted Imperial Preference in 1932, a protected trading bloc within the empire. World trade in 1930s never recovered to 1929 level, and bilateral trading agreements came to replace liberal system of multilateral trade and exchange**. Germany in 1934 adopted a 'New Plan' for state-regulated trade, and in 1936 a programme of 'autarky', or self-sufficiency.... **Declining trade encouraged domestic sources of economic revival. In Britain, Sweden, Germany and the USA; experiments in state work-creation projects soaked up some of the millions of unemployed**. In 1933 **Roosevelt** introduced a package of recovery policies known as the **'New Deal'**, which extended state regulation of the economy in a country with little tradition of such government activity. In Britain and Germany state regulation by the late 1930s had produced what was called a 'managed economy', a forerunner of the mixed economies of the post-1945 era.... **In more vulnerable economies such as Germany or Japan, radical nationalist groups argued for economic empire-building and an end to the old liberal capitalist order. In France and Spain economic crisis stimulated communism or anarchism, sparking prolonged social conflict. Only in Britain, with its large empire markets, did democracy survive and a modest prosperity set in during the 1930s. In much of Europe and Latin America various forms of dictatorship came to replace parliamentary systems now irrevocably associated with economic disaster**." (Overy, 2008, pp. 280 & 281) (*As we may pause a little and peep deeper into the above quoted lines; we may very clearly be able to discern* ***the then highly unsettling and fast emerging states of both the world-affairs as well as, the world-economy****. The above discussed multiple-aspects of the then global 'economic depression' apart; consequentially the highly alarming and fast manifesting—the many kinds of indeed the overwhelmingly sweeping geostrategic and politico-military upheavals and turbulence were soon going to be much more menacing as well as, violently destructive for the entire mankind, in fact without any such earlier precedence in the pages of man's multiply vast living-history itself.*).

During the above discussed highly turbulent times leading to the 'Great Depression, during the post-1914-18 global war in the first half of the twentieth century and thereby the setbacks triggered by the above discussed numerously momentous setbacks tragically perhaps, the otherwise much sought after peace itself happened to be the mostly as usual, no less elusive for the mankind. It was indeed tragically a queer situation again during that precise timeline when the initiatives in this regard, had almost completely been passed-on into fast-moving domain of the variously manifesting historic developments and events and consequently man happened to be broadly looking like with hardly any 'tools' left in his hands, to steer the world to perhaps prevent any further war. Hence, dark-clouds of another much more extensive and destructive global-war had already begun gathering, to again engulf the mankind with much greater tragedies. **So come 1939, and Second World-War came so menacingly to visit the unfortunate humanity and consequently for the next five to six years, unleashing multiply a chain-reaction of the all sorts of violence, deaths, and destructions in the world**. It can be said that due to various kinds of the ironical omissions and commissions on the part of political as well as, military leaders and their advisers coming from the different fields of expertise, in the leading and also the then imperialistically obsessed nations in Europe, the fast incubating and the fundamentally causative-seeds—of this unprecedentedly momentous war may perhaps quite naively, had been already sown during the mankind's highly chequered recent past itself. And these highly momentous follies on the part of man himself immediately after the so-called 'botched-up' end of the 1914-18 First World-War and also, during the callous and hotchpotch process of the very formulation of the terms and clauses of the 'Treaty of Versailles' onward, any worthwhile space for even somewhat meaningfully influencing the multiply continuing processes of since the on-going making of man's living-history, had already slipped from his hands. And from then on, the numerously manifesting and historically far-reaching developments and events had perforce gathered some kinds of their own uncontrolled directions and momentum subsequently, and indeed so tragically injurious for the man himself.

As also brought out by Richard Overy in his above quoted paragraphs, consequent to the already discussed unsettling fallouts of 'Great Depression', due to their own perhaps perceptively fixated reasons, both **Germany in Western Europe and Japan in the Far-East** felt quite alarmingly threatened politico-economically and at the same time, they also felt both geostrategically and politico-militarily isolated as well. **On top of it since the somewhat ancient as well as during the whole of medieval ages, they too had their traditionally different but, diehard and feudalism-oriented cultures of the very strongly fixated nationalism, imperialism and historically the extreme kinds of militarism too**.

Now precisely coming to the highly complex fallouts of the 'Great depression', we do find that due to their own kinds of perceived reasons, both Germany and Japan felt threatened and isolated and at the same time various kinds of already operative diplomatic channels between principally the then imperial Britain and France on one hand, and those two other nations didn't happen to be essentially functional in any vibrant and also the much needed mutually trustworthy manner to reduce the information-gap and also, rather speedily cultivate the essential and wider environment for the maximum possible co-operation, so that the otherwise already fast manifesting urges for going into the recklessly extensive military-conflicts are kept within any tangible control by the Parties concerned. But historically in any case, these highly complex and fast moving situations were understandably no simple affairs, and hence the emerging challenges were indeed extremely daunting. Moreover, Germany was still unstoppably sulking under the avoidable humiliations, since meted out to her by the victorious Powers at Versailles in the year 1918 and that too, immediately after the First World-War. Moreover, in many nations in Europe and South America the different kinds of dictatorship, too got planted and thereby, turning the already messy situations around, into their highly dangerous proportions. During such extremely turmoil-ridden scenario then prevailing in Germany, **Adolf Hitler** came onto the already problematic German political scene and with his no peaceful but hidden ideas, got himself quite craftily elected as the Chancellor of Germany in 1933 and then subsequently soon, he also managed to become Fuhrer too in 1934, which ultimately transformed him into the all-powerful, unquestionable as well as, ruthless dictator. And we may recall that broadly in similar contexts, and also during few years earlier, we have since briefly discussed—perhaps the no less craftily as well as quite manipulative rise of another eccentric and ruthless dictator in Germany's southern neighbour too, i.e. Italy; in historically the highly disturbing persona of **Benito Mussolini**. And therefore, it may appear to be equally befitting and also relevant to come to certain questions, related to precisely the human-level personality traits of **Hitler**, which were instrumental then, to soon transform him into so dominating and strong a dictator; as be able to engulf almost the whole of world into a kind of indeed the firestorm of war. **What sort of man Hitler was and since, what were the dangerously decisive 'grains' in his psychological make-up**?

Concerning Hitler as a man and his personality traits, **E.L. Hasluck** has mentioned that, "**From long before the armistice Hitler, as an obscure N.C.O., had been more uncompromisingly militarist and authoritarian than the most aggressive Generals and *Junkers*. He was that rare phenomenon, a poor and humble working man whose mind lived perpetually in the realm of high politics of the most blatant imperialist type**. The sabre-rattling colonels naturally looked with sympathy upon Adolf's little National Socialist Party in Bavaria, since alone of the popular organization it appeared to stress the martial and militant aims of the Kaiser's Reich, and some officers supported the Nazis with subscriptions; but Hitler never allowed any army organization to control his movement. Rather he accepted the help that the army imperialists could give him and courted the Reichswehr (*i.e. German Armed Forces*) leaders as potential commanders of his future all conquering army. Nevertheless, without Hitler's Nazi movement it is highly improbable that the German imperialists could have been able organize for the next war in anything like so short a time, and they certainly would not have done it so comprehensively or with such enthusiastic support from the masses.... Even Hitler's success in capturing the allegiance of the masses was due to his programmes of social reform than to his denunciations of the **Diktat** of Versailles or his belief in an expansionist future for Germany. **All political Parties in a democratic State tend to modify their programmes for the purpose of vote-catching, and Nazi Party was by no means exceptional in this respect**." (Hasluck, pp. 12 and 13)

We learn about Hitler's personality traits, his mind, thoughts regarding his home country and the external world, and also his highly obsessed geostrategic and politico-militarily extra-ordinarily aggressive ambitions to get some valuable clues towards his future actions, and course of the soon coming Second World-War. In fact, besides his not so common dictatorial ruthlessness, his real politico-strategic genius lay in his crafty socio-political designs to keep his well-organised and multi-dimensional war-preparations, the exceptionally well-guarded secrets; both from his ordinary fellow-citizens as well as, from the eyes and ears of the very foreign powers, with whom he was soon going into this prolonged and bloodiest war in the man's living-history. It is also equally true that the very recently bygone Hitler's era, didn't belong to any period of rather the long-past medieval-age, and needless to say that by that time so many differently far-reaching and also revolutionarily transforming fruits, consequent to the **'Industrial-Revolution'** in the multiple-areas of transport, communication and also, so many other socio-economic-cum-political systems too, had since already begun manifesting fast onto the man's expanding civilisational-canvas and thereby, bringing-in new kinds of tools and capabilities for the intelligence-gatherings in his hands. And still his country's next door neighbours, no less techno-industrially advanced and further developing; and in addition also since the archrivals, in the forthcoming war Second World-War, couldn't have any worthwhile knowledge of Hitler's multiply momentous plans, and the feverishly proceeding—military-centric preparatory activities on the Nazi Germany's part; which could have shaken them up little earlier, from their deep slumber and geopolitical lethargy in this regard.

Often, strange are the ways of even otherwise not so naïve and also not so ill-aware people, since authoritatively stationed in the highest corridors of powers in any concerned societies/nations, rather amazingly turning out to be so clueless and also showing some kind of blankness or hollowness, in their thoughts as well as, in the decision-making processes and thereby losing their geopolitically valuable strategic-space and, seriously harming their own nations in many ways. Historically in a way precisely, this happened ironically, with respect to France vis-à-vis Nazi Germany, during the year 1936 and, i.e. just prelude to the 1939-45 war. In our present contexts at this precise point the particular incident, which relates to the amazingly easy occupation of the strategic Rhineland on the west of River Rhine, by Hitler's just a couple of Infantry Units and also since located between the eastern and western borders of France and Germany respectively, during the month of March 1936. Another militarily important aspect prevailing momentously in favour of France, happened to be that of the well-prepared as well as heavily reinforced—but rather the 'obsolete' **French Maginot Line** was also passing through the France's eastern borders, since facing the Nazi Germany. As we may begin to delve little deeper into these historic aspects, we do find that the precise point in this regard, stands out that during 1936, Hitler was not yet substantially prepared to launch any such terribly risky military misadventure and still with his rather highly reckless audacity, Hitler issued orders to his **war minister and Chief of the Nazi Armed forces General Werner von Blomberg** to march into the Rhineland for the German occupation. Hitler was rather audaciously indulging in a gamble and also, playing a great bluff. But both geostrategically as well as politico-militarily France, in spite of her tremendously stronger military position on the spot remained in the dark and thereby, Hitler could have a smooth walk into the highly strategic Rhineland, and occupy it without any resistance. The amazing fact was that when just two or three Nazi units were walking into the Rhineland, France had already brought-in thirteen divisions towards the Maginot Line to man and reinforce the same; but they remained idle, as none of them was ordered by the French High command to move forward and chase away or apprehend the intruding Germans. **No wonder that sometimes, the hard-facts, since as already recorded into the relevant pages of man's living-history; do appear to be stranger than the fiction.**

What an ironically amazing situation it was then in Rhineland as, if the French would have remained up to the mark in their intelligence gathering system with its own vitally needed institutional dynamism, France with just one division then, would have easily moved forward and smashed the German units with hardly any cost to herself—both in equipment as well as, the valuable military lives. It was historically a typical far-reaching situation for France, during her this vitally crucial tryst with destiny; when multiple-aspects of the French **'essential strategic-culture and thoughts'** were found to be so much derailed. On the other side, knowing all these highly dangerous and disturbing ground realities then prevailing, it was no wonder that the General Staff of the German High Command in fact mortally worried and shaky, General Blomberg was almost literally shivering, while passing-on Hitler's these audacious orders to his troops. But Hitler was adamant and quite surprisingly perhaps, he was sure in his belief that, France was certainly not going act against his troops in Rhineland; and so he was going to have an easy walk over there. His luck then was favouring him and at the same time; Hitler's immediate larger aim happened to construct the formidable **'West-wall' i.e. the famous German Siegfried Line, to effectively stall any future military assault against Germany from the West**. It was all a very sad commentary on France's the then prevailing strategic thoughts and vision. During the late 1930s ironically, France like her other Allies in Europe, was still living in the past and thereby, almost all members of the Western Allies barring somewhat, the USSR in the East, remained callously 'disjointed' in their rather mutually supportive and complementary—in the essential areas of their geostrategic-cum-politico-military oriented visionary thoughts and preparations. And in contrast on the other side, Nazi Germany was since feverishly engaged with maximum possible secrecy, in her well-planned and ruthlessly aggression-oriented preparations.

Specifically in these larger contexts, in *Reader's Digest: Illustrated Story of World War II, 1969* (p. 52) it has been very precisely mentioned that, "And with reason Minister of War, who five days before had issued in his own handwriting the order to march, was losing his nerve. But the French never made the slightest move to oppose the Germans. **The most Gen. Maurice Gustav Gamelin,**

Chief of the French General Staff, would do was to concentrate thirteen divisions near the German frontier to reinforce the Maginot Line...Even this was enough to throw a scare into the German High command. Blomberg, backed by most of the officers at the top, wanted to pull back the battalions that had crossed the Rhine. As one of them testified at Nuremburg, 'Considering the situation we were in, the French Army could have blown us to pieces.'.... It could have—and had it, that almost certainly would have been the end of Hitler, after which history might have taken quite a different turn, for the dictator could never have survived such a fiasco.... It was Hitler's iron nerve alone which now, as during many crises that lay ahead, saved the situation and, confounding the reluctant generals, brought success. Confident that the French would not march, he bluntly turned down all suggestions by the wavering High Command for pulling back.... 'What would have happened' he exclaimed on March 27, 1942, in recalling the Rhineland coup, 'if anybody other than myself had been at the head of the Reich? Anyone you care to mention would have lost his nerve. I was obliged to lie, and what saved us, was my unshakable obstinacy and my amazing aplomb.'.... Hitler's successful gamble in the Rhineland brought him a victory more staggering and more fatal in its consequences that could be comprehended at the time.... At home it fortified his popularity and his power, raising them to heights which no German ruler of the past had ever enjoyed.... Finally, and above all, the Rhine land occupation, small as it was, opened the way to vast new opportunities in a Europe whose strategic situation was irrevocably changed...For France, it was the beginning of the end. Her allies in the East—the Soviet Union, Poland, Czechoslovakia, Rumania and Yugoslavia—began to realize that France would soon not be able to lend them much assistance because of Germany's construction of a Westwall behind the Franco-German border. The erection of this fortress line, they saw, would quickly change the strategic map of Europe, to their detriment. They could scarcely expect a France which did not dare, with her 100 divisions, to repel three German battalions, to bleed her young manhood against impregnable German fortifications while *Wehrmacht* ('Armed Forces') attacked in the East. (*Indeed ironically, how momentously true as well as prophetic, all these historically came out to be later.*)."

In spite of the above elaborated incident of Hitler's aggression as well as, the easy gains in 1936 into the strategic Rhineland; even the leading Western European Powers like Britain and France, rather 'refused' to wake-up from their deep slumber, causing highly dangerous geostrategic-cum-politico-military shortsightedness and all these were going to shortly cost them very dearly in many ways so much so that after 1945, the entire geopolitical situation was henceforth going to globally change forever. Hence in these larger perspectives, it has been rightly said that the man's living-history since always continuously in the making, favours or fears none and accordingly spares none whatsoever, in its perpetually on-going march onto the 'expanding civilisational-canvas' of the mankind in this world.

Hitler was ruthlessly clear and focused in his thoughts, planning and actions in the beginning; but subsequently as the 1939-45 wartime situations began to turn adversely for him principally since 1943 onward, any decisive control of events and also the geostrategic space for any valuable initiatives too, fast began to slip from his rather dictatorial hands. As we may begin to objectively observe these momentous developments and events, since then taking place in Europe during the mid twentieth century, we with our contextual criticality may also observe that the man's living-history precisely in this regard, had already begun turning-over to repeat itself rather dispassionately as well as distantly in its own way, towards facilitating the impending downfall as a fait-accompli, for the concerned dictator. And consequently, countdown for Hitler and his globally unsetting regime, without indeed any such historic precedence earlier; had begun and, the clock for him had already started clicking too. Historically, it is a different matter that in spite of all these fast and momentously emerging scenario then, Hitler regressively took much time to realise these hard realities and that too, when he found himself inescapably surrounded and perforce confounded in his secret underground accommodation bunker in Berlin, along with his consort. In these contexts precisely at this point, man's living-history was rather decisively taking-over command of its own since ongoing making and ironically but rather inevitably in the process; Hitler found himself reduced to a mere 'pawn' in the history's unsparing hands.

Hitler was a master-expert in the games of bluff and deceit. And he quite unscrupulously in a rather unashamed manner used his skills, to the best of his and also his regime's advantages almost throughout the 1930s decade towards secretly relentless military preparations for war, to avenge his country's humiliation at Versailles. Precisely in these perspectives, the earlier referred book (*Reader's Digest Illustrated Story of World War II*, 1969, pp. 50 and 51) has recorded that, "To talk peace, to prepare secretly for war and to proceed with enough caution to avoid preventive action by the Versailles powers—such were Hitler's tactics during his first two years in power. He built up the armed services with unflagging energy. The Army was ordered to treble its numerical strength, from 100,000 to 300,000 by October 1, 1934. **Joseph Goebbels, Minister for Propaganda and Enlightenment, was admonished never to allow the words 'General Staff' to appear in the press, since Versailles forbade the very existence of this organization. The official rank list of the German Army ceased to be published lest its swollen roster of officers give the game away to foreign intelligence.... Submarines, which Versailles prohibited, were under secret construction in Finland, the Netherlands and Spain. As Minister of Aviation—supposedly *civil* aviation—Hermann Goering put manufacturers to work designing warplanes, and training of military pilots began immediately under the camouflage of the League for Air Sports. By the end of 1934 rearmament in all phases, had become so massive that it no longer be concealed.... Then, on Saturday, March 16, 1935, the Chancellor decreed a law establishing universal military service and provide for a peacetime army of thirty-six divisions, roughly half a million men. That was the end of the military restrictions of the Treaty of Versailles—unless France and Great Britain took action. As Hitler had expected, they protested but did not act.**" (What indeed, a great irony of history and hence; we find that here too, it took two to tango!).

In the above quoted paragraphs, we find that the generally hidden or not so well-known facets of Hitler's personality, character and, his thought-processes' fixations, have indeed been very revealingly narrated, which were then soon going to push almost the whole world into the furnace of human-history's the most destructive global war and thereby, thrusting upon the humanity—the most extreme and also so far unprecedented levels of deaths, destructions and devastations. Hence whatever could be the historic reasons for all these; it would perhaps be historically erroneous to even contend that, somehow Hitler was pushed into war by the then leading European Imperial-Powers; but rather he knowingly asked for it and consequently, he himself turned out to be the most stubborn chief architect of Second world-War. In fact, every treaty or the so called 'pious' understanding that he arrived at, with any of his western or eastern neighbours was ultimately meant to be violated by him on the slightest pretext later or, at any earliest kind of the emerging politico-diplomatic-cum-military opportunity coming in his largely evil-sponsored ways and thereby, displaying his rather most unscrupulous and stealthily aggressive intentions, regardless of the any kinds of tragedies to other affected nations or even to his own fatherland. In fact broadly speaking, the 'love of power' anywhere possibly in its all kinds of manifestations is known to be universal and hence in the ruthlessly rivalry-ridden, highly challenging and, the always intoxicating areas of power-politics beyond certain degree of limits, is bound to be rather uncontrollably undisciplined and in the process thereby ultimately, destined to finally devour the concerned personality himself or herself.

Therefore in these larger contexts, Bertrand Russell is known to have rightly said that, 'power corrupts and absolute power corrupts absolutely'. Here the 'corruption' has much larger connotations and doesn't mean only financial but, it may even pertain to anything or everything conceivable within the humanity's mental-power's reach and dimensions. Well interestingly in these very specific contexts, in the otherwise highly complex socio-political domain of the games of power-politics within any concerned Nation-State, which is rather ruthlessly as well as even unscrupulously played by the concerned politicians/dictatorial-power seekers; all possible kinds of the selfish-motives-cum-falsehood-based activities, are not only indulged-in exclusively in any kind of the dictatorial-regime; but at times and under certain kinds of unsettling circumstances (*even at times perhaps, willfully created or otherwise*), the same could also be possible in any democracy as well. It may sound rather amazing normally but in reality at times, somewhat even in quite reassuringly functioning democracy; some such very disturbing derailment of the constitutional norms as well as, rather even temporary subversion of democracy has been recorded by the man's living-history as a source of warning-cum-lesson for the days ahead. Specifically at this precise point we may appropriately render sovereign democratic India's example—when during 1975-77, even a democratically elected but rather overwhelmingly power-hungry Prime Minister, for protecting her official chair and satisfying her greed for political-power, did not hesitate to subvert the Constitution itself and even, went so far as towards imposing the most uncalled-for 'state of internal emergency' upon the entire nation. Man's living-history since continuously in the making and also from the beginning itself, has been showing that most of the powerful dictators anywhere and historically during any timeline, have been known to unquestionably wield absolute power over his or her regime, while staying in power. Hence precisely in this regard, Hitler too comes within the same overall historic framework and thereby, as his time rolled-on he gradually became destined to be devoured by the overwhelmingly devastating combination of his unquestioned dictatorial power, as well as massive follies. We may recall that little earlier in this chapter itself, we have since very briefly discussed the making of man's living-history's own and ground-work, in the post-First World-War Italian Nation-State which soon historically became instrumental in ultimately pushing-up **Benito Mussolini** into the seat of the Prime Minister of Italy. Thereby with his all sorts of the extremely unscrupulous as well as, Fascism-triggered machinations, Mussolini could very craftily spiral himself soon, into his new incarnation of Italy's all powerful dictator. Broadly and in somewhat similarly parallel manner and nearly one and half decades later in Germany as well—**Adolf Hitler** could succeed in occupying the chair of the German Chancellor and thereby soon with his no less Nazism-triggered manipulations, he too became Germany's unquestioned dictator and quite interestingly again even substantiating that at times, facts do seem to be stranger than fiction. In these specific contexts, so far as the then Benito Mussolini's Italy and Adolf Hitler's Germany were concerned; during the unprecedentedly momentous 1939-45 global war, both Mussolini and Hitler became partners-cum-allies and comrade-in-arms—for jointly taking-on the Western democratic Powers' Allied military-might, in almost all theatres of war except in the Pacific. But their principal areas of operations were in and around the European continent itself.

On the Allied side and, as subsequently the events turned out to be and especially after the

Pearl Harbour incident in December 1941 in the Pacific, the USA was the mightiest as well as leading Military-Power to jointly take-on the Axis Powers and, with the maximum possible quite visionary joint-creation of the geostrategic-cum-politico-military synergy; globally in the Allied camps. So far as the vast theatre of war in the Pacific and, including the multiple-aspects of geographically—quite a politico-diplomatic-cum-militarily engaging cluster of the numerous island nations in both the South-East and the Far-East regions of the Asian continent were concerned, the main clashing military-might consisted principally, of the then Imperialist Japan on one side, i.e. belonging to the Axis Powers' bloc and then on the other side; principally of the USA and UK jointly on the side of Allied Powers. **Historically it is also true that the unprecedented Pearl Harbour incident since so suddenly triggered by Japan on the 7th of December 1941, with almost total geostrategic-cum-politico-military surprise and thereby within a couple of hours only causing massive naval devastations onto the US fleet since anchored over there and along with, also to the vast shore-based support installations/systems, did not happen all of a sudden out of blue.** Tensions between USA and Japan were mounting and had already begun reaching to dangerous proportions. We have already touched upon some important aspects of the Japanese background, their mindsets and also, their some 'real/imaginary grievances' against European and American seafaring powers, and at the same time Japan's imperialistic greed against the Northern Chinese territories, were getting more pronounced by the day. And accordingly again since 1930s itself, Japanese warlords had already begun all round feverish military preparations for war and understandably in the overall geostrategic-cum-politico-military terms, USA was taking such Japanese initiatives as aggressively hostile acts against not only weaker China then but in long terms, as no small threats to the world peace as well.

As anticipated perhaps, the 'balloon of war' in the Pacific, went up with the Historic Pearl Harbour incident. **Specifically with respect to this the *Reader's Digest Illustrated Story of World War II*** (p. 25), **Capt. Mitsuo Fuchida of the Japanese Navy, who led this momentous attack on Pearl Harbourwrites that,** "The attack on Pearl Harbour had been planned by the Japanese as early as January 1941. The enemy fleet consisted of 6 aircraft carriers bearing 414 airplanes, 2 battleships, 3 cruisers, 9 destroyers, 3 submarines and numerous auxiliary craft. It sailed from the Kurile Islands on November 25, 1941 (U.S. time), under strict radio silence and complete blackout. The northern route to the objective with its prevailing bad weather and high seas was deliberately taken to lessen the chance of detection by U.S. aircraft and merchant ships. Well ahead of the main Japanese force was a group of submarines, 25 to 28 in number. Five of these carried midget two-man submarines fastened to their decks.... At 6 a.m. on December 7 (*1941*) the signal was given for the attack. The assault was made in two waves: the first by 50 horizontal (level) bombers, 54 dive-bombers and 45 fighters; the second by 54 horizontal bombers, 81 dive-bombers and 36 fighters.... There were 94 ships of the US Navy in Pearl Harbour on that Sunday morning of December 7, 1941. Of these 70 were combat vessels, including 8 battleships.... The Japanese planes met no opposition. While their fighters and dive-bombers attacked the airfields at Wheeler, Hickam, Kaneohe, Bellows and Ewa, torpedo and horizontal bombers concentrated on Pearl Harbour. **Surprise was complete**.... The double file of great U.S. battleships was struck again and again. When the first wave of planes had spent its fury, the second wave tore in on the ripped and blazing targets.... **In less than two hours the winged raiders were gone and ships of Japanese striking force were withdrawing into the northwest. The savage assault had taken a serious toll**—7 of the 8 U.S. battleships had been either sunk or badly crippled; 3 cruisers had suffered severe damage, as had 3 destroyers. Of a total 394 U.S. planes, 188 were demolished and 159 damaged; 2403

Americans were killed and 1178 were wounded. The Japanese lost 29 aircraft, 5 midget submarines and 1 full-sized submarine."

Numerous books on the subject-matters pertaining to the various historic battles between different Navies along the civilisational dawn and the continuing advancements, including the matters pertaining to the man's gradually developing wherewithal—concerning precisely the multiple-aspects of man's seafaring adventures and voyages are already available to the students of military-history, research-scholars as well as also, to the military professionals. **And accordingly, among so many of the historically prominent naval engagements down the ages; the Pearl Harbour incident during the 1939-45 global war, does stand out indeed as a case apart.** There could perhaps be no other naval-battle in the vast annals of man' living military-history like that of Pearl Harbour, where within so short a spell of just two hours only, so massive occurrence of deaths, destructions and devastations could have been possible. It is accepted that one prime reason in this regard then, had been the post-Industrial Revolution era's multiply continuing the substantial scientific and technological advancements including the developments in the then new areas of aviation technologies—leading to the far-reaching developments of military technologies but at the same time it is also true that, these specific aspects were no less taking place on either side of the military-divide, since affecting the battle-lines having already drawn between them.

Hence as already aided and abated by the multiple aspects of fast fructifying RMA, i.e. 'Revolution in Military Affairs'; the phenomenally successful application of one of the age-old 'Principles of War', i.e. 'Surprise', by the Japanese military commanders carried the historic day in Pearl Harbour then, with such great success for the Japanese Forces. In a way from the multiple-aspects of the 'profession of arms'—since the dawn of various human civilisations down the ages and, contextually in broader perspectives, we may also opine that almost each of the 'principles of war', in its own specifically situations-based circumstances; can also turn out to be no less and valuably indeed, far-reaching **'force-multipliers'**, in any particular battlefield scenario within an overall warlike situation. Well at the same time, broadly it does also remain a different matter that **the final outcome of any particular major war, during any specific point of the man's living-history depends on numerously other but no less decisive factors, since pertaining to the 'geostrategic and politico-military' related multiple-aspects as being fruitfully perceived and also proactively made use of, in the overall processes of formulating the highest strategic-level war-plans of the rival multi-national alliances**. And needless to say that as we proceed to critically analyse further the same historically timeless aspects, happened to be equally true specifically, in case of the 1939-45 Second World-War as well. In further and even larger conceptual-framework in this regard, we may also say that any fruitful conceptualization of these precisely warfare-related multiple-aspects, always remain proactively linked to again the fruitful degree of the critically continuing phenomena of cultivation, evolution, development and, imbibing of the **multiple-aspects of 'essential strategic-culture and thoughts'**; by the concerned Nation-States and their military alliances. And as since also borne out by the man's living-history, and specifically concerning the above mentioned multiple-aspects in this regard, in the long-run during the Second World-War, finally the Western Allied Powers did decisively score over the Axis-Powers.

Now again coming back to historically the phenomenal surprise—so decisively achieved by the Japanese forces over the US Naval fleet anchored in the strategic Pearl Harbour, this was not any one-sided affair; as generally it is never so in any war since, the dawn of human-civilisations. Man's living military-history, since also always in the making within the much larger domain of the living-history itself; has always shown that if one side is found to be rather more operationally smart and thereby, being able to achieve any relatively minor or even at times phenomenal surprise over its rival forces; then it is also found that the other side has not been adequately alert and quick enough to foresee the enemy's intentions and precise moves well in advance, to be able to effectively counter those moves in the desired self-defense. At times, even some crucial piece of information little in advance—if taken rather casually or somewhat callously; when time is at premium, can turn out to be very tragic and costly indeed. And needless to say that ironically, that's what precisely happened with respect to the US Forces, on the morning of 7th December 1941 in their strategically valuable naval-cum-air base in Pearl Harbour; so strategically important in the Pacific.

Precisely with respect to the above, **Louis Snyder, a leading historian,** in his well-informed article (*Reader's Digest Illustrated Story of World War II,* pages 30 and 31) has mentioned that, "The man who engineered the attack on Pearl Harbour was **Adm. Isoroku Yamamoto, Commander-in-Chief of the Imperial Japanese Navy**. In the belief that 'the fiercest serpent can be overcome by a swarm of rats', the Admiral carefully amassed an armada of aircraft carriers long before December 7, 1941, and supervised the training of a force designed to obliterate the American power in the Pacific.... In mid-October 1941 **Prince Fumimaro Konoye**, who had suggested for moderation, resigned as premier. He was succeeded by **Gen. Hideki Tojo, the 'Razor Brain'**, whose new Cabinet of Army and Navy officers 'smelled of gunpowder'. The anti-American campaign went into high gear.... Not long before, there had been indications that something was radically wrong. But by a combination of unfortunate circumstances the attacking Japanese were able to achieve complete surprise.... There came about an elaborate piece of playacting by the Japanese militarists, planned so that Japanese naval units could have time to reach Pearl Harbour for the surprise attack. On November 14, 1941, Tojo's special envoy, Saburo Kurusu, arrived in San Francisco on his way to Washington to assist the Japanese Ambassador, Adm. Kichisaburo Nomura, in a last ditch effort to 'maintain peace'. 'Kurusu announced to the press that he had come 'to make a touchdown'. It is highly probable that neither he nor Admiral Nomura was aware that they were being used as decoys in a fixed game.... **At 6:45 a.m. the American destroyer *Ward*, on routine patrol duty off the naval base, identified and sank a Japanese midget submarine. No one on the destroyer dreamed that it might have been a part of a large task force.... Shortly after 7 a.m. on December 7 two U.S. soldiers were watching their mobile radar set on the northern slope of Oahu. An oscilloscope signal began to blip wildly. The screen showed what seemed to be a swarm of aircrafts approaching at a distance of 137 miles. One of the soldiers immediately informed an officer by phone. The officer's answer was disastrous for many men at Pearl Harbour. He told them to ignore the signals, assuming they indicated friendly B–17s expected from the mainland. It was a human mistake but a costly one**. Less than half an hour later, through the fleecy clouds over Kahuku Point, came the roar of the first phase of Japanese planes."

In retrospect it may appear to be matters of conjecture only, but specifically regarding the then US fiasco in Pearl Harbour and contextually also from matters-military considerations; these do historically have substantial weightage indeed i.e., had the Commanding Officer of USS Ward and his other key officers had been little more farsighted as well as professionally imaginative; they could have immediately communicated emergency-signals to their superiors back in the base, for the fastest possible operational alertness and going-in for all kinds of counter-measures to keep the base and important assets in Pearl Harbour, as far as possible safe from the impending attack. Moreover on top of all these little later, the officer concerned in the base itself totally neglected the crucial information since passed-on by his men who were engaged in watching their mobile radar, ironically meant for these very purposes only. And tragically the result was so disastrous then, for the US strategic forces and their valuable military assets in the Pacific and interestingly for the Western Allied Forces, then already engaged in fighting grim battles with Hitler's invading Nazi herds in the Europe itself, this Pearl Harbour incident turned out to be rather more than enough to shake-up the whole US Nation from its deep slumber and thereby, getting the USA proactively engaged on the side of Allied Partners; in fighting the Axis Powers simultaneously, both in the West and also in Pacific in the East. Historically in fact, this was indeed no small development then, for subsequently shaping-up ultimately the final outcome of the 1939-45 Second World War. We may recall that little earlier, we have briefly delved upon as to how the Nazi Germany under Adolf Hitler's dictatorship, secretly indulged in the feverish preparations for an all-out War principally with France and in addition, not only with Britain but also with the whole of British Empire with great deceit and geostrategic-cum-politico-military cunningness, since the early 1930s itself. Well interestingly, as a somewhat parallel phenomenon in the Far-East, Japan too was preparing in no less feverish manner for a major war since the 1930s itself to first expand her Asian Empire by militarily occupying as much territories as possible in the eastern part of the Chinese mainland, by swiftly launching her invading forces across the Sea of Japan and also, across the Yellow Sea as well. From matters-military points of view in a way historically, we may also term these swift and surprise attacks by Japan onto the then weaker China, as an eastern example of the Japanese 'Blitzkrieg' as well. It is a historic fact that during those times, the imperialist-cum-militarist Japan was very sinisterly looking for, or even very aggressively trying to invent some sort of excuse to attack through the eastern coastlines of China to occupy more territories for expanding her empire. And accordingly as per her some ulterior motives in 1937 itself, first the Japanese forces were ordered to intrude into China towards her eastern railroads passing through Peking Junction (i.e. now, Beijing) and then accusing China that her soldiers had attacked the Japanese forces. Japan's extremely aggressive military designs became quite clear as, massive Japanese forces invaded eastern China towards Nanking, which was already vacated by the retreating army of **Generalissimo Chiang Kai-shek**. Japanese forces then very ruthlessly unleashed a reign of terror and cruel human-slaughter on massive scale; raping and killing many thousands of innocent Chinese civilians, for spreading their reign of terror.

Obviously then, the USA had a soft corner and friendly relations with the nationalist Chinese leader Chiang Kai-shek and his forces; who were since actively opposed to the Chinese communists —who were in turn then, quite menacingly struggling to come to power in China by any means. So during those years even before the beginning of Second World-War, China was already weak, geopolitically divided and also, quite destructively trapped in an extreme kind of the civil-war between the Mao's communist and Chiang's nationalist forces. It was very bad and trying times for the poverty-ridden as well as, already ruthlessly exploited since centuries—the Chinese masses. Militarily during the siege of Nanking, the Japanese militarism-oriented imperialists didn't even hesitate to attack the US gunboat *Panay*, since anchored in the Yellow river where nearby, the USA had some support-cum-logistic facilities. **In these and also in numerous other contexts then, Japan was looking at the USA, as a powerful hostile power to herself and geostrategically-cum-politico-militarily therefore; in order to clear her hegemonic paths to fulfill her imperialistic-ambitions, both in the Far-East and South-East Asia. Japan quite aggressively wanted the USA to completely clear-off from the entire vast Pacific region at the earliest. So far as China was concerned then, Japan had conceived three point strategic aims: first to slaughter the Chiang's nationalist army; secondly to bloc all supply-routes into China from anywhere outside; and thirdly, compel Chiang Kai-shek to unconditionally surrender before the Japanese High command, accept her hegemonic supremacy over China as being part of the Japanese Empire in Asia**.

Historically it remains a different matter that in spite of the all sorts of aggression into the Chinese mainland, Japan could not achieve any tangible military success in the vast Chinese lands and; as Chiang proved to be no 'pushover' and, his nationalist soldiers duly supplied by the USA fought vigorously against the Japanese invaders. All these obviously led to some frustration among the Japanese fanatic militarists. And as per the above narrated scenario and also Japan's larger strategic ambitions Pearl Harbour incident too, was meant to coercively display the range and extent of the wherewithal of her military-might to the big Uncle Sam, with a somewhat devastating shock and as a clear message to her. In order to pursue her blind and untamed military-cum-imperialistic ambitions and thereby, Japan while feverishly increasing her forces with relentless drive for armaments; even domestically within her own homeland the Japanese warlords since already in power, just did not hesitate to commit any heinous crimes against their own fellow citizens, as early as the 1920s itself. In a sort of the wild pursuits of their aggressive intentions, the Japanese militarists then were found to be even surpassing Hitler in Germany and Mussolini in Italy. Historically therefore during those times and due to all these unsettling happenings; even more than a decade before the start of Second World-War, all was not too well and socio-politically healthy inside Japan for the Japanese themselves then and consequently; the masses were mostly kept quite terrorized and always ruthlessly exploited as well, ironically by their own militarism-obsessed and extremely feudalistic rulers.

Specifically with respect to the above, **famous naval historian Rear Adm. Eliot Morison, USNR (Retd)**, in his paper on 'How and Why Japan Prepared for World War' forming part of the book *Reader's Digest Illustrated Story of World War II* (pages 94 and 98) has written that, "[*In the 1920s a movement closely resembling Hitler's National Socialism was gaining strength in Japan. It envisioned a totalitarian state under Army control. Numerous secret societies, of which the Black Dragon was the most notorious, furthered its cause. It was sometimes called Showa ('Enlightened Peace') Restoration, sometimes Kodo-Ha ('the Way of the Emperor'). At home, its zealots aimed to destroy the power of big capitalists, liberalism and Western influences. Abroad, they wished to achieve absolute Japanese domination over the Far-East, with a dream of Hakko Ichiu ('the eight corners of the world under one roof'), the leading position among the nations, as the final goal. The movement was fanned to white heat by limitations on Japanese naval expansion attained by Great Britain and the United States in the treaties of 1921 and 1930, as well as the U.S. Immigration Act excluding Orientals. The Showa and Kodo fanatics were openly committed to desperate measures*—Ed.].... In the meantime, the Japanese militarists had consolidated their power by the simple means of getting rid of inconvenient persons; during the first half of 1932 the Minister of Finance, the chief manager of the Mitsui interests and the seventy-five-year old Prime Minister, Ki Inukai, were assassinated. It was even planned to kill U.S. Ambassador Joseph C. Grew and Charlie Chaplin (then visiting Tokyo), in the hope of provoking a war with the United States and thus riveting military control on the government.... Government dominated by the military were in during the next few years. When opposition began to develop, terrorists struck again. The refusal of Japan's Finance Minister in February 1936 to sanction a further increase of Army appropriations was a signal for his assassination by Army extremists, together with that of Adm. Makoto Saito, Lord Keeper of the Privy Seal, and Gen. Jotaro Watanabe, the Inspector General. The Prime Minister, Adm. Keisuke Okada, was also marked down, but the murderers shot his brother-in-law by mistake.... The Genro, the elder statesman, chose the next Cabinet, Prime Minister Koki Hirota's, which took office in March 1936, and at once began preparations for war with China. When the Army was ready, it invaded China proper. That was the famous China incident of July 7, 1937, a cooked-up clash between Japanese troops on maneuvers and a Chinese outpost on the Marco Polo Bridge near Peiping (now Peking—*well, even Beijing.*).... For two years the U.S. Government concentrated on diplomatic efforts to 'bring Japan to her senses' and restore peace in China." (*but to no avail as, all these initiatives by the US President then, fell on arrogantly deaf ears of the Japanese Warlords since already firmly entrenched in the seat of power inside Japan, and hence turned out to be the exercises in futility only.*).

So we have now seen as to how since the 1920s as well as 1930s only, Germany and Italy in Europe and then Imperialist Japan in the Far-east; had already momentously begun rather independently but in parallel manners, their preparations to fight again an all-out war against the leading democratic western European and North American Powers. They were not only expanding their Armed Forces majorly and arming them with all sorts of the then modern offensive weapon-systems for their in a way shortly forthcoming future war; but also at the same time, they were very seriously engaged in updating, refining as well as, fine-tuning their matters-military-oriented strategic thought-processes of their 'General-Staff outfits' within their military-command systems, for the impending multiple kinds of warfare as well. **Whereas rather ironically, so far as the leading nations on the so called Allied side, such as the UK and France in the Western Europe and the USA across the Atlantic in North America were concerned; they had not even begun paying any kind of the worthwhile minor attention to their any short or long terms multi-dimensional geopolitical and matters-military related strategic imperatives, for the sake of their own defenses both in the general terms and obviously thereby; certainly not against the menacingly aggressive designs of the hostile Axis Powers, both in Europe and Asia.** France had still not come out of the groove of obsolete military thoughts, of the trench-warfare-oriented static defenses. Great Britain was basking in her victory over Germany during the 1914-18 First World-War and at the same time, was also quite obsessively occupied in enjoying the fruits of her vast and global British-empire. In addition, so far as the giant USA was concerned, in a way she was rather understandably then considering herself somewhat reassuringly safe being geographically located far-off on the other side of the globe, hence much beyond the capacity-cum-reach of any kind of the military-forces of both militarist Germany and the imperialist Japan.

And moreover on the top of all these their intelligence-services too, were then functioning somewhat casually and perhaps even on a 'half-sleep' mode. Hence no wonder that, even if indirectly but substantially in a way all these then on the part of Western Allied Powers, were coming as being unintentional but, certainly of great strategic help to the Axis Powers. These are precisely the reasons in this regard and, the multiple-aspects of **'essential strategic-culture and thoughts'**—do also show us that some weighty faults or errors committed in the past cannot be ignored and, if not taken care of at the earliest consequently then they may at times, cause even substantial harms in the coming days. **Needless to say that these are also some of the many precise geopolitical-cum-matters-military related reasons as to why it is generally said and that too, all the more so for democracies that, 'the eternal vigilance is the price of liberty, peace and the national well-being'—always needed to be paid attention to, both ungrudgingly and perpetually by any nation or group of nations concerned**. In these precise perspectives of always paramountcy importance and then, even at the risk of being out of context here, it is considered of some substantially pragmatic necessity that today in the present modern and super-age of the multiply fast-moving high-tech era, any Indian citizen with some geopolitical as well as, 'geostrategic-cum-politico-military awareness do very painfully realise that the sovereign and democratic India's irresponsibly short-sighted and callous 'Politician-cum-bureaucratic combine', persistently keep-on refusing to see prime reasons and, imperatively wake-up to comprehensively understand and learn to evolve their grips onto the multiple-aspects of the overall regional security-environment around India and thereby, to get desirably geared-up towards fulfilling their responsibilities to indeed—multi-dimensionally brace-up the Indian-Nation's both short and long terms' defence preparedness in the mandatory as well as, visionary manners. Even the today's fast-growing Indian youths are asking on their arrogant politicians' largely insincere faces, as to be these in any manner, any kind of the 'tall orders' for them. In fact, it is unquestionably mandatory on the part of India's rather always sinisterly scheming 'Politician-cum-bureaucratic-combine' to prove their worth of their salt and, get rid of their regressively fixated callousness and arrogant ignorance in order to come out clean, take the nation's Armed Forces into sincere confidence and at the same time then to comprehensively begin making their efforts to get hold of the India's these national-security related paramountcy issues and thereby indeed, reassuringly satisfy their fellow-citizens without any delay whatsoever. Already more than half a century has passed since India gained her political independence in 1947 and hence, in today's modern-era it is indeed amazing that still India's 'politician-cum-bureaucratic-combine', needs repeatedly so much of coaxing and chasing by their fellow-citizens to get woken-up towards doing their national duties, and for which they are being regularly paid for and also, along with their substantially expensive perks as well.

Now specifically with respect to the essentially needed preparedness for the impending war then and on the part of both Great Britain and France, **Field Marshal Montgomery** has emphatically mentioned that, "Neither Britain nor France was satisfactorily prepared of war in 1939. How about Germany? In number of trained men she was better off, since Hitler had introduced conscription early in 1935, and in the theory of warfare the Germans had moved far ahead of the victors of the 1914/18 conflict. **It was realized in Germany that a new warfare of vastly increased mobility and striking power was possible with the use of tank and aeroplane. The German army had developed a new tactical doctrine, the *Blitzkrieg*—the essence of which was to achieve a breakthrough and deep penetration by an armoured force, supported from the air**. Aircraft would create havoc among the enemy communications and installations, would assist the field artillery during the advance by attacking ground targets, and would keep the attacking force supplied with men and materials. Tanks, supported by infantry, would achieve the break-through on the ground. **The tactical principles in the *blitzkrieg* were concentration, surprise and speed—the enemy being overwhelmed and shattered with ruthless thoroughness**." (*Hence, no wonder that when the war began, the speedy German blitzkrieg found no difficulty in smashing the French 'Maginot Line'; ironically supposed to be impregnable then, as the France's main strategic defenses against the German invasion.*). (Montgomery, 1983, p. 310)

So Germany along with her ally Italy in Europe, was not only preparing since years ahead of the start of this most consequential and devastating global war in 1939 but also at the same time, had made sufficient efforts to learn from the past, for further honing-up her massive preparations both at strategic and tactical levels in the German Armed Forces with in fact, focused intentions for creating the maximum possible synergy during peace and war. Further in the Far-East as well, the extremely militarism-obsessed Imperialist Japan too was all the more relentless in making all sorts of military preparation with in fact, focused ambitions to menacingly enhance the multi-dimensional offensive capabilities of the Japanese Imperial Armed forces. German High Command since the 1930s only, had realised that during the 1914-18 Great War, it was in a way their great strategic blunder to make all sorts of amendments at the last minute, in the original concepts of the ***Schlieffen Plan*** with certain perhaps half-baked ideas which in fact, brought many avoidable problems later in the field. We may recall that little earlier, while discussing certain contextually important aspects of the 1914-18 First World War, we have also very briefly touched upon some salient features of the ***Schlieffen Plan*** as well. So in Europe the Nazi Germany in combination with the Fascist Italy and then, the aggressive Imperialist Japan in the Far-East Asia, put together as a powerful line-up of the Axis Powers had already begun to emerge fast, as great danger to the world-peace as well as, to the already somewhat fragile economic stability of the world economy. Historically in fact, these powerful constituents of the Axis Powers had become so much obsessed with their fanatic and war mongering Armed Forces, that they had no time to pay any attention towards providing even some of the basic succor to their own suffering and already ruthlessly exploited peasants, workers and other poor peoples among their masses, always struggling to keep their bodies and the souls together. But at the same time, it is also emphatically borne-out by the man's living-history that—howsoever any military under any kind of the tyrannical dictatorship regime may appear to be very menacingly powerful and aggressive, in the end they mostly remain destined to again be militarily destroyed by the democratic or any other similar somewhat people-caring kind of Powers, since coming forward decisively as that concerned dictatorship's geostrategic-cum-politico-military adversaries. And simultaneously perhaps, any concerned dictatorship also appears to be mostly destined to commit even some far-reaching tactical-cum-strategic blunders during the war in a foolhardy manner and also thereby, bringing-in its ultimate doom. Understandably, some may perhaps quite perceptively, even call these as matters of historic conjecture only. Interestingly therefore we may also opine that contextually in the larger perspectives these have also been in fact again, some of the perhaps destiny-guided systems or ways of the further far-reaching making of the man's vast, highly complex as well as, the multi-dimensional living-history onto his expanding civilisational-canvas in the world.

In the above discussed multiple-aspects, **as we further begin to critically delve into certain geopolitically important dimensions of the 1939-45 global war; we just cannot miss to note in a praiseworthy manner, the British Nation's greatly history-making 'tryst with destiny' during the initial years of this war and also, in its extremely adverse and trying times. It is definitely a great saga of the British peoples and their both political-cum-military leaders' resolute courage, dogged determination, wits and great vision; which enabled them to change the very course of the subsequent making of their history. It is indeed one of the all-time great lessons for any truly democratic nation and hence; while considering the overall and quite menacingly persisting geopolitical-cum-security environment-related scenario—around now the sovereign and democratic Indian-Nation and hence, all the more so for the modern and not so well reassuring Indian political-class.**

And accordingly with respect to the above and precisely concerning the initial years of Second World War and the British Nation as well, Field Marshal Montgomery has written that, "The German conquest of the Low countries and France was due to superior tactical methods and leadership; the allies actually had a slight superiority in numbers and equipment: 146 to 126 divisions. **Manstein's (*the German Field Marshal*)** overall plan and the panzer tactics by which it was carried out, particularly **Guderian's (*i.e. another German Field Marshal*)** strong westward thrust, altogether constituted a remarkable military achievement.... **The British people prepared for invasion; for nearly 900 years no foreign invading force had set foot on the soil of Britain, and although nobody in the country knew how the Germans were to be beaten, nobody thought of giving in**. Even Winston Churchill did not see at that time how it was to be done, as he told me in June 1940; but he was in on doubt about the final result; nor was I, **because I saw in him the leader we needed, the man who could rally a defeated nation, and would lead it through further troubles....**

Churchill had been Prime Minister since May, and after the disaster of Dunkirk he called for the spirit of the British people by his courage and oratory:

> **'We shall not flag or fail. We shall go on till the end. We shall fight in France, we shall fight in the seas and oceans, we shall fight with growing confidence and growing strengths in the air; we shall defend our island whatever the cost may be. We shall fight on the beaches, we shall fight on the landing grounds, we shall fight in the fields and the streets, we shall fight in the hills; we shall never surrender'.** (*These are historically indeed very far-reaching and superb examples—as to what all differences and good a brilliantly capable and sincere national-level political leadership can bring-in, even during any very serious crisis situation; especially when the very survival of the concerned Nation-State is at stake. Well and though historically in much limited sense; aren't we seeing some similar top political-level brilliant leadership traits and understandably in the altogether different contexts, in the* ***Indian Prime Minister Late Lal Bahadur Shastri, too****, during the 1965 India-Pakistan border-war?*).

(*And Field Marshal Montgomery goes on further,*): A Home Guard sprang into being. Britain's productive capacity was turned over to the war effort, and much material aid came from neutral America. The re-equipment and enlargement of the army and the adaptation of training to the up-to-date methods got under way. Great things from now on were achieved in the organization of the British army by **Sir Alan Brooke, C-in-C Home forces and later C.I.G.S**. But there was much to be done, and I myself, commanding a corps on the south coast of England, how long we had before Hitler would strike. The armed forces of Germany were in occupation of the whole of Western Europe from Norway to France; Churchill told us that Hitler knew he must conquer Britain or lose the war.

We shall now examine the war in the air in Europe between the summer of 1940 and May, 1941. To prevent the Germans in 1940 from crossing the channel, Britain had to rely on her strength in the air and the sea. The essential preliminary for a German invasion against the island shores of an enemy who was immensely superior at the sea was in command of the air." (Montgomery, 1983, pp. 313 & 314)

And of course as expected by the British High-Command the above narrated extremely challenging wartime scenario concerning the British Nation, soon resulted in the **historic and so devastating aerial-battle of Britain**. Nazi Germany's Air Force—**Luftwaffe**, under the command of the **German Air-Chief Goering**; put all its skills and resources to cripple and force capitulation of the British Nation. But the British Royal Air force (**R.A.F.**) withstood the German onslaught and courageously geared-up fast to retaliate, by causing serious attrition on the German pilots and flying machines and thereby, tide soon began to turn in favour of the RAF: as the aerial-saviour of the British Islands. No. II Group RAF, was engaged in the battle of Britain then, under the command of **Air Vice Marshal Keith Park** gradually but effectively countering the Luftwaffe's menacing air-offensives and thus, decisively winning the early aerial-battles of Britain. During those times, the entire RAF Fighter Command was placed under the brilliant leadership of a **legendary RAF Officer, Air Chief Marshal Sir Hugh Dowding**, and the rest is indeed history.

We have seen that wars have always periodically been—inseparable parts of the human societies and their various kinds of political entities since dawn of the history itself. We have since discussed and also quite ironically found with considerable dismay that, **in spite of his variously enchanting civilisational advancements and achievements—having their glorious excellence down the ages—man so far has not at all been successfully able to effectively subjugate the always inevitably menacing urges for going into the wars on any pretexts, against none else but**

his fellow human-beings. We have justifiably termed them as the human-blood soaked dark-corners and also the no less frightening dark-spots onto the expanding civilisational-canvas of the mankind.

With the dawn of twentieth century itself and, **one after the other, the 1914-18 First and 1939-45 the Second World-Wars** came violently visiting the humankind in the world and thereby what all were the highly demonic consequences for the man himself, we have since earlier very briefly tried to paint them through the words, in course of our continuing discussions. But at the same time, we may also recall and since contextually discussed too by us earlier, that even few years before the historic commencement of the larger First World-War there have been two significant wars and though these being smaller in range and hence confined into their respective regions only—and these had been **the South African Boer-War (1899-1902) and then, the Russo-Japanese War (1904-5)**. Between the above mentioned two world-wars we find just a brief gap of hardly 21 years only but, within this relatively very short-span of few years; historically lot of waters had already flown down the Rhine, Thames, Mississippi and many other rivers in various continents of the world. The victor-nations of the World-War I, had then gone-in for constituting the **'League of Nations'** to preempt further wars and also thereby, make some kinds of the institutional efforts through negotiations, mediations and dialogues towards solving various kinds of the international disputes and also diffusing any apparently complex conflict-prone situations between different nations. But there were many in fact, inherent flaws in the making of this international Body, i.e. the 'League of Nations' and hence ironically in retrospect we may say that it was no wonder that, this multi-national League turned out to be nothing substantial, but just a 'Lame Duck' only. The gross failure of this in fact non-starter Body too, was quite instrumental in keeping Allied Powers in the dark and also, clandestinely encouraging the Axis Power nations to menacing prepare fast for an all-out next world-War. So the 'die for this soon forthcoming war' was already getting cast and thereby little earlier, we have already very briefly delved upon certain since geostrategic and politico-militarily important aspects as to, how the Axis Power nations got very obsessively engaged in multiply preparing for the fast and hence, for the next major and much more devastating war. In all these efforts, not only the matters-military oriented tactical-cum-strategic levels thinking were getting highly revolutionized with momentous advancements in this regard but also by **making far-reaching uses of the multiple-fruits of the already more and more fructifying the—Industrial-Revolution**—substantial advancements in the fields of military-technologies as well as weapon-systems and weapon-platforms too, were getting revolutionized and thereby the entire gamut of warfare was then almost totally changing. In fact contextually in the larger perspectives in this regard, we do also find that these multi-faceted developments were indeed taking place somewhat differently, on both sides of the dividing-lines and thereby, resulting in the many kinds of the then wartime queer situations, as since borne out by the earlier relevant paragraphs.

After the end of Second World-War in 1945, perhaps some better sense prevailed among the victor Allied Powers and also while somewhat learning from the inherent lacuna in the making of the earlier 'League of Nations'; they soon proceeded to constitute a much better equipped and also, multi-dimensional another International Body since superseding the earlier one—in the name of the **'United Nations Organisation (The U.N.O.)'**. But has this one too, been substantially effective as desired? The experience of more than half a century in this regard, shows again not so reassuring mixed results and hence; there have been consistent demands from various international platforms, to initiate processes for its essential reforms. Interestingly and also in a very much positive manner, immediately after the end of Second World-War, indeed quite unprecedented new phenomena having sincere and far-reaching and globally macro-level and also with genuine humanism-triggered touch, soon dynamically surfaced—for achieving fast-track and multi-dimensional results internationally at the geopolitical-cum-socio-economic development and well-being of the humankind. These were the **'Marshal-Plan'**—as conceived and formulated by **General George Marshal**, the wartime Chairman of the US Joint Chiefs of Staff and soon after, appointed as the US Secretary of State for the speedy reconstruction of war-devastated Europe and in addition, the phenomenal launching of the well-conceived and well-executed plans for indeed comprehensive reforms and reconstructions under the supreme command of the American **General Douglas MacArthur**, of the so much devastated, but still feudalistic as well as imperialism-obsessed: the defeated Japanese Nation-State. We may recall that earlier in this chapter itself, we have since very briefly touched upon some aspects of the very fast and indeed so successful reforms achieved within a span of few years in Japan. Both these phenomenally international efforts were then, launched under the overall leadership of the industrial-cum-economic Giant—none else but the USA. We would be discussing them more relevantly in some detail, in the next eighth chapter.

Now again coming back to the some critically important and also decisively far-reaching aspects during prelude to the momentous beginning of the Second World-War, we many recall that we have since already contextually delved upon indeed the very focused and comprehensive geostrategic-cum-politico-military oriented preparations for war by both the Germany in Western Europe and then also, by the militarism-obsessed Imperialist Japan in the Far-Eastern Asia and thereby under the no less fanatic but at the same time also quite charismatic leaderships of **Adolf Hitler in Germany** and then, under the similar combination but having altogether different textures and culture of the Japanese **Emperor Hirohito and Prime Minister General Tojo in Japan** respectively. Now at this precise point a very interesting question arises too—as to **why the man's living-history has not been able to record the phenomenally somewhat similar efforts by the Italian Nation-State then, under the almost no less fanatic, charismatic and also quite dictatorial leadership of Benito Mussolini**? As we may begin to critically delve into these aspects during those prewar years concerning the Italian Nation-State; we may notice that the aspects of the national-level leadership apart, the paramountcy difference happened to be that, Italy almost totally lacked the other essential wherewithal; being always both economic-cum-militarily and strategically, so inescapable in this regard. In addition, Mussolini was also subsequently found wanting in having his essentially desired grips on the multiple-aspects of **'essential strategic-culture and thoughts'** as well.

Specifically with respect to the above and precisely concerning the Italian nation-State, during prelude to the World-War II era; **E.L. Hasluck** has mentioned that, "However much the Italian people might be roused by propaganda suggesting that they should emulate the achievements of the ancient Romans, modern Italy does not possess the material basis for the organization of heavy military enterprises. In normal times Italy imports 95 percent of the coal needed for her industries, a similar proportion of her oils and copper, and nearly all her raw cotton. Nine-tenth of her wool supplies comes from abroad; she imports two-thirds of her supplies of iron-ore. The most optimistic champion of 'autarky' could not reasonably expect Italy to produce within her own borders more than a quarter of the products needed for a war of any appreciable duration. Famous as some of them were, her steelworks produced only some 10 percent of Germany's pre-war output, and iron production was even smaller in proportion to that of the German Reich. Rubber, tin, phosphates, chromium and wolfram—all had to be brought in from abroad.... It was difficult to envisage an Italian war of expansion which was not guaranteed the freedom of the seas by at least the benevolent neutrality of the great naval Powers, for her normal routes of supply depended on sea-borne transport to the extent of 85 percent. Of her imports, 60 percent came through the Straits of Gibraltar, and about 10 percent through the Suez Canal. If Mussolini's dream of the Mediterranean as a Roman lake were to come true, the expansion must be largely at the expense of Britain and France, whose naval power was a force to reckon with." (*Hence no wonder that the same author has also mentioned and, since also contextually quoted by us earlier too that, it was when Mussolini began to follow the will-o'-the-wisp of imperial expansion that the Duce started down the slippery slope to ruin. And this highly ill-advised and totally ill-conceived any such imperial misadventure on the part of Mussolini as the unquestioned Dictator of Italy then; historically shows quite clearly as to that—in spite of the some important aspects of the charismatic leadership on the part of Mussolini, he rather ironically proved himself to be considerably naïve in the perpetually paramountcy domain of the multiple-aspects of* ***'essential strategic-culture and thoughts'****. And hence here too in case of the Italian Nation-State during the times of World-War II, indeed the rest is again history. But again it is rather contextually necessary to mention in this regard that, if the outcome of the 'Battle of Britain' during the beginning years of World-War II, turned out to be a shocking British rebuff to the Germany, then at the same time initially the Italian Naval Forces in Mediterranean and also, the sizeable Italian Army divisions in Libya along the North African coast—indeed also turned out to be no 'pushover' for the Allied High Command; both at the political level under President Roosevelt and Prime Minister Churchill and along with militarily, under General Eisenhower in the Supreme Headquarters of the Allied Expeditionary Forces in Europe.*). (Hasluck, pp. 113 and 114)

Contextually earlier, we have since already noted that during the so-called somewhat lull-period of around 21 years only, between the two World Wars, many momentously far-reaching changes took place in almost all the areas and hence, no less or rather all the more so in the vast and highly complex areas of the matters-military and thereby, in the fields of warfare too. All these resulted in momentously transforming the multiple-aspects of warfare into indeed an altogether three-dimensional phenomenon and thereby, numerous resultant effects to it. One more militarily very important development then took place as to—the new kinds of mechanization too, was brought into the armed forces on both sides of the diving-lines. All these together led to the formation of armoured divisions in the armies; as the powerful spearheads of the armies' massive strike-forces, with tremendous shock and destructions against their adversaries in the battlefields. In these areas too, Germany was substantially ahead of the allied Powers in really action-oriented formulation as well as, fine-tuning their strategic-thoughts and doctrines and also, coupled with the essential refinements of their almost 'failsafe' offensive tactical-concepts to begin with; meant precisely for the shockingly speedy conquest of almost whole of the Western Europe—touching its southern Mediterranean coasts and also, its western coasts touching the English-Channel across France with amazing military success and thereby for the time being, putting the Allied Forces' High Command, off-gear in a quite worrisome manner.

Arthur Birnie has mentioned that, "The tactical problem left over from the last war was solved by mechanized troops. Break-through became practicable, and trench warfare was shown to be a passing phase, peculiar to the conditions of 1914-18, not a permanent military tactic. As usual, the German General staff showed itself, most alive to the new developments. It worked out a new offensive technique, by which masses of tanks assisted by flights of bombers were launched against fortified lines, feeling for weak points. If they broke through, they hurried on, spreading confusion among the enemy's communications, while the gap they created was widened by motorized troops coming up behind. Finally the ordinary infantry marched through and spread out fanwise, taking the enemy line in the rear. These original methods enabled the Germans to win striking and unexpected successes in the first stages of war." (*As since brought out earlier too; these concepts precisely form the concepts behind* ***the then remarkable German doctrine of 'Blitzkrieg'****, during the Second World-War. In addition to all these; the German High Command was very particular towards the focused training as well as, putting their units and formations and their commanders at all levels—through rigorous field-rehearsals for successfully participating in these impending operations.*). (Birnie, 1942, p. 240)

Hence no wonder that when the 'balloon' of war finally went-up in the year 1939 the above narrated remarkably historic developments in Germany and also then, still prevailing very much hotchpotch thinking as well as military preparations in the allied camps, decisively facilitated the German forces in ensuring their remarkable victories in the western Europe, during the initial phases of war. Though contextually from the points of views of the man's living military-history precisely in this regard, it may perhaps appear to be some thoughts of conjectures only but still, we find some deep strategic prudence on the part of **Hitler** in the sense that even before momentously embarking on his campaign to speedily conquer the whole of Europe's western continental territories, he first wanted to be doubly sure of making the Germany's eastern frontiers quite safe and militarily secured. Historically it is also true that with his great politico-diplomatic vision and skill, **Hitler** constructively heard the call of **Stalin** and thereby even before 1939, first entered into a 'no-war pact' with the East-European giant the USSR; which in any case little later, was ultimately going to be violated by the Nazi Germany herself.

And in spite of all these and also to militarily garner some valuable strategic space in the further east of the eastern German borders, Hitler first and foremost proceeded to attack the neighbouring Poland and hence, Nazi Germany's this military operation for the conquest of Poland is also historically known as—**Hitler's September 1939 Polish Campaign**. For this campaign to begin with, Hitler had already earmarked a very large body of troops consisting of not less than 1.50 million men, almost 4000 tanks and nearly 1000 fighters and bombers—for the extraordinarily fast subjugation of Poland because he since knew that the then Poland too, had comparatively almost no smaller armies and also the vast bodies of the conventional cavalry troops. But when Germany invaded Poland on the 1st September 1939 the Polish forces and their large horse cavalry troops soon proved to be no match to the well-trained, well-led and highly mobile attacking German infantry and also the Panzer armoured divisions, along with decisively getting close air-support by the Luftwaffe formations. In addition, there were hardly any natural or man-made obstacle systems coming in the way of fast advancing Nazi forces to crushingly push the Polish armies further into the east. The already weak Polish Air force fighting formations and ground installations too, soon got wiped out by the advancing Nazi forces. It became very clear that it was just a matter of a very short time that, Poland was overwhelmingly soon going to be militarily conquered. And hence on 17th September itself, USSR too attacked Poland with their massive forces already launched from the East. Warsaw was already captured by Germany and also due to massive Russian attack, the Polish Government elements became panicky and hence fled the country into Romania leaving Poland in the hands of Germans and the Russians. And so within a month, the territories of Poland got divided between conquering Nazis and the Soviets. And to all these the Allied European Powers then, remained apparently helpless and moot spectators to the military conquest and division of Poland.

Now in addition to Nazi Germany then, even the USSR had already tasted blood. And hence, the Russians soon proceeded to launch an attack onto the North European nation of Finland resulting in a violent **Finnish campaign of the USSR during World-War II**. In comparison to the giant Russia, Finland is a very small nation, having much smaller resources, manpower and also the military strength. But the smaller Finnish territories are gifted with very harsh winter weather and also, a network of difficult natural obstacles. Though the smaller Finnish military forces were no match to the giant Soviet Red Army but in spite of all these, the Fins proved to be very brave and tenacious fighters and also guerilla soldiers. The massive first wave of Russian attacks on Finland, resulted in great failure and indeed embarrassing military-setback for the powerful Soviets. When the USSR actually began invasion of Finland on 30th November 1939, severe winter had already started setting-in over there. In fact as per the natural availabilities of access into Finland from the Russian side in the East the easiest entry point happened to be through the open gap between the Gulf of Finland and Lake Ladoga, but here the strategically prudent Finnish Military Command had already prepared the strongly fortified defenses as their **'Mannerheim Line', (*like the German 'Siegfried-Line'.*)** to plug this gap and thus blocking this entry for the Russians; in addition, the Britain was helping the Finns with supplies. So the Russians entered into Finland all along the borders from their side, wherever it became feasible for the Red Army. The second time the soviets launched their more massive attacks on to the Finnish territories, properly supplied and equipped and also, with duly learnt from the lessons of the fiasco of their first wave of attacks. This time the Finnish forces got overwhelmed and, so on 10th March 1940, the Finnish Government was forced to enter into negotiated settlements with the USSR.

Immediately after this, Nazi Germany too again got active in the **Northern European countries of Denmark, Norway and Sweden in Scandinavia**. It was a queer situation since, both Denmark and Sweden then so much keen to maintain their neutrality and perhaps, to save themselves from any direct Nazi onslaught onto them and thereby did not permit some British army and naval forces to use timely and easy passages through their territories and sea-lanes, to come fast into Norway for actively helping her against the Nazi Germany's invasion; now since historically known as the **'Norwegian Campaign'** of this war. But soon both Norway and Denmark too got conquered by the masquerading herds of the Nazi forces. In case of Norway, and also with the support of powerful Royal naval units; the British forces were able to occupy strategically important Norwegian Ports on Norway's North Atlantic coast-lines, for launching offensive operations against the German forces and their installations in the Norwegian territories. But the panic of the Norwegian Commander-in-chief came on the way, as he began then desperately asking for the diversion of these allied troops towards the West to rescue his threatened forces. All these hasty actions and also some chinks in the British armour already prevailing then, for the time being prevented even their any partial success in Norway against the well-equipped and stronger German forces.

Now with his typically more and more wartime confidence, Hitler very prudently planned to turn to the West with his principal aim to conquer France at the earliest, to become the unquestioned 'Supreme-Lord' of almost the whole of continental European landmass, and thereby geopolitical-cum-militarily to put him in a strategically commanding position to be able to directly confront the Great Britain—for forcefully enabling him to set eyes on the whole of British Empire itself. During the early years of the war these were precisely the nodal points of Hitler's grand strategy of this global war. To move westward as per his above mentioned strategic aims **Hitler first proceeded to move fast as per his blitzkrieg-oriented concepts, to quickly overwhelm both Holland and Belgium and in the process to also preempt any Anglo-French joint initiatives to confront his Nazi forces from successfully proceeding with their further campaigns in the Western Europe**. Hitler had the information that Britain had already placed a sizeable portion of the British forces in France to extend the already in fact, in a way 'overmanned' the French 'Maginot Line' since, already prepared to stall the German invasion into the French heartland. The British forces consisted of four infantry divisions and one British armoured brigade. Further in the interior France, the French High Command had already mobilized almost eighty divisions and in addition the British also had already ten more infantry divisions in France by May 1940 itself. But ironically both the British and French High Commands were still rather somewhat 'regressively' fixated in a sort of the static defensive-mode only and thereby, not being able to see through Hitler's larger game-plan, and consequently to think with the desired boldness for fast offensive action against the Nazis and as dictated by the then prevailing situation-based essential strategic-dynamism.

These were precisely their typical loss of their otherwise mandatory grip, especially that of the British themselves over the always paramountcy multiple-aspects of the **'essential strategic-culture and thoughts'**. It was indeed a very sad commentary on them and even all the more so as the same soon turned out be definitely very costly and unsettling with great tragedies for their Nation-States. Therefore no wonder that, we have otherwise too in various contexts earlier, already delved upon the numerous aspects of always of the paramountcy weightage and also reaches—the **multiple-aspects of 'essential strategic-culture and thoughts'. Coming precisely to the situations prevailing in France then in 1940 the essentially visionary strategic prudence perhaps demanded that, when Hitler was already occupied with his sizably large and multiply constituted forces in the invasion of Poland in the East, both Britain and France could have advantageously attacked Germany from the West with their ready and already mobilized large forces, and thus achieving great surprise and also indeed, substantially throwing the German High Command and its military wherewithal off-balance. Had the Allied Powers then pressed on boldly, with further movement and concentration of forces; the subsequent history of the World-War II itself, could perhaps have been quite different.** Well, these points now in retrospect may understandably be arguable indeed and also, may be called matters of perhaps the conjectures only. **But in any case, taking any such wartime initiatives by the any concerned top politico-military-level leadership do always require extraordinarily bold strategic-thinking as well as, iron-will for taking rather uncommon risks and also timely bold decisions.** But it is perhaps also true that the earlier talked about—the then prevailing strategic-thoughts, based on the statically fixated defensive-mode, on the part of the then allied Powers in the Western Europe in any case, squarely prevented any such indeed far-reaching and opportunity-based possibility, for imginatively bold initiatives against the Nazi Germany's western borders.

Hence as the events turned out to be, Hitler had a field-day in the beginning and thereby, no difficulty in quickly subjugating Poland in the East, then speedily conquering Norway and Denmark in the North and then boldly turning to the West and then within a week or so, be able to decisively overwhelm both Holland and Belgium as well. With all this Hitler's confidence and boldness got a boost, also both geostrategic and politico-militarily even outflanking the French 'Maginot-Line'—tremendously enabled him to very menacingly knock at the doors of France, both from the North and Eastern directions and thereby in a way entering into rather the concluding phase of his **West European campaign**. In his these moves, **Hitler very advantageously adopted the masterly conceived and formulated plans, since prepared by German General von Manstein for attacks and superbly swift military manoeuvres for the extraordinarily fast and highly successful results, leading to his victories in Europe then**. All these also resulted in a very embarrassing **Dunkirk-Fiasco** for the British, in France across the English Channel. Field Marshal Montgomery has indeed painted a very vivid picture of this historic episode concerning Dunkirk in his already earlier referred book.

Specifically in the above discussed multiple-contexts and also the **Nazi campaign in France,**

Montgomery has mentioned that, "By 23rd May (*1940*) the Germans reached Boulogne. At midnight on 27th May the Belgians capitulated. This was awkward for me and my division, because by that time we had become the left division of the Franco-British forces, with a Belgian army prolonging our left to the sea. At dawn on 28th May I learnt that the king of Belgians had surrendered the whole of his army to the Germans. I decided in my own mind that it is not suitable in the mid-twentieth century for kings to command their national armies in battle. There was little which could be done for the British forces. **By 4th June the British troops and 120,000 French soldiers had been evacuated from Dunkirk with their personal weapons, all vehicles and other equipment were left behind.... The evacuation of the British forces from the Dunkirk beaches lasted nine days, and saved 338,000 men. This was a remarkable achievement by the British navy which assembled 887 vessels of all sizes, and the R.A.F. which in one period of four days shot down 179 enemy aircraft for the loss of 29. The relief in Britain at the extrication of the troops countered the shock, but the truth was that the British army was thoroughly defeated in battle. So much of its equipment had been abandoned in France that in the summer of 1940 only one division in England was respectably equipped.... The German had not pursued the British up to Dunkirk; this was on orders of Hitler, who decided to turn south and finish off the French. French morale cracked, and on 16th June the French Government under Petain surrendered. The Germans occupied northern and western France. At this stage Italy declared war on Germany's side....** In the following years of the French resistance movement and agents of the British Special Operations Executive harassed the German forces of occupation, and organized escape routes for soldiers who had escaped from prisoners-of-war camps and for airmen shot down over France. Commando raids were a nuisance for the Germans, and Free French units served in Britain and Africa. But the fact had to be faced that in June 1940 Britain and her empire stood alone against the Axis powers, Germany and Italy." (Montgomery, 1983, pp. 312 & 313)

The above narrated momentous historic facts, leading to subjugation of the French Nation-State by the Nazi Germany then, rather decisively paved the way for Hitler's intense **'Battle of Britain'**. We may recall that we have since already briefly delved upon some of the historically important aspects of this battle of Britain, little earlier in this chapter itself. In this campaign for the military subjugation of the much tenacious and also, geostrategic-cum-politico-militarily indeed more visionary: the British Nation-State, the principal strategic aim of Hitler's this **'Operation Sea Lion'** was—first to possibly annihilate the strong Naval and Air Arms, positioned in and around the British Islands and secondly then to enable the massive Nazi herds to land onto the British shores across the English Channel, to speedily proceed deep into the British heartlands for the final and militarily decisive conquest of the British Nation-State itself and thereby to take-on the rein and convert the then unprecedentedly transcontinental British Empire, rather automatically into a sort of the readymade and perhaps, Hitler's own kind of the global German-Empire. These overwhelmingly fixated Nazi geopolitical ideas and dreams on the part of Hitler, obviously happened to be too far-fetched and perhaps even wild at its face-value itself. But that's the way the man's living-history, since continuously in the making onto his expanding civilisational-canvas has been showing down the ages that—the dictators anywhere and in any particular given situation, are mostly very much and uncontrollably at times, more prone to think and also plan regardless at times, rather in quite wild manners indeed. Anyway, in the 'Battle of Britain'; Hitler put his massive Luftwaffe to begin bombing and strafing the numerous and strategically vital British cities, industrial and other infrastructural centres, located in the British Islands. And precisely here, the Nazi Luftwaffe met its no less stronger match—in the shape of British R.A.F. Fighter Command. Hence as the days passed then, Luftwaffe got more and more decisively check-mated with indeed massive losses. Consequently Hitler's Massive Air-Arms began to face shocking defeats, in the hands of superbly tenacious and professional R.A.F. Fighter Command and its brave pilots, over the British Skies across the English Channel.

Specifically with respect to the above and indeed, with great historic precision; the then **British Prime Minister, Sir Winston Churchill**, in his 'contributory' article on 'The Battle of Britain' in *Reader's Digest Illustrated Story of World War II* (page 129) has very perceptively written that, "Yet the Battle of London was still to be fought out. Although invasion had been called off, it was not till September 27 (*1940*) that Goering gave up hope that his method of winning the war might succeed. In October, London received its full share; the German effort was spread by day and night in frequent small-scale attacks on many places. Concentration of effort gave way to dispersion; the battle of attrition began. Attrition. But whose? In cold blood, with the knowledge of the aftertime, we study the actual losses of the British and German Air Forces in what we may well be deemed one of the decisive battles in the world.... No doubt we were always over-sanguine in our estimates of enemy scalps. In the upshot we got two to one of the German assailants, instead of three to one, as we believed and declared. But this was enough. **The Royal Air Force, far from being destroyed, was triumphant. The stamina and valour of our fighter pilots remained unconquerable and supreme. Thus Britain was saved. Well might I say in the House of Commons, 'Never in the field of human conflict, was so much owed by so many to so few."**

From then on, tide, though slowly in the beginning, had already begun turning adversely for Hitler and his Nazi military forces. And it was then a matter of time consisting of few years only that, the final end of Hitler's menace and total defeat of his Nazi forces, in various theatres of this war in the different continents; were going to be the end results of these highly tragic and, massive blood-bath and also, all round destructions and indeed, the unprecedented tragedies for almost the entire humanity across the world. **Further on, after his failed attempts to militarily overwhelm Great Britain in the intense and ruthless aerial-battles in the skies over the British Isles in the North Atlantic Hitler turned east and, committed a far-reaching blunder of rather recklessly invading Russia too**. Indeed living, and both military as well as, the political histories of World-War II are so multiply vast that no wonder; far many volumes have already been written in numerous languages of the world, during more than the last half a century and, they are since always available in the libraries or otherwise too for the interested readers, research-scholars and the professionals for their perusal and the needful. In this book, as already contextually spelled out earlier during our continuing discussions, that here we are precisely engaged in attempting at the strategically critical appraisal as well as, some kinds of the focused reviews; as we continue to gradually move along the timelines and pages of the man's vast, highly complex and indeed, multi-dimensional living-history; since always perpetually in the making too, and hence thereby, within the framework of the specifically defined perspectives in this regard.

It remains always historically true that, any kinds of the critically perceptive and even brief discussions on some importantly crucial aspects of the 1939-45 World-War II, cannot be perhaps of any substantial worth, without equally delving on some of the no less paramountcy and also indeed equally pivotal aspects of the simultaneous as well as intensely ruthless—the Allied Powers' all-out war in the Pacific and the Far-East with principal aim to militarily defeat the Japanese imperial forces over there and also, finally destroy since the centuries-old regressively lingering and extremely feudalism-fixated ruthless militarism, in the Japanese societies themselves. At this precise point of our continuing discussions we may also recall that in this chapter itself, we have since already earlier touched upon these historically far-reaching aspects concerning Japan and also highlighted **as to how after her decisive conquest by principally the USA, Japan was again reconstructed and also almost totally reformed and got reincarnated into a dynamic, modern and vibrant democracy within a short span of few years only, under the overall and the then continuing supreme command of an indeed legendary US General of the Army, Douglas MacArthur**. At that point we may also recall that, we have since already quoted some historically valuable and also contextually very informative passages too, from the General's well-known 'Reminiscences' (his 'Autobiography') as well.

So far as the geostrategic and politico-militarily multiple-aspects of the World-War II are concerned, historically it indeed remains true that at the highest political-levels of Allied Powers in the beginning itself, it was jointly decided that, it was rather very much imperative to first bring the war in Europe against the Nazi Germany—to totally successful conclusion and meanwhile, to also keep Japan militarily onto somewhat hold and subsequently then come to the Pacific with full strategic-cum-military might and thus, facilitating General MacArthur's multi-national forces and also his Pacific Command—to decisively conquer Japan and thereby, receive unconditional surrender of Emperor Hirohito and his Japanese imperial forces. In overall historic perspectives in this regard, it indeed appears strategically sound decisions, but as usual mostly in any warlike situations, the contents of this the then highest-level decisions did carry its own no less weighty and worrisome riders and hence consequently thereby, ruthless Japanese then did extract some very heavy prices both in human and material terms, from the indigenous peoples of both the South-East and Far-East island-nations and hence wherever they could succeed in their invasions, since spread over almost whole of the Pacific regions over there. Well, while no less momentously continuing to remain strategic-cum-militarily engaged in Europe with overriding priority to first finish-off Hitler and his Nazi-regime, but at the same time how much even some additional and extra token kinds of rather clearly visible aids and supplies could have been spared by the Allied Powers, for their extremely hard-pressed Pacific Command, do also remain historically some matters of conjectures to perhaps even this day, for the scholars of the man's no less living military-history in the world. Since during those extremely trying times of late 1941 and very early years of 1942, the Governments and peoples of those Island Nation-States in the Pacific, along with General MacArthur and his multi-national Pacific Command, were all positioned in the sharp-end and hence, facing the heavy Japanese onslaughts; it was indeed understandable: the perceptive causes of their dismay and also disappointing bewilderment, over the so-called apparently visible—some sort of strange callousness on the part of both President Roosevelt and the Prime Minister Churchill and their highest-level military headquarters and the Chiefs of Staff, placed in their capital cities even if they do remain otherwise broadly debatable as well.

In precisely the above mentioned multiple-contexts, General Douglas MacArthur has given detailed accounts of the historically momentous aspects in the Pacific ('Reminiscences'), and hence in also the earlier referred book *Reader's Digest Illustrated Story of World War II* (pages 160 and 161) some passages from this has also been quoted and thus making General MacArthur's contributory article on 'Disaster—And Glory—In The Philippines'. And therein, in the Reader's digest book's (pages 160 and 161), General MacArthur has been quoted as saying that, "A top-level decision had long before been reached that the Atlantic war came first, no matter what the cost in the Far East. **President Roosevelt and Prime Minister Churchill**, in a Washington conference after the attack on Pearl Harbor, reaffirmed a policy to concentrate first, on the defeat of Germany. Until victory was won in Europe, operations in the Pacific would be directed toward containing the Japanese with the limited resources available. **General George C. Marshal, the Army Chief of**

Staff, supported this policy. Unhappily I was not informed of any of these vital conferences and believed that a brave effort at relief was in the making.... To allow my main bodies to be compressed into the central plain in the defense of Manila by an enemy advancing from both directions could only mean early and complete destruction. By retiring into the peninsula, I could allow maneuverability of my full forces and gain our only chance of survival.... The Japanese themselves realized the important effect of the protracted resistance in the Philippines. Their historical records state: 'There was influence, a spiritual influence, exerted by the resistance on Bataan. Not only did the Japanese at home worry about the length of the period of resistance on Bataan, but served to indicate to the Filipinos that the Americans had not deserted them and would continue to try to assist them.... [*To the Filipinos, and to many mainland American fighting men, the U.S.; failure to provide the beleaguered islands with even minimal munitions and supplies was utterly bewildering. They felt themselves abandoned,—Ed.*].... If only help could have reached the Philippines, even in small form, if only limited reinforcement could have been supplied, the end could not have failed to be a success. It was Japan's ability to continually bring in fresh forces and America's inability to do so that finally settled the issue... It seemed incredible to me that no effort was made to bring in supplies. I cannot overemphasize the psychological reaction of the Filipinos. They were able to understand military failure, but the apparent disinterest on the part of the United States was incomprehensible. Aware of the efforts the Allies were making in Europe, the feeling ranged from bewilderment to revulsion.... He [**President Quezon *(of the Philippines)***] told Col. Charles Willoughby, my versatile G-2 [military-intelligence officer]: 'For thirty years I have worked and hoped for my people. Now they burn and die for a flag that could not protect them. I cannot stand this constant reference to Europe. I am here and my people are here under the heels of a conqueror. Where are the planes that they boast of? America writhes in anguish at the fate of a distant cousin, Europe, while a daughter, the Philippines, is being raped in the back room."

The above quoted paragraphs, since penned by the legendary US General who was then himself on the spot and also in command of the Allied forces deployed in the Pacific region in South-East Asia and the Far-East, and also mandated to counter the relentless Japanese onslaughts on the island countries over there, do give us a vivid picture of both—the then prevailing socio-political scenario as well as, the Allied Powers' indeed precarious military conditions in Philippines and also in the other neighbouring countries there. The inside rather somewhat specifically compulsive reasons for all these and, as then prevailing in the Allied Camps both in Europe and the USA during those extremely momentous post-Pearl Harbour early years of late 1941 and the 1942 years, have also been touched upon and these are indeed understandable as well. But there remain other historic dimensions to those the then existing compulsions too and the same have precisely been and also very perceptively highlighted by the Commanding US General MacArthur over there in the Pacific then and thereby, he also very professionally brings out as within the then prevailing geostrategic and Politico-militarily larger framework, the apparent possibility of indeed crucially desirable and also since as per the prevailing situations-guided necessities, for rather judiciously fine-tuning of the highest level responsibilities concerning the overall gamut of logistics-management towards little more accommodativeness for perhaps the no less precarious needs of the Pacific Command. Historically the specific point in this regard remains that, any such highest-level strategically imaginative and also, the geopolitically far-reaching timely initiative taken so proactively perhaps, by the Allied-Camps' concerned decision-makers would have gone a long way in heralding the so much expected succor in the Pacific region. And that's what General MacArthur has very precisely brought out, in his above quote paragraphs. Further, we may recall that precisely like the above historic episode during our ongoing discussions from time to time, we have come across numerous 'ifs and buts' into the pages of man's living-history since, continuously in the making onto his expanding civilisational-canvas and also, differently in many contexts since the very advent of various human-civilisations themselves. And hence in other words, we may also say that these aspects too, concerning the very vast, highly complex and also multi-dimensional living-history of the mankind, do constitute indeed its another very engaging dimension, for man's rather continuing explorative studies in this regard.

Contextually as per above and in the larger perspectives as well, as we again come back to some historic aspects leading to the war with Nazi Germany as early as in 1936 only, when Hitler very audaciously proceeded to occupy the strategic Rhineland in the east and, across the river Rhine, it has indeed been very perceptively observed that if at least even a couple of the French divisions already deployed along the Maginot-Line, would have promptly moved ahead to challenge the Nazi aggression over there German battalions would have retreated fast in panic and thereby, perhaps handing-over many German prisoners too, and that would have also been a kind of the politico-military end of Hitler. And it has also further been conjectured and needless to say that, this in turn would have indeed triggered—numerously path-breaking and historically very positive fallouts for the entire humanity in the world. These observations do remain very logical and also, geostrategic-cum-politico-militarily appear to be rational as well. But as we may critically delve deeper into these momentously historic aspects and then, the no less weighty moot point begins to rather confront us that would these otherwise definitely decisive developments at that precise point of historic timeline, have really ensured the final relegation of Hitler into the dustbins of history? In fact, considering the type of the typically charismatic man that Hitler was during that moment of perhaps comparatively appearing to be some kind of a minor 'fiasco', for the then so much hysterically war-fixated Nazi Germany, Hitler with his extremely manipulative lies-based game-plans and deceit, could have again managed to soon spring-back to his dictatorial powers. These too are some of the many conjectures in this regard. Though opinion on these counts do vary but, one thing historically perhaps, do remain quite clear that—nothing short of a kind of physical elimination of Fuhrer—would have ensured his final demise with his so menacing and extreme war-mongering kinds of politics and thereby, his relentless efforts to create war-hysteria in the then Nazi-Germany. Hitler was in fact psychologically such a character and that too, with his already overwhelming as well as, deeply entrenched dictatorial powers over the whole of Germany that nothing could diplomatically be taken for granted and also internationally reassuring, with respect to his thoughts and actions. To historically substantiate this point, we may have to contextually recall his pure bluff and crafty deceit in 1938 itself, when Hitler managed to scare the quiet-mannered **Austrian Chancellor Herr Schuschnigg** by inviting him for a 'conference' in his mountain retreat on the Obersalzberg, and with his some wild outburst, he very craftily cornered him for signing an already some kind of the pre-drafted 'document of agreement' and thus; very quickly facilitating the Nazis to take over the Government in Austria.

But later-on during wartimes, as tides began to turn adversely against the Nazi Germany during the early 1943 itself some historic personalities especially in Britain, rather hastily came to the opinion that, to end the war in favour of Allied Powers, some diplomatic 'face-saving' avenues could be offered to Hitler in Europe and then to Japan in the Far-East to work out some 'conditional-surrender'. But historically rather a cool point does remain that, would Hitler have sincerely co-operated as being wished for in some quarters? **Somewhat interestingly in these contexts and in some kinds of indeed the historic 'ifs and buts', Field Marshal Montgomery (page 322) has**

written that, "Confident of victory, Churchill met Roosevelt at Casablanca in January 1943 to formulate strategy. Together they declared that the policy of the Allies was to force the 'unconditional surrender' of Germany, Italy and Japan. I have always considered this decision to have been a tragic mistake. What it led to was well summarized by **Lord Hankey**:

> 'It embittered the war, rendered inevitable a fight to the finish, banged the door on any possibility of either side of offering terms or opening up negotiations, gave the Germans and Japanese the courage of despair, strengthened Hitler's position as Germany's 'only hope', aided Goebbels' propaganda and made inevitable the Normandy landing and the subsequent terribly exhausting and destructive advance through North France, Belgium, Luxemburg, Holland and Germany. **The lengthening of the war enabled Stalin to occupy the whole of Eastern Europe, to ring down the iron curtain**. By disposing of all the more competent administrators in Germany and Japan, this policy retarded recovery and reconstruction. Unfortunately also, these policies, so contrary to spirit of the Sermon on the Mount, did nothing to strengthen the moral position of the Allies.'

But from the time of the war turned in favour of the Allies, the unconditional surrender of their enemies was their strategic objective."

In the above quoted paragraphs, in the last line (*as highlighted above*), Montgomery has indeed quite perceptively concluded that finally nothing short of an unconditional surrender of the enemies, was geostrategic-cum-politico-militarily agreed upon as the principal aim of the Allied Powers. It is a historic fact that the Allied Powers conclusively did succeed in successfully achieving their principal aim, as since already agreed upon in advance and thus—bringing finally the end of World-War II. But at the same time, it is also equally true that all these were finally achieved at indeed unprecedentedly tremendous costs, in the forms of colossal deaths and also the all kinds of destructions and thereby, rather uncontrollably bringing-in great tragedies, miseries as well as, sufferings onto almost the entire humanity in this world so far unknown to the humankind. Now principally the pivotal point in this regard, emerges as to were all these perhaps avoidable then in any manner at all? As since already quoted above that, some expertly perceptive thoughts have been historically rendered that, these were indeed possible if some geopolitical-cum-diplomatic due care would have been exercised for timely taking some requisite visionary-initiatives in this regard. Well, weren't such thoughts although scholarly perceptive in retrospect, just well-meaning wishful thinking or mostly rather flights of imagination only? And precisely here perhaps, lie the various kinds of differences in opinions. **Contextually, we may recall here that little earlier in this chapter itself, we have already discussed and emphasised the different kinds of unchallenged dictatorial regimes socio-politically so much overwhelmingly entrenched in each of the then Nation-States of Italy, Imperially militaristic Japan and the Nazi Germany and in turn thereby, this trio were then having their differently distinct kinds of regressively fixated die-hard 'culture of extreme militarism'**. In addition, Japan being from Orient, was still socio-psychologically rather submerged in her deep-rooted medieval culture of 'Emperor-worship', duly soaked in her age-old and also powerfully lingering the multiply Japanese kinds of feudalism. No wonder that even after somewhat earlier capitulation of Italy and then later-on, the totally unconditional surrender of Germany, Japan still rejected an offer of her similarly unconditional surrender to the Allied High Command in the Pacific. And hence, the unprecedentedly cruel and also indeed totally inhuman droppings of two 'atom-bombs' one after the other on two Japanese cities, finally forced the Japanese Emperor and his Imperial Command; to unconditionally surrender before the Allied Powers' Pacific Command, since led by the US General Douglas MacArthur.

After all in these historic contexts of momentous significance, indeed globally for almost the entire humanity we may need to bear in mind that **none of these countries then constituting the Axis-Powers, i.e. Italy, Germany and Japan, had any kind of meaningful democratic systems of governance, which always provides some kinds of the constitutional 'checks and balances' and thereby, ensures free debates to thrash out important points, before arriving at any national-level decisions in the concerned corridors of power**. Needless to say that any kind of dictatorship and also more so with some powerfully overriding militaristic-culture and value-systems, provide no room for any such socio-political as well as, well-meaning endeavour on the part of concerned citizens under any dictatorial-regime. It is true that both Italy and Germany—did have some sort of democracy with some systems of Parliaments, before their dictators came in power; but once **Mussolini** came to power earlier, heralding fascism as a tyrannical system of dictatorial regime in Italy and as subsequently after a gap of some years, **Hitler** almost similarly came to power bringing-in Nazism in Germany, the differently flickering candles of democracy, in both the countries were extinguished and thereby these two extremely militarism-fixated leaders, but in their two differing models, became all-powerful dictators in their respective countries. In both Germany and Italy, press-censorship was also strictly enforced. Both had their agendas for their respective nations' imperialism, towards securing their colonial empires duly triggered by their differently ingrained fixations of jealousy, towards Britain and also the British Empire. Historically it was a queer situation then prevailing in Europe and hence, the requisite groundwork was getting completed for the fast incubation of the seeds of World-War II.

Then so far as **Japan in the Far-East** was concerned, in fact the pre-war situations were all the more momentous, and interestingly had some rather additional 'Asiatic' dimensions as well. In spite of Japan being the first Asian Nation-State—to be able to very imaginatively garner numerous benefits, possibly from both Europe and North America to trigger very rapid progress towards multiple-advancements in the fields of technologies and sciences on her home-turf, after the momentous advent of **'Industrial-Revolution'** and thereby indeed, achieving accelerated industrialization but both socio-culturally as well as socio-politically, Japan continued to largely live in the bygone medieval-era, still fixated with her extremely feudalistic value-systems. It was again a somewhat differently queer situation, inside Japan. Industrially on her home-turf itself, Japan was getting modernized fast, but socially she had not come out of her cocoon of the past. **There was her old cult of the 'Emperor-worship' still prevailing and even no less revered and in addition naturally, the extreme forms of the age-old Japanese 'multiple-culture of feudalism' too, continued to be deeply entrenched. In the domain of matters-military, in spite of the massive modernisation of the weapons, ammunitions and weapon-platforms; at the socio-psychological level in the military, the old kinds of Japanese 'Samurai-culture' remained powerfully prevalent and hence, always determining the value-systems within all aspects of soldiering in Japan.** No wonder that during those times, the Japanese people were yet to get exposed as to what the democracy and different democratic ideas were all about. **Hence, in spite of the momentous industrialization as well as, massive modernisation of the Armed Forces; Japan mostly remained a 'closed-society'.**

Now so far the pre-war era's Germany and her dictator **Adolf Hitler** were concerned; historically we already do have a very revealing book, *Mein Kampf* from the pen of Hitler himself, which does give quite in detail, a vivid picture of his inner self and thoughts on various subjects. Quoting some passages from this book originally written in German, E.L. Hasluck *The Second World War* (pp. 35 and 36) has mentioned that, "In that extraordinary and revealing book, *Mein*

Kampf, Hitler had clearly outlined the methodology of his political system, To Many Germans the serious relaxation of the military restrictions of Versailles seemed impossible of achievement until the generation that remembered the Kaiser's war should have passed away. Hitler scoffed at this pessimism. If the German Government would content itself with gradual, piecemeal encroachments on the Treaty provisions, the divided Allies of 1914, particularly pacific democracies of the West, would abstain from intervention. A Government that is disinclined to undertake a war, he says, will yield a comparatively small concession to avoid trouble; and when another concession is demanded, that Government, having brought itself to depart from the principle of maintaining the old system, will find it easier to make another retreat, whilst if yet another concession is demanded, the stage will then have been reached when there is so little left of the original situation that little is not worth fighting for. Each concession, in fact, paves the way for another of the same kind. The successful application of this principle became apparent very soon after the accession of the Nazis to power. The rearmament of Germany was not carried through at one fell swoop; it was developed in clear-cut stages. Later, when the manoeuvring for strategic positions in Central Europe began, each German advance was accompanied by emphatic protestations that the aggression was of purely local importance and that Nazi Germany had no more territorial ambition whatever." (Hitler, 1968, Indian Edition).

Paradoxically in these momentously far-reaching and history-making events during the times before the start of World-War II indeed, indicate as to how much both the political and military leaders then in the Allied-Camps, still continued to suffer from their strategic short-sightedness and also refused to come out of their typical naivety-based cocoons. Whereas Hitler since the beginning, both militarily as well as geopolitically; was craftily shrewd enough to read the Allied Powers' systems of functioning and even, see through the psyche of their leaders whereas; both the Allied political leaders and top military commanders were not caring **to closely read in-between the lines of '*Mein Kampf*'** and thus, were unable to see through Hitler's mind and his game-plans. The irony was precisely here and hence the Allied Powers, in spite of their first-hand momentous experiences in the First World-War, did not remain geostrategic and politico-militarily visionary and even tactically alert enough, **to proactively see-through the secretly but fast developing and indeed—very menacing situations both in Germany in Europe and, Japan in the Far-East** and thereby to preempt them in requisite manners. And here comes precisely the timeless and paramountcy import of the multiple-aspects of **'essential strategic-culture and thoughts'**, since always having their numerous nuances, subtexts and, dimensions.

Hitler's ***'Mein Kampf'*** reveals numerous things about him—his ambitions and aspirations, socio-cultural-cum-political ideas, many kinds of his rather confused thoughts on some aspects of human-civilisations and, his militarism-fixated psychological stubbornness and so on, and very amusingly been revealed by he himself, through his writings in this autobiographical work. Interestingly Hitler had rather very forcefully written that, "I am not one of those people who will begin something today and just give it up the next day for the sake of something new. That was the main reason which made it so difficult for me to decide in joining something newly founded (*i.e. Nazi Party*); for it must become the real fulfillment of everything I dreamt, or else it had better not be started at all. I knew that such a decision should bind me forever and that there could be no turning back. For me there could be no idle dallying but only a cause to be championed ardently. I had already an instinctive feeling against people who took up everything, but never carried anything through to the end.... **The so-called 'intellectuals' still look down with infinite superciliousness on anyone who has not been through the prescribed schools and allowed**

them to pump the necessary knowledge into him. The question that what a man can do is never asked but rather, what has he learned? 'Educated' people look upon any imbecile who is plastered with a number of academic certificates as superior to the ablest young fellow who lacks these precious documents. I could therefore easily imagine how this 'educated' world would receive me and I was wrong only in so far as I then believed men to be for the most part better; than they proved to be in the cold light of reality.... Just as many varieties of prehistoric animals had to give way to others and leave no trace behind them, so man will also have to give way if he loses that definite faculty which enables him to find the weapons that are necessary for him to maintain his own existence.... Thus the indispensable prerequisite for the existence of a superior quality of human beings is not the State but the race, (*what a strange mindset and thinking.*) which is alone capable of producing that higher human quality.... The capacity is always there, though it lie dormant unless external circumstances awaken it to action. Nations, or rather races, which endowed with the faculty of cultural creativeness possess this faculty in a latent form during periods when the external circumstances are unfavourable for the time being and therefore do not allow the faculty to express itself effectively. It is therefore outrageously unjust to speak of the pre-Christian Germans as barbarians who had no civilization. They never had been such. But the severity of the climate that prevailed in the northern regions which they inhabited imposed conditions of life which hampered a free development of creative faculties. If they had come to the fairer climate of the South, with no previous culture whatsoever, and if they acquired the necessary human material—that is to say, men of an inferior race—to serve them as working implements, the cultural faculty dormant in them would have splendidly blossomed forth, as happened in case of the Greeks, for example." (Hitler, 1968, pp. 190, 191, 329 & 330)

The pivotal purpose behind quoting the above paragraphs from **Hitler's 'Mein Kampf'**, has been to facilitate a critically detailed profile of Hitler as a man and his personality, his mental make-up, his even civilisational-cum-racial and, ideological fixations and also thereby of his die-hard intolerance as well as, his blind militarism towards other peoples in the world. At this point, if we may also care to refer back to our already earlier given some critically penetrating but brief descriptions of the many crucially important aspects of the then highly militarism-fixated Italy in Europe and then, those of the Imperial Japan, her Emperor and his both war-mongering politico-military leaders in the Far-East, then we may find them too, having many similarities with Hitler and, understandably though in their own distinctly different kinds of make-up and 'models' as well. In other words we may also say that, **all these three nations of Germany, Italy and Japan and also their rulers, politicians and military-leaders, in fact were so much overwhelmingly entrenched in their ruthlessly ambitious imperialism, and also in their blind militarism-cum-war-oriented geopolitical aspirations; that any kind of the midway efforts by the democratic Allied Powers—towards somewhat mutually satisfying attainment of the genuinely sincere and lasting peace—would have largely been futile and nothing but sadly some kind of a 'kite-flying' only**. Hence in retrospect, as we may try to rather objectively look into almost all the important aspects even before the start of 1939-45 World War II hostilities, and then unfolding of the numerous event and episodes during the war, we do find the above given historic scenario, as the plain and cold reality in this regard.

Then there is another rather extremely momentous dimension to Germany's so obsessive and relentless war-potential as well as in fact, always ongoing and totally focused war-efforts as well. And this so much far-reaching dimension, concerning the then Nazi Germany—pertains to the untiring work towards the earliest possible development of **V-weapons**. It was then in fact Nazi

Germany's most 'top-secret' project, concerning the most dedicated and fastest possible development of a totally new kind of offensive weapon and weapon-platform. Then Hitler's blindly dedicated German scientists and engineers were also working (*or even perhaps compelled to work*) hard to develop and perfect new kinds of the **'Poison-Gases'** and, even the high-technologies concerning the use of **'Nuclear Energy'** as perhaps the greatest source of the most devastating explosions; to be used at the earliest possible, against the Allied Powers. Hence geopolitically, even after some kind of the somewhat successful fructification of any midway and mutually agreed-upon termination of war won't have stopped Hitler, the kind of man that he was, from relentlessly continuing further with the development of these indeed, extraordinarily lethal weapon-systems and hence needless to say that—these 'demonic' developments in the then Nazi-camps, would have been too menacing and traumatic situations for the mankind.

But no sooner the tide of war turned in favour of the Allies and to begin with, the Allied Air Forces resorted to the relentlessly continuous bombings of the German industrial centres, cities, their vital and strategic installations and in turn; totally crippling the Nazis towards an enforced collapse, before they could develop and perfect the above mentioned strategic weapons and their delivery-systems. The point is that if Hitler and his herds would have been lucky enough to get through those weapon-systems, then being readied for use; there would have been perhaps nothing like the most decisive and massive offensive operation then, launched by **the Allied Supreme Commander in Europe, General Eisenhower** with his multi-national seven Army-groups across the English Channel to France and then on-way to the German heartlands and thereby leading to the final defeat of the Nazi military's forces and the end of Hitler. In other words perhaps in the first place itself, there won't have been anything like indeed the **historically momentous 'Normandy-landings on the D-day'** for finally defeating the Nazis in their homelands. But it may now be reassuring to say with great relief that, indeed the world was lucky enough and, during those historically momentous times; Providence did bestow kindness onto the mankind so that, the Axis-Powers were finally vanquished and thereby paving the paths for the 'Free-World' to survive and also at the same time, democracy in the world to gradually expand into newer countries and get strengthened.

During the period of almost six years of 1939-45, **when the Second World War was in fact so devastatingly in full swing in its numerously vast theatres of the all kinds of military operations on lands, in seas/oceans and also in the skies above a number of continents in the world there were indeed momentous military-phenomena of a series of the history-making 'Clashes of the Titans'—in fact, belonging to both sides of the dividing-lines**. Now historically to begin with in this regard, from among those 'Titans' from the both sides of the dividing-lines and thereby from the relevant pages of the man's momentous living military-history, we do very fascinatingly find in fact, a number of names of the military men and women of the all ranks and positions as well as, no less important those of civilians from probably all walks of life, then engaged in different professions and capacities really making momentous contribution in this historic World War. We would perhaps understandably be able to record names of just few of them here, to contextually substantiate the perceptive point concerning the above mentioned 'clashes of Titans'.

Now so far as the **Nazi Germany** was concerned, while tapping the ***Hitler's War directives* edited by H.R. Trevor-Roper and then also referring to the book *The Last Battle* by Cornelius Ryan,** we could possibly name Field Marshals—Fedor von Bock, Walther von Brauchitsch; Admiral Wilhelm Canaris, Grand Admiral Donitz Karl, Hermann Goering: Chief of German Air Force, Field Marshal Guderion; Generals Nicolaus Falkenhorst, Hellmuth Felmy, Franz Halder, Froehlich,

Gause; and also, General Alfred Jodl, Field marshals Wilhelm Keitel, Albert Kesselring, Wilhelm List, Erich von Manstein, Friedrich Paulus; Admiral Erich Raeder, Field Marshals Erwin Rommel, Erich von Rundstedt, Ferdinand Schorner; Generals Ludwig von Schroder, Kurt Student, Walter Warlimont, August Winter and, General Gustav von Zangen. In addition then, we may also note the names of Generals Froehlich, Gausz, Cruewell; Col. Gen. Gotthard Heinrici, General Martin Gareis, Lt. Gen. Wolf Hagemann, General Hasso von Manteuffel, Major Generals Franz Reuss, Hellmuth Reymann, Hans Speidel; Lt. Gen. Friedrich Sixt, SS Gen. Martin Felix Steiner, Rudolf Hess, General Heinrich Himmler—Head of Gestapo and the SS, General Reinard Heydrich—Deputy Chief of Gestapo and also; General Walther Wenck and General Kleeman as well. Then some of them got killed during the war and, some of them were captured by the Allied Forces and also, many of them surrendered too and thereby were tried for many kinds of war-crimes and sentenced to death or various lengths of prison-terms. These post-war famous trials are now historically known as the **'Nuremberg Trials'**.

Out of the above mentioned Nazi Germany's historic legends, most appropriately considered to be Titans in this regard—**Field Marshal Rommel was in fact, both—a born writer as well as no less a born fighter too**. A book, since produced as a compilation of Rommel's valuable writings by **B.H. Liddell Hart** and along with **Rommel's son—Manfred Rommel, titled as *'The Rommel Papers'* would always remain a valuable source of information on Rommel's personality and, his thoughts on warfare as well. In his 'Introduction Chapter' itself, Liddell Hart (pages xiii, xiv and xviii) has written that, "The impact that Rommel made on the world with his sword will be deepened by his power of the pen. No commander in history has written an account of his campaigns to match the vividness and value of Rommel's—which for the most part has now been retrieved from its various hiding places and put together in this volume**.... Rommel's narrative is remarkably objective, as well as graphic. In drafting it he certainly had, like most men who have made history, a concern for his place in history. But while he shows a natural desire for justification in his explanation of events, it is subordinate to his burning interest in the military lessons of the campaigns.... The clarity and the high degree of accuracy which distinguish Rommel's picture of the operations are the most notable because of the confused impressions that are to be produced by fast-moving tank battles, especially in the desert. The clearness of Rommel's picture owes much to his way of command—his habit of getting right forward and seeking to be near the crucial spot at the crucial time. It also owes much to his prolonged self-training in observation, highly developed eye for spotting what was significant in the scene, and knack of registering it. His passion for taking photographs at every step of advance was a symptom of this characteristic—as it was with Lawrence in the Arabian theatre of World War I.... **In most fields, genius is associated with originality. Yet it has been rare among those who are usually acclaimed as the great masters of war. Most of them have gained their successes by using conventional instruments speculatively well, and only a few have sought new means and methods. That is strange, since history shows that the fate of nations has been repeatedly decided, and the most epoch-making changes in history determined by change in weapons and tactics—especially the later**.... In the history of war great ideas have been less numerous than great generals, but have had a more far-reaching effect. The distinction between the two is a reminder that there are two forms of military genius—the conceptive and the executive.... In Rommel's case they were combined. While the theory of ***Blitzkrieg***—the new super-mobile style of warfare with armoured and motorised forces—had been conceived in England, long before he came on the stage, the quickness with which he grasped it and the way he developed it showed his fresh-mindedness and innate conceptive power. He became next to **Guderion**, the leading exponent of the new idea." ('The Rommel Papers', Natraj Publishers, Dehradun, 1977).

(a) Battlefield

We note that therein, and even while facing an extremely hot and challenging situation on 26 May 1942 during the second year of war in Africa (p. 208), **Rommel** with his rare coolness of mind and also clarity of thoughts has very perceptively as well as with great professional precision, has written that, "Certainly we had seriously mauled the brigades which the British had thrown against us in the south-east of Bir el Harmat. The **3rd Indian Motor Brigade** had suffered such heavy losses that it was unable to make any further appearance during the whole of the battle. It was also to be a long time before the 7th Armoured Division recovered from the blows it has been dealt that day.... But I will not deny that I was seriously worried that evening. Our heavy tank losses were no good beginning to the battle (far more than a third of the German tanks had been lost in this one day). The 90th Light Division under General Kleeman had become separated from the Afrika Corps and was now in a very dangerous position. British motorised groups were streaming through the open gap and hunting down the transport columns which had lost touch with the main body. And on these columns the life of my army depended.... However in spite of the precarious situation and difficult problems with which it faced us, I looked forward that evening full of hope to what the battle might bring. For Ritchie (*i.e. British Lt. General Sir Neil Ritchie*) had thrown his armour into the battle piecemeal and thus given us the chance of engaging them on each separate occasion with just about enough of our own tanks. This disposal of the British Armoured brigades was incomprehensible (*So who says that it is crucially not important for a General in war, to also be no less lucky as well.*). In my view the sacrifice of the 7th Armoured Division south of Bir el Harmat served no strategic or tactical purpose whatsoever, for it was all the same to the British whether my armour was engaged there or on the Trigh Capuzzo, where the rest of the British armour later entered the battle. The principal aim of the British should have been to have brought all the armour they had into action at one and the same time. They should never have allowed themselves to be duped into dividing their forces before the battle or during our feint attack against the Gaza line. The full motorisation of their units would have enabled them to cross the battlefield at great speed to wherever danger threatened. **Mobile warfare in the desert has often and rightly been compared with a battle at sea—where it is equally wrong to attack piecemeal and leave half the fleet in port during the battle**." (*It is certainly a historically timeless and also an outstanding piece of writing from the 'mighty' pen of indeed a legendary military genius, during the momentous times of Second World-War; and precisely that too, in the midst of an extremely hot and fluid battlefield scenario itself.*).

In fact, the famous **'Nuremberg Trials'** were indeed path-breaking and also great history-making events in many ways. In these historic perspectives; **Professor Mark Raginsky**, in his article on 'Before the Tribunal of the Nations' and since published in the magazine 'New Times' (47-85) and therein has very thoughtfully mentioned that, "The Nuremberg trial played a major role in the progressive development of international law. The records of the trial, the judgment, and the principles of Nuremberg are in this respect of inestimable value. But it is not enough to recognize this fact; it is imperative to work for the full and undeviating implementation of these principles. This end is served to no small extent by the recognition by the UN General Assembly in December 1946 of these principles as standards of operating international law. In August 1979 these principles were endorsed by the 11th World Congress of the International Political Science Association. At a meeting of one of the sections of the congress this writer had occasion to present the position of Soviet political scientists on this question. Despite some differences of opinion, all the participants agreed that observance of the principles of the Nuremberg trial could effectively help safeguard world peace.... It was not only the war criminals, albeit the main ones, that were tried in Nuremberg. And the case was not closed when the last of the men sentenced to death by the international Military Tribunal was executed. In the dock of Nuremberg was fascism, a political system of undisguised terroristic dictatorship of the imperialist bourgeoisie, an ideology expressing the interests of the most chauvinistic and aggressive circles of monopoly capitalism.... The Nuremberg trial has gone down in history as an anti-fascist, anti-imperialist and anti-militarist trial. It helped to reveal the real motive forces of aggression. The international Tribunal passed judgment not only on the fascist top men, but also on all those forces whose sights were and are still set on war. And herein lies its vast significance for the present.... What **Robert Kempner**, US Prosecutor of the trial and now an eminent legal scholar, said in his speech at Nuremberg is highly appropriate today too. The 'communist danger', which the Nazis invoked as the pretext for a reign of terror against anti-fascists and all the progressive forces of the German people, was, he said, 'a mere subterfuge and was one of the things which finally led to the second world war."

Now again coming back to the various aspects of above mentioned 'clashes of the Titans'; on the side of **Mussolini's Fascist Italian forces**, we may note the names of **Generals Grazani, Soddu** and, that of **General Cavellero** as well. In addition to them, there were other Italian Army, Navy and Air Force commanders too; who answered the extremely militarism-oriented imperialistic calling of the Italian fascist dictator **Benito Mussolini**, to fight his war against the Allied Powers in Europe, as well as in North African lands. So far as Mussolini's geostrategic-cum-politico-military thoughts and ideologies were concerned, which were instrumental in shaping his 'war-plans', even if somewhat 'half-baked' we may contextually recall that, we have since already briefly delved upon them in the earlier pages in this chapter itself.

Now specifically in the above perspectives so far as the then **Imperial Japan**, being on the side of Axis-Powers, was concerned Japanese too had a substantial line-up of their Army, Navy and Air force commanders for this war to protect their imperial-cum-economic interests as well as their hegemonic colonial-empire in the far-east and hence to fight the allied forces since deployed in the vast Pacific region. Among them, we may note the names of Baron General Giichi Tanaka—author of the 'Tanaka document'; General Hideki Tojo—wartime Prime Minister of Japan, General Hejime Sugiyama—Japanese Army Chief in 1937, Admiral Isoroku Yamamoto—Chief of the Imperial Japanese Navy; General Umedzu, Rear Admiral Raizo Tanaka—the mastermind behind the 'Tokyo Express', i.e. he had the responsibility to transport the troops and supplies to Guadalcanal in the Solomon, since situated in the somewhat north-east of the Australian continent in the Pacific. Then **General Yamashita**—the overall commander of the Japanese forces since deployed in the Pacific region and, Colonel Shuttu Matsumura—Chief of Information Section in the Imperial Headquarters during the war. In addition to the above and since also, momentously triggering the entry of Japan in this war and hence thereby; the big-four principally responsible for the surprise Japanese attack on the US Naval Base in Pearl Harbour, happened to be—Prince Konoye, Mr. Matuoka, Admiral Yoshida and, General Tojo as well. And at the culmination of second world-war, **these and many other Japanese military commanders too, who were captured or surrendered before the Allied commanders were tried by the Allied Powers' Military Tribunal in Japan, under the overall control and supervision of the US General Douglas MacArthur in Tokyo and also, broadly on the lines of 'Nuremberg Trials' in Europe**. We may also recall that in the earlier paragraphs in this regard, we have already briefly touched upon the multiply brilliant activities then, launched by General MacArthur for indeed the momentous reconstruction, rehabilitation and also, the far-reaching socio-cultural and socio-political-cum-economic reformation of the then militarily defeated as well as, the devastated Nation-State of Japan.

In the 'clashes of the Titans' Great Britain too, indeed had a highly competent line-up of the British Armed Forces' military commanders, professionally taking-on the relentless onslaughts of the menacing Axis Powers in the all theatres of this war across the world. There was a galaxy of highly focused British commanders and staff officers at all levels of the military command-systems then, since engaged in their various kinds of the highly challenging tasks in this war. Some names out of so many of them, may contextually be mentioned such as: Field Marshal Sir Alan Brooke—then Chief of the Imperial General Staff (CIGS), **Field Marshals Sir Bernard Montgomery, Wavell, Claude Auchinleck;** Admiral Cunningham, General Sir Harold Alexander, Air Chief Marshals Sholto Douglas, Tedder; General Sir Humfrey Gale, Admiral Morse, Air Chief Marshal Sir Hugh Dowding—the legendary AOC-in-C of the Fighter Command in the British Isles and thus, successfully defending the British Nation from the relentless air-assaults of the Nazi Air force Luftwaffe; and in the end, resulting in the decisive British victory in the 'Battle of Britain' itself. Further, we may also record the legendary names of **Major General Orde Wingate**—commander of the famous Chindits Brigade, then operating behind the enemy lines in the historic Burma campaign; **Admiral Lord Mountbatten**—Commander of the Allied Forces under the South-East Asia Command; Field Marshal Hull, Admiral Luce, General Cassells, Air Chief Marshal Elworthy, Admiral Somerville, General Ismay, General Sir George Giffard, Air Chief Marshal Sir Richard Peirse, Admiral Power, **Field Marshal Viscount Slim**, Lt. General Wheeler and, Air Chief Marshal Park as well. In addition to the above, there happens to be numerous other names of the British soldiers, sailors and airmen of all ranks—other than officers too, and in addition to the so many names of officers of all ranks, who have magnificently contributed with their lives and blood—in writing the immortal living-history of this unprecedented global war and hence their names also, would always remain written in golden letters in the pages of that history. Indeed, the fruits of victory in the end, is always sweet for all concerned in any war; but the multiply momentous challenges, menacing Axis-Powers' threats to the lives and properties across the world, had been magnificently faced, endured and also successfully countered by the soldiers, sailors and airmen and their officers since belonging to the Armed Forces of Allied-Powers; and especially much more so, during the initial years of this war; when the Axis-Powers happened to be rather unstoppable indeed, in their well-planned as well as, the all sorts of offensive operations on numerous fronts; and also particularly then, the Nazi Forces' 'blitzkrieg' in Europe.

Then as this most extensive, intensely devastating, truly intercontinental and, indeed multi-dimensional Second World-War progressed from the year 1939 onward and immediately after the Pearl Harbour incident on December 6, 1941, the **United States of America** decided to actively join the war, with great intensity against the Axis-Powers, with all her massive resources and military might. Under the momentous presidential orders of the US President Roosevelt then; USA feverishly commenced her multiply massive mobilization as well as, increasingly pushing-up her indeed unmatched and also multi-dimensionally unprecedented war-efforts. Consequently then in the vast industrial complexes in the USA, production of the all kinds of weapon-systems, weapon-platforms, and munitions and, also mind-boggling—the multiple kinds of wartime goods as well as accessories too had advanced, without any kinds of generally the peacetime systemic-cum-routine delays at all. In addition and besides all these, at the highest level of the newly combined operations concepts in Europe against the Axis-Powers in fact lot of matters-military oriented as well as, multi-dimensional planning-cum-staff-work too, had commenced under the overall supervision and guidance of **US General Dwight D. Eisenhower** without delay immediately after his appointment as the Supreme Commander of the Allied Expeditionary Force in Europe.

In fact, the colossal range-cum-dimensions and the multi-dimensional complexities of the huge work involved in the matters-military oriented operational planning-cum-staff-work in this regard, has very vividly been painted with great historic precision, in a book by **Joseph P. Hobbs**. Concerning the multiple aspects of World-War II, this book also is of great historic value and that too, when it pertains to the direct correspondence-papers between two key US military legends of those times, who were then manning rather two of the most crucial appointments. Especially in case of multinational military forces and their command-systems, the otherwise generally prevailing common matters-military as well as also the personality-oriented complexities, do perforce assume many extra dimensions and thereby, throwing uncommon kinds of challenges as well. This book reveals many such things and also many kinds of the otherwise rarely known crisscross currents of professional ideas, opposing thoughts and so on in this regard.

Find the very first letter of **General Eisenhower to General George Marshal** and therein he writes that, "Dear General: This is the first of my informal reports to you, written exactly twenty-four hours after I landed here. It will be longer than those to follow, because I would like to give you the highlights of the current picture.... One interesting thing I immediately encountered is the state of mind of Lord Louis Mountbatten. He has been waging a rather serious even if friendly argument with General Paget about a matter in which I think Lord Louis is quite correct, but in which we have no direct concern. It involves Paget's conceptions of training for the assault. Instead of designating a corps, with its commander and attached troops, to begin training and organizing *now* for this particular task, the British Army officials persist in viewing the cross-channel movement as merely another advance guard exercise. They therefore apparently propose to send in two or three corps abreast, expecting that the leading detachments of each corps will have received some special training before that time landing from boats. This of course diffuses responsibility and is almost the exact antithesis of our idea, under which we have designated General Clark as the assault corps commander to head the American Army on its great venture. It seems to me that our method is so logical, simple and sound that I cannot perceive any reason for attempting to do the job otherwise.... I fear that some inter-service jealousies or politics may be involved. At any rate, realizing that I am powerless to intervene, I hope to demonstrate such successful results from our own method that the British will adopt it speedily. For this reason I am more than ever determined to be as forward, in the matter of time, as is humanly possible.... I shall try to apply your dictum—'Persuade by accomplishment rather than eloquence." (Hobbs, 1999, p. 25)

Historically, those were indeed unprecedentedly momentous times in the Allied Camps. And understandably therefore, there were larger dimensions to all these massive historic developments and where since the very beginning of 1942, the USA did assume the multiply vast and also indeed, very decisive roles—both in ensuring the proactive maintenance of unmatched and unobstructed systems of logistics in almost the all theatres of war the world over and, simultaneously at the same time in all those theatres of war massive formations of the US Army, US Naval Fleets and the Air Force fighters and bombers Fleets too, were launched in coordination with the military forces of the then British Empire, to effectively counter the Axis-Power forces and thus, advancing towards decisively winning this war. **It is a plain historic fact that indeed then, only the USA had these kinds of economic, industrial and also the manpower resources—to venture upon such gigantic kinds of the multi-dimensional wartime global initiatives and efforts**. Now so far as the already above discussed **multiple-aspects of the 'clashes of Titans'** are concerned, USA too had her own magnificent line-up of the brilliantly professional military commanders, who could indeed magnificently ensure to provide the successfully proactive leadership-inputs at all levels to the US military forces in all sorts of the extremely challenging situations during this war. The lists of their names are no wonder long but in order to substantiate these aspects we may record few of the names here, such as: **General Douglas MacArthur in the Pacific, General Dwight D. Eisenhower in Europe, General George Marshal as Chairman of the US Joint Chiefs of Staff, in the US Armed Forces Headquarters in Washington; Admiral Nimitz**, Admiral Leahy, Admiral Ernest J. King; General Thomas Holcomb—US Marine Commander; General H.H. Arnold, of the US Air Force; Lt. General Jonathan M. Wainwright in the Philippines; **General Joseph Stillwell in China;** General Clark, General Patton, General Bernard C. Freyberg—Commander of the New Zealand Expeditionary Force, who was then preparing to send his troops to take on the German defenses along the mountain, and so on. In addition to the above, in US Army Air Force (as the USAF was known then); there was indeed a galaxy of names such as: General Laurence Kuter—the Assistant Chief of Air Staff for planning and analyses; General Elwood R. Quesada—Head of the XIX Tactical Air Command; General Nathan F. Twining—Commanding General of the Fifteenth Air Force; General Anderson—Commanding General of the US VIII Bomber Command; General LeMay, General Norstad—Commanding General of the XXI Bomber Command; Colonel Cecil E. Combs—General Norstad's Deputy for Combat Operations; General Mason Patrick—US Chief of the Air service; General Carl Spaatz—Combat Commander and Manager of large-scale air operations and the man whose forces planned and repeatedly went for air offensives for some of the most effective-cum-decisive and also, controversial air actions during the Second World War; General Doolittle and then, General Groves as well.

The massive quantum of the US Air Force fighters and bombers assets which in fact were launched by the USA during World-War II, against not only the military might but simultaneously also against the socio-economic, human resources, industrial as well as, the scientific and technological wherewithal of the Axis-Powers both in Europe and the Far-East, were indeed unprecedentedly devastating. But these too were executed by the USA, as per the comprehensive as well as in fact no less ruthless US Air-War Doctrine-based operational plans and also, in strategic-cum-tactical co-ordination with other Allied Partners with pivotal aims to break the enemies' will to fight anymore. It is true that these were finally achieved in the end, but indeed with great costs especially in terms of colossal human-lives and sufferings, and also along with unimaginable collateral damages. But that's the way war is largely fought especially when the human-level multiple-aspects of chivalry as well as also, otherwise the common humanitarian considerations are overwhelmingly subjugated finally, by the far-reaching, multiply decisive and also no-less captivating—new developments in the technologies concerning the air-war related weapon-systems, weapon-platforms and their delivery-systems. Hence, in the entire gamut of the specifically this kind of warfare and, even before the war begins in fact, the first casualty possibly becomes the all kinds of humanitarian considerations in this regard. Even then at the time of such extremely pressing situations, when necessity to win the war itself in the first place cannot in any case be compromised at all and that too irrespective of the any kinds of costs involved, the highest-level political clearance especially in any truly vibrant democracy, is never easily obtained. Before arriving at any such final decision in a democratic set-up, lot of debates and arguments from various points of views, are resorted to—before arriving at the final positive decision for going ahead and that too, substantially with heavy hearts. Hence, it is rightly said that uneasy lays the head that wears the 'crown'. Precisely similar things did substantially happen during those history-making momentous times in the USA too, when multi-dimensional as well as multi-layered comprehensively offensive air-war plans against the no less tenacious Axis Powers, were finally given the green-signal. These indeed multiply historic aspects have been discussed in lot of valuable details and also with great geostrategic-cum-politico-military insight by **Professor Ronald Schaffer**.

Professor Ronald Schaffer while delving upon the **multiple-aspects of the American Air-War doctrine and the bombing of civilians** has very reflectively written that, "Officers like Hugh Trenchard of the Royal Flying Corps, Billy Mitchell of the Air Service and the Italian General Giulio Douhet believed that in the next war the airplane could prevent this kind of stalemate, in making war again an effective instrument of national power. **They envisioned fleets of warplanes heading towards the vital centres of an enemy nation to paralyse and destroy them with poison gas, incendiary bombs and explosives, terrorising civilians until they begged their government to surrender. In the future war, bombers would spare the lives of soldiers by moving the battlefields from the trenches to the cities**.... The most original and influential of these theorists was Douhet, who before World War I had begun to formulate a fairly coherent doctrine of air strategy. Novel as it is in major respects, Douhet's theory had roots in earlier military thought. Like the nineteenth century Prussian strategist Karl von Clausewitz, author of *On War*, Douhet considered war an act of force used by one nation to compel other nations to do its will. He tried, as Clausewitz had done, to view war without illusions, to examine the most extreme form it might take.... That form, as Douhet conceived it, was conflict of whole peoples, employing the entire human and material resources of society, a struggle to which the distinctions between combatant and noncombatant vanished. In modern war everyone took part, 'the soldier carrying his gun, the woman loading shells in a factory, the farmer growing wheat, the scientist experimenting in his laboratory.' **In this total war the primary object of military action would no longer be the enemy's armed forces, as in Clausewitz's time, but the vitals of the nation itself, the source of the enemy military power, now exposed by technology to attack from the air**.... In some respects General Douhet's vision of war resembled the view of the American naval theorist Admiral Alfred T. Mahan. Admiral Mahan had recommended far-ranging offensive action as the best way to defend its trade, protect its homeland, and secure its other objectives. Douhet likewise considered **strategic attack by far the best form of defense**. Mahan wrote of the necessity to secure command of the sea, and idea paralleling Douhet's belief that a nation must dominate the third dimension by seizing the command of the air. This would be done not through the inefficient method of individual air combat, but by striking, at the airports, supply bases, and production centers. For Mahan the chief means of war was a long-range battle-fleet. For Douhet it was an air force independent of other military arms.... Once it had secured command of the air, this independent air force could disrupt the mobilization of the enemy army and hamper the enemy's navy by bombing oil supplies, naval bases, arsenals, and ships at anchor. Or it head directly for the chief objective, the population centres: Rome, Paris or London. Bombers would terrorize the nation, breaking its physical and moral resistance. Douhet felt civilian workers were more susceptible than the troops to a collapse of morale and that unlike an army they would be unable to resume their operations after an attack. 'How could a country go on living and working,' he asked, 'oppressed by the nightmare of an imminent destruction and death?'.... The future war, as Douhet imagined it, would begin with warplanes bombing a city, dropping high explosives, poison gas, and incendiaries. The attack would erupt suddenly and last just a few minutes." (*So contextually in our broader perspectives we do find that during World-War II in fact, both clashing Powers-Groups had applied the above quoted strategic air-war concepts against their adversaries, in their own differently formulated strategies and war-plans.*). (Schaffer, pp. 20 & 21)

Much before the commencement of the Second World War in fact, the **massive Communist Russian Empire in the geopolitical shape of the USSR**, had already come into being and thereby, overwhelmingly both geostrategic and politico-militarily covering a vast and continuous Eurasian landmass. It is historically an altogether different matter that during the post-World War II era the USSR along with her East European Satellite States, for many kinds of the then globally altered geostrategic-cum-politico-military reasons, turned out to be diehard adversary of the Western democratic Nation-States—principally in Europe as well as in the Northern American continent. In fact objectively speaking in their now historic perspectives, both the massive 'Camps-cum-Power Blocs' had their own and altogether mutually opposing compulsive geopolitical-cum-multiply-strategic reasons in this regard, and now interestingly in retrospect it is a different matter altogether that all these bygone cold-war era historic aspects—do now obviously appear to be even naïve and rather comically futile too. But historically these are different matters and, in any case we would contextually be briefly discussing them in the next eighth chapter. But during the early years of momentously history-making Second World-War times the USSR was in fact first, very sinisterly betrayed by Hitler and here, we may also recall that we have since briefly discussed these aspects little earlier, in this chapter itself. In any case, **Joseph Stalin** of the then USSR too, was no 'saint' in any manner at all but when Hitler's attacks on USSR clearly appeared to be imminent, Stalin perforce prudently joined the war from Allied Powers' side and thereby; with all round Soviet efforts for the massive mobilization and, as the wartime momentous events began unfolding onto the Eastern fronts in Europe, obviously the Soviet forces' intensive participations too, began enlarging manifold. **Now in these specific contexts and while again coming back to the multiple-aspects of the already above discussed historic 'Clashes of the Titans' during this momentously global war; the USSR too had her own line-up of the famous military commanders to lead their forces to relentlessly fight with the Axis Powers' military-might, with some rather largely military-oriented understandings with the other Allied Powers, to finally achieve decisive victory in this war**. But there was indeed always a world of difference in multiple areas of the 'matters-military' as well as, the multiple-aspects of militarily professional ethos; between the then Soviet military and, military-forces of the Western democratic nations and their respective close-allies.

Some of the now historic names in this regard are—**Marshal of the Soviet Union G.K. Zhukov**, who on behalf of the USSR, formally signed the act of Nazi Germany's unconditional surrender; Commander of the 37th Army Lt. General M.N. Sharokhin, Commander of the 46th Army Lt. General I.T. Shlemin, GHQ representative Marshal of the Soviet Union S.K. Timoshenko, Commander of the 3rd Ukrainian Front General of the Army F.I. Tolbukhin, Commander of the 57th Army Lt. General N.A. Gagen and, Commander of the 5th Striking Army LT. General N.E. Berzarin as well. In addition to them, we may also note the names of the Military commandant of Berlin and Hero of the Soviet Union and 2nd Far Eastern Front General of the Army M.A. Purkayev; Member of the Front Military Council Lt. General G.S. Leonov, and then; Soviet Pacific fleet Military Council—Lt. General S.R. Zakharov, Fleet Commander Admiral I.S. Yumashev and, Chief of Staff Vice Admiral A.S. Frolov and then; General K.N. Derevyanko, who signed the Act of Japan's Unconditional Surrender as the official representative of USSR, on board the USS Missouri—then anchored off the Japanese coastlines and since, acting as the official venue for this formal ceremony, and then; General of the Army I.D. Chernyakhovsky, Commander of the 3rd Byelorussian Front and his Chief of Staff Colonel General A.P. Pukrovsky; Members of the Military Council of the 3rd Byelorussian Front: Lt. General V.Y. Makarov, Front Commander Marshal of the Soviet Union A.M. Vasilevsky and Chief of Staff Colonel-General A.P. Pokrovsky; Members of the Military council of 2nd Ukrainian Front: Lt. General M.M. Stakhursky, Chief of Staff Colonel-General M.V. Zakharov, Front Commander Marshal of the Soviet Union R.Y. Malinovsky, Colonel-General I.Z. Susaikov;

and then, Marshal of the Soviet Union P.I. Tolbukhin, his Staff Officer Lt. Colonel S.S. Birynzov, and so on. **In addition to all these Soviet Commanders and their numerous soldiers, sailors and airmen; there were other USSR's Allies too whose significant efforts, resolute resistances against all odds and their sacrifices, in conjunction with the Soviet Armed Forces, in facing the ruthlessly relentless Nazi Germany's assaults onto the Eastern European Fronts during those extremely trying times, just cannot be historically marginalized**. In fact again the historically irrefutable point in this regard, that remains well accepted is, that without their sustained, disciplined, highly dedicated and also, brave services what the then Soviet Armed Forces did significantly endure and factually as well as indeed, so magnificently achieved in finally turning the tides against the Nazi Forces and hence (*in any case, minus the all sorts of officially sponsored Soviet propaganda therein and well, it is also true that all these the then lingering propaganda inputs from the Soviets, have now been rather quite befittingly put in the dust-bins of history.*), leading to the conclusive victories all along the vast Eastern European Fronts could perhaps never have been so decisively possible.

So long as the USSR existed on this planet Earth, as the massive Communist Empire covering huge territories in continuity, in both the continents of Europe and Asia, the Soviets had their own historic version of the Second World-War. For any objectively discerning reader of the Man's living military-history, the problem with the official accounts of any war had been that these remained obsessively intermingled with their then rather overwhelmingly prevailing some kinds of their diehard communist ideologies and also their political propaganda, against the Western Democracies. In fact the then Soviet Armed forces too, were not at all apolitical as well as, purely professional as in case of the Western democratic nations' Armed forces in this regard. Moreover, the then Soviet Armed Forces were trained and also politically groomed on the totally unchallenged ideologies of the ruling Communist Party of the USSR and thus for all kinds of the policy-lines laid down by the Supreme Soviet as well as, according to their extremely rigid ideologically-fixated and also, together with their equally rigid geostrategic-cum-politico-military purposes the Soviet Armed Forces were always supposed to dedicatedly act as their powerful Military-Wing in this regard. Hence, these the then prevailing Soviet-aspects in any case just could not even marginally be reconciled with the military-norms, systems of functioning as well as, the liberal democratic ideals of the Western Allied Powers. Hence, in the Western democracies' domain of the essentials of their matters-military during those historically momentous times, their Armed Forces just could not be operationally integrated with the Soviet Military for any joint-operations, in any workable manner whatsoever. Of course now in retrospect, we find that the USSR is no more in existence and so these aspects are contextually just of some historic relevance only and hence broadly keeping them in mind, we may find it relevant to refer to **one of their many Soviet-era publications**.

A Soviet Marshal has written that, "Almost sixty years ago a new type of soldier appeared on the historical scene. He had the same old grey greatcoat but there was a big red star on his cap. He was a Soviet soldier who not only did not threaten the toiling masses, but, on the contrary, was prepared to endure the greatest privations and give his life for their interests. Since then the Soviet Armed Forces have been vigilantly and reliably guarding the peaceful labour of the Soviet people and the cause of world peace.... The Second World War was a rigid test of the moral and material forces of many countries and peoples. It was the most hard-fought and devastating war history had ever known. Like the First World War it was an imperialist war which had been prepared over a period of many years by the imperialists of many countries. It broke out as a result of a fresh sharpening of contradictions in the capitalist world, and the responsibility for it devolves on the entire capitalist system. It was started by Nazi Germany which acted as the shock force of the world capitalism....Yet, as distinct from the First World War, it erupted when capitalism ceased to be a world-wide social system, when besides the capitalist states there was also the Soviet Union, the first state of workers and peasants in the history of mankind.... The Soviet people value highly the role played by the armies of the USA, Britain, France and other countries in smashing the common enemy. But almost for three years the Soviet Union bore the main burden of the war as it faced the combined forces of the fascist aggressors single handed (*Look at this: what?*).... The main battles of the Second World War were fought on the Soviet-German front where 607 enemy divisions were destroyed, routed or taken prisoner—almost three and a half times as many as on all the other fronts taken together (*Wah: what a distortion of the truths of history in the then Soviet Lands.*). The Anglo-US troops in Africa and Western Europe smashed or taken prisoner 176 divisions. The Wehrmacht sustained 80 percent of its casualties on the Soviet-German front. There the enemy lost the bulk of the weapons of his ground and air forces and also a considerable part of his navy (*In any case, what more to comment on all these purely propaganda stuff?*)." (Grechko, pp. 5, 6 & 9)

So in the above few pages, during the momentous beginning of the 1939-45 Second World-War, we have recorded a few of indeed the famous names of mostly top-level military commanders since, got lined-up as professionally historic 'Titans'—on both sides of the warring Power-Groups—to take to battlefields in the different no less history-making theatres of this war and needless to say that, theatres/warfronts were also located mostly in the continents of Europe, Asia, Africa and also, Australia as well. In addition to them, numerous extensive battles were also fought in the seas and vast oceans surrounding these continents and then in addition also, in the vast skies over these continents as well. And obviously therefore, very many different kinds of the weapon-systems as well as, numerous verities of weapon-platforms too and since belonging to the different Wings of the Armed Forces of both, the mutually opposing Axis-Powers and Allied Powers, were extensively employed to fight this war. Before the war actually began, it does perhaps historically appear that—neither side really had any substantial idea in any case, as to what this war so far so unprecedented in both its range, dimensions and also in deaths, destructions and devastations was soon going to be, and in fact so momentously colossal and obviously that, not even ever imagined by the humankind earlier. Hence in other words, **we may also say that irrespective of any perceptions in this regard, even before the first shot of this war was fired by either side the first and foremost rather largely invisible casualty in war, turned-out to be nothing else but, the widely accepted and valued—the mankind's civilisational-cum-spiritual humaneness in the world**. Consequently, as the weeks, months and years kept on passing-by gradually as per the ever unchanging rate of the so-called forward movement of the timeline in this world, especially pertaining to and also later came to be known as—nearly the six years' period (i.e. 1939-45) of this war—there indeed remained nothing credible in the hands of mankind—to end midway this most tragic as well as ruthlessly inhuman saga of a series of such deaths, destructions and devastations in almost the whole of Europe and the Far-East, then so relentlessly taking place during this war. During those bygone years of war, it appeared that the whole world was rather inescapably trapped in a kind of unending violence and destruction. But it is also true that, onto the man's expanding civilisational-canvas; any rather appearing to be even unending and frighteningly bad-patch also, has to come to its end and hence this 1939-45 World-War too, did end finally to the great relief for the mankind, the world one.

Historically in any case, when the die for the World-War II was finally cast and as usual mostly in any previous minor or major war **this war too, but in a totally unimaginable as well as unprecedented scale, was also fought simultaneously and in addition also in mutually**

complimentary manners at two distinct but inseparably overlapping levels, i.e. political-cum-politico-diplomatic and then the military as well. In retrospect, as we may care to critically delve into specifically the momentous living-history of this 1939-45 global war, we may notice that first at the political-level and, in spite of certain amount of at times, the obviously understandable surfacing of the passing-by heat, dust and even confusions among the **nations in the Allied-Powers' camp since mainly being democratic had distinct long-term advantages as well as also, ensuring the socio-political-cum-cultural kinds of the force-multipliers; in comparison to the Axis-Powers**. And clearly the pivotal reason in this regard, being that all the leading nations since lined-up in the Allied-Camp, were having well-assured and vibrantly healthy-system of democratic governance and in any case, these were missing among the nations in the Axis-Powers' camp. At this precise point and also in broad historic perspectives, the point to be contextually borne in mind remains that in any case this was no small matter in this war and thereby this too went decisively far, in ensuring the final victory for the Allied-Powers.

Now, so far as the matters regarding exclusively the military-level of fighting this war were concerned, situations were certainly not so easy for either side. But during the beginning phase of this war in fact, the Axis-Powers had distinct and indeed menacing edge over the Allied Nations due to their more focused and well in advance preparations under their autocratic and ruthlessly militarism-oriented regimes. During the series of momentous 'clashes of Titans', many kinds of the ups and downs as well as, even rather astonishing and also decisively history-making kinds of the shifting of fortunes in the numerous theatres/warfronts of this war did also take place as no less momentous turning-points, within the ongoing phenomenon of that history-making itself. But by the time tides of war started to gradually turn in favour of the Allied Forces, the decisive moments of reckoning too, began to arrive and thereby, the principal players in the Axis-Powers' camp, i.e. the military forces of the Nazi Germany in Western Europe and then, those of the militarist Japan in the Far-East as well, were compelled to unconditionally surrender before their opposing Allied Forces' top commanders in the field and thus, finally bringing the curtains down on this most menacing-cum-devastating war. We may contextually recall and already briefly discussed earlier that Italy had already been decisively subjugated by the Allied-Forces even earlier in this war. **These indeed were great moments of history and also at the same time, it was becoming absolutely clear that the world too was soon going to undergo many irreversible changes as well from then onward after the decisively successful but also simultaneously, no less highly chequered and devastating end of this war, during the mid-twentieth century**.

Now so far as the historic aspects of the momentously unconditional surrender of Nazi Germany were concerned, a well-known British Conservative Party member, politician and statesman in his book, *The Blast of War, 1939-45*, has along with numerous other politico-diplomatic as well as strategic-cum-military episodes of this war very vividly described the historically decisive turn of some significant events finally leading to the Nazi Germany's unconditional surrender. **Harold Macmillan** in the rank of a Deputy Minister and, who was then a kind of Political-Assistant to the **Prime Minister Sir Winston Churchill** during the war and later-on, he himself became Britain's Prime Minister in the sixties. In his above referred book (pages 586 and 587) Macmillan has written that, "On the next day (*i.e. during the first week of May 1945 itself*) we heard that Berlin had fallen. Still, we were, and Alexander had a right to be proud. In spite of our anxieties, the actual carrying out of the surrender proceeded in an orderly way. But this was still only a military surrender.... **The German plan is now becoming pretty clear. It is to obtain *de facto* what they failed to obtain *de jure*, viz., surrender to the British and Americans instead of to the Russians.**

They are now making and will go on making these piecemeal surrenders—holding the Eastern line with rearguards and trying to save their men, not their territory.... Naturally the German knew that if their people became prisoners of war of to us or the Americans, they would be looked after, fed, cared for and ultimately returned to Germany. But if they fell into Russian hands, they would probably never be seen or heard of again.... Kesselring (through Wolff) has sent Alex, a message asking to be put in touch with Eisenhower—a very interesting development.... Although from the German point of view this was their best chance, the British and Americans stood resolutely by their alliance and in spite of all Stalin's suspicions, Churchill and Truman kept both the letter and spirit of their undertakings.... On 7 May, the following message was circulated on behalf of the Field Marshal to Fifteenth Army Group and repeated to the many organisations, missions, rear headquarters, delegations and bases under his command.... **At 0241 hours 7 May, a representative of the German High Command, signed the unconditional surrender of all German land, sea and air forces in Europe to the Allied Expeditionary Force and simultaneously to the Soviet High Command.... All forces under German control are to cease active operations at 0001B hours 9 May.... It seemed, therefore, that 8 May was now to be V.E. Day.... It seemed almost impossible to imagine that it is all really over.... So ended six year of the most terrible war in our history**. We had surmounted many perils by air and land and sea. We had been buffeted by fierce tempests, but were now moving into calmer waters. **Yet who could tell what future might await our Empire and our people after the storm**." (Macmillan, 1967-68). (*Well, interesting isn't it? But at the same time, in the broader terms concerning its contextual historicity it is understandable too. At this precise point, while going back into the broader historic aspects in this regard, we are indeed able to see through that era's since, British diehard imperialism-oriented policy-lines as well as, thought-processes on the part of leading the then British leaders and decision-makers and at the same time, we do also interestingly be able to have some glimpses of their highly fixated obsessions for their once gloriously global Empire and that too in spite of the such unprecedentedly colossal destructions, deaths and devastations in the British homelands themselves during this war. We may recall that we have since already touched upon those imperial-era fixations on the part of British-Nation, in different contexts earlier and thereby; we would be discussing them further in some more details little later*).

During the 1939-45 World-War II, the legendary **US General Dwight D. Eisenhower** was the Supreme Commander of Allied Expeditionary Forces in Europe, with his Headquarters based in Britain and obviously. It was under his wartime supreme command—that the Nazi Germany's various top-level field commanders along with their sizeable military formations in the month of May 1945—eventually surrendered almost all over Europe, to their counterparts in the Allied Forces over there. Hence, it may now appear to be appropriately relevant here, to historically bring-in his personal observations and perceptive comments too in this regard. In his well-known book on ***'Crusade in Europe: A Personal Account of World War II'*** (pp. 425 and 426), the General has written that, "Field Marshal von Kesselring, commanding the German forces on the western front, also sent me a message, asking permission to send a plenipotentiary to arrange terms of capitulation. Since von Kesselring had authority only in the West, I replied that I would enter into no negotiations that did not involve all German forces everywhere... When Admiral Friedeburg arrived at Reims on May 5 he stated that he wished to clear up a number of points. On our side negotiations were conducted by my chief of staff, General Smith. The later told Friedeburg there was no point in discussing anything that our purpose was merely to accept an unconditional and total surrender. Friedeburg protested that he had no power to sign any such document. He was given permission to transmit a message to Doenitz, and received a reply that General Jodl was on his way to assist him in negotiations... To us it seemed clear that the Germans were playing for time so that they could transfer behind our lines the largest number of German soldiers still in the field. I told General Smith to inform Jodl that unless they instantly cease all pretense and delay I would close the entire Allied front and would, by force, prevent any more German refugees from entering our lines. I would brook no further delay in the matter.... Finally Jodl and Friedeburg drafted a cable to Doenitz requesting authority to make a complete surrender, to become effective forty eight hours after signing. Had I agreed to this procedure the Germans could have found one excuse or another for postponing the signature and so securing additional delay? Through Smith, I informed them that the surrender would become effective forty eight hours from midnight of that day; otherwise my threat to seal the western front would be carried out at once.... Doenitz at last saw the inevitability of compliance and the surrender instrument was signed by at two forty-one in the morning of May 7. All hostilities were to cease at midnight May 8.... After the necessary papers had been signed by Field Marshal Jodl and General Smith, with the French and Russian representatives signing as witnesses, Field Marshal Jodl was brought to my office, I asked him through the interpreter if he thoroughly understood all provisions of the document he had signed.... He answered, '*Ja*'.... I said, 'You will, officially and personally, be held responsible if the terms of this surrender are violated, including its provisions for German commanders to appear in Berlin at the moment set by the Russian high command to accomplish formal surrender to that government. That's all.' He saluted and left."

So these were, indeed, some inside views into the phenomenal historic details—of dramatic surrender ceremony—concerning the final capitulation of Nazi Germany's High-Command and its forces, since deployed in the various operational-theatres/war-fronts; all over Europe during the first week of May 1945. It was a great historic moment, as finally the end of human-history's most perilous, tragic and devastating World-War II in Europe and these in turn came as a great relief to the entire humanity in the world and, all the more so to the peoples of the entire Europe. But even in a historic retrospect, it would be rather naïve to suggest that all was over for the good all over in the whole world. The Far East region was still in ferment, where any tangible-cum-enduring peace was still to be achieved. The Allied Forces under the supreme command of rather all the more legendary **American: General Douglas Mac Arthur in the Pacific Command** were still fighting the fanatic Japanese Forces for decisive military superiority in the region. Both sides had their own even interestingly rather well-perceived but diametrically opposed geopolitical and politico-military strategies for fighting this war. But even much before the final surrender of Nazi Germany in Europe, tide had already momentously begun turning in favour of the Allied Forces in the Pacific too and consequently on the military fronts, the massive Allied Air Forces of Mac Arthur's Pacific command, had commenced devastatingly crippling bombings of the Japanese homeland islands, including their strategic resources all over the Pacific region. It cannot be denied that it was very massive and extremely ruthless affair but wars have always been like that—full of all sorts of human tragedies, sufferings, deaths and destructions and hence, what to talk of any massive and global war: like the bygone mid-twentieth century's 1939-45 Second World-War!

In spite of some kinds of the rather peculiar similarities between their socio-cultural-cum-geopolitically guided as well as imperialistic-cum-militarism oriented extremely rigid fixations on the parts of both the Nazi Germans and the then fanatic Japanese; there indeed were many historic differences in their socio-psychological make-up and value-systems. In fact as it turned out to be—handling of the Japanese high-command by all possible kinds of geostrategic and politico-military means and then, finally bringing them to the table meant for signing the documents of the unconditional surrender and thereby to end the war in the Pacific; turned out to be relatively much more complicated and difficult affairs. In fact tragically as well as ironically, for finally subjugating the Japanese both militarily and otherwise, droppings of even few atom-bombs on the two most unfortunate cities in Japan had to be resorted to, for decisively ending the war on all fronts in the world. Interestingly for some informed discussions sake, we may also say that—as during any such massive global war, we come across various kinds of the clouds of war, then at the same time, we do also come across no less engaging as well as conflicting 'clouds' in the vast domain of its politico-military living-history too—where-in accounts and perceptively considered opinions of the wars on their various fronts and theatres, do vary historically and thereby, making their explorative studies and research—all the more engaging and challenging as well.

Andrew Roberts has clearly mentioned that, "Not knowing that the Americans had no more atomic bombs to drop, and shocked by Russia's interventions in the Pacific War on 8 August, which they were unable effectively to counter, the Japanese did finally surrender on 14 August, with the **Emperor Hirohito** admitting to his people to a broadcast at noon the next day that the war had gone 'not necessarily to Japan's advantage', especially in view of 'a new most cruel bomb'. Even as he prepared to broadcast, a group of young officers invaded the palace grounds in an attempted coup intended to prevent him from doing so.... A fortnight later, on Sunday, 2 September, six years and one day after Germany had invaded Poland, General Douglas Mac Arthur and Admiral Nimitz and Sir Bruce Fraser and representatives of other Allied nations took the formal Japanese surrender, which was signed by the one-legged Foreign Minister Mamoru Shigemitsu and the Army Chief of Staff General Umezu, aboard the battleship USS Missouri, by then moored in Tokyo Bay. (She was chosen because she had served in Iwo Jima and Okinawa and was Nimitz's flagship; it was mere coincidence that she was named after President Truman's home state.) Mac Arthur concluded the ceremony by saying: 'Let us pray that peace be now restored to the world and that God will preserve it always. These proceedings are closed." (Roberts, 2009, p. 577)

The above quoted paragraphs from one of the books of a well-known author, and if read even in-between the lines, do indeed give a historically interesting as well as socio-cultural-cum-politico-military insight into the extent of extreme militarism-fixated psyche of the Japanese peoples, their feudal societies and military forces and, the constitution of their Nation-State along with rather strangely 'divinity-bestowed' Monarchy and their systems of governance—incorporating the still unchanged culture of now the obsolete or somewhat illogical medieval-era military chivalry—where only either of the two paths were supposed to be available, i.e. **either total victory in war or hara-kiri**! We may recall that contextually in different perspectives in the earlier chapters, we have since already discussed these as well as some other no less momentous aspects of the making of even the ancient and medieval-era Japanese peoples, their psyche and evolutionary phenomena of their societies and their value-systems—where-in mostly the various dimensions and aspects of their extreme militarism-soaked matters-military, did mostly dominate and thereby always, leading to the fanatic contempt for their adversaries and subjugated colonies. **All these multiply socio-cultural aspects historically dating back to the Shogunate-era Japan and even earlier and then, further nurtured by the Samurai-culture and hence, always remained so valued in the extremely feudalistic Japanese societies.** In fact much later, with the advent of historically unique Industrial-Revolution, Japan being the singular Asian nation to first dynamically learn, adopt and imbibe multiple aspects of the ongoing advancements in the vast and complex fields of the sciences and technology, and thus became the first Asian nation to get substantially industrialized. But Japan failed or perhaps refused to socio-culturally modernize in the contemporary modern sense and also as per the ongoing dictates of the mankind's expanding civilisational-canvas, unlike the other

European industrialized nations and thereby, even created a regressive drag or a kind of darkness around her thoughts-cum-decision-making processes.

In other words ironically, in spite of these far-reaching developments and praiseworthy industrialization of Asiatic Japan, the centuries-old and extremely fixated socio-cultural aspects of feudalism as well as the extreme kinds of fanatic militarism too lingered and, remained dominatingly present then, even in the modern-era Japan till her total military defeat and the final end of Second World-War in the year 1945. We may contextually also recall that earlier in this chapter itself, we have since discussed these aspects and thereby, significant methodologies for bringingin so many kinds of changes, reconstruction, modernisation as well as democratisation of Japan **under the historically charismatic leadership of American General Douglas Mac Arthur,** who was later asked to stay-on in Japan for few more years after the Japanese surrender, for ensuring the needful. **By the time Mac Arthur left, Japan was a changed and democratic nation and in addition—even in rather historically a brilliant kind of continuity—of healthily retaining the changed and constitutionally ceremonial system of Monarchy as the Royal Head of State**. It does appear and even if as a matter of historic conjecture only that in all these processes of revolutionary changes then taking place—since the already earlier achieved a significant degree of the industrialization of Japan, must have been a far-reaching and yet apparently hidden source of assistance-cum-facilitation—in these well-crafted, well-led and well executed processes of multi-dimensional transformations of Japan and that too, immediately after the extremely violent and also highly chequered end of the 1939-45 World-War II. Whatever may be rather differently perceived matters of opinions in this regard—these were indeed historically momentous and far-reaching developments in the then war-ravaged Japan—leading to her so unprecedented, indeed total and also, enduring transformation, with soon to follow: the multiple kinds of many positive as well as, reassuringly contributory consequences for now the entire modern world.

Hence it would be contextually naïve to even imagine that there were no major differences in the overall makings as well multiple value-systems of the Nations-States of the then Nazi Germany and the militaristic-cum-imperialistic Japan. But at the same time, though the Nazi Germany since being a Western European Nation-State; was definitely not so feudalistic as per the age-old Asian socio-cultural sense but, she too was no less imperialism-cum-militarism soaked in their own kinds of the socio-cultural and politico-militarily fixated value-systems where-in, the Nazis fanatically believed that—going to the war or even giving various kinds of the coercive threats for war—are the only ways to come to terms with others or to establish and maintain peace in the world and that too, largely to their own benefits alone. So in other words in retrospect, we may also say that in the multiple contexts of the Second World War; both the then Nazi Germany and the imperialistic-cum-militaristic Japan, in spite of being geographically located so far-off, turned out to be ideally made for each other and hence, became menacing threats to the Allied Powers representing entirely different kinds of ideologies. Therefore in the overall global perspectives, and also multi-dimensionally comprehensive politico-military strategic prudence; there was no midway for the any kind of negotiated enduring peace and thereby, the only option was to decisively defeat these two western and eastern Kingpins of the menacing Axis-Powers once for all by the overwhelming military-might, so necessarily mobilized by the Allied Powers. Now coming back to the unconditional surrender of Japan, it would be relevant to refer to **General Mac Arthur**'s highly informative and valuable autobiographical work, wherein, he has mentioned that, "The formal ceremonies of surrender aboard the Missouri were fixed for September 2, 1945. I had received no instructions as to what to say or what to do. I was on my own, standing on the quarterdeck with only God and my own conscience to guide me.... I have read several accounts of what occurred and what was said that morning, but my favourite of all is that which was officially rendered to the Emperor after it was over by a Japanese member of the surrender party. Mr. Toshikazu Kase, an alumnus of Amherst and Harvard, and a Japanese diplomat for twenty years' service with the foreign office. It has been published but I would like to include it here: 'It was surprisingly cool day for early September. The sky was dull gray with clouds hanging low. We left Tokyo at about five o'clock in the morning. There were nine of us, three each from the foreign office, the War and Navy Departments, besides the two delegates, Shigemitsu, the Foreign Minister representing the government, and, General Umedzu, the Chief of Staff of the Army representing the Supreme Command.... To begin with there was much ado in selecting the delegates. Nobody wanted to volunteer for the odious duty. The Prime Minister, Prince Higashikuni, was the Emperor's uncle and was considered unsuitable on that account. Next choice fell on Prince Konoye, who was Vice Premier and the real power in the government, but he shunned the ordeal. Finally the mission was assigned to Shigemitsu, the Foreign Minister. On accepting the imperial command to sign the surrender document as the principal delegate, he confided in me what an honor he felt it, since it was the mark of the sovereign's confidence in him.... Not so with General Umedzu, who reluctantly accepted the appointment as the second delegate. He had opposed the termination of hostilities to the last moment and was, moreover, a soldier born to command and not to sue. When he was recommended for the mission he grew, so it is reported, pale with anger and laconically remarked that if it was forced upon him, he would instantly commit hara kiri in protest. It required the Emperor's personal persuasion to make him execute the duties with good grace.... It may sound somewhat silly, but as precautions were deemed necessary, the appointment of two delegates was not intimated to the press until the last moment. The names of nine persons who accompanied them were not published at all as the service officers were against this, though these names had been communicated to and approved by the Allied authorities. Such, indeed, was the temper of the times.... Mac Arthur walks quietly from the interior of the ship and steps to the microphone: 'We are gathered here, representatives of the major warring powers,' he said, 'to conclude a solemn agreement whereby peace may be restored.... But rather it is for us, both victors and vanquished, to rise to that higher dignity which alone befits the sacred purposes we are about to serve, committing all our people unreservedly to faithful compliance with the obligation they are here formally to assume.... The terms and conditions upon which the surrender of the Japanese Imperial Forces is here to be given and accepted are contained in the instrument of surrender now before you'.... In a few minutes' time the speech was over and the Supreme Commander invited the Japanese delegates to sign the instrument of surrender. Shigemitsu signed first followed by Umedzu. It was eight minutes past nine when Mac Arthur put his signature to the documents. Other representatives of the Allied Powers followed suit in the order of the United States, China, the United Kingdom, The Soviet Union, Australia, Canada, France, the Netherlands and New Zealand.... When all the representatives had finished signing, Mac Arthur announced slowly: 'Let us pray that peace be now restored to the world and that God will preserve it always. The proceedings are closed.'.... Mac Arthur broadcast to the American people: 'Today the guns are silent. A great tragedy has ended. A great victory has been won. The skies no longer rain death—the seas bear only commerce—men everywhere walk upright in the sunlight. The entire world is quietly at peace. The holy mission has been completed. And in reporting this to you, the people, I speak for the thousands of silent lips, forever stilled among the jungles and beaches and in the deep waters of the Pacific which marked the way. I speak for the unnamed brave millions homeward bound to take up the challenge of that future which they did so much to salvage from the brink of the disaster."... (MacArthur, 1965, pp. 311, 312, 313 314 and 315) ***(This war indeed had been mankind's unprecedented and truly global in scale and also unsurpassed in scales and quantum of deaths, destructive devastations and multiple human-tragedies.)*** The foremost momentous development that soon began to surface were that the curtains—on the Western European nations' socio-economic, geostrategic and politico-military culture of the extremely fixated imperialism-based restlessness and greed towards building of the various kinds of intercontinental empires, began to come down heavily and irreversibly as well. The Western Imperial and seafaring-Powers with theirs the above mentioned centuries-old as well as, unchallenged cultures of the pursuits of imperialism and acquisitions of their vast overseas colonies, soon found that their huge geopolitical-cum-economic overseas colonial assets, began to slip-out of their now helpless imperial hands, as new and independent political-entities in the world. With the end of World-War II, time and also the overall global geopolitical situations had already begun to rapidly change and thus; heralding new kinds of the international power-equations along with the momentous emergence of the new kinds of 'great games' in the arena of international power-politics. One kind of the momentous event led to another kind of the totally new development and thereby also—numerous kinds of the new developments triggered completely unforeseen kinds of the newer geopolitical games, international rivalries and, the geostrategic-cum-politico-military alignments.

It was historically indeed an unexpectedly queer situation—since the older Western Imperial-Powers along with their homelands back in Europe, stood almost cripplingly devastated at the end of World-War II, and at the same time, when they perhaps needed their empires the most, they helplessly found that their numerous and vast colonial assets in Asia, Africa and elsewhere; had already begun to slip out of their 'imperial hands'. Since the vast and overseas British empire was premier and the most extensive one, these developments came all the more as severe blows to the British nation, who had traditionally then believed in some kind of a perpetuity of their vast and overseas colonial-empire. We may recall that earlier in this chapter itself as well as, in the last sixth chapter too, we have since already discussed in differently numerous contexts these vary aspects and also many other historically allied and relevant subject-matters, concerning the British Nation-State and her truly magnificent global empire in some substantial details. And hence while briefly discussing in our specifically broader contexts here, we do also find it to be largely true that historically since centuries, the socio-cultural-cum-psychological and geopolitically—the gradual evolutionary making of the British societies as well as the Nation-State, has been such that they have indeed been able to rather substantially develop their inbuilt geostrategic and politico-military resilience—to be able to mostly adjust and reorient themselves quite advantageously and also somewhat smoothly as per the historic phenomenal dictates of the momentously changing times.

First and foremost at the earliest and, after the end of World-War II, the leading victor nations did compulsively display some kind of higher and larger geopolitical wisdom—by sincerely striving towards the phenomenally historic establishment of the somewhat visionary and so far unprecedentedly global **United Nations Organisation (UNO)**, having its principal seats in city of New York in the USA. Then the concerned people constituting the global leadership in this regard too, did care to learn from the mistakes of its **earlier somewhat ineffective and poorly set-up the 'League of Nations'**, which was quite hastily as well as insincerely constituted then, after the end of 1914-18 First World-War. It is true that the UNO too is not as effective as it was initially intended to be and hence, it has ironically numerous shortcomings. It is also true that many such shortcomings are preventable and hence not insurmountable, and at the same time in these respects, onus principally lies at the doors of the present-day leading World-Powers. Well, it is hoped that in course of time many more reforms in the make-up and more fundamentally, in the Constitution of this world-body, would be affected for larger good of the humanity. As the years have been rolling-by, UNO has been expanding and in addition, been serving the world-community in many areas and also, in multiply sincere ways. Further with the ongoing passages of time, numerous kinds of changes in the already many complex-cum-problematic spheres in the world, have continuously been taking place in the overall geopolitical as well as already otherwise too—very volatile international military situations, and consequently thereby many kinds of the multiply far-reaching roles on the part of UNO, have also been expanding and in the process, triggering more and more—the internationally somewhat focused inclinations of the concerned nations to increasingly gravitate towards the UNO—to perhaps compulsively search for the desirably more tangible solutions to the many such daunting international problems, in the larger interests of world-peace and cooperation. In these multiple contexts, always the pivotal point has been that—with respect to any particular international or bi-national problematic 'logjam' haunting the concerned nations in any particular region—for taking any geopolitical-cum-diplomatic initiatives, the today's comity of nations do always compulsively need some kind of a vibrantly functional world-body and its multi-dimensional forum for the needful in this regard. At this precise point here, to substantiate these presently ongoing modern-era perceptions, many obviously well-known illustrations from the recent past to the present times, may be cited and hence just as a couple of examples, we may contextually note the numerous already existing and complex as well as fluid Middle-Eastern highly volatile problems and then in addition also now, the globally 'burning'—numerous kinds of the environmental-issues too, do readily come in our minds.

While contextually keeping the above discussed points in our minds, we may also say that as the times have been passing-by; in the international geostrategic and politico-militarily triggered multi-dimensional-cum-multi-layered games of the international power-politics in today's world, have perforce simultaneously been multiply enhancing both the importance and reach of the UNO, in the modern-era comity of nations. This does not mean that now in the fast-moving and high-tech world, the so-called modern nations have become more sagacious and also, geopolitically more wisdom-oriented. In fact far from it, as the plain realities are—that consequently with the dawn of the times of 'Industrial-Revolution' and then the 'Post-Industrial-Revolutions' as well as, their multiple-fallouts, did ensure indeed the mind-boggling multi-dimensional advancements in the multiple fields of modern sciences and technologies and thereby in the process, perforce made the nations, especially the leading and more powerful ones, to realise that in their any concerned areas, they no more can always have a field-day and hence, will not be able to do anything and everything as per their sweet-will. In our specifically larger perspectives, we do find these modern-era's multiple developments onto the mankind's expanding civilisational-canvas, as phenomenal and thereby hence, making the subsequently highly-chequered birth, on-going development-cum-expansion and also the international reach of the UNO too as indeed phenomenal—for multiply the larger good of the humanity at large. It is true that the paths of UNO as a whole, have never been smooth and easy and results also have been mostly mixed and in addition, they would perhaps never be otherwise in the future too, but that is basically civilisational in its reasons and content and hence in the larger perspectives, these do remain understandable as well.

Now contextually coming back to the post-Second World-War scenario, historic processes of the peaceful and largely somewhat gradual-cum-friendly winding-up of the once so vast and gloriously trans-continental and mighty British Empire, we do indeed find that these unprecedentedly historic processes did display magnificence of the earlier already noted multiple-aspects of the British nation's **'essential strategic-culture and thoughts'** multiply in action, with their visionary and long-term geopolitical-cum-strategic prudence; towards perhaps the best possible larger good of both the British people as well as that of once their subjects in their erstwhile overseas colonies. Well historically in largely identical contexts, the same cannot be so emphasised in case of the similarly other no less ambitiously obsessed erstwhile European seafaring Imperial-Powers who then happened to be geopolitically as well as politico-economically the die-hard competitors of British imperialists, during now the bygone era of the relentless pursuits of their overseas militaristic colonialism. During the historic phase of the largely gradual-cum-friendly winding-up of once her so vast and mighty British-Empire, with the British peoples' typically vibrant and flexibly well-stationed the multiple-aspects of **'essential Strategic-culture and thoughts'** and also, coupled with their no less visionary geopolitical strategic-prudence, they hit upon a novel idea of persuasively setting-up the fast moving modern-era's international **'Commonwealth of Nations'**—having most of their erstwhile imperial colonies as the sovereign member-nations and then, along with also, politico-diplomatically putting-up the British Constitutional Monarch as the ceremonial head of this quite unique international Institution in the today's so much changed world. This Institution of the 'Commonwealth of Nations' too, has been making some noteworthy contributions at different levels, in the socio-economic developments of its member-nations and also thereby, promoting in its own ways—the multiple aspects of world-peace and also well-being of the larger humanity as well.

Historically it is true that, end of the era of highly exploitative Western European colonialism in the vast continents of Asia and Africa and in other places too, soon after the so much chequered end of the Second World-War, came as a great relief to the native populations in these vast continents, who were now able to breathe in the free air of their political independence. But at the same time it is also true that most of them found themselves trapped in their own indigenous but totally unexpected kinds of socio-economic and strange political turmoil and all these new developments in turn, ironically but understandably with their typical centuries-old historicity, left them totally unprepared to face such complex and highly challenging situations. **It was another kind of a queer situation** for them, as they no more had their colonial masters to look up to them, ironically as their alien guardians and also, many of them even had quite strange as well as uneasy relationship with them, but whatever they had earlier at least existed over their heads—though mostly very much exploitative and also largely unconcerned with the traditional local problems in the lives of these natives. But at least they earlier happened to be there, even if for their own greedy imperial motives and thereby, readily available to face the brunt and take some sort of charge of the situations. But after the friendly or not so friendly departures of those European colonial-masters, most of these native populations of Asia and Africa, became free with their all kinds of large, medium and even smaller politically independent Nation-States mostly as 'problematic-cum-crying babies in their laps' and on top of it, many of them had not had their any tangible historic tradition or some past civilisational legacy to live in some kind of any unified and sovereign political entities to fall back-on for some kind of now a modern-era socio-cultural-cum-political anchorage or moorings, even under a 'geopolitical ceremonial umbrella' of today's UNO, among the modern comity of independent nations.

In the present era's these multiple contexts, the pivotal point is that a large number of those newly independent-cum-newly born Afro-Asian nations, due to their various internal and/or even externally induced socio-economic and politico-military reasons, soon began to be seen as landing in the categories of various kinds of the totalitarian, despotic, and/or tin-pot military dictatorships, with hardly any escape routes apparently available to them. Moreover tragically, and some of them inevitably began to even fall in the 'abyss of failed nations'. In today's larger global perspectives, such scenario have now been staring as internationally complex and menacing socio-economic and also highly explosive geopolitical problems in the eyes of the whole world. And also ironically in addition to these, during those recently past decades **all kinds of rivalries and the mutually opposing the then twin Super-Powers' ruthless Cold-War era games of the international power-politics then**, began to constitute as a 'menacing-icing' over the above mentioned extremely complex multiple problems of the many of those newly independent Afro-Asian nations. In fact, the ground realities turned out to be that, in such ruthless big-powers' games, many of those poor and hapless Afro-Asian nations, got forcibly sucked-into the numerous kinds of extremely adverse consequences for their unfortunate peoples. These poor and suffering peoples in any case, had not bargained for them at all and so, these were forcibly thrust-upon them by their corrupt and most unscrupulous leaders whether civil or military, and also by the ruthless machinations of the leading big-powers and hence consequently, multiplying their miseries.

In fact in these precise contexts, it is important to note that once internationally the so much menacing cold-war between the then ruthlessly rival twin Super-Powers, i.e. the USA and the USSR from almost immediately after the so much chequered end of the World-War II, to almost up to the end of twentieth century, lasted for a number of years, wherein the world witnessed many kinds of the geopolitical and military upheavals and also, experienced a number of the 'clashes of arms' as well as, **almost irrational arms-race between the two Super-Powers** and in addition, leading to the numerous kinds of traumatic developments across the world too. The multiple-aspects of the so much both geostrategic and politico-militarily obsessed cold-war between these two super-powers during those recently bygone decades and their highly chequered history in fact, are too recent to even somewhat slip into some remote corners of the mankind's otherwise not so very enduring memories. It is a different matter now that this is all over and hence enabling the larger humanity to feel a great sigh of relief. But the above mentioned the recently passed decades of the cold-war, had indeed menacingly caused highly burdensome times with heavily suffocating environment for the mankind. **So far as the massive line-up of the nuclear-arsenals and their various kinds of delivery systems and also, their logistics-cum-support as well as command and control systems—since then pertaining to each of the then twin Super-Powers were concerned, they were indeed unprecedentedly large and also in fact, had mind-boggling capacities to even completely destroy the world many time over**. Hence during those highly surcharged and dangerous times, in overall global geopolitical-terms, the state of panic as well as the prevailing tensions across the world, remains even now understandably amazing too.

And after the Second World-War and within few years when the massive heat and dusts of this war were yet to settle, the above mentioned cold-war had just begun to unfold that during its early years, it even triggered the start-up of another war i.e. **the unfortunate Korean-War** in the geographical meeting-areas of South-East and Far-East Asia, in the year 1950 itself. During those eventful-cum-highly turmoil-ridden times in Europe, Africa and Asia as well; the Western-Bloc now under the globally new leadership of the USA, was then hardly prepared and geared-up both at geopolitical and the military levels, to effectively face this Korean-War, in fact so suddenly thrust upon them. **Precisely in these larger perspectives concerning the almost always very fluid geopolitical gamut of the globally geostrategic and politico-military dynamics of international power-politics and, all the more so with strongly overwhelming-cum-highly-conflicting as well**

as politically, the totally irreconcilable ideological contents; one thing remains always heavily present—that for the strategically beneficial solutions towards any such problems-causing issues—whether regional or global—do have both geopolitical-cum-diplomatic and the direct/ indirect military avenues. The conceptually central point, to always bear in mind in this regard, is that sometimes the geopolitical-cum-diplomatic avenues become more dominant and decisive and where, the direct/indirect military-avenues are supposed to function prudently in synergy-creating supporting-roles; and at the same time in certain other kinds of the emerging situations, they are mutually made to reverse their strategic-roles as well. In addition also, these are in fact principally, part of the larger tenets of again the **multiple-aspects of 'essential strategic-culture and thoughts'** too, and hence, their paramountcy importance in this regard, cannot be overemphasized. And further as we critically delve deeper into these perceptions, one more dimension too also, does become visible—and precisely that is—if on one side, the acutely obsessed ideological tenets of communism as the political-contents—do perforce occupy the central space; then in any such scenario, the above mentioned multiple-aspects concerning the geopolitical-cum-diplomatic and the direct/indirect military avenues, along with the clear decisiveness of the either of them, as per any specifically emerging situations, do become all the more pronounced.

(b) Korean-War: 1950-53 (President Truman's blunders of not listening to General MacArthur then – further, 'Vietnam' won't have happened, if 'Korea' would have been won decisively)

Keeping this precise conceptual-framework critically in our minds, we may clearly perceive that so far as the 1950-53 Korean-War was concerned in fact the proactively direct military-avenues were singularly decisive to victoriously settle the concerned conflict-prone issues and thus in turn, effectively check the then menacing advancements of communism.

Further, the politico-military history, of this war then fought in the Far-East-cum-South-East region of Asia, shows that geopolitically the Communist-Bloc happened to be absolutely clear that the keys to victorious solution lay in their highly focused and the decisively proactive military-avenues and hence, they had massively prepared for this eventuality, even before they had started the war. Whereas just opposite the Western-Bloc and that too principally the USA, remained strategically unclear as well as, geopolitically largely confused, in their crucial policy-formulations and also in the areas of resource-mobilizations and hence; unable to effectively respond to this war since, thrust upon them by the Communist-Powers. The unfortunate reason being that **the then US President Harry Truman** perhaps lacked the long-range geostrategic-cum-politico-military vision to effectively provide some kinds of the statesmanship kind of the highest-level political leadership to the Western-Bloc (*like that of President Roosevelt, earlier during the Second World-War.*), in this war. **And needless to say that these top-level historic shortcomings at the highest-levels of the then US political-level leadership aspects, during this Korean-War, had indeed been a kind of geostrategic and politico-military set-back for the USA, with heavy costs to her and even to the larger world, and all the more so now, this remains a menacing problem and also, with the continued passage of time; even today the communist North Korea is becoming more and more irresponsibly problematic and, continue to indulge fanatically in more and more irrational sabre-rattling.**

With respect to the above and to precisely identify the main causes of this Korean-War we may prefer to again look into the relevant pages of the already earlier referred the historic autobiographical book of the **US General Douglas Mac Arthur** wherein he has very reflectively written that, "With the formation of the Republic of Korea and the withdrawal of the American forces, my official connection with Korea ceased. From August 15, 1948, the doomed little country was under the sole charge of the State Department. But my intelligence section was increasingly aware of the distinct menace of an attack by the North Korean Communists in the summer of 1950.... The Joint Chiefs of Staff reiterated the Administration's unwillingness to commit itself to the defense of South Korea and had recently drawn up a plan of strategic defense in Asia which was based on the assumption that under no circumstances would the United States engage in the military defense of the Korean peninsula.... In vain were my attempts to expose the growing Communist threat in the Far East. From June 1949 to June 1950, constant intelligence reports of the increasing urgency were submitted to Washington, advising of a possible North Korean thrust. But little impression was made against the general apathy and the inspired 'agrarian reform' propaganda. One of these reports even suggested that June 1950 would be the likely time for North Korea to cross the 38th parallel.... On June 19th, John Foster Dulles, as the personal representative of Secretary Acheson, visited Korea. He had come to Tokyo to discuss a Japanese Peace Treaty with me, the framework of which I had submitted to Washington. In Korea, he apparently reversed the previous policy enunciated by the State Department, by stating his belief before the Korean legislature that the United States would defend Korea if she was attacked. It made me wonder just what United States policy in Asia....was Dulles took a brief motor trip from Seoul to the demarcation line between North and South Korea, and what he saw alarmed him not at all. He noted that the South Korean forces appeared quite ready if any attack should come from north of the border. With his tactical inexperience and possible lack of accurate information, Dulles clearly did not realize the inferiority in both troop strength and material of the forces he had seen in comparison with their Communist kinsmen above **the 38th parallel**.... (*In fact, after the balloon had gone-up on the Korean frontlines.*) Even then it was evident that it was far more than a 'police action', as President Truman was to euphemistically characterize it, far more than any localized clean-up of border-raiding North Koreans. **In Korea, Communism had hurled its first challenge to war against the free world. Now was the time for decision. Now it was as clear as it would ever be that this was a battle against imperialistic Communism. Now was the time to recognize what the history of the world had taught from the beginning of time: that timidity breeds conflict, and courage often prevents it.**" (MacArthur, 1965, pp. 370, 371 & 375) (*In historically similar contexts and though perhaps in more localized geopolitical domain, during 1962 the then Indian Premier Nehru too, had very callously refused to meaningfully learn the precisely above enunciated the timeless lesson of history and thus, causing some humiliating embarrassments to the democratic and sovereign Indian Nation-State and also, to her totally apolitical, loyal and gallant Armed Forces as well.*).

So far as the various aspects of **1950-53 Korean War** are concerned, now they are all part of history but at the same time it is also equally true that due to some kinds of the very callous and far-reaching 'goof-up' on the part of highest-level US political-leadership during prelude to this war, unfortunately and in addition also, as **a great menace to almost the whole world the regressively lingering Korean-crisis (*courtesy: almost the 'failed' and communist North Korean Nation-State in the Far-East.*), still remains malignant on the world's larger geopolitical horizon**. It is hoped that like the earlier unification of Vietnam in South-East Asia and, that of Germany in Europe, one day Korean peninsula in the Far-East too, would get united, to a great sigh of relief to the entire humanity. Till it doesn't finally happen, this so far menacing Korean-crisis would continue to remain as a '**lingering pain in the neck of mankind**'. **General Mac Arthur** in his autobiography's chapter on 'Frustration in Korea', has very vividly narrated these historic aspects of the Korean-

War and also all the more so; his writings assume great historic significance since, he was the highest-level military-commander from the Allied-side and hence, functioning then on the spot in the Korean peninsula. **In our specifically larger perspectives and also, in a totally different geopolitical historic contexts; shall we say that in South-Asian region again, in a some different kind of the so-far lingering 'Kashmir-problem' since, so sinisterly created by another pseudo-democratic, religiously fanatic and somewhat 'failed' Nation-State of Pakistan too; is another similar 'pain in the neck' for the mankind?** So far as the multiple-aspects of this so-called ***'Kashmir-problem'*** are concerned we would contextually be discussing in some relevant historic details, in the last sub-chapter of the next eighth chapter. Ironically, so far as the Korean-War was concerned, this was not the first geostrategic and politico-military 'fiasco' in fact, on the part of unchallenged Super-Power the USA, but interestingly to the great geopolitically historic embarrassment to not only the USA, but also perhaps to the whole democratic and free world, and as further times have been rolling-by, we would be coming-across more of such but also contextually different kinds of 'fiascos' since, periodically committed by the USA and that too, in different parts of the world.

As we move-on in our continuing discussions and, contextually in our larger contexts here, as we further critically delve deeper into the turbulent and yet fast-changing times of the post-Second World War era of the latter half of the twentieth century onwards,, we do notice that with the demise of the older and colonial-era slow-moving international geopolitical equations, the world also saw the emergence of altogether new kinds of the international geopolitical situations and the consequent 'great games' of the international power-politics. It is true that the **far-reaching Bolshevik revolution in the older Imperial Russia** since having taken place even little before the 1914-18 First World-War and which in turn, had already triggered emergence of the communism over there and thereby consequently, **the world saw the momentous rise of a massive and powerful communist Soviet-Empire encompassing in continuity, almost the whole of the vast Eastern European and the Central Asian landmass**. We may recall that earlier in this chapter itself contextually, we have since already briefly discussed some aspects of this **Russian Bolshevik Revolution** and thus also, momentous rise of the vast Soviet-Empire. Both socio-culturally as well as geopolitically, there was nothing in common and hence, no love-lost—among the Western European nations, their overseas colonial empires, the USA and others across the Atlantic on one side, and then the USSR on the other. Hence their uneasy military alliance during the World-War II, became a possibility just as a marriage of convenience and, just for successfully taking-on the commonly menacing and powerful Axis-Powers. And needless to say that even before the war finally ended, there in fact obviously irreconcilable as well as also inevitable parting of their ways too, had already begun and thus thereby, dividing almost the whole world into two opposing and massive **Power-Blocs: wherein, one side was led by the then USSR and the other Western-side was led by the USA**—as an unprecedentedly unknown kinds of the dangerously rival **'lethal combinations'** and thereby, triggering a totally new kind of 'war' between these two camps since, now historically known as the **Cold-War**—and that too little later-on, with again a totally new kinds of the then already known **nuclear-weapons' (i.e. weapons of mass-destruction)** arsenals in both camps, so menacingly lined-up against each other. During those years it had rightly been said that the entire humanity seemed to be helplessly sitting on top of a huge mountain-like massive stock-piles of dynamite in the world! In any case now in retrospect, whether any unimaginably traumatic ignition of its fuse was prevented by the mankind's overall wisdom-cum-geopolitics-based strategic prudence, or ultimately by the divine mercy of the Providence, will always remain in this regard, the matters of historic conjecture only!

Well by now even if, since the massively imposing existence of the Soviet Empire is already a matter of the past, has the world become reassuringly and also reasonably safe? Has the world become free from all the more numerous kinds of irrational threats of new and also multiple kinds of even not so scientifically designed and 'fail-safe' nuclear-weapons and their delivery-systems—mostly so irresponsibly-cum-dangerously flaunted by some rogue-cum-failed States and/or their crazy kinds of the politico-religiously fanatic as well as uncontrollable, non-state actors, mostly so menacingly operating from within their geographical territories? All these dangerously serious menaces and many other kinds of the geopolitically very complex problems, have had their births in the basic genesis of the cold-war. Now, as the nightmarish era of the cold-war is over but, the many lingering geopolitical, politico-economic as well as the militarily disturbing fallouts, still proactively survive and thereby, disturbing the peace and trans-national tranquility in various regions across the world. Some of them are too complex to be linearly understood and thus, defying their even any rational and politico-diplomatically well-intended solutions across the board. In fact, there are numerous kinds of localized, regional and even international problems as well as, 'flashpoints', originating from the times of cold-war era—in Africa, Asia, South-American continent and then even in some places in Europe as well.

In the above multiple-contexts, **Richard Overy** has very aptly mentioned that, "The roots of the Cold War lay in the Russian Revolution of 1917. Communism, with its belief in its own domination of the world, was seen as a profound threat to the world capitalist system, of which the USA was taking the leadership. The United States and the other western allies had found themselves allied to the Soviet Union in the Second World War through forced circumstance: the imperative of defeating Hitler overrode all other considerations, though this did not prevent persistently strained relations throughout the war. But with Hitler beaten, the underlying tensions resurfaced. The west saw in Soviet communism the spectre of a second expansionist authoritarian system, while for its part the USSR believed it had helped defeat one form of capitalist imperialism only to be confronted with an even more powerful one in the form of the USA.... **With both sides eager to avoid open conflict, the cold war was fought out in the world of spies and secrets, political defeat and subversion. In large measure it was a war fought by proxy. Both camps exploited or entered local conflicts in which they armed, equipped and trained the opposite sides, each seeking to extend their spheres of influence without coming to blows directly**.... Tension was most acute in the period between 1947 and 1963, as the new political order in Europe and Asia was being fashioned, above all in central and Eastern Europe where the USSR sought to impose itself on those territories it had liberated in the war. The consolidation of communist regimes in the region combined with the success of the communist revolution in China in 1949 under **Mao Zedong** created an apparently a solid communist bloc from Europe to the Pacific, though by 1960 an open rift developed in Sino-Soviet relations.... **America in 1947, committed itself to 'containing' communism (the Truman Doctrine) and support to countries fighting wars against the communist threat**, in Greece (1947), in Korea (1950-3) and then in the long and draining conflict in Vietnam (1961-73). At the same time, the United States tried to *bolster* these efforts by creating security blocs—NATO in Europe in 1949, SEATO in Southeast Asia in 1954 and SENTO in the Middle East in 1959. The Soviet Union and China retaliated by giving military and the political support to anti-colonial or nationalist struggles against 'world imperialism' throughout Asia, Latin America and Africa. In 1955 the Soviet Union also sponsored a military alliance of its own—the Warsaw Pact.... **The most important factor preventing the shift to 'hot war' was the existence of nuclear weapons. First used in 1945 against Japan, America's monopoly of the weapon was broken by the Soviet Union in 1949. In 1952 the USA developed thermonuclear weapons, with still more destructive**

power, and by 1953 the Soviet Union had them too." (Overy, 2008, pp. 291 & 292) (*All these numerous kinds of the geopolitically, mutually opposing ideologies-based highly-competitive conflicts were able to ultimately trigger the unprecedentedly ruthless and almost the blindly overwhelming kind of the 'arms-race' in the world and thereby in the process, threatening the very survival of life on this planet including that of the human-beings. Such was the larger menace of the cold-war era!*).

We may recall that interestingly and contextually in these larger perspectives, we have since already noted and discussed at various stages that, in the multi-dimensional paths of the continuous making of the man's living-history onto his always 'expanding civilisational-canvas' and, also as per the basically inherent evolutionary systems of the mother-Nature, anything and everything remains directly or even indirectly, inter-related and mutually influencing too. Similarly what kinds of the both positive-cum-negative new developments, we have been witnessing during the many decades of Post-Second World-War era now, have been having their seeds sown during the war-years and even prior to them in continuity as well. In spite of the so many kinds of indeed irrefutable advantages-cum-plus-points of the democracy as a system of political governance anywhere, rather ironically peoples and their leaderships in democracies as Allies, somehow are mostly found wanting in geopolitically seeing through the 'fogs of future', in geostrategic-cum-politico-militarily visualizing as well as, multi-dimensionally estimating the perhaps inevitably forthcoming—all sorts of the menacingly 'nefarious' designs and proactive planning of their always diehard non-democratic adversaries. And again quite ironically, the democratic nations with their comparatively slow-moving decisions-making systems, are at times put to some bothersome disadvantages, vis-à-vis their militarily or otherwise totalitarian and/or communist adversaries, in the paths of somewhat successfully countering their adversaries' 'nefarious' ambitions. As mostly borne-out by the man's living-history, it is equally true that in the end as well as, also historically in the long-term perspectives the success mostly comes to the democratic world but in the overall strategic bargain in this regard, costs, both in human-cum-material terms too, mostly come out to be avoidably rather very high and hence contextually, similar phenomena have mostly been witnessed regarding the western-bloc since, led by the USA in its struggles against her communist-bloc adversaries, in different regions of the world.

In precisely the above multiple-perspectives, ironically the first historic scenario since unfolding before the world was, immediately after indeed the heavily chequered end of World-War II, i.e.—some sort of rather preventable and highly crucial geostrategic-cum-politico-military 'goof-up' on part of the already victorious super-power: the USA, during the above discussed geopolitically momentous years of the 1950-53 Korean-War itself. It remains important to note that during those years as since, General Mac Arthur was then at the highest military-level and on the spot, in Japan/Korea, historically in the fitness of things, he has been able to give an analytically very penetrating as well as well-perceived geopolitical-cum-military accounts of this complex Korean-crisis, and hence that remains even today, a very precise and learned source for somewhat comprehensive understanding of the very genesis of this crisis. Various commentaries on several dimensions of the still lingering Korean-crisis, have since already appeared across the world and hence these too certainly do constitute valuable treasures, for the man's continuing studies and research in this regard. And hence at the same time, as we may care to further delve deeper into these writings contextually, we may perhaps come to the conclusion that while the USA and her other Western-Allies no doubt then, momentously remained engaged in the historic Marshal-Plan sponsored reconstruction-cum-rehabilitation of the war-torn Europe in the West, as well as similarly, the massive reconstruction-cum-democratically-reshaping of the devastatingly ruined Japan in the Far-East! USA perhaps did not care to pay enough strategic-cum-militarily visionary, effectively preparatory and also multiply-proactive attentions while there was still invaluable time of few very crucial years after some troublesome but also unconditional surrender of Japan in September 1945 for some militarily successful pre-emption of the sinister communist-plans in the Korean peninsula, where the USA and her Allies were perforce soon going to face the massive armour and other military wherewithal of the Soviet and Chinese forces. It is said that in these contexts, **any strategically crucial timeline lost, prior to or even during any emerging crisis/conflict situations consequently may also turn out to be at times, geopolitically-cum-militarily indeed, no less crucial and historically invaluable opportunities lost as well.**

These precisely have also been somewhat ironical saga of the typical 'confusions-cum-goof-up' on the part of USA historically, not only during handling of the Korean-crisis, but also during her quite embarrassing Vietnam-episode as well as, in addition much later even during the Middle-Eastern Iraq-crisis in the 1990s and beyond, and similarly during the Afghanistan-crisis too. Further, the USA has also been victim of her own some sort of avoidably quite costly and amateurish handling of the so-called 'failed' and almost perpetually crisis-ridden and rather artificially created Nation-State of Pakistan in the South-Asian region, and at the same at time, this has also been turning-out to be again quite embarrassingly counter-productive as well as, wasteful in terms of dollars and also, in many kinds of the costly military equipment and assets, concerning the larger US strategic-interests in the region, along with her continuing war against fanatically Islamic-cum-Jihadi kinds of terrorism. We would again be contextually discussing these aspects critically in brief and more appropriately in the next eighth chapter. Hence, globally in the larger geopolitical contexts, concerning the USA and her close Allies and since all the more so, USA being today the sole super-power in this relentlessly fast-moving as well as, hyper high-tech modern-era international scenario and at the same time now, she also being the uncontested leader of the larger free-world, it is imperative that especially in the long-term perspectives, she is always able to keep her strategic feet and thought-processes firmly anchored onto the ground realities and in addition no less important, be able to fruitfully keep the crucial lessons of history in her minds. Irrespective of others, these cannot be termed as tall-orders for the USA. And that precisely remains—geopolitically the establised point here.

Needless to say that wearing the crown of the sole super-power is never easy and also, it perforce always carries its own geostrategic-cum-politico-militarily-oriented price-tag too and the same in any case perhaps, cannot be avoided in any manner. **Now coming back again to the 1950-53 Korean-War which the USA since historically, could not have been able to avoid militarily fighting directly, the long-term and multi-dimensional essentials of the fruitfully conceived geostrategic-cum-politico-military prudence precisely in this regard, demanded that after the unconditional surrender of Japan in 1945, the USA should not have whole-hog demobilized her forces totally, i.e. her the then massive army, marine, naval and air-force assets in the Pacific, but had kept a strategically sizeable portion of them intact and operationally deployed there too as crucial strategic reserves, for soon to be victoriously employed in the Korea-War, perhaps with long-term strategic-aims, to effectively check any further possible communist hegemonic incursions in the vast Pacific-region, possibly once for all.** Now in retrospect, these matters may appear to be largely conjectural-thoughts of some geopolitical-cum-military 'flights of imagination' only, but if we multiply delve deeper into these perhaps 'out of box' perceptions in this regard, man's living-history as well as his destiny too in the making then, expected from the then top-level US political and military leadership since, proactively engaged then at the helm of affairs in Washington to indeed fruitfully display competently visionary statesmanship with strategically brilliant sense of history, in the meaningful long-term interests of the larger humanity. In perhaps no way, these were tall-orders for the then historically so victorious super-power the USA, to whom the entire free-world was so expectedly and even dreamily looking upon, as their saviour and also geostrategic-cum-politico-militarily—as their singularly capable guardian too. In any case for the USA as super-power then, who happened to be so historically instrumental in achieving the Allied Powers' decisive victories both in Europe and in the Far-East in Second World-War, it was not at all in the fitness of things to think, plan and act as the way she did during the Korean-crisis and thereby, unfortunately created rather all the more menacing and costly problems during the years to come, to herself and for the larger humanity as well. **In these larger perspectives and precisely at this point of our continuing discussions, we may also contextually add that if the Korean-crisis earlier would have been handled differently and hence achieving the final victory in Korea, then hardly after a decade later perhaps, all the more bothersome and costly: the Vietnam-crisis, won't have cropped-up in the first place in South-East Asia, and there too in any case, the USA couldn't avoid her another all the more embarrassing and costly war.** Whereas ironically at the same time, it is also equally true that in those momentous times during now the bygone cold-war era, and in each of the above mentioned crisis-situations, the then communist-bloc, as determined adversary of the Western powers, acted precisely in a diametrically opposite manner, to its best possible and numerous geopolitical-cum-military interests, and the rest is now history.

Now precisely with respect to the above as well as, in terms of many other larger perspectives, the historically timeless and also multi-dimensional importance of the **multiple-aspects of the Intelligence Services** and, the **concepts of Special Forces** cannot be overemphasized. Historically, it is true that both these concepts are quite ancient—as since valuably versatile tools of war, coercive diplomacy as well as, for the maintenance of peace—possibly geostrategic-cum-politico-militarily, in somewhat dominating manners and thereby, in the best possible both short and long-term perspectives, of the any kingdom, empire or Nation-State concerned. These noted concepts have also been changing in their multiply-conceptual incarnations and consequently, more and more strategically and also, as per any timeline situation and emerging need-based scenario—do also get fine-tuned as well. It is obvious that, these multiply complex aspects in this regard, cannot be linearly visualised and in addition also, their organizational set-up as well as, proactive policy-guidelines to get-going on these challenges-cum-risks ridden fields too—cannot fit-in within any preconceived rigid framework. In fact, any such kinds of ill-conceived notions and ill-advised planning are sure to lead to the very costly and sometimes, even very dangerous failures for any concerned Kingdom or Nation-State. So these broad-based conceptual-thoughts, in their own differently specific moulds and also, with their inherently different flexibilities, equally happen to be applicable to the multiple aspects of both **'Intelligence Services'** and also, the **'Special Forces'** as well.

So far as specifically the multiple-aspects of the Intelligence-Services and their basic 'conceptual-framework' are concerned and, in our specifically larger perspectives, one scholarly subject-specific book does first come in our minds and, that happens to be ***The Craft of Intelligence*, written by Allen Dulles who being once, also a well-known expert, as one of the earlier Directors of the USA's CIA**. This is a short and scholarly book providing a very captivating bird's eye-view—broadly across the genesis and development of the art and crafts of the profession of intelligence as well as, gradually increasing their multi-dimensional complexities for the mankind.

Allen Dulles writes that, "In the fifth century B.C. the Chinese sage **Sun Tzu wrote that foreknowledge was 'the reason the enlightened prince and the wise general conquer the enemy whenever they move.**' In 1955, the task force on Intelligence Activities of the **second Herbert Hoover Commission in its advisory note to the government stated that 'Intelligence deals with all the things which should be known in advance of initiating a course of action.**' Both statements, widely separated as they are in time, have in common the emphasis on the practical use of advance information in its relation to action.... **The desire for advance information is no doubt rooted in the instinct for survival**. The ruler asks himself: What will happen next? How will my affairs prosper? What course of action should I take? How strong is my enemies and what are they planning against me? From the beginning of recorded history we note that such enquiries are made not solely about the situation and prospects of the single individual but about those of the group—the tribe, the kingdom, the nation.... Mythology and the history of religion contain countless instances of the revelation of the divine intention regarding man, solicited or unsolicited by men themselves. But not many of them have to do with the practical affairs of state, with the outcome of military ventures and the like. Yet there are some and I look upon them as the earliest recorded instances of 'intelligence gathering'.... Cassandra's prediction that the rape of Helen would spell the ruin of Troy and her warning about the famous Trojan Horse—one of the first recorded 'deception' operations—were disregarded.... In a chapter of the *Art of War* called the 'Employment of Secret Agents', Sun Tzu gives the basics of espionage as it was practiced in 400 B.C. by the Chinese—much as it is practiced today. **He says there are five kinds of agents: native, inside, double, expendable and living**. 'Native, and 'inside' agents are similar to what we shall later call 'agents in place'. 'Double', a term still used today, is an enemy agent who has been captured, turned around and sent back where he came from as an agent of his captors. 'Expendable agents' are a Chinese subtlety which we later touch upon in considering deception techniques. They are agents through whom fake information is leaked to the enemy. To Sun Tzu they are expendable because the enemy will probably kill them when he finds out their information was faulty. 'Living' agents to Sun Tzu, are latter-day 'penetration agents.' They reach the enemy, get information and manage to get back alive.... To Sun Tzu belongs the credit not only for this first remarkable analysis of the ways of espionage but also for the first written recommendations regarding an organized 'intelligence service'. He points out that **the master of intelligence would employ all five kinds of agents simultaneously**." (Dulles, 1964-65, pp. 11, 13 & 14)

It is true that **Sun Tzu's 'Treatises on the Art of War'**, now indeed recognised as the earliest, timeless and also, even today, a very valuable beacon-light for the mankind, for any further modern-era conceptualizations in this regard. But it is also equally true that after nearly one and half to two hundred years after Sun Tzu, appeared; a similarly outstanding scholar, true genius and a thinking sage came-up in India too, known as **Chanakya or Kautilya**. He too in his equally and also probably much more versatile-cum-multi-dimensional manners, has given his scholarly treatises as the, **'Kautilya's Arthashastra'**, as another timelessly valuable source for the multiply on-going enrichment of human knowledge-cum-guidance—on the multiple-aspects of almost the entire gamut of governance, warfare, Intelligence Services as well as contextually, in certain manners on the then ancient-era Indian concepts on Special-Forces too. In fact in these multiple-contexts at this precise point, we may recall that in previous chapters, we have since already critically discussed and analyzed in some detail, the timeless, indeed remotely ancient and invaluably scholarly 'Treatises' of both Sun Tzu of China and also Chanakya of India. Though it is true that Sun Tzu preceded Chanakya by probably 150 to 200 years but, Chanakya was then not at all aware of the so historic appearance of the Chinese Sun Tzu and also at the same time, he never had any knowledge of his conceptually brilliant writings, such as the mankind's oldest and timeless treatise on multiple-aspects of 'The Art of War'. It is also true that both China and India have been two of world's most glorious and very ancient human-civilisations, but given the then earlier ancient-era's existing state of affairs across the humanity's, even though somewhat then very slowly expanding and enriching civilisational-canvas, as well as, the overall wherewithal of communications and the connectivity, it couldn't have been possible otherwise. **And in addition to all these, it is also no less historically true that the original 'manuscripts' of both Sun Tzu in China and Chanakya in India and, as they might have initially existed in either case, are no longer available to the mankind and hence as a greatly adverse civilisational consequence, got lost long time back, during the so remote antiquarian fog of the ancient-era. Whatever that be; in detailed forms and conceptualizations in either case—these treatises as available now to the man and for the future—are principally based-on the saga of the then marvellous traditions of the oral-journey of human-knowledge, through the concerned seers down the historic timeline and accordingly, from some later periods' writings of the scholarly sages—coming onto the then historic-scenes, separately in both China and India respectively.** We may also further recall and, as these being otherwise contextually no less historic aspects too, we have since critically delved upon them quite relevantly, in the earlier chapters. In the already earlier referred book on, **'Kautilya: The Arthashastra'**, we indeed find a very scholarly-cum-interpretive analyses as well as, a comprehensively ready source of reference concerning **Chanakya** and his multiply invaluable and timeless thoughts on a range of subjects—related to the entire gamut of the statecraft and multiple-aspects of the governance.

Hence precisely concerning the timelessly multiple-aspects of the art and crafts of intelligence in the above referred book while reproducing Chanakya's thoughts; Rangarajan writes that, "*In a well-organized and well-governed State*: a king can reign only with the help of others, one wheel alone does not move (a chariot). Therefore, a king should appoint advisers (as councilors and ministers) and listen to their advice.... **No enemy shall know his secrets. He shall, however, know his enemy's weaknesses**. Like a tortoise, he shall draw in any limb of his that is exposed....(*On covert operations*), a king shall have his agents in the courts of his enemy, the ally, the Middle and the Neutral kings to spy on the kings as well as their eighteen types of high officials.... Miraculous results can be achieved by practicing the methods of subversion.... A single assassin can achieve, with weapons, fire or poison, more than a fully mobilized army.... *On creating the secret service*: after verifying the integrity of high officials by means of seven tests, the king shall create a secret service consisting of the following types: (i) Agents based in one place: the intelligence officer, spies under the cover of monks, householders, merchants or ascetics. (ii) Roving agents: the secret agent, the assassin, the prisoner and the woman mendicant.... Any information corroborated by three different spies shall be taken to be true.... Any agent who is constantly unreliable shall be eliminated....(*On Intelligence Officer*): A courageous and sharp student shall be trained to be an intelligence officer. [After training] he shall swear to be loyal to the king and the Chancellor. He shall be well rewarded with honours and money and sent out with instructions to report back any activity against the king. (*On using different roles as cover*): Honest and intelligent persons shall be recruited to act as spies according to the cover specified below: (i) Monk—a monk who has relinquished his vocation but pretends to be one after recruitment. (ii) Householder—an impoverished farmer. (iii) Trader—an impoverished merchant.... These agents having been provided with plenty of money and assistants, shall pursue a [profitable] occupation at the places assigned to them. Out of the profits, they shall give others [monks, farmers, or traders, as applicable] food, clothing and shelter. Among those thus helped, ones who want to earn money shall be recruited as spies using the appropriate cover. They shall swear loyalty to the king and be promised payment of salaries and expenses. With the new recruits also recruiting others, a network of spies shall be set up.... **The intelligence gathered from roving spies shall be collected together in the establishments of spies based in one place and shall be transmitted by code. The transmitters shall not know who the gatherers were**. [Sometimes transmission of intelligence gathered by agents with the house may become difficult if nuns are prevented from entering it.] In such a case, the intelligence gathered shall be transmitted by means of songs, speech, signs or messages in code hidden inside musical instruments or vessels, to other agents appearing at door in disguise (as parent of a servant, an artist, a singer or a slave). Alternatively, the agent inside the house may attempt to get out either secretly or pretending to be ill, mad, or feigning escape from fire or poison." (Rangarajan, pp. 177, 498, 503, 504, 505 & 506)

The above quoted lines from the treatises of both Sun Tzu of China as well as, Chanakya of India, since both belonging to the remote ancient era during the few centuries preceding Christ, are indeed the ever alive and firm conceptual foundations for the subsequent developments of the multi-dimensional thought-processes as well as, the various doctrine-formulations concerning the entire and also, multiply-complex and vast gamut of the art, crafts and sciences of the intelligence, and at the same time, being almost equally relevant and important even to the fast-moving and the so high-tech oriented super-age of the presently on-going modern-era. In other words, we may also say that, these unique and vast storehouse of the timeless and multi-dimensional wisdom—are indeed beyond any degree of obsolescence for the mankind.

But it is also broadly true that the fundamental tenets of the so ancient conceptual thoughts and strategic-doctrines of both Sun Tzu of China and Chanakya of India in this regard; do constitute the pivotal fulcrums even today, in their own different ways—for the continuously ongoing developmental processes and expansions of the thoughts and activities in since the always so hush-hush domain of intelligence services across the world. In addition and as already substantiated during the ages back, by both the above mentioned Chinese and Indian seers in this regard; the indeed extraordinary world of the intelligence, has always been highly secretive, challenging and even risk-prone fields of since the State-sponsored human activities, and hence everybody and anybody just cannot be suited for such special kinds of work of the very crucial national interests. Further, as we continue to delve deeper into these indeed extraordinary aspects of the statecrafts; we do also notice another dimension to all these and that being—that precisely these State-sponsored activities of any two rivalry-cum-enmity-ridden kingdoms/nations down the continuing historic timelines—have always been inescapably becoming mutually competitive, ruthless and also, even violently clashing as well.

In today's ultra-high-tech and also the accelerating as well as, very fast moving modern-era times; the multiply-complex and all the more highly challenging areas of intelligence activities across the world, are getting more and more mindboggling and high-tech oriented too. **What is intelligence by itself and also, what's the basic purpose of intelligence, if any**? Well to say the least, intelligence gathering since the dawn of man's living-history itself; has always indeed been, a crucially important component of the entire gamut of the statecrafts, as practiced by any ruler/ruling-class or the Government of any kingdom/nation; to protect her security by thwarting any kind of the hostile designs of any adversary/enemy and thereby, to be able to always protect the strategically vital and also, her other multiple kinds of the geopolitical-cum-economic interests. Broadly intelligence can be of two kinds; i.e. **(i) Human intelligence** and, **(ii) Electronic intelligence**.

Human intelligence has historically been of the oldest method of intelligence gathering, and even in the present super-age of modern era; any systemic methods of human-intelligence will remain always no less crucially important and hence perhaps, can never be dispensed with altogether, irrespective of the any possible levels of the marvels of sciences and technologies, ever to be achieved by the mankind itself. And as the years have been rolling-by and, along with continuing advances in the multiple-fields of sciences and technologies, both the above mentioned kinds of the intelligence gathering in this regard, have been continuously acquiring numerously more forms and dimensions and also, all the more complexities too. Since the very beginning, communication in any form—as timelines have been moving ahead between any two individuals or a smaller or much bigger groups of individuals, or even any extraordinarily larger societies/States has always inherently been a basic civilisational feature of the mankind. And interestingly therein itself, we do also notice—the always historically as well as, proactively stationed—the very secretive and hush-hush world of the all kinds of highly complex as well as, multi-dimensional interplays of the activities of various kinds of intelligence-agents, so deployed wherever needed and thereby, giving desired effects to the intelligence gathering strategic-plans, of any concerned kingdom/nation.

We all are since aware that communication is resorted to: for the exchange of ideas, thoughts, comments, information, and then, even to facilitate the routine or any vital business of statecrafts, at any level or in any segment of the entire gamut of governance of any concerned Nation-State. Well in larger perspectives, we may also say that at times, any communication for any routine/ casual purpose, anywhere, at any level or in any manner, looking otherwise normal and very innocent, could also carry a vital piece of intelligence. And these are all the more so now and since, also acquiring newer dimensions as well as more and more complexities, concerning the domain of intelligence, as regularly being triggered by the super fast-track developments continuously taking place in multiple areas of electronics, other dynamic technological areas of the computer hardware and software and in addition, also in the otherwise numerous areas of sciences and technologies, and thus in totality thereby, phenomenally expanding indeed the multi-dimensional frontiers of the human knowledge and capabilities. But at this precise point, one timeless and universally applicable fundamental truth also needs to be borne in mind that, every coin always has two sides. Hence, if the man is continuously getting more and more equipped with newer kinds of capabilities-cum-wherewithal to continuously enhance his potential of intelligence gathering then at the same time, inherently in a parallel manner, newer as well as more menacing kinds of challenges and threats too, are perhaps unendingly cropping-up and consequently thus making things far from any easy for him. In fact, these multiple-fruits of the continuing advancements of sciences and technologies are mostly like two edged swords as wherewithal or 'weapons' since, if these are available to any one nation then, the same do also get identical places on the shelves of concerned adversaries as well.

Contextually in other words, we may also say that in the presently on-going super-age of the modern-era, in the man's multiply expanding and also the highly complex domain of the intelligence gathering the above mentioned multi-dimensional advancements have also since resulted in the numerously challenging kinds of the 'cat and mouse' games. And at the same time, as the time is passing-by different kinds of the inherent intensities of these games, are also increasing by the day, and hence thereby making the multiple kinds of now, all the more challenging business of the statecrafts-cum-governance of any concerned Nation-State and, as these being always so multiply situations-specific too are becoming geopolitically-cum-militarily more and more demanding for the concerned rulers/decision-makers and also menacingly perhaps, putting them more in increasing tight-corners as well. **And contextually, as we continue to critically delve deeper into these far-reaching issues in this regard, we do clearly notice that otherwise too, always being crucially important—the timeless multiple-aspects of the 'essential strategic-culture and thoughts'—do tend to acquire rather a state of 'critical-mass' within the overall geostrategic-cum-politico-militarily centric conceptual-framework of any concerned nation's larger gamut of the statecrafts and systems of governance**. Hence in today's modern-era world, these somewhat contextually analytical perceptions in turn, do substantially point towards the paramountcy importance of the entire gamut of intelligence gathering activities, and whose negligence or even taking any vital intelligence inputs in a somewhat casual manner by any concerned nation, may turn out to be matters of some strategic disasters. Now, where does the modern sovereign and democratic Indian Nation-State stand in this regard, in the comity of nations and, how far her multiply strategic-moorings are firmly secured? Well, we would more relevantly be discussing these aspects in some detail later on, in the last sub-chapter of the next eighth chapter.

In the above multiple-contexts, we do at the same time also notice that since the beginning of man's living-history itself, the entire gamut of intelligence has been having the numerous dimensions and areas concerning the human-activities related secret-operations. What again the continuous multi-dimensional advancements in the vast frontiers of sciences and technologies now, have been doing that: these are providing new impetus, the means and potential into the all kinds of activities concerning the art, crafts and sciences of the highly secretive intelligence activities of any concerned Nation-State. Each nation has its own largely distinct kinds of her socio-psychological-cum-cultural and political ethos and also along with these, their specifically distinct historicity as well. So when it comes to the subject-matters of evolving-cum-formulating any nation's Intelligence-cum-Secret Services' strategies and thus, establishing the concerned Institutions in this regard, finally each Nation-State does naturally end-up with her own distinctly different kinds of the institutional mechanisms, methodologies for their functioning, administration, as well as, their command and control-systems. Hence, obviously if these are up to the maximum possible extent—professionally having their synergy-creation-oriented concepts and thereby, having perhaps least of hassles into their interactive systems of the institutional-functioning—then they are most likely to achieve maximum optimization of their allotted resources—and also remain mostly effective in serving their most of the paramountcy national-interests.

Based on the above discussed perceptions perhaps, and in addition also, bringing-in any particular nation's own and specifically different essential inputs, that concerned nation, has designed and established her distinct kinds of the Intelligence-cum-Secret-Services Institutions for fulfilling her geopolitical as well as, the other kinds of requirements. Requirements in these respects, can be both internal as well as external and at times, they may even be somewhat overlapping too; and hence different nations have mostly two distinct Institutions, to take care of the internal and external needs, as much separately as possible and at the same time also, both having their in-built institutional mechanisms for their mutual cooperation and in addition also, for their mutually helpful professional facilitation as well. **Contextually in our larger perspectives in this regard we find that the USA has her FBI and CIA; UK has her MI-5 and MI-6; Israel has her Mossad; India has her IB, NIA and RAW; Pakistan has her FIA and ISI and so on; for serving their own specifically different kinds of covert and overt needs.Mossad of Israel is unique in the world in many ways, in its operational capabilities and super effectiveness.** Well interestingly, these needs could be: (i) espionage—geopolitical and military, (ii) then as parallel to it, the counter-espionage, (iii) all kinds of the needs-cum-specific-situation-based intelligence gathering activities, both internally and externally as well, (iv) spreading disinformation as well as, false propaganda too, for misleading the potential adversaries and also at times; for creating all sorts of confusions in the enemy camps and lastly, (v) as per the geostrategic and politico-military needs, to go in for sabotage and subversions to harm the adversaries and serve one's own vital interests. How India's above mentioned agencies have been faring in these respects; more suitably and contextually in some greater details, would be dealt with, in the last sub-chapter of the next forthcoming eighth chapter itself.

After the highly chequered end of the (1939-45) Second World-War; the so much historic, multi-dimensional and, ruthlessly competitive intelligence gathering activities and also, their mutually rival institutional mechanisms, on the part of both Western-bloc and Communist-bloc groups of nations can be compartmentalized into two distinct segments: (i) the Cold-War-era period and, (ii) the post-Cold War-era period, respectively. These periods now historically relate to—a highly chequered part of indeed still, a very fresh chapter concerning few end decades of the quite recently bygone 20th century—and concerning once so menacing, the Cold-War itself. In these multiple-contexts, many scholarly as well as, historically very engaging books in different languages across the world are already since available for the interested readers and research-scholars and in addition, many historically memorable films too, have been made in this regard. **We have already since earlier referred above, one very historic book on. 'Craft of Intelligence'; authored by Allen Dulles, one of the earlier Directors of the CIA, in the USA.** Similarly, another book indeed gives a very interesting insight in the modern-era hush-hush world of the espionage/counter-espionage; and also the very murky, extremely dangerous and ruthless and also, the other multiple-aspects of the activities of Secret-Services on both sides of the iron-curtain; during the highly chequered Cold-War years and also, since belonging to the leading powers of that historic era. **During those historic times, another Super-Power, the USSR also had a highly feared Secret-Service as KGB, and that being a great rival then, of the USA's CIA. KGB too, had a no less vast organisation and also, no less efficient institutional mechanisms and the requisite wherewithal once, with their truly global reach as well.**

The author **Stephen Dorril**, has written that, "For MI-6, the early sixties proved to be the best and worst of times as its 'special' relationship with the CIA went through highs and lows. There were the occasional outstanding intelligence-gathering successes which drew the admiration of the Americans, who were awash with money and technical advances, but these were largely overshadowed by security lapses which fed the almost continual mole hunts that ripped into the heart of the Service, colouring relations with the Americans and undermining operations.... A complete breach of trust was probably avoided only as a result of the high quality intelligence that the Service was able to supply from the few sources it developed behind the Iron Curtain; though in truth, MI-6 had no hand in the recruitment of these sources as they were 'walk-ins', volunteering their services.... From late 1958, MI-6 had been running three agents—the principal one code-named Noddy—inside the Polish security intelligence service, the Urzad Bezpieczenstwa (UB), controlled by the former chief of station in Berlin, Robert Dawson, 'a solid fatherly figure with the air of a countryman' who responsible for eastern Europe as head of the Directorate of Production 4, and run through the Warsaw embassy where John Quine and a future chief of the Service, Colin Figures, were stationed. The UB officers proved to be MI-6's first major sources in the Soviet orbit of the postwar period.... Close liaison between the Poles and the KGB meant that intelligence handed over included unprecedented information about the Russian service, and profiles of military and political personalities in Moscow. There was also information on operations against French and American citizens and diplomats living in Warsaw. Realising the value of the intelligence, Dick White made sure that it was quickly passed on to a grateful CIA, which assessed it as 'the best we ever had'. According to Soviet sources, which were probably based on revelations from George Blake, who was then in the Russian section of DP4, the **Americans undertook to finance some of MI-6's operations; in particular, they offered $20 million for espionage in Poland**. The paradox was that the operation's success exposed the extent to which MI-6 was reliant on the American support." (*So interestingly, those were the days in the hush-hush world of intelligence in Europe then; during the historic Cold-War era decades, even at times sounding stranger than the fiction itself.*). (Dorril, 2000, pp. 703 & 704)

Now while contextually discussing the multiple-aspects of intelligence services quite briefly in the above few pages, since almost the dawn of man's living-history; we clearly notice one historic fact in its on-going continuity that, the **'Concepts of Special Forces'** are, inseparably as well as inherently, linked to those of the intelligence services. **In these precise contexts, we may also say that for the special-forces (SF) units in any specifically given situational-tasks; the vital and well-timed intelligence-inputs are in fact, always the most crucial sources of oxygen, the beacon-lights and also, some kinds of the compass; in their dark, dangerous and highly challenging paths, towards accomplishment of any concerned SF's assigned mission.** Historically, the operational ideas and, concepts of Special Forces, for ensuring one's security and for achieving victories or even cornering some kinds of the upper-hand over the enemy, both in war as well as during the coercive diplomatic-encounters have been in force in different manners, since the very ancient of times. That is inevitably and also obviously a different but quite logical matter that, as the centuries have been rolling-by over the continuously expanding and also dynamically altering civilisational-canvas of the mankind, multiple-things in almost all the civilisational-segments of man's facilities, multiple wherewithal, and ways of life in many ways, have been changing and improving; and hence consequently thereby the special-forces too, have been changing across the world. Universally, change has been the law of nature and hence for the man in this world, and in almost the all segments of his life as well as also, in the numerous areas of human-activities, changes have always been perhaps unalterably the order of the day, anywhere and everywhere. Hence, keeping pace with all these changes—with essential strategic prudence and visionary wisdom and also along with the well-stationed 'sense of history', are mandatory prerequisites, towards ensuring one's security and multiple well-beings in the world.

SF units are no substitute for the conventional military forces in any way; but they are always militarily, very valuable complementary assets and also at the same time, any particular situation-specific force-multipliers as well. Like personnel in any regular and conventional Armed Forces, men in the SF units too, are equally soldiers of their any concerned nation. But, there is a remarkable difference here. The SF soldiers are especially selected from among the regular military forces, through very strict screening-processes, of the uncommon kinds of exceptionally tough selection-boards. The prerequisites are exceptionally high and rigid norms have no room for any flexibility at all. Their professional job-profiles are exceptionally demanding and hence, their physical as well as mental fitness-norms are uncompromisingly demanding too. They are rigorously trained for special kinds of operations and hence; mostly required to penetrate deep into the enemy territories, for carrying out the numerous varieties of tasks of some special tactical/strategic significance and hence, bringing-in some high strategic-cum-military gains for the home forces. They are especially trained and soldierly groomed to be able to operate both in groups and even at times individually in some very adverse of situations without any panic and also, not to always look for orders from the above. Hence, they have to be professionally capable of tactically assessing and reading any given extremely risk-prone and challenging situations, for carrying out some crucial strategic-cum-politico-military mission successfully, with soldierly focused and professional tenacity; both in his groups of identical companions and even at times, individually as well. With his personally inherent as well as, professionally acquired some extraordinarily exceptional attributes and soldierly qualities through his uncommonly rigorous training-programmes and thereby, ensuring his all these unique qualities getting well honed-in together; he becomes capable of bearing any kinds of extreme stress and hardships, and yet be professionally capable of thinking logically and taking quick decisions, always with sole aim to accomplish the assigned mission.

Any SF unit or sub-unit's task in any given situation may include anything, such as concerning—any conflict situation, any conventional war, special kinds of deep and extensive reconnaissance behind the enemy-lines, deep infiltration into the enemy territories—for any sabotage, subversion, assassination of some very important or top-grade social, political, or military personality in the enemy-camps for creating panic in the enemy country, or for some crucially important kinds of the intelligence gathering, for counter-espionage or for counter-infiltration, or for carrying out any similar or other kinds of special assignments, or for carrying out some **psychological warfare (PSYOPS)** missions against the enemy and thereby, to perforce put him in some off-balance situations; by some kinds of the attrition-oriented tasks and even at times, to create civil-disturbances in the enemy territories. In other words in this regard and as per orders, the concerned SF unit/sub-unit proceeds to accomplish it's the very special kinds of the some situation-specific given mission/missions, towards serving one's own larger aims to subvert, sabotage, disrupt, damage or, even considerably smash the enemy's morale, his will to fight and thereby, finally derail his overall war-efforts—against one's own nation and the military-forces in any emerging conflict situation(s). Similar to the already above discussed hush-hush world of one's Intelligence-Services any specific situation-based given-tasks and thereby, the activities regarding highly focused operational planning and professionally tying-up all the essential details, training, rehearsals, etc., to be effectively carried-out by any concerned SF unit/sub-unit too, are equally no less secretive and the hush-hush world; as per its also, the extra-ordinarily demanding profession. And at the same time contextually in our larger perspectives; as we continue to critically delve deeper into these multiple-aspects of both the Intelligence-Services and also, those of the Special Forces; we may clearly notice the vital multiple-threads of mutually supporting and always essentially proactive linkages and professionally very far-reaching and always dependable inter-relations, between the respective domains of their multi-dimensional activities.

Hence no wonder that, as in case of the highly secretive and hush-hush professional domain of the Intelligence-Services; many scholarly and also professionally very valuable books on the above listed multiple-aspects, concerning the equally secretive-cum-hush-hush domain of the Special Forces as well, and also at times, since very vividly depicting the numerous bone-chilling accounts of many SF stalwarts and historic legends across the world too; are available, to meet the further needs of the interested readers in this regard. As contextually just specific examples we may cite: (1). 'Shadow Warriors—Inside the Special Forces', authored by **Tom Clancy** with **General Carl Stiner (Retd)** and **Tony Koltz**, and published by G.P. Putnam's Sons, New York in the year 2002; (2). 'A Special Forces and CIA Soldier's Fifty Years on the Frontlines of the War against Terrorism—Hunting the Jackal', by **Billy Waugh,** Master Sergeant of the U.S. Special forces; with **Tim Keown** and, published by Avon Books (An Imprint of Harper Collins Publishers), New York, in the year 2004/2005; (3). 'No Easy Day—The Only First Hand Account of **the Navy Seal Mission**

that Killed Osama Bin Laden', written by **Mark Owen** with **Kevin Maurer**, and printed and published by Michael Joseph—An Imprint of Penguin Books, London (England), in the year 2012 and then; (4). 'Mission to Pakistan—An Intelligence Agent in Pakistan'; in fact, a somewhat fictionalized, but very fascinating account of an Indian Secret Agent in Pakistan, based on indeed, some substantial historic facts as well; and since published by the Manas Publications, New Delhi (India) and, authored by **M.K. Dhar (IPS) Retd**; for our quite informative references in this regard. (*So far as specifically any real-life exploits of the Indian SF units anywhere are concerned, unlike the USA; their historically factual and precise accounts are yet to come to the public-domain, for many obvious geopolitically very fluid and rather different kinds of the malignantly lingering situations, almost all around the Indian borders.*). Today in almost all the leading nations of the present modern-era world; all the three principal wings of Armed Forces: the Army, Navy and Air Force, have their distinctly separate kinds of the SF Units meticulously organised, trained and equipped in their distinctly special manners for their likely assigned roles.

India too, has her own SF outfits on the similar grounds, to take care of their specialised tasks whenever they are called upon to do so, as per her paramountcy national interests. **Specifically concerning India, recently one valuable book in this regard has appeared namely, 'Special Forces: Doctrine, Structures and Employment across the Spectrum of Conflict in the Indian Context', and published by Knowledge World, New Delhi, in the year 2006, with also a quick reprint in 2007 itself.** It is basically a compilation of some quite learned and professionally, very focused 'papers' since authored and presented by some of India's leading experts and top-level seasoned professionals and, who quite befittingly are having the well-honed expertise-cum-experience-based scholarly grips over the highly challenging multiple-fields in the vast and complex domain of the Special Forces, with precise references to the Indian contexts. The papers do have precision in thoughts and also, valuably futuristic vision, covering multiple-areas and the daunting challenges, since always emerging onto the highly complex, mostly very fluid, and also, the no less volatile contemporary geopolitical scenario in South-Asia as well as, into the other adjacent regions in the present times. Particularly the first listed 'Approach Paper' in the above mentioned book, and since titled as, 'Special Forces: Doctrine, Structures and Employment across the Spectrum of Conflict in the Indian Context and, since presented by **Lt. Gen. Vijay Oberoi**, PVSM, AVSM, VSM (Retd); does give a very informative panorama of thoughts and also professional expertise on the concerned subject-matters and thus, setting the tone and scholarly momentum for the further paper-presentations, deliberations and discussions in this regard. Hence, this book is indeed a very valuable contribution on these subjects specifically covering the presently prevailing overall geostrategic and politico-military scenario, confronting now the sovereign, fast-developing and democratic Indian Nation-State in today's overall geopolitical environment as well as, the multiple kinds of the quite menacing games of international power-politics in the world.

While broadly substantiating the thoughts-lines of our above discussions on some principal dimensions concerning the multiple-aspects of Special Forces, in the above referred book and in the 'Approach Paper itself, **Lt.General Vijay Oberoi** says that, "**Throughout history, success by a small force against a strategic or operational level objective has usually required units with a combination of special equipment, training, manpower, or tactics that go beyond those found in conventional units**. This is what has driven militaries to form special units, with specialised, highly focused capabilities. The SF outfits are small, elite military units, with special training and equipment that can infiltrate into hostile territory through land, sea, or air, to conduct a variety of operations. The SF personnel undergo rigorous selection and lengthy, specialised training. The SF are not a substitute of for conventional forces, but a necessary adjunct to existing conventional capabilities.... **Rapid advancements in technology, in all spheres, are changing the concept of warfare. In the field of IT, digitization, networked communications, electronic warfare (EW), as well as information warfare (IW), have highlighted the importance of systems integration and the concepts of 'system of systems'.... The nature of war has changed, and, in many respects, it continues to change on account of environment and situational factors.** Low intensity operations, a general reluctance to change existing boundaries by conventional military means, increased cooperation and engagement between nation-states to counter terrorism; countering non-state actors; and similar sub-conventional operations, have assumed much great importance than hither to fore.... Today, joint and combined operations are the norm. Concepts and methodologies of waging war are also changing. It is now difficult to divide operations into conventional and **LIC (*low-intensity conflicts*).** The modern battlefield demands that war be fought in all six dimensions—space, cyber, air, land, sea and sub-surface.... All these changes and those in the offing have a direct impact on how armies are going to carry out their tasks of deterrence and waging war. In such a dispensation, the SF too have to conform and integrate themselves to the overall structure.... India's military strategy needs to be effective across the entire spectrum of possible conflicts, from a high intensity conventional war under the backdrop of nuclear threat, to LIC, which may include high intensity insurgency, militancy and terrorism.... These monumental changes and the continuing **RMA (*revolution in military affairs*)** demand that India tailors its SF to meet these new challenges. For this, a holistic view needs to be taken, not only about the size of SF we ought to have but more importantly how they should be organised and structured. **The most important of an SF unit is the individual, called 'commando' or 'special warrior' or 'special operator'.** These personnel require a particular type of personality that may not be fully in tune with the more conventional, live-by-the-book, military personnel. Commandos should possess a certain confidence to their capabilities and independence of character. They should also be much more intelligent. Unlike regular military personnel, they must have the ability to analyse a situation and do what it takes to resolve it, instead of waiting for orders. They need to believe in 'speed of action' and 'violence of aggression'. This contrasts with a regular soldier, who waits to receive his orders from a long chain of command and flounders if none are received.... The SF are the nation's penetration and strike forces, which are able to respond to specialised contingencies across the conflict spectrum with stealth, speed and precision." (*Interestingly, in these multiply far-reaching contexts, we would briefly be coming across more of the SF units deployed in their actions across the world; in a number of contextually different and also demanding situations, during our discussions in the next eighth chapter.*). (Oberoi, pp. 13, 15, 16, 17, 18 & 20)

*

CHAPTER 8

'Super-age' of the Present Modern Era

In the multiple-contexts of our on-going discussions and, also to put the numerous contextual perceptions related to the very many subject-matters in this regard, in their some, appropriate perspectives; it may appear somewhat befitting to note that, **the timeline of the historically momentous as well as, very far-reaching advent of the "Industrial Revolution' along with its vastly numerous fallouts,** may be taken; broadly as some sort of the conceptual demarcation-phase—between the finally passing-away of the ending centuries of now, since the already bygone medieval-age and consequently—a kind of the somewhat accelerated beginning of the present modern-era.

Based on the ongoing discussions **Professor E.H. Carr** has quoted a passage from **Professor Powicke**'s inaugural lecture as Regius Professor in Modern History in Oxford as: '**The craving for an interpretation of history is so deep-rooted that, unless we have a constructive outlook over the past, we are drawn either to mysticism or to cynicism.'.... Like the ancient civilisations of Asia, the classical civilisation of Greece and Rome was basically unhistorical. As we have already seen Herodotus, as the father of history, had few children; and the writers of classical antiquity were on the whole as little concerned with the future as with the past**.... Acton, in the report of 1896 on the *Cambridge Modern History* which I quoted in my first lecture, referred to history as a 'progressive science'; and in the introduction to the first volume of history wrote that, 'we are bound to assume, as the scientific hypothesis on which history is to be written, a progress in human affairs. In the last volume of the history, published in 1910, Dampier, who was a tutor of my college when I was an undergraduate, felt no doubt that '**future ages will see no limit to the growth of man's power over the resources of nature and of his intelligent use of them for the welfare of his race.' (*Well, is it so and, are these completely true so far as; any total reliance on the man's perhaps, so-called 'unfaltering' wisdom is concerned? We would contextually be discussing these important aspects subsequently, in some of the forthcoming relevant sub-chapters.*)**.... The conception which I have put forward in these lectures of history as a constantly moving process, with the historian moving within it, seems to commit me to some concluding reflections on the position of history and of the historian in our time. We live in an epoch when—not for the first time in history—predictions of world catastrophe are in the air, and weigh heavily on all. They can be neither proved nor disproved. But they are at any rate far less certain than the prediction that we shall all die; and, since the certainty of that prediction does not prevent us from laying plans for our own future, so I shall proceed to discuss the present and future of our society on the presumption that this country—or if not this country, some major parts of the world—will survive the hazards that threaten us, and that history will continue.... **History begins when men begin to think of the passage of time in terms of not of natural processes—the cycle of the seasons, the human life-span—but a series of specific events in which men consciously involved and which they can constantly influence. History, says Burckhardt, is 'the break with nature caused by awakening of consciousnesses'. History is the long struggle of man, by the exercise of his reason, to understand his environment and to act upon it**. But the modern period has broadened the struggle in a revolutionary way. Man now seeks to understand and to act on, not only his environment but himself; and this has added, so to speak, a new dimension to reason, and new dimension to history. **The present age is the most historically-minded of all ages. Modern man is to an unprecedented degree self-conscious and therefore conscious of history. He peers eagerly back into the twilight out of which he has come, in the hope that its faint beams will illuminate the obscurity into which he is going; and, conversely, his aspirations and anxieties about the path that lies ahead quicken his insight into what lies behind. Past, present, and future are linked together in the endless chain of history**." (Carr, 2008, 109, 111, 133 & 134) (*Yes indeed; it has been so and, it would continue to remain so always, for the mankind, within the overall on-going creation of the mother-Nature and hence thereby within it, consequent to the all sorts of ongoing human endeavour; as well as simultaneously no less, the multiple-kinds of the ongoing natural phenomena onto this planet-Earth too.*).

(a) Post-Industrial Revolution, Multiple Advances in Electronics, Advent of Computer-Age and, the Beginning of Automation

Perceptively in these multiple contexts we may broadly say that with the advent of computer and consequently, the scientifically mesmerizing birth of the computer-science soon to be multiply-used for the revolutionary furtherance of the man's highly complex and multi-dimensional knowledge-pool, concerning almost the all disciplines of the modern sciences and technologies as well as, their multitude of applications in the man's service, for enhancements of the overall quality of his life; meant largely in physical-cum-material terms only, in this world. Hence we may also say that, the birth of computer was in fact, instrumental in heralding the dawn of indeed the unprecedented as well as, broadly the unexpected momentous era of now, the generally understood **'Post-Industrial Revolution'**. But prior to that in these respects earlier, the continuously on-going relentless research, developments and the many path-breaking advances in the multiple-fields of since the already highly engaging ***discipline of the electrical engineering areas***; first led to the birth of **Electronics** in this regard.

And consequently, further multi-dimensional research and advances in the multiply fascinating fields of the sciences of electronics led to the many new kinds of breakthroughs: such as knowhow concerning the sciences of semi-conductors, printed-circuit boards and systems and further on, leading to the scientific and technical developments in the engineering concepts of the **'transisterisation'** and the multiple kinds of their applications. Then further-on, within some years only, the above mentioned concepts led to the far-reaching break-through in the momentously multiple-aspects of the **'technologies of digitisation'** as well. Hence, as per very nature of the phenomenal processes of since the continuously on-going multiple advancements, in the numerously vast and highly complex fields of the sciences and technologies, onto the mankind's expanding civilisational-canvas itself, we do interestingly find that one thing leads to another and thereby, the multiply continuing scientific-cum-technological advances do keep-on progressing and hence consequently, do also keep-on registering newer and newer scientific breakthroughs too; into the vast and highly-complex and, also the expanding store-house of the humanity's knowledge in these regards. All these multiple-aspects concerning indeed, the paths-breaking advancements in the past, during the beginning years of so momentous-era of the 'Post-Industrial Revolution', during the latter half of the last twentieth century itself; have been very knowledgeably summarized by **Professor Vannevar Bush** of the M.I.T., USA in *Science is Not Enough: Reflections for the Present and Future* (pp. 80, 81, 83 and 91). **Professor Vannevar Bush** has very perceptively written that, "And an excellent example of how the advance goes forward is the history of the thermionic tube and the transistor. The thermionic tube was, at its inception, largely a matter of ingenious tinkering, without much reliance on science. **Edition**, who was no scientist, noted a current from the filament of one of his electric lamps to a plate he put in, but did nothing about it. **De Forest**, who probably knew still less science, added a grid between them, and the thermionic tube was born. For many years, it was erratic in operation and likely to fail at any moment. Then engineers learned to make a really good vacuum and it became much better. Finally, it became so reliable that it could be installed in an amplifier of a submarine cable at the bottom of the sea and expected to last for forty years. It became so rugged that, in the proximity fuses of the war, it could put into a shell, and fired out of a gun, and still be expected to work as a sensitive electronic detector.... The transistor which can be as small as the eye of a fly, requiring extreme precision of construction, rugged and long lived when once built, is perhaps the most versatile device man has yet produced. With the use of very little power, and in a small space, it will amplify, modulate, rectify, and do dozens of other things. It is one of whole family of devices based on the use of semiconductors: photoelectric cells, rectifying valves etc. When the transistor is combined with other elements, resistors and capacitors sealed in a resin, an assemblage the size of thimble will do all that used to be done by a radio receiver as large as a suitcase. More than this; by some very method of depositing very thin layers of material in a vacuum, the whole thing can be reduced to a thin wafer the size of a flyspeck, and a thousand such can be produced identically in a single manufacturing operation.... A very great advance—possibly the greatest so far—as we look forward the future memex, is magnetic tape. We have known it for some time in dictating machines. The idea is a simple one, and the tape is merely a plastic strip covered with magnetic material in finely powdered form. As it passes over an electromagnet, the voice, picked-up by a microphone, causes the strength of that magnet to vary, and these variations become impressed on the tape in the form of its magnetization. A wavy form of air vibrations from the voice become an identically wavy form of magnetism on the tape. Then, when the tape is run in front of a coil, the voltages there produced can be amplified and fed into a loudspeaker, and a replica of the original voice appears as sound waves in the air.... All this is now ancient history. But it is relatively new to put on the tape the variation of light impinging on a photocell as an optical system sweeps its view over a scene, and to do this so that all the details of a complex scene are thus recorded in as small fraction of a second. And then to reverse the process and that is, to reproduce the scene to a viewer a thousand miles away. (*And that is Television.*).... Beyond this the laser renders possible an exciting process called holography, which may render it possible to project the record so that it is three-dimensional.... Logic can become enormously difficult, and it would undoubtedly be well to produce more assurance in the use of it. **The machines for higher analysis have usually been equation *solvers*. But we now have equation *transformers*, which will rearrange the relationship expressed by an equation in accordance with strict and rather advanced logic**. Progress here is a bit inhibited by the exceedingly crude way in which mathematicians express their relationships. They employ a symbolism which grew like Topsy and has little consistency; a strange fact in that most logical field." (*Very engagingly knowledgeable and so informative as well, indeed providing panoramic views in these regards; isn't it? These too are the ways and means, for the growth of Science.*).

In these multiple-contexts, precisely the above depicted scenario, do give us systematic glimpses of the interesting scientific phenomena concerning indeed, the paths-breaking advancements in the fields of electronics as well as, the fundamental concepts of the design and fabrication aspects of the various electronic devices, having vast potential for their multiple-applications and thereby also leading to the subsequent processes of **automation** and also, technical formulations of their mass-scale manufacturing technologies as well. And at the same time in fact, all these pertained to the beginning years of the momentous 'Post-Industrial Revolution' times. But compared to those otherwise indeed quite enchanting times and since then, within less than even four decades or so time in this momentous era, has indeed moved quite fast and thus in these regards we do also find today another enchanting emergence of the successful development and mass-scale production of the **'digital micro-chips'** and their multiple applications in so many ways. In our contextually broader perspectives, as we may begin to critically delve deeper into these aspects we do also interestingly visualise that perhaps suddenly, time is appearing to have begun moving in an overwhelmingly very fast manner. That apparently looks to be so, but in fact the speed of the perpetual movement of time is always unalterably fixed by the mother-Nature herself and hence, that is in a way, just a superficial illusion only. **The far-reaching and also, quite captivating truth in these regards, precisely is that the rate of the highly complex and multiple advancements in the so many vast fields of the sciences and technologies have been accelerating so fast and thus, leading to indeed the mesmerizing births of the all kinds of the mostly unexpected as well as, the newer and newer technologies, electronic and other devices, tools, equipment and machines with mind-boggling applications, leading to their substantially advantageous uses and hence consequently, the mankind across the world finds itself totally overwhelmed, by these earlier unknown technological marvels.** And therefore consequent to all these phenomenal developments in multiply accelerating manners, since continuously materializing onto the man's expanding civilisational-canvas, and hence so interestingly the man himself rather quite illusively feel that, the time during this super-age of the present modern-era, has begun moving faster and faster.

Hence, as since also apparently clear now, that the other basic reasons towards causing such apparent illusions in this regard, reside somewhere else; and i.e., now indeed in the overwhelmingly accelerating pace of the highly complex as well as, multi-dimensional break-through and their far-reaching impacts, onto the overall ways of life of the mankind, since always being relentlessly triggered by now the expanding as well as, extensive continuation of the highly complex-cum-multi-dimensional research and development efforts, in the vast fields of the modern-day sciences and technologies.

We may contextually recall that in the earlier chapters, and in course of our on-going discussions, **while critically delving upon the advent and subsequent advancements of the various human-civilisations across the world; we have briefly covered some of the very far-reaching and also, the multiply developments-heralding historic 'Revolutions' too which were in turn, indeed instrumental in triggering the numerous and also highly consequential multi-dimensional advancements of these very human-civilisations along with their different timelines and also, in the different geographical areas of the world**. These 'Revolutions' were largely apolitical to begin with and hence, substantially civilisational in nature and thus, leading towards being the socio-economic and, scientific-cum-technological aspects in their contents and at the same time also, the multiply developments-oriented directional thrusts; indeed in their phenomenally paths-breaking as well as, further new history-making manners too. Broadly we may say that in these respects, the first one was the quite ancient and historic **'Agricultural-Revolution'**; the second one now being the so well-known and, the multiply-momentous **'Industrial-Revolution'**; and then the third one being the presently on-going—the **'Post-Industrial Revolution'**—with its new kinds of the previously unknown, highly complex and also, numerously vast scientific and technological dimensions. All of the above mentioned 'Revolutions' in these regards, during their historically momentous and different timelines and then further on have in their own specifically different, multiply far-reaching and at the same time also, with their paths-breaking turning-points in their multi-dimensional manifestations and since thereby, have also provided the new impetus from time to time, to the on-going civilisational advancements of the mankind. And in addition consequently, these were also instrumental in substantially altering from then on the subsequent continuous making of the man's living-history, onto his expanding civilisational-canvas itself. Hence in other words in fact, we may also say that, consequent to the advent and then, the multiple-fallouts of each of the above mentioned 'Revolutions', during their specifically recorded timelines, have been momentously instrumental in bringing-in, the numerous paths-breaking changes in their own distinct manners, in the subsequent multi-dimensional making of the man's living-history in the world. Needless to say that, **the consequent to each of the above mentioned and so unique kinds of the historic 'Revolutions' during their specific timelines; the man's subsequent history has never remained the same and thus, heralding multiply far-reaching changes in the lives of, as well as, in the overall environments around the mankind**.

In these precise and multiple-contexts we may in addition, recall that during our earlier discussions on the multiple-aspects of the 'Industrial revolution' in the last seventh chapter itself we have since also noted another kind of perhaps, no less far-reaching the then historic emergence of the global **'Market-Revolution'** as well. And then consequently in-turn, we did also notice that, the systems of both national as well as, international trading activities too, got totally altered for each of the vast comity of the mutually interacting nations; **heralding totally new kinds of both, domestic and international market-economies in the world. Hence, these new kinds of the market-centric developments in the multiple-areas of trade and commerce in turn, perforce acquired altogether newer dimensions, having momentous socio-economic-cum-socio-political contents in both domestic as well as, international spheres and then in addition also specific to the international-levels, bringing-in numerous geopolitical contents too and thereby; their fallouts spilling onto the multi-dimensional arena of the complex-games of international power-politics as well**. At this precise point of our on-going discussions in these regards, as we may critically delve deeper into these no less important aspects, concerning the larger humanity in some global perspectives; we do find that **in international spheres, the simple economic considerations mostly get momentously superseded by the powerful geopolitical compulsions of international power-games and also, due to the numerous other geostrategic and politico-military considerations, on the part of today's powerful and economically leading nations and then; even for somewhat regressive and perhaps, unnecessary multiply-fixated few other nations of the world**. Interestingly, to contextually bring-in an analogy in this regard; we may also say that, in the vast and highly complex arena of the games of international power-politics; the various kinds of the interactive relationships, between the present modern era's big and smaller and also, between the stronger and relatively weaker Nation-States; are mostly akin to those of the bigger and smaller fishes in the world's vast oceans. In fact in these specific contexts, we may also say that, these interactive relationships between various nations in the world; are mostly piloted, as per the any internationally emerging situation-specific compulsions of the international diplomatic as well as geostrategic-cum-politico-military power-games, since always being played among the nations concerned, and hence precisely there; it is mostly the big and powerful nations who call the shots and also thereby, shape-up the decisions required and, so dictating the further actions to be taken by the concerned hapless nations. Hence, these are the precise ways and means available to the comity of nations, to compulsively let the global economic-order to function and to thrive or otherwise, in the today's world.

In the above paragraphs, we have indeed deviated from precisely the present subject-matters in hand, as it is needed to be emphasised that, the numerously multiple-kinds of the fallouts of the so many newer and also mesmerizing scientific-cum-technological marvels, since now continuously fructifying onto the man's expanding civilisational-canvas and specifically so, during **the presently on-going 'Post-Industrial Revolution' era** are as usual, not confined to the technological domain only, but are always spilling-over, not only into the any particular nation's internal spheres, but also into the much larger and the highly complex international geopolitical as well as, other domains as well. The reasons being that as usual these are always interrelated as well as, mutually-influencing too. Hence, in our contextually larger perspectives, we do very momentously find, that the above discussed so far latest scientific and technological marvels and their indeed mind-boggling fallouts are not only affecting the man, his life-patterns and his overall environment or so; but simultaneously his national and, international dimensions—concerning the much more important, the numerous kinds of geostrategic-cum-politico-military as well as, the much larger geopolitical aspects too—get greatly impacted, with numerous far-reaching consequences in these regards. So, these indeed looking quite different but, no less important aspects too perhaps, are needed to be discussed and put in their desired perspectives. Now as we move-on, we do also notice that in some altogether different perspectives the ground-level realities across the world, do remain far from being identical when other factors too, are taken into critical considerations along with the multiply positive and indeed, reassuringly far-reaching contributions of the very briefly above discussed and also, the much more numerous kinds of the scientific and technological developments, since being relentlessly achieved by the mankind, across the world. In these precise contexts and also, during any somewhat comprehensive as well as, critically analytical assessments therein, we do perforce have to include other dimensions too, such as globally always, the very fluid geopolitical and international power-politics situations, and then with respect to the any specific nation's own socio-cultural and the socio-economic-cum-socio-political factors as well. In fact, in order to facilitate the acceptance of science and then, positively imbibing their multiple-fruits for ensuring the optimal benefits all round, at both the human and social levels, each nation's both the socio-psychological as well as, the socio-cultural ethos too, do have major roles to play, for making them a substantial success or perhaps, towards regressively creating some obstacles or social-inertia-based resistance to change in these regards.

Precisely with respect to the above and accordingly, while looking into the above discussed specific aspects especially concerning the modern sovereign and democratic Indian Nation-State we do come across a very relevantly informative article, titled as, 'Science of the Times: From Abstract Concepts to Daily Life'; and published in the *Times of India* newspaper, dated 1st. March, 2000; and since written by **Professor J V Narlikar**, a very well-known Indian Scientist and learned Astrophysicist. And he in his this article has said that, "In late January (*i.e. in the year 2000*), an international meeting took place that brought scientists and science communicators together to discuss how science and technology can be integrated with the life of the common person. Although organised on a much more modest scale than the high profile Indian Science Congress that had been held in Pune three weeks earlier, this meeting nevertheless addressed practical issues related to reducing the gap between S&T and the prevailing ethos. As the participants from different countries shared their experiences, it became clear that despite the sweeping advances of technology, human societies are still a long way from achieving the scientific ethos. As far as the Indian ethos is concerned, the following experience of mine speaks for itself....when I was eating in the hotel coffee shop I was recognised again, this time by a diner in an expensive suit wielding a cellular phone. He sought my confirmation of what he already believed was going to happen, namely the catastrophic end of the world because of planetary alignment.... The contrast illustrates the fact that scientific literacy or the lack of it, does not depend on the social or economic strata. Nor does it seem to be related to education.... February 28, 1928, was the date on which C.V. Raman made the important discovery that was to be immortalised as the Raman Effect. February 28 has been used since 1987, for nationwide celebration of Science Day.... In short, students should be your prime target on Science Day.... Science Day also provides opportunities to scientists to reach out to public and the media. An open house organised by a typical research centre is means of showing the taxpayers how their contributions are put to use in R&D programmes. Popular lectures, films, demonstrations of equipment, and question-answer programmes are serious attempts to increase public awareness of the dual and interrelated forces of science and technology which are determining how we live. Media print as well as electronic can provide very effective means of communication. However, because Science Day falls a day before the annual budget speech every year, media too preoccupied with speculation over rupees and paisa, to do justice to an important occasion.... **Sure, one should project science seriously and factually, informing people that its benefits and pitfalls depend on how it is used. But all means should be tried to project it to the masses, for it is as much part of our daily existence as soft drinks and detergents. Actually it is much more so."** (*Hence, the central point in these regards remains that it all depends upon as to how prudently, the science and the multiple-fruits of the scientific and technological advancements are put to use, by the mankind. Now at the same time, with respect to these multiply far-reaching aspects, concerning any specific Nation-State as well as, even considering the entire comity of nations; there are many more and also, quite complex dimensions into all these, than just largely and indeed multiply, the scientific and technological considerations only; and at the same time always, they are all interrelated and also, mutually influencing too; in both their national/domestic and also, international perspectives.*).

(b) Dawn of Information Technology

Contextually in the earlier paragraphs, we have since been somewhat critically striving to survey; as to how and in what all manners, the various kinds of the history-making and, multi-disciplinarily on-going scientific-cum-technological progresses since indeed, the unprecedentedly momentous breakthroughs as well as advancements; have been greatly instrumental in triggering the **gradual dawn of the phenomenal 'Post-Industrial Revolution' Era**, broadly since the 1950s decade onward. Accordingly we find that, during these indeed, far-reaching and the phenomenally on-going multi-disciplinarily relentless, scientific and technological R&D efforts, the first and foremost **scientific knowledge-based breakthrough happened to be the multi-dimensional technological concepts of the engineering sciences of Electronics** and also, their subsequently very fast-track multiple applications in almost all the modern and vast areas of human endeavour.

Hence perhaps, it won't be any exaggeration to say that specifically in these regards, the scientifically phenomenal advent of the multiple-sciences and engineering aspects of the Electronics, was indeed a kind of the developmental watershed, which was centrally pivotal in heralding the dawn of the multiply-momentous Era of the 'Post-Industrial Revolution'. We have already since conceptualised that, universally always one thing as a kind of some new development in any field of the multiply on-going human-endeavours leads to another all the more newer development, or even a number of such breakthroughs and developments; and phenomenally these in turn, trigger their numerously far-reaching fallouts. Hence, all these are naturally many kinds of the developmental processes within the overall highly complex as well as, multi-dimensional phenomena of the mankind's continuously on-going scientific-cum-technological advancements onto its always expanding civilisational-canvas in the world.

Precisely in these regards and also, as already briefly discussed above; we do notice that, as the earlier on-going R&D efforts in the multiple fields of sciences and technologies; were instrumental in the far-reaching breakthrough and the consequent advent of the sciences of electronics and thereby, triggering another scientifically path-breaking birth of the computers and the computer-science as the mankind's multiply momentous technological equipment as well as, new kinds of the scientific-concepts with tremendous potential, and therefore similarly further the computer and the allied fields of the computer-science, soon turned out to be the centrally pivotal in the phenomenally very fast and subsequent birth of indeed, the highly complex and the mind-boggling scientific discipline of **'Information Technology'** and thereby heralding the dawn of the **'I.T. Age'**, along with the then approaching 21st century in the world. And therefore, this quite unprecedentedly phenomenal as well as indeed, the very mind-boggling coming of the 'I.T. Age' has totally transformed the world into its altogether a new kind of incarnation, being so overwhelming for the man; both technologically as well as, environmentally in the overall human-terms too; and thus tremendously altering his ways of life, the situation-based various kinds of his new and emerging opportunities and challenges and thereby; revolutionarily transporting him into his absolutely new and technologically enchanting arena of the **'Cyber-World'**. Interestingly, as we may delve further deeper into these I.T. related aspects; we may notice that consequently, **the mankind finds its multiply vast world indeed, transformed into a very captivatingly sensitive kind of the 'Global-Village'**. No wonder that, the man finds himself unprecedentedly as well as, multiply overwhelmed; by his own indeed, such scientific breakthroughs and thereby, his multiply far-reaching and the on-going technological innovations and also advancements, in the today's super high-tech world.

And further consequently in these regards **more the man strives to gain some kinds of the master-control over his own now, these so enchanting scientific advancements as well as, the superbly ongoing technological innovations/achievements, the more he finds himself overwhelmed and even perhaps at times, 'digitally-conquered', by these relentlessly fast manifesting advents and at the same time along with, their further refinements-cum-fine-tuning of such multi-dimensional breakthroughs in the already above mentioned highly complex-cum-gigantic and also, continuingly enlarging scientific-cum-technological spheres; during the present super high-tech age of the modern times**. And at the same time, these do have always their multiply numerous dimensions too. Now specifically in the above multiple-contexts, as we may precisely delve deeper into these unprecedentedly momentous aspects, we do find that these multiply numerous dimensions, do also appear to be grouped into two principal segments; and that being: (1) largely the **civil areas**, concerning the mankind's multiply on-going endeavours and, (2) the no less consequential as well as, highly complex and multi-dimensional areas, concerning the **matters-military** of the mankind.

Now in the above precise contexts, first coming to the highly complex as well as, multi-dimensional **civil-areas** we do notice that other than the contextually obvious—the multiply scientific and technological considerations—these do have numerous other dimensions as well and hence these being pertaining to the socio-psychological, socio-cultural, socio-economic and then, even socio-political aspects too. And then at the same time in these highly complex and multi-dimensional civil-areas the above noted precise aspects, do also have always the proactive dimensions concerning, the nationally domestic as well as also, the geopolitically international connotations as well. And these connotations happen to be since being mostly positive as well as negative too, depending upon principally the triggering intentions behind various kinds of the uses and applications of these super high-tech marvels, now so fast manifesting onto the mankind's continuously expanding civilisational-canvas itself. Hence in other words, we do notice that such multiply overwhelming as well as, the relentlessly fructifying numerous scientific-cum-technological fruits and fallouts are always affecting and influencing the lives and also, the overall multi-dimensional environments around the mankind, in so many far-reaching and also the multiply history-making manners; across the accelerating-cum-fast-moving times in the present modern-era world.

Further in these precise I.T. related areas, the **momentous advent of the various kinds of communication satellites and their appropriate stationing, into relatively the little outer-space and then at the same time, multiply on-going advancements in the highly complex domain of the software-developments** have always been decisively the central fulcrum, towards now, since the ongoing manifestations of the above mentioned as well as much more and, so numerous and also enchanting scientific-cum-technological marvels on the part of mankind itself. Anything and everything concerning the satellite-based connectivity and also, proactively establishing the fail-safe as well as, multiple kinds of the far-end communications, spanning over the continuing '24×7×365' timeframe have been rendered possible by indeed, **the multi-dimensional advent and also, relentlessly on-going advances in the areas of space-technologies**. And accordingly in these multi-dimensional areas of the present super high-tech era of the continuing human endeavour some knowledge-based and indeed, gigantic data-storehouse kinds of the techno-commercial Companies, such as the **Google**, **Yahoo** and so on, have also come-up contextually, on the present global-scene in these regards. Contextually, more of these aspects will be briefly dealt-with, in the forthcoming sub-chapter on the **'Coming of Space-Age'** in these regards.

Precisely and also, very informatively in these multiple-contexts, **Professor N.D. Batra**, author of since the *Digital Freedom* and also, who teaches communications and diplomacy at the Norwich University, USA has very reflectively written on these aspects, in his two short and indeed, intellectually very crisp articles titled as: 'When Our Daily Lives Become Hyper-Personal' and, 'Global Civic Management Through Technology'.

And accordingly therein **Professor N.D. Batra** has written that, "The idea behind surf and search record-keeping is to create patterns that anticipate the searcher's behaviour and provide the information quickly next time he looks for an item. More than that, based on searching patterns, Google, according to CEO Eric Schmidt, could answer hypothetical questions, for example: Where should I invest my bonus, in mutual funds or a savings certificate?.... Freedom in cyberspace is becoming 'curiouser and curiouser', as 'Alice in Wonderland' would have cried. The more you have it, the more you lose it.... **The internet has created a new media environment, that not only enables people to communicate, discuss and exchange information, give and receive feedback, but it also provides an interactive collaborative experience, doing things together, when words**

become deeds and speech becomes action. Networked computers, the building blocks of the internet, are much more than mere productivity tools and informatics systems. Unlike the traditional media, **they are capable of creating cyber-environment** that can be designed to be persuasive, and can motivate people to disclose personal information and change their social behaviours. Social engineers, especially those who serve the authoritarian governments, **know the potential of the internet for creating an illusion of boundless freedom only to entice people to digital chains**.... The challenge for software programmers whether they work for Google, Yahoo or some authoritarian regime is **to build a network architecture that enables the designing of the virtual environment** to enable people, for example, to buy, sell, invest, obey the supreme leader; or even induce good social and personal habits: such as not to drink and drive, not to have unhealthy social behaviours. **Computer codes have the potential force of law, according to Stanford Law professor Lawrence Lessing, and programmers could bypass the government and, in a manner of speaking, take law into their own hands. Or the government could also use the programmers to stifle dissent**.... So many teenagers go astray in cyberspace because it lets them assume fake identities and gives them freedom to pretend what they fancy themselves to be. Some of them became victims of con-men and predators, who too assume identities desirable for their teenage victims. **An avatar can be very deceptive.... The playfulness of virtual environment, an environment of 'Be what you want to be', creates a pleasurable experience, a sensuous flow, in which we feel in control of our environment that real life might deny us. Those who spend too much time in Second-Life might forget what real life is. In search of a fake identity, we are in danger of losing our real identity** (*No wonder that, the 'Internet addiction' is capable of transforming citizens into the '**netizens**' and hence, it is rightly perceived that, this addiction is no less powerful and also, could be no less damaging than the alcoholism or even drug-addiction, for the mankind.*).... **The use of technology for a purpose different from its original intent is called bricolage. It is the practice of tinkering and using the available material to create something new or solve a problem in a unique way**. Americans are not the only bricoleurs, though they are most fertile and creative.... **If somehow we could know a way of tuning up and enhancing the nervous system to a higher level, the reality would change**. Think of the time when zero was discovered and, the subsequent development of decimal system; and how that might have changed the perception of reality, by subjecting the physical phenomena to measurement.... Measurement is actuality, which can be enhanced by technology. When Galileo's telescope was embedded into the human nervous system, the view of the universe changed and the earth ceased to be a stationary planet. Or consider some modern political events. Had satellite pictures revealed the existence of the Soviet's missiles in Cuba in the 1960s, Americans wouldn't have perceived the threats to the country. **If the remote sensing technology could pick-up the sights and sounds of human sufferings (*Well, tomorrow sometimes most probably, this would also become equally possible for the mankind.*)**, it could become possible to know how the Chinese have been brutalising the Tibetans and their culture. The Tibetan cultural pogrom will find full description. Description either leads to action or generates guilt (*Well, can it do so everywhere and, under the all kinds of situations*?). Guilt demands action and atonement. **That's how human consciousness has been evolving perhaps.... Communication technology—satellites, radio, television, cell phones and the internet—which made free flow of information possible**, has castrated communism.... **History teaches a lesson that it is through the control of narratives that the powerful, the ruling classes, exercise their hegemony**. I have always wondered how Britain ruled so successfully over India for about two hundred years. The British succeeded to a great extent by supplanting the native stories and narratives with their own literature and legends, so that knowing the legend of King Arthur was deemed more civilised than knowing the legend of King Ashoka.... China has begun to transplant its own stories and historical descriptions upon the young Tibetans, and by the time they grow up into adulthood, their reality will clash with the truth held by their parents.... By denying the Tibetans access to their language and culture and access to communications technology, China is doing unto Tibetans what Americas have done unto Native Americans.... **It is time for global civic engagement, through all available means of diplomacy and technology**." (*Yes it is indeed so, and hence undeniably now; an unavoidable imperative as well, for the entire mankind in the world.*).

Since man **entered into this so phenomenal the present I.T. Age—his life as well as his 'world' too, have now got incomprehensively transformed, in so much far-reaching manners that he is now continuously struggling to be able to acquire some sort of a 'controlling-grip', over these highly complex and dynamic phenomena, around his life and also, around his multi-dimensional environments as well**. In any case and in these precise contexts, as per the mankind's overall civilisational-template; all these have broadly been multiply consequent creations and the fructifications of his, on-going endeavour and the fallouts of his relentless R&D efforts in the highly complex, multi-dimensional as well as, interdisciplinary-oriented, vast areas of the sciences and technologies; now since pertaining to super-age of the present modern-era. And yet he finds himself so outflanked by the sudden as well as, phenomenally very fast manifestations of the all kinds of surprises in these regards that perhaps, he sees himself totally overwhelmed and interestingly in a way and, though at times, being somewhat unaware too even finds himself 'digitally-conquered' as well.

We may recall that in the different multiple contexts from time to time, in course of our continuing discussions in the previous chapters, we have since been contextually emphasising that, within the overall perspectives of the man's highly complex as well as, multi-dimensional living-history perhaps anything and everything in these regards, have been interrelated, even if at times quite remotely and so, mutually-influencing as well. Hence, the so much **paths-breaking dawn of 'Information-Technology'** as well as, indeed the phenomenally multiple fructifications of its immensely far-reaching scientific-cum-technological fallouts too have momentously together and also, simultaneously been equally consequential; for the both civil applicability areas, and the domain of matters-military as well. In fact, as we may care to contextually delve little deeper into these aspects, we may also recall that from the very beginning of our multiply continuing discussions, we have since been quite discerningly noticing that, along with the dawn of the various human-civilisations themselves; the mankind's so many numerous kinds of the creative developments as well as new innovations from time to time, onto its expanding civilisational-canvas have always been equally consequential and applicable too, in both the civil and military areas. The somewhat comprehensive and chronological listing in these regards would understandably be colossal and hence, there appears to be no need to go into these data and also the allied vast historic facts. In order to contextually illustrate the relevant points here, we may prefer to somewhat critically revisit the multiply complex and vast aspects of the momentously historic, the earlier **'Industrial-Revolution'** since already briefly discussed by us, in the beginning of the last seventh chapter itself.

And accordingly, as we may proceed to critically delve into these aspects, we would also be very interestingly observing that the so many unprecedented, phenomenally paths-breaking-cum-multiply-facilitating as well as, bringing-in the mankind's far-reaching civilisational-enrichments too—the immensely numerous kinds of the fructifying fallouts and the very many socio-economic, socio-psychological-cum-political and also, the geopolitical consequences—had in fact, always both civil and military dimensions, concerning their eventual applications in the multiple areas of human-endeavour. Hence naturally, the vast politico-economic, geopolitical as well as, the military histories of the both (1914-18) First World-War and (1939-45) Second World-War and then, of the so many other wars too, do substantially and also historically quite in detail, corroborate these always valid and also, specifically precise civilisational-perceptions in these regards. And in addition to all these, **as we may proceed further and hence, specifically encompassing the domain of matters-military, we in a way fairly and overwhelmingly, do also notice certain and indeed, unprecedented as well as momentously altogether the new dimensions having been added into since, the already highly complex and multiply numerous aspects of the modern warfare; consequent to the above discussed numerously far-reaching fallouts and also, the momentous manifestations of the multiply versatile I.T. tools of the presently continuing 'I.T. and also, quite accelerating fructifications of the expanding cyber-age'; in the today's super high-tech world**. Hence, while critically looking into all these multiply consequential aspects; we may also say that, God knows as to what all lies in the future ahead, for the mankind.

At this precise point, we quite obviously do also notice that the scientific-cum-technologically inevitable but also, highly consequential as well as, indeed very far-reaching dawn of now the **'Cyber-Age'** has multiply brought new dimensions and challenges, into already the highly complex areas of the matters-military. Hence it is quite natural that, the learned experts in the multiple-fields of the electronics, I.T. and Cyber-Age, modern defence technologies' related R&D scientists and engineers space-technologies' related learned specialists and also, the scientists working in the multi-disciplinary fields of the classical sciences, are since relentlessly engaged in achieving the R&D oriented breakthroughs in the multi-disciplinary basic sciences as well as, the material-sciences throughout the world; in taking care of the military-threats and challenges of the concerned nations. Consequently, many leading military professionals and technical experts too, from time to time have been having their own and rather some sort of the 'closed-doors' techno-scientific 'brain-storming' as well, and also publishing highly informative articles in their various well-received Journals in these regards.

Precisely in these contexts therefore, **Brigadier Abhimanyu Ghosh** (Retd) of the Indian Army, has written that, "History has shown that wars are won not only by the courage and strength of the men and women on the battlefield, but also, by the evolving technology adopted by the military. In the cyberspace, two distinct regimes emerge for the military. The first regime is the open network with its inherent risks and vulnerabilities which is essential for collaboration, information sharing, logistics etc. The second regime of closed network meets the demands of security, speed of operation and availability of information at the right time and place. **To maintain confidentiality, integrity and availability, closed network is air gapped from open network. In addition to these networks, the military is also exposed to commercial off the shelf (COTS) hardware and software products for wireless, cellular phones, computers, networking equipment etc**. Therefore, increased dependence of the military on civilian cyberspace capabilities have inherent risks and make them vulnerable to cyber-attacks by attack vectors, which are prevalent in commercial/open cyberspace.... **Cyberspace includes not only the internet but also telecommunication, cellular phone technologies and wireless data services. The technologies involved such as, bandwidth, interoperability, processor speed, functionality and security vulnerabilities have evolved over time**.... Cyberspace has emerged as a major new environment for political and military competition, and would necessitate political and military intervention to protect economic and informational interests vital for national security. The challenges for military, in the era of on-line connectivity and information flow, are unique and require a great amount of coordination among the nations." (Ghosh, October-December 2011, pp. 517 & 526)

But precisely in the above multiple-contexts in these regards and, also as per the presently prevailing and internationally ruthless and highly competitive geopolitical conditions onto the man's expanding civilisational-canvas, are hardly conducive to even imagine some preliminary amounts of trusts among the concerned nations and thus, any meaningful fructification of the above mentioned 'any great amount of coordination among the nations' appear to be just 'tall-orders' in the presently prevailing overall international geopolitical scenario. In fact to begin with, essentially the basic prerequisites are first and foremost, the mutually some tenable trusts among the nations to enable them to speedily cultivate the necessary geostrategic-cum-politico-military vision, to be able to collectively promote the much desired—larger well-being of the mankind across the world. **But the emerging scenario is quite different and hence, always the all sorts of lingering trust-deficits among the nations concerned; have perforce induced almost all of them, to get obsessed with multiply enhancing their cyber-warfare related techno-military capabilities. Though at the same time, given the overall and also the always scientifically improving and advancing dimensions of the cyberspace-related technologies it is also equally understandable that, the enhancements of such techno-military capabilities of any particular Nation-State do also fall short of fully ensuring the geopolitical-cum-military security, against any possible hostile cyber-vectors and their sudden attacks. Hence it is no wonder that, these multiply emerging aspects in these regards; have thrown-up altogether now, a totally new kinds of the previously even unimaginable security-risks and challenges for any concerned Nation-State in the present ultra-high-tech world.**

Hence understandably as rather a fait-accompli, the military professionals and also, the matters-military related scientists and technocrats all over the world in fact, geostrategic-cum-politico-militarily have also accepted the cyberspace, as a new and quite mid-boggling arena of modern warfare. Further quite appropriately therefore, one of the national-security experts of the USA; **Richard A. Clarke, the author of *Cyber War* has defined 'Cyber-Warfare' precisely as, 'the actions by a nation-state to penetrate another nation's computers or networks for the purpose of causing damage or disruption'**. Understandably therefore, for fighting any Cyber-War; the concerned specialised technical-branches of the military of a nation, and also along with the active cooperation from the concerned, both the Government and private-sector Institutions too; need the **'Cyber-Warfare tools', i.e. the Cyber-War weapon-systems and weapon-platforms** to rather promptly and effectively neutralise any such hostile action against own nation and then at the same time, also launch any further befitting counter-offensives, since as deemed mandatory in any given situation. Strategically therefore in addition to all these, the properly as well as proactively well-stationed—the 'alert-cum-vibrant systems' in these regards; are simultaneously meant to multiply act also—as the much needed effective deterrent too. Hence at this precise point, we now may recall **the very ancient-era, but also indeed the timeless strategic perceptions of Sun Tzu of China—that always the best victory in any war—happens to be the one since achieved by without actually fighting that war**.

And needless to say that the above mentioned since **the timeless and also universally, the so well-known strategic-thoughts of the legendary Sun Tzu, do remain equally applicable to now rather the somewhat 'frighteningly' challenging as well as, the highly complex, ultra-modern, and also the super high-tech domain of the Cyber-War**. Further at this precise point of our ongoing discussions; it may appear to be strategically imperative to also highlight another no less consequential ground-reality that, the above discussed domain of the geopolitical-cum-militarily so 'threatening' Cyber-War; is no longer remaining something akin to the recently bygone era's

'Cold-War' since, so ruthlessly fought between the erstwhile two global Super-Powers. But the multiple-aspects of Cyber-War now, have already got quite menacingly transformed into a kind of the 'Hot-War'. And there are some recorded examples of the clashes of Cyber weapon-systems of two or more hostile nations since, having already taken place from time to time, in the different geographical locations of the world. Hence in these multiple-contexts, it is also equally true that in terms of the present-era's overall fast-moving global geopolitical-cum-politico-military scenario these are now indeed the highly dangerous war-like developmental trends onto the mankind's expanding civilisational-canvas itself. Therefore quite befittingly, **Lt. Gen. Harbhajan Singh**, a retired Signal Officer-in-Chief of the Indian Army, in one of his recently published article, on 'Cyber Warfare: Dangerous Trends' have already given two of the some important Cyber-Attacks and also in geopolitical and military terms, briefly discussed them too, as great warnings since specifically meant for the modern democratic Indian Nation-State. He has cited the first example as 'the cyber-attacks on Estonia in May 2007, targeting Estonian Government and the private web-sites, on a much larger scale and which happened to last for nearly a month. And in addition then as a second example in his article he has briefly and also very relevantly recorded some important details concerning the Stuxnet worm attack on Iranian nuclear plants/establishments, in September 2010 and then, these attacks were able to intrude into some other countries as well. Hence, scientifically as well as geopolitical-cum-militarily now, we cannot deny that in the larger strategic perspectives in these regards; the '*Cyber-War chickens*' have already since quite menacingly, 'come home to roost'.

In the above mentioned article the author, **Lt. Gen. Harbhajan Singh**, has written that, "International and political turbulences have at times led to hacking/defacing of website across the world. Israeli and Palestinian hackers have launched attacks on websites of each other and, India and Pakistan hackers have done the same.... The Chinese have been suspect for a number of cyber-attacks in the USA, India and some other countries. These attacks, however, have been rather limited in scope and for short periods, i.e. interruptive.... 'Kill Switches' and 'Backdoors' secretly installed in chips can disable, betray and even blow up the equipment in which such (*doctored*) chips are used. 'Backdoors' provide access to the equipment for malicious actions. Chips, microprocessors and printed circuit-boards (PCBs) on which these are embedded, contain millions of components and circuits. There is, therefore, ample scope to slip-in secret codes.... It is possible to make chips which after specified usage, become ineffective or on external/programmed command, carry out malicious actions. Devices in equipment can be remotely switched on and off whether connected or not to internet. Even a soldier on a PCB can act as an antenna, making possible intrusions, from mobile phones/drones/satellites and aircraft.... Considering the above, the world is now looking at a new era of **'weaponised cyber attacks'**. This is likely to multiply the power of cyber-attacks to much higher and dangerous levels.... **Cyber-attacks will not only be able to shut down power grids, air traffic control, banking systems, nuclear facilities and other critical infrastructure, but also cause damage to electronics and other hardware and, corrupt the software controlling them. Such attacks will therefore become more lethal and destructive, and corruptive of data and programmes. The scale will vary depending on the resources with the attacker**. Most of the preparatory work is being done by such countries on 24x7 basis.... **As there is nuclear deterrent, similarly there could be cyber/electronic deterrent too, because electronics are all pervasive and nothing works without electronics. No doubt, nuclear attacks mean tremendous physical casualties and damage, but electronic attacks will immobilise functioning of a country i.e. cause paralysis of the nation, and seriously damage its electronic infrastructure. Any deterrent has to be plausible and demonstrated**. Spurts of cyber-attacks between nations having inimical relations are efforts towards this end and also to test their techniques and responses from other side. Efforts are being initiated to reach international consensus, and may be agreements on some rules on use of cyber weapons. **The biggest problem is that, it is very difficult to locate the source of cyber-attacks.... The USA is seriously thinking of declaring cyber-attack as an act of war, depending on its severity, as it can cause destruction/disruption comparable to a hostile conventional attack and would take retaliatory actions—weaponised or others. This is a very significant development and, shows how seriously the US takes cyber warfare.**" (Singh, April-June 2011, pp. 228, 229, 230, 231 & 233)

In addition to the above in the highly complex and also, now the already very fast expanding domain of this totally new kind of the warfare i.e. 'Cyber-War'; another mindboggling and new dimension has been added to it. And that dimension contains the **'Anti-Satellite Weapon-Systems'** and thereby possibilities of the highly complex and dangerous war-fighting in the relatively an outer space, with main geostrategic-cum-politico-militarily strategic thrusts towards destroying the adversary's satellites in the space itself. No wonder that to begin with, even any possible imagination of this sort of any possible scenario appears to be quite frightening; to any person who does have some basic ideas concerning the **'I.T. and Cyber-Age technologies'** and also, all the more so for all those who are since actively as well as professionally engaged in the I.T-cum-Cyber related activities, both in the civil and military spheres. All the leading and powerful nations in the world, both globally as well as regionally too who have any worthwhile since, multi-disciplinary scientific wherewithal as well as the technological infrastructure, so essential in these regards, have been relentlessly working in developing, updating, and fine-tuning their multiple capabilities to forge the shields for ensuring the defences of their own multi-utility satellites and simultaneously also, strategically perfecting the arsenals of their 'Anti-Satellite Weapon-Systems' to act as some kind of the deterrence or if the need arises ever, then even to undertake effective counter-offensives against one's any indiscreet-cum-mischievous adversary. And India too understandably, remains focused and active in these regard.

In the above multiple-contexts, **Group Captain RK Singh** of the Indian Air Force and presently a research fellow with the USI, New Delhi; in his article on 'Indian Anti-Satellite Weapon: Necessity, Urgency and the Way Ahead' (pages 85 and 86) has written that, "When so many satellites are there in space, their safety and security becomes crucial for India's progress and sustained growth, especially when India is facing a hostile adversary Pakistan in the West and not so friendly a neighbour in China in the North and East. **The security of Indian satellites have become a matter of concern, ever since the shooting down of a defunct Chinese weather satellite by China's anti-satellite (ASAT) test in January 2007.** The test opened a new flank of vulnerability in India's $12 billion (Rs 60,000 crore) space infrastructure. China's alarming test spurred India's quest for a satellite-killing system.... Offence is the best form of defence. Apart from the military, even a common man understands this axiom. However, when transposed to the national level (*i.e. regarding India's.*), it is surprising that the 'killer instinct', which pumps the adrenaline and justifies one's belief in 'offence is the best defence', appears to be amiss. Perhaps that's the only justification for the country's political leadership to withhold the decision of developing its own ASAT programme, in spite of the fact that the country, as per the top defence scientists, does have the capability of developing an ASAT weapon for India, as and when the situation demands. In this regard **Dr VK Saraswat**, the Director General of Defence Research and Development Organisation (DRDO), has gone on record while briefing the reporters at Thiruananthapuram in January 2010, 'India is putting together building blocks of technology that could be used to neutralise enemy satellites. We are working to ensure space security and protect our satellites. At the same time, we are also working on how to deny the enemy access to our space assets."

Well, some peaceniks understandably perhaps, may even say that these extremely menacing developments across the world and that too, onto the vast, multi-dimensionally complex, and also the mankind's very fast expanding and advancing wherewithal of the sciences and technologies, since relentlessly being used as platforms for the developments-cum-perfecting of the multiple kinds of **'Cyber-War weapon-systems' and also their delivery-systems**, onto the intended targets are nothing but rather an altogether new kind of the indeed, much more hyper-dangerous—the ongoing phenomenon of the **'maddening arms-race'**, manifesting now more prominently, with dawn of the 21st century itself. These well-intended and also quite perceptive kinds of arguments in these regards, are indeed understandable and well-taken; but at the same time, while confronting the overwhelmingly powerful and since always prevailing ground realities; the essentially cold strategic prudence does compel us to realise that, neither the world or the human-beings are essentially and evolutionarily made that way, within the always ongoing evolutionary phenomena as well as command, of the infinite mother-Nature. **And hence rather ironically, we may also say that collectively in any manner perhaps, the man does have certain inherent 'limitations/ shortcomings' which perforce make him to fall quite short of sincerely accepting and imbibing the above noted indeed, well-intentioned and perceptive arguments—in his principally the ruthless trade-cum-commerce oriented as well as in addition also, the overwhelmingly matters-military oriented too—the multiply-competitive thoughts and actions**.

Thus we may contextually recall **the famous saying of General George Washington that, 'if you want peace, be prepared for the war';** and hence herein remains the universally practicable 'wisdom' of keeping and maintaining the requisite kinds of the need-based viable deterrence in these regards, by perhaps quite compellingly not disturbing the indeed, very fragile 'applecart' of some sort of the overall geostrategic-cum-politico-military 'balance of power' in today's world. And contextually therefore in the broader perspectives, it does remain quite an enduringly 'queer situation' for the mankind in today's world. But that's what realistically it is and at the same time it is also equally true and also, as borne out by our above discussions in these regards; that the highly complex and, multi-dimensionally as well as, inter-disciplinarily ongoing advances in vast domain of the ultra-modern sciences and technologies as since relentlessly being triggered by the man's so multiply continuing and focused R&D activities, have brought such overall emerging scenario concerning the vast domain of 'matters-military' that the entire gamut of warfare appears to be somewhat 'overturned', and thus opening new avenues and the techno-military wherewithal for fighting any modern-cum-futuristic war. Precisely in the above multiple-contexts, and also specific to the Indian parameters; **Lt Gen PC Katoch** (Retd) of the Indian Army, in his article has written that, "The regional security environment surrounding India today, includes failed and failing states. We are faced with nuclear threat, missile threat and cyber-threat, cross border terrorism, infiltration, demographic assault, conventional threat and insurgencies. Besides, we are battling continuing asymmetric wars waged by both, China and Pakistan. The 27 odd terrorist organizations operating in India, including the Maoist insurgency are open to exploitation by our adversaries. The US thin out from Af-Pak region and increased Chinese forays into POK, Pakistan, South Asia and the Indian Ocean Region (IOR) will enhance collusive threat from China and Pakistan.... Listing out all the technological advancements with military applications by say year 2025, would require many pages, but existing improved assault rifles will perhaps include phased plasma guns. Plasma weapons already reported in Russia, focus beams of electromagnetic energy produced by laser or microwave radiation into upper layers of atmosphere to defeat targets flying at supersonic or near-sonic speeds, bumping the targets out of trajectory. Laser weapons would be fielded on land, sea and air. Our DRDO is developing a Laser Dazzler for police forces that will impair vision temporarily to control unruly crowds. DRDO's Laser Science and Technology Centre (LASTEC) is developing a **vehicle mounted gas dynamic laser-based Directed Energy Weapons (DEW) system**, named 'Aditya' as a technology demonstrator. A 25-kilowatt laser-system is also under development for hitting a missile in terminal phase at a distance of 5-7 kms. DRDO identifies DEWs, along with space security, cyber-security, and hyper-sonic vehicles as future projects. MoD's 'Technology Perspective and Capability Roadmap' identifies DEW and anti-satellite (ASAT) weapons as thrust areas over the next 15 years. However, given the track record of DRDO, how much they will actually deliver by 2025 is anybody's guess.... **While there was much talk of stealth helicopters in the recent US Navy Seals raid in Pakistan to kill Osama bin Laden, the US has already developed a Reusable Space Plane (X-37B)**,—another step in weaponising space. In addition, a powerful conventional weapon (Prompt Global Strike), as alternative to nuclear warhead, is under development, which can travel halfway round the world from launch to target in less than 30 minutes, using missile launch, release of hypersonic gliders and eventual release of 1000 pound deep penetration bombs." (*Need we say more on the entire gamut of the projected—overall futuristic scenario—concerning the fast-moving and global geostrategic-cum-politico-militarily oriented 'power-games', among nations during the present highly chequered times?*). (Katoch, July-September Issue, pp. 305-308)

In our multiple-contexts interestingly here, as we continue to somewhat critically review these overall so far-reaching and multiply prevailing scenario in these regards, we also quite engagingly notice the **powerfully decisive and also altogether new dimensions that are highly secretive and also, the no holds barred ruthless domain of the 'Intelligence world'. Contextually now, as we continue to delve little deeper into the multiple-aspects of Cyber-War; we find equally overwhelming cyber sciences and technologies-triggered manifestations of some new dimensions, related to the intelligence-world.** Needless to say that, these do throw-up altogether new challenges too in these regards. As usual, the today's sole super-power the USA again, does remain far ahead of her any possible rival and at the same time in spite of all sorts of the multi-dimensional superiority perhaps, in any conceivable area of the ongoing human-endeavour and also since, so far commanding some kinds of an unassailable decisive edge, as per any critically evaluating parameters, she with her typical and also any developing situation-based strategic prudence, is just not taking these issues and the related challenges lightly.

In precisely the above multiple-contexts, one news-item as since appeared in an Indian Daily, The Sunday Times of India, Ranchi, dated 16th June 2013 has reported that, "***(the name of the Chief of US NSA, i.e. National Security Agency), is General*** **Keith Alexander**. He is one of the most powerful men in Washington and possibly one of the most dangerous men in the world. He is the Director of the US National Security Agency (often called the 'No Such Agency' for its paranoid secrecy), the chief of the Central Security Service and commander of the US Cyber Command. He also commands the US Navy's 10th Fleet. The 24th Air Force and the Second United States Army.... James Bamford, profiling the secretive Alexander in *Wired* Magazine, writes, 'The military has for years been developing offensive capabilities, giving it the power not just to defend the US but to assail its foes. Using so-called **cyber-kinetic attacks**, Alexander and his forces now have the capability to physically destroy an adversary's equipment and infrastructure, and potentially even to kill.' **It was Alexander's team in the NSA, in partnership with the CIA and Israeli Intelligence, that was behind the 'Stuxnet virus' attacks on Iranian centrifuges in 2010—something that the Obama**

administration acknowledged in 2012 (*So that's that.*). More Stuxnet like attacks are expected to take place in the future.... Stuxnet, says Bamford, was just the beginning. The NSA has recruited thousands of computer experts to give an edge in cyber warfare. The Pentagon's budget request of $526.6 billion allocation for cyber ware, up $1 billion from last year.... Bamford likens Alexander to legendary FBI Director J Edger Hoover, a man who was suspecting of maintaining secret dossiers on a wide-ranging group of people, from politicians to civil-right activists, from media personnel to movie stars. 'We jokingly referred to him as Emperor Alexander—with good cause, because whatever Keith wants, Keith gets', says one former CIA official who agreed to speak on condition of anonymity. 'We would sit back literally in awe of what he was able to get from Congress, from the White House, and at the expense of everybody else,' quotes Bamford.... Alexander's control centre is Fort Meade, Maryland. The budget indicator of the growing scope of the NSA's operations is the growth of private security contractors' offices around Fort Meade. General Dynamics, the defence contracting company opened up a 28,000 sq foot facility close by. Science Applications International Corporation (SAIC), another defence contractor providing engineering and systems integration services, set up a Cyber Innovation Centre there. CSC established a 'virtual cyber-security centre' that included its labs at Hanover, Maryland. Two years ago, Boeing also jumped on the bandwagon with Cyber Engagement Centre.... Alexander was appointed the head of the NSA by Donald Rumsfeld, George W Bush's Secretary of Defense, in 2006. Since that time, he has seen the NSA grow in capability and power. **He is expected to retire in 2014, but by that time, he would have changed the face of modern warfare through the NSA, and completed his mission**." (*Well, personalities apart, what all so overwhelming matter in these substantially paths-breaking as well as indeed, so momentously history-making—the 'Cyber-World' related and also, matters-military-oriented phenomena—are specifically the policies, programmes and also the institutions in the US since multiply and also so proactively, engaged in the Pentagon and elsewhere in these regards*).

Well in any case, that is not all since at this precise point; we may also contextually prefer to note as to, what the Ex-US President himself had to say on these indeed very weighty matters. And accordingly thereby, we do further find an interesting as well as very informative news-item appearing as '*Obama defends NSA surveillance as transparent*' as dateline Washington in *The Times of India*, 19 June 2013, therein, it is reported that. "**President Barak Obama**, under attack for allowing the controversial National Security Agency Surveillance programme, has defended the policy by saying the covert operations were 'transparent' and had disrupted multiple terrorist plots, not just in the US but overseas....'The one thing people should understand about all these programmes, though, these have disrupted plots, not just here in the United States but overseas as well,' he said. He added that while other factors were at work, 'we are increasing our chances of preventing a catastrophe alike that through these programmes.'....'My job is both to protect the American people and to protect the American way of life, which includes our privacy. And so every programme that we engage in, what I have said is 'Let's examine and make sure that we're making the right trade-offs,' Obama told the popular 'Charlie Rose' show on PBS channel.... Obama has strongly defended the effectiveness of the surveillance programmes in heading off threats to the US."

Consequent to the above quoted news-items at this juncture, we have now interestingly been brought to the geopolitically, two very important and relevant points and that being: (a) **multiple-aspects of the 'heavy burdens' of being the sole ultra super-power in today's so fast-moving and super-high-tech world** and then also, (b) **the multiple-aspects of the prevailing menace of all kinds of international terrorism and the extremism**. In these respects since centuries one thing remains clear and irrefutable that, irrespective of any other considerations, the highest national interests of any particular Nation-State always come first and supreme. This holds good for any big or small, poor or rich; developed, developing or even underdeveloped; global super-power, regional-power, or even any very small or no power—Nation-State in the world; quite interestingly similar to any particular individual—for whom, the self comes first and foremost and every time. But at the same time, concerning the comity of nations in the world, there are certain apparently hidden other dimensions too. In fact, what may not be perhaps of any worthwhile considerations of some smaller nations would at the same time in some other respects indeed, be of substantial considerations for any particular super-power in the world. Further, these aspects do also assume all the more weightage and complex textures in view of also the multi-dimensionally very fast-manifesting advancements, in the so far-reaching disciplines of the ultra-modern sciences and technologies as well as, their multiply complex fallouts, in the already very vast domain of the matter-military in today's world. And at the same time, these specifically national-cum-international defence-oriented aspects in case of any particular super-power Nation-State, do perforce assume lot of extra weightage and also, the extra-dimensional significance too. And precisely that's what **Barak Obama** was trying to spell-out and geopolitically emphasise, during one of his news-conferences and, as since contextually reflected in the above quoted newspaper report.

And needless to say that, we have not seen that since the present fast-advancing and also, so momentously flowering cyber-age has ironically indeed, manifested the dawn of cyber-war related cyber weapon-systems and weapon-platforms too. Any concerned nation's multi-dimensional cyber-security aspects as well as, the protection of its cyber assets since already functionally deployed and stationed on the land, in the sea, in the air and then, also in the outer space have quite agreeably become very sensitive and hugely bothersome too. If all these are agreeably so problematic state of affairs in general terms then contextually one may imagine the multiple-dimensions and their momentous reach too, of the overall prevailing scenario in the USA in these regards! Now interestingly at this precise point of our multiply continuing discussions we may even perhaps in a lighter vein, add that the contextually above discussed indeed the momentously extra-large dimensions-cum-burdens concerning the sole super-power's cyber-related threat-perceptions as well as, the consequential wherewithal of the cyber-security aspects, may not be confused or even perhaps inadvertently bracketed with erstwhile the globally colonial-era, then prevailing mind-set concerning the **'white-man's burden'** among the European imperialists and also, as since historically portrayed in **Kipling**'s novels and literary story-writings. But what indeed very menacingly gets highlighted here is that, the multiply so bothersome dangers of the internationally lingering all sorts of terrorism and extremism have entered into the present modern-era's ultra-high-tech 'Cyber-World' too. These are naturally bringing-in signals of great dangers, to mankind across the world. These further enlarge the multiple aspects of since the always very complex—internal, regional and even globally mandatory tasks of threat-perceptions and then, ensuring the necessary counter-measures. And that's what we do contextually find, and as emphasised in the above quoted paragraphs and the news-items from different sources that, any nation with some substantial consequence and also having the requisite scientific-cum-technological wherewithal, are now found to be relentlessly occupied with the above mentioned activities. And obviously therefore contextually in the global perspectives, the USA is again found to be momentously in the decisive lead in these regards.

In these indeed momentously multiple-contexts again, another news-item concerning President Obama's press-conference, along with the German Chancellor **Angela Merkel** in Berlin while defending the USA's NSA, has very emphatically said that the NSA has been able to quite effectively foil more than 50 potential terrorist plots in as many as 20 countries, **including India**. Further, informed sources however, have said that India was figuring rather somewhat prominently in the list of those 20 countries or so and at the same time, one such India-related plot indicate that, it was around the successful Commonwealth Games in Delhi in 2010. While geostrategic-cum-politico-diplomatically rather, sharply defending the NSA's such comprehensive cyber-surveillance ongoing programme, President Obama said that it saved precious and innocent human lives in more than 20 nations, by effectively neutralizing some highly dangerous evil-designs of the various internationally deployed terrorist groups; and also then added that, 'this is not a situation in which we are rifling through ordinary e-mails' of the huge number of citizens in the US or elsewhere. In addition then, he also clarified it as a 'circumscribed narrow surveillance program' adding that it was a program closely supervised by the courts to ensure that any encroachment of privacy is strictly limited. On the other side, another very interesting news-item also reveals that, some Australian scientists are working on **Quantum Mechanics**, which is being explored for its possible applications to computing and thus, to develop the most powerful and impenetrable cyber-security system, ever scientifically conceived. Well at the same time, it is also very appropriately said that anything which floats, can indeed be sunk. Rest we may prefer to leave to the future, to unfold to us.

Now we return to deliberate upon **now the so engaging new and emerging dimensions of the 'Globalisation'** and accordingly we do also notice that historically so paths-breaking phenomena of the globalisation, in any case being never simplistic and linear and so, do also have their own complexities as well as, numerous dimensions too. Hence, in these precise contexts, **Robert L. Lindstrom, the founder president of C60 Communications, USA**, during an interview few years back in Singapore (*Ref: The Times of India, Patna/Ranchi, dated 12th November 2003*) has very reflectively said that, "Globalisation is not like communism or capitalism. A globalised environment is neutral and today we are connected like never before. Of course, there is a downside to it—so what do we do, cut our wires, throw out the satellites, withdraw flights? The digitisation of information—has transformed everything into zeros and ones, the binary language of computers....there is a proposal to link Chennai and Singapore through underwater fibre optic cables. So don't underestimate the potential of connectivity. Also, remember that digital codes are neutral—it's how we as human beings react and relate to technology that reality matters. We have to put science in context. Technology causes people to converge. Letters (snail mail) connect you very slowly. Digital technology connects you instantly—this is what rich media and broadband is all about. The outcome of all this convergence is predictable—but as happens in any new situation, there is a natural and automatic self-organisation process that is set in motion.... **It is true that a globalised world is perfect for terrorist networking. It is equally true that it enables us to fight anti-social networking with the same neutral tools**.... As technologies converge, so do people, creating chaos and complex systems. But just as rhythm emerges from diverse music patterns, technology convergence produces cogent new systems of communication.... Most of us like things to be predictable, controllable, secure,—we fear the unknown. But the fact is we don't know what will happen in the future—we never have known. In a less connected world change is gradual but in a globalised world, the pace of change is astonishing.... Now, we are completely wireless. And the new technology allows us to sit in a park or forest in the world of nature, and still stay connected with work or home. **The relationship between people, the planet and technology is amazing. The planet today is in distress with all the abuse being heaped on it by insensitive development. People too, have become increasing lonely, carried away by the freedom afforded by the new technology. But**

the same technology is also making it easier for us to reconnect with nature and strengthen relationships with people."

In these highly consequential and multiple-contexts therefore we do also notice that almost every aspect of the man's, not only the overall but even almost the every moment of his daily life too; has indeed been overwhelmingly affected by the multiply phenomenal dawn of the I.T. Age, as well as also, by the amazingly fast and, multiply continuing fructifications and, along with also the increasingly on-going sophistications of the I.T. technologies-cum-I.T. tools, all around the man across today's modern world. Hence, these precisely have been the truly powerful, indeed multi-dimensional and also, all-encompassing impacts of this I.T. Age on the mankind and its lives, across the world. Man's both short and long terms as well as, his daily lifestyles too and, almost everywhere across the world, has totally changed and so his whole world and his overall environments too, have also been irreversibly changed. Things till yesterday, being part of largely the areas of science-fiction as well as also, since belonging to rather the fictitious world of the man's some regularly off and on manifesting, 'flights of his imagination' only; and at the same time even probably much more so, have now not only come out to be true but also, even become indivisibly essential segments-cum-components of the chores of his daily-life everywhere and also, under the all kinds of situations.

There is another quite new and, earlier somewhat unforeseen and also, no less influencing dimension to all these unexpectedly fast-paced and, the on-going I.T. age related developments, i.e. the rate of the technological-cum-systemic **'phenomenon of obsolescence'** in these regards. There is another all the more amazingly invisible component to the any and every kind of the I.T. related functioning system and, that happens to be precisely the continuously advancing **'systems of software'**.

The software is mostly specialised and, exclusively given to any particular job-cum-application and hence conceptualised, developed and thus, especially designed for that specifically intended task; by the concerned **'team of software-experts'**, in the I.T. related software-development Companies, since engaged in the those specialised areas of business. And these new and highly specialised areas of business activities are since related to the multiply-complex jobs of software-development assignments, capable of data-processing, developing systems for the data-storage and safe preservation, establishing multiply different kinds of the I.T. based communication-links and allied systems and also in addition any other kinds of the specifically needed job assignments, as given by any concerned customers or their Companies, since engaged in altogether very many different areas of the trade, commerce, and also the industrial activities. One phenomenally voluminous off-shoot concerning these new, since already well-established and fast expanding I.T. related business areas; has been the often talked about: the techno-commercial **'systems of the B.P.O.'**, i.e. 'Business-Processing Outsourcing'. India is one of the lead players in this particular I.T. related business world, which has now acquired huge volumes not only in jobs, but also in plain monetary-terms and turn-over's, and at the same time phenomenally manifesting its international dimensions too. These dimensions now, not only pertain to the international trade and commerce but also; highly sensitive and multiply geopolitical-dimensions as well. We would more relevantly be discussing these momentous aspects in the last sub-chapter of this eighth chapter itself.

Thomas L Friedman, a well-known contemporary American thinker, author and, three times Pulitzer Prize winner has written a very thought-provoking book, on precisely this subject and very appropriately, the title of this book is, *The World is Flat: The Globalized World in the Twenty-*

First Century. This book has been crafted in a very engaging and lucid style and also therein the highly consequential as well as, multi-dimensional phenomena, concerning the multiply strategic and also, technological-cum-commercial 'flattening of the world' in today's super high-tech I.T. Age with great detail and thus, providing the readers, indeed very informative-cum-fascinating panoramic views in these regards.

Richard L. Friedman (pages 9, 10 and 11), "I found that trail again on my journey to Bangalore in February 2004. Once I did, I realized that something really important had happened, while I was fixated on the olive groves of Kabul and Baghdad. Globalisation had gone to a whole new level. If you put *The Lexus and the Olive Tree* and this book together (*interesting, isn't it?*) the broad historical argument you end up with is **that there have been three great eras of globalisation**. The first lasted from 1492—when Columbus set sail, opening trade between the Old World and the New World—until around 1800. I would call this era **Globalisation 1.0**. It shrank the world from a size large to a size medium. Globalisation 1.0 was about countries and muscles. That is, in Globalisation 1.0, the key agent of change, the dynamic force driving the process of global integration, was how much brawn—how much muscle, how much horsepower, wind power, or, later, steam power—your country had and how creatively you could deploy it. In this era, countries and governments (often inspired by religion or imperialism or a combination of both) led the way in breaking down the walls and knitting the world together, driving global integration. In Globalisation 1.0, the primary questions were: Where does my country fit into global competition and opportunities? How can I go global and collaborate with others through my country? The second great era, **Globalisation 2.0**, lasted roughly from 1800 to 2000, interrupted by the Great Depression and World Wars I and II. This era shrank the world from a size medium to a size small. In Globalisation 2.0, the key agent of change, the dynamic force driving global-integration was multinational companies. The multinational went global for markets and labor, spearheaded first by the expansion of the Dutch and English joint-stock companies and the Industrial Revolution. In the first half of this era, global integration was powered by falling transportation costs, thanks to the steam engine and the railroad, and in the second half by falling telecommunication costs—thanks to the diffusion of the telegraph, telephones, the PC, satellites, fibre cable, and the early version of the World Wide Web. It was during this era that we really saw the birth and maturation of a global economy, in the sense that there was enough movement of goods and information from continent to continent for there to be a global market, with global arbitrage in products and labor. The dynamic forces behind this era of globalisation were breakthroughs in hardware—from steamships and railroads in the beginning to telephones and mainframe computers toward the end....that right around the year 2000 we entered a whole new era: **Globalisation 3.0**. Globalisation 3.0 is shrinking the world from a size small to a size tiny and flattening the playing field at the same time. **And while the dynamic force in Globalisation 1.0 was countries globalizing and the dynamic force in Globalisation 2.0 was companies globalizing, the dynamic force in Globalisation 3.0—the force that gives it its unique character—is the new found power for *individuals* to collaborate and compete globally**. And the phenomenon that is enabling, empowering, and enjoining individuals and small groups to go global so easily and so seamlessly is what I call the *flat-world platform*.... **The flat-world platform is the product of a convergence of the personal computer** (which allowed every individual suddenly to become the author of his own content in digital form) with fibre-optic cable (which suddenly allowed all these individuals to access more and more digital content around the world for next to nothing) with the rise of work flow software (which enabled individuals all over the world to collaborate on that same digital content from anywhere, regardless of the distances between them). No one anticipated this convergence. It happened—right around the year 2000.... But Globalisation 3.0 differs from the previous eras in not only in how it is shrinking and flattening the world and in how it is empowering individuals. It also is different in that Globalisation 1.0 and 2.0 were driven primarily by European and American individuals and businesses. Even though China actually had the biggest economy in the world in the eighteenth century, it was the Western countries, companies and explorers who were doing most of the globalizing and shaping of the system. But going forward this will be less and less true. Because it is flattening and shrinking the world, **Globalisation 3.0 is going to be more and more driven not only by individuals but also by a much more diverse—non-Western, non-white—groups of individuals. Individuals from every corner of the flat world are being empowered.** Globalisation 3.0 makes it possible for so many more people to plug in and play and, you are going to see every color of the human rainbow take part." (*But at the same time, as we continue to further delve deeper into these multiply phenomenal, history-making and indeed, substantially incomprehensive and also, superbly fast the I.T. Age-related developments; we do also very fascinatingly notice that, all these multiply on-going phenomena are not only concerning the individuals now perhaps, appearing to be quite a bit masters of their own destinies and dreams; but these phenomenally advancing scientific and technological marvels in the I.T. areas, are also revolutionizing the entire gamut of national governance and, those in vast arena of the international-relations as well.*). To precisely illustrate this particular perception and thus, rendering an example; Thomas Friedman in his already above quoted book, and therein, in the page 248; has written that, "Before Colin Powell stepped down as Secretary of State, I went in for an interview; which was also attended by two of his press advisers, in his seventh floor State Department Suite. I could not resist asking him about where he was when he realized the world had gone flat. He answered with one world 'Google'. **Powell said that when he took over as Secretary of State in 2001, and he needed some bit of information—say, the text of a U.N. resolution—he would call an aide and have to wait for minutes or even hours for someone to dig it up for him. 'Now I just type into Google—UNSC Resolution 242—and up comes the text', he said. Powell explained that with each passing year, he found himself doing more and more of his research', at which point one of his press advisers remarked, 'Yes, now he no longer comes asking for information. He already has the information. He comes asking for action."** (*So that is that, just some glimpses of the so far-reaching and, also the, so phenomenally paths-breaking as well as, history-making: the multiple-kinds of the on-going fructifications-cum-fallouts of the present super high-tech I.T. Age. Well, as the earlier 'Industrial-Revolution' changed the world forever; similarly the I.T. Age too, has now again momentously so much, changed the world forever. Precisely in fact, these are few of the numerous phenomena of mankind's continuing civilisational advancements.*).

(c) Scientific and Technological Multiple-Advances; Nanotechnology and Cryogenics

The briefly above discussed and the phenomenally advancing as well as, multiply fructifying aspects of the I.T. Age in turn have also been further, triggering again, the continuously on and on, advances in the so many other areas of now, the ultra-modern breakthroughs in the multiple-fields of sciences and technologies. At present now, perhaps the most fascinating aspects regarding the latest scenario in these regards; in the present horizon concerning the levels of the continuing advances in the multiple fields of the sciences and technologies in fact, have been even beyond the various kinds of principally the multiple utilities-cum-applications-oriented developments and thereby, further triggering the still expanding—the vast and also multi-dimensional domain of the I.T. Age itself. So in other words, **the always relentlessly ongoing multi-dimensional as well as,**

the high-end R&D endeavours in the so many scientific-cum-technological disciplines have enabled the mankind to indeed, phenomenally advance through the multiply fascinating paths of the I.T. domain and thus, in their evolutionary continuity now, to also enter into even the much more mesmerizing and mind-boggling corridors of the Nanotechnology and the Cryogenics as well. Hence, in this ultra-modern era's multiply overwhelming and, also the latest kinds of the fast advancing and fructifying fruits and fallouts of the multi-disciplinary sciences and technologies, the larger humanity is certainly not expected to get somewhat 'statically-fixated', only with always the well-accepted—super high-tech I.T. Age and thus also required to look beyond all these and thereby, towards the no less phenomenal dawn of the Nanotechnology and Cryogenics onto the man's expanding civilisational-canvas. Interestingly at the same time in fact it is also equally true that little earlier, the multiply fast emerging advances in the momentously far-reaching as well as, the highly consequential history-making—the Computer Science and Information Technologies together—have been playing decisive-roles in triggering the above mentioned dawn and also, the fast-paced and highly consequential advances now, concerning further-on—the super high-tech disciplines of Nanotechnology and the sciences of Cryogenics, as these aspects are always interrelated and mutually influencing as well. Hence, these phenomenal aspects truly now in fact are no less fascinating and at the same time, scientifically engaging in the multiple domains of relentless endeavours of the vast fraternities of scientists, technologists and the engineers across the world.

Scientifically at the same time it happened to be apparently logical that the phenomenal advent and thereby, **the very fast-paced advances in indeed, the highly complex disciplines of the Nanotechnology also soon triggered the superb technological skills towards developing as well as, the fine-tuning of technologies of the Nano-manufacturing techniques and processes concerning the multiple kinds of Nano-tools and Nano-machines for their various kinds of the previously unimaginable applications and far-reaching utilities, towards the multiply valuable services of the mankind**. We may contextually note here that, earlier the conventionally quite broad-based processes of the manufacturing of various kinds of goods, tools and other technical items in different kinds of industries across the world were considered to be principally belonging to the engineering fields, and thus no wonder that consequent to these developments, the then relatively new disciplines of the Industrial and Production Engineering came into being. But since then, it is also equally obvious that the time has begun 'to move still faster' and thus, many kinds of the newer and newer scientific-cum-technological advancements, have since also been relentlessly fructifying onto the man's expanding civilisational-canvas. **And the latest developments in these regards perhaps, have been the multi-dimensional scientific knowhow of Nanotechnology and the Cryogenics.**

At the same time in addition to these, we also quite fascinatingly notice that unlike the industrially traditional manufacturing activities, the Nano-manufacturing techniques and processes are not confined to engineering only, but these are truly much more multidisciplinary as well. Hence in addition to the mechanics of industrial, mechanical and electrical engineering, metallurgy, sciences of ferrous and non-ferrous materials, production-processes' optimization and so on, these also momentously concern with the scientific disciplines of chemistry, biology, physics, nuclear-sciences and the knowhow regarding bio-medical engineering too. But contextually at this precise point, we do also no less engagingly find that in these multiple respects indeed the pivotal contributions of the computer science, the computers and the super-computers, and also the knowhow of Information-Technology too cannot be minimized. Hence in our broader perspectives; while critically delving into these, we may somewhat visualise up to some extent, the enormities of their complexities as well as, the aspects of indeed, their scientific-cum-technological multi-disciplinary dimensions. Now at this precise point of our ongoing discussions we may also notice that and, specifically in these regards too; that the seeds of all these indeed, the unprecedentedly mesmerizing scientific-cum-technological developments onto the mankind's expanding civilisational-canvas; were historically sown-in during somewhat the mid-twentieth century times and even little earlier, when the truly paths-breaking as well as indeed, **the scientifically phenomenal 'Theory of Relativity'** of the legendary scientist **Albert Einstein**, saw the light of the day and thereby, triggering a series of the very fast-paced breakthroughs in the domains of nuclear-physics, nuclear-chemistry, the quantum-physics, and also the many other allied sciences as well.

These timeless, universal as well as, the momentously historic aspects indeed in contextually different manners, systems, and methodologies—are also equally true in case of now the scientifically so mesmerizing and, the ultra-modern domains of the still fledgling, but also the quite accelerating and the fast advancing—multi-disciplinary Nanotechnology and the sciences of Cryogenics as well. In these multiple contexts, it is important to note that just few years back a National Conference on the 'Emerging Technologies: Nanotechnology and Cryogenics' was held on 30-31 October 2009, at the G.L. Bajaj Institute of Technology and Management in Greater Noida (India).

Professor, **Dr. S. Prasad** of the same Institute, was then the senior and presiding member of the Committee of Editors responsible for the Proceedings of this Seminar; titled as, 'Advances in Nanotechnology and Cryogenics'. Dr. Prasad, who has been a mentor of this author in many ways like an exceptionally affectionate elder brother, has been very kind to share a printed copy of the Proceedings of this Seminar for this book. Hence in the 'Preface' of this book itself the Committee of Editors, consisting of **Dr. S. Prasad, Professors B. Sahay, A.K. Pandey and, S.K. Mishra**; with indeed their precise and the futuristic-vision, have very perceptively said that, "Nanotechnology is one of most significant emerging technologies in today's world. In India too, premier institutions such the Indian Institutes of Technology (IITs) and central government funded research organisations are taking leading roles in bringing nanotechnology applications in engineering, technology, agriculture and medical sciences. Cryogenics has already taken up various important roles, such as cryogenic engines, industrial use of LH2 and LN2 (liquid Hydrogen and liquid Nitrogen), cry coolers etc. In fact, the field of material science has been enriched by both nanotechnology and cryogenics. In the future, this field will have more meaning for engineers and scientists to meet various challenges facing our world. It is, therefore, essential to create awareness amongst the students and faculty of institutions both at the graduate and postgraduates levels, since they form the bulk human resource to take up the future revolutions."

It is true that the multi-dimensional and indeed, the phenomenally fast emerging sciences of Nanotechnology and the Cryogenics are too complex and specialised for the laymen and, also for conventionally the general kinds of scientists and technocrats to be able to even preliminarily acquire some grasp and basic understandings of these latest kinds of the ultra-modern breakthroughs, in the multiple domains of altogether the latest frontiers of technologies and the sciences; in the presently ongoing super high-tech era, in our fast-changing world.

It is also contextually quite interesting to note that the other paragraphs and pages of these highly specialised articles, since contained in the above referred book, are plainly meant for the concerned scholars and experts in these latest scientific areas and hence obviously, fall outside the conceptual framework of our continuing discussions as well as, our principal subject-matters. Further, in these larger perspectives, it is also generally clear that any fructification of the intended specialisation in any particular field of the expanding human-knowledge of the sciences and technologies the sharply-focused and exploratory studies with specifically mandatory academic backgrounds too, are always the necessary perquisites to be able to proceed further with unwavering perseverance. These are precisely the reasons, but without any kind of the intended or otherwise any negative connotation in this regard as perhaps generally perceived—as to why any kind of the academically steadfast process of **specialisation is known as, 'knowing more and more of less and less'**. Understandably therefore, we may also note that this academic methodology for achieving specialisation in a systemic as well as scientifically formulated manner, in any specifically identified/ defined field or branch of the human-knowledge, being always so vast and complex is considered to be relatively of much later origin, onto the expanding civilisational-canvas of the mankind.

Now coming back to indeed, the quite engaging subject-matters of the Nanotechnology and Cryogenics, **Professors R. Sharma, R. Aggrawal and A. Sehgal** in their technical paper (G.L. Bajaj Institute of Technology and Management, pages 103 and 104) have said that, "The atomic structure of the matter was quantitatively revealed by X-ray diffraction in the early part of the 20th century. It led to the currently much-celebrated structure determination of DNA by Watson and Crick in 1953, which was based on the stunning X-ray diffraction patterns obtained by Rosalind Franklin. However, diffraction patterns, for all the quantitative information they contain, are not 'real space' representations of the matter. They reveal ordered structures through which one cannot pinpoint the position in space of a given atom, molecule or cluster. To achieve this, one requires new developments in microscopy. In the second half of the century, a range of microscopes were developed, which are capable of producing atomic-resolution images of atom at surfaces. The most innovative of these developments, was the scanning tunneling microscope, which not only imaged individual atoms but can also be individually manipulated. Synthetic chemists and scientists have demonstrated a remarkable ability to synthesize large quantities of desired products ranging from tenth of a nanometer including metal clusters, antibiotics, pigments, esters, polymers and a wealth of others. However building a single molecular structure atom-by-atom is impractical and expensive, but it excited the thought-process of many scientists around the world. This is how a new buzzword appeared in science and, nanotechnology was born.... Nanotechnology is the new arena of science and engineering. Nanotechnology is being taken as the new technological revolution. It is so profound that it will touch all aspects of the economy and society. Technological optimists are looking forward to transform the world to a better place, using nanotechnology. As it will reduce the production costs of all the goods and services and also, permit the developments of new products; along with self-assembly modes of production. Further, it also allows miniaturization of the control systems. These social effects are inherent part of such revolution. In such 'Nano-society', energy will be clean and abundant, the environment will be repaired to a pristine state, and any kind of material artifacts can be made with minimum cost. Space travel will be cheap and easy, distances will be a thing of the past, and we can all expect to life for 100 years.... We should also distinguish between Nano science and nanotechnology. Nano science is a convergence of physics, chemistry, material science and biology, which deals with the manipulation and characterization of matter on length scales between the molecular and the micron size. On the other hand, nanotechnology is an emerging engineering discipline that applies methods from Nano science to create new and advance products."

Further in continuity, **Professors N. Kumar, A. Yadav and V.K. Yadav** in their scientific paper on 'Nanotechnology and its Applications' (pages 116, 118, 119, 120 and 121) have brought out that, "Nanotechnology is the essence of present time.... Nanotechnology is defined as fabrication of devices with atomic or molecular scale precision. Devices with minimum feature sizes less that 100 nm (*i.e. nanometer.*) are considered to be the products of nanotechnology. A nanometer is one billionth of a meter, and is the unit of length that is generally most appropriate for describing the size of single molecule. The nanoscale marks the nebulous boundary between the classical and quantum mechanical worlds; thus realisation of nanotechnology promises to bring revolutionary capabilities. Fabrication of nanomachines, nanoelectronics, and other nanodevices will undoubtedly solve an enormous amount of the problems faced by mankind today.... Computer chips are rapidly shrinking according to predictable formula (by a factor of four every three years—Moore's law). According to SIA's extrapolation of formulas such as this one, it is expected that the size of circuits within our chips will reach the size of only a few atoms in about 20 years (*That's it and precisely, this aspect of continuing scientific-cum-technological advancement in this regard; happens to be one of the phenomenal contributions of the nanotechnology to computer science.*).... The other field in which molecular scale manipulation of matter is receiving abundant attention is medicine. Since all living organisms are compound of molecules, molecular biology has become the prime focus of biotechnology. Countless diseases have been cured by our ability to synthesize the small molecules commonly known as 'drugs' that interact with the protein molecules that make up the molecular machinery that keeps us alive. Our understanding of how proteins interact with DNA phospholipids and other biological molecules in what allows such progress.... Living systems are able to live because of the vast amount highly ordered molecular machinery from which they are built? The central dogma of molecular biology states that the information required building a living cell or organism stored in the DNA. This information is transferred from the DNA to the proteins by the process called transcription and translation. These processes are all executed by various bimolecular components, mostly proteins and nucleic acids.... Molecular biology is a field in which the study of these interaction has led to the discovery of numerous pharmaceuticals that have been enormously effective in curing diseases.... Nanotechnology is expected to have impact on nearly every industry. The research community is actively pursuing hundreds of application in nanomaterials, nanoelectronics and bionanotechnology. Most near term (1-5 years) applications as nanotechnology are in the form of nanomaterials. These include material such as lighter and stronger nanocomposites, antibacterial nanoparticles, and nanostructured catalysts. Nanoscience is an inter-disciplinary field that seeks to bring about mature nanotechnology."

As we move along, we find that **Professor J. Husain** of Aligarh Muslim University, Aligarh (India) in the same seminar, (pages 129, 130 and 132); has also said that, "Nanotechnology in India is receiving adequate support from the Government. Academia has also been quick to jump on the band-wagon. Already dozens of M.Tech degree programmes have been started by various Indian universities, both in the government as well as private sectors. However, Indian business has been slow at capitalizing on the nanotechnology catalogs on the internet (http://www.nanotechproject.org/inventories/consumer) lists 24 countries, including a few developing ones, which manufacture nanotechnology products, but India is absent from the list.... Why is Indian business slow in catching on to nanotechnology? Part of the reason is that the minds of the Indian business community are still under the glamour of the information technology revolution. Another important reason is the lack of entrepreneurship courses being embedded within the M.Tech. Courses on Nanotechnology—that have been started throughout the country. Contrast this with the education of students of other engineering branches, which predominantly consists of B.Tech, followed by MBA. The two degrees combined help to expose the students to both the technology as well as, the entrepreneurship sides.... **Nanotechnology** is the technology at the sub-micron level. Milli is one thousandth and micro is one millionth. Nano is one billionth. At the nano level we are speaking in terms of large single molecules or a group of a few dozen atoms. Not only is the nano level very small, but the properties of matter at the nano level are quite different from the microscopic level.... Until micro level, laws of classical physics are good enough, but at nano level, quantum physics only can explain phenomena.... Swords manufactured in Damascus from 10th century to as late as 1750, have been found to contain carbon nanotubes (strongest and stiffest material known). The sword makers knew nothing about nano-science, but were accidently utilizing nanotechnology. Nanomaterials are not difficult to produce, but their characterization is hard.... Nanocity—http// www.nanocity.com—Nanocity spans 11138 acres of flatland located less than 25 kms east of Chandigarh.... It is public/private partnership between Sabeer Bhatia (52%), Parsvnath Developers (38%) and Haryana State Industrial and infrastructure Development Corporation (HSIIDC) (12%). Similar PROJECTs have been announced in Himachal Pradesh and Gujarat, in collaboration with Nanobioysm USA, whose founder is an NRI by name Anita Goyal."

Now, while futuristically visualizing the prospects concerning the scientific possibilities of further advancements in areas of technologies of computing and multiple data-processing, along with their global-connectivity too, **Professor N. Cheshan** of the Department of MCA, Ideal Institute of Technology, Ghaziabad (India); in his paper, since titled as 'Emergence and Development of Cloud Computing' (pages 137 and 152) has said that, "Computers have become an indispensable part of life. We need computers everywhere, be it for work, research or in any such field. As the use of computers in our day-to-day life increases, the computing resources that we need also go up. For companies like Google and Microsoft, harnessing the resources as and when they need it is not a problem. But when it comes to smaller enterprises, affordability becomes a huge factor. With the huge infrastructure, come problems like machines failure, hard drive crashes, software bugs etc. This might be a big headache for such a community. I think that Cloud Computing offers a solution to such problems and overcome to this situation.... According to me, the Cloud computing is a paradigm shift in which computing is moved away from personal computers and even the individual enterprise application server, to a 'cloud' of computers. A cloud is a virtualized server pool which can provide the different computing sources of their clients. Users of this system need only be concerned with the computing services being asked for. The underlying details of how it is achieved, is hidden from the user. The data and the services provided reside in massively scalable data centers, and can be ubiquitously accessed from any concerned device all over the world. Cloud computing is the style of computing, where massively scaled IT related capabilities are provided as a service across the internet to multiple external consumers and are billed by consumption.... Cloud computing is a powerful new abstraction for large scale data processing systems which is scalable, reliable and available. In cloud computing, there are self-managed server pools available which reduces the overhead and eliminates management headache. Cloud computing services can also grow and shrink according to need.... Also it is new emerging architecture needed to expand the internet to become the computing platform of the future."

Now coming to the no less scientifically mesmerizing aspects of the sciences of Cryogenics, for the general readers like this author; we do also find that a research scientist, **Dr. R.B. Saxena**, Director of the national Physical Laboratory in his paper on 'Application of Cryogenic Technology with a Focus on Recent Research and Development in the area of Cryogenics and Superconducting Magnets at National Physical Laboratory, New Delhi', (pages 171, 172, 175, 176 and 180 of the same published proceedings) has said that, "Cryogenics is a rapidly developing technology embracing the production, control, and applications of very low temperatures. Below –100 Degree

C, i.e. between the temperature of liquid oxygen (90 K) and absolute zero (*–273 degree C.*), numerous phenomena occur, which are of immense importance to science, industry and medicine. Cryogenics is the means of exploiting these phenomena. Since cryogenics is mainly concerned with temperatures below 100 K, the gases encountered are those with boiling points below this level. Although temperatures within a minute fraction of absolute zero have been achieved in laboratory equipment, it is necessary in many applications to drop below 4.2 K—boiling point of liquid helium. Amongst these gases with suitable properties from the engineering point of view are those, found naturally occurring in the atmosphere at ground level.... At cryogenic temperatures, physical properties of many materials undergo dramatic changes. For instance, the thermal conductivities of some metals and alloys change considerably and specific heats of all solids become progressively smaller as temperature is lowered.... At temperatures near absolute zero, the electrical conductivity of some metal and alloys abruptly becomes infinitely high, so that an induced current is self-sustaining. Superconductivity is one of the most important of cryogenic phenomena. As temperatures get very low, the thermal motion of atoms within a material is greatly reduced. The noise level in electronic circuits drops. Transient molecules or compounds occurring in chemical reactions can be frozen.... (*In the multiple-areas of the medicine and biology*): **Cryobiology** is the science, which deals with the study of living organism at temperatures below the normal freezing point of water. The physical events, which occur when individual cell or group of cells are low in temperature until they freeze, are very complex. A particular cell has a so-called survival signature with an optimum cooling rate for maximum cell viability and an otherwise considerably increased probability of cell death on either side of this optimum rate. In relation to the survival rate, the different research areas in cryobiology can be characterized as follows:

(a) **Cryopreservation**—the freezing of cells or organs within the narrow band of cooling rates on either side of optimum rate;

(b) **Cryosurgery**—in which cooling rates exceed the optimum rate as much as possible, in order to destroy infected tissue and;

(c) **Cryofixation**—the principal objective of which is the retention of the morphology within a cell by freezing it as rapidly as possible; cooling rates at the upper end of the survival rate.

Concerning the Cryogenic Engine

The various rocket propulsion systems namely, solid, storable liquid, semi-cryogenic and cryogenics are used successfully in the launch vehicles. Among these, the cryogenic engines using Liquid Oxygen (LOX) and Liquid Hydrogen (LH2), gives maximum propulsion efficiency. Initially, cryogenic propulsion system was used for the upper stages of satellite launch vehicle, where the propulsive energy is used to impart sufficient velocity to the payload to achieve the mission. Currently cryogenic engines are used both as upper stage and also used as booster engines. Cryogenic propellants provide highest performance in terms of specific impulse, which is defined as kilogram of thrust, per kilogram per second of propellant." (*In our specifically larger perspectives, we would be further discussing these aspects too; when we come to forthcoming sub-chapter on the 'Coming of Space-Age, little later*).

In the last, in these indeed magnificently ongoing multi-dimensional R&D efforts in the scientifically high-end areas, concerning the so fascinating Nanotechnology and Cryogenics, **Professor (Dr) S. Prasad**, in his paper titled, 'An Overview of Salient Landmarks in the Applications of Cryogenics as well as Historic Achievements' and (pages 186 and 187) of the already above referred published proceedings in these regards has mentioned that, "The author was in Nagpur, having close interactions with IITs, defense research organisations as well as industries. During that period, there was a big tragedy in 'Nagpur Engineering Company', manufacturing huge size of water valves, used to be treated at –80 degree C, in a room size compartment (*i.e. Environment Chamber*) being cooled by liquid nitrogen. Cryogenics are obviously produced at very high vacuum. Due to some leakage, the chamber exploded like nuclear bomb, killing 2 DGM and 3-4 engineers at the spot. Due to some reason or other, news was suppressed; but the author got intuitive decision in his mind that, it was high time that a serious awareness needed to be created in engineering students, faculty, serving engineers, industries, as well as society at large.... 'Nanotechnology and Cryogenics'.... Wider applications of cryogenics may get limited to a few highly advanced areas, of course they are indispensable. **Superconductivity is much awaited accomplice, if energy-environment is viewed from 'Vision – 2020', authored by the living legend, Dr. A.P.J. Kalam**.... Nanotechnology is more in focus today and that's why, large number of papers were presented during this seminar, and the cryogenics took back-seat, except NPL (*i.e. India's National Physical Laboratory*) headed by Dr. R.B. Saxena.... It is because of such situation, the author offers to venture this paper for the benefit of readers and researchers: (i) **Nanotechnology**: Particles of the size of one billionth of a meter.... How do we get such particles? Fragmentation to that extent is going beyond molecule size as well as atomic size. We have mentioned earlier that Einstein theory 'E = MC square' is more relevant today. Free electron as in nuclear science; leads to conversion of thorium to plutonium and, uranium—by involving mass. The author does not have authority to talk more about the nuclear physics and chemistry. But one thing is clear, as to how we are getting carbon nanotubes, by impinging nanoparticles from agricultural wastes in the polymers. (ii) **Cryogenics**: Unlike nanotechnology, which is still at the laboratory scale, cryogenics have taken up very important ventures in the space ships, cooling by LH2 (*i.e. Liquid Hydrogen*), in a number of sophisticated power generating units, cryogenics treatment of metals, cry coolers; preservation of cells, human organs, blood samples etc. As said earlier, production of cryogenics is by applying Stirling Cycle. It is significant to note that we have been able to get to a temperature, as low as –270 degree C (approx)."

We may also recall that while discussing the various scientific and technological developments; we have since noticed that most of the advancements thereof have been having dual applications, i.e. both in the civil and military areas. And hence in the similar manner, the fruits of both Cryogenics and Nanotechnology too though presently at their beginning stages only, do also have the same dual uses. **It fact, it is also true that the sciences and technologies of Cryogenics have precisely been instrumental in developing and perfecting the technological wherewithal for sending satellites into the space through launching of the space-rockets, and then in addition also towards developing the on-going mastery over the multiple kinds of missiles since, equipped with different kinds of the warheads. Broadly in the same manner, the multiple-aspects of Nanotechnology too in the years to come, would soon be contributing towards the ongoing furtherance of matters-military as well.** Here too we find that, these being some of the man's always ongoing multiple progresses onto his expanding civilisational-canvas; are also interrelated and mutually influencing with so many other multiple-kinds of earlier accomplished developments and in addition, simultaneously encompassing both the civil application-areas and then also, those in the matters-military domain as well. The security, wellbeing, and the urges for multiply continuing creativities of the mankind along with also, those of the various human-civilisational aspects too are fully as well as reassuringly need to be ensured, against any possible threats emerging in the world.

(d) NBC Warfare and Fast Emerging-cum-Magnifying Phenomena of Multiply Both, the Intra-national and International Terrorism and Extremism

In fact, the 'NBC' warfare means: the Nuclear, Biological and Chemical warfare. Hence, **these specifically much more lethal, devastating and even, in a way abnormally inhuman means of waging, whether limited or otherwise; wars can also quite justifiably be termed as demonically great negations of the human-civilisations, onto the mankind's expanding civilisational-canvas itself**. Contextually, one may also quite emphatically say that due to whatever kinds of perhaps, any pressing reasons or compulsions at any given point of time in any warlike situations; whosoever resorts to the first use of these indeed draconian weapon-systems, anywhere against any kind of adversary does commit totally inexcusable sins, against the entire humanity in this world. We all are aware that especially, soon after the end of so far the most violent and devastating (1939-45) Second World-War; there have been seminars and conferences of the people from various walks of life from across the world, from time to time under different 'banners' or in different contexts to put some kinds of somewhat effective checks on this menace, but nothing much has effectively been achieved in these regards. Even United Nations too, are quite seriously seized of these highly sensitive, volatile and also extremely dangerous subject-matters but with hardly any substantially tangible gains for larger good of the humanity as a whole.

And consequently in these specific contexts even now, the overall scenario doesn't remain encouraging at all and thereby all sorts of perhaps irresponsible proliferations of these so far most inhuman weapon-systems, are going-on regardless of any geopolitically civilisational considerations, on the part of anybody who—regionally or internationally—matters in some manners in these regards. And today internationally perhaps, the greatest irony remains that geopolitical-cum-militarily; irrespective of commanding—bigger the reach and also some decisive influence of any concerned Nation-State on these matters—she fixatedly remains all the more short-sighted as well as obsessed, with her some kinds of rather a queer geostrategic-cum-diplomatic callousness. And further at the same time, now the fast fructifying and expanding—**another draconian culture of both the intra-national as well as international terrorism and extremism**—are adding newer but all the more menacing dimensions to the already much worrying aspects of the 'NBC warfare. Hence needless to say that today, in mankind's larger perspectives in these regards; all these don't constitute any kind of perhaps, even a reasonably encouraging scenario at all.

In the above multiple contexts, while first taking-up certain aspects of the **'chemical weapon-systems'** for our brief discussions at this precise point, we do historically notice that this entire gamut of the some kinds of **'chemical warfare'** in the specific areas of the mankind's living 'military-history', it is found to not be a new phenomenon. In fact, even willfully poisoning the left-over food, water and fruit etc, with sole intention to trap and harm one's enemy forces have been rather quite common practices, since earliest of times related to the mankind. And then in addition, even poisoning the tips of one's arrows, swords, daggers, lances and so on, to kill or incapacitate one's enemy soldiers for winning any specific battle or war have been prevalent in the expanding domain of matters-military, since the dawn of human-civilisations varyingly and also, in different and initially unconnected geographical regions of the world and, during some remote corners of the very ancient-era itself. During those historically bygone times of the war-fighting, man used to mostly get such poisons and toxic items/chemicals etc, from the mother-Nature itself by extracting juices from the poisonous roots/branches/leaves, etc. of some wild trees/plants for these purposes, before man could master some knowledge of chemistry for chemically synthesizing certain desired poisonous chemicals, for their uses in war. While somewhat critically delving into the historicity of these phenomena during the man's ancient past, we interestingly also notice that while man had then begun to make fast advancements in the areas of developing medicines and other elixirs, to be used for fighting the diseases and his different kinds of sickness; simultaneously the above mentioned, though initially quite primitive to begin with, the 'chemical-weapons-based' his warlike activities too, were developing no less faster.

Expanding our these ongoing discussions little further, we find that later and since the very early medieval ages and even earlier too; when to begin with, some technically primitive know-how concerning the construction and erection of forts and fortifications had developed; from the top of the fort-walls and also through their some later periods' construction of suitably designed battlements along the top edges of the forts' walls, throwing off the boiling-hot water and/or oil, sometimes even mixed with poison/toxic materials; onto the invading enemy warriors down below, may also be taken as different kinds of relatively the primitive-era manifestations of the then some applications of some practicable 'chemical weapon-systems'. As the centuries kept-on rolling-bye, newer and newer inventions/innovations too kept-on manifesting themselves; and so historically in different warlike situations from time to time, sabotage and subversions against the enemy forces were also resorted to, by possibly somewhat innovatively making use of some kinds of the 'chemical weapon-systems'. Then at times based on the direction and speed of the winds too, some suitably toxic/poisonous items were burnt from favourable vantage points; to spread the emitting poisonous gases/smokes into the enemy camps or even among the formed-up enemy ranks and files, to kill/incapacitate them even before the actual battle had begun. **Interestingly, the military-history or some prevailing historic legends/stories of ancient India, gives even some glimpses of the '*vish-kanyas*' (i.e. 'poison-girls'); employed for luring and killing enemy-kings and their top military commanders, by resorting suitably to the different acts of sabotage/subversions in war. And so contextually; the historic stories in these regards, even sometime with some 'juicy' inputs; go on as peoples' folklores, having some historic facts and some captivating myths too.**

And thereby, as further centuries kept on rolling-bye and along with, the man's living 'military-history', since always being integrative-cum-inseparable segments too; of his much larger as well as, multi-dimensionally highly complex overall living-history itself; kept-on multiply manifesting and marching ahead and thus, giving new impetus and progress into the mankind's the then prevailing know-how regarding the 'chemical weapon-systems'. Hence contextually, the man's living-history does also informs, us that the real breakthroughs in these regards; substantially came along with the historically momentous advent of 'Industrial-Revolution', onto the mankind's expanding civilisational-canvas. We may recall that in the beginning pages of the last seventh chapter itself; we have since already discussed in some detail, **the various aspects of this historically so momentous and indeed unprecedented 'Industrial-Revolution'; along** with its multiple-fallouts and thus thereby, manifestations of its so many kinds of the multiply empowering fruits for the mankind and its enduringly further-on, larger well-being as well. But at the same time, it remains also equally true that, 'history' since in course of its unending march; does not fear or favour regardless of any possible considerations and thereby, if the man himself becomes sinisterly greedy, shortsighted, irresponsible or callous; the same 'history' does not spare him, or show any mercy at all. Hence, it is needless to say that, the overall welfare and wellbeing of the mankind, as since already again touched upon above; always depends upon its sincerely individual-cum-collectively qualitative display and also, the mandatorily judicious uses of its 'analytically creative and thinking mind', as per the any situation-specific developments around.

Now coming back to the recently passed timelines of earlier fast fructifications of the historically momentous **'Industrial-Revolution'**, which resulted in the multi-dimensionally fast-paced advancements in the multiple areas of the sciences and technologies; in turn also facilitated the mankind to substantially enhance its multi-disciplinary knowledge of the physical chemistry, both inorganic and organic chemistry and also the biochemistry as well and thereby; man could acquire newer wherewithal and also more lethal formulations, with respect to the various kinds of **'chemical as well as, the biological weapon-systems'**. And then, as the opportunities came during the two World-Wars and further-on; during the subsequent relatively lesser war/conflicts, from time to time and in different regions of the world; these were so lethally as well as even quite inhumanly used, during extremely violent conflict-situations. At the same time, during the highly dangerous advents of these the then relatively modern types of the **'chemical and biological weapon-systems'**; the ongoing multi-disciplinarily engineering-cum-technologies related developments consequent to the continuing manifestations of the multiple-fruits of 'Industrial-Revolution' too, enabled the mankind to ensure far-reaching developments in the highly consequential areas of these weapon-systems' more and more effective delivery-systems, with fast-paced enhancements in their ranges and accuracies as well. So there was no looking-back and contextually, we may also call them as various kinds of their **'weapon-platforms'**. So historically here too; we find that these multiple developments in the multidisciplinary areas of the sciences and technologies happen to be almost equally encompassing both the vast civil-sector as well as also, the matters-military domain as well.

In some larger perspectives, concerning the mankind's these indeed crucially important issues; **Colonel R. Rama Rao** (Retd) of the Indian Army and later, a Research associate at the Institute of Defence Studies and Analyses (IDSA), New Delhi; in his a very deep researched 'Books' Review' article, titled as, 'Chemical and Biological Warfare' and, covering the following books—(a) 'Chemical Warfare: A Study in Restraints' (Frederic J. Brown; Princeton University Press, 1968), (b) 'We All Fall down: The Prospect of Biological and Chemical Warfare'; (Robin Clarke; Allen Lane, the Penguin Press, 1968), (c) 'Chemical and Biological Warfare: London Conference on CBW', Edited by Steven Rose (George G. Harrap and Co. Ltd, 1968), (d) 'SIPRI Year Book of World Armaments and Disarmament', 1968-69 (September, 1969) and, (e) 'Report of the U.N. Secretary General on Chemical and Biological Weapons'—(A/7575 July 1969); and since published in the January 1970 Issue of the IDSA Journal, and therein (pages 354 and 355) he has written that, "The devastating power of nuclear weapons was brought home to the common man throughout the world following the destruction wrought by their use in Japan towards the end of World War II. The results of subsequent tests of more powerful weapons, in which a number of fishermen, several miles away from the test area, fell victims to the nuclear 'fall out', and the continued testing and development of yet more powerful devices have served to remind mankind that should a nuclear war be unleashed accidentally or otherwise, casualties could run into hundreds of millions and the survivors may envy the dead.... The recognition of the dangers of unrestricted chemical war led to the evolution and general acceptance by the majority of nations, of the Geneva Protocol of 1925. The Protocol was, no doubt, not perfect. Some nations, among them the United States, chose not to sign the Protocol. Eighteen nations including Britain, France and most countries of Eastern Europe, who formally signed the Protocol, did do so with reservations which left them free to use the weapons against a non-signatory to the Protocol, as well as against any country in retaliation or as a measure of self-defense. Further, Protocol only sought to ban the use of chemical and biological weapons; not their manufacture and storage. In spite of these reservations, chemical weapons were not used in World War II, which was global in character and was in other respects also an unlimited war.

Both sides had ample stock-piles of chemical weapons, which in the inter-war years had been considerably improved; protective clothing and respirators for combatants as well as good proportion of non-combatant population had also been laid by. Experiments with biological agents had also been carried out, but both sides were uncertain of the possible consequences of the release of such weapons on a large scale—especially in the context of the uncertainty of meteorological conditions over battle zones, and the nature of the immediate and long-term epidemic effects of the micro-biological agents on man, beast and vegetation." (*In their contextually larger perspectives in fact, all these do appear to be pointing towards the subsequently ongoing highly complex as well as, the multiple games of international power-politics and hence; we would also be somewhat critically delving upon them in the next 'sub-chapter itself*).

As we move-on further in our these specifically multiple-contexts, to critically delve deeper into these historic aspects; we find that in many ways the highly chequered and indeed ruthlessly human-blood soaked living military-histories of both the (1914-18) and (1939-45) two World-Wars, though their some important and also allied aspects have since already been briefly discussed in the last seventh chapter itself. But at the same time, the contextual history in these regards, do also show that during those conflicts and warlike situations; many kinds of the 'chemical weapon-systems' were in fact, also in a quite limited manner, used to create and spread some highly toxic and lethal gases, like **'Mustard-gas'** and so on, against one's enemy forces. **In fact, the most horrifying example during the Second World-War, may be cited as the Nazi Germany's infamous 'Gas-chambers'; for the most demonic kinds of elimination of the Jewish men, women and children; as per Hitler's war-directives to kill as fast as possible and also, as many of the Jewish population as feasible**. These warlike activities in turn, resulted in another no less fast-paced developments and their actual application-oriented fine-tuning of the many kinds of protective-gears and effective counter-measures; to protect own troops and other strategic-cum-tactical assets, from any adversary's the above mentioned hostile actions. And further on, various kinds of the warlike **'biological-agents'** too like **'Anthrax'** etc, have been developing in many tailor-made scientific laboratories and establishments across the world. But, have the nations across various kinds of the political-divides in the world; have even now cared to refrain from these activities, to rather mindlessly enhance their so called CB 'arsenal', precisely in these regards? It is true that, internationally in numerous geopolitical-cum-diplomatic ways from time to time; many initiatives have been taken and also are being taken, but to what avail and at the same time, how far those nations who do really matter and also command the latest kinds of the techno-military wherewithal; are really sincere about it for making these so called well-intentioned efforts, for making the world more safe from the menace of the nuclear, chemical, and the biological weapon-systems? It is true that, largely the efforts towards improving one's many kinds of the defensive measures against these weapons; do remain justifiable but again, are these measure always remain confined to only the defensive measures? **In fact in these specific contexts, it remains rather an irrefutable fact that, all these highly worrisome weapon-systems are first conceived and developed since, driven by different kinds of the geopolitical-cum-military reasons, by the different international 'State-actors' and only then; the various 'Non-state actors', always very clandestinely as well as sinisterly and, for their own totally unforgivable nefarious intentions, are able to lay their hands on them**. Well at this precise point, there appears to be no need to so much, otherwise rather internationally already identified one particular 'Nation-State' today; as in any case, we would be both naming and also contextually discussing about her little later; as being both a leading sponsor as well as, a kind of the nerve-centre of international terrorism. But in addition, it would perhaps suffice to say here, that there is another irony too precisely in these regards that, this particular

'Nation-State' interestingly, due to her own all types of the extremely nefarious designs and initiatives; she herself now appears to be trapped in a typical kind of the 'Catch-22' situation, principally of her own making and thereby also, looking like perhaps most regressively proceeding towards an 'abyss'. These and also many more questions specifically in these respects now, do keep-on haunting the mankind and hence today, all these appear to be something like a much larger 'catch-22' situation in this 'super-age of the high-tech era' world.

In the above multiple-contexts as well as, precisely concerning the present not at all any comfortable scenario in these regards; **Lt Col Ravindra Singh** has very thoughtfully mentioned that, "Terrorism has spread its deadly tentacles rapidly in the various parts of the world, bringing in its wake untold miseries, loss of life and considerable economic damage. Various measures have been initiated to prevent these attacks and to keep ahead of the perpetuators. Terrorists however are seeking newer methods to push their agenda of causing mayhem and death. Use of Chemical, Biological, Radiological, or nuclear (CBRN) material offers them the opportunity to not only to kill, but generate mass fear psychosis in the minds of potential victims.... Past use of CB materials by terrorists; threats/attempts to use CB Agents have been made by terrorists in the recent past across various parts of the world. A few of these are listed as: (a) Threat of use of CB Agents against civilian/military targets by Left-wing and Right-wing extremists in Europe and North America. (c) In March 1995, Aum Shinri Kyo carried out a Sarin attack (a nerve agent) in a Tokyo subway leaving 12 dead and 5,500 injured and use of CB agents in Matsumoto resulting in death of seven and over 270 affected. (d) In 1984 the Rajneesh sect in Oregon successfully used Typhoid bacillus for poisoning food affecting over 750 people. (g) Al-Qa'ida documents recovered in Afghanistan in summer 2002 indicate crude procedures for making sarin, VX (Video tapes containing evidence of experiments using an apparent nerve gas against dogs), mustard agents and other typical agents as well plague, Cyanide and botulinum toxin. Trace amounts of ricin and anthrax was found and anthrax was found at five or six sites. (k) Anders Breivik (the perpetrator of the 22 July 2011 massacre in Norway) in his treatise—the *2083: A European Declaration of Independence* addresses the issue of acquisition, weaponization, and use of CBRN agents. (n) In 2002 an Al Qa'ida plot by Moroccans and Algerians to poison the water supply of the US Embassy in Rome, and London using cyanide was disrupted.... These incidents are presumed to be precursors to the ominous intent of terrorists.... Reports indicate that extremist groups are making an endeavour to obtain CB agents and delivery means. The end goal is being to cast mass casualties. However, most attacks will be small scale, incorporating relatively crude delivery means using easily produced or obtained chemicals, toxins, or radiological substances. The success of any attack and the number of ensuing casualties would depend on many factors, including the technical expertise of those involved, but most scenarios could result in panic and disruption.... **It is evident that terrorists are committed to the use of non-conventional CB weapons. Given the appeal to terrorists of non-conventional weapons, we obviously need to focus urgently on preventing these attacks and equipping the emergency services appropriately**." (Singh, October 2012) (*And here lies the rub and thus, the larger scenario of the above mentioned 'Catch-22' situation too, in these regards.*).

The highly menacing and also, so far perhaps the most inhuman as well as now, somewhat regularly recurring phenomena of **terrorism** and **insurgency** in the varying degrees and almost all across the world; have turned out to be rather biggest nightmare and also, geopolitically-cum-socio-economically the most unsettling situations for the entire humanity in today's in fact, no less turmoil-ridden world. On top of it, to further increase the multiple botherations in these regards; these problems are denying so far, any tangible solutions and irrespective of their any genuineness too. Today in post cold-war as well as now, all the more multiply fast-changing I.T. Era and so on; the enlarging menace of the **terrorism and insurgency**, may appear to be much more pronounced and indeed mind-blowing; but the historic facts are that since the ancient times, these have been associated with the mankind in some forms or other. So far as the various aspects of **extremism** are concerned, these are principally related to the all sorts of dissenting thoughts and socio-cultural-cum-geopolitical, mostly the regressive 'ideas', since powerfully lingering with some sections of societies anywhere and in differently emerging situations; ultimately leading to some kinds of the somewhat calibrated responses from those dissenting sections of any particular societies, in the different forms of terrorism and insurgency; to ultimately force others or the concerned State finally, to agree to their viewpoints and ideologies. Many a times, their grievances may be based on perhaps their some preconceived notions or even, propagated by some sinister propaganda of certain vested-interests only. And at the same time, the pivotal methodology in these regards, are principally based generally, on all sorts of the **'irrational cults of violence'**. And in addition at times, even hardly any patient and tangible 'space for any genuinely purposeful dialogue' too, is seldom allowed to exist. Hence the State too, in response, is compelled to resort to force naturally to suppress the menace of any such terrorism and insurgency. And as since already mentioned in some contexts, all these do perforce lead to some sort of the 'Catch-22' situation. And paradoxically here now, almost the whole world is finding itself getting stuck, in different manners and also in different geographical regions; multiply in different kinds of the above mentioned 'Catch-22' situations. Hence, we may also say that, as it is difficult to fully understand the multiply menacing phenomena of the terrorism and insurgency and so similarly, it is all the more difficult to find any substantially tangible and any 'quick-fix' solutions in these regards.

In these specific respects, we also do need to keep in our minds that the numerous aspects of terrorism and insurgency; are of broadly two distinct types, i.e. intra-national and international and further at times in some ways or other; they are found to be both somewhat interrelated and mutually influencing as well. And at the same time, these aspects in fact; do substantially tend to add to their complications. As befitting examples here in the present times, and that too in the modern and democratic India's contexts; we may cite the violent Maoists sponsored disturbances all along the so-called ***'Red-Corridor'*** since, largely spread over the central heartlands of the quite vast Indian peninsula in fact, as intra-national (i.e. substantially home-grown, due to various indeed far-reaching reasons, such as—all kinds of corruption, political favouritism and nepotism, lack of performance and culture of unhesitatingly cheating the local inhabitants mostly rural and also then, overall lack or extremely poor governance, and so on.); and then, the prevailing terrorism-cum-insurgency in India's north-western State of Jammu and Kashmir, since largely aided and encouraged by **the multiply ailing and also, religiously fanatic State of Pakistan:** as international, and further on at the same time; all kinds of the militancy-ridden socio-cultural-cum-political turmoil, as prevailing in the Indian States situated in its north-eastern region, since India's pre-independence times: can be cited as both international and also, intra-national in these regards.

In fact, after attainment of India's political freedom in 1947, from the British colonial rule spanning over nearly two and half centuries, **the Chinese Communist Nation-State, due to her all kinds of the geopolitical-cum-militarily vested-interests; quite unhesitatingly began to 'fish in the troubled waters' of the India's geographically very difficult north-eastern, and also somewhat isolated region, mostly through the north-eastern Himalayan mountain-passes as well as also, through the remote, largely isolated and the hills-cum-jungles ridden age-old Indo-Burmese corridors.** In all the above cited three examples, pertaining to the present times and since concerning the Sovereign and democratic Indian Nation-State, broadly the different kinds of mountains-hills-jungles-seasonal or otherwise rivers-cum-streams ridden broken as well as, extremely difficult terrains and topography both strategically-cum-tactically in fact, in many ways obviously, do substantially favour the terrorists, insurgents and their masters located in the neighbouring foreign soils both in West and the East who in quite immune manners, manage and exercise command over them, through different kinds of the 'remote-control mechanisms'. But, both in the north-west and north-east, as spelled out above since the Indian Armed forces have both directly and indirectly, got involved in professionally focused manners, problems have quite dominatingly remained under firm control of the Indian Nation-State and then that since not being the case in the above mentioned India's so-called the 'Red-corridor, the problems are much beyond the capacities of both the local police and some Para-military units. **We would more relevantly and also in some more details be discussing these now crucially important issues, when we come to the highly consequential as well as, multiply India-specific matters, especially in the last 'sub-chapter here, at the end'**. Now in the present era, all these highly complex issues in fact geopolitically, do form some segments of the various kinds of internationally ongoing 'great-games'; and no wonder that, volumes have already been written and also, are being written by the numerous experts since, belonging to the different disciplines all across the world. We may clearly notice that, these extremely violence-ridden problems are not confined to any one particular nation or just a few nations only; but in fact now, have certainly now acquired global proportions and logically therefore, these do need to be effectively tackled internationally, by the much larger comities of nations, in more focused and also, synergy-creating manners; if the mankind wishes to solve these highly complex and perhaps, presently the most menacing problems, affecting the entire humanity in some ways or other.

In precisely the above multiple-contexts **Shri S.K. Ghoshhas written that,** "Insurgency is as old as organised state. There are innumerable instances in history where regular bands of insurgents, usually operating in mountainous and forest terrains and adopting what is now known as guerilla tactics, achieved success over regular army. They were able to compensate for their deficiency in numbers and traditional military skill by bravery, superior mobility, elements of surprise and intimate knowledge of the countryside. (*Well, both historically as well as contemporarily, these precise aspects are not so simplistically linear, as Mr. Ghosh is perhaps, attempting to elucidate; since the various aspects of the success or otherwise of any professionally highly trained, motivated and well-led regular army, as engaged in militarily tackling any insurgents/terrorists do depend upon various other kinds of the larger geopolitical variables, such as legitimacies of the concerned State's causes and policies etc. as well as, concerning the many qualitative and other wherewithal-related aspects of the internally ground-level prevailing systems of governance and so on.*). These tactics are even now the forte of the insurgent movements throughout the world, although the character and dimensions of the movement have gone a marked change in course of time. **While medieval insurgency was simple in character, modern insurgency is a complicated affair, which calls for understanding of the military, political, social and economic aspects. Probably the advent of communism in the early decades of current century (*i.e. in fact, the last 20th century*) significantly contributed to this change in the character of insurgency. Lenin recognised the utility of insurgency—terrorism, subversion, infiltration, propaganda and sabotage—and defended its 'Marxist character', in his treatise on 'Partisan warfare' (*Ironically, what indeed an historic negation of the superbly magnificent: the humanity's other much larger civilisational thoughts, ideas and also, marvels of the multiple-achievements therein.*); and Mao Tse-Tung, one of the pioneers of modern guerilla warfare, developed the thought on this subject during the anti-Japanese civil wars in the thirties and forties (*Well,***

there are many otherwise negative dimensions too, in Mao's largely 'regressive' thoughts; which now stand considerably exposed, and discarded as well.*). More than even communism, nationalism has been the prime fuel, on which insurgency has been able to feed and sustain itself**.... South-East Asia is the most vulnerable part in Asia or, for that matter in the world, susceptible to the so-called 'subversive insurgency', in American parlance, or the 'people's liberation war' as the communists call it. Insurgency in this area has become complicated with foreign intervention—overt as well as covert.... (*In these precise contexts historically and also, particularly concerning the recently bygone indeed, extremely violent and turmoil-ridden times of the second half of the last twentieth century in the South and South-Eastern Regions of the vast Asian continent; Mr. Ghosh in his same article, further mentions that*).... In Vietnam a local 'limited civil war' has (*had*) assumed international dimension as a result of open intervention and counter intervention by the Super Powers—USA and USSR—as also by China. In other areas of South-East Asia, China's verbal support for 'liberation movements' led by local insurgents and covert intervention in certain cases, have complicated the situation and focussed international attention. The presence of a strong Chinese minority in this area has considerably added to the anxiety of the local regimes about the possible Chinese attempts at subversion. China's interest in the area has recently (*the then*) been reiterated by **Lin Piao** in his statement to the Ninth Party Congress that, 'the armed struggles of the peoples of Southern Vietnam, Laos, Thailand, Burma, Malaya, Indonesia, India, Palestine, and other countries...are steadily growing in strength'. Another Chinese document published recently has described Asia as 'one of the storm centres of world revolution'. (*Well, much water has already flown down the numerous rivers and streams since then, in these South-East Asian nations and far-reaching changes too have also taken place in these nations, consequent to a number of major and indeed, history-making world-events, principally such as, the historic demise of the 'Cold-War' etc. But these geopolitically far-reaching and historic changes are never isolated since, as they were linked to the past they are also connected and, very influencing to the present and then, as well as to the future events too. In different context earlier, we have been noticing these phenomena and similarly, we would be coming across them later as well onto the perpetually multiplying pages of the man's multiply vast living-history.* ***What contemporarily are the matters of great worries to the mankind here, that China is still not refraining herself from rather sinisterly meddling in others' affairs, in these Asian regions with her many geopolitically hegemonic designs *and hence; China should be more careful and understand that now, in the present high-tech era and also, so substantially changed-cum-changing situations; her own such reckless obsessions in many ways, may ultimately boomerang on herself only.*).... *And now in similar historic perspectives, Mr. Ghosh further goes on to contextually add that*.... The above account illustrates the complexity in the pattern of insurgency in various parts of the world. There are cases in which the Soviet Union has been assisting the established regimes, while the Chinese and in some cases Western Powers and their agencies have been assisting insurgents. Except in Portuguese colonies and white minority regimes in South Africa anti-colonial nationalism is not the primary impulse behind insurgencies, even though there have been cases in which insurgencies are being sustained by the urge to preserve sub-national identities. There have also been encouraging developments of diverse experiments in political organisations which accommodate the aspirations of ethnical minorities in an effort to counter some insurgencies. A large number of insurgencies are directed towards securing social, economic and political rights in countries ruled by oppressive and traditionalist elites. What is most striking about insurgencies in various countries in the world is the difficulty in fitting them into a common framework. As has happened with the international communist movement, an insurgency tends to acquire its own national characteristic and there does not seem to be any more chance of the development of a central inspiration for insurgency than there is to have a central direction to international communist movement." (Ghosh, October 1969, 150, 151, 152 & 178)

The above quoted long paragraph, along with certain comments given therein too, for further somewhat clarifying the various aspects involved in these regards have been able to put the numerous issues concerning the presently menacing phenomena of the intra-national, international or transnational insurgencies-cum-terrorisms in their geostrategic as well as, politico-militarily indeed, in some comprehendible perspectives. In fact, every conceivable factor since as relevant and also applicable in these regards such as, the multiple-fruits accrued due to the phenomenally ongoing advancements, **during the Post-Industrial-Revolution times;** in the multiple disciplines-cum-fields of the modern era's, ultra high-tech sciences and technologies and thereby also, the far-reaching and multi-dimensionally continuing fructifications in the vast areas of the **RMA (i.e. Revolutions in Military Affairs)** all in their own different manners, have been contributing towards substantially enhancing the capabilities and also the operational wherewithal of the insurgents/terrorists, in any given situations across the world. And then, as if these are perhaps not enough, the various nations too whether directly affected or not, have their own problems within, and also between them in different manners; and in addition also, their differently uncomplimentary-cum-conflicting fixations with respect to others; and thereby these in turn, further add to the complexities of these problems concerning the insurgencies/terrorisms. All these otherwise highly complicated and mind-boggling issues in turn, almost all across the world perforce throw-up **both the Non-State and State Actors** into, the mostly no-holds-barred arena of the international power-politics, to 'play their balls' respectively, in their own different ways and also, in their different geographical locations as well as, having different geopolitical ambitions too. So the problems concerning the insurgencies/terrorisms are indeed increasingly menacing and obviously, there are no easy or quick-fix solutions to them anywhere.

In fact in their broader perspectives, we may even find that mostly at the roots of the different kinds of manifestations in different geographical regions of the world, of the so worrisome menace of insurgencies/terrorisms either of the two principally different kinds of 'driving-forces, comprising of i.e. (a) wherewithal of rather the **extremely radical and regressively fixated forms of the so called 'communism'** and, (b) All the more **different forms of the blindly rigid, uncompromisingly fanatic as well as also, extremely intolerant 'militant Islam'** are proactively involved. Then at times, we have other reasons too, such as: (c) so there are in fact, other than 'communism' or 'militant Islam' oriented, but mostly triggered by specifically the any case/example-based ideological aspirations, some particular situation-based geopolitical, 'racial' or, even due to money-making criminal motivations—such as drug/arms-trafficking, money-making etc—or any other kinds of reasons. We may cite examples such as: the then LTTE in Sri Lanka, earlier IRA terrorisms in the British territory of Northern Ireland, drug-trafficking from some Latin American and other countries and thereby; promoting money-laundering, arms-smuggling, printing and smuggling of fake-currencies of some nations, with the nefarious designs to destabilise their economies and so on; and these too in turn, sponsor and promote different kinds of terrorist/insurgency activities as well. Internationally quite menacing consequences of the specifically identified motivational-cum-triggering reasons in turn, have given birth to **some hyper-active and highly menacing arms-smuggling, drugs-trafficking as well as, the money-laundering 'corridors' too,** with which different international and multi-national security and the intelligence agencies are trying hard to grapple-with, and smash these crimes-cum-criminals-outfits' corridors and their sinisterly established and also hidden 'strategic-bases'. Therefore for many nations, these aspects do add additional dimensions and complexities towards their both internal and external security environments and geopolitical-cum-military compulsions. There are numerous reports since originating from time-to-time, from the different UN bodies and, other Agencies/Study Groups and Think-Tanks, across the world in these regards and hence, we do also all know as to which quite irresponsible and terrorism/insurgency-obsessed nations are active in such nefarious pursuits and they too, need to be thoroughly exposed before the comity of nations and then effectively reined-in for the larger well-being of the global humanity and no less important, the world-peace as well.

Hence, after the collapse of both the vast USSR-cum-China led communist-empire and thereby also, eclipse of the earlier so much motivated and focused international communist movement, the world began to experience quite far-reaching changes but even then, all sorts of 'pockets' of the regressively lingering communism in different regions of the world, though largely even corned continued to survive, bringing-in at times, the different kinds of the quite serious socio-political-cum-economic unsettling effects in their areas of influences and these in turn, perforce brought mostly quite negative kinds of the fallouts onto the larger societies within the concerned Nation-States, wherein the above mentioned 'pockets' existed or even now continue to exist in some forms or the other. Further some new kinds of 'great-games', concerning the international power-politics, gradually began to be again played in their highly complex 'international arena', by not only the great-powers but all others too, having some differently specific situations-based, 'stakes' in them.

Consequent to the fall of communist empire especially after historically former Soviet Union's geostrategic and politico-militarily costly fiasco in Afghanistan wherein, in those great-games then being played and, wherein also, the **fiercely fanatic Islamic-Jihadi Afghans and the so-called Pakistani Mujahidins** since duly aided, abated and also liberally armed by almost with the unrestricted dosages of the US Dollars in fact, became very far-reaching and instrumental, particularly after withdrawal of the once, so mighty Soviet military-forces in **giving birth and the accelerated expansion of 'radical-cum-militant Islam', and the fanatic Islamic-warriors now since, known as the Jihadi or Islamic terrorists/militants the world over**. So consequent to all these mind-boggling 'recipes' for the multiple-kinds of intra-national, transnational or the international, insurgencies/terrorisms in these regards, in fact, were fast getting fructified too, the world over. And interestingly further to these highly turmoil-ridden developments as we continue to delve deeper into these aspects, we also notice that since the already earlier identified and defined above as: **(a) the wherewithal of rather the extremely radical and rigidly fixated forms of the so-called 'Communism' and, (b) all the more different forms of the blindly rigid and uncompromisingly fanatic as well as also, the extremely intolerant 'militant-Islam',** further continued to multiply flourish, expand and magnify, in their reach and capabilities wherever they could possibly consolidate their roots. In addition to these, they have also appeared to be 'tacitly collaborative', not geostrategic-cum-politico-militarily but, by means of their multiply unsettling cumulative effects in these regards; in the affected regions or the Nation-States concerned. In fact, some sovereign Nation-States now, like the modern and democratic India as an example are clearly seen to be victims of the both above mentioned kinds of the insurgencies/terrorisms and, along with their collaborative tendencies. (In 2017 onward, State of 'Kerala' here, is the latest example!).)

Hence, onto the wider socio-psychological-cum-cultural, and also the socio-economic-cum-political canvases and, their multiple connotations in different manners in these regards and thereby, in any particularly concerned/affected Nation-State their such 'combine' since as contextually elucidated above, do also mostly remain both interrelated and, mutually influencing as well, and at times, even mutually-supportive too. And then, they do in fact also get manifested, mostly as quite a proactively menacing 'combination' and thereby throwing-up some serious kinds of the socio-economic-cum-political and, in addition also above all, the multiply security-related problems and challenges for the concerned governments and her peoples all across any affected Nation-State. So obviously in these precise contexts, we may also say that, the above mentioned specific 'combination' has all the requisite recipes to transform itself into some kind of a **'lethal combination'** as well; if not checked and decisively taken care-of both geopolitical-cum-militarily, while still some requisite space and time are available to the concerned highest-level political and military leaderships, for the needful.

We have since quite befittingly deliberated upon the paramountcy importance of the well-cultivated, well-groomed, well-imbibed and also then, proactively-cum-vibrantly, the well-placed—**multiple-aspects of the 'essential strategic-culture and thoughts'**—which in any case always universally remaining crucial in every aspect of any sovereign Nation-State's healthy-cum-assured existence, the entire gamut of her effective governance and statecrafts, the ongoing cultivation and upkeep of the higher human-values as well as also no less important, for the truly successful management of her both internal and external security dimensions. In fact with respect to any particular Nation-State if perpetually, the paramountcy multiple-aspects of the 'essential strategic-culture and thoughts' are so-much effectively and fruitfully kept vibrant and secured in place, as emphasised above, almost all aspects, necessities-cum-dimensions of her assured survival, multi-dimensional developments as well as, the ongoing civilisational advancements and, in addition also, her multiple security-aspects too; are always desirably looked after and, kept geopolitically under her control in the world. **The pivotal reasons being that specifically, the various dimensions of the above mentioned mandatorily well-stationed, 'multiple-aspects of the essential strategic-culture and thoughts' in turn ensures any concerned nation's, the following since no less crucial aspects in these regards and that being: (a) The irrefutable success of the systems and multiple conducts of the nation's governance, (b) Always both imaginatively and proactively; being in possession of the multiple-skills concerning since, the numerous dimensions of the statecraft, as applicable in home as well as abroad, (c) Both geostrategic-cum-politico-militarily, the futuristically imaginative and consistent all round composite developments and also, fruitfully nursing of the concerned nation's essential aspects of the 'matters-military', (d) Successfully always, the multi-dimensional conduct and the fine-tuned administration of the art and crafts; of the highly complex, challenge-ridden and also the rivalry-infested games of geopolitics and the diplomacy; since perforce being played, in the ruthlessly competitive as well as, the 'pitfalls-cum-diplomatic-minefields' ridden and also in fact at times, truly mindboggling 'arena of the international power-politics'. (e) And then, prompt and the effective management-cum-handling of the any otherwise quite unexpectedly at times, and/or any situation-specific and both internally and externally the sudden manifestations of the any further kinds of challenges/threats, since confronting one's Nation-State**. Hence, while multiply delving deeper into the above mentioned numerously different dimensions precisely in these regards; we may both contextually or otherwise too, be able to have the always timelessly ongoing and also, the universally panoramic glimpses of almost the all-encompassing comprehensiveness as well as, the paramountcy importance of **those multiple-aspects of the 'essential strategic-culture and thoughts'**. Well in these so precise and important contexts as to where precisely does India stand, and how she happens to be now faring, both in her internal and external perspectives; we would since more appropriately be discussing these aspects in some substantial details and along with some other important subject-matters too, in the last sub-chapter of in fact, this eighth chapter itself.

What precisely are the various modules of insurgency/terrorisms and also, what are the methodologies for their so much menacing manifestations in any given situations anywhere? In these perspectives, **Major General Afsir Karim** (Retd), of the Indian Army; in his 'Lancer Papers–5' (pages 1, 2 and 4) has written that, "Terrorism and insurgency are warlike activities which can be successfully employed by a group of adversaries to destabilise an established order by subverting the loyalty of its people, by sowing the seeds of ethnic divisions, and by encouraging sub-national separatist movements. Eugene Walter, a noted authority on the subject, writing on the use of terror by a nineteenth century Zulu King says, 'The aim was to deny the people any resistance potential, i.e. to deny their ability to participate in the power process, by withdrawing their empowering responses that would collapse the established order....' Such methods have been used in the past by various fascist leaders to 'disrupt key elements' in an established society. Those methods aim to take political decision-making away from the Parliament to the streets, through violent demonstrations, assassinations, kidnapping and murder of innocents. A general breakdown of trust in the established order is the first objective of such an assault or war. **R Gaucher**, a French authority on terrorism notes, 'The truth is, at a time when it is difficult to mobilise great masses of people without provoking a global conflict with irreparable damage, terrorism tends to become a substitute for war'. (*Precisely in these multiple-contexts interestingly, we may recall our discussions concerning the overall prevailing socio-political-cum-military scenario and consequently; the then fast manifesting and highly disturbing historic developments during the pre-Second World-War years, since taking place in Germany and Italy in the Western Europe, and also in Japan in the Far-East simultaneously; and wherein we have seen initially indeed, the so fast and far-reaching fructifications of the above quoted highly perceptive thoughts of R Gaucher as well.*).... However, after the advent of nuclear and other weapons of mass destruction, social warfare has gradually become an independent means of waging proxy wars; the aim remains the same as in a conventional war, destroying the basic structure of a state. According to **Barry Buzan**, a state has the following basic components which may be attacked:

(i) ***Physical component:*** The territory and the people of the state subject to its authority.

(ii) ***Institutional component:*** The institutions of Law and the Government under which the government functions.

(iii) ***Metaphysical Component:*** The concept or the idea of the State acknowledged or accepted by the populace.

If any of the above components cease to exist, the state is likely to collapse as a sovereign entity. If the state is already experiencing internal instability, its social and political structure, are easier to attack. According to Buzan, 'The distinguishing feature of weak states is their high level of concern with domestically created threats to the security of the government." (*Well, specifically in the above multiple-contexts, it doesn't appear to be necessary to look into the various aspects of few of the somewhat gradually 'failing' or perhaps, already looking like substantially 'failed' states; but at the same time, it would be appropriate as to where and in what all manners, the modern, sovereign and democratic Indian Nation-State does stand in these regards, as she quite justifiably claims to be fast developing as well as an emerging regional 'Power' in the present 21st century? As since already stated above, we would also be discussing these aspects too in some detail, in the last sub-chapter of this eighth chapter itself*).

The above quoted broad-based definition, encompassing almost all kinds of the terrorism/insurgency, does enable us to see through their making as well as, their aims, ambitions and aspirations. We have also come across the three vital components of a Nation-State, as since perceived by Barry Buzan and which always remain the likely strategic targets for any concerned terrorist/insurgent groups, which happen to be active in any particular state. Hence, the focal point is that the multi-dimensional menace of such ultras and outlaws are in no way any simplistic or routine law and order problems, for any concerned government anywhere as perceived by the many naïve politicians and bureaucrats in India, or perhaps may be anywhere else in today's world. Whatever may be anyone's any ideological, socio-economic-cum-political beliefs or lines of thoughts anywhere and irrespective of them we may even say that in many ways, these now extraordinarily menacing phenomena are indeed great civilisational-negations of the mankind and that too, onto its expanding civilisational-canvas itself. Within these broader perspectives, it is equally true that to begin with, at the roots of so much menacing manifestations of the many kinds of terrorisms/insurgencies almost all across the world and also, in differently situation-based intensities-cum-reach there, always happen to be **many kinds of the political and administrative corruption, nepotism and in addition, shamelessly continuing 'culture of fooling-around' to hoodwink the toiling masses, both rural and urban, and thus leaving them hopelessly impoverished and frustrated.** Thereby; offering the lurking ultras/outlaws, ready and even at times deeply fertile socio-psychological and the extremely poverty-infested 'fields' to enable them to sow their seeds for the fast fructifications of their desired kinds of terrorisms/insurgencies.

It is also true that the ultimate enduring solutions to these menacing 'diseases', have to be socio-economic and constitutionally political, through the state-sponsored multiple developmental-cum-empowering processes, but in order to enable the state-machinery and its wherewithal to proceed ahead with these assigned programmes and tasks perhaps, first it may be strategically necessary to politico-militarily neutralise the all kinds of such highly dangerous outlaws and ultras and their outfits, with the concerned state's unwavering kinds of the military decisiveness. And now fortunately during the year 1947 onward in India, Narendra Modi Government, alongwith—the concerned State government within, has exactly been doing the same, so much decisively, for the needful. Hence those states who, have been able to solve this menace with indeed acceptably substantial success, have done so by largely proceeding on these lines and also as per their different situations-based strategies. The multi-dimensionally momentous complexities in these regards in any case, cannot be wished away, but it is also no less true that any state infected with any forms of this menace of terrorisms/insurgencies and, if she sincerely aspires to ensure its geopolitically sovereign existence then it has to be ready to pay the necessary price for decisively solving this menace once for all. Moreover it is since true that, every nation-State has to solve her problems herself and also fight her own wars; and thus be always ready to pay the requisite price, with her own vision and strategic prudence. These aspects may also be termed as few of the numerous 'keys' towards in fact, fruitfully understanding the geopolitical complexities and 'rules' of the always continuing in some ways other; the 'great-games' of international power-politics since being played by the nations and, onto its geostrategic-cum-politico-militarily 'minefields-infested' international-arena. Hence, there is never any room for any nation, both internally as well as externally, to be naively callous and thus keep-on making all sorts of rather laughable false-moves; if she sincerely wishes to avoid any possible geopolitical-cum-military 'calamity' visiting onto herself. We may recall and as already emphasised earlier in different contexts that, there are no permanent friends or enemies in the above mentioned 'arena' of the always ongoing great-games of the international power-politics. Hence, all these realities-based perceptions in turn do also constitute some segments of the multiple aspects of the paramountcy **'essential strategic-culture and thoughts'**.

Precisely in the above multiple-contexts in fact, the presently ongoing international menace and its enormous magnitudes in these regards, can be gauged by the following quoted facts and the data, which can make any geostrategic-cum-politico-militarily perceptive person to sit-up and also, indeed get alarmed. The well-known Indian News-Weekly *India Today*, dated September 29, 2008, and therein in the page 36; has brought out that, "The 2007 Report on Terrorism by the National Counterterrorism Centre in the US tracks terror strikes across the world and looks for trends and changes. The report says that 'tracking attacks can help us understand some important trends, including the geographic distribution of incidents and information about the perpetrators and their victims. 'Year-to-year changes in the gross number of attacks across the globe, however, may say something about the international community's effectiveness in preventing these incidents, and thereby reducing the capacity of terrorists to advance their agenda through violence against the innocent.... **Approximately 14,000 terrorist attacks occurred in various countries 2007, resulting in over 22,000 deaths. Compared to 2006, the attacks remained approximately the same in 2007, while deaths rose by 1,800, a 9 percent increase over the previous year's number**. As was the case in the previous two years, the largest number of reported attacks and deaths occurred in Near East and South Asia. These two regions accounted for about 87 percent of the 355 casualty attacks, each of which killed 10 or more people. **As the chart shows, India ranks fourth in the world in terms of the number of terror attacks, after Iraq, Afghanistan and Pakistan**:

(a) Columbia: 364.
(b) Democratic Republic of Congo: 178.
(c) Somalia: 767.
(d) Sudan: 403.
(e) Chad: 368.
(f) Algeria: 192.
(g) Pakistan: 1,335.
(h) Iraq: 13,606.
(i) Afghanistan: 1,966.
(j) Thailand: 859.
(k) India: 1,093.
(l) Russia: 150.
(m) Philippines: 209.
(n) Sri Lanka: 241.

....'To combat terror', as Director IB (India), P.C. Haldar pointed out in an 'ongoing' Asia-Pacific Regional Seminar on Terrorism, 'traditional counter-terrorism strategies require a quick, dynamic and uniform response conforming to national systems.' He said that the paradigm of terrorism has seen a major change with 'the terrorist remaining invisible and exhibiting a cross-national spread, spawning intricate networks, with the Internet being used as a force multiplier and a resource base for terror cells."

The above quoted paragraphs in fact, do convey some very worrisome data and important information in these regards and hence it is all the more important that the present sovereign-cum-democratic Indian Nation-State does take them seriously with mandatory strategic-vision and geopolitical prudence so that she is able to effectively take care of her both internal and external security-interests, peace and well-being, in short as well as long terms perspectives. **These are the**

issues not at all to be taken casually and thus, quite naively initiate some half-hearted and even at times, haphazard and thereby counterproductive measures. And that's what the peoples sitting in the Indian 'corridors of power' in New Delhi, have been so hopelessly indulging in and, regardless of the very serious and adverse multi-dimensional consequences for the modern democratic Indian Nation-State. Hence, the presently prevailing overall scenario in India, are not all that encouraging and, both the Indian masses as well as the Paramilitary security-forces, do not seem to be even reasonably confident and reassured towards the 'Indian political-class and bureaucrats combines' at different levels in the entire gamut of India's national-governance, who are in any case inescapably charged at the policy-formulation levels, to be able to do their assigned jobs always with highest national-interests; and not get callously carried away by their different kinds of vested-interests and, and all sorts of their 'selfish-fixations'. In the present super-age of high-tech world, nothing remains hidden and India also, is always under the regional-cum-global scanners of a number of foreign powers in many ways and, with different motivations in these respects. And hence, she is not expected to look like a soft-state and thus contextually be rather 'laughable' too, as during the recent Manmohan Singh's Government years, so sadly, in most unbecoming manners.

At this precise point of our continuing discussions, it perhaps appears to be quite important and also contextually relevant to delve further and endeavour to identify as to, what all happened to be in fact perhaps, the historically momentous triggering reasons towards the **globally manifesting militant-Islam oriented sponsoring of the so violent and menacing international Jihadi terrorisms/insurgencies?** Contextually therefore in the mankind's continuing historic perspectives, **the momentous birth and the fructifications-cum-expansion of the many types of Islamic Jihadi intra-national-cum-transnational terrorisms/insurgencies** too, as the man's great civilisational negations happen to be within the above mentioned ongoing phenomena of the continuous making of mankind's highly complex-cum-multidimensional living-history itself.

We may recall in our multiple perspectives that historically, the apparently sudden invasion of Afghanistan in December 1979 was in fact, part of the then ongoing geostrategic-cum-politico-military great-games, being played by the vast Eurasian Soviet Empire to capture new territories for the further expansion of this once Super-Power Soviet Empire. But nobody on either side of the then **Iron-Curtain**, was in a position to remotely foresee and even sketchily evaluate the subsequent indeed, so path-breaking and historically momentous fallouts-cum-effects onto the entire global humanity itself. **Contextually, one specifically larger historic reality here remains that no war has ever been conclusively won by the any invading military forces, in the rugged and barren mountains, high plateaus, semi-deserts, inhospitable, and hostile terrains inhabited by the deeply proud and fiercely independent peoples of Afghanistan**. In fact, the man's living-history precisely in these regards, does show that whenever any alien army has entered into Afghanistan with military aims for territorial conquests the rugged-cum-barren and hostile Afghan mountains and plateaus have always been found to turn into some kinds of the 'quick-sands', where the invading soldiers and their military-hardware tend to frustratingly get stuck and thereby, the final victory becomes rather embarrassingly elusive. Quite ironically, history keeps repeating in Afghanistan in these regards, as in case of earlier invasions of the British imperial forces with the undivided colonial-era Indian subcontinent as their strategic-bases and then, for the much later Soviet forces and now, for the subsequent American-cum-British invading military forces too.

As any war in Afghanistan remains unwinnable and so, garrisoning her for militarily subjugating would always remain futile, costly and even counterproductive; and geopolitically therefore, the only option remains some sort of prudently formulating workable settlements for some socio-political-economically workable arrangements for the Afghan peoples; both internally and externally for them and thereby also, for the sake of the enduring larger peace in the world. But in these broader perspectives interestingly rather, the much bigger irony has been that mighty Empires and their strong and proud military forces have seldom been known to pause and carefully read the finer-prints of man's living-history to learn and hence the rest again repeatedly become parts of the same living history itself. These multiply historic as well as the presently ongoing aspects concerning Afghanistan, have been narrated with great analytical precision and comprehensive perception, by a British Author and adventurous Journalist, **Jonathan Steele** who has very perceptively mentioned that, "What I have tried to reveal the 'ghosts'—the specters of the past mistakes that haunt the American and British in Afghanistan today. The ghosts of the Soviet misadventure in Afghanistan point the way. **The Kremlin should not have taken military action to achieve regime change in Kabul. It acted from a desire of revenge and a false reading of how best to protect national security. It failed to consider the consequences of a non-Muslim state, inserting troops into a country with a long history of defying invaders**. What was intended as a quick in-and-out invasion soon suffered mission creep. Regime change turned into nation-building and an occupation that lasted nine years. The comparison with U.S. policy is clear, except that Washington's occupation remains unfinished after ten years.... It was not as if American policy makers were ignorant of Afghanistan before they went in. The United States played a key role in exploiting the Kremlin's blunders by arming rebels during the Soviet period, many of whom—including Osama Bin Laden—wanted to impose a fundamentalist version of Sharia law that was not part of Afghan tradition. In 2001, hubris led Washington to think it would be able to do a better job of providing stability and progress than the Soviets had done.' (*Look at this, and this too, coveys numerous things concerning the Afghan societies and cultures.)*.... *No wonder that concerning the Soviet forces then 'struggling in Afghanistan; Jonathan Steele in the page 55 of his book has very interestingly mentioned that,* 'What particularly embarrassed the diplomat (*i.e. Soviet*) was my remark that it was odd that Hafizullah Amin, the Afghan leader who had invited the Soviet troops, was by his guests on their first day in Kabul. This smelled more like regime change than fraternal help. The point produced a joke that circulated in Moscow some months later: Question: Why are we still in Afghanistan? Answer: We're still looking for the people who invited us.' (*Wonderful, Isn't it?*).... *Then on page 215, we find a quote that,* 'Even the nomads don't know where to go these days'.... *Further on page 391, we do come across strategically a very perceptive comment from a former U.S. Defence Secretary, Robert Gates saying that,* '**Any future Defence Secretary who advises the President to again send a big American land army into Asia, or the Middle East or Africa, should have his head examined**.' (*How geopolitically precise and visionary indeed, isn't it? Well, better be late than never in these regards.*)." (Steele, 2011, pp. 393 & 394)

If we may care to pause and try to read 'in-between the lines' in the above quoted paragraphs, we find that Jonathan Steele while commenting that **the US while in Afghanistan; tried to geopolitically exploit the earlier Soviet blunders over there, by arming the rebels, to her strategic-cum-military benefits; but what it really turned out to be in larger long-term perspectives concerning the US strategic interests in the region and even, regarding the overall enduring world-peace for well-being of the global humanity? Who were the peoples the US was arming then in addition also, liberally pumping-in dollars into theirs as well as, into their masters' hands? They were Afghan Muslims so-called 'freedom-fighters' and also, the Pakistani fanatic Mujahidins. And their Masters were Taliban, born out of al Qaeda, a ruthless transnational Islamic terrorist-outfit. Their masters were Osama bin Laden (now already quite befittingly**

killed by the US Navy-Seals.), and Mullah Omar in Afghanistan and then, General Zia ul Haq and the ISI in Pakistan respectively. It is anybody's guess now, as to how then the USA was being led through their sinisterly fabricated 'garden-path' and thereby, how the US so naively was being trapped into a rather expanding 'cobweb' of trans-nationally the fast manifesting Islamic terrorisms/insurgencies. Now the sole hyper Super-Power and along with her, the numerous other countries too are yet to fully recover and get out of this so menacing a transnational 'cobweb' in fact by no other means but to completely smash it into pieces and also conclusively thereby, to geopolitically 'sterilize' the larger grounds involved, with some kinds of the all-inclusive multi-dimensionally visionary strategic-prudence, so that any further seeds of the so demonic Islamic-militancy are just not able to sprout in any manner whatsoever in future, towards the larger well-being of the global humanity. These are not tall orders for the world community, but in any case, that remains largely a wishful thinking for the time being. So let us hope for the truly democratic world's earliest possible victory in the presently ongoing wars against all kinds of the Islamic and other terror/insurgents and their outfits in a near future, to make the world free of the any kinds of such demonic threats anywhere.

So when the US geopolitically-cum-militarily entered Afghanistan, rather first to indirectly counter the Soviet occupation the texture and rules of the 'great-games', then being played over the rugged and violently hostile Afghan 'chessboard', had drastically changed and also acquired some all the more threatening new dimensions. Every player involved in these **geostrategic and politico-military great-games in Afghanistan**; had their different aims and methods to achieve their sole 'victory', by outsmarting all other adversaries engaged in Afghan arena. The **erstwhile Soviets** wanted to outflank the **USA** and thus be able to geopolitical-cum-militarily enter into the **warm waters of Indian Ocean** through her shortest route, to further expand the communist hegemonic empire and thereby, even be able to threaten the US long-term and far-reaching strategic interests over there and also, the African coastal belts, the vast South-East, the **vital Malacca-Strait** and, the Far-East Asian regions. Further comparatively in some very limited perspectives; the **Taliban led by Osama and his associates** wanted to get firmed-up in Afghanistan to begin with, for establishing and promoting fanatically fundamentalist forms of Islam in the regions all-round the Afghan mountains and valleys, i.e. the Middle-East, North Africa and the South Asia. Thereby, be able to substantially harm the USA and her strategic-cum-economic interests, by the all possible Islamic terrorist/insurgent means. And then comes **Pakistan**, the strategically so-called 'frontline' State, rather readily available to be multiply used by the USA for promoting her many strategic interests in the region even further and also, improving her hold over the **'oil-well in the Middle-East'**. General Zia and his ISI in Pakistan, had altogether totally different aims and sinister geostrategic-cum-politico-military designs, which happened to be **largely India-centric,** for even militarily acquiring the **Indian State of Jammu and Kashmir'** in her East, and also for some **hegemonic domination of Afghanistan** for gaining some **'strategic-depths'** in her West. But in the process Pakistan too, was rather naively biting much more than she could geopolitically chew; and the rest is again a highly chequered history.

But contextually in their larger perspectives all these were not adding-up to the complete story then as consequent to the then somewhat hidden but the fast incubating and fructifying **fanatic culture of terrorisms-oriented militant-Islam,** that was ironically not going to be friendly for any of the above discussed so-called 'players' in the then extremely turbulent Afghan 'chessboard', for these highly complex and ruthless 'great-games'. It was soon going to manifest into rather **an unprecedented demonic monster, which was going to be friendly to none and**

enemy to all and, totally uncontrollable by any kind of its central authority and then in addition, being capable of devouring its own limbs and cadres and simultaneously thereby, also acquiring all the more menacing new ones to replace them. In other words, we may also say that this monster was soon going to be like self-actuating as well as, self-multiplying 'clusters of deadly viruses' with unprecedented lethality for the mankind. So the highly demonic recipe in these regards, began fast getting complete with extreme lethality for the world-peace and tranquility and in the process, even all the above mentioned so-called highly ambitious and fixated players too, were soon going to be caught in a state of 'napping'. And these precisely are the ways the **Islamic fundamentalism-cum-terrorisms/insurgencies,** since so suddenly born, got trans-nationally fructified and also expanded as arguably now, the greatest menace for the mankind in this world. And then the biggest paradox here is that, all those so-called rather 'strategically-shrewd' players of the erstwhile 'Afghan great-games', now find themselves no less facing and suffering this demonic menace of Islamic terrorisms/insurgencies and thereby, somewhat pathetically struggling to crush this menace. But in any case, this monstrous menace has to be crushed and finally wiped-out from the face of the world, as a great service to the global humanity itself. Precisely at this point, we may also add that here too, the paramountcy significance of the **multiple-aspects of 'essential strategic-culture and thoughts'** cannot be ignored. Now a word for the present-era sovereign and democratic Indian Nation-State as in addition to this menace, she finds herself also additionally trapped in another kind of no less threatening menace of the **Maoist 'red-corridor'** passing through India's heartlands and, also being dangerously restless to further expand its this 'corridor' inclusive of Delhi and thus, painting the whole of India into red, which is nothing but senselessly violence-oriented and an extreme form of, otherwise now the **defunct and discredited, communist ideologies**. But these India-centric aspects can wait to be subsequently taken-up and since also spelled-out earlier too, in the last sub-chapter here.

How the leading nations, particularly the USA, has already geared-up and also developed some appropriate kinds of the professional-cum-institutional mechanism with truly global reach, which she only remains capable of, with their geopolitical-cum-military concepts and strategies for firmly catching the 'bull' of this menace of the Islamic fundamentalism-cum-terrorisms/ insurgences and then, 'turn its neck round and round' to finally break it into pieces? And for all these, the USA has to herself first get rid of at times, her own certainly not at all befitting **'smallness of minds as well as, the strategic short-sightedness'**; for the larger long-term good of the global humanity, and for herself too. Particularly in these specific contexts, it may be very relevant to recall that during the last decade of the twentieth or perhaps during the early years of the first decade of the present twenty-first century itself; the then Indian Prime Minister **Shri Atal Bihari Vajpayee**, while addressing US Congress in Washington had emphatically made it clear that, the inhumanly violent culture of terrorism has no limitation and also, does not recognise any international boundaries. Hence all the affected nations inescapably need to sincerely come together, and fight this demonic menace jointly with total geopolitical synergy. But how far sincerely the USA was listening then? And to everybody's surprise, within few years the **most dreaded and indeed tragic—'9/11'**—happened in the USA itself. So ironically, even the mightiest and the singular hyper super-power: the USA, perhaps needed the most demonic 9/11 to wake her up from a deep 'geostrategic and politico-military slumber'. In case of the USA, this menace has numerous international dimensions and hence, she is much more required to fight it with the requisite geostrategic vigour, military decisiveness and also, in the all-inclusive manners—more onto the both near and far-off foreign soils—rather than within her own domestic turf only; but internally she has to be multi-dimensionally vigilant to timely neutralise any emerging threats for her national security and the safety of her peoples. And precisely for her all these internal and external compulsions rather, she is expected to substantially get geared-up and also be proactively occupied-with, in various ways. Concerning some dimensions in these regards, we have since already contextually discussed briefly, in the end of last seventh chapter itself.

Needless to say therefore, that in these present-era conflicts for any nation, the multiple-importance of the both **internal and external professionally institutionalised intelligence-services** as well as also, that of the any concerned nation's **special-forces under the professional and direct military-command**, can never be over-emphasised. The keys to success lie precisely in that nation's political-will, unwavering decisiveness, geostrategic-cum-politico-militarily: the well-tuned strategic vision and perseverance and also no less in addition to these—the need to sincerely discuss and listen to the professional advice and thoughts—in the one's highest national-interests since no slipshod and half-hearted approach and plans can bring the intended success and, which can even be rather dangerously counter-productive too. **To begin with, the concerned top-level political-leadership of the any concerned nation, must first understand and accept that, these are no simplistic police or civil-administration related some kinds of the law and order problems if one desires to avoid being a non-starter and thereby possibly, even land into the 'fire' directly from a 'frying-pan', and in the process be instrumental in creating very dangerous and embarrassingly very costly mess, for peoples of that concerned nation. Further, it also needs to be noted that, these are not at all any kinds of the classical military-problems requiring any conventional and direct military action. These are totally different and special kinds of the extremely menacing problems and hence, do need special kinds of the geostrategic-cum-politico-military solutions in fact—revolving around the well-coordinated and professionally skilled: the multiple intelligence-services and simultaneously with due strategic prudence, launching of the specifically formed, suitably equipped, well-trained and also professionally led: the special-forces—with all levels' synergy and sincere involvement**. And it is important to note that here too, it remains paramountcy that all the likeminded and truly democratic nations in all the continents of this world, need to proactively come together with utmost geopolitical sincerity, for decisively winning now this unavoidable war against all forms of the transnational terrorism/insurgencies.

In fact, there is now no escape from decisively fighting this demonic menace on all fronts, with geopolitically the utmost international togetherness, synergy and, sincere geostrategic-cum-politico-military-oriented 'strategic-prudence'. Well needless, it is essential to say that here too, the USA is expected to positively get robust, far-sighted and, strategically be visionary to come forward to fruitfully galvanize the world opinions and take the requisite lead for the needful. Is it a 'tall-order', beyond the geopolitical capability of the USA, or is she more inclined to withdraw herself in her self-made and short-sighted 'geopolitical-cocoon'? USA first needs to cut-down on some very costly, wasteful, useless and, outdated kinds of conventional military-budgets and also, the colossal expenditures on her still bloated 'nuclear-arsenal' and then with due strategic-vision, gear-up for her some new international roles as per due dictates of the time. And at the same time, it could perhaps never be any saner person's 'brief' to even suggest that, the USA should refrain from taking proactive note of the **ongoing RMA (i.e. *Revolutions in Military Affairs*)** across the world; and in addition also, not be going ahead strategically with the continuing up-gradation of her **military technologies**, in her best short-cum-long terms national security interests. The pivotal point here is, stressing towards the fruitful needs for strategically visionary optimizations of any leading democratic nation's defence allocations and allied security-related expenditures. Times are changing and new challenges are emerging fast and hence, there is no use living in the past and thereby, remain fixated with the obsolete strategic ideas and concepts, but to listen to the call of times and meaningfully undertake the new global leadership role, for the much larger and truly enduring world-peace and tranquility. In fact, the man's a new living-history is in the making and hence, the USA has altogether unprecedentedly new and decisive roles to play, to render much bigger and long-term services to the mankind and hence at present, it is no less a **'tryst with destiny'** for the USA herself.

The US Special Forces and also, the highly checkered and historic evolution of her JSOTF (i.e. Joint Special Operations Task Force), as they are organised, professionally perform and exist as the US Nation's pride. Anywhere in a democracy, it is no easy way to think 'out of the box' and do something substantial with some brilliant vision and that too, in a relatively short spell of time. But where there is a sincere will with patriotic perseverance, there is always a way and in these specific contexts too at times, it is some committed individuals' leadership acumen in the nation's overall best interests that takes the lead in augmenting the prevailing thought-processes in the concerned 'corridors of power' for overcoming any inertia, towards achieving the desired goal. And precisely in these paramountcy context too, India has many important lessons to learn in her greatest national and security-related interests and also, how the Indian political-class must wake-up as since we would soon find as to how, even the USA had very imaginatively taken note of certain outstanding attributes of Israel and was quick to learn some far-reaching lessons in her best national interests, and thereby and as an example out of so many, we may cite the brilliant accomplishment of US Navy Seals—in finally eliminating the perhaps most dreaded terrorist Osama bin Laden inside Pakistan itself.

In the above multiple-contexts and the US Special-Forces and also the JSOTF, **Tom Clancy, with General Carl Stiner (Retd) and Tony Koltz** have written that, "The evolution of JSOTF: The traditional Function of wars is to change an existing state of affairs. In the early 1970s, a new form of warfare emerged—state sponsored terrorism. Nations that were not militarily powerful learnt to use terrorist tactics to obtain objectives and concessions they could never win through diplomatic or military means.... When this new form of warfare broke out, the United States quickly showed itself, unprepared to cope with it. It had neither a national policy nor intelligence capabilities aimed at terrorism, nor any military forces adequately trained and prepared to respond to terrorist provocations, Although, the United States was the most powerful nation in the world, its military capabilities were focused on the Soviet Union and not on something like this.... In 1972, Israeli athletes at the Munich Olympics were massacred by Black September terrorists. This outrage might have been avoided if German snipers had the ability to hit the terrorists as they led the hostages across the airport runway to their getaway plane.... **The Israelis took this lesson to heart**, and on July 4, 1976, eighty-six Israeli paratroopers landed at Entebbe Airport in Uganda. Their mission was to rescue the passengers from an Air France airliner hijacked eight days earlier. In the matters of minutes, the paratroopers had rescued ninety-five hostages and killed four terrorists—though at the cost of the lives of two hostages and the paratroop commander. News of the raid flashed all over the world—and pointed out even more sharply America's inadequacies in fighting terrorism.... **To correct this shortfall, in the mid-70s, three farseeing people began lobbying for the creation of a special 'elite' unit to deal with the unconventional threat: Lieutenant General Edward C. 'Shy' Meyer, Director of Operations for the Army; Major General Robert 'Bob' Kingston, Commander of the Army's Special Forces; and Robert Kupperman, Chief Scientist for the Arms Control and Disarmament Agency, who was managing the government's studies on terrorism....**

The three initially made little headway. Scant support for the 'elite' unit could be found among the services and even within the Army (*typical of the human nature anywhere and perhaps in any field: having even mindless 'resistance to change'; and moreover especially among the government institutions/ departments, typical shortsightedness, traditional bureaucratic habits to even arrogantly refuge to come out of one's 'dark-groove' and think 'out of the box' and also, senselessly refusing to get one's real or imagined 'turf/empire' even marginally tampered or altered for the larger-cum-all-inclusive good.*), **even though it was devastatingly clear that the technology in which the Army was investing so heavily--tanks, helicopters, air defence missiles, armored personnel carriers, and all the other machinery of the modern-day battlefield—was of little use against the terrorists**. (*Well, these strategic perceptions— are somewhat equally true in any situation-specific calibrated manner anywhere--in any kinds of the prevailing scenario concerning the menaces of internal-terrorisms too.*). The opposition stemmed from two sources: a bias against elite units as such—elites have never been popular in the U.S. Army— and the perception that the unit would rob resources and available funds from the existing force structure.... On June 2, 1977, Lieutenant General Meyer presented the concept of this special mission unit to Army Chief of Staff General Bernard Rogers.... **This unit was to be premier counterterrorist force. Because it was expected to deal with the most complex crisis situations, it would have capabilities like no other military unit. It would be organized with three operational squadrons and a support squadron; and it was to be composed of handpicked men with demonstrated special maturity, courage, inner strength, and the physical and mental ability to react appropriately to resolve every kind of crisis situation—including imminent danger to themselves**.... On November 19, 1977, the Army officially activated the unit, but it took another two years to develop the tactics and procedures required for unit's projected mission.... The unit's final exam and validation exercise was held at Hunter Army Airfield at Fort Stewart, Georgia, and ended in the early morning hours of Sunday, November 4 (*1979*). It was now certified for its special mission requirements. (*Well, based on the overall prevailing national-cum-global realities any specifically comprehensive threat perceptions concerning the USA, would understandably be always different from others in these regards. But any multi-dimensionally different, massive, menacing and simultaneously threatening—internal terrorisms/insurgencies, from different directions and in different regions and, duly aided and abated quite sinisterly by very hostile and highly militarism-oriented neighbours—**of any quite large sovereign-cum-democratic Nation-State like India. Moreover, that too, having numerous diversities, traditions, and socio-cultural complexities** should perforce be having appropriately, very different kinds of the comprehensive threat-perceptions towards activating her proactive multiple-planning and preparations, raising of the few specifically trained and special kinds of the **'Army Light Infantry Divisions', now so essential for India** for decisively annihilating these menaces of terrorisms/insurgencies finally, in her both micro-cum-macro levels' long-term national-security interests.*).... *On* making Professionals: Or almost ready to go—As previously noted, one of SF's problems (*specific to the US Military*) was that many of the generation that emerged from Vietnam, or who came into the force after Vietnam, failed to attain the high levels of professionalism expected of men who make up a force that calls itself elite. In Vietnam, they'd operated out on the end of a string without much supervision. Others, recruited after the Army drawdown in the 70s—were not the best group of men to begin with. Some of them had simply been looking for greater freedom and intrigue than they could get in conventional units, and had found their way into SF. Meanwhile, back then SF did not give a strong enough professional orientation to its younger officers. As a consequence, some of them picked up 'outsider' attitudes, simply because that was in the air.... On the other side of the coin, it was hard for them to get promoted, and that also didn't help their attitude any. Normally, if you were good, you moved through key positions in a variety of conventional units.

Your performance and potential were recognized by people who counted, and in due course you were selected for promotion and for attendance at Leavenworth and later the War College, Carlisle, Pennsylvania, or one of the other high-level service schools. Promotion and Selection Boards were composed exclusively of officers with conventional backgrounds.... Back in the 70s and the early 80s however, most officers were dead-ended in Special Forces. The personnel assignment people in Washington were content to drop them there and 'forget' about them. In many ways, assignment to SF was career suicide, so it was small wonder that some officers just figured: To hell with it. Such people simply reinforced the perception that special operators were not 'real' army." (Clancy, Stiner & Coltz, pp. 3, 4, 5 and 221) (*But, things were not going to continue like this for long and, soon during the tenures of Generals Maxwell Thurman, the Vice Chief of Staff of the Army TRADOC commander, i.e. 'Training and Doctrine Command of the US Army', General Bill Richardson others like Generals—Mike Spigelmire, tom Fields, Fred Franks and Ed Burba Jim Guest as head of the Special Warfare Centre and School, subsequent entry of General Karl Stiner into the SF and so on historic and indeed path-breaking changes, reforms and improvements were made into the U.S. SF units.* ***In case of India, the main problems are revolving round the regressive, callously 'hard-hearing', ignorant and arrogant—the 'Indian political-classes-cum-bureaucrats combine': sitting in New Delhi and, in the capitals of the different Indian states' 'corridors of power' who have certain strange 'fixations', since not being helpful in the wise and professional handling of India's national-security compulsions and also, fulfilling the Indian defence-apparatuses' multiple-requirements; and these would more relevantly be discussed little later*).

(e) Nations' Ongoing Complex 'Great-Games' of International Power-Politics

Since the advent of various human-civilisations as well as, dawn of the man's living-history in this world, these highly checkered phenomena of the **'great-games'** have been continuing in some forms or other. We may also say that, these momentously ongoing and, history-making phenomena of the 'great-games' are in fact, proactively inherent into the fundamental psyche of the man as well as, inherent to the various kinds of initiatives and activities of his so many socio-political and even other groupings, manifesting and 'flourishing' in the different locations/regions of the world. Hence, these since momentously history-making phenomena have been continuing in the mankind's world, since the remotely ancient of times. A great British philosopher and also, the humanity's one of the finest of minds—thinker, commentator and also, well-known author, **Sir Bertrand Russell** has in fact, very scholarly delved into the man's these phenomenal aspects; for our mental illumination and, some meaningful understanding in these regards.

Bertrand Russell with his unique intellectual penetration, has mentioned that, "Between man and other animals there are various differences, some intellectual, some emotional. One of the chief emotional differences is that some human desires, unlike those of animals, are essentially boundless and incapable of complete satisfaction.... The activities of animals, with few exceptions, are inspired by the primary needs of survival and reproduction, and do not exceed what these needs make imperative.... With man, the matter is different. A large proportion of the human race, it is true, is obliged to work so hard in obtaining necessities that little energy is left over for other purposes; but those whose livelihood is assured do not, on that account, cease to be active. Xerxes had no lack of food or raiment or wives at the time when he embarked upon the Athenian expedition. Newton was certain of material comfort from the moment when he became a fellow of Trinity, but it was after this that he wrote the ***Principia***. St. Francis and Ignatius Loyola had no need to found Orders to escape from want. These were eminent men, but the same characteristic, in varying degrees, is to be found in all but a small exceptionally sluggish minority.... **Imagination is the goad that forces human beings into restless exertion after their primary needs have been satisfied.... Of the infinite desires of man, the chief are the desires for power and glory**. These are not identical, though closely allied: the Prime Minister has more power than glory; the King has more glory than power. **As a rule, however, the easiest way to obtain glory is to obtain power**; this is especially the case as regards the men who are active in relation to public events. **The desire for glory, therefore, prompts, in the main, the same actions as are prompted by the desire for power and the two motives may, for most practical purposes, be regarded as one**.... When a moderate degree of comfort is assured, both individuals and communities will pursue power rather than wealth: they may seek wealth as a means to power, or they may forgo an increase in wealth in order to secure and increase in power, but in the former case as in the later their fundamental motive is not economic.... **It is only by realizing that love of power is the cause of the activities that are important in social affairs that history, whether ancient or modern, can be rightly interpreted**...**Propaganda**, if it can create an almost unanimous opinion, can generate an irreversible power; but those who have military or economic control can, if they choose, can use it for the purpose of propaganda. To revert to the analogy of physics: power, like energy must be regarded as continually passing from any one of its forms into any other, and it should be the business of social science to seek the laws of such transformations. The attempt to isolate any form of power, more especially, in our day, the economic form, has been, and still is, a source of errors of great practical importance.... To understand our own time and its needs, history, both ancient and medieval, is indispensable, for only so we can arrive at a form of possible progress not unduly dominated by the axioms of the nineteenth century." (Russel, 1948, pp. 7-12 & 14)

At present in our these multiple-contexts; the above quoted paragraphs indeed do show us, as to the inherently continuing phenomena of the self-actuating manifestations of the various kinds of 'seeds' of the above mentioned 'great-games', the ongoing processes of their incubations and fructifications onto the man's expanding civilisational-canvas and thus thereby—the various kinds of the phenomenally continuing and also history-making 'great-games'—which are since always being played by the nations, in the highly complex as well as, the multiply turmoil-ridden 'vast arena of the international power-politics'. And as Bertrand Russell, since very thoughtfully has tried to 'educate' us in these regards, through some of the numerous books and writings and thus showing that it is the man's multiple-desires for very many kinds of the 'power and glory', that are behind powerfully coaxing and multiply motivating the man, his various groups and also, his numerously sovereign kingdoms/nations, towards begin playing different kinds of the 'great-games', onto the already above identified 'vast international arena', for fulfilling those multiple-desires. Moreover, among the man's numerous nations/kingdoms/empires since the beginning of history, one more allied fact has also been seen that **there are no permanent friends or enemies but always, one's paramountcy and various in fact, situations-specific, national-interests**.

And then we do also come to the fact that, power is of many kinds in the world like: the money-power, scientific, technological and industrial power, military-power, political-power, geographically sizeable and also multiply strengths-giving territorial-depth and geopolitical power, human-resources power, and then indeed always of the paramountcy importance—having the meaningful wherewithal concerning the **multiple-aspects of 'essential strategic-culture and thoughts'**—since as a kind of the perennially great fountain-head of the numerous kinds of greatly valuable and unique sources of power, and so on. But at the same time, it is also true that, in order to channelize and fruitfully activate any or the combination of a number of the above mentioned kinds or any other kinds of power, to one's desirously advantageous uses, another prerequisite essentially remains that the availability of the first-class systems of the leadership-aspects, being vibrantly in place and also having or even cultivating the desired quality of followers since already motivated by the concerned leadership and thus, fired with a zeal towards achieving the intended power and glory. Hence in a way, we may also find that these indeed paramountcy and also far-reaching aspects in these regards, as the very genesis of the already discussed variously ongoing, the numerous kinds of the 'great-games'. As in this and so in the earlier chapters too, especially from the fourth chapter onwards, we have since already discussed in differently allied contexts, the timeless importance and the multiply-fruitful knowledge of 'History', from its beginning itself and thus enabling the mankind to be able to acquire and also multiply-groom the **always paramountcy 'sense of history'**, for the furtherance of its aims and ambitions towards attainment of the various requisite kinds of 'power and glory'. Bertrand Russell in his even quite earlier book, *Philosophical Essays*, had already elaborated upon these aspects of the timeless importance of the man's living-history in the following paragraphs, as to be now quoted from the above cited wonderful book of **Bertrand Russell**; we would be coming across this great scholar and thinker's thoughts on the man's engaging and multi-dimensional living-history and thus; so analytically authenticating, in fact, the timeless importance of the **'multiply-unique lessons of history'** to be learnt, and which are so invaluably stored in the history's irreplaceable treasure-house; and thereby enabling us to cultivate, develop, and also fruitfully imbibe—the meaningfully vibrant **'sense of history'**—into our ever **creative and analytically-thinking human-minds**. Precisely in these multiple-contexts, we may recall that in the earlier chapters; we have already quoted and discussed some passages from an eminent **Professor E.H. Carr**'s book, *What is History*, containing the collection of his 'The George Macaulay Television Lectures and, as delivered in the University of Cambridge (January-March, 1961). Thereby he too in his distinctly stimulating manner, has also emphasised the 'multiply-unique lessons of history', and thus stressing upon the values of the 'sense of history' for the mankind.

Bertrand Russell has written that, "**History is valuable, to begin with, because it is true; and this though not the whole of its value, is the foundation and condition of all the rest**. That all knowledge, as such, is in some degree good would appear to be at least probable; and the knowledge of every historical fact possesses this element of goodness, even if it possesses no other. Modern historians, for the most part, seem to regard truth as constituting the whole of the value of history. On this ground they urge the self-effacement of the historian before the document; every intrusion of his own personality, they fear, will involve some degree of falsification. Objectivity before all things is to be sought, they tell us, let the fact be merely narrated, and allowed to speak for themselves if, they can find tongues. It follows, as a part of the position, that all facts are equally important; and, although this doctrine can never be quite conformed to in practice, it seems nevertheless to float before many minds as an ideal toward which research may gradually approximate.... That the writing of history should be based on the study of documents is an opinion which it would be absurd to controvert. For they alone contain evidence as to what really occurred; and it is plain that untrue history can have no great value.... **Nevertheless, history has a function in regard to current affairs, but a function less direct, less exact, and less decisive. It may, in the first place, suggest minor maxims, whose truth, when they are propounded, can be seen without the help of the events that suggested them**. This is largely the case in economics, where most of the motives concerned are simple. It is the case also, for a similar reason, in regard to strategy. **Wherever, out of the facts, a simple deductive argument from indubitable premises can be elicited, history may yield useful precepts. But these will only apply where the end is given,**

and are therefore of a technical nature. They can never tell the statesman what end to pursue, but only, within certain limits, how some of the more definite ends, such as wealth, or victory in war, are to be attained.... Another and a greater utility, however, belongs also to history. **It enlarges the imagination, and suggests possibilities of action and feeling which would not have occurred to an uninstructed mind. It selects from the past lives the elements which were significant and important; it fills our thoughts with splendid examples, and with the desire for greater ends than unaided reflection would have discovered. It relates present to the past, and thereby the future to present. It makes visible and living the growth and greatness of nations, enabling us to extend our hopes beyond the span of our own lives. In all these ways, knowledge of history is capable of giving to statesmanship, and to our daily thoughts, a breadth and scope attainable by those whose view is limited to the present....** Against this spirit, whereby all leisure, all care for the ends of life, are sacrificed to the struggle to be first in a worthless race, history and the habit of living with the past are the surest antidotes; and in our age, more than ever before, such antidotes are needed.... **The record of great deeds is a defeat of time; for it prolongs their power through many ages after they and their authors have been swallowed by the abyss of non-existent. And, in regard to the past, where contemplation is not obscured by desire and the need for action, we see, more clearly than in the lives about us, the value, for the good and evil, of the aims men have pursued and the means they have adopted. It is good from time to time, to view the present as already past, and to examine what elements it contains that will add to the world's store of permanent possessions, that will live and give life when we and all our generation have perished. In the light of this contemplation, all human experience is transformed, and whatever is sordid or personal is purged away. And as we grow in wisdom, the treasure-house of the ages opens to our view; more and more we learn to know and love the men through whose devotion all this wealth has become ours. Gradually, by the contemplation of great lives, a mystic communion becomes possible, filling the soul like music from an invisible source.**" (Russel, 2003, pp. 60, 61, 64, 65 & 66)

We find as to how lucidly and yet intellectually with deep penetration, the greatly thinking as well as, so scholarly mind of Bertrand Russell has propounded for us and also, for the future generations; the greatly continuing multiple-values of the storehouse of knowledge of history, since always containing numerous dimensions and, indeed their multiply prudential reach as well for good of the mankind and also, its perpetually ongoing advancements onto its expanding civilisational-canvas itself. In addition therefore, we may also say that since always in any creative sphere of the multiply-ongoing human endeavour; any such fructifications of the enchanting marvel, can only possibly be achieved by a great mind having its intellectually sharp prowess and also, with its brilliant craftsmanship towards the fine weaving of indeed always a **'magic-of-words'**. And so at the same time, **it also remains true and more so precisely here, in our broader perspectives concerning the principal subject-matters of our since ongoing discussions that, without having any student like learning-eyes continuously onto indeed the enchanting storehouse of the mankind's multi-dimensional knowledge of history, can any worthwhile 'Strategically Critical-Journey' in any specifically defined contexts, be ever possible**? Though it remains also understandable in any contexts, that any prudential-aspects concerning any average mind too, would always be largely average; but in any case prudence demands that, in order to fruitfully cultivate the desired 'sense of history' as already delved upon earlier and also, to avoid any pitfalls of the any possible 'confused-state of thoughts', it remains timelessly paramountcy to keep diving into the same storehouse of the knowledge of history, specifically contextual to one's lines of the continuing studies at any stage and also, since as one's engaging pursuits for the intended mental enjoyment and the resultant intellectual enrichment-cum-satisfaction as well.

Now coming back to the historically continuing multiple-aspects of the 'great-games', we may also recall that the phenomena of interestingly both any intra-national as well as international-level 'great-games' in some forms or the other, have been continuing onto the man's expanding civilisational-canvas, since the dawn of man's living-history itself and that too regardless of the consequences both good and bad, to the man, the multiple-aspects of his overall environment and, also to the advancing or even perhaps at times declining—various human-civilisations as well. If the man's multi-dimensional living-history, is the story of the various advancing and/or gradually declining human-civilisations and also, the chronological records with their historic datelines as well as, their critical interpretations-cum-narrations of the various events and developments, then it is also said that the man's same living-history is no less historically momentous records of the man's continuing legacies of the warfare and also, the differently intended other kinds of struggles too. It has been rightly said that the war or perhaps some sort of the 'culture of war' itself, and although howsoever it may quite appropriately be considered always to be undesirable or even beastly in the humanity's terms; but it remains historically also true that, war had remained rather unfortunately, an unavoidable part of the almost all primitive societies too. And as since already talked about earlier in a number of different contexts, naturally when even the remotely various ancient human-civilisations gradually began to sprout and, slowly but consistently commenced to multi-dimensionally fructify and flourish in the different geographical continents/locations these historically captivating phenomena thereby, triggered the momentous emergence of the man's indeed all encompassing **'civilisational-canvas'** as well. Simultaneously therefore, as these civilisations began to gradually advance, that 'civilisational-canvas' too, commenced expanding within the supremely infinite 'canvas of time', and thus contextually we have also said that, all these different kinds of the human-civilisations have been phenomenally advancing or even at times some of them seen to be declining as well, in fact nowhere but, onto the man's that **expanding civilisational-canvas** itself. Historically it remains factual that, these multiply interrelated phenomena have always been manifesting within the perpetually ongoing march of time but at the same time also, they can always be interpreted and elucidated in different manners. We also find the phenomenal fructifications-cum-interplay of the various kinds of 'great-games', being always centrally instrumental in the ongoing multiple-making of the further momentously numerous historic developments and events. At the same time, **we do also notice that in these momentously history-making developments and events, since consequential of the various kinds of 'great-games and their interplays', the main instruments for playing them, by the powers that be have been first towards: the geostrategically-cum-politico-militarily preparing for the any intended wars and then fighting them, as per their larger strategic-cum-tactical plans and also secondly; indulging in the all sorts of coercive-cum-crafty diplomacy too, in all possible manners**. In precisely the above highlighted contexts; we may quite appropriately recall the well-known and since universally applicable **thoughts of Clausewitz** that, the war is nothing but an extension-cum-fulfillment of policies/strategic-aims by other means. Since the ancient of times, the most powerful urge for going into wars and also resorting to the other means, has been the irrepressible desires for the power and glory, and also if the need be, then to further expand one's empires or territorial possessions. To begin with first we do also contextually find that all these historic aspects have been quite true in case of the ancient Egyptian, Mesopotamian, Chinese, Indian, Greek, Roman, Western, Islamic, or any other ancient and medieval-era human-civilisations all over the world. Then further as we critically continue to have a panoramic survey of the advancing or at times, even declining human-civilisations, we indeed continue to witness momentous manifestations of the same phenomena with many kinds of the changed textures and fallouts. In other words we may also say that, while critically interpreting-cum-analyzing these ongoing multi-dimensional saga, we do find that the same saga in turn, do continue to facilitate us to have panoramically captivating glimpses of indeed these multiply history-making phenomena, which in fact remain the principal constituents of the man's living-history, since always continuously in the making, onto his expanding civilisational-canvas itself.

In all these different kinds of the manifesting and ongoing phenomena, since fructifying during different ages, we find the **same irrepressible urges for achieving power and glory and also, for acquiring all possible kinds of the riches, further territorial gains/domination and so on**, in rather unparalleled manners. **'Industrial-Revolution'** indeed brought very many paths-breaking developments, in these specific contexts of the 'great-games' too, with all kinds of their subsequent fallouts, which in turn, were decisively instrumental in almost totally altering the multiple-courses of further ongoing making of the man's living-history itself. We may recall that, these historic aspects have since already been discussed in different contexts, including also the multiple-aspects of these 'great-games' too; in the previous chapters as well and hence, these may not be repeated again. Well, all these may not mean that the different kinds of 'great-games and also their differently resultant fallouts'; since continuously being played as the time has been passing-on, from the beginning of history and, in their differently emerging and situations-based manifestations—of the newer and newer kinds of the 'arenas of the international power-politics'—onto the man's expanding civilisational-canvas and thereby, triggering further making of the man's living-history have always been bad and undesirable for the mankind and its overall well-being in the world. As we look back into the beginning period of the already bygone medieval-age, we may take as an example—the momentous and also historically the far-reaching advent of the **'Magna-Carta'** in 1215 A.D. It phenomenally triggered the multiply gradual fructification of the edifice of the socio-cultural-cum-political traditions of the English liberty. Historically, this remains a noted intra-national example of the 'great-games' to begin with, and also though initially not at all intended to be so far-reaching and in fact, only to put some kind of an effective rein onto the then cruel and highly exploitative tyranny of the ruling British Monarchy by the cornered but no less feudalistic English barons. This subsequently soon triggered a path-breaking phenomenon of the first and unique dawn of the tenets of modern democratic-thoughts and thereby establishment of the British Parliament, into the British Kingdom and thereby rightly being granted the status of being the mother of all Parliaments in the world. We may also cite another great example of the **'Monroe-Doctrine'**, as an effective counter to the newer kinds of the further European imperialistic overseas 'great-games' into the American continents since momentously formulated by the **5th US President James Monroe**, during his presidential terms for eight years, from 1817 to 1825. And needless to say that, these indeed visionary thoughts of an US President then so much both geopolitical-cum-militarily prevented the then overwhelmingly imperialism-cum-greed soaked obsessions of the seafaring European Powers, towards launching fresh adventures into the South American continent. We may recall that all these historically great developments and events, have since been already discussed in the earlier 6th chapter on the 'Medieval Ages, Birth of Empires and Dawn of Democracies in the West'; including their fallouts, which indeed subsequently triggered the far-reaching and also, altogether newer kinds of the history-making: socio-cultural-cum-political revolutions and developments, onto the man's expanding civilisational-canvas and, all across the world.

Further, we have seen historically very violent and extremely unsettling emergence of **Two World-Wars**—one after the other and also, within a short-gap of just 21 years only and that too, within the first half of the recently bygone 20th century itself. Historically in fact, starting the wars with whatever real or perceived compulsions, were 'great-games' as indeed also, being perhaps the most demonic civilisational-negations for the mankind. But thereby, soon the geostrategic-cum-politico-militarily global mobilizations of the so-called 'free-world', against interestingly the so-called 'evil-powers' were indeed some contextually perhaps, shining examples of the different 'great-games' having far-reaching positive fallouts soon for the mankind. And then lo and behold, immediately after the Second World-War in 1945 came the historically unprecedented and also the multi-dimensional manifestations of the **'Cold-War era'**, during the second half the 20th century itself. Then during the cold-war era's ongoing 'great-games', in early 1960s came the highly volatile face-off between the world's the then two Super-Powers—the USA and USSR, then led by President Kennedy and Premier Nikita Khrushchev, during the **Cuban Missile-Crisis days** bringing the whole world on the brink of almost total nuclear annihilation. But the mankind was lucky and, the total nuclear war-mongering soon cooled-down and the rest is history.

Understandably, due to perhaps the unexpected and somewhat sudden emergence of the above mentioned **'Cuban Missile-Crisis' in the 1960s and later-on, concerning the massive deployment of the US cruise-missiles in Europe against the USSR and the Warsaw-Pact nations in the 1980s** the whole world came to a state of panic and thereby, in the entire Western Europe, including the UK, panic-buttons were pressed raising the alarm-bells to get alert and be prepared with whatever possible means, for the impending **global nuclear-war or to begin, even attacks by 'chemical-weapons'**. Recently a news-item pertaining to these historic crises as well as, since precisely concerning the then far-reaching panic-ridden environment all over the Western Europe, has now appeared in *The Times of India* (Ranchi Edition); and in retrospect now, this indeed gives a glimpse of that climax of the 'Cold-War Era' in Europe. This historically interesting 'News-Item' reads thus, "**Queen Elizabeth II** was prepared with a public broadcast in the eventually of a nuclear exchange as part of an imaginary war game exercise that provisioned the Soviet Union launching a chemical attack on the UK, at the height of Cold War back in 1983.... The speech devised by senior Whitehall officials to announce Britain's involvement in what was feared as the Third World-War, was never recorded.... (*Now the declassified speech*) 'Not a single moment did I imagine that this solemn and awful duty would one day fall on me,' reads the imaginary broadcast, released here on Thursday (*i.e. 1st August 2013*) as part of Britain's 30 year de-classification rule for classified documents.... The speech was written for broadcast at midday on March 4, .983 and adds, 'Now this madness of war is once more spreading through the world and our brave country must again prepare itself to survive against great odds. I have never forgotten the sorrow and the pride as my sister and I huddled around the nursery wireless set listening to my father's (King George VI) inspiring words on that fateful day in 1939 [at the start of Second World War]'....'But whatever terrors lie in wait for us all, the qualities that have helped to keep our freedom intact twice already during this sad century will once more be our strength' the speech said.... It was part of an imaginary war game exercise, Wintex-Cimex 83, designed to seem as realistic as possible during which Orange Bloc forces—representing the Soviet Union and its Warsaw Pact allies—launch a chemical weapon attack on the UK.... Blue forces—representing NATO—retaliate with a 'limited-yield' nuclear strike, forcing Orange to initiate a peace process.... The exercise came in the year **that US President Ronald Reagan denounced the Soviet Union as the 'evil empire' and deployed US nuclear cruise missiles to Europe.**"

Next we may analytically look into now the historically multiple-aspects of some kinds of the hegemonic initiatives and thereby soon, leading to the quite frustratingly entangled involvement of the erstwhile USSR in Afghanistan, since beginning with the Soviet armed invasion into that ruggedly mountainous as well as, both geopolitical-cum-militarily and also socio-culturally unwinnable landmass, since situated along the Hindu Kush mountain-ranges and valleys. And hence, within few years of the Soviet invasion of Afghanistan quite unexpectedly the newer kinds of 'great-games' got triggered inside the then Superpower USSR and, for the entire volatile region around Afghanistan as well and thereby, also involving another Superpower the USA too, into this complex 'cobweb'; first to begin with indirectly and subsequently then even directly, with far-reaching consequences interestingly, for both the Superpowers. The far-reaching and indeed substantially, the new history-making fallouts of all these still simmering 'great-games' came as, the ultimate disintegration of USSR and also trapping USA—into the 'haunted and loose sand-ridden battlefields' of geopolitically the always scary Afghanistan—out of which the USA is still struggling to somehow credibly extricate herself.

We may recall that during earlier years of the turmoil-ridden times in Afghanistan and thereby in the then USSR, after **Brezhnev, Mikhail Gorbachev** came to power. During Gorbachev's reign in fact, the withdrawal of Soviet military forces from Afghanistan was initiated, to the great relief to his regime. And as we continue to contextually delve deeper into all these historic aspects as well as, those of the 'great-games' involved therein, we do notice that consequential to all these and also in addition to the then ongoing 'Cold-War era', and its much costly arms-race between those Superpowers, the Soviet economy got rather quite unbearably stretched too far. Not only for the Soviets, but even for the USA too; the then economic burdens of these in fact clearly fruitless 'great-games', were turning out to be quite back-breaking and hence both of them were rather forced to come to the high-table for their **Strategic Arms-limitation Talks (SALT)** and thus, leading to the substantial reduction in their nuclear and other arsenals as well as, their extraordinarily costly delivery-systems. On the home-front during his terms in office, the Soviet President Mikhail Gorbachev, took some visionary initiatives for the socio-economic and also some far-reaching political reforms, since called ***'Perestroika and Glasnost'*, i.e. the restructuring and reforms**. And thereby, he as the then Soviet President did trigger lot many far-reaching economic as well as, other reforms and in addition to some kinds of the partial democratization of the Soviet political-system, its governance models, the bureaucracy and also, even attempting to begin reforms in the entrenched socio-cultural ethos and value-systems. All these multi-dimensional initiatives/actions did put once the mighty Superpower USSR, initially in a state of the daunting 'flux'; which soon appeared to be magnifying as well as, becoming rather unmanageable too, for both the declining Soviet economy and, also the Government and its President. Were it so that perhaps, each the fast collapsing Soviet political-economy, the systems of governance as well as, the Soviet President in a hurry had bitten substantially more, than they could possibly chew? Now, as the time has been passing-by since Gorbachev's highly eventful innings as President and also later, his passing away too, the Soviet Union has already disintegrated into a number of the East European as well as, the Central-Asian sovereign Nation-States and out of their new 'Federation' the principal one, largest and the strongest, is **Russia** and thereby also, all of them have now got reasonably stabilized and geopolitically consolidated and in addition, multiply-progressing in their respective manners. For posterity sake, **President Gorbachev has also written a book, titled *Perestroika—New Thinking for Our Country and the World,*** giving lot many insights into his thoughts and his numerous historic initiatives/decisions and actions while in power there. And understandably therefore, the historic heat and dusts raised by Gorbachev's initiatives and reforms in his home country and also around the world took some time to settle down; and to allow some sort of contextually clear scenario to emerge and thus, enable some relatively immediate assessments of the multiple-fallouts and also globally their after-effects as well. Today, we find that 'Cold-War' is a thing of the past and also, the erstwhile USSR as a Superpower then, stand geopolitically liquidated and thereby heralding a new era in Europe as well as, the whole world. Historically in any case, the USSR was not formed by the erstwhile communist Russia to begin with, by any peacenik means or by exchanging any bouquets of flowers and was destined to wither away; as in case of the various ancient-cum-medieval eras' earlier big and powerful empires in the world.

In his above cited book **Mikhail Gorbachev**, with his precise and visionary perceptions; has written that, "The **concept of restructuring** with all the problems involved had been evolving gradually.... Their analysis then became the basis for the documents of perestroika. Using the recommendations of scientists and experts, our entire potential, all the best that social thought had created, we elaborated the basic ideas and drafted a policy which we subsequently began to implement.... **The first question to arise was one of improving the economic situation, stopping and reversing the unfavourable trends in that sphere**.... To be sure, we saw that these means alone would not impart a great dynamism to the economy. The principal priorities are known to be elsewhere—in a profound structural reorganization of the economy, in reconstruction of its material base, in new technologies, in investment policy changes, and in high standards in management. All that adds up to one thing—acceleration of scientific and technological progress.... **The economy has, of course, been our main concern**. But at the same time we have **set about changing the moral psychological situation in society**. Back in the 1970s many people realized that we could not do without drastic changes in thinking and psychology, in the organization, style and methods of work everywhere—in the Party, the state machinery, and upper echelons. And this has happened, in the Party's Central Committee, in the government, as well as elsewhere ...**People, human beings with all their creative diversity, are the makers of history**. So the initial task of **restructuring—an indispensable condition**, necessary if it is to be successful—is to wake-up those people who have 'fallen asleep' and make them truly active and concerned, to ensure that everyone feels as if he is the master of the country, of his enterprise, office, or institute. This is the main thing.... But we should also see that dishonest people try to exploit these advantages of socialism; they know only their rights, but they do not want to know their duties; they work poorly, shirk and drink hard.... It is essential to look ahead. We must have enough political experience, theoretical scope and civic courage to achieve success, to make sure that perestroika meets the high moral standards of socialism.... **Perestroika means overcoming the stagnation process, breaking down the braking mechanism, creating a dependable and effective mechanism** for the acceleration of social and economic progress and giving it greater dynamism.... **Perestroika means mass initiative**. It is the comprehensive development of democracy, socialist self-government, encouragement of initiative and creative endeavour, improved order and discipline, **more glasnost**, criticism and self-criticism in all spheres of our society. It is utmost respect for the individual and consideration for personal dignity.... **It is true to say that post-revolutionary development underwent difficult stages, largely due to the rude meddling of imperialist forces in our internal affairs** (*?*); policy miscalculations also occurred.... Industrialization in the twenties and thirties was really a very hard trial. But let's now, with hindsight, try to answer the question: Was it necessary? Could such a vast country such as ours have lived in the twentieth century without being an industrially developed state? (*Indeed, very moot points.*) There was another reason that very soon made it clear that we had no option but to speed up industrialization. As early as 1933 the threat of fascism began to grow very swiftly. And where would the world now be if the Soviet Union had not blocked the road for Hitler's war machine? Our people routed fascism with the might created by them in the twenties and thirties. (*Accepted, but it is wrong to ignore, the no less bigger and mightier multiple-efforts of the Western Powers too, in these regards.*) Had there been no industrialization, we would have been unarmed before fascism (*quite true.*).... Lively debates and sometimes polemics started in the media. The country's past began to be assessed critically. Thousands of people—workers, farmers, and intellectuals—eagerly took part in these debates—at meeting of their collectives, in the press and in letters to the highest Party and government bodies which contained both criticism and suggestions. Different, and sometimes directly opposite points of view were expressed on many specific problems, and an eager search for a way out of the existing situation was openly launched. We consider such plurality of opinion both natural and useful.... **A new thinking is not an improvisation, nor a mental exercise. It is a result of serious reflections on the realities of today's world, of the understanding that a responsible attitude to policy demands scientific substantiation, and that some of the postulates which seemed unshakable before should be given up**....(*And in conclusion pages*) Great are the dangers facing the mankind. There are enough elements of confrontation, but the forces wishing and capable of stopping and overcoming that confrontation are growing in strength and scope before our very eyes.... Moving from suspicion and hostility to confidence, from a 'balance of fear' to a balance of reason and goodwill, from narrow nationalist egoism to cooperation—this is what we are urging. This is the soul of our peace initiatives, and for this we shall continue tirelessly to work.... **There is a great thirst for mutual understanding and mutual communication in the world. It is felt among the politicians it is gaining momentum among the intelligentsia, representatives of culture, and the public at large**. And if the Russian word 'perestroika' has easily entered the international lexicon, it is due to more than just interest in what is going on in the Soviet Union. **Now the whole world needs restructuring, i.e. progressive development, a fundamental change (*How true, but even now, including the Post-USSR Russia too.*).**" (Gorbachev, 1987, pp. 27, 28, 29, 31, 34, 38, 39, 61, 145, 253 & 254)

To be contextually precise here, while undertaking a specific example concerning since the visionary thoughts-cum-programmes expressed by Gorbachev, in the above quoted paragraphs during his indeed history-making Presidency, as he initiated the multiply far-reaching action-plans towards the then USSR's momentous *'perestroika and glasnost'*, he could not have intended or even remotely visualised, the historically so path-breaking as well as fast manifestations of the subsequent developments and events such as—the sudden disintegration of once the unchallenged Superpower USSR—and thereby, the speedy withering away of the Soviet communism. **These momentous developments in turn then, pulled down the final curtain on the globally menacing culture of 'cold-war' and also indeed, geopolitically changed the subsequent texture of the modern world. And needless to say that, the rest might have looked for the time being, some reassuring history for the mankind, but the future is still appearing to be somewhat in a state of flux.**

Precisely in these multiple-contexts, we may undertake another historic example from the medieval-era Mughal India, to further contextually substantiate these internationally valid perceptions, i.e. when historically the most glorious **Mughal Emperor Akbar** was ruling the Indian subcontinent he had initiated a **new religion as *'Din-e-Ilahi'***, and also no less sincerely and persuasively, tried to propagate his new religious thoughts. But it is a different matter that his new and indeed, so humane and large-hearted 'religion' could not take roots in his empire and flourish, but due to his tolerant and humanistic spiritual thoughts and, also consequent to many other socio-politically harmony intended initiatives and *'Shahi-firmans'* i.e. 'Royal-writs'; he was indeed able to substantially develop and socially firm-up the culture of religious tolerance, brotherhood and, even promote inter-religious marriages and that too, without any compulsory religious conversion as a precondition, by sincerely setting-up his personal examples. But again diametrically opposite of the above cited historic example and that too, during the India's medieval-era Mughal period itself, as we look into the thoughts and policies of the historically last and noted **Mughal emperor Aurangzeb**, contextually the then emerged picture was totally different. Much of the Akbar's such far-reaching and indeed praiseworthy socio-cultural-cum-political achievements and marvels were rather 'foolishly' undone, with not only serious fallouts for Aurangzeb himself but, even adversely affecting the Mughal Empire itself and thus leading to its fast decline and withering away. Now, **in their historically broader perspectives we may also say that, the eras of both Emperors Akbar and his great-grandson Aurangzeb respectively, and both during the medieval India's Mughal period itself, their 'great-games' during their ruling years, were indeed the 'two sides of the same coin'; one being so fruitful for the empire and ironically, the other being even so destructive for the same empire**. While again coming back to the Gorbachev-era in the then USSR, and also concerning the post-Gorbachev period in the present non-communist Russia, we have come across interestingly a recorded interview of Late **Gorbachev**, as conducted by **Mary Dejevsky** and, as since published in *The Statesman'*; therein, Mary records that, "It is almost 19 years (*i.e. with respect to the 2010 only*) now since Mikhail Gorbachev announced his resignation as President and, with it, the dissolution of the Soviet Union.... How things have changed. Foros, the small Crimean resort where he was placed under house arrest by the conspirators, is now in Ukraine. Today's Russian Presidents make do with more stolid residences further east at Sochi. **Gorbachev out of office for the best part of a generation, has been elevated to the ranks of the global great and good.... Like so many Soviet and Russian reformers, Gorbachev traces the first major change in his outlook to Khrushchev's 1956 denunciation of Stalin and the cultural 'thaw' that followed**. He says he gravitated to the ranks of Russia's Shestidesyatniki (sixties generation), craving change, and came to realise that **serious reform was not 'just a matter of people, but of the system'**.... Today, Gorbachev says he would turn to social democracy because of its ability to 'bring together capitalism—and social priorities. But he is far from dismissive of capitalism. 'It acts as a stimulus, it has serious achievements to its name and shouldn't be rejected just like that.' Deep down, Gorbachev also retains something of the Leninist he was as Soviet leader, while admitting that Lenin's relationship with democracy left something to be desired. The way he seized power was not exactly democratic, to put it mildly'. But among Lenin's great merits, from his points of view, was the 'wisdom and decisiveness' with which he recognised the early mistakes of the **Bolsheviks** and changed course.... For Gorbachev, it is clear, the battles he had to fight in pursuit of Perestroika and glasnost—with the conservatives of the party hierarchy to one side, and the **bombastic Russian leader, Boris Yeltsin** to the other—rankle almost as much today as they did then, in the late 1980s. He cannot forgive Yeltsin, not just for his stubborn refusal to compromise as the Soviet Union collapsed, but for the 'chaos' that, in his view, the Russian President unleashed on the land once he held power, the 'wild capitalism', the 'privatisation at a stroke' that led to what Gorbachev still deplores as the 'plundering' of the country.... But for the collapse of the Soviet Union, he also blames himself. In the end, he says, 'we'—and I think he means his small band of Kremlin reformers here—should have prevented it'. He goes on, 'Mostly though, I reproach myself even today. **We called free elections, carried out political reforms, tried to build a modern Parliament—and the reality was that the Communist nomenklatura could not withstand the test of democracy and**

freedom'.... Self-critical Gorbachev may be, but some more recent judgments by outsiders seem to wound. 'People criticise me for indecision,' he says, 'But, in general, I think we went too fast. A country with our history should have taken an evolutionary course. I said reforms should need 20 or 30 years—but such passions were raging, as glasnost and perestroika gained pace, and the calls were mostly to go faster, faster—he thumps the table in time—'to go on, go on'.... Does he then agree with today's Russian Prime Minister **Vladimir Putin**, who as President, described the dissolution of the Soviet Union as 'one of the greatest geopolitical catastrophes of the 20th century'? Well, says **Gorbachev**, 'I think his language was a bit extreme, but the destruction, the dissolution of the Soviet Union is something worth discussing'.... So is he disappointed with how things have turned out, especially the slow progress in modernising Russia? 'No, that's not what disappoints me', he replies. 'I'm disappointed with the way the next administration (Yeltsin's) abandoned evolutionary change and launched 'shock therapy' with the result that everything disintegrated: state structure, the army, culture, medicine, education, everything. Destabilisation became the number one problem'.... It is only just now, he says that Putin has acknowledged that nothing can be done without democracy. 'If he acts accordingly, then he'll be on the right track'. But he is not underestimating the difficulty. 'We can't modernise the economy in isolation from all else, we have to modernise society generally, but that's very difficult, very difficult to do."

In these far-reaching multiple-contexts and also, while critically delving deeper in the above quoted paragraphs, we find that there are indeed geopolitical-cum-strategically very valuable lessons in their differently specific contexts, for the numerous European and Afro-Asian Nation-States and also, for **no less including India as well**. Now, again coming to the above discussed so far, the multiple-aspects of the always ongoing phenomena of both intra-national and the international 'great-games' in the world; by picking-up just few of the examples from pages of the man's living-history, since continuously in the making ; we may also say that all these do belong broadly to the one side of this 'coin' of the 'great-games'; hence, ironically then on the other side, the various contextual stories continue to remain sadly very different. But in any case these too, deserve the equally engaging attention in these regards, and perhaps all the more so even if they continue to be quite unsettling in many ways and also, no less being the nauseating negations of the various human-civilisations as well. **Moreover as time has been perpetually passing by the multiple-kinds of civilisational negations, the certain kinds of developments-cum-events belonging to since the other side of that 'coin', have been dangerously enhancing in the multiply worrisome manners for the mankind. And that remains an irony for the human-civilisations the world over and thus, giving births to such unprecedented situations and challenges of the Islamic Jihadi-terrorisms and also of the other kinds too with which the man is now sadly found to be struggling to come to terms with, so that he remains himself safe and keeps his civilisations also secured and thereby, these remain in a state of multiply-thriving as well**. But at the same time, the irony of ironies is that, many a times the earlier discussed multiple-aspects of again the multiple kinds of these 'great-games—since being played onto the highly complex, fast altering, and also multi-dimensional 'arena of the international power-politics'—for the differently focussed aims/ambitions to achieve more power and glory, additional territorial gains, as well as also for, cornering more and more resources and wealth from some geopolitically alien lands, or may even at times, be apparently for some laudable purposes too. And if on one side, these have been throwing-up new opportunities and the multiple-wherewithal, for neutralizing the above mentioned menaces/ dangers, then at the same time perhaps paradoxically, some of them have also been instrumental in creating new menaces and dangerous problems along their paths as well.

And these menaces-cum-problems then, incubating and at times fast fructifying along the way and also based on any particular situation-specific emerging factors may have some kinds of the intra-national fallouts for the concerned nation(s) and/or also, their some international manifestations. As we may proceed contextually exploring into these additional aspects and dimensions of the 'great-games', we may find that our emerging analytical perceptions in these regards substantially tend to become conjectural in their nature and narrations, and that in any case are understandable too. Contextually in these multiple perspectives and also precisely in the any particularly given situation, the centrally pivotal point may desirably needing to be borne in mind is that, whatever may be the futuristically articulated aims/ambitions and thereby, the geopolitical-cum-strategically fine-tuned and proactive 'plans of actions', that might have been prepared by any relatively more powerful nation or even by any Superpower concerning that situation-specific 'great-games' to be played, and since a number of the consequential-cum-new developments and events soon begin to unfold, it has also interestingly been noticed that many a times some unforeseen fallouts begin to surface fast, and some of them not so favourable too. And thereby as a further quite weighty consequence, they may even at times go out of the any geopolitically tangible control of that very 'powerful' nation or even that of the concerned Superpower as well. In other words and in their multiple-contexts, we may also say that conjecturally, these happen to be some of the perhaps geopolitically inherent 'ifs and buts' in the always multi-dimensional domain of since the already above elaborated continuing phenomena of the 'great-games'. But in addition, at the same time it is also equally true that, those consequentially unfolding developments and events and their other allied fallouts too are all interrelated and mutually influencing as well, in their some forms or other and also, in their further history-making manners onto the man's expanding civilisational-canvas itself.

Now in these larger perspectives, while coming back to the post-World War II era of the then so overwhelming cold-war times, and as discussed earlier, the unchecked arms-race scenario between the two erstwhile Superpowers of the 20th century, an extremely obsessive culture was getting manifested—in multiple-forms of the unprecedented phenomena of the so fast setting-up of massive **'military-industrial complexes'**, their allied R&D outfits, marketing as well as their highly focussed management-systems and, their installations—strategically almost all over the USA which in any case since its inception, has always remained a free-market-cum-private-sector oriented the most powerful and also unassailably the largest capitalists-systems' based economy in the world. These multiply vast and also highly complex aspects concerning the USA have also since been critically analysed and discussed in many different perspectives the world over, by numerous scholars, academicians, political-economy experts, and the strategic-thinkers. Hence these are befittingly for the concerned experts to further delve deeper and in any case thereby, these do fall outside our duly specified and also relatively limited domain of the ongoing discussions. Having said that, we may have to always contextually bear in our minds and as also already emphasised earlier that all these and everything else concerning almost all kinds of developments and events taking place anywhere in the world since ancient times, have mostly been closely, directly or indirectly or, even remotely, interrelated and mutually influencing as well. The **differently evolving high-potential seeds of the numerous kinds of 'great-games' do get sown-in**, and then in course of time sooner or later they get incubated and begin to sprout and hence further-on also, phenomenally begin to fructify with multiple consequences in both short and long terms perspectives for the mankind. The inherently basic human socio-psychological-cum-cultural level desires for the any multiply feasible 'great-games', with always changing situations-based ambitions have been rather overwhelmingly there, onto the man's expanding civilisational-canvas, since the very ancient of times. Hence as since seen down the always multiplying pages of the man's 'living-history' itself their changing patterns, gradual fructifications as well as, their game-changing manners themselves, do get dictated by other simultaneously manifesting factors such as, the ongoing advancements of the sciences and technologies and thereby—since, as the time keeps-on moving ahead—triggering of the multiply gradual increase in the mankind's various kinds of the wherewithal and resources too.

Concerning the world's most powerful and also the largest economy the USA and, especially in the multiple-contexts of her multi-dimensionally massive 'military-industrial complexes'; **Professor John Kenneth Galbraith has mentioned that,** "*The New Industrial State* pictures the world of the large corporation as the outgrowth of the neoclassical world of monopoly and oligopoly. At least by implication what was left behind was the world of the competitive market. That also I here correct; what is left behind is, in fact, something resembling the neoclassical admixture of entrepreneurial monopoly, oligopoly and competitions. The consequence of so seeing matters is a better explanation of the behavior of the entrepreneurial firm and what here I call the market-system. It shows, also, the world of the large corporation is something new—that it is a clear break with what is described by traditional doctrine.... The practical effectiveness of a particular monetary action is enjoyably debated. No one can tell what the effect of a given tightening of the money supply—with accompanying increase in the interest rate—will have. Often there can be considerable such tightening with no visible effect, and there comes a sharp shrinkage in borrowing, investment and demand with, on occasion, disconcerting effect on output and employment. There is some uncertainty as to the borrower's response to lower interest rates and easier borrowing terms. Unless this is accompanied by seemingly good prospects for selling goods or houses, nothing much may happen. Partly because of this uncertainty central bank actions always reflect the collective judgement of men who, individually, are in ignorance of the immediate consequences of their action. This and the intense, priestly and deeply lugubrious discussion which surrounds central bank policy serve, on the whole rather effectively, to disguise the uncertainty as to its effects.... Given what was not foreseen, the future of economics could be rather bright. It could be in touch with the gravest problems of our time. Whether this is so—whether economics is important—is up to the economists. They can, if they are determined, be unimportant; they can, if they prefer a comfortable home life and regular hours, continue to make a living of the infinitely interesting gadgetry in disguise. They will, as was the case in the summer of 1971 when price controls were imposed in the United States, or a year later when they were put into effect in Britain, have few and no words of guidance or advice on great issues. They will be socially more relevant than Keynes's dentist, for he would feel obliged to have a recommendation were everyone's teeth, in conflict with all expectation, suddenly to fall out." (*Interesting, isn't it? Well, learned economists understandably, have their own logical methodologies of putting across their perceptive views. But needless to say that, whatever; the economists' intellectually innovative thoughts and expert analyses, even if they may be generating many contextually conjectural debates across any concerned democratically vibrant and liberal Nation-State, excluding any totalitarian, dictatorial or thereby, even 'failed/failing' nations, the economists quite discerningly, are always able to command important relevance and power, in the larger strategic-domain.*) (Galbraith, 1973, pp. 22, 23 & 312)

In any case and quite understandably, almost all noted nations, or any geopolitical conglomerate of nations including the EU (i.e. European Union) and hence, what to talk of India, do get always strategically concerned and begin to closely study and examine, every important perception and published thought, whether economic or otherwise and since related to the USA in their different contexts and strategic imperatives, for setting their own multiply strategic 'compasses' right, to serve their needs. The universally pivotal-point in these regards and, in today's super high-tech era modern world anything of consequence, whether intra-nationally or internationally or, both in varying combinations, happening or being published in the USA, whether one may relish or not, always do perforce have different kinds of the global ramifications. Well, that's one of the numerous dimensions regarding being a Superpower; and that too, being a singular one and, the largest economy, as well as, the ultra high-tech industrial nation in today's world. Hence no wonder and though in a lighter vein; it has been quite rightly said that, **'if the USA sneezes in any manner, the whole world appears to be catching cold'**. Hence the highly consequential and also, the all-encompassing-cum-magnifying manifestations of the modern times' super-connectivity across the world have in turn, multiply multiplied the intensities of the economic linkages amongst almost all nations in today's world, and wherein the USA for the numerously obvious reasons, has emerged as the singularly most decisive and dominant player in these regards.

In the above multiple-contexts, **Professor Galbraith** has mentioned that, "**Wealth is not without its advantages and the case to the contrary, although it has been made, has never proved widely persuasive. But, beyond doubt, wealth is the relentless enemy of understanding** (*Interesting, isn't it?*). The poor man has always a precise view of his problem and its remedy: he hasn't enough and he needs more. The rich man can assume or imagine a much greater variety of ills and he will be correspondingly less certain of their remedy. Also, until he learns to live with his wealth, he will have a well-observed tendency to put it in the wrong purposes or otherwise to make himself foolish. As with individuals, and so is with nations. And the experience of nations with well-being is exceedingly brief. Nearly all, throughout all history, have been very poor. The exception, almost insignificant in the whole span of human existence, has been the last few generations in the comparatively small corner of the world populated by Europeans. Here, and especially in the United States, there has been great and quite unprecedented affluence.... **Armies would still cross national frontiers and with considerable capacity for doing damage where they fought and trod. But the economic consequences of national armies in the age of nationalism have been almost infinitely small as compared with the damage wrought by feudal, marauding, or crusading armies in the centuries before**.... The people of the so-called backward countries have frequently heard from their presumptively more advanced mentors in the economically more advanced lands that they should be patient about social reform, with all its disturbing and even revolutionary implications, and concentrate on increasing production. It can be remarkably inappropriate advice. Reform is not something that can be made to wait on productive advance. It may be prerequisite to such advance.... However there was never any dramatic assault on the older confidence that, rising prices and inflation would cure themselves. Keynes led the attack on the conviction that depressions were self-correcting by picturing the possibility, indeed the probability, that the economy would find its equilibrium with an unspecified amount of employment. Dozens seized his point and sought to persuade the politicians and the public. The notion that, in peacetime, prices might as a normal thing rise continuously and persistently has had no Keynes.... The research programme will be built around the need to advise the 'selling points' and 'advertising pegs' or to accelerate 'planned obsolescence'. All this suggests that the incentive will be to allocate research resources to what, in some sense, are the least important things. The quantity is more impressive than the way it is allocated. Still one would wish to suggest that the American economy is delinquent in the attention it devotes to change and improvement in consumer goods. Clearly it is not.... These incentives, however, operate over but a small part of total scientific and research activity and, indeed, over but a small part of that potentially applicable to the production of goods. Thus a very large amount of very useful research cannot be specialized to be sustained by any marketable product. This is most obviously true of much so-called basic research. But it is also true of a large amount of applied effort. **The modern air transport is the stepchild of the military aeroplane**. It would never have sustained the underlying research and development endeavour on its own. The same is true in even greater degree of the non-military uses of the nuclear-energy, and with the communications satellite and the electronic computer." (Galbraith, 1958, pp. 31, 47, 104, 176 & 226)

Interestingly in our broader perspectives here, as we may delve little deeper into the above quoted paragraphs by Professor Galbraith, we do also contextually notice that our earlier rather repeatedly emphasised perception that, along with the unendingly ongoing march of the time itself, almost all of the history-making developments and events since continuing to occur from time to time and, onto the man's expanding civilisational-canvas; happen to appear to be always closely, directly, indirectly, or even perhaps remotely, interrelated as well as, mutually influencing too. In fact, precisely substantiating the above mentioned multi-dimensional realities concerning the ongoing making of the man's 'living-history' onto his expanding civilisational-canvas itself, that Professor Galbraith is seen to be elucidating in his so distinguished scholarly manner, when he says that **'the modern air transport is the stepchild of the military aeroplane'**. Hence no wonder that, similarly any of all the areas/disciplines of the continuously advancing and multiply thriving sciences, medical-sciences and the technologies oriented consistent 'R&D' efforts in both civil and military fields and wherein, best scientific-cum-technical brains all across the world are since relentlessly engaged and working to achieve newer and newer scientific-knowhow and cutting-edge technologies and also thereby, acquiring expertise over the breakthroughs towards their numerously new uses.

It is quite befittingly true that, the military's systems of command have to be perforce 'militarily-regimental' in its functioning and, understandably these may appear to be perhaps quite 'autocratic' to any concerned larger civil-societies, but at the same time wherever it may be applicable it would rather be a social-fallacy to assume that the concerned military in any manner, remains quite aloof or unaffected by the tenets and the already functioning systems of the democracy in its own sovereign Nation-State. **Soldiers in military-uniforms, and that too in today's fast advancing and high-tech-oriented super-I.T. Age are much more multiply-aware, educated, tech-suave, and with free thinking minds; and thereby, they can't possibly be by any kind of the politico-bureaucratic or even any military-level gimmick and hence obviously, the last thing they would relish or expect is, any kind of the 'culture of fooling around'**. It is needless to say that in any meaningful and vibrant democracy the larger societies are quite liberal and thereby, the peoples are quite confident about their since constitutional rights to the freedom of thoughts, speech, writings, and their free media (both print and the electronic). Accordingly at the same time the people in their military-uniforms too therein, and especially after their off-duty hours as well as, during their leave-days interestingly get rather no less conscious about their constitutional rights and obligations. Hence, the always multiply-conscious as well as, professionally proactive-hierarchy of the military-command systems therein too; remain no less rather appreciatively aware of these dimensions, as since now in the modern democratic India and, similarly in other liberal democracies across the world. But at the same time, it remains equally true that concerning these important dimensions of the 'matters-military', the military officer-corps do quite proficiently and also effectively always ensure that these aspects, on the part of their men and women in uniforms are always befittingly peppered by the inviolable norms-cum-soldierly-compulsions, regarding the necessarily mandatory maintenance of the healthy military-discipline, professional decorum, and the always uncompromisingly expected gentlemanly thoughts-cum-behaviours; irrespective of the any kinds of challenges and emerging situations anywhere during their stay, movement/journey, and their deployment. Therefore any slip-up whether minor or major, on the part of any person(s) at any level in these regards are always frowned-upon and taken very seriously by the concerned superiors and thereby wherever possible, severe punitive actions are promptly taken as per the applicable military legal-provisions.

But in no circumstances all these do ever mean that, these brave and disciplined military-men, outside their cantonments or the areas of duties are ever expected to quietly tolerate any kinds of the excesses or uncivilised conducts from any other quarters. In spite of all these, it needs to be appreciated and that too in any liberal democracy that even uniformed people can't be kept socio-cultural-cum-politically insulated and isolated in any manner and hence, any developments or happenings, whether good or undesirable are bound to affect the military-men as well. **In addition, there is always and otherwise no less desirable too that the military in any vibrant democracy, constitutionally, willingly, and healthily, is supposed to function under the civilian supremacy at the political-level, and thereby it remains always a matter of paramountcy importance, that the politico-bureaucratic combine must always be multiply-cum-befittingly conscious of their important duties, obligations, and responsibilities; towards the perpetually crucial 'matters-military', and the military-personnel, in everyone's collectively highest-level national security, and other vital interests of the concerned democratic nation**. Well, since now already to everybody's knowledge not only internally, but ironically perhaps in abroad too, that it remains a 'million-dollar question' as to how much the modern Indian Nation-State, her political-class and the bureaucracy have meaningfully been able to cultivate their always duty-bound awareness in these regards and, as also elucidated above, in the areas of their official-cum-social functioning at their various levels, in the best interests of the nation? And needless to say that here too precisely, comes-in, the always paramountcy importance of **the multiple-aspects of 'essential strategic-culture and thoughts'**. In the above multiple-contexts, we may need to bear in our minds that, with respects to the any meaningfully vibrant-cum-liberal democracy and then, different models of totalitarianism-cum-dictatorship-oriented country, the entire system of constitutionality of either of these two kinds of the any concerned Nation-State, its systems of governance, socio-cultural-economic-cum-political ethos and value-systems are always totally, as well as irreconcilably rather 'poles-apart'. Interestingly if that be the case, and that's what the facts are then it is naturally logical that, in their respective vast and highly complex domains of the 'matters-military' too numerous things both at human as well as, military professional-levels, would also be largely different and in addition, each one's both socio-cultural and professional ethos would also be no less 'poles-apart'. In these quite engaging perspectives, whatever may be the many differing perceptions and opinions and also the lines of contextual debates, one thing within the man's expanding civilisational-canvas itself, remain an established fact that, so far the mankind is yet to evolve and develop any system of the socio-cultural, as well as political governance of the any concerned Nation-State—even slightly superior to that of the democracy. It is contextually agreed that, democracy anywhere whether in any relatively older or more developed or, in somewhat newer and thereby, a fledgling democracy, there are mostly any differently situations-specific interestingly, lots of the all kinds of 'heat and dusts' getting raised but at the same time, all these have their own no less fascinating 'political-flavours', which always keep the concerned peoples fairly engaged and amused too. Well at the same time, it is also all the more interestingly true that in any liberally vibrant democracy, the intensities of the above noted mostly 'multi-coloured' aspects, do always get magnified manifolds, when the periodically regular 'election-fever' grips the entire any concerned democratic Nation-State, in multiply different manners, since located in any far-off continent of the world. It is also a no less interesting matter that, symptoms and the quite 'colourful' complexities of the such 'election-fevers' would be of quite different kinds in the different democratic 'Nations-State', depending on her specific historical, socio-psychological-cum-cultural, politically systemic-ethos and also since, similarly other variables too. **Now coming back to specifically the multiple-aspects of the already earlier discussed gamut of the 'great-games' we may contextually say that in any liberally vibrant democracy, the multi-dimensionally ongoing systems of the functioning of that democracy too appear to be no less fascinating and also, 'flavoured' example of that particular nation's intra-national 'great-game'; and at the same time, any such 'example' does always remain an excitingly welcome development, onto the man's expanding civilisational-canvas in the modern world**.

Precisely in the above multiple-contexts, **Bertrand Russell** has very thought-provokingly mentioned that, "Democracy, in a modern great State, has certain disadvantages, not, indeed, as compared with other forms of government over the same area, but inevitably owing to the immense population concerned. In antiquity, the representative system being unknown, the citizens assembled in one market-place voted personally on each issue. So long as the State was confined to a single city, this gave to each citizen a sense of real power and responsibility, the more so as most of the issues were such as his own experience enabled him to understand. But owing to the absence of an elected legislature, democracy could not extend over a wide area.... The geographical difficulty was overcome, in the modern world, by the practice of choosing representatives. Until very recently, the representative, once chosen, had considerable independent power, since men living at a distance from the capital could not know what was happening soon enough, or in sufficient detail, to be able to express their opinion effectively. Now, however, owing to broadcasting, rapid mobility, newspapers etc., large countries have become more and more like the City States of antiquity; there is more personal contact (of a sort) between men at the centre and voters at a distance; followers can bring pressure on leaders, and leaders reciprocally can exert influence on followers, to an extent which was impossible in the eighteenth and nineteenth centuries.... **Democracy, when it is new, arises from resentment against the previous holders of power; but so long as it is new it is unstable**. Men who represent themselves as enemies of the old Monarchs or oligarchs may succeed in restoring a monarchical or oligarchical system: Napoleon and Hitler could win public support when the Bourbons and Hohenzollern could not. **It is only where democracy has lasted long enough to become traditional that it is stable**. Cromwell, Napoleon and Hitler appeared in the early days of democracy in their respective countries; in view of the first two, the third should be in no way surprising. Nor is there reason to suppose him more permanent than his predecessors.... There are, however, some serious reasons for doubting whether, in the near future, democracy is likely to recover the prestige that it had in the latter half of the nineteenth century. **We have been saying that, in order to become stable, it must become traditional. What chance had it of becoming sufficiently established in Eastern Europe and Asia to begin the process of becoming traditional?Government has at all times been greatly affected by military technique**. In the days when Rome was tending towards democracy, Roman armies were composed of Roman citizens; it was the substitution of professional armies that brought about the Empire. The strength of the feudal aristocracy depended upon the impregnability of castles, which ended with the introduction of artillery. **The large almost untrained armies of the**

French-Revolution, by defeating the small professional armies opposed to them, showed the importance of popular enthusiasm for the cause, and thereby suggested the military advantages of democracy." (Russel, 1948, pp. 201, 202, 208 and 209)

Also, precisely in the above multiple-contexts, **William Greider** has very perceptively stated that, "The common wisdom of politics has therefore settled on a much narrower idea of what elections are for—election are a search for good character. Issues and ideas may provide the fodder of campaigns, but these are mainly useful in illuminating the temperament of the men and women who are running. Are they wise and honest people? Or do they have some hidden flaw? This definition allows conscientious observers to retain their faith in the electoral process and defend its efficacy.... If the search is mainly for good character, then the only 'issues' that matter in electoral politics are the ones that may help identify the candidate's personal qualities (or denigrate his opponent's). Thus, campaigns turn naturally to irrelevances that provide quick emotional attachments, but have no connection to the real sources of popular discontent.... Electoral politics in the age of mass communications serves as an elaborate mask—concealing what goes on in government from the untutored mass of voters. But, if the voters have only weak influence over those governing decisions, who does influence them? That is the question neither political party will discuss with any candor, but citizens at large have inferred the answer. In the 1060s, surveys found that 28 percent of the public was convinced that 'the government is pretty much run by a few big interests looking out for themselves. 'A generation later, this resigned view of politics was held by two third of the people. (*Well, what about now? could be anybody's guess.*).... **The gleaming temples of democracy that tourists visit in Washington, the marble shrines to great leaders and great ideals, are no longer an appropriate emblem for the nation's capital. Washington now is more aptly visualized as a great bazaar**—a steamy market place of tents, stalls and noisy peddlers. The din of buying and selling drowns out patriotic music.... **The art of governing—making laws for the nation and upholding them—has been reduced to a busy commerce in deal making**. (*How true. Interestingly in India, the story is much more diverse, oriental and 'colourful', noisy and chaotic; at times the Indian Parliament may resemble like a 'fish-market', or even similar to a 'rudderless ship in the midst of a choppy sea'.*). Thousands and thousands deals are transacted every day in diffuse of corners of the city. The rare skills required for the politics at the highest level are trivialized as petty haggling, done with the style and swagger of rug merchants.... The government, however, is a principal violator itself. As the *Times's* exposure of the nuclear weapons scandal suggested, major scandals of lawlessness often involve not just private companies, but installations of the federal government itself. **During a generation of enacting ambitious environmental protection legislation, the government has been without doubt, the single worst polluter in the nation**. Private corporations can always plead that they were merely following the example set in Washington.... The White House apparatus killed, stalled or watered down hundreds of regulations and boasted of the billions it had saved private industry. Warning labels on children's aspirin were abandoned for the drug companies. Safety rules for the underwater divers were weakened for the offshore oil drillers (George Bush's line of business before he entered politics). Cotton-dust controls were blocked for the textile industry. Industrial air pollution standards were set aside for the steel industry. And so on across dozens of fields of federal law enforcement. The White House interveners, intent on secrecy, usually communicated their demands orally to the regulatory agencies. As one said the absence of any written memos or letters, 'leaves no footprints'.... The quality of democracy is not measured in the contentment of the affluent, but in how the political system regards those who lack personal advantages. Such people have never stood in the front ranks of politics, of course, but a generation ago, they had a real presence, at least more than they have now. The challenging conditions they face in their daily lives were once part of the general equation that the political system took into account when it decided the largest economic questions. Now these citizens are absent from politics—both as participants and as the subjects of consideration.... The citizens are not the idle poor, though many hover on the edge of official poverty and virtually all exist in a perpetual condition of economic insecurity. These are working people—the many millions of Americans who fill the society's least glamorous yet essential jobs and rank at the bottom of the ladder in terms of compensation. A large segment of working-class Americans have effectively become invisible to the political debate among governing elites. They are neither seen nor heard nor talked about. Their absence is a crucial element in the general democratic failure of modern politics.... Increasingly people buy a product not because of its benefits but because they identify, or strive to identify, with the kind of people they think they use it.... Rational benefits are vulnerable, because with today's technology it's easy to knock off a competitor's innovation quickly or play on his marketing turf. Emotional bonds, on the other hand, are hard to break.... **The same logic has become the prevailing rule for political competition in the media age. Campaign consultants and managers describe the electoral process in the same dispassionate—and amoral—terms. Elections are for selling, not for governing and certainly not for accountability. The selling depends, not on rational debate or real differences, but on concocting emotional bonds between the candidate and the audience.... The human yearning for democracy is profound and universal**, as the upheavals in Communist nations have most recently confirmed. **It is driven fundamentally by the impulse for individual self-realisation, a desire to discover and establish one's one worth in the collective context of the surrounding world**. The struggle to achieve this is so difficult that the search has proceeded fitfully even in the most congenial settings, such as the United States.... The history of nations has reflected this search for over four or five hundred years as different societies sought to discover democratic meaning and establish its terms. Some places have advanced not at all; some created the forms of democracy but without the substance. Other places—most notably the United States—embraced the idea in their national character and have worked fairly steadily toward improving upon it. **Democracy is a presumption, not that the majority of the voters will always be right, but that real choices will be put before them**." (Greider, 1993, pp. 23, 105, 117, 142, 183, 271, & 417)

Indeed, as we delve deeper into the above quoted paragraphs from William Greider's book, we do begin to face the waves of his exploring and yet so precise thoughts, enunciated in such systemically logical manners that your own minds begin to get overwhelmingly charged on the many aspects in fact, irrefutable sharp-end realities of our democracies as since we continue to practice. What all William Greider might have said so thoughtfully as well as, also so captivatingly engaging for his readers, might have principally been elucidated by keeping in mind the prevailing overall US scenario in these regards, but their such 'stormy' connotations do get no less realistically and also indeed, thought-provokingly echoed in almost all democracies across the world. Democracies, as we do comprehend and practice in the polity of our respective nations and, as since conceptually elucidated in their Constitutions, are not always supposed to be but also, actually meant to be totally above board and transparent so that in any given situation as well as, under any kinds of both internal and external challenges, the peoples concerned are always reassuringly confident to retain their unshakable and proud faith in their democracies. **Interestingly an elephant has two sets of teeth: one for the showing-off in different situations for differing purposes and similarly, the other one for munching and eating his staple foods**. Sadly what we do in fact, witness today in any meaningfully vibrant democracy and of course, in varying proportions and also with different politico-economic-cum-cultural textures at times, are not that what it was meant to be initially and also what with a highest sanctity to begin with, it was enshrined in the concerned nation's Constitution itself. If that be so and then at the same time, it does never mean that a democracy since deeply rooted as a matter of continuingly long tradition in any concerned nation, could even perhaps remotely be in some danger; but certainly as some kinds of the inviolably sanctified tenets, for the any concerned nation's systems for governance and its well-being, do rather quite ironically appear to be somewhat losing their socio-politically dazzling shine as well as, their systemic credibility too. And in any case these too, do always remain capable of giving some worrisome shockwaves as well as, at the same time do also create some socio-cultural-cum-politically all possible kinds of fault-lines; and then these are sadly seen to be running all over the places, in any concerned nations' body-politic and also its democratic-institutions.

Well in any case, each concerned nation being a meaningfully vibrant democracy does also remain uniquely different—with different historical backgrounds, differing socio-psychological-cum-cultural make-up and also thereby—with differing kinds of the politically evolved its Constitution as well. **In fact, these matters are so differently complex and dissimilar in their multiple manifestations that, no standardised template can be possible, and hence into which different democratic nations' differing 'conceptual-framework' in these regards, could fit-in**. But broader symptoms of concerning any substantial success of the democratic functioning or possibly, some kinds of the both macro and micro-levels systemic malaise setting therein, could perhaps have lot many similarities as well. Precisely in these larger perspectives, we do contextually notice indeed the globally broader applicability of William Greider's the above noted and also so perceptively analytical observations, concerning the factual functioning of a democracy in the present modern-era world. And though his above noted thoughts have principally been meant for the American democracy in these regards; but in any case, the universality of any such so broad-based as well as, conceptually deep penetrating thoughts if they be so, cannot remain confined to any political boundaries. Thought-processes and also the contemplatively evolving and knowledgeable concepts concerning any discipline of the human-knowledge as well as, any such intellectual endeavour on the part of mankind all across the world, are indeed perceptively similar to the 'Sun rays' and hence, always capable of being all encompassing and all-pervading as well.

Keeping in our minds the above noted thoughts of **William Greider** we do in addition, also get driven towards looking for some apparently hidden but also no less specific and, always powerfully manifesting in different manners and in differently emerging situations the 'sprouting-seeds' of the triggering reasons, towards the multiply varying situations-based and also momentously history-making materialisation of the numerous kinds of 'great-games' in their both intra-national as well as, transnational incarnations. If we may care to conceptually visualise them in their broader perspectives and, on the perpetually moving-ahead timelines of the man's multiply-vast and highly complex 'living-history' since always in the making, we may very fascinatingly notice that all these are truly indeed, equally inclusive segments of the much larger ongoing-cum-history-making as well as, all-encompassing phenomena continuing to manifest onto the man's expanding civilisational-canvas itself.

Nobel-Laureate **Rabindranath Tagore** has this to say, "A nation, in the sense of the political and economic union of the people, is that aspect which a whole population assumes when organized for a mechanical purpose. Society as such has no ulterior purpose. It is an end in itself. It is a spontaneous self-expression of man as a social being. It is a natural regulation of human relationships so that men can develop ideals of life in co-operation with one another. **It has also a political side, but this is only for a special purpose. It is for self-preservation. It is merely the side of power,**

not of human ideals. And in the early days it had its separate place in society, restricted to the professionals. But when with the help of science and perfecting of organization this power begins to grow and brings in harvests of wealth, then it crosses its boundaries with amazing rapidity. For then it goads all its neighbouring societies with greed of material prosperity, and subsequent mutual jealousy, and by the fear of each other's growth into powerfulness. The time comes when it can stop no longer, for the competition grows vaster, and selfishness attains supremacy. Trading upon the greed and fear of man, it occupies more and more space in society, and at last becomes its ruling force. It is just possible that you have lost through habit consciousness that the living bonds of society are breaking up, and giving place to merely mechanical organization. But you see sign of it everywhere. It is owing to this that war has been declared between man and woman, because the natural thread is snapping which holds them together in harmony; because man is driven to professionalism, producing wealth for himself and others, continually turning the wheels of power for his own sake and for the sake of the universal officialdom, leaving woman alone to wither and to die or to fight her own battle unaided. And thus there where co-operation is natural has intruded competition. The very psychology of men and women about their mutual relation is changing and becoming the psychology of the primitive fighting elements, rather than humanity seeking its completeness through the union based upon mutual self-surrender. For the elements which have lost their living bond of reality has lost the meaning of their existence. Like gaseous particles, forced into a too narrow space, they come in continual conflict with each other till they burst the very arrangement which holds them in bondage." (Tagore, 1937, pp. 9-11)

Ironically whether peoples inhabiting this planet Earth, since the very beginning of the gradual evolution of man himself may or may not like it, 'change is the law of Nature', and hence the same law is perpetually under endorsement regardless, with all its continuingly multi-dimensional dynamism. Tagore is also able to give the intellectually very subtle hints of the various kinds of situations to follow in the social-lives of mankind in this world, consequent to the soon to manifest indeed, so much overwhelmingly continuing and enchanting advancements in the multiple areas of sciences and technologies, onto the man's expanding civilisational-canvas itself. All these phenomena in turn, do keep-on throwing-up the various kinds of the fast manifesting situations of such kinds, which continue to trigger any possible kinds of conflicts and the warlike situations and thus, bringing the men face to face and begin fighting among themselves.

Precisely in the above multiple-contexts as well as, very interestingly highlighting the historic birth, phenomenal rise of now the sole Superpower USA, **Thomas P.M. Barnett** has been able to discuss many dimensions of the so many history-making developments and events and thereby, also the multiple-aspects of the 'great-games', numerous geostrategic and politico-military opportunities, compulsions and challenges; since coming the ways of USA, in a sharply focussed and engaging manners for his readers. And obviously while doing so, he has very critically but indeed, with his brilliant intellectual objectivity has also done very precise appraisals of the USA's geopolitical thoughts and history-making initiatives-cum-actions, as her responses to them from time to time. Thomas Barnett has highlighted, with their both great historicity as well as their far-reaching and contemporary relevance, many history-making aspects of some of the 'great-games', as since dictated by their respective timelines and thereby; as having been played or being played by the mighty USA, onto the man's expanding civilisational-canvas in the world. (Barnett, 2009, pp. 83, 84, 93, 94, 95, 119, 120-122, 131, 132, 191, 192, 225, 379, 397-399)

(i) And hence with respect to the above and also, precisely concerning the American War of Independence and the birth of a new nation the USA, Thomas Barnett of his above referred book, has mentioned that, "In colonial America, the proximate trigger for such revolution was British crown's illegitimate taxes, but the ultimate rationale was that the colonies simply felt they could do better on their own and that Britain was holding them back from their natural potential to dominate the continent, regardless of the geopolitical grief it might cause distant London with Europe's other colonial powers. **As such, George Washington's retiring admonition to avoid 'foreign entanglements' spoke less to avoiding meddling in European affairs than to concentrating all attention on America's westward advance at the expense of those same European powers. We didn't revolt from Britain to become something unlike it but ultimately to grow beyond it in strength and world influence**. (*Interestingly, we do see here the historically momentous, as well as since very fast sprouting and fructifying, seeds of the forthcoming and, soon to flourish US culture of the transnational 'great-games' with truly global reach and the multiply history-making American influences all over the world.* ***And at the same time, in the process; America was also able to, though not nearly totally but substantially at least; cultivate, evolve, groom, and quite fruitfully imbibe the all-important multiple-aspects of the 'essential strategic-culture and thoughts'***.). Not surprisingly, our great divorce from Britain lasted until British needs for protective alliance against Germany early in the twentieth century made a few arrangements possible and necessary. Only then did the 'special relationship' take root. Before that, America was at once dependent on British investment and resentful of British power and prestige—a situation the United States would experience from the other side with many countries in the twentieth century.... Intimidated by the loss of Britain's protection on the high seas and subject to its economic sanctions (*well, once those were the days too; and obviously now, since buried in the perpetually accumulating sands of the time.*), America had the grand strategic goal from the start, as Kagan notes, to become a 'great world power' by establishing an inland empire. (*And thereby, ensuring the multiply fast fructification of the geostrategic, politico-economic-cum-industrial, and the massive military assets-cum-wherewithal in the hands of the USA, to unprecedentedly proceed ahead with her geopolitically unrelenting 'great-games', onto the truly global 'arena of the international power-politics'.*). On this point, all our Founding Fathers were in clear agreement. The real question, however, was the best method for achieving that power, and here is where the matter of grand strategy shaped the future course of American politics by ultimately encouraging the competing political parties, each centered on a particular answer to the question, how best to make America a world power? This development was neither preordained nor desired by the Founding Fathers, virtually all of whom feared divisive 'factions' as a source of internal weakness leading to failed statehood that would invite foreign meddling. Indeed Washington himself saw the possibility of secessions and civil strife as a result of political parties, believing they were inimical to the long-term health of the nation." (*No wonder, the historically unintentional and apparently hidden seeds of the forthcoming 'American civil-war' even within less than a century of the indeed, phenomenal birth of a huge Nation-State of the USA had in fact, perhaps quietly begun to sprout all over the American hinterlands.*). (Barnett, 2009, pp. 83 & 84)

(ii) And thus contextually concerning the very genesis, as well as the final outcome of American civil-war Thomas Barnett has written that, "**From Jackson's election in 1828 until Lincoln's shocking rise to the presidency, in 1860**, the South had effectively achieved political 'safe harbor'. But with Honest Abe's stunning victory, the Whigs' grand strategy of consummating America's potential as a global power was reborn in the newly formed Republican Party, the main stalwarts of which all made it into **Lincoln's original cabinet**, including the party's original standard-bearer, **William Seward, as Lincoln's Secretary of State**. Lincoln and Steward, **Kagan** notes, 'had remained consistent supporters of the American system throughout the Jacksonian era, favoring federal support for internal improvements and a strong federal bank.... **Lincoln's grand plans** came to nothing, thanks to the national economic crisis of 1837, but when later granted the extraordinary powers of a wartime president, he would take up those plans on a continental scale.... **Lincoln believed deeply in the essential equation of American democracy, seeing its truest test in its ability to 'elevate the condition of men'**. In this manner, **historian Doris Kearns Goodwin links Lincoln, Steward, Treasury Secretary Salmon Chase, and Attorney General Edward Bates, 'a team of rivals' for the Republican presidential nomination in 1860, to a new generation of Americans eager to leave behind the eighteenth century and its ills—specifically slavery**.... Of course, Lincoln's very election, leading to the secession of **a slew of Southern states that immediately reformed themselves into a confederacy committed to remaining forever independent, constituted the single greatest crisis the still young nation had ever known. The first order of business was salvaging the Union through war**. Faced with fielding a 'B team' because so many of the country's best flag officers had stuck with their native South, including **General Robert E. Lee**, to whom Lincoln had officially offered command of the federal forces, the new president was forced to run through numerous less talents before finally locating the primary architects of his victory, **Ulysses S. Grant** and **William Sherman, two previously undistinguished prewar officers who rose to the rank of general, and proved capable of the impossible task of not merely defending the North but of subjugating the South.... Beyond that, the primary strategic challenge of the war lay in the ability of Lincoln and Steward to forestall any entry by European powers in support of the Confederacy, whose economy Union naval forces slowly strangled with a very effective blockade**." (Barnett, 2009, pp. 93-95) (*So that was the historic significance as well as, momentous magnitude of the American 'Civil-War: 1861-65' and in fact the ultimately victorious end of this extremely menacing war for President Lincoln, finally enabled him to successfully maintain the geographically massive USA's magnificently federal and also, socio-cultural-cum-political unity; indeed reassuringly for the generations of Americans who would gradually keep-on coming, with the unending forward march of the timelines and thereby in turn, also enabled Lincoln to fruitfully implement his 'grand-plans'. This one specifically may be taken as a grand and historic example—of an intra-national good as well as, so successful a 'great-game'—onto the man's expanding civilisational-canvas itself*).

(iii) Even now the historically rather unforgettable **'Great-Depression'** of 1929-30s in fact, shook the economies of the almost entire world and thus, its severe tremors were also very frighteningly felt across the otherwise mighty citadels of the US economy too. In fact, 'chicken had come home to roost'; in the forms of massively unsettling fallouts of the all sorts of blunders, so callously and arrogantly, made at Versailles by the US and her other European allies, i.e. all combined victorious powers at the end of WWI. And the result was that 'Great-Depression'. But here too, FDR's visionary statesmanship in the USA came into play as his historic intra-national 'great-game', with indeed its many positive international repercussions which went a long way to finally tide over the consequent difficulties, for the concerned people all across the world and thereby the man's 'living-history' since perpetually in the making, kept-on marching further regardless, onto his expanding civilisational-canvas. In these precise contexts, concerning the multiple-consequences of the 'Great-Depression'; Thomas Barnett in his same book has written that, "The great Depression, of course, changed all that, empowering FDR's administration to become, as Keynes put it, 'the economic laboratory of the world'. Keynes himself was fascinated by Roosevelt, who, on the basis of one brief face to face encounter, appeared to return the complement to the man whose theories would encapsulate much of FDR's own instinct that **America needed to 'spend to save' itself from the clutches of the Great Depression's monetary trap**. As my own Depression-era mother always describes the times, 'Everybody wanted work but nobody had any money'. Many economic historians have argued that FDR's new Deal did little on its own to lift America out of depression, because that hard time was not simply limited to the U.S. economy but rather reflected a global capitalist crisis extended—seemingly *ad infinitum*—by the trade protectionism to which many governments resorted across the 1930s. FDR could reverse that policy tide at home, but he couldn't force larger architectural changes on international trade—until WWII gave him the opportunity.... Almost immediately once the war started, the nation's best and the brightest, an entire generation of New Dealers with years of experience in economic stimulus packaging, institution-building, and state-building in general, were put to work thinking ahead about **how the New Deal could be translated into a new international economic order**. Paramountcy in their thinking, as Keynes's biographer Robert Skidelsky notes, was 'one overriding aim: to do better than last time'. As policymakers on both sides of the Atlantic knew, America was—in Skidelsky's description—the 'great unknown' in the equation, having chosen withdrawal after the last world war. As long as Roosevelt remained in power, America's commitment to follow through seemed clear." (Barnett, 2009, pp. 131 & 132) (*Understandably, it seems logical that volumes have been written but one thing remains greatly established and that being, FDR's great vision, rare qualities of political-leadership and also, his unwavering statesmanship; even during the gravest of challenges, then facing not only the USA but the whole of world during those not so common trying times. After all such indeed history-making greatness is never common; and that precisely being the reason as to why the history bows in respect and as a tribute before their persona, onto the man's expanding civilisational-canvas in the world.*).

(iv) In these multiple-contexts, now coming to the such historic times of the 1914-18 and then 1939-45, the first and Second World-Wars respectively, and then, specifically concerning some unique aspects of the US 'great-games', Thomas Barnett has written that, "By making such concessionary loans, FDR did two things: (1) he established the precedent that made **Marshal Plan** possible; and (2) he so indebted the British to U.S. economic aid that during subsequent postwar liberal trade order, **Roosevelt**'s administration was able to hold the British to their promise, made in Article 7 of the Lend-Lease agreement, to end their system of imperial trade preferences and submit to the Bretton Woods Agreements—the 'real price' of **Lend-Lease**.... As we look from today's perspective, it's interesting to note how private-sector loans and harsh economic reparations defined WWI, while WWII is best known for Lend-lease and the Marshal Plan. As is for today's 'global war on terrorism'? So far we cite **higher oil prices** and an unprecedented amount of government postwar reconstruction contracts granted to private contractors, but no overarching economic agenda proposed or implemented by the **Bush administration, which, like the Wilson administration, seemed to overestimate our home market's power to force them to our way of thinking**—to wit, Bush's admonition to citizens to keep shopping despite the war on terror. But we are already seeing with the rise of a global middle-class, flush with insatiable desire and enough disposable income to warrant market respect, America no longer wields the same immense—if passive—demand power within globalisation. (**Build your domestic constituency from the start and keep it bipartisan**).... **Wilson, whose 1885 scholarly book *Congressional government* set the standard for describing as Clements termed it, 'the American political system not as it was supposed to function but as it actually worked'**.... After all, when Wilson landed in France in 1918, he was arguably the most popular leader in the world. If Wilson's arrogance and ambition blinded him during this crucial period, the danger was apparent to others.... America was liable 'to win a war but lose the peace', unless the President's rhetoric was quickly reconciled with the reality that, as one academic involved in the U.S. government's secret planning for the region's postwar structure put it, 'most of Wilson's European counterparts at Versailles viewed it as the **'great loot of the war'**. In the end, Arabs who had believed Wilson's 14 Points had promised them freedom were sorely disappointed by the colonial mandates produced by the **Versailles Treaty**.... Franklin Roosevelt would likewise be blamed for disappointing postwar outcomes, particularly in his apparent accession, at the infamous **Yalta Conference of February 1945**, to Soviet demands for political hegemony over Poland and the rest of Eastern Europe. Given the military facts on the ground at that time, Roosevelt had little choice, in my opinion especially since he was eager to gain Russia's entry into our Pacific War, which, at that time, appeared that it could drag on for a significant amount of time and at great human cost. (*Well, the point is well-taken, but in retrospect we may also say that by bringing-in Russia onto the East, China began getting some substantial arsenals for their combined* ***Sino-Soviet sinister designs in Indochina and subsequently further in the South-East Asia, towards the unhampered spread of communism wherever feasible****; and thereby within few years, more severe* ***'menace of the Korean-War'*** *came visiting onto the U.S. and which was further facilitated by the shortsightedness of* ***President Harry S. Truman*** *as well as, quite naively hasty withdrawals of the mighty US Forces from the Pacific, since then under command of the* ***legendary American General, Douglas MacArthur****, who was called upon to hold the 'hot-potato' of the Korean Peninsula with his no more his thick and powerful military gloves to put-on'. And then sooner than later, the disturbing fallouts of Korea then; did trigger the equally menacing as well as, all the more embarrassingly costly the* ***'Vietnam-War'****—with more tragic and sadly, much bigger 'price-tag'—for the USA. And now, the rest is history with its malignant scars only. Hence, in any such highly chequered and also fast incubating situations, since anywhere in the world precisely comes the interplay, successfully fruitful or otherwise, of the* ***multiple-aspects of 'essential strategic-culture and thoughts'****. The may also be taken as some examples of the messing-up one's own history, by some, prejudiced, and naively shortsighted leaders since, chanced to occupy the top-level decision-making seat of power.*). Nonetheless, it was exactly that sort of balancing between external goals and internal cost considerations that made FDR's management of war so impressive. Beyond the strategic choices made, Roosevelt was simply a master at shaping public opinion through the frequent 'fireside chats' with America, as well as a slick political manipulator of his political opposition through his consistent willingness to 'reach across the aisle' for bipartisan support. Once the war began, FDR also benefited greatly from the public's strong historic awareness of the mistakes made by politicians following WWI. Senator Arthur Vandenberg, for example, an ardent isolationist prior to WWII, came to realize during the course of the war how history seemed to vindicate Woodrow Wilson and vilify his main Republican opponent in the Senate, Henry Cabot Lodge.... As for the Bush administration since 9/11, it was not known for generating any long-term bipartisan support for his 'global war on terror'—just the opposite." (Barnett, 2009, pp. 121 & 122) (*It is universally true that in a truly vibrant liberal democracy, the spirit of bipartisanism is always desirable and crucially valuable too; but so far as the lapses leading to the so tragic occurrence of 9/11 and the, subsequently ongoing developments and events and thereby, making of the man's 'living-history', since contextually both in the USA as well as, across the world still appears to be in a 'fluid-state' and continuing too and the 'dusts are yet to settle-down finally, as all these in any case are always time-taking and hence understandably the history's judgement and its final crystallization for the posterity, will follow sometime in the near future.*

(v) After the end of World-War-II, and even much more so after the highly relieving demise of the erstwhile 'cold-war' times, lot many far-reaching changes took place almost all over the world and thereby, even the multiple-paths of since the further ongoing making of the man's 'living-history' too, got totally altered. All these in turn, brought many kinds of the totally new situations concerning the entire gamut of the 'great-games', into since the always highly challenging as well as, **all sorts of the turmoil-ridden: 'arena of the international power-politics', in the world**. In these present times' as the totally new kinds of the multiple-situations now prevailing in the world, and hence concerning the **USA's newly formulated and contextually the 'great-games-oriented' designs for her otherwise quite expected, 'initiatives-cum-responses' in both Africa as well as, towards China,** Thomas Barnett respectively, has very thoughtfully mentioned that, "So can greed save Africa, as one *Business Week* story asked a while back? Only if it's African entrepreneurial greed properly incentivized and channeled. Because if it's America pushing new security rule sets, China simply seeking raw materials, and investors everywhere looking for the newest 'frontier market' (the emerging buzz phrase), then we're arguably looking at no better outcome than the European colonial powers achieved previously (*What an irony indeed.*). A gold rush mentality is fine, so long as the infra-structure built up—both 'hard' connectivity and 'soft' institutions—increases African long-term resilience as well as their shot-term profits. But it also doesn't make much sense for American politicians to start lamenting China's economic 'penetration' of Africa, even when authoritarian states with poor human records are involved, because **China's growing economic connectivity becomes an asset to be leveraged, not discouraged, and nothing pushes the human rights agenda better than a rising middle-class, something Africa needs desperately. Plus, the more China pumps in Chinese nationals, the more China gets on the hook for protecting their rights and property, not because America says so but because those nationals' relatives back in China are making themselves known and pressing their demands upon China's Ministry of foreign Affairs. In short, anything that increases China's exposure as a rising great power forces it to develop a foreign policy more in line with ours. If we do it well, we help Beijing explore what it means to be responsible great power on globalisation's frontiers**. (*Aren't these for the moment, geopolitically 'tall-orders'? But in any case, these in fact, new kinds of the paths-breaking thoughts, are indeed soaked with contemporarily very relevant strategic-prudence and also, geopolitically great futuristic-vision as well.*)..... Last year's volume noted that 'improvements in China's strategic capabilities have implications beyond the Asia-Pacific region' and that the lack of transparency regarding Chinese intentions 'will naturally and understandably lead to hedging against the unknown' by the international community. Care to guess who constitutes the 'international community' on this one? Because if you can't, let me introduce you to 535 prime suspects, the vast majority of them have a major Defense Department 'program of record' in their home state or congressional district.... Okay, that's a bit harsh. **The truth is that several Asian countries (e.g. South Korea, Japan, Taiwan, Australia, and India) also produce similar reports on Chinese defense spending**. **Why? Because China goes out of its way to hide what it procures and then slyly trots out its big-ticket items every so often to pointedly give our satellites a few snapshots**. We have within our power to alter this dynamic, but for now, it is simply too useful to too many **members of the military industrial complex** on our side. Much as in the case of Iran, so long as we don't link our demand for their transparency to some larger understanding of what they shouldn't fear from us, we're unlikely to change this situation. **But remember this: We choose to force this military balancing dynamic on others—yet another example of why America's grand strategy matters more than any other great power's**." (Barnett, 2009, pp. 191, 192 & 225)

We rather quite fascinatingly also, do notice some of the transcontinental and newly emerging patterns of the 'great-games', as per the USA's own all the more newer kinds of the fast manifesting prime-cum-long-term national-interests, vis-à-vis the African continent and the fastest developing giant economic-cum-military power: the Chinese Nation-State in the North-Eastern Asian continent. In any case these by themselves today, are the far-reaching as well as, momentously consequential development in both the African and Asian continents. And hence it is to nobody's surprise that today's Superpower, the USA, with her own globally as well as, the multi-dimensionally and so dynamic intercontinental kinds of high-end geopolitical, economic and the military interests was ever in a position to ignore these overseas developments and events, and look the other way, regardless of the multiple kinds of their internationally impending and the history-making fallouts. **Therefore as expected, the USA too, began to geopolitically-cum-technologically, activate her close strategic surveillance over these fast manifesting multiple developments and events in both Africa and the China.**

(vi) In the presently continuing modern-era of the super-age of ultra high-tech world; concerning multiple-kinds of the ongoing US sponsored 'great-games' since, precisely concerning the multiple-aspects of her **'the inescapable realignment: re-blending diplomacy, defense, and development'**; Thomas Barnett has written that, "**President's George W. Bush's 2002 national security Strategy is best known for its enunciation of the military doctrine of preemptive war. More lasting and important was its more subtle argument that development and diplomacy are the equals of defense when it comes to victory in the long war against violent extremism**. Taking that high-level cue, the U.S. Agency for international development began **promoting the so-called 3D approach of blending each element** for more synergistically than was done in either the Afghanistan or the Iraq occupation. Based on my interviews locally for a 2007 *Esquire* story on the rise of **U.S. Africa command**, ground zero for this new way of thinking appears to have been **the USAID mission in Nairobi, Kenya, one of USAID's larger missions with responsibility for much of the agency's programs across the Horn of Africa**.... Of course, USAID wasn't the only government entity calling for such a blend. **General James Mattis** had been arguing, right from the start of his nation-building exercise with the Marines' 1st Infantry Division in southern Iraq that 'jobs, jobs, jobs.' was the primary method of diminishing popular support for an insurgency. The COIN field manual, in arguing for an 80/20 blend of non-kinetics and kinetics, was stealing a line from that great progenitor of twentieth century 'revolutionary war', **Mao Ze-dong, who argued, along with his generals, that successful insurgencies were 20 percent military in nature and 80 percent political**. As such, throughout the Iraq and Afghanistan occupations official Washington has been barraged by blue-ribbon-panel recommendations, op-eds by experts, and calls from the field for some sort of new, blended organizational approach that was not quite State and not quite Defense but somehow made these two antagonistically opposed bureaucracies work with each other seamlessly. While most nongovernmental voices called out something akin to my proposed Department of Everything Else, government-sponsored 'group-thinks' favored expanding the State Department, a solution few familiar with our bureaucracy's perfectly preserved nineteenth century style of operations found acceptable. Both Defense and State have created new senior-level positions to focus on postwar operations, but beyond that, the level of cooperation between the two remains stunningly limited. (*My God, we may on our own, no less stunningly, look at these some rather situations-based 'fractured-functioning' in the US Government Departments' fault-lines within their structural-scenario; and that too, concerning precisely the present-era's sole Superpower.*). A good example? In Afghanistan, the United States has sought to hand over coordination of economic and political reconstruction to the United Nations—a sure sign of lack of coordination leading to bureaucratic capitulation.... Again U.S. history suggests that we've approached this challenge better in the past. **By most historians' judgement, the U.S. occupation of the Philippines following the Spanish-American War was quite successful both in terms of counterinsurgency and state-building**. To create unity of command and effort, **President William McKinley** created two successive temporary occupational authorities, the first to deal with the immediate aftermath of the war and a second, led by future **President William Howard Taft** (who later performed a similar role for **President Theodore Roosevelt** in Cuba), to oversee the transition from military to civilian rule. **General**

MacArthur's postwar rule in Japan is likewise often cited as an example of successful transition, as is the Allied coalition rule over postwar West Germany. The difference in all these cases would seem to be the underlying logic that everyone involved (military and civilian) brought to the effort—nation-building was a government-wide effort involving skill sets reasonably transferred from recent American experience back home.... **The problem we face today is that America is so far distant—historically speaking—from such nation-building experiences at home that both its leadership and its people have an innate aversion to the concept**. (*What an ironically, both disjointedly as well as, geostrategically-cum-politico-militarily confused state of affairs now, prevailing in the US minds; precisely with respect to the sole Superpower's naturally; the unrivaled geopolitical domains—both within the huge US homelands and also, concerning the unimpeachable overseas—US obligations and compulsions in these regards.*). **What was once viewed as a noble endeavor is now viewed with real distaste, as if it's morally wrong to 'impose' a better form of governance upon a society typically lacking a functioning version altogether. The downside, of course, is that we know what the libertarian paradises of the Gap usually produce: widespread violence, disease, and starvation. As soon as America gives up on a country, expect globalisation's vultures to start circling**." (Barnett, 2009, pp. 281 & 282)

In the above identified and so penetratingly elaborated multiple-aspects, concerning now the 21st century's globally so momentous and far-reaching 'great-games'; which only the today's sole Superpower: i.e. the USA, is both capable of, as well as obliged to, play onto the presently all the more violently turmoil-ridden and also, unpredictably fast-altering arena of the 'great-games' and, since always being full of 'shifting-sands'; and at the same time, do also project some no less compulsive obligations-cum-responsibilities in these regards, exclusively now with respect to no one else, but precisely the USA. In these multiple-contexts, the geopolitical point of paramountcy importance is that, being a Superpower and that too exclusively the singular one does always carry some unique and also extraordinarily weighty 'baggage', which the concerned Superpower just cannot afford to ignore in any manner as there is always some exclusive kinds of the 'price-tag' attached to the entire gamut of being a Superpower, and hence from the very ancient of times, though in the changing forms and parameters, as since the man's 'living-history' has been marching ahead onto his expanding civilisational-canvas it has remained quite valid that, it has never been any kind of a 'cake-walk' for being a Superpower in the world. In the above quoted paragraphs, we do also find the mention of the historically very valid point concerning quite successful US occupation of the Philippines with its multiply-glorious as well as contextually also, the American 'great-games-oriented' some indeed praiseworthy achievements as well. But it is also no less true that, the then geopolitically prevailing times and situations regarding Philippines vis-à-vis the USA were quite different. And as of now, since the ongoing phenomena of the making of newer and newer kinds of the 'sand-dunes' from the more and more accumulating 'sands of time' too, have gone-ahead quite far, especially in today's fast-advancing high-tech era of indeed, the unprecedentedly globalised world. Hence, the above desired and even contextually, all the more differently required 'great-games' within newer and geopolitically much more innovative kinds of the geostrategic-cum-politico-military conceptual-framework, towards bringing-in suitably some 'better forms of governance' into a society typically lacking any worthwhile functioning version, would mandatorily be requiring altogether different kinds of the diplomatically very fine-tuned skills as well as, politico-military acumen to safeguard against any possible kind of the 'logjam', in today's much globalised and yet, much more dicey world; and thus, fruitfully achieve the desired all round successes in these regards.

In today's the highly-fluid, fast-changing and also, quite unpredictable global geostrategic-cum-politico-military scenario these are indeed very dauntingly frightening challenges for any other leading Power and hence, even for a Superpower too, it does always remain a tough-calling. But it does not mean that with due strategic-vision and prudence and along with the well stationed—multiple kinds of the wherewithal as deemed necessary and also, with flexibly fine-tuned and Superpower-oriented geopolitically robust willpower; the well-crafted and no less desired multi-dimensional ambitions since, always being good and peace-bringing for both the Superpower concerned and also, for the larger humanity, cannot be realised. **The paramountcy needs for the concerned Superpower are to be flexibly far-sighted and also, geopolitically sincere-cum-large-hearted and certainly never to be small-minded; so that while triggering the already above discussed 3Ds-oriented initiatives, the geopolitically prudent realignment-cum-re-blending of the multiple-aspects of diplomacy, defense and the development too, are equally fruitfully ensured.** In these contextually broader perspectives, as we may care to critically delve deeper into the relevant pages of USA's 'living-history', concerning her successfully historic occupation of the Philippines we may also notice that, the above mentioned prerequisites were prudently taken care of, with due geostrategic and politico-military vision by the then concerned US personalities, both military and civilian, since occupying the decision-making positions then, at the helm of affairs in the USA. Even during those times, both challenges and the tasks were no less daunting but the strategically brilliant-vision as well as, the leadership qualities of the then concerned American personalities made all the difference and thereby, made a glorious history for not only the USA, but indeed for the entire world.

Hence, the above discussed thoughts in these regards contextually, do also highlight the universally critical importance of the personalities—including the political, military and bureaucratic—since, occupying the top-levels' driving-seats both in the national-capital in home country of the Superpower concerned and also, in the fields of the target areas/country as well as their intellectual-cum-professional vision and thus, generating the crucial synergy along with their well-coordinated strategies and plans and so on, cannot be over emphasised in any manner. In fact, we may contextually bring back and also stress-upon that, the fountain-heads for all these paramountcy prerequisites and, as all these happen to be the: well-groomed, well-imbibed and also, flexibly well-stationed—the **multiple-aspects of 'essential strategic-culture and thoughts'**—which do always remain the proactive 'incubators' for them to vibrantly fructify and thereby, these do always remain in some kinds of the auto-programmed to get multiply proactive well in advance, in case of the any impending or even suddenly fast developing and, relevantly challenging situation for the concerned Superpower anywhere in the world; and thereby, geopolitically demanding to proactively get-in for the needful. Therefore, howsoever any justifiable or even otherwise laudable intentions, on the part of any concerned Superpower by themselves are not enough since along with them, various kinds of the critical prerequisites as now contextually discussed and highlighted in the above paragraphs too, are equally important for the any desired successes, with respects to any specifically concerned 'great-games' in these regards. In fact, it is needless to say now, that the absence of any of the above mentioned crucial aspects or even in the situation of some kinds of the disjointed or incoherent relationships among those crucial aspects are bound to result in some kinds of the 'failure' or 'embarrassment', or even at times as per any specific kinds of situations, may lead to some acutely costly and awkward 'fiasco' too.

Well, a very precisely relevant example in the above multiple-contexts, happen to be the USA's adventure or perhaps, *'misadventure'* in Iraq in 2002 and hence, these do need to be looked into.

Accordingly, we have come across a book titled as *Fiasco: The American Military Adventure in Iraq*, written by **Thomas E. Ricks.** This highly analytical and absorbing book does give us a panoramically inner glimpses into the entire geopolitical-cum-military gamut of the American invasion of Iraq in 2002, which since so clumsily conducted by the US political and military establishments then, with all sorts of the avoidable fault-lines in almost anything and everything conceivable that, far from successfully achieving any long-term strategic aims—and thereby consequent to this, Iraq is still in a state of turmoil and whose tremor is still being felt nearly all over the world. And hence, the USA is still not able to fully extricate itself from the dicey 'quick-sands' of the post-Saddam era Iraq. In fact, the 'risk-factors' in these kinds of overseas 'great-games' concerning a Superpower; if one or a couple of crucially important things gets loose and go wrong and still, the concerned Power remains bent upon acting naively then, almost all the relevant things tend to go wrong simultaneously and thus, bringing-in a very embarrassing 'Catch-22' situation.

(vii) Hence, precisely in the above multiple-contexts; **Thomas Ricks**, in his above referred book has very perceptively written that, "It is a tragedy in which every major player contributed to the errors, but in which the heroes tend to be anonymous and relatively powerless—the frontline American soldier doing his best in a difficult situation, the Iraqi civilian trying to care for a family amid chaos and violence. They are the people who pay everyday with blood and tears for the failures of high officials and powerful institutions.... The U.S. invasion of Iraq, Army **Lt. Col. James Scudieri** wrote later, 'may be the most planned operation since D-Day on 6 June 1944 and Desert Storm in 1991.' **The irony is that in eighteen months of planning, the key question was left substantially unaddressed: What to do after getting to Baghdad**. **Franks, Rumsfeld, Wolfowitz, Feith and other top officials** spent well over a year preparing to attack Iraq, but treat almost casually what would come after that. (*It was amazing that no lessons from history were learnt and, no worthwhile* ***'Plan-B'*** *was prepared to effectively address this question in fact, of no less paramountcy importance.*). 'I think people are overly pessimistic about the aftermath', Wolfowitz flatly stated in an interview in December 2002.... The combination of all these moves—a prolonged foreign occupation that was built on **de-Baathification**, dissolution of the military, and economic upheaval—radically undercut social stability and built opposition to the American presence. 'What we have done over the last six months in al Anbar has been a recipe for instability', **Keith Mines** would write in a November memo. 'Through aggressive de-Baathification, the demobilization of the army, and the closing of factories the coalition has left tens of thousands of individuals outside the economic and political life of the country'.... Taken together, **Bremer**'s approach had for many Iraqis a punitive feel, a result that was a key misstep, Wolfowitz's old mentor **Fred Ikle** would later observe: **'Democracies that have achieved a military victory ought to refrain from seeking revenge'**, Ikle wrote.... Taking revenge is a Neanderthal strategy. Instead of giving priority to policy that can transform a defeated enemy into an ally, the revenger helps the hawks on the enemy's side to recruit angry fighters who will undermine the peace settlement. **During the critical weeks following the collapse of Saddam Hussein's rule in Baghdad, the emphasis on punishment and revenge clearly harmed America's long-term objectives**. Obsessed with a desire to punish and revenge, the U.S. managers of Iraq's occupation delayed this [taking a conciliatory] approach for more than a year, at which time the United States was confronted by an organized, hostile insurgency.... The cost will go beyond that initial bill of blood and treasure; Iraq is likely to dominate American foreign policy for years. As the 'National Strategy for Victory in Iraq', the document released by the White House in November 2005, put it, **'What happens in Iraq will influence the fate of the Middle East for generations to come, with a profound impact on our own national security'**. (*Iraq's multiply negative fallouts soon later, did come onto all the more dicey affairs in the*

Afghanistan.).... **In Iraq, the U.S. position also suffers from the strategic problem of the fruit of the poisoned tree--that is, when a nation goes to war for faulty reasons, it undercuts all the actions that follow, especially when it won't concede those errors**.... Even if the United States stays, there is no guarantee that Iraq won't slip into civil-war. That threat hovers constantly, discussed quietly by American officers as a possibility and more openly by many Iraqis. Americans tend to remember the horrors of their own civil-war and so assume that all parties would do their best to avoid it, a perspective that obscure the fact that there is a considerable pro-civil-war lobby in Iraq, Essentially, there may be more people in the region who want to see the United States leave Iraq than want to see it stay, from Sunni Islamic extremists to their Shiite foes. The quickest way to achieve that ejection of the U.S. presence may be to start a civil-war, on the calculation that the U.S. public wouldn't stand for seeing American troops die trying to keep apart the warring factions." (Ricks, 2006, pp. 4, 78, 165, 166, 431 & 436) (*This is a portrayal of a depressing, and a highly explosive scenario now, concerning all sorts of extremely violent situations prevailing in Iraq. It is now somewhat encouraging that, so far the civil-war has not started; but, the overall present situation in Iraq is far from being capable of displaying any faint or even perhaps, a 'tattered' picture of the so-called socio-cultural-cum-politico-military normalcy over there. Tragically for the world at large, Iraq still continues to remain heavily entangled in the highly complex 'cobweb' of the all sorts of mindless terrorism and insurgencies.*).

We may recall that in various places and in a number of different contexts in the earlier chapters and as well as, in the present sub-chapter of our this eighth chapter too, we have been highlighting that in course of the numerous kinds of 'great-games' having been played since the very ancient of times, and are even now being played almost everywhere in the world. As the many unforeseen or quite unpredictable kinds of the subsequent developments and events get strangely manifested on the part of those actively participating powers or even any Superpower involved, and as the times pass by, these powers are hardly in any tangible position to exercise perhaps, even any semblance of the strategic and/or politico-military control over those very developments and events. These too are also some of the numerous ways-cum-systems of the perpetually ongoing making of the man's so vast and highly complex 'living-history', onto his expanding civilisational-canvas in the world. In these presently multiple-contexts concerning the variously numerous aspects of the mankind's 'great-games'; we may also need to bear in our minds and, as also emphasised earlier that, the man's 'living-history' never favours or fears anyone in any manner whatsoever and hence, does continue to keep-on marching ahead regardless, onto its multi-dimensional paths of the continuous as well as, highly chequered making in these regards.

We may recall that the various Kingdoms/Empires/Nation-States, since almost the very beginning of the human-civilisations in world, have been indulging in the all possible kinds of 'great-games' principally to look-after their paramountcy strategic/national interests, to achieve more power, glory and also no less, to garner more wealth and territories and, even to expand their imperial holds and thus, to more effectively indulge in the all kinds of the dominating power-projections and thereby, enhance their geostrategic and politico-military reaches in the all possible geopolitical directions in the world. Hence all these multiple-aspects in these regards always in their multiple-manners, do remain integrative-cum-interrelated, as well as mutually influencing too. While precisely keeping in minds their broader perspectives, we may also contextually recall that in the fifth chapters itself earlier, we have quoted few paragraphs from a lecture delivered in the USI, New Delhi by one of India's leading Journalists and a Strategic Analysts, Mr. **M.J. Akbar**, concerning now the bygone late medieval-era British imperial-conquests of the erstwhile extremely feudalistic Indian subcontinent. In it he has very perceptively said that, nobody takes anywhere so much hazardous sailing through thousands of miles of rough seas and oceans, just to conquer the vast territories of desert-wastelands and squander-away so much wealth and energy like fools. **India has been through the ages both geostrategic-cum-politico-militarily, a very richly bestowed 'golden-bird', being always irresistibly attractive for any daring, strong and adventurous imperial-power to undertake some kinds of the highly ambitious 'great-games', and the rest is now well-known history**.

And subsequently later, geopolitically on similar lines, during the medieval-era and all the more so after the discovery of another kind of since the no less valuable the **'black-gold' i.e. the 'Oil'**, in almost the entire Middle-Eastern huge desert-mass, these regions became the highly chequered and also the extremely turmoil-ridden 'arena of all kinds of the great-games-oriented international power-politics', for the competing overseas imperial powers and then much later during the post-WW-II times of the twentieth century, between the two ruthlessly rival Superpowers. Well, in the similar manners, we may contextually bring-in very many highly chequered examples from all across the world as well as, since the very ancient of times, substantiating these momentously history-making numerous aspects of all kinds of 'great-games'; as having been played and being played even now, onto the man's expanding civilisational-canvas itself. But that is not all and now, geopolitically all the more new, extensively overwhelming as well as, geostrategic-cum-militarily very far-reaching fast developments and events have been taking place all over the Middle-East, and other oil-producing-cum-exporting countries in today's world; and thus newer kinds of challenges and opportunities are getting thrown-up for the concerned players and oil-rich nations and thereby, triggering all the more complex and daunting 'great-games-scenario'; onto the same multiply fluid-kind of the 'arena of since the relentless international power-politics'. The Man's 'living-history' since continuously in the making, has now quite amazingly begun to acquire many unexpectedly newer dimensions in the multiple-paths of its ongoing making itself.

(viii) Precisely in the above multiple-contexts **Thomas Barnett** has written that, "The underlying reality on oil is more complex, than simply noting global supplies are concentrated in unstable areas, which, of course, is absolutely true. The core used up most of its oil a long time ago, forcing global economy's narrow embrace of several gap regions—most significantly the Persian Gulf. Because the global economy's demand was relatively light during the Cold War, **it wasn't until its radical expansion to include 3 billion new capitalists that we entered during this period of dangerously tight production capacity, meaning no matter how big our global oil reserves are, our current worldwide oil industry infrastructure is dangerously close to being maxed out in its capacity to extract, refine and delver oil around the planet**.... With oil stuck the $100-a-barrel mark for much of last year, there's little surprise that we're once again inundated by call for more regulation of so-called super-major companies (e.g., Exxon-Mobil, BP, Royal Dutch Shell, Chevron) and an end to our military presence in Persian Gulf, in addition to the usual demand for a **Manhattan Project on alternative energy**. The assumptions are that America's energy demand drives prices, the super-majors determine supply, and instability in the Middle East explains recent spikes. So if this is all our doing, then it can be all our undoing as well. Would that Washington were so eminently in control of global energy markets. **The driver on oil prices today is rising Asia's increasing demand. At the Cold War end, Asia accounted for 10 percent of global oil demand. Today it gobbles up double that share. Before 2025, Asia will become the oil-market's global demand centre, dislodging North America. So as I've noted before, it's basically our blood and their oil, not surprisingly, when Americans are confronted with that strategic reality, we want out—and fast**..... A Second big driver is that many key producers are themselves becoming significant consumers, shriveling their capacity for exports. Indonesia became a net importer years ago, while Mexico, a crucial supplier for the United States, heads down the same path unless it soon opens up it national oil company to foreign direct investment. Meanwhile, strongman **Hugo Chavez diminishes** Venezuela's future export capacity by scaring off investors and—in essence—eating his seed corn: funneling today's windfall profits into socialist programs designed to buttress his popular standing. His latest scheme involves Venezuela's national oil company using its logistical network to import foodstuffs currently in short supply. Fire breathing **Mahmoud Ahmadinejad** has pursued a similarly short-sighted course in Iran to predictable outcomes. **In general all the big producers are seeing their oil consumption skyrocket far above global growth, with Saudi Arabia, Russia, and Norway leading the way**. So while rising Asia demands a lot more oil, the biggest sources have trouble boosting their production.... What can the super-major oil companies do about any of this? Not as much as you think. **Increasingly, it's the home team national oil companies, or NOCs, that control the bulk of reserves. Today, these state-owned firms directly manage 40 percent of the world's production. Because the NOCs own roughly 90 percent of known reserves, any increase in production naturally favors their increased control over the industry. By 2030, NOCs are expected to control as much as three-quarters of global production, meaning ExxonMobil, BP, and others will be squeezed out of new big projects as national oil companies grow rich enough to finance these on their own**. Already, a chunk of the super-majors' current production is locked into sharing agreements that favor national oil companies as prices rise. **As the national oil companies grab more control over future production, the super-majors face more restricted access to existing oil-fields, pushing them towards increasingly complex ventures to discover new supplies: tougher geology, deeper sea-beds, harsher arctic environments, and thicker oil. Those tougher conditions drive up production costs just as the worldwide shortage of rigs and workers emerges. Toss in years of under-investing by everybody in refining capacity, and this situation won't be remedied anytime soon, even as a global recession depresses demand temporarily**." (Barnett, 2009, pp. 379-381) (*Indeed globally, what a frighteningly depressing scenario. If some reasonably viable and globally well-assured wherewithal concerning the altogether new sources of energy, don't multiply emerge, along with their techno-economic-cum-industrially vibrant and sustainable bases all across the world the entire humanity and their crucial well-being, would come under deep threat towards its very survival.*).

(ix) In the above quoted paragraphs and, specifically concerning the developing Asia's fast increasing demands for oil, **Thomas Barnett** has very analytically said that, **'it's basically our blood and their oil'**. Well, this overall scenario doesn't pertain only to the multiple-aspects of the extremely crises-ridden global 'oil-supply and demands' situations, but has some other much wider dimensions as well; and which in turn, have much greater ramifications in the vast and, highly complex arena of the—geostrategic-cum-politico-military-oriented 'great-games'—concerning the 'international power-politics' in today's, though indeed super-high-tech-oriented, but no less turmoil-cum-conflicts-ridden world. Precisely in today's contexts, **Thomas Friedman** (the author of his famous book *The World Is Flat*), in his one of very perceptive articles, titled 'The US and Its Double Game'; and also since published in an Indian news magazine, *The Sunday Indian*, dated 15 August 2010 and therein (page 18) he has mentioned that, "In case of the Great Game of Central Asia, that's us... Best I can tell you from WikiLeaks documents and other sources, we are paying Pakistan's Army and intelligence service to be two-faced. Otherwise, they would be just one-faced and 100 percent against us. The same could probably be said of Afghanistan's President, Hamid Karzai. But then everyone out there is wearing a mask—or two.... China supports Pakistan, seeks out mining contracts in Afghanistan and lets America make safe Afghanistan for Chinese companies, all while smiling at the bloody nose America is getting in Kabul because anything that ties down the US military makes China's military happy. **America meanwhile, sends its soldiers to fight in Afghanistan at the same time it rejects an energy policy that would begin to reduce our oil consumption, which indirectly helps to fund the very Taliban schools and warriors our soldiers are fighting against.... So why put up with all this duplicity**? Is President Barak Obama just foolish? It is more complicated. **The 9/11 terrorist attack was basically planned, executed and funded by radical Pakistanis and Saudis. And we responded by invading Iraq and Afghanistan. Why? The answer is because Pakistan has nukes that we fear and Saudi Arabia has oil that we crave**.... So we tried to impact them by indirection. We hoped that, by building a democratizing government in Iraq would influence reforms in Saudi Arabia. After expelling al-Qaida from Afghanistan, we stayed on to stabilise the place, largely out of fears that instability in Afghanistan could spill into Pakistan and lead to Islamist radicals taking over Islamabad and its nukes. The strategy has not really worked because Pakistan and Saudi Arabia are built on ruling bargains that are source of their pathologies and our fears.... **Pakistan, 63 years after its founding, still exists not to be India. The Pakistani Army is obsessed with what it says is the threat from India and keeping it alive is, what keeps its Army in control of the country**. The absence of either stable democracy in Pakistan or a decent public education system only swells the ranks of the Taliban and other Islamic resistance forces there. **Pakistan thinks it must control Afghanistan for 'strategic depth' because, if India dominated Afghanistan, Pakistan might be wedged between the two.... So we pay Pakistan to help us in Afghanistan, even though we know some of that money is killing our own soldiers. (*Well, all the money doesn't end-up there only, but a substantial portion of it also ends up in running various Jihadi terrorist-camps, located inside the Pakistan-occupied Kashmir and in addition, to lethally indoctrinate and equip them to enable them to illegally infiltrate into India, for indulging into all sorts of the unpardonable terrorist-activities and thus, cause serious harms and injuries to the innocent Indian civilians. The USA needs to be fully aware of these no less disturbing facts, for pragmatically coming out of her old-era 'cocoons' as well. First, she needs to discard her rather 'regressive fixations' concerning the so-called 'frontline nation', i.e., the 'failing-nation' Pakistan.*). And we send Saudi Arabia money for oil, even though we know that some of it ends up financing the very people we are fighting, because confronting the Saudis over their ideological exports seems too destabilizing. (Addicts never tell the truth to their pushers. *'Yes, indeed'*.)**.... Is there another a way? Yes. If we can't just walk away, we should at least reduce our bets. We should limit our presence and goals in Afghanistan to the bare minimum required to make sure that turmoil there doesn't spill over into Pakistan or allow al Qaida to return. And we should diminish our dependence on oil so, we are less impacted by what happens in Saudi Arabia, so, we shrink the funds going to people who hate us and we make economic and political reform a necessity for them, not a hobby.... Alas, we don't have the money, manpower or time required to fully transform the most troubled states of this region. We do, though, have technology, necessity and innovators to protect ourselves from them—and to increase the pressure on them to want to change—by developing alternatives to oil. It is time that we started that surge. I am tired of being the sucker in this game." (*What an irony—like a 'Catch-22' kind of the trap—into which, the today's solitary ultra-high-tech-Superpower: the USA, has pushed herself so naively and thereby very embarrassingly making herself, so much a 'laughing-stock', even in the eyes of these 'failed or failing' senselessly fanatical Islamic-States.* ***It is high time, especially after unnecessarily paying so much in terms of American blood and money that the USA, geopolitically 'wakes-up' and thereby both geostrategic-cum-politico-militarily teams-up with Israel, U.K., India, Japan, South-Korea, and Australia, and also including any other similarly trustworthy nations, whether***

in Europe or elsewhere indeed, in a prudently far-reaching and fruitfully synergy-creating manner for crushing the al Qaida and its other Jihadi-outs anywhere, including in Pakistan, once and for all and thereby, successfully reining-in these so much menacingly nuisance-creating fanatic Islamic-States*. The greatest needs of the hour for all these nations, since as proposed 'to be led by the USA' to be geopolitically focussed, strategically prudent and also, proactively-cum-sincerely, to gather enough political-will, for this far-sighted strategy towards the paramountcy common-purpose. Are these just wishful thinking and hence, only 'tall-orders'? Well, certainly not. And, with respect to these proposed multi-national 'grand-plans for the joint-actions' if China in any manner dares to unnecessarily throw her weight around and, begins to poke her nose in anyway then the team of these nations should not have much of a difficulty 'to even cut China's nose', for the larger common good in these regards. In fact, if China dares to threaten then, she can all the more seriously be threatened too. Since the principal geopolitical reality in these regards, remains that, in the highly challenging, as well as so much turmoil-ridden 'arena of the international power-politics' and thereby, the multiply consequent 'great-games the paramountcy necessity for the desired success, is to always talk and act from positions of strengths and never with the any kind of weakness. It is high time that, the USA begins to 'start thinking out of the box', in befittingly far-sighted and multi-dimensionally robust manners, as quite justifiably expected from today's this singular Superpower; since similar to the US President Roosevelt and the British Prime Minister Churchill, during the much more dangerously menacing and challenging, the World-War II years. Precisely, in these much larger and the highly momentous perspectives, in fact, the pivotal point boils down to that, all the above discussed—the presently so much paramountcy geopolitical aspects and thereby; the suggested geostrategic-cum-politico-militarily, the alternative 'plans of actions'—will always decisively be driven and successfully steered by the vibrantly, as well as proactively flexible, and hence the well-placed:* ***multiple-aspects of the 'essential strategic-culture and thoughts'****, with respect to those concerned nations. Man's highly complex and the multiply vast 'living-history', since continuously in the making never fears, favours or disfavours anyone and also waits for none and hence, it depends upon the mankind itself or any concerned Nation-State(s), to timely grab the opportunities coming on the way and thereby, advantageously change the momentous courses of the subsequent making of the mankind's same 'living-history' itself.).*

Hence globally, both geopolitically, as well as techno-industrial-cum-economically there seems to be no feasibly effective answers to the above depicted—**such dauntingly complex problems concerning the 'cobweb' kinds of the oil-supply and demand aspects**—in any foreseeable future in the world. On the contrary as the time keeps-on rolling-by, these already scaring problems are perhaps, unstoppably going to magnify and thus, fast going to turn into more and more menacing scenario for the mankind; and hence needless to say that, the whole world unfortunately appears to be sliding quite quick-paced: into an **'abyss' of the dangerously deep 'energy-crisis'**. There are many overall environmental factors as well as, numerous kinds of mankind's inexcusably callous, and shamelessly self-destructive 'omissions and commissions' over the years, since the momentous and otherwise definitely well-meaning advent of the **'Industrial-Revolution'**, which has perforce been made responsible for now the so menacingly manifesting and also magnifying, this deep energy-crisis. It is also true that the **man's demonically uncontrollable 'greed' and thereby, birth of his irrationally expanding-cum-insatiable kinds of the blind and vulgar 'culture of consumerism' too, are no less responsible for so dangerously pushing the man himself, into the above mentioned 'abyss'**.

All these globally in turn, have unfortunately led to the **man's many kinds of civilisational-negations in the world, like: ransacking of the Nature's globally sensitive and very fragile ecological balance, global warming, 'Greenhouse' effects, such fast depletion of the global**

natural-resources, frequent and increasingly destructive occurrences of the all sorts of natural-disasters, all round drinking water-crisis, diseases and epidemics, all sorts of conflicts and wars, many even barbarian kinds of the religious fanaticism, mindless arms-race in many areas of the world and then, now the most dangerous emergence of all possible kinds of extremism, terrorism and the insurgencies almost all across the world. Well the list can go on even longer as one thing leads to another and hence, many things give birth to many more things; at the same time in these multiple-contexts, and as history's timelines keep on marching ahead the man's all these and much more many different kinds of depressing civilisational negations also, keep on emerging and expanding regardless and that too, in spite of the mankind's all kinds of civilisational-glory and the magnificently numerous achievements as well as, the multi-dimensional marvels in the multiple-areas of the relentless human-endeavours in the world. We would more relevantly and also in some greater details be discussing these aspects, when we come to the forthcoming next to next sub-chapter, on 'Population-Explosion and Environmental Degradation' shortly, within this eighth chapter itself. As since already emphasised many a times earlier and in numerous other contexts too; anything and everything in the world, is directly, indirectly or even remotely perhaps, always remains interrelated as well as, mutually influencing as well. At the same time and, onto the man's expanding civilisational-canvas, it has also been seen down the ages that for the man himself, **'necessity has also been found to be the mother of inventions'**. Hence, the man's inherent mental-cum-intellectual faculties with his built-in 'sense of optimism', multiply relentless and dedicated, of the mankind's many brilliant minds the world over do indicate that as the 'energy-crisis' and also, the other kinds of crises have been so menacingly emerging, he should be in a position to devise and throw-up some credible solutions too, with his continuously enriching and expanding knowledge-base in the numerously multiple-areas of the sciences and technologies and thereby, his enlarging very many kinds of the wherewithal in these regards; would indeed, further facilitate his ongoing search for the many newer kinds of the credible solutions in these regards. The desired fructifications of all these solutions, are likely to be seen in the near and possibly, in not so distant future as well.

Now coming back to the above discussed multiple-aspects of the various kinds of the US-sponsored, or even otherwise triggered 'great-games'; we do also find that things have not remained static or getting stopped at the doors of Iraq and Afghanistan only, but moved ahead into the territories of Libya, Egypt, and in Syria too, in many different ways. **And now quite amazingly, we find that, even Iran is now strategically and rather quite actively 'blipping', onto the US geopolitical radar since already deployed for the numerously multiple-purposes, in the Middle East.** Whatever may be, but there is always 'two sides of a coin'; and Iran too, all the more irrationally perhaps, trying to rather clandestinely develop its unnecessary nuclear-weapons; and Iran's reckless programme, is bound to be quite unsettling in the Middle-East, in many ways; and thus, could be no less problematic for the whole world. USA may have her own geostrategic-cum-politico-military justifications, for so proactively entering into one after another country in the Middle East or even at times quite obsessively be desirous of rather offensively coming onto any of them militarily, through aerial means. But the moot point remains, that have the US political-cum-military establishments, multiply done their homework; with their always mandatory strategic prudence as well as, with the requisite geopolitical vision in these regards? Have the necessary cool-calculations been resorted to, inescapably in far-sighted manners or, are they leading their mighty and great nation into the murky quick-sands of another 'fiasco'? Famous thinker **Jodi Picoult** is learnt to have said that, "**It was one thing to make a mistake; it was another thing to keep making it**".

At this precise juncture of our ongoing discussions, it may contextually be said that the above one-line 'quote' conveys it all. The **former US President George W. Bush** is understood to have said that, **"Champions of democracy understand that democracy is not perfect...but it is the only path to national success and dignity"**. The point is well-taken but at the same time, it is also equally true that, in no country anywhere and in any continent of the world and, at any given timeline of the ongoing making of man's 'living-history' itself; any external power, through any means of the geopolitical-cum-military kinds of the 'transplant-surgery'; has successfully been able to transplant 'democracy' as a political-system of governance, onto the 'body-politic' of any particular target-nation or political entity. And at the same time, it is substantially even much less possible, in any Muslim-society, since being multiply as well as, overwhelmingly fixated with the **medieval-era type of an 'extremely conservative-Islam'** and irrespectively, located anywhere in the world. Therefore it would perhaps be certainly a limit of the geopolitical naivety to ever think and, even by any Superpower too, that any such 'naïve-adventure'; in any conservative Islamic-nation, is ever going to succeed. **'Seeds of democracy' anywhere and in any given situation; have to be indigenously planted by the indigenous population and its leadership themselves and then, gradually and peacefully be allowed to incubate, as well as fructify; and thereby also at the same time, it is always a time-taking process in quite a prolonged manner**. In fact, it is true that the **'urge for democracy is universal'** but at the same time; it is all the more true that in any fanatically conservative religious society, and more so in an Islamic one; and also from the broader human civilisational-perspectives as well, there can never be any 'quick-fix' solutions. And in any case, any shortsighted and quite hastily planned military-action, with all sorts of half-baked geopolitical ideas and strategies is bound to lead to quite an embarrassing 'fiasco' and, may even turn out to be counter-productive and hence, sheer wastage of human as well as, other kinds of the valuable resources in these regards. As it is said that, where even 'angels fear to tread'; will it be strategically prudent to just hastily enter with some short-sighted ill-planning and also, quite perforated kinds of the preparations itself? One always needs to be not only strategically visionary but also, equally requires to go back into the relevant pages of the man's 'living-history', since continuously in the making onto his expanding civilisational-canvas itself, to get some more 'light' and learn valuable lessons, so that any setback or logjam is prevented, for the overall good of the larger humanity, especially in the present times' high-tech-cum-super-connectivity-oriented world where the fast improving and expanding 'awareness-levels' too, are getting more and more instrumental in making the man more impatient and intolerant as well.

These 'great-games' in the Middle-East region and, precisely in the specific-contexts of Syria itself, as since already briefly discussed above; it may particularly be very relevant to mention here that, the democratically very focussed responses of the Great Britain, as since being always both the geostrategic-cum-politico-militarily, the closest-Ally of the USA, do highlight some additional dimensions in these rather ironically, quite confused state of affairs. Down her no less checkered history-lanes, both within the European continent, and also almost globally later, during the vast British-Empire times, of the later centuries, during now the already bygone medieval-age, as well as even during the modern-era the UK has always been found to be both geopolitical-cum-militarily, much more prudent and strategically far-sighted as well. Though being a much smaller nation and in spite of earlier, forced to directly face the extreme brunt of the German onslaughts both during the First and Second World-Wars, along with her other European Allies she has seldom acted naively. Against all sorts of multiply heavy-odds, she has almost never been found to be losing her strategic-coolness and betraying her geopolitical emotions. That remains the premier reason as to why, she remains successful in maintaining her close friendly relations even now, with almost all her erstwhile colonies. These unique geopolitical dimensions, do demonstrate the multi-directional marvels always vibrantly-cum-flexibly well-placed, the multiple-aspects of her **'essential strategic-culture and thoughts'**. And that makes for most of the praiseworthily unique: the geopolitical, political-militarily strategic, and the economically different; so far as her position, in the global comity of nations is concerned. It is a different reason and justifiably so, in the globally post-colonial-era modern world, that UK is quite prudently always ready to play the role of a 'B-team' of the USA, and as also her closest strategic Ally, but the conclave of US politico-military leadership, is seldom found to be willing or be strategically far-sighted enough, to closely discuss their ideas and plans with the British politico-military leadership establishments for being able to come to better, cool, and multiply far-sighted decisions. In other words, being mostly obsessed with her 'blind-arrogance', since duly triggered by the unchallenged supremacy of her mighty dollar and also, by the industrial-cum-militarily towering multiple-kinds of the American wherewithal, the USA perhaps thinks below her dignity, to bring-in the UK into her 'multiply-closer confidence' in these regards. But whatever these may be, the level of maturity of the British-democracy, commands altogether an unmatched excellence and stature. Hence a recent Indian Newspaper-Editorial since as quoted below, would very reassuringly be substantiating these precise perceptions, concerning the British Nation-State. And in no way these are any newly evolved dimensions in the overall making of the British Nation-State, but the dawn of all these multiply unique, living and vibrant, and wisdom-oriented as well as, the well-gelled British socio-cultural-cum-political attributes and their roots; may be found to be continuing right into the historic advent of the centuries-old **'Magna-Carta'** itself.

(x) In specifically the above multiple-contexts; the Editorial-column of the *The Statesman* (Kolkata, dated 01 September 2013) very perceptively and also, in a sharply focussed-manner; has commented that, "A NEW chapter in British constitutional history was scripted on Friday with British Prime Minister **David Cameron**'s defeat in the House of Commons on a matter of foreign policy. The parliamentary vote has rejected the idea of Britain joining the concert of invaders—notably America and France—to teach Syria a lesson in the aftermath of the alleged chemical attack. It is rare for the government to be defeated on foreign policy; for Mr. Cameron, who had pitched for Western invasion ever since the visuals were aired on YouTube, the vote has been a political catastrophe.... It was 200 years ago that a Prime Ministerial war motion was defeated, pre-eminently Lord North over the American War of Independence. Such ironies of war and peace shape history as do judicious initiatives by leaders. Not that the landmark repetition of military history will affect the warlike designs of **Barack Obama** or **Francois Hollande**, unlike in Iraq, Britain shall not be part of the attacking forces, however, Iraq was a war that the average Brit didn't approve of yet **Tony Blair** had bashed on regardless in the company of **George Bush**. Friday's robust disapproval of the legislature signifies that enough is enough. The commons have spelt out the difference between 2003 and 2013: in the span of a decade it has influenced international power-games by setting Britain's terms of engagement. Across the Atlantic, President Barak Obama's inaugural rhetoric to end the 'decades of wars' has been fulfilled by the British legislature though he himself, increasingly egged on by Secretary of State **John Kerry**, displays an impatient desire to strike.... Britain's legislative rejection must rank as a classic case-study of how parliamentary democracy ought to work...free from rumbustious scenes and logjam as in this part (*i.e. Eastern or the Asian.*) of the world. Did Mr. Cameron misjudge the mood of the people? Neither he nor his ministers could convince the nation with their claim that there could be no analogy between the planned limited strikes on Syria and the invasion of Iraq. Arguably, the price of misjudgment would have been still more disastrous than the legislative setback. The Prime minister has been humbled as has his Lib-Dem deputy, **Nick Clegg**, who too had shrilled for War. **Britain's international stature is now diminished, while its democratic credentials are reinforced**. The resounding note of the Commons vote resonates across the Atlantic." (***Indeed, both historically, as well as contemporarily the contents of the above quoted editorial, are how precise and true. But, it is more important, as well as being multiply far-sighted too, to always keep-on reinforcing and nursing one's democratic credentials back home rather than unnecessarily get fixated, with some highly shaky kinds of the 'false-notions', regarding the largely make-believe 'edifice of one's fictional prestige' in the foreign lands. As earlier very sadly, the then newly independent India's first Prime Minister Pundit Jawaharlal Nehru, during the years between 1947 to 1962, got so much obsessively fixated with largely similar kinds of the much more serious 'wrong-notions'; which in turn, were able to drive him into a very embarrassing 'fiasco', in the form of '1962 Indo-China War'; and thus he so naively pushed India into very costly humiliations, in full view of the whole world. Hence it has been rightly said that, the man's 'living-history', since always in the making onto his expanding civilisational-canvas never fears or favours anyone, waits for none and also at the same time, spares no one too, in any manner whatsoever.***).

(xi) Thomas Barnett has written that, "Despite Americans' current angst about globalisation and the challenges it presents, what we face over the next couple of generations are **problems of more success—not failure**. By introducing, codifying, and protecting our international liberal trade order across the decades of the Cold War, America not only set in motion Europe's recovery from its disastrous great-power wars (WWI, WWII), but also expanded the global economy to encompass effectively more than the four-fifths of the world's population (*Yes, facts are facts.*). **By doing so, America has made possible the global correction of the so-called Great Divergence that began in 1800, an unprecedented historical period in which the West's income grew twelve-fold over two centuries while the rest of the planet's income dropped by half. The Great Convergence of income and consumption that this century will witness will change everything in global affairs yet again. The projected twelve-fold have/have not divide by generating globalisation's first truly dominant middle-class—for the first time in human history.... This is America's gift to the world**: the chance for a global middle-class to finally escape the sort of Malthusian growth limits that the advanced West left behind in the nineteenth century. **Our success in spreading our liberal trade order is without parallel in human history**. At its height, the European colonial world order encompassed a mere third of the earth's surface; our globalisation model encompasses virtually all of it. What that means is this: **Long before the end of this century we will see the bulk of the planet move into the same sort of middle-class lifestyle (not to be automatically confused with consumption) that America has known over the past half-century-plus**. The European colonial-era globalisation simply does not compare with what America has achieved in the past several decades. The European colonial order created no widespread wealth and no global middle class, and came nowhere near encompassing the bulk of the planet's geography or population. **If globalisation signifies America's 'empire', then it's the first empire in history ever to enrich those integrated**.... Here is the America's still revolutionary impact upon the world is clearly seen: By expanding the liberal international trade order of our American system-cum-globalisation, we now elevate—over the next several decades—the vast bulk of the rest of the world's population out of the Malthusian resource equation and into a **planetary age of abundance**. By 2020, the global middle class should expand from just under one-third of the world's population to just over half, with much of that increase found in **China (the world's largest middle-class by 2025) and**

India (a tenfold increase in a generation's time). The problem is that, culturally speaking, much of this newly and soon-to-be-elevated chunk of world population remains wedded to the 'virtues' of their not-so-distant Malthusian past. Need a good example? The West's world of security now tries to incorporate **New Core powers (e.g., Russia, India, China)** with decidedly nineteenth century views of great power politics, along with Gaps whose style of mass violence remains as primordial as ever—with great local pride in the obvious (to them at least) 'virtue'. **Thus, numerous profound social shifts will occur as we successfully extend globalisation around the planet, consolidating vast levels of new connectivity and suddenly elevating entire societies from sustenance to abundance, often turning their definitions of vice and virtue on their heads**." (Barnett, 2009, pp. 397, 389 & 399) (*Aren't the above, reflected specific perceptions indeed, contextually so fascinatingly engaging in these regards, concerning the multiply numerous aspects of since the always momentously history-making 'great-games'? But as we begin to read in-between the lines of the above quoted paragraphs and thereby also commence delving much deeper into the perceptively, as well as no less momentously contemporary thoughts contained in the above quoted paragraphs, as so gratefully taken from Thomas Barnett's already earlier referred book, we do also witness that these inherently proactive thoughts, since now in multiple kinds of the action-mode have momentously begun to change the entire face and also the overall making of this world, in their unprecedentedly far-reaching manners. Mankind's civilisational facets, specifically concerning here—now the so fast fructifying, as well as expanding and, the vibrantly dynamic middle-classes almost all across the world—since as usual manifesting onto its expanding civilisational-canvas itself, may also be termed as nothing but the man's enchanting civilisational marvels, being always consequent to his highly focussed ongoing quest in the multiple areas of human-endeavours and thereby, also causing, the relentless flow of human sweat, tears, and the blood onto the advancing timelines of the man's 'living-history' itself, and thus soaking with these, the history's perpetually multiplying pages as well.*).

Contemporarily, while continuing our multiply continuing with these indeed multiply-important, and the far-reaching discussions in these regards, we may now quite relevantly prefer to switch-over our critically delving attentions from North-American and the European Continents, contextually to the no less turmoil-ridden Nation-State of Iran, as well as territories of the South-Asian Region with their own different kinds of the highly chequered historic past and then also, hardly any settled-cum-peaceful kinds of the present state of affairs. In both the places ironically we, again, become quite sad to witness the presently ongoing geostrategic-cum-politico-militarily-oriented 'fiasco' in fact rather so naively triggered by the geopolitically, historically, as well as socio-culturally, short-sighted and ignorant Americans so far as the regions of Middle-East, and that of the South-Asia are concerned. It is indeed a very negating commentary and certainly, not at all befitting to a Superpower Nation-State i.e. the USA. In fact, regarding any concerned leading power and then much more so, specifically concerning any Superpower; if she is not historically conscious and careful, as well as geopolitically prudent and flexibly far-sighted enough, with respect to the always paramountcy**multiple-aspects of the 'essential strategic-culture and thoughts',** then she is bound to land in a quite embarrassing kind of the geopolitical 'logjam' and thus, resulting in many kinds of the costly 'fiasco' for the larger world too.

We may recall that just after the highly chequered end of the World-War II and, while considering the long-term the geopolitical interests of the Western democratic nations, then onwards to be led by the USA—as the once so glorious era of the intercontinental British-Empire got finally confined to since the multiplying pages of the man's 'living-history' only; the UK divided India before quitting, by creating rather being an artificial and thus geopolitically untenable—the so-called Nation-State of Pakistan—to facilitate the new Superpower leader of the Western-Bloc i.e., the USA to geopolitically halt the expansion of the then menacing international-communism and thus, also by providing her a strategically valuable springboard to enable her to successfully checkmate the Soviet-Bloc and thus, prevent it from entering into the warm-waters of the Indian-Ocean. And consequently the desired success was achieved, without fighting any 'hot-war'; and so the Soviet-Union got finally dismantled, and in addition also the Chinese communism too, got almost liquidated and now, the rest is since well-known history.

These historic aspects substantiate the multiply-ongoing traditional enrichment of the British Nation's **multiple-aspects of the 'essential strategic-culture and thoughts'** in this world. **Though political division of the Indian subcontinent, in such ruthless, uncared for, and rather a somewhat avoidably cavalier manner rendered a greatly violent disservice to the newly independent Indian Nation-State and that too, India being a big and fledgling democracy but as it is already said earlier, there is no room for any such geopolitically moral and ethical considerations in these 'great-games', in the always very dicey 'arena of the international power-politics' and hence, prudently speaking, there is no space for the any kind of geopolitical 'sentimentalism' and any 'humane' behaviour, from any quarter in these regards**. Two historically very recent, as well as contemporarily very relevant examples, since specifically concerning the precisely above discussed multiple-contexts in these regards; one pertaining to the Iranian Nation-State and, the other one geopolitically pointing towards the so-called 'failing' Nation-State of Pakistan in South-Asia may appear to be very far-reaching and substantially reinforcing the above delved perceptions, concerning the US-sponsored short-sighted-cum-ill-conceived 'great-games' and thereby, the resultant 'fiasco' in these regards. More important, so far as the Middle-East countries are concerned; Russia too, with her own kinds of the ongoing 'great-games', have not lagged behind in 'fishing in these trouble waters'. **The paramountcy need in these regards, remains that both the USA and Russia essentially need to be geopolitically more pragmatic and thus, geostrategic-cum-politico-militarily more cooperative and sincere to each other; in the larger interests of the entire humanity and thereby, restrain themselves in getting fixated with their short-sighted 'great-games'**. At a first glance itself, these do appear to be rather 'tall-orders'; especially while somewhat closely looking into the numerously turbulent 'cesspools' of the above discussed and, even rather irrationally fixated 'great-games' within the globally much larger 'arena' of since, the always continuing in various textures of the different nations' situations-based and thus, altering geopolitical priorities; with respect to the intensely ruthless, and the competitively ongoing 'international power-politics'. But needless to say that in spite of all these, it is precisely here that, 'keys' for solving these international 'logjams' are to be found, by the concerned leading nations and also, by today's the singular Superpower, provided they begin thinking 'out of the box' in requisite cooperative manners, while still the fruitful time remains available.

(xii) Professor **Subrata Mukherjee**, from the Faculty of Political Science, University of Delhi in his very thought provoking, as well as highly informative article on 'Coup That Changed Iran' and published in *The Statesman* (Kolkata, dated 23 September 2013) writes that, "The operation by the CIA and the MI 6, was basically misguided. And the long-term effect on Iran in particular and on West Asia in general was disastrous. It strengthened the Shah's dictatorship and Iran drifted towards Khomeini's Revolution of 1979, which demonstrated **that Islam could also be politically potent**.... Sixty years ago—**on 19 August 1953—the popular democratic regime of Dr. Mohammad Mossadeq in Iran was overthrown in a coup d'etat that lasted for nine hours and with the support of America and Britain**. On 20 August, Dr. Mossadeq and three of his followers: **Dr. Abdullah Moazami**, former speaker of the Majlis (Lower House of Parliament) and member of

Dr. Mossadeq's oil-nationalisation team. **Dr. Ghulam Hussain Sadiqi** and **Dr. Sayed Ali Shayegan,** another member of the oil-team, were arrested. Dr. Mossadeq's wife and children were ejected from Tehran with the protection of imperial troops. As many as 300 Iranians died in the coup. **Had the USA and Britain not lent their support, the history of Iran and the region as a whole would have been very different. When the coup took place, President Eisenhower had vehemently denied any involvement. However, papers released very recently by the CIA testify to the contrary**.... Mossadeq was an anti-British nationalist who became the Prime minister at the age of 70. He was for many years a member of the Iranian Parliament (Majlis) and was a champion of the constitutional movement.... Commentators generally agree, that he was not corrupt. His outlook was deeply influenced by the virtual occupation of Iran twice during the preceding 50 years by Britain and Russia. **He was a democrat, and wanted the Shah to reign and not to rule and in a model of constitutional monarchy similar to Great Britain's. He believed that the army and police ought to be under civilian control**. (*That was very far-sighted and geopolitically so visionary that, these paramountcy imperatives concerning the very much genuinely, as well as indeed the democratically desired—the overall structuring of the entire gamut of the governance of Iran—were in the best long-term interests of not only the Iranian people, but the entire Middle-East region and then no less of the larger world, i.e. including those industrialized Western Nations too. In fact and in retrospect we may also quite perceptively maintain that historically then,* ***it was not Iran, indeed as quite a fledgling and even highly unsettled democracy then, who was being somewhat regressively short-sighted; but it were the two great and also, undeniably so 'nakedly' greedy Western-Powers, i.e. the USA and UK, who were being so obsessively short-sighted and also geopolitically so naïve; and that too, for the unjustifiably quite short-term 'loot' of the Iranian oil, that they became sadly blind and uncaring for their own long-term larger interests and at the same time obviously, for those of the global humanity as well.*** *It is a sad commentary on these two leading Western-Powers and also, ironically displaying such naïve mess-up of their indeed, so ill-planned 'great-games' based upon their half-baked ideas; and the rest is a very recent saga of a highly preventable and, the extremely unfortunate and at the same time, a no less consequential, and still continuing 'chapter' of the man's 'living-history' since continuingly in the making, onto the same global humanity's expanding civilisational-canvas in this world.*). He had tried to establish parliamentary supremacy and was opposed to foreign intervention, and eventually to end the domination of by the British controlled Oil Company.... **It is ironical that the 1979 Islamic revolution was initiated in Mossadeq's name (*look at this. Strange are the ways of these self-seeking, regressively-fixated, and the callous Mullahs, religious Pundits and Priests in this world.*),** and then during the Carter Presidency the Islamists detained 52 diplomats for 444 days. **This demonstrated the inherent animosity of the Islamists towards the USA for having toppled Mossadeq and installing the Shah. The rise of Hindu fundamentalism, East Asian Confucianism and China's blatant violation of human rights is equally unethical to the Western liberal creed. However, it does not elicit the same denunciation as Islamic fundamentalism.** (*The points are well-taken, but in these contemporary contexts, Professor Mukherjee needs to also realise that, the so-called 'Hindu-fundamentalism', though certainly not justifiable in any way and also, as per the modern and indeed, the fully acceptable democratic liberal thoughts, has largely been as a kind of the desperate reaction to the rise of foreign sponsored quite demonic, as well as the irrationally violent 'Islamic fundamentalism-cum-Jihadi-terrorism' and hence, these can't judiciously be put into the same 'bracket'.* ***And moreover, it is also no less important for the Western-World, as well as for India's the so-called 'pseudo-secularists' too to honestly and fully understand that, the modern, sovereign, and liberally democratic India, remaining very much unchallengeable in her so praiseworthy—the socio-culturally-cum-politically far-reaching, and also the geopolitically***

reassuring, her such vibrant and healthy secular-credentials—because, even more than eighty percent of her teeming Indian population, have traditionally been the liberal and tolerant Hindus.). This is because of the special interest the West has in the Arab world, indeed one that it wants to perpetuate at any cost. **The affluence of West is based on the cheap supply of oil**.... There is an increasing realisation of the fact that there is nothing so unique about any religious belief, including Islam. There are different versions of war, peace, pluralism and authoritarianism in all the major religions of the world—Buddhism, Christianity, Confucianism, Hinduism, Judaism, Shintoism and Taoism. The different views are manifest in different periods and circumstances. (*Yes, these are universally quite true and hence, need to be fully understood, and even rationally revised; for their harmony, further enrichment, as well as for better understanding, in their present contexts as well.*). **The idea of secular democratic state is very recent phenomenon even in the United States. The egalitarian concept of democracy—one man, one vote—attained fruition only after the Civil rights movement led by Martin Luther King Jr. in the 1960s. The task is to create more and more democratic institutions in the Muslim majority countries as was done after the Second World War in the Catholic countries of Europe and Latin America, to avoid what had happened in Egypt very recently**... Democracy was imposed in Germany and Japan by the victorious Western powers after the Second World War. The same process of democratisation has to be initiated in West Asia and the best opportunity to begin would be by acceding to the demand of an independent democratic state of Palestine. (*Yes indeed but for that, first Jordan has to surrender the traditional major chunk of the Palestinian territory, which she continues to occupy unjustifiably and at the same time Israel too, has to be assured of her permanently peaceful, strong and democratic existence. All these have to be both ways, on 'give and take' basis and hence, can't be one-sided only. All the major powers honestly need to come on the board, in the larger interests of the entire humanity and thereby no more any kind of the turmoil, may be permitted in the Middle-East. Solutions have to be geopolitically viable, enduring, all inclusive and also, clearly justice-oriented for all concerned and hence, with no room for any kind of the 'greasing' of any 'vested-interests' from any quarters.*). It is also ironical that the world had to wait for 60 years to know the truth though the facts were known more than a decade ago through newspapers and the national Security Archives of George Washington University. The activities of **Edward Snowden** and **Glen Greenwald**'s courageous attempts to bring these secrets to light assume legitimacy because in a sense, they are crusaders of truth. It is because of such acts of bravery that we may not have to wait for another 60 years to know the truth behind the war on terror." (***Whatever, but these are justifiably, as well as very intensely—the multi-dimensionally debatable issues—in numerous global contexts and also, containing differently conflicting views and perceptions in these regards.*** *Hence in order to patiently and gradually, to arrive at a globally harmonious reconciliations, on these indeed apparently irreconcilable issues having the high potentials for triggering all sorts of the seriously unsettling fallouts, and serious crises in various regions and so, being somewhat hastily judgmental on them perhaps, may turnout to be even largely counterproductive as well.*).

(xiii) Adhering to the above covered multiple-contexts and, concerning the all kinds of ongoing 'great-games' in the various regions of the world and precisely, with specific reference to the almost continuously turmoil-ridden and the so-called 'failing'—the USA's 'client' Nation-State of Pakistan—a quite knowledgeable Pakistani commentator, **Mr. Nasser Yousaf**, in his considerably informative and a quite thought-provoking article on, 'Choosing Not To See The Real Enemy'; (since republished in *The Statesman*, Kolkata, dated 23 September, 2013; and courtesy, the Dawn/ANN); and therein, he has written that, "Many politicians and arm-chair Fata analysts keep referring to the situation in the tribal hinterland as an uprising by tribesmen whereas ground realities indicate that the composition of the opposing forces is very different to what they imagine....ironically the controversial drone strikes have given us several leads regarding the true identity of the real enemy but vital information gets lost in heated debate surrounding the legality of the lethal technology. (*Indeed these are very interesting and, revealing too.*). A majority of the casualties of the two most recent drone strikes this month were non-local militants who had come from as far as Central Asia to the Arab lands.... At any given period of time over the last ten years, perhaps one in every three persons from the seven tribal agencies has been found to be living as an IDP (*i.e. internally displaced persons*) in camps across the settled areas of Khyber Pakhtunkhwa.... The living conditions in the camps have been so deplorable that it has moved even the otherwise stolid civil servants.... An officer who appeared to have been greatly touched by the tragic events, recently disclosed that tribal women rendered homeless by the unrest in the Bara region had been forced into begging, 'This is a catastrophe', the officer was heard to remark.... How could the so-called valiant and formidable tribesmen of yore be forced to leave their hearths and homes? The question needs to be probed and not only from a strategic point of view. It is one that also requires a meticulous study by students of anthropology.... If anything, the recent phase in the tribesmen's life could also expose a docile nature as he is found to have forsaken his home under pressure from outsiders, and done so with little or absolutely no resistance. (*Well, are these perceptions really true totally? It perhaps appears that those intruding 'West-Asian and Arabian aliens', being followers of the same religion and broadly of the same racial-stock, could get merged into the local crowd and thereby, were able to also establish the very fast-paced kinship-relationships through marriages and so on.*). This would have been unimaginable until recently.... This narrative over the years suggests that the authorities tasked with looking after Fata seem bent on keeping the residents of the tribal agencies in the shackles of a primitive tribalism as otherwise there remains very little that distinguishes the so-called tribesmen from their Pakhtun brethren in the settled areas.... The design appears to be to keep the tribal land as jinxed territory to further the ambitions of some who cannot think beyond the level of the dreaded lowest grade revenue official called patwari. (*Interesting all these appear to be, but so humanly tragic and also, totally unjustifiable too.*).... With the approach of the withdrawal of US-led NATO forces in 2014, tribalism will need to be revisited. No one can afford to let matters be as there are today.... An officer who has served as a political agent in two tribal agencies deeply regrets his two stints. He holds the political agent and the malik system responsible for the underdevelopment of the tribal areas and the subjugation of its people by brute force resulting in inexorable corruption.... In the days to come Pakistan will have to decide between keeping tribalism in place and permanent stability across the country....**The enemy in the tribal lands stands identified, it's not the tribesmen but the disgruntled zealots from other parts of the country as well as from abroad.... Successive governments in KP (*i.e. Khyber-Pakhtunkhwa*) have, without mincing words, blamed the unrest in the province on lawlessness in the bordering tribal areas. Fata's educated and law-abiding people, alone hold the key to security and development in their area and not the artful political agent.** (*It is true and this remains the only way out, towards the long-term and viable solutions in those remote, highly troubled, and also ruthlessly exploited tribal-lands.*).... ***Sun Tzu has said that,*** **'If you know the enemy and know yourself, you need not fear the result of a hundred battles. If you know yourself but not the enemy, for every victory gained you will also suffer a defeat. If you know neither the enemy nor yourself, you will succumb in every battle'.**" (*It is timelessly how true indeed. And that's precisely, what happened in 1962 to the Indian Prime Minister Jawaharlal Nehru in India.*).

In the above recorded as well as, briefly commented upon two last 'case-studies', concerning the globally ongoing and also, principally the US-sponsored and indeed, the multiply-ambitious 'great-games', what we broadly notice is that, the eventual 'success or failure', or even at times, somewhat partial or not so substantially gainful kinds of some 'success', almost entirely depends upon the concerned leading powers' and/or the concerned Superpower's, the multi-dimensional tangibility, as well as the vibrantly proactive stances; of theirs already cultivated, multiply developed and evolved and then also, flexibly imbibed—**the multiple-aspects of 'essential strategic-culture and thoughts'**—within those Powers themselves. Hence, any persisting shortcomings in these regards may lead to the whole efforts going quite ironically tangential and thereby, resulting in some kinds of the ironically, very costly 'logjam'; as well as at times, these may even result in no less embarrassing 'failures' too. These overall 'great-games-oriented' phenomena are always multi-layered, as well as multiply-complex and at times, due to the various kinds of even many the unknowable reasons, or developments-cum-events, in advance, may even manifest into their numerously unexpected ways. But, if any concerned leading Powers or the Superpower, have remained sincerely focussed, prudently consistent, and multiply-conscious enough in their always ongoing cultivation, evolution, development, as well as the flexibly imbibing of the numerously different contexts, the already earlier discussed—**the multiple-aspects of 'essential strategic-culture and thoughts'**—then the just above mentioned 'pitfalls' in these regards; can quite substantially be prevented and indeed, in a numerously beneficial ways, as well as, in their far-reaching geostrategic-cum-politico-military manners too. And hence, precisely in their larger perspectives; these very multiple-aspects have already been discussed and critically analysed appropriately, since in the above cited examples—as some contextually, a number of the 'great-games-oriented' 'case-studies'—and thus, highlighting always **the universal paramountcy of the multiple-aspects of 'essential strategic-culture and thoughts'**. Hence in other words, what we may apparently be able to visualise, with respect to the any leading Powers' or any concerned Superpowers,—the 'keys' towards their any contextual success, and also specifically, in any particular situation-driven 'great-games'—happen to be their geostrategic-cum-politico-military acumen, as well as their multi-dimensional economic, industrial and military wherewithal—being in fact, their uniquely valuable assets since vibrantly fructifying from nothing else, but precisely from theirs—the already well-cultivated, multiply-developed and then, flexibly imbibed—**the multiple-aspects of their 'essential strategic-culture and thoughts'**.

(xiv) Quite understandably, as the saying goes that any strongly built and multiply entrenched Establishment, both civilian and the military, and at the same time all the more in a truly vibrant-cum-healthy, and liberal democratic setup are mostly resistant to change and hence, always slow in adapting any crucially desired and so strategically mandatory changes, since as deemed necessary and, being consequent to the fast changing times and situations. The overall conventionally prevailing, and the geostrategic-cum-politico-militarily manifesting any 'conceptual-framework', in the Establishments of the any concerned Nation-State, and at any given point of its any situation-guided timeline are inherently slow to discerningly and also quickly see through, into the freshly emerging and also the largely changed and quite different overall geostrategic-cum-politico-military scenario needed to be confronted and managed, as per the dictates of one's vital national, as well as the larger geopolitical interests. And needless to say that herein, the proactive roles of the **multiple-aspects of 'essential strategic-culture and thoughts'**, and since also differently delved-upon earlier; assume much greater importance, precisely with respect to the above spelled-out the paramountcy wider aspects, concerning the newly emerging overall geostrategic-cum-politico-military scenario and thereby, the different kinds of challenges. **And at the same time, it is also no less important to note that these overall geopolitical aspects assume much greater, truly global as well as, certain extra-dimensional challenges as and when, these do come into the much larger and more complex domain of any particular Superpower; and accordingly now in fact, in the so fast manifesting, far-reaching, and also the totally changed global scenario; concerning**

the sole Superpower—the USA and thereby, in its numerous kinds of the differently ongoing—globalised aspects of the US-sponsored 'Great-Games', in various regions across the world. But paradoxically and as somewhat critically discussed above; the USA too, has been displaying its own kinds of the 'regressively-fixated', as well as the no less Establishments-oriented resistance to change, otherwise so visibly common all across in most of the nations and thus thereby, not being able to flexibly see through the overall changing scenario which in any case, are imperative to fruitfully realign its wider 'conceptual-framework' and then, reshape the new strategies and further plans, for the requisite geopolitical initiatives. Changes everywhere and in almost in every respects, are the inherent laws of Nature and hence, they cannot be wished away. So the imperative strategic needs are, to timely adapt changes with prudent flexibilities, for reaping maximum possible advantages. These as well as, numerous other wider but broadly allied aspects, have since been very thought-provokingly dealt with and also critically delved upon since having quite different global-perspectives and, concerning now, the altogether newly altered and the wider geopolitical scenario by **Thomas P.M. Barnett**, and he has indeed very perceptively written that, "In the early 1990s, the conventional wisdom around the Pentagon said we were living in a 'more dangerous world' than the one we had just left behind in the Cold-War. That's right. Even though the threat of global nuclear war had basically passed, many national security strategists were touting the notion that the world was a far more chaotic place, because the Cold-War's 'bipolar order' had collapsed. Because Washington and Moscow were no longer able to manage their respective camps, the global security environment was evolving into 'chaos,' the new Pentagon buzz-word to describe the lack of a global security rule set to replace the one that had defined the 'stable' Cold War standoff between America and the Soviet Union—you know the good old days.... **I have a lot of problem with the notion of chaos as a guiding strategic principle. It just seems like such a cop-out. So like any good Roman Catholic, I obsess instead about 'rules', or all the procedures, laws, treaties, rules of thumb, and conventional wisdom that seem to guide the actions of individuals, corporations, governments, and the international community at large. I focus on rules because wherever I find them in healthy abundance (*read* equity, not quantity), I know the U.S. military's role in enforcing them will be small, because once you have rules, you typically find rule enforcers already built into the system (e.g., our very robust and distributed U.S. law enforcement network of federal, state, and local police)**. (*These do constitute in fact, the pivotal core of the present-era's, now prevailing 'multiple-aspects' of the global-geopolitics; especially concerning the sole Superpower: the USA. And hence, any lack of concern for these aspects, on the part of US Military-Establishment, does somewhat quite fixatedly manifest into a kind of 'fog over the strategic-ideas' in Pentagon.*). **Likewise, wherever rules are clear because most players in that system agree they're good, there is not so much reinforcement required, because most participants simply decide on their own that playing by the rules is the best course of action. But where you don't find generally agreed-upon rules, or where rules are out of whack or misaligned across social sectors, then you are talking about the future of instability, the potential for misconceptions leading to conflicts, and the clash of competing rule sets.... It's as simple as that: The fewer the rules, the more war you have**. Back in the early 1990s, it was becoming clear to me that the world was not a dangerous place, just one in which the rules concerning war and peace were not as clearly defined as they had been during the Cold War. So, yes, the Pentagon's security blanket (the Soviet threat) may have been torn to shreds, but a new global security order was clearly in the works—just not one defined by a military superpower rivalry. Figuring out what the new global security rule was set was and what larger forces were shaping it would determine the future strategic posture of the U.S. military.... **I know it would not be easy to convince anyone in the Pentagon that there was some underlying method**

to the seeming madness of U.S. military crisis responses in this yet unnamed 'post-Cold War era'. Containment is coherent, chaos is not. Throughout the 1990s, the Pentagon lurched from Somalia to Haiti to Bosnia to Kosovo, and it did so without the slightest understanding *why*. Each engagement had its merits, but taken as a whole, they did seem to spell chaos is no one's idea of a strategic paradigm, especially if you are the world's only superpower.... According to most people in the defense community, if you want to understand the future of conflict, you figure out the future of weapons technologies. Then you plot out who's likely to get their hands on these technologies. Then all you need to assume that anyone who wants to get their hands on those technologies must inevitably fear or hate America and thus seek to do her harm and—*presto*. You have identified the future threat. Unless, of course, they are already allies and it was you who sold them the weapons in the first place. (*What these happen to be ironically indeed, certain 'hilarious-cum-confused' lines of the geostrategic and politico-military thought-processes and that too, on the part of the today's sole Superpower the USA.*).... An even easier calculation is who's an ally? Answer: Basically anyone who agrees with your definition of the future threat, or anyone to whom you have already sold weapons. (*Shall we comment anything more, on this kind of geopolitically transnational 'strategic naivety'; especially during the presently ongoing super high-tech modern-era and that too again, from now the sole Superpower: the USA?*).... **This sort of approach is the reason the defense community was so flabbergasted by the rise globalisation in the 1990s**. As far as most security experts were concerned, it was a complete negative, for all it did was facilitate the spread of dangerous military technologies around the world. So if you described globalisation as a good thing leading to the integration of the economies, the development of emerging markets, and the 'shrinking' of the world, you were almost immediately labeled naïve or a wide-eyed optimist. As far as the Pentagon was concerned, the post-Cold War era was one in which constant warfare raged around the planet and dangers to U.S. security grew at every turn.... I've been reading Pentagon and CIA projections about future threats for years, and each report I've perused has always said the same thing: anything is possible and eventually *everything will happen*. That sort of brave futurology means you'll never take one in the end, but it doesn't mean you'll be pointed in the right direction or that you'll be able to move any faster than the flow of current events. You will always end up being reactive, even as the inevitable ex post facto congressional hearings 'prove', yet again, the 'disturbing' and 'alarming' failures of your intelligence community to predict the unpredictable—like a 9/11. Doesn't it seem weird that the same senators who prattle on during Sunday news programs about how the world is a chaotic, unpredictable place still always seem to show up on C-SPAN following some security disaster or to decry yet another 'intelligence failure'? Who are these people kidding? **Rather than dwell on the unpredictability of future threats or attacks, our strategic vision for national security needs to focus on growing the community of states that recognize a stable set of rules regarding war and peace, as in 'These are the conditions under which it is reasonable to wage war against identifiable enemies to our collective order'. Growing that community of like-minded states is simply a matter of identifying the difference between 'good' and 'bad' regimes, and rallying the former to work collectively to encourage the later to change their ways, applying military power when diplomacy alone does not do the trick. But changing 'bad' states to 'good' ones requires that we generate some broadly accepted definition of what a 'good state' is, meaning that a government that plays by the security rules we hold dear—like 'Don't harbor transnational terrorists within your territory' and 'Don't seek weapons of mass destruction. Enunciating that rule set is the most immediate task in this global war on terrorism, and promoting the global spread of that security rule set through our use of military force overseas (e.g. preemptive war against regimes that openly transgress the rule set) is our most important long-term goal in this struggle.**" (Barnett, 2004, pp. 22, 23, 24 & 25)

We may contextually recall that earlier in this sub-chapter itself, and broadly in the same lines of thoughts, as reflected in the above quoted paragraphs, we have since also delved upon the contemporarily very crucial issues concerning the multiple-necessities of first: the USA needs to come out of its rather the 'isolationism-fixated' as well as, the no less 'self-created' some sort of the geopolitically counterproductive 'cocoon'. And then with always the universally desirable, and the far-sighted geostrategic-cum-politico-military prudence to be able to team-up with the matured, peace-loving, trustworthy and the truly democratic Nation-States, like Israel, UK, India, Japan, South-Korea, Australia, and also similarly the more, like-minded States in Europe and elsewhere to collectively, as well as effectively, combat and then decisively subjugate the all sorts of extremism-cum-terrorism all over the world. **And thereby all the more importantly, the first and foremost, in globally the similar manners; to do the needful with the most menacing and also, now the always so much dangerously lurking, the transnational Islamic Jihadi extremism and terrorism; while no less sincerely, adhering to the collectively arrived at the rule-sets in these regards. Now in these very precisely, the broader geostrategic and politico-military perspectives first and the strategically foremost step perhaps, will have to be internationally, to identify and map the variously menacing epicentres of Islamic extremism and terrorism, with all kinds of the necessary details in each case; so that always the most valuable synergy-creating comprehensive geopolitical-cum-military plans of action are drawn-out in advance and that too, without any kind of fear, favour, 'drag', or even bias; for quickly and prudently achieving the desirably decisive results, to save the entire humanity from this extremely demonic menace in today's world**. Ironically in the globally broader and also, being the highly chequered geopolitical domain, especially in the post-Cold War era today's Superpower the USA, has mostly been stuck with its only very 'limited and short-term vision', as per the region-wise, and also with mostly having the piece-meal urgencies only; and thus needless to say that as per the larger geostrategic and politico-military imperatives, the USA is stuck with only *'counting the trees while missing the wood'*. And these precisely are the contemporarily crucial as well as, very far-reaching aspects; on which the above quoted paragraphs since as picked-up from Thomas Barnett's book, do expertly reflect upon, and contextually in the so much penetrating manners.

It is said that in any truly vibrant and liberal democracy, if politics is a game of possibilities and then at the same time, the strategy too, happens to be the prudently flexible as well as, the multi-dimensionally visionary gamut of the always ongoing exercises of 'policies-formulations' in any given situation and thereby accordingly, the proactively shaping-up of the responses to any emerging threats/challenges, as per the dictates of one's overall highest national-interests. But the no less 'dicey-catch' in these regards, happens to be that the just mentioned 'dictates of the overall highest national-interests' are seldom very clearly, or even at times apparently, visible to the decision-makers in the any concerned 'corridors of power'; and thus ironically, resulting in the formation of a kind of the ***'fog over any desirably emerging strategic-ideas'***. It remains essential, and since **Sun Tzu** has said that, it is always important to know your enemy and also, know yourself too. And another perhaps, no less important aspect is that, the antidote to the above noted 'fog'; happens to be the proactive as well as, always the flexibly well-placed **multiple-aspects of the 'essential strategic-culture and thoughts'**.

Interestingly, as well as in a very thought-provoking manner; Thomas Barnett, in his already above referred book has discussed the broadly similar to but, all the more comprehensively and in much more the geopolitically far-reaching manners the above delved-upon 'thoughts and concepts', with his always the so well-known—the intellectually captivating master craftsmanship—and he calls them precisely, the multiple-processes of the **'horizontal thinking'**. What precisely is this horizontal thinking and, how it facilitates an interested person's thoughts-processes and thereby, be able to effectively remove since the already above discussed, some kind of a *'fog over any desirably emerging strategic ideas and concepts'*? For conceptually clearing these points in these multiple-contexts, Thomas Barnett has written that, "When I left Harvard in the summer of 1990 with my spanking-new Ph.D. from the Government Department, I was pretty sure I had all the intellectual tools to cut it as a policy analyst in Washington, D.C. I knew a lot of fellow students and professors looked unfavorably on the choice, believing it was quite a step down from a career in academia, but I was eager to find out how things worked in the real world.... Harvard as it happens, is the perfect finishing school when it comes to working for the government. Because in Washington, the only way to stand out, when you are surrounded by thousands of people just like you; to tear down their ideas. **The problem is, in that vision-hostile environment it is impossible to come up with any stories with happy endings—as I like to call strategic visions. That is because no one in Washington is really interested in your happy ending; all they obsess about is preventing, what they are certain will be the disastrous outcome of your ill-conceived plan**. (*And interestingly here, we may also see some glimpses of precisely the above discussed 'fog', in these multiple-contexts; and that too ironically, in the US 'Corridors of Power' as well as, at the policy-level decision-making 'Establishments' in Washington D.C.)*.... **Henry H. Gaffney Jr.**, as everyone calls him, is probably the smartest man alive when it comes to understanding how the Pentagon and national security decision making actually work. But what really marks **Hank** as unique in the defense community is the combination of his wide-ranging but detailed knowledge of security issues, his innate ability to ferret out weak assumptions in strategic logic, and his crankiness and lack of respect for authority.... During my eight years in Washington, Hank Gaffney and I did a lot of research together and coauthored a number of significant reports, and most of the big ideas I have hatched over my career either began in or were intimately shaped by my time with him. **But what Hank really taught me was how to think horizontally. By that I mean thinking broadly across subject matters versus drilling down deep into a particular subject, which I call vertical thinking. In both the Pentagon and Washington in general, the system of debates awards points almost exclusively to those who think vertically, because intense subject-matter expertise allows you to poke holes in everyone else's thinking. Inside the Beltway, vertical thinkers are expert at telling why something will never succeed, and little else. Horizontal thinkers tend to be the exact opposite.... The most important advantage of horizontal thinking is the ability to see the future unfolding in realistic stages, never becoming too invested in any one particular pathway. Most of what passes for strategic planning in the Pentagon involves the acquisition (or buying) of future technologies in the form of weapons systems and platforms (e.g., ships, aircrafts, Tanks). Within those narrow confines, most vertical thinkers do well. But these same vertical thinkers are incapable of making similar forecasts about the messy world lying outside the Pentagon**. Here their tendency to define the future as straight-line projections of current trends consistently betrays them, because **you can never mechanistically extrapolate tomorrow from today**." (Barnett, 2004, pp. 22-25)

It is true that you can never succeed anyway and in any kind of given situation anywhere, in mechanistically extrapolating tomorrow from today, even somewhat correctly in any manner. But it remains a great paradox that in spite of all these indeed, the concerned people sitting in the corridors of power and so thereby, commissioned with the onerous responsibilities to preside over the destiny of a Nation-State remains ironically, so rigidly fixated with their obsessive systems of 'vertical-thinking'. No wonder, it has very thoughtfully been said that since the very ancient of times, mankind is rather universally found to be displaying its rigid-cum-obsessive 'culture of resistance to change'; and paradoxically in spite of the so much multi-dimensionally ongoing the humanity's progress as well as, all the more no less multiply continuing advancements-cum-broadening of now, the highly complex 'store-house' of the human-knowledge, onto the man's expanding civilisational-canvas itself he has not substantially succeeded in breaking the shackles of 'resistance to change' and thus, remain unable to come out of 'dark cocoon' of the 'regressive' culture of vertical-thinking. **Thomas Barnett,** indeed very perceptively, has been able to intellectually dissect the strategically very 'dangerous' as well as, any nation's multiple-kinds of the economic-cum-military resources-wise, very tragic-cum-wasteful 'arrogance' of their squandering-away consequent to the—professional but no less 'regressive' intellectual-culture of the higher strategic-cum-policy-planning-levels', the 'malignantly lingering culture'—of *vertical thinking*. And hence, we wish to say that as we may proceed to further critically delve deeper into these aspects, the modern-era's indeed extraordinarily bothersome, and also the highly fluid and, the presently prevailing—the overall international security-scenario all around us—we may clearly be able to see through the far-reaching quantum, as well as the extremely menacing dimensions of its 'fallouts', not only for us but also, to our future generations. And needless to say at this precise point here that with respect to all these multiply-complex and also indeed, the international security-related scenario their multi-dimensional complexities, reach and also in addition, their depths and dimensions, assume their multiply very large-scale enlargements, in case of the present sole Superpower: the USA and especially, in the today's super high-tech era world.

But ironically indeed, such 'regressive' fixations regarding the—professionally somewhat lingering 'intellectual-culture' of the vertical-thinking—is certainly not with the sole Superpower USA only; but at the same time no less and, with differently varying limited dimensions and reach, the same 'regressive' obsessions do also remain true with many other Nation-States as well; and all the more paradoxically, with the modern, fast-developing and also, the liberal democratic Nation-State of **India** too. And thereby rather quite comically at times for India especially, these do result in the 'foreign-policies-related' embarrassing 'fiasco' for her; and no less due to her own numerous kinds of the 'omissions and commissions', in differently emerging situations in the highly slippery 'international arena' of the nations' since always ongoing—the different kinds of **'great-games'**—in the world. We all are since aware that in today's Asian continent and north of India, another very giant Nation-State of **China** is situated and it has a very different kind of the political-system of national governance, so unlike India or any other liberal democratic nation in the modern world and thereby obviously, many things, as well as their multiply consequent fallouts too, are quite different for the peoples in China. In any case, we would be more relevantly touching upon these aspects with their different dimensions, when we would soon be coming to the last sub-chapter of this eighth chapter itself.

It is a matter of common knowledge that, 'change is the law of nature' and hence thereby, almost nothing in this world remains static and unaltered for any kind of the somewhat substantially longer-period; onto the ongoing timelines of the man' 'living-history' since, continuously in the making, onto his expanding civilisational-canvas itself. And the same remains indeed equally true, in case of the any concerned nation's overall security-situations as well and at the same time also much more so, along with any presently manifesting timelines-related, the overall prevailing international security-scenario in the world. It is indeed the widely known very current historic fact that, during the recently passed-on initial decades of the post-Cold-War era in the world, many kinds of the very far-reaching as well as, unprecedentedly the new kinds of and also, the no less momentous history-making developments and events, have now been taking place; and that too interestingly, in the presently ongoing new and totally altered kinds of the globally economic, geostrategic and politico-militarily—the overall international scenario—and wherein surprisingly today, almost all the leading nations, including the sole Superpower USA too, have been caught off-guard and thereby, geopolitically not all that desirably found to be prepared as well. And if that be so then, what to talk of the other so-called third-world countries in these regards. Among the above mentioned new kinds of the recently manifested developments and events, the two perhaps most far-reaching and also the no less globally situations-altering phenomena have historically been found to be manifesting so far, and thereby, first the most menacing phenomena of the **transnational Islamic extremism and terrorism** on one hand; and then at the same time, multi-dimensional manifestations of indeed, the globally so much economically-uplifting phenomena of again the **ongoing globalisation** almost all across the world. Coupled with all these, the indeed mind-boggling multiple advancements in the so fascinating fields of the modern sciences and technologies too, are adding many unprecedentedly new dimensions-cum-impetus to now the overall since fast altering global scenario, for the humanity at large. We contextually may recall here, some very relevant and also, since the already earlier quoted indeed, certain intellectually penetrating perceptions of **Thomas Friedman**, that the I.T.-Age, coupled with the processes of globalisation; had made the world 'flat' and thus, resulting in some 'level playing-fields' for the all concerned. Well, these thought-lines may also appear to be the no less contextually debatable too.

In precisely the above multiple-contexts; **Thomas Barnett** again has mentioned that, "**Hank Gaffney** taught me how to put my natural inclination for horizontal thinking to good use in diagnosing security situations in the world around me. Hank used to present these wildly complex vugraphs in the 1990s that displaced a number of bad security situations around the world as horizontal scenarios, meaning he would measure the 'pain or 'instability' in any situation and plot it out over time as it rose and fell—'peaks and valleys', he would call them. His argument was that the Pentagon tended to ignore these situations until they reached some critical point where they magically tripped our 'national interest' alarm and then the White House would call over to the Pentagon and say, *'Do something about that now'*. Of course, the Pentagon would respond with complete shock: How could anyone have predicted this? Then the military would rapidly muster an interventionary force and response to the 'crisis', afterward bragging that it handled the situation with virtually no warning time. (*Amazing, isn't it. But in any case as some credible counter-measures there must always remain as rather strategically, some inbuilt professional systems of futuristically visualizing any possible forthcoming developments and events in these regards, and expertly see through them and then simultaneously, make the multiply military-levels: the geostrategic and politico-military preparations and thereby, render professionally effective advance advice to the White House while there is still some valuable time available to the constitutional 'Head of both the Executive and the State', and also Commander-in-Chief of the US Armed Forces; to be able to undertake other kinds of the geopolitical initiatives, possibly first to save the valuable lives of few of the likely frontline US military-men.*).... It was those vugraphs that eventually pushed me to defining what I call the Post-Cold War Horizontal Scenario, which I describe as follows: **First, there is no clear beginning or end to the conflict, meaning it feels as if it has been around forever and that it will continue well past our lifetimes.... Unlike the classical vertical scenario, these horizontal ones feature loads of adaptive planning for the military. Rather than one giant battle, it is strike after skirmish after ambush after.... Planning is done on the fly. It is not chess but something close to soccer. The ball is always moving, and substitutions are constantly changing the composition of both your team and your enemy's. But worse still, your political leadership's definition of the 'problem' you are trying to solve keeps changing, making**

your attempts to keep score almost meaningless. You want to know what today's definition of the problem is? Try reading the op-ed pages; you will have plenty to choose from.... If we think globalisation as the ultimate horizontal scenario, then we immediately move beyond the simplistic notion that globalisation is a binary outcome—*first it was not there, and then it was*. Globalisation is not a *yes* or *no*, despite the all-encompassing outcome suggested by its name (by definition, globalisation must be everywhere, right?). Rather globalisation is a process, a pathway, a *what*, combined with a *where*. **Understanding where globalisation has taken root and where it has not is the first step toward mapping the international security environment of the twenty-first century. For the military, the importance of this demarcation cannot be over stated. It is like the difference between land and water, jungle and desert, mountains and grasslands. Knowing where globalisation begins and ends essentially defines the U.S. military's expeditionary theater. It tells us where we will go and why. It tells us what we will find when we get there, and what we must do to achieve victory in warfare.... Thus, within an amazingly short time of less than two decades, Russia effectively synchronized the bulk of its internal rule sets with that globalisation's emerging rule set. By doing so, Russia was transformed from America's number-one threat to country deemed worthy of limited membership of NATO**. Think about that for a minute.... **Gorbachev** began this process hoping to save the Soviet Union, only to bury it and the Cold War in the process. **But what he gave the world was a Russia reborn**. (*Yes indeed, and we may also contextually recall that, while dealing with these, historically so momentous aspects and also, while quoting some paragraphs from Gorbachev's well-known book on 'Perestroika', and certain other observations too, earlier; we have since quite befittingly, emphasised his contemporary greatness in these regards.*). Russia, while embedded within the Soviet Union, was one of the most isolated, disconnected countries in the world. Not surprisingly, it was a huge source of danger and threat to the United States. When Russia ended its disconnectedness, it ceased to be a military threat.... **A Chinese friend of mine who had been active in the democracy movement explained how this rule-set shift changed his outlook: 'Before Tiananmen, we believed that freedom is 90 percent political and 10 percent economic. A few years later, we came to realize that real freedom is 90 percent economic and 10 percent political**.' You may find my friend's change of heart troublesome, but think about your own daily life and then try to tell me that second formula isn't better description of how things really work for the vast majority of Americans. Anyway, my friend's rebalanced equation makes perfect sense to me when I think about China's sequencing rule set reform.... **Is China a democracy yet? No. In terms of economic freedom, though, it has synchronized its internal rule sets with those of globalisation to a great degree**, and as we witness the birth of a litigious society there, I become increasingly confident that China will accomplish the same in terms of its legal rule sets. Nowhere is this seen more clearly than in China's joining the World Trade Organization, from which, in effect, Beijing's leadership is importing legal and economic rule sets they could not otherwise generate effectively on their own—a sure sign of their weakening authoritarian grip over the political system as a whole. China will be a democratic society within a generation, because by synchronizing their internal economic and legal rule sets with those of globalisation's Functioning Core, China will end up accepting far more internal change that its Communist leadership ever bargained for ...**So if Russia's path was political-to-economic-to-legal rule-set change, China's is economic-to-legal-to-political rule-set change, can a country begin with legal rule-set changes first? It can, and frankly must do, although most observers would be hard-pressed to describe the internal rule-set shift as constituting what we like to call rule of law**.... So if these are my fundamental definitions of what constitutes functioning within globalisation's Core, which countries can be considered in? My list of the Functioning Core of globalisation consists of North America,

Europe both 'old' and 'new', Russia, Japan, China (although the interior is less so), **India (in a pockmarked sense)**, Australia and New Zealand, South Africa, and the ABCs of South America—Argentina, Brazil, and Chile. That is roughly four billion out of a global population of six-plus billion." (Barnett, 2004, pp. 118-121, 128-132)

So in the last few lines of the above quoted paragraphs while clearly bringing-in his thoughtful definitions as to what all constitute functioning within the overall gamut of globalisation's Core, as well as while also, listing certain nations, being both geopolitical-cum-economically, fit to be included therein; Thomas Barnett has indeed, included India *'in a pockmarked sense'* and hence contextually whatever if any, some commentators in India, may question or perhaps even criticize him, for his rather not so charitable remarks but while objectively speaking, he appears to be 'above board' and thus, can't be faulted at all in these regards. Similarly, his intellectually so focussed comments on China too, appears to be no less precise and fair. In fact, in any intellectually penetrating as well as, equally comprehensive, both in its range and dimensions concerning any critically objective analysis-cum-explorative studies of any globally far-reaching and also, the momentously history-making developmental phenomena, one is not supposed to waver in any manner, while fairly calling 'spade a spade, spot a spot or even, a stripe a stripe' possibly, in any specific context whatsoever. While it also remains true and broadly acceptable too, that in these specific contexts; contemporarily any sovereign-cum-democratic Nation-State, thus geopolitically-cum-economically getting engaged in embracing the ongoing processes of globalisation in today's global-scenario, is also required to take care of its legitimately paramountcy national-interests, and indeed compulsively above the all other considerations. But in any case understandably, these broader aspects can never be justified as excuses for any kinds of the otherwise, no less avoidable internal failures or any kinds of the functional shortcomings in its processes of the expectedly pragmatic statecraft for its policy-formulations, any limitations in its diplomatic acumen as well as possibly, the unbecoming 'cracks' in its entire gamut of the democratic governance. Now as in various contexts since, already emphasised earlier for any truly democratic and also, all the more consistently developing Nation-State in order to considerably-cum-fruitfully take care of the above identified, some critical shortcomings and challenges, what always remains paramountcy in fact—are the prudently, as well as the no less simultaneously cultivating, developing, grooming and also, flexibly imbibing the—**multiple-aspects of 'essential strategic-culture and thoughts'**. And hence interestingly, in all these our presently ongoing discussions, and their multiple-contexts it is precisely here that, lie both the 'key' as well as, the proverbial 'rub' in these regards. **So if India's presently ongoing indeed, the somewhat 'cocky' processes of globalisation, is externally viewed in a 'pockmarked' sense then precisely here, that lies the 'rub' and needless to say that, in spite of India being a gloriously so ancient a civilisation; she has not yet substantially succeeded in coming to the always desired-cum-fruitful 'grips', over the above mentioned multiple-aspects of the 'essential strategic-culture and thoughts'**. Precisely in the above multiple-contexts, Professors **Jean Dreze** and **Amartya Sen** have written that, "The recent achievements of modern, democratic India are not inconsiderable, and have been widely recognized across the globe over the last decade or more. **India's record in pioneering democratic governance in the non-Western world is a widely acknowledged accomplishment, as its basic success in maintaining a secular state, despite the challenges arising from its thoroughly multi-religious population and hugely problematic history of violence around the ending days of the Raj**. To this can be added the achievement of rapid economic growth.... And yet—despite these great achievements—if the much talked about glory of today's India is deeply uncertain, it is not because an unblemished sunny day stands in danger of being ruined by a freshly arriving shower, as was feared by Proteus of Verona. The uncertainty arose, rather, from the fact that together with the sunshine, there are dark clouds and drenching showers already on the scene. It is important and urgent that we try to evaluate both the achievements and the failures that characterize India today. What remains to be done? And are there new problems that India has to address? In historical perspective, the accomplishments are large indeed, especially in light of what the country was at the time of independence in 1947. India emerged then from an oppressive colonial rule, enforced by dogged imperial rulers; there was little devolution of power until the British actually left, and it was not unnatural at that time to doubt India's capacity to run a functioning democracy. A second challenge was to avoid the danger of chaos and conflict, or even a violent break-up of the country. There is long history—stretching over thousands of years—of cultural affinities across India, and the struggle for independence generated a great deal of popular unity. And yet the diversities and divisions within India—of many languages, religions and ethnicities—gave skeptics good reason to worry about the possible break-up in the absence of authoritarian rule. More immediately, the chaotic partitioning of pre-independent India into two countries—India and Pakistan—gave justified cause for anxiety about whether further violent splintering might occur.... And indeed in 1943, just four years before colonial rule ended, India did actually have a gigantic famine in which between 2 to 3 million people died.... **India had not always been a symbol of poverty and hunger—far from it....how the country became so poor. What is not in doubt is that the economy of British India was remarkably stagnant, and that living conditions around the time of independence were appalling for a large proportion of the Indian population, and not just in famine years**." (Dreze and Sen, 2013, pp. 1 & 2)

(xv) Now coming back to the certain crucially important dimensions concerning the multiply vast and complex aspects of the indeed, numerous kinds of **'great-games'** since relentlessly being played by the various powers across the world and then, much more so, by the sole Superpower: the USA, onto the highly dicey, as well as, rather some kinds of the 'shifting-sands-oriented' and also altering—the geopolitical international arena—the pivotal-cum-decisive dimension pertains to the military wherewithal of any Power concerned in these regards. Interestingly to substantiate these perceptions, as we may look into the various dimensions of even during the times predating the holy-birth of **Jesus Christ** we do contextually notice that, when the ancient India's **Emperor Ashoka**—*c*268 to 233 BC (grandson of Chandragupta Maurya—*c*321 to 297 BC) commenced sending the then, momentously history-making peace and love for the fellow human-being-oriented delegations to both South-East and the Far-East Asian regions, to propagate the spiritual thoughts of Lord Buddha; he could desirably do so, from the then far-reaching, as well as indeed, with his almost unchallenged positions of the geostrategic-cum-politico-military strengths only. And needless to say that, the rest is unique, monumental as well as, momentously enduring, and indeed, overwhelming civilisational-history of the captivating peoples of the above mentioned Asian regions. And hence for all these, what the first and foremost since essentially needed, are—the multi-dimensionally requisite matters-military-oriented wherewithal—for the concerned Power; whoever wished to trigger, indulge-in, as well as maneuver with some credible successes, with respect to its any intended 'great-games'. But at the same time, any substantially credible matters-military-oriented wherewithal in any case by itself, won't bring-in any desired success, if the already conceptually and since differently emphasised the universally paramountcy—**the multiple-aspects of 'essential strategic-culture and thoughts'**—have not been prudently taken care-of, and also both proactively-cum-flexibly, kept in place with its multiple dimensions and the desired effectiveness, since years well in advance. And hence even at the risk of being repetitive, we may say that precisely here, lie both the 'rub' and the 'key' in these regards. Now specifically in these multiple-contexts, we do also need to bear in mind that, the always multi-dimensional matters-military-oriented wherewithal has many kinds of both strategically 'cold-cum-hot-centric' applications. Besides the commonly understood launching of one's military-forces in different kinds of war/conflicts; any substantially credible military wherewithal, is also inescapably needed for maneuvering the different kinds of any situation-specific coercive-diplomacy and then at times, since also quite effectively needed to be applied for the different kinds of 'power-projections'; as well as even at times, for some highly ceremonial kinds of displays, for indeed the greatly strategic and also, for the far-reaching geopolitical effects and accrual in these regards.

Indeed a highly ceremonial Presidential review of only some representative kinds of 'Naval-Fleets' since, belonging to a number of the somewhat recognized Naval-Powers of today's world, **Thomas Barnett** has very perceptively commented that, "Why is this shared history so important? Most Americans just do not understand what a huge disparity exists between our military capabilities and those of the rest of the world. We really own the only global military, whereas everybody else pretty much has the national or, at best, regional militaries. If we are not out there using our military to export that security, these bonds will simply never be created.... This fundamental reality was made greatly apparent to me when the President and Prime Minister of India conducted their 'review of the fleet' in Mumbai harbor, zigzagging past all eighty ships in rapid order in their 'presidential yacht', which was really an Indian Navy frigate gussied up for the event. The bigwigs sat up front on the bow, I sat in the cheap seats at the stern. **In such a review, it is impossible not to notice that, compared with the United States, the rest of world's navies are simply the equivalent to our Coast Guard. The United States owns the only blue-water navy in the world**. The size disparity in ships was so great that when we finally passed by the sole U.S. Navy ship in the review, a mere guided missile cruiser (one of our smaller surface combatants), I almost fell off my chair. It was substantially more impressive--not to mention simply bigger—than every other ship in the review. It was as if we had been reviewing a line of Coast Guard cutters and then all of a sudden this glooming bruiser loomed over us. Everyone on board took out cameras at that moment and snapped picture.... My car, adorned with American flags on the hood, along with a sign on the side describing me in my entire titular splendor, followed that of a British naval expert in the existing parade of limos. When the crowd saw the British flag flapping smartly in the breeze, a nice cheer rose. But they spotted the American flag, the cheer turned into a frenzy as people jumped up and waved their arms ecstatically, shouting at the top of their lungs.... At first, I was simply too stunned to do anything but stare out the window. Of course, it had nothing to do with me. A trained monkey in the backseat would have gotten the same response. This was the biggest party in Mumbai's history, and average citizens were simply thrilled that America decided to show up. So after about a minute of watching rolling wave of cheering Indians, my handler, a young Indian Navy commander, leaned over in the back seat and showed me how to wave my hand like I was the Queen of England." (Barnett, 2004, pp. 233 & 234)

We all are also fully aware that, in almost most of the ways today; the USA happens to be the sole Superpower, especially after demise of the yesteryears' Cold War era. Now being the sole Superpower, brings many kinds of responsibilities too and thereby, the Superpower is never expected to rather quite unexpectedly shy-away from them. **One prudently geopolitical and also, no less far-reaching method may even be: the multi-dimensionally 'exporting' security, across the continents as well.**

Precisely in the above multiple-contexts; **Thomas Barnett** has said that, "There are many different ways a superpower like the United States goes about making other states feel more secure. Sometimes exporting security means training their future military leaders at our schools, like the Naval War College. Other times it means having our ships visit their ports, to let them know—like the cop walking the beat—that, we are there, when they need us. Still other times, it means setting up shop in their neighbourhood, stockpiling some of our equipment and supplies, telling them we are ready to help if anything really bad happens. Sometimes those really bad things do happen, and then we use all our influence with the involved countries to deal with the situation as best we can. **In this way we stop wars from happening.... That is exporting security. It consists of America giving the world something we have in abundance: a belief in future. It is a wonderful gift, and frankly, only the United States has either the wherewithal or the generosity to actually provide it. It is one of the best things we provide the planet, and it has changed the course of human-history for the better.... This exporting of security is, in large part, is nothing more than a by-product of the U.S. military's continuous worldwide operations. We are the only military in the history of the world to possess a planet-spanning command scheme**. But exporting security is also done on a person to person basis, across countless exchanges, meetings, and conversations. **In short, it's a people business as much as a platform business**—as much face to face as aircraft fly-over and carrier port calls." (Barnett, 2004, pp. 231 & 232)

While contextually reviewing the above recorded continuing discussions, we do indeed find that, Thomas Barnett does have a lot may many far-reaching, as well as contemporarily, some very important geopolitical points, having truly global dimensions. And at the same time, it is no less irrefutably true that in today's world it's only the sole Superpower USA, which has all the essential geostrategic-cum-politico-military wherewithal in these regards. But in the larger geopolitical perspectives, the pivotal moot points do remain that, how far the sole Superpower USA has in fact contextually succeeded to indeed live up to, as per even the somewhat prudently fair expectations of the larger world? The larger humanity quite legitimately expects and also commands—on the part of USA—to always be able to display the no less crucial, the multi-dimensional-cum-visionary 'sense of history' and thereby, with strategic prudence-cum-farsightedness **enable the entire humanity at large, to always feel the no less sense of being reassuringly secured, from the already looming threats of any anticipated dangers or security-risks, principally due to the tin-pot military dictatorships or even, due to any kind of the 'failed and fanatic' leadership getting somehow regressively entrenched onto the seats of power, in any similarly other 'failing States'**.

To be contextually precise here, the ongoing contemplative-deliberations in fact do geopolitically point towards the **'artificially-created' Islamic State of Pakistan in South Asia, including the Af-Pak region and thereby, the al Quada; and then also towards the no less bothersome and all the more turmoil-ridden—a number of, though mostly 'oil-rich' but even now, being socio-culturally-fixated into the already long bygone medieval era; the States/ Kingdoms in the Middle-East**—where the Western-Powers since led by the USA are in a so naïve hurry to rather untenably transplant the somehow 'performing' tenets of the some kinds of modern constitutional democracy. And that too ironically in fact, without facilitating the making of their socio-political 'soil' ready for the purpose. And thereby, it appears that there are certain kinds of the 'lingering lack of inner synergy and synchronizations'; at highest-levels of the leadership in fact, into the highly complex arena of both the political, as well as, the military establishments within the sole Superpower USA itself. The larger reasons in these regards could even be the somewhat absence of the no less crucial flexibility, as well as the essential fine-tuning, within the national-levels' multi-dimensional domain of **the multiple-aspects of 'essential strategic-culture and thoughts'** in the USA, since their universal paramountcy in fact, cannot be just wished away in manner whatsoever. But it is also true that at times and, in the differently manifesting situations and also, in the different geographical regions in the world, USA as per the above, have indeed been expectedly successful too, in its equally far-reaching transnational initiatives and efforts; whenever the above noted lacunae in these regards, have prudently been attended to well in advance, as per the multi-dimensional dictates of the 'essential strategic-culture and thoughts'. And hence, it has indeed been quite appropriately said that precisely here are always, the relevant 'rub' and the 'keys'.

Man's 'living-history' has been showing numerous such historic examples in fact, of the varying dimensions, consequences and connotations, since ancient of times. But it would be perhaps of more contemporary interest and relevance, if we may pick-up some of the very recent examples, of the presently ongoing modern-era and all the more so if some of them do happen to be still 'alive' and thus, continue to guide and influence even now, the further making of the mankind's 'living-history' on its expanding civilisational-canvas in the today's world. **Then at the same time ironically, all the more bothersome, the no less quite regressive, and also the equally unbecoming phenomena as seen in the today's world; are getting duly triggered by even some kinds of the 'failed' or 'failing' nations in fact, since seen quite counterproductively masquerading as some kinds of the 'make-believe Powers' of some consequence, towards its neighbours; mainly to mislead its own suffering and restless masses and in addition also at times, to wish-away its own lingering 'identity-crisis', mainly in futile manners**.

As a first example to substantiate our these perceptions, we may wish to contextually look into the extremely human-blood-soaked and also, the no less futile—recent 'Syrian-Wars'—first pertaining to the still smothering 1954-62 Algerian-French War of Independence and then, the still later Algerian War between the Islamists and the Government in the 1990s. In these specific contexts, **Robert Fisk**, in his recent article titled, 'Guarding Secrets of the Syrian War' republished in *The Statesman* (Kolkata, dated 22 October 2013; *courtesy: 'The Independent', London*); and therein, he has mentioned that, "**Major General Jamaa-Jamaa** was not a popular man in Beirut. One of Syria's most senior intelligence officers in Lebanon until the withdrawal of **Bashar al-Assad**'s troops in 2006, he was headquartered in the run-down Beau Rivage Hotel in west Beirut and also in the Bekaa town of Anjaar, where Lebanese men would be taken for interrogation and later emerge—or not emerge at all—sans teeth or nails. He was loyal, ruthless apparatchik for **Bashar's father Hafez**, and his mysterious killing last week in the Syrian war provoked no tears in Beirut.... The UN had interviewed Jamaa about the murder of former Lebanese Prime Minister **Rafiq Hariri** whose 2005 assassination brought about the Syrian withdrawal from Lebanon. But how did Jamaa die? Syrian state television would say only that he was 'martyred while carrying out his national duties to defend Syria and its people and pursuing terrorists (sic) in Deir el-Zour'.... All kinds of rebel groups including, of course, **the equally ruthless al-Qa'ida affiliates**—wanted to add his name to their 'kills'. (*Look at this, limit, isn't it?*). He was shot in the head by a sniper in the eastern Syrian oil town. Jamaa was also killed, we were informed, by booby-trap, and blown up by a suicide bomber. All that we can be sure of is that he remains, as they were, were taken for burial in the village hills above Lattakia where he was born. **How long before we know the truth? I am brought to this question by the secrecy which still smothers the 1954-62 Algerian-French war of independence where a cruel French regime of occupation fought a war against an equally cruel and determined Algerian resistance, primarily led by the National Liberation Front, the FLN.**

French officers indulged in an orgy of torture while their Algerian opposite numbers slaughtered each other—as well as the French—in a Stalinist purge of thousands of their own followers suspected of collaborating with the French occupation. For decades, the French refused to discuss this most dishonourable of wars—censoring their own television programmes if they dared talk of torture—while the subsequent FLN dictatorship only published infantile accounts of the heroism of their 'martyr' cadres. (***There appears to be no limitations regarding, the all kinds of suppression of truth, as well as, the 'war-crimes'. Well, these bizarre aspects too at times, do form parts of some 'great-games', with certain motives.***). The French, you see, were fighting 'terrorism'. The FLN were fighting a brutal, Gaullist regime.... The parallels are, of course, not exact. But over the past months, a remarkable phenomenon has made its appearance in Algeria. Dozens of elderly Algerians maquisards from the conflict that ended just half a century ago have turned up at small publishing houses in Algiers and Oran with private manuscripts, containing frightening accounts of the savage war in which they fought and in which their officers tortured and massacred and assassinated their own comrades. Rival Algerian resistance groups—not unlike the 'Free Syrian Army' and their Islamist rebel enemies in northern Syria today—also slaughtered each other.... The French investigative journalist, **Pierre Daum**, has spoken of the 'extreme timidity of Algerian historians', and recounted how one Algerian publisher said he lacked the courage to print a book on the infiltration of the FLN.... In 2005, this guy came to see me, 'I refused his manuscript because it was filled with names, X tortured Y, and so on. Imagine the children of a martyr—who believe their father died under French gunfire—discovering that he perished under Algerian torture' ...**The real story of the much more recent Algerian war—between the Islamists and the government in the 1990s (total deaths 250,000, a hundred thousand more than in Syria today)—still cannot be told by Algerian historians**. How long must we wait, then, for the secrets buried beneath the rubble of the Syrian War?" (*Could anything be more bizarre here than, the negations of both—the mankind's 'living-history' and also, that of the human-civilisations—themselves? Well quite strangely, that's the way at times even today, the things are.*).

Any particular kind of perhaps, the perceived to be a 'victim State', or even otherwise a 'State', seem to still remain fixated with some medieval-era times' strange socio-cultural-cum-political 'value-systems, norms and laws'; due to their own inherent and also its some powerfully irreconcilable contradictions and hence thereby, quite incapable of reconciling with those of the presently ongoing modern-era, finds itself in a bizarre kind of the geopolitical logjam. In the process, inside any such 'State', the mostly quite irrational and violently manifesting—the all kinds of turmoil, as well as, its obvious inability to chalk-out any internationally integrative place in the today's modern, fast altering and also, the multiply interconnectivity-oriented comity of nations—geopolitically pushed it towards a kind of a state of perhaps; the already *'failed'* or fast *'failing'* state. This state of affairs in any such kind of state in any geographical region of the today's world, and due to its own contradictions, and various other reasons they begin to quite relentlessly slip into such a dangerous state that there becomes every possibility sooner or later, to trigger any probable kind of its 'implosion or explosion', which in any case could be of no less danger to its immediate neighbours, but also may turn out to be quite risky to even any little farther-off nations too, especially in today's super high-tech era.

Precisely, this kind of a situation anywhere in turn may sooner or later, geopolitically turn out to be a, though quite turmoil-ridden but also in fact, some overwhelming kind of a 'ripe-canvas', for the any concerned Power or the Superpower; to make highly consequential and, with some kind of mostly the any particular 'great-game' oriented entries; mostly accruing some 'mixed-results' or even at times, resulting in an embarrassing 'fiasco' too. All these always turmoil-ridden, as well as the history-making, developments and events, in totality do at times, create some highly complex and also, a 'cobweb' kind of the strangely 'trapping situations'. Interestingly any such menacingly developing circumstances anywhere, for the concerned Power/Powers or even any Superpower; mostly turn out to be a kind of the 'Catch-22' situation wherein, any state of non-initiative or non-entry may quite temptingly perceived to be geopolitically instrumental, into being even a greater risk or imminent dangers to one's own State. And in these multiply specific contexts precisely, lies indeed the proverbial 'rub'. In other words, we may also say that, onto the mankind's expanding civilisational-canvas, during the presently ongoing super high-tech-oriented modern-era, these too, happen to be some highly complex and newer kinds of the problems-cum-challenges for the larger world. What precisely are the basic symptoms-cum-features of any *'failed'* or *'failing'* state, in the today's modern world? In these specific contexts, **Noam Chomsky**, presently being a Professor of Linguistics and Philosophy and, one of the leading and most influential thinkers of the USA, has some very interesting, as well as intellectually perhaps, even somewhat 'shattering' comments in these regards.

Noam Chomsky, in his one of the highly thought-provoking, as well as well-received books ***Failed States: The Abuse Of Power And The Assault On Democracy'*** (pages 1 and 2) in the 'Preface' itself has written that, "The 'system' is coming to have some of the features of failed states, to adopt a currently fashionable notion that is conveniently applied to states regarded as potential threats to our security (like Iraq) or as needing our intervention to rescue the population from severe internal threat (like Haiti). Though the concept is recognized to be 'frustratingly imprecise', some of the primary characteristics of failed states can be identified. One is their inability or unwillingness to protect their citizens from violence and perhaps even destruction. Another is their tendency to regard themselves as beyond the reach of domestic or international law, and hence free to carry out aggressions and violence. And if they have democratic norms, they suffer from a serious 'democratic deficit' that deprives their formal democratic institutions of real substance.... **Among the hardest tasks that anyone can undertake, one of the most important, is to look honestly in the mirror. If we allow ourselves to do so, we should have little difficulty in finding the characteristics of 'failed states' right at home. That recognition of reality should be deeply troubling to those who care about their countries and future generations. 'Countries', plural, because of the enormous reach of US power, but also because the threats are not localized in space or time**.... The issues are closely interlinked, and arise in several contexts."

If one may try to closely read 'in between the lines' and, delve deeper into the above quoted passage from Noam Chomsky's book, one may find the author's certain perceptive thoughts beneath it, quite disturbing and indeed debatable as well. But whatever, whether one may agree, or partially agree, or even totally disagree with him and his thoughts; most likely one would find that there is some weighty truth, in what he intends to convey. But at the same time, it is also true that the world cannot correctly be 'divided' into two distinct shades of 'black' and 'white' only. In any kind of the pragmatically realistic and, globally geopolitical assessment with any maximum comprehensiveness possible with any contextually discerning eyes on the world-map today, one would notice that there are indeed, quite a few 'fast failing', or already 'failed' countries among the third and/or even perhaps 'fourth' world countries. Even if, any such thoughts may appear to be matters of some conjectures only, but while both objectively and realistically speaking; as we geopolitically begin to move contextually closer and closer to the presently prevailing—the global ground realities—we find that there are some countries going so substantially astray, as well as, quite untenably remaining so much stuck with the multiple kinds of regressive-fixations, that in the greater interests and well-being of the larger humanity, some kind of effective and timely intervention by any credible-cum-capable Power, is mandatorily called-for, before it is too late to prevent wider and the multiply much more serious tragedies for the mankind. And again precisely here too, the universal paramountcy of the multiply well-groomed, as well as flexibly-cum-proactively, well-placed and indeed always holding the 'keys'—the **multiple-aspects of 'essential strategic-culture and thoughts'**—do remain as unassailable as ever. Now precisely in the above noted multiple-contexts here, two very recent and momentously historic examples with truly global reach, dimensions and consequences; and though also having altogether very different circumstances, as well as, global connotations, come in our minds, to indeed quite argumentatively substantiate these perceptions. The first one being even just before the actual start of the second World-War, when France though having enough of military wherewithal, totally lacked both the courage, as well as historically very crucial and quite expected geostrategic-cum-politico-military vision; to effectively take-on then, only the few intruding Nazi German military units on the spot, when they had just so audaciously crossed the Rhine and hence; prevented them from crossing-back to their bases, without getting severely mauled and totally disarmed. If France would have displayed this crucial courage and vision then perhaps, even the soon to come much bigger war would have been it not totally then partially, prevented. At least Hitler won't have dared to trigger little later, such all the more audacious beginning of the second World-War and thereby rest of the subsequent history too, would have been substantially different and much less tragic and cruel for the larger humanity as well. And as second example at times, in order to prevent some unimaginable global tragedy it becomes geostrategic-cum-politico-militarily essential, to even maintain a rather some otherwise frightening 'balance of terror' to prevent any globally annihilating war as the USA, and the then USSR geopolitically did so for years together in Europe and elsewhere during the Cold-War era, for judiciously preventing a nuclear-holocaust and thus perhaps leading to the almost total global destruction, duly saturated with extreme kinds of nuclear-radiation.

Again during the Cold-War times only, the USA then under **President John F. Kennedy**, did not lack the necessary courage and vision, as in case of France earlier and as noted above; and did very firmly carryout a strategically effective Naval-Blockade around Cuba, against the then USSR to successfully prevent it from installing nuclear missiles, in the backyard of USA itself. Much later then, during the 1982 'Falklands-War' in the South-Atlantic Ocean across Argentina in South America; when the British Joint-Military forces, under the Prime-Ministership of **Margaret Thatcher** then, did successful landings on the Falkland islands and victoriously reclaimed them from the Argentine forces before it was perhaps, geopolitically too late for the UK! **Hence no wonder, as to why precisely the man's 'living-history' since continuously in the making is always called multi-dimensional, highly checkered, and also no less complex as well.** Now again, in these multiple-contexts only, the next example coming in our minds precisely at this point of our continuing discussions is in fact, much more momentous and indeed also unprecedentedly paramountcy but, of presently the urgent concern and also, futuristic in its contents too. And that specifically pertains to the **multiple-aspects of the presently indeed, so menacingly ongoing 'population-explosion and environmental degradation'** and they would be discussed in the next to next subchapter, in more details, as a very relevant part of our larger discussions.

Now contextually, while looking little differently to our specifically continuing discussions on the **'great-Games'**, we may now prefer to also pay some critical attention to the highly checkered history of the **ill-famed 'Vietnam-Episode'**, which first resulted in a kind of a very embarrassing military 'defeat', on the part of erstwhile one of the European colonial-power, France, and then subsequently later all the more geopolitical-cum-military 'fiasco' for the Western Superpower, the USA. Volumes have been written on 'Vietnam' in various countries during all these years and naturally, giving quite differing versions for posterity. It was indeed the extremely menacing Cold-War era, when Vietnam took place and obviously therefore, each Bloc was rigidly fixated with a kind of the extreme norms of geopolitically 'painting' the other side as nothing but a total 'black'. And hence, it took some years to pass-by even after the end of Cold-War, before some fog could wither away and thereby, some objectively discerning look into the highly checkered, as well as the no less violent and tragic history of the 'Vietnam-years' could be possible, even by the free world, including the USA. Anyway this too, has been one of the inherent ways of the mankind's 'living-history', to get finally manifested. But the history can't be suppressed forever, and finally sooner or later, it always finds its own inherent ways and means towards its place in the domain of the mankind's 'living-history', through always critically discerning manifestations.

On the passing away of indeed the **legendary General Giap of Vietnam**, **Professor Subrata Mukherjee**, retired professor in the Department of Political Science, University of Delhi, in his article on 'Unforgettable Soldier' published in *The Statesman*, Kolkata, dated 24 October 2013, has written that, "Giap's dictum was that it was the people, and not weapons, who won wars. He was a genius who popularised the phrase that every inhabitant was a soldier and every village a fortress. He always reminded his followers that surrender did not exist in his vocabulary as it was a fight for freedom.... **Vo Nguyen Giap (1911-2013) has earned a distinguished place in history by his strategy of a protracted battle to be fought jointly by the peasantry and the army trained to fight a far superior opponent with guerrilla tactics combined with conventional army operations. He had a reasonably wealthy middle class background. Giap was school teacher and journalist in the 1920s and 1930s and had a no prior military training**. In his reckoning, French colonial rule was exploitative and intolerable, and he had joined **Ho Chi Minh** in China, to avoid detection by the French colonial authorities, who were working out a strategy to fight the Japanese occupation of Vietnam during the Second World War. Viet Minh began its journey to achieve it under the leadership of **Diem**, with 31 men and women.... France was the colonial master of Vietnam before the rise of Japan. In September 1945, Ho Chi Minh declared independence and became President. He appointed Giap as interior minister and Chief of the Army and subsequently the defence minister. But when the French reasserted their control over Vietnam, the revolutionaries were forced to go to the jungle as the French troops were reinforcing colonial authority. For nine years the conflict continued till the French army suffered a humiliating defeat in 1954 at Dien Bien Phu at the hands of Diem and his army.... However, post-French Vietnam was to face a larger, longer and a more savage battle. By the Geneva Agreement (1954), Laos and Cambodia were made independent, but nullifying the earlier declarations of Ho Chi Minh that entire Vietnam was independent, it was temporarily divided into two states at the 17th parallel. Ho Chi Minh's government was recognised as the legitimate government of North Vietnam, whereas for the time being South Vietnam would have a separate government with a promise of an election to be held in 1956 for the entire country, which then would be united. The partition disappointed Ho Chi Minh, but he was confident that the communists would win the national election. **But the elections were never held and it seemed to be repetition of the Korean situation**. The USA, which had helped the French, was backing the South Vietnamese regime and did not press for elections as its defeat at the hands of the communists was certain.... Both Kennedy and Johnson followed this lines of (*Truman's policies*) action and enhanced economic and military aid to South Vietnam.... But the real reason was to keep the country outside the Communist bloc as a continuation of the containment policy. The intensity of the US involvement changed course when it became common knowledge that Vietcong (the name of the guerillas) was receiving substantial assistance from North Vietnam. Ho Chi Minh defended the policy because of South Vietnam's refusal to agree to the earlier agreement to hold elections. It was under such circumstances that Giap's master strategy was to maintain the supply chain through the 2000-mile-long Ho Chi Minh trail, which was the most bombed area in order to retain vital connectivity from North Vietnam to Vietcong. He followed Mao's guerilla tactics with innovations, with his vast experience of using it against the French.... **Giap argued that 'guerilla war is the war of the broad masses of an economically backward country standing up against a powerfully equipped and well-trained army of aggression'.... Giap's last great act was in 1978 when under his stewardship the mighty Chinese which had invaded Vietnam, suffered a humiliating defeat....** Officially he left politics in 1991 but **remained vocal on issues of corruption and environment.... He also took note of Vietnam's efforts to enhance US cooperation to balance China's growing clout in the South China Sea**. He added, however, that 'we can put the past behind but we cannot completely forget it'. The architect of the most defining phase of post-war history has reminded us that his deeds may perhaps fade, but he will never be totally forgotten." (*And interestingly, as the history's timelines do continue to move ahead onto the mankind's expanding civilisational-canvas, numerous kinds of the 'great-games' continue to be played in all over the places in the world and thereby, the man's 'living-history' too, do continue to be made further in all possible manners.*).

(f) Coming of the Space-Age

Since from the very beginning of man's 'living-history', being always in its continuous making and also, quite appropriately remaining in tune with the history's advancing timelines, and thereby, multiply-consequent to the man's relentless quest in his always magnifying, as well as enriching human-endeavour, in the numerous fields/disciplines of the sciences and technologies, the mankind has been able to achieve and master many path-breaking scientific-cum-technological marvels, onto his expanding civilisational-canvas; and thus these in turn, have been rather magnificently instrumental in multiply-triggering the subsequently, as well as a kind of the gradual dawn of indeed, the man's further enchanting civilisational-marvels. All these do remain always interrelated, as well as mutually-influencing: the multiply ongoing phenomena; since the very beginning of the mankind on this planet-Earth and hence, continuingly from even the remotely ancient-era itself. We may recall and since already noted earlier that quite surprisingly, the man was able to cultivate his some initial awareness of the mining, metals, metallurgy and alloys etc; and hence copper was discovered as early as 6000 B.C., and Bronze come to his knowledge around 3000 B.C. But the greatly path-breaking development came, as the **man's highly phenomenal discovery of wheel** during 4000 B.C. and then, **discovery of iron** fructified around 1500 to 1600 B.C. And as the further centuries did keep-on rolling-by and then, another great path-breaking discovery and, i.e. the **gunpowder**, came to the man's knowledge around 1260 A.D. And thereby; subsequently in the domain of matters-military, fructified the momentous advent of the Artillery as well as, its continuing improvements too, have continued since the 1415 A.D. onward.

And all these combined and then appeared another path-breaking, as well as the unprecedentedly momentous advent of the **'Industrial-Revolution'**, during 18th and 19th centuries in Europe. Broadly the triggering-point in this regard, was the scientific identification and the initiation of efforts towards harnessing the steam-power got a head start by the scientific genius of **James Watt** in 1769 A.D., and then that of **Stephenson** in the year 1814 A.D. We may contextually recall that in more details, we have since already discussed in the earlier seventh chapter itself; the phenomenal advent of the "industrial-Revolution' with all its numerous dimensions and also, its multiple-influences too; and then subsequently in some logical continuity, in the first sub-chapter of this eighth chapter itself. And then we have since also discussed, the phenomenal times of the **'Post-Industrial-Revolution'**; advent of the **Electronics**, indeed the multiply path-breaking beginning of the Computer-Age and thereby; also the beginning of multi-dimensional manifestations of the **technologies of automation** and then; another path-breaking dawn of the **Nano-technology and Cryogenics** as well. And needless to say that the mankind's, all the above very far-reaching, as well as the paths-breaking fructifications of its so many numerous kinds of indeed, the enchanting civilisational marvels; down the thousands of years of since the continuing timelines of the man's 'living-history' and then onto his expanding civilisational-canvas, and all these in turn have very cumulatively enabled another so momentously path-breaking fructifications of the fast developing **scientific-cum-technological marvels, principally consequent to the present 21st century's fast advancing 'Space-Age'**. Presently, these very momentous, multi-dimensional, as well as so fast accelerating advancements of the Space-Age related sciences and technologies in turn have unprecedentedly enabled the man to expand his mental-horizon and thereby consequently, his many new and all the more daring urges, triggering his scientific-cum-physical initiatives to commence proceeding into the outer space. Along with his numerous kinds of ambitions towards exploring the outer space for expanding, improving, and also strengthening his many civil and also matters-military oriented modern wherewithal, man also wants to serve his numerous kinds of fast emerging: strategic, military, and the economic compulsions. Hence we may now prefer to contextually put-in, another indeed, very unique dimension of the mankind's since multiply continuing advancements in the numerous fields of the sciences and technologies and all the more interestingly, this phenomenal marvel, may earlier being in somewhat very primitive in nature but no less very captivating too, **did manifest in India first, as the 'science of rocketry'** in the areas of matters-military; even before so far-reaching, and the history-making, civilisational advent of the man's **'Industrial-Revolution'** itself. To cut the long story short, we may now prefer to very appropriately quote few paragraphs from the autobiography of a former President of India, **Dr. APJ Abdul Kalam**. Dr. Kalam has, indeed very informatively mentioned that, "From LRC (*i.e. Langley Research Centre in Hampton, Virginia, USA*), I went to the Goddard Space Flight centre (GSFC) at Greenbelt, Maryland. This Centre develops and manages most of NASA's earth-orbiting science and applications satellites. It operates NASA's tracking networks for all space missions. Towards the end of my visit, I went to the Wallops Flight facility at Wallops Island in East Coast, Virginia. This place was the base for NASA's sounding rocket programme. Here, I saw a painting prominently displayed in the reception lobby. It depicted a battle scene with a few rockets flying in the background. **A painting with this theme should be the most commonplace thing at a Flight Facility, but the painting caught my eye because the soldiers on the side launching the rockets were not white, but dark-skinned, with the racial features of people found in South Asia. One day my curiosity got the better of me, drawing me towards the painting. It turned out to be Tipu Sultan's army fighting the British. The painting depicted a fact forgotten in Tipu's own country but commemorated here on the other side of the planet**. I was happy to see an Indian glorified by NASA as a hero of warfare rocketry.... **My impression of the American people can be summarised by a quotation from Benjamin Franklin. 'Those things hurt instruct.' I realised that people in this part of the world meet their problems head on. They attempt to get out them rather than suffer them**.... My mother had once narrated an incident from the Holy Book—after God created man, he asked the angels to prostrate themselves before Adam. Everybody prostrated themselves except Iblis, or Satan, who refused. 'Why did you not prostrate yourself?' Allah asked. 'You created me of fire and him of clay. Does not that make me nobler than Adam?' Satan contended. God said, 'Be gone from paradise. This is no place for your contemptuous pride.' Satan obeyed, but not before cursing Adam with the same fate. Soon Adam followed suit by becoming a transgressor after eating the forbidden fruit. Allah said, 'Go hence and may your descendants live a life of doubt and mistrust.'.... **What makes life in Indian organizations difficult is the widespread prevalence of this very contemptuous pride.' It stops us from listening to our juniors, subordinates and people down the line. You cannot expect a person deliver results, if you humiliate him, nor can you expect him to be creative if you abuse him or despise him. The line between firmness and harshness, between strong leadership and bullying, between discipline and vindictiveness is very fine, but it has to be drawn**. Unfortunately, the only line prominently drawn in our country today is between the 'heroes' and the 'zeroes'. On side are a few hundred 'heroes', keeping nine hundred and fifty million people down on the other side. This situation has to be changed." (Kalam, 1999, pp. 37 & 38) (*Yes indeed, and all the more so because, there is no alternative to it; if India really wishes to come out of her self-imposed some crippling shackles of the presently prevailing situation, of the very embarrassing state of her 'uncertain glory'. We would more relevantly and in greater details, be discussing these indeed very crucially important issues, and their numerously complex aspects; when we come to the last sub-chapter, concerning India and the Indians, in this eighth chapter itself.*).

In these substantially broader respects and, precisely concerning India and the Indians, but having different parameters and some equally facts-based storylines, we may contextually prefer to also quote here, another interesting episode so that even modern, democratic and also now, the multiply quite fast developing Indian Nation-State and the modern-era's Indian citizens are suitably awakened and, coaxed to discard their old colonial-era's some absolutely uncalled-for and, unbecoming notions as well as, no less regressive fixations. This particular true story in fact, belongs to the times of post-1947 and pre-1962 years, when the independent and democratic India's fledgling Armed Forces were trying to militarily firm-up their professional moorings as well as, no less dynamically struggling to prepare and stand-up, as per dictates of the sovereign India's supreme constitutional-framework; and without any kinds of the British geostrategic-cum-politico-military crutches and that too, under the befitting and also no less constitutional civilian-supremacy of the duly elected political leadership. This interesting matter in fact, specifically pertains to the then **General J.N. Chaudhuri**, who became India's Chief of the Army Staff immediately after the India's very embarrassing but quite avoidable military debacle in the 1962 India-China border clashes, under the highest national-level political leadership of India's first Prime-Minister **Jawaharlal Nehru**.

General J.N. Chaudhuri, in his book has written that, "In 1951, I was a Major General, commanding the 1st Armoured division.... There was plenty of time to soldier seriously and get the division absolutely tuned up. There was also time to study and think. At the end of the month (*i.e. May 1951*) I did a routine trip to Delhi during which I dropped in to a cocktail party at Sinclair's hospitable house. Normally cocktail parties are an unprofitable excursion but this one was different. Sinclair introduced me to Evan Charlton of *The Statesman* and after a bit Charlton asked me if I could suggest the name of someone who would write a fairly regular column for them on military affairs. The articles were to be non-political, easily understandable, to leave Pakistan alone and to arouse an interest in military matters. Charlton felt this was necessary in a newly independent country where, in the past for political reasons, the intelligent citizen had, for generations, been kept out of anything concerning defence. We talked it over and finally I offered to do them myself.... Over the next ten years, as *The Statesman*'s Military Correspondent, I wrote a number of articles, forty-five of which are reprinted in this book.... My anonymity was very well-kept. Even on, *The Statesman*, apart from the Editor and one other, no one knew who was writing these articles. **Excerpts were often quoted to me as examples of foreign military thinking for, of course, the Military Correspondent of *The Statesman* had to be a foreigner. Once in Jullundur, a learned civilian with an article of mine in his hand, lectured me on our army's inadequacy in producing military writers. The soldiers liked the article and occasionally I saw quotations in prize essays and/or in the answer papers for the Staff College**. The fact that someone read and even remembered what I was writing about naturally gave me a lot of pleasure. I also quite enjoyed writing the stuff. If inspiration was there it came straight from a typewriter. If I hadn't thought it out properly, then the birth pangs were repetitious, long-drawn-out and time-consuming.... In this book I have put this article down in the order in which they were written. Not a word has been changed and even the titles given by the various *Statesman* sub-editors remain unaltered.... I am grateful to the Editor of The Statesman for permission to publish these articles as a book." (Chaudhari, 1966, pp. v, vi & vii)

The earlier quoted scientific facts above, from one of the indeed modern, sovereign, and the fast developing-cum-maturing **democratic India's, very few illustrious Presidents, Dr Kalam's earlier referred Autobiography, the '*Wings of Fire*'**; though may now appear to be largely insignificant, especially to those somewhat 'arrogantly snobbish' with different motives—the so-called intellectuals, rather regressively western-world-fixated scientists and members of the academia, bureaucrats, the short-sighted politicians, and even many of the rather amusingly 'rank-fixated' senior-officers of the Indian Armed-Forces—but irrespective of any such thing in India or anywhere else, the above quoted historic facts and that too, during the medieval-era times of **Tipu Sultan** in India were no less path-breaking, as well as history-making developments in these regards. **The 'science of rocketry', then having been mastered by the Army of Tipu Sultan during the seventeenth century Medieval India and that too, even before the unprecedentedly momentous-cum-multiple fructifications of the 'Industrial-Revolution' may be required to be viewed indeed, in the above mentioned broader perspectives.** And needless to say that now, precisely in these very broader perspectives the USA's NASA, has been able to view and commemorate Tipu's historically fascinating aspects concerning his rockets as some kind of the then quite lethal projectiles, in his battles against the British imperialists, who were since so obsessively busy due to their overwhelmingly far-reaching imperialistic ambitions and also then needless to say that, for their some very powerful geostrategic, economic, as well as politico-military compulsions in establishing and expanding their colonial-rule in the Indian subcontinent and also in other far-off nations and continents, making different segments of the then fast fructifying, but now the historically bygone global British-Empire. Naturally then, since always the mostly strategically prudent and also scientifically very imaginative British, after their final victory over Tipu Sultan in their battle for the strategic Srirangapatnam Fort near Mysore, in the so vast South-Indian peninsula were certainly not going to quite naively overlook this path-breaking knowledge of the 'science of rocketry'. And finally the NASA of USA, as a globally unbeatable Institutional Giant, has multiply emerged as the lead-player in these regards.

Precisely, in the above indeed very captivatingly multiple-contexts the former Indian President, **Dr. Abdul Kalam** has very informatively written that, "The development of Indian rockets in the twentieth century can be seen as a **revival of the eighteenth-century dream of Tipu Sultan. When Tipu Sultan was killed, the British captured more than 700 rockets and subsystems of more than 900 rockets in the battle of Turukhanahally in 1799**. His army had 27 brigades, called Kushoons, and each brigade had a company of rocket men, called Jourks. **These rockets had been taken to**

England by William Congreve and were subjected by the British to what we call 'reverse engineering' today. There was of course, no GATT, IPR Act, or patent regime. With the death of Tipu, Indian rocketry also met its demise—at least for 150 years.... Meanwhile, rocket technology made great strides abroad. **Konstantin Tsiolkovsky in Russia (1903), Robert Goddard in USA (1914) and Hermann Oberth in Germany (1923) gave rocketry new dimensions. In Nazi Germany, Wernher Von Braun's group produced V2 short range ballistic missiles and showered fire on the Allied Forces. After the war, both USA and the USSR captured their share of German rocket technology and rocket engineers. With this booty, they started to run their deadly arms race with missiles and warheads.... Rocketry was reborn in India thanks to the technological vision of Prime Minister Jawaharlal Nehru. Prof. Sarabhai took the challenge of giving physical dimensions to this dream.** Very many individuals with myopic vision questioned the relevance of space activities in a newly independent nation which was finding it difficult to feed its population. **But neither Prime Minister Nehru nor Prof. Sarabhai had any ambiguity of purpose. Their vision was very clear: if Indians were to play a meaningful role in the community of nations, they must be second to none in the application of advanced technologies in their real-life problems.** They had no intention of using it as a means to display our might." (Kalam, 1999, pp. 42 & 43)

As the further years kept rolling-by, there was never any kind of 'break' in indeed the multi-dimensionally relentless pursuits of the highly talented and dedicated Indian scientists and engineers in the continuingly fast flowering, as well as multiply expanding—the super high-tech oriented Indian Institutions of the ISRO (i.e. Indian Space Research Organisation) and, the DRDO (i.e. Defence Research and Development Organisation)—along with also a vast network of the Indian Scientific and Industrial Research Establishments, as strategically located all across the country and, functioning under the Central Government's Council of Scientific and Industrial Research (i.e. CSIR); and jointly always remaining teamed-up, along with the leading Scientific-cum-Technical Teaching Institutions/Universities. And much more so in today's super I.T. age and also, the present times of instant connectivity; the above mentioned Indian Scientific-cum-Technical Institutions, since always engaged in their various kinds of the high-end research pursuits do also multiply remain connected with the similar Institutions abroad too. In these very fascinatingly broader perspectives, it is obvious that unlike the research-work in any other highly specialised domain of any basic sciences-subjects, the multi-dimensional, as well as multi-disciplinary research assignments, in any high-end technological fields, such as the today's 'Space-Age related R&D pursuits', are always very complex-cum-vast teamwork activities, consisting of a very large number of the scientists and engineers since belonging to the numerous scientific-cum-technical disciplines and also, along with their latest and different kinds of the cutting-edge specialisations.

Modern India's Dr. Kalam, is always very be fittingly known as the 'father' of India's IGMDP (i.e. Integrated Missile Development Programme) and no wonder, that the democratic Indian nation-State owes much to his relentlessly focussed efforts and leadership for years together, as himself being a dedicated scientist but much more so, as a brilliant and highly inspiring scientific team-leader of the Indian scientists in these regards. And consequently what the Indians find today, is a reassuringly matters-military-oriented and also, the multiply fructifying strategic-assets; since consisting of a range of the different kinds of missiles in the arsenals of Indian Armed Forces, principally meant to geostrategic-cum-politico-militarily act as effective deterrents towards fruitfully maintaining peace in the extremely turmoil-ridden geopolitical dimensions of the today's world. Understandably, these do always remain the highly debatable-issues. **Hence, so far as the mankind's**

enlarging civilisational-vision and also, the geopolitically ongoing enrichment of its multi-dimensional acumen are concerned, all these are still not able to throw-up indeed, any pragmatically realistic-cum-practical, all-inclusive as well as geopolitically robust and enduring alternative directions-cum-solutions—in these regards.

Needless to say that under the presently prevailing circumstances, the geostrategic-cum-politico-militarily effective maintenance of the different kinds of deterrents; happen to be the only ways to keep preventing, even any remotely possible and also, the extremely senseless nuclear-holocausts. **But, there are many other no less bothersome dimensions, to these indeed very maddening possibilities; since still lingering in different manners and also due to different reasons, involving various kinds of the irresponsibly fixated some state and non-state actors being active in different geographical regions of the world.** In fact, what we do very sadly notice today is—that the overall scenario in these regards and that too, onto the man's expanding civilisational-canvas itself—have become so depressing and also, quite mind-bogglingly so complex that even the angels fear to tread over there. But all said and done, some tangible solutions in these respects have to come out sooner or later.

At the present moment, in order to have some magnificent glimpses of **Dr. Kalam's** imaginative and the highly inspiring traits, concerning his since well-gelled acumen and thoughts towards the multiple-aspects of the scientific-leadership, he writes that, "Anyone who has taken up the responsibility to lead a team can be successful only if he is sufficiently independent, powerful and influential in his own right to become a person to reckon with. This is perhaps also the path to individual satisfaction in life, for freedom and responsibility is the only sound basis for personal happiness. What can one do to strengthen personal freedom? I would like to share with you two techniques I adopt in this regard.... First, by building your own education and skill. **Knowledge is a tangible asset, quite often the most important tools in your work. The more up-to-date the knowledge you possess, the freer you are. Knowledge cannot be taken away from anyone except by obsolescence**. A leader can only be free to lead his team if keeps abreast of all that is happening around him—in real time. **To lead, in a way, is to engage in continuing education**. Abroad it is normal for professionals to go to college several nights every week. **To be a successful team leader, one has to stay back after the din and clutter of a working day to emerge better equipped and ready to face a new day**....The second way is to develop a passion for personal responsibility. **The sovereign way to personal freedom to help determine the forces that, determine you**. **Be active. Take on responsibility**. Work for the things you believe in. If you do not you are surrendering your fate to others. **The historian Edith Hamilton wrote of ancient Greece, 'When the freedom they wished for most was freedom from responsibility, then Athens ceased to be free and was never free again**.' The truth is that there is a great deal that most of us can individually do to increase our freedom. We can combat the forces that threaten to tame us. We can fortify ourselves with the qualities and conditions that promote individual freedom. In doing so, we can help to create a stronger organisation capable of achieving unprecedented goals....**To succeed in your mission you must have single-minded devotion to your goal**. Individuals like myself, are often called 'workaholics'. I question this term because that implies a pathological condition or illness. **If I do that which I desire more than anything else in the world and which makes me happy, such work can never be an aberration. Words from the Twenty-sixth Psalm come to mind while I work: 'Examine me, O Lord and prove me'**.... Technology, unlike science, is a group activity. It is not based on an individual intelligence, but on the interacting intelligence of many. I think the biggest success of IGMDP is not that in record time the country acquired the capability of making five state-of-the-art missile systems but that through it, some superb teams of scientists and engineers have been created. If someone asks me about personal achievement in Indian rocketry, I would pin it down to having created an environment for teams of young people to put their heart and soul into their missions.... In their formative stages, teams are much like children in spirit. They are excitable, full of vitality, enthusiasm, curiosity and the desire to please and excel. **As for children, however, these positive attributes can be destroyed by the behaviour of misguided parents. For teams to be successful, the environment must offer the scope for innovation. I confronted many of such challenges during the course of my work at DTD and P (Air), ISRO, DRDO and elsewhere, but always ensured my teams of environment which allowed innovation and risk taking.**" (Kalam, 1999, pp. 77, 90 & 169) (*While calmly reflecting through the above quoted lines, we may even tend to readily agree that if they are read closely and with an open-mind, as well as with requisite pragmatic-optimism, they can even stir a sick person and make him or her sit-up with renewed vigour.* ***Dr. Kalam's writings are always inspiring, especially to the children and youth all across the world.*** *If we do pause a little and may care to delve deeper into these aspects with some criticality,* ***we may visualise that it is far more difficult to inspire a team of young scientists and technocrats than perhaps the similar kinds of teams/groups constituting as specific fraternities, in other areas of human endeavour.***).

Now Coming back to the presently ongoing super high-tech-oriented modern-era's, the multiple-aspects of the so accelerating-cum-fast advancing—the mankind's scientific-cum-technologically, the multi-dimensional explorative initiatives and adventures; towards some deeper and deeper domain into the infinite outer-space all around the planet-Earth—with the numerous unprecedentedly different aims and ambitions, **also have both the civil, as well as the matters-military-oriented dimensions**. In fact, in these interestingly larger perspectives **we may also contextually say that, without the multi-dimensionally far-reaching advents of the 'sciences of rocketry; along with the technologies concerning electronics, computers and super-computers digitization, I.T. Age, and the Cryogenics and so on now the so momentous and mind-blowing fructifications of the Space-Age, won't have been thinkable, or even scientifically imaginable in any context, or in manner at all.**

Contextually at this precise point of our continuing discussions, with regard to the multiple kinds of Space-Age-related sciences and technologies, we may recall that little earlier, we have since mentioned that all these globally far-reaching, and geopolitically-cum-economically very crucial aspects and, their multi-dimensionally very fast fructifying, as well as relentlessly enriching techno-scientific wherewithal have both in fact, multiply-important and civil-cum-matters-military-oriented dimensions, with immense possibilities. Some scholars and strategic experts, coming from both civil and military fields, the world over have indeed very discerningly been commenting that, some quite unjustifiably reckless pursuits of some very alarming kinds of the matters-military-oriented Space-Age-related technologies are in fact, resulting in some highly dangerous kinds of the **'militarization of the space'** all around this planet-Earth. While very critically reviewing these specific aspects in their all sorts of the present, as well as in their futuristic perspectives; quite worriedly we do notice both leading and the other fast-developing nations in fact, have very actively constituted and also activated their **strategic-level 'Space Commands'**, within the gamut of their geostrategic-cum-politico-military-level military command and control systems. In these multiple-contexts, while critically considering and talking about the globally far-reaching as well as, domestically always perhaps the most crucial aspects, regarding any particular Nation-State specifically from the economic angles, we do very clearly notice that the cost-factors are indeed very high in these regards. Hence, if any nation with considerably limited financial resources and reach and also, no less rather overwhelmingly ridden with its all sorts of internal-problems too, is unnecessarily allowing itself to rather irrationally get trapped into these kinds of the extremely dicey and indeed abnormally costly—the 'Space-Age' related and also, the always cutting-edge-cum-super-high-tech-oriented: matters-military-related 'great-games'—then these kinds of the undeniably regressive obsessions, may even push that concerned nation into a dark, destructive and a 'financial-cum-geopolitical kind of an abyss'. In order to keep any of these kinds of the 'abyss' at bay, the need is to always exercise an objectively cool-cum-calculative, and also very befittingly pragmatic strategic prudence, so that the future of the coming generations in any such similarly concerned nation, is not allowed to be put in a jeopardy, as in case of the any of an already any 'failed or the fast failing-nation'. **The hardcore strategic prudence, concerning any such nation in these regards to keep away from the matters-military related pursuits but, to take initiatives and thereby, advantageously team-up with any concerned leading nation in one's neighbourhood, for solely the very sharply focussed purpose towards availing the maximum possible and, the multiply very valuable benefits, in the civil-sectors. Needless to say that the fallouts of any such proactively strategic prudence-cum-initiatives for any such concerned nation would be always multiply advantageous, first within the domestic societies of that concerned nation itself.**

Now contextually speaking in the wider techno-scientific perspectives and the momentously history-making coming of the space-age, we have already seen that its presently ongoing multi-dimensional flowering in fact, was very competitively triggered during the peak-years of the historically bygone 'Cold-War-Era' itself. And hence that at the outset itself, the then both ruthlessly rival Superpowers, i.e. USSR and the USA had already begun their own top-secret and also, since the relentlessly multi-dimensional R&D programme to race ahead of the rival in these regards. These geostrategic-cum-politico-militarily far-reaching and the relentless pursuits of these kinds of the cutting-edge 'R&D Programme', were in fact kept under the so top-grade secrecy-cum-security wrap; that the wider world-media, even including the leading ones in both the Superpowers themselves, were not at all aware of such historically path-breaking developmental events since taking place then in such vast scales, within those two Superpowers. And consequently when the 'big-bang' did take place during the late 1950s itself and so with great surprises, the whole world suddenly woke-up to the so far, the unprecedented dawn of the coming of a new age, i.e. now so commonly known as the present 'Space-Age'. And then within a couple of decades only the modern, democratic and the fast-developing Indian nation-State too, jumped in the fray with widely encouraging results for itself. Precisely in the above multiple-contexts regarding the space-age, **Prof (Dr) N.R. Bandyopadhyay**, Editor of an Indian Technical-Journal, *Technorama*, in the December 2012 issue (p. 3, as part of his 'Editorial') has written that, "The launch of Sputnik 1 in 1957, marked the advent of Space Age and on July 20th 1969, Apollo 11 achieved the first manned moon landing. In 1981, the launching of space shuttle 'Columbia' begins the era of regular manned access to the orbital space. A sustained human presence in orbital space started with 'Mir' in 1986 and is continued by the 'International Space Station' ....**21st century witnessed the sea-change in the development of space technologies. The technologies like weather forecasting, remote sensing, GPS systems, long distance communications systems, astronomy etc are the most notable outcomes from space technology**. Moreover, since 1970, each of the NASA Research Centre is continuously working to facilitate the transfer of Space Programme Technology to the medical industry. This reduced the cost of medical technological research and development.... Those were the days when computers and telemetry were the leading cutting edge technologies, which have been considered as critical to 'Space Technology' because of their applications in spacecrafts. Enormous developments have taken place over the last few decades in our country (*i.e. India*) in space infrastructure, which manifested themselves as Indian National Satellite System (INSAT) for Meteorological and communication service, and Indian Remote Sensing Satellites (IRS) system for resources monitoring and management service. The Indian researchers have developed two satellite launch vehicles, namely Polar Satellite Launch Vehicle (PSLV) and Geosynchronous Satellite Launch Vehicle (GSLV), to place INSAT and IRS satellites in their required orbits.... PSLV-XL has been developed for launching of first Indian lunar orbiter called Chandrayaan-1, whose grand success encouraged the Chandrayaan-2 mission which has better orbital configuration. This new mission has an objective to improve the understanding of evolution of the moon by using orbiter and in-situ analysis of planetary samples and studies of its properties through robotic technologies."

But as the saying goes that in almost any field of human-endeavour if the one important and at times, even indeed any path-breaking accomplishment leads to a kind of the quite encouraging opening of the doors towards another all the more fulfilling achievement, onto the man's expanding civilisational-canvas, then needless to say that at the same time, new problems and all the more difficult challenges too in these regards, do begin manifesting in no less momentous manners. In our present multiple-contexts, while critically looking deeper into these indeed quite weighty aspects in fact techno-philosophically we may even say perhaps, that the man's civilisational marvels may also be looked into as the differently situations-specific since, the ongoing evolutionary developments within the much wider, highly phenomenal as well as, the proactive manifestations of always the all-encompassing laws of Nature. So in the similar manners and also, as per since the same always proactive laws of Nature as the man has already begun his multiply scientific-cum-technological forays into the orbital space, by indeed repeatedly launching different kinds of the manned space-vehicles, for regular scientific access into the orbital space; various kinds of the technical, medical, and also the environmentally problematic challenges and hazards too, have been manifesting and staring into the eyes of scientists and engineers in these regards. Hence interestingly in the multiply vast, as well as highly complex fields of the sciences and technologies, the always continuingly ongoing human-endeavours are no 'cake-walks'; and similarly the same happens to be no different, regarding the man's all the more challenging and adventurous initiatives into the outer orbital-space as well. But as we may somewhat further delve deeper, into these indeed very interestingly—captivating scientific-cum-technological aspects—we do also quite discerningly notice that, the various dimensions and also their complexities in these regards are much more daunting and demanding, concerning the man's ventures into the orbital-space; than the same with respect to his similarly explorative, and the path-breaking—numerous techno-scientific endeavours—within the planet-Earth itself.

Hence precisely in the above noted multiple-contexts, concerning the space-age, **Prof. V. Adimurthy**, Satish Dhawan Professor and Senior Adviser (Interplanetary Missions) ISRO HQ, Dean (R&D), Indian Institute of Space Science and Technology, Thiruananthapuram, has mentioned that, "However, wherever human beings venture, microorganisms follow. Microbial risks in space are all pervasive, because the closed quarter of a manned spacecraft or habitat, are ripe for infectious diseases. The problems accentuate because of the recycled water and air, because of food spillage and material degradation. In space, bacteria do not behave as they do on Earth; differences in growth rate and metabolism and in antibiotic susceptibility are reported. This Russian study indicates that a typical microsphere of a space station contains as many as 234 species of microorganisms, 108 species of bacteria and 126 species of fungi. An aggressive growth of microorganisms in space as compared to earth conditions in reported. Microbial risks in space can be medical or technological. Medical risks are related to auto infections and cross infections. Technological risks are due to resident colonization on structural material of the interior and equipment by potentially pathogenic bacteria and fungi which can cause initiation of corrosion of metals, biodegradation of polymers, bio-interference which can lead even to failure and malfunctioning of equipment.... On a two year trip to Mars, a crew of six humans will generate more than 6 tons of solid organic waste. For long-term exploration, one wants to recycle it, because it holds resources that astronauts will need. It will provide pure drinking water and fertilizer. Using Geobacter, a microbe discovered in 1987 by **Derek Lovely** of the University of Massachusetts, recycling may also provide electricity. Geobacter microbe feed on and can decompose organic material. They also have the unexpected ability to move electrons into metal. Thus, under the right conditions, Geobacter microbe can both process waste and generate electricity. This will be a great boon in long-term human-space-operations.... The last five decades of international space initiatives have left behind a multitude of space objects, which no longer serve any useful function, but pose risk to space operations. Thus, space debris becomes an important subject for all space faring nations in particular and humanity in general. Space debris includes non-operational spacecraft, spent rocket bodies, material released during planned space operations, and fragments generated by satellite and upper stage breakup due to explosions and collisions. After 45 years of the international space operations, more than 29,000 objects have been officially catalogued, with more than 15,000 objects are still in orbit. Catalogued objects are larger than 10 cm-20 cm in diameter in Low Earth Orbits (LEO) of up to 2000 km altitude and 1 m in diameter in geosynchronous orbits; and they are regularly tracked and their orbital elements maintained in a database.... One may recall that following the loss of Columbia in February 2003, the debris situation in the International Space Station worsened without the removal of tons of waste matter, generated inside ISS by regular space shuttle sorties which was the normal practice. In such situations, the waste products from spacecraft may be dumped into space creating more space debris.... **Hazards from asteroids can be another factor one should reckon with in space expeditions.** There is however an interesting positive aspect in this situation, namely, the extraordinary resources that are contained in asteroids. Several thousands of asteroids are our companions in the outer space. Majority of them are in asteroid belt between Mars and Jupiter orbits. Some groups of asteroids are very close to Earth and they pose collision danger in future.... **It is widely recognised that there is a small but definite probability of large Near-Earth Asteroids (NEAs) impacting on our planet. The Earth has experienced several asteroid strikes in its 4-5 billion year history, and about 170 impact craters have been identified on its surface.** The 1.8 km impact in Lonar, India is estimated to have occurred about 60,000 years ago.... In 1908, an asteroid, perhaps 30m-60m across, exploded in the sky above the Tunguska River of Siberia, producing an air blast that flattened about 2000 square km of forest. The Peekskill impact near the suburbs of New York in 1992 was video graphed by about dozen onlookers. More recently, the September 15, 2007 impact of meteorite in Peru caused a crater of about 4.5 m deep and 14 m wide.... A large steroid can survive passage through atmosphere and create a crater. The initial impact creates a shockwave that travels through the crust. Due to the intense heat, both the crust and the asteroid melt locally and fuse together. Material is ejected out and can cause secondary impacts. Depending on the energy imparted, simple or complex crater shapes can result." (Adimurthy, December 2012, Technical Journal of the Institution of Engineers)

Concerning the presently flowering 'Space-Age' as well as, the mankind's numerously planned and designed explorative initiatives into the outer-space all around the planet-Earth—what all challenges and problems are being generated for the man himself—have since been very precisely discussed in the above quoted paragraphs. Wherever mankind exists and conducts its own multi-dimensional living as well as, pursuits of its needs, desires and ambitions; all kinds of garbage and waste-materials are bound to get generated and hence, are thrown around or sometimes, scientifically disposed-off in some desirable manners. Interestingly, any phenomena of the natural-cum-automatic bio-disintegration leading to their final disposal or 'melting-away' into the mother-Earth itself do need the earth, water and oxygen, for triggering and sustaining such processes, which in fact, we are so used to witnessing daily in our lives in the world. But the same phenomena in the similar manner are just not thinkable even, into the outer-space where first the Earth's gravitational-force are not similarly strong and decisive to keep the wastes confined to one place and at the same time, the solid earth-mass, water and oxygen too, are not available for any similarly desired purpose. In nutshell in fact, the entire physical-cum-biochemical situations and thus; the overall environmental scenario since prevailing into the outer-space, are totally different. And obviously therefore, the entire basic as well as, complex-cum-advanced physical-cum-chemical scientific laws, always being so irrefutably remaining proactive onto the mother-Earth are just not relevant into the so expansively vast domain of the outer-space and hence as per every possible techno-scientific consideration, the whole scenario along with its many allied factors, over there become different and totally out of this world. Therefore understandably, the various kinds of the biological processes of germination, multiplication, growth, survival of the bacteria and micro-organisms, and also the patterns of their behaviours too, get totally altered while into the outer-space. **In addition, we have also been told by the above quoted paragraphs that certain so-far unknown kinds of the new bacteria and micro-organisms, since not existing onto the planet-Earth have also been found to being capable of getting materialised and thrive over there.** Thus, bringing-in new problems and also opportunities for the mankind, while venturing into the outer-space. Then techno-scientifically it is also no less true that, while critically delving deeper into the various multi-disciplinary engineering aspects therein into the outer-space, we do find that the most crucial stage in these regards, remains breaking through the Earth's gravitational-barrier and thus, getting in a way 'liberated' to be able to venture into the outer-space.

Precisely in the above contexts, **Professor S. Ramakrishnan**, Director of the India's Vikram Sarabhai Space Centre has said that, "To reach space from earth, climbing out of the deep gravity 'well', transcending the dense atmospheric shield is the most energy intensive challenging task and the rocket propulsion based satellite launch vehicles are the currently viable means of accessing space. The launch vehicle technology encompasses a wide spectrum of engineering and science disciplines. While propulsion systems or the 'rocket' constitutes the backbone for the multi-stage launch vehicle, to perform a precise mission of deploying spacecraft into a specified orbit, the control/steering, guidance, navigation and staging systems are equally vital requirements.... For a satellite to remain in orbit, it has to travel at great speed in the range of 7 km/s to 8 km/s (25000 kmph). The task of satellite launching essentially boils down to taking an object at 'rest' on the earth's surface to the required altitude and imparting to it sufficient tangential velocity to sustain it in orbit. The means of achieving this function can range from the most elementary concept of shooting it out of a giant gun barrel to carrying it in the cargo bay of a 'Space Shuttle' and gently releasing it in the orbit. As on date, expendable multi-stage rocket launchers continue to be the most practical and economic approach to perform the above task, while Single-Stage-To-Orbit/reusable space planes may become reality one day.... A satellite launch vehicle is a complex space transportation system, which is essentially a flying machine capable of transcending the atmospheric regime and beyond to deliver the payload into a desired orbit. Since, it has to cross the atmosphere, it is aerodynamically streamlined and its basic structural frame is designed to withstand dynamic loads experienced in flight. Like any flying machine, it has to be of light weight to maximize the performance.... A launch vehicle has well defined functional sub-systems (such as, Propulsion, Guidance and Control, Navigation, Staging/Separation, Vehicle Avionics, Major Vehicle Modules, and so on).... The first step in accessing space is to reach a payload to an orbit around earth, from where one can navigate the spacecraft for a rendezvous with Moon, Mars or any other planetary body. The remarkably versatile PSLV which has so far accomplished an unbroken string of nineteen successful missions provides India an assured access to space and enables the country to undertake more complex planetary missions. The operationalisation of GSLV and the development of LVM3 launcher will further enhance this capability in the coming years." (Ramakrishnan, Technorama, pp. 21 & 24)

In the larger techno-scientific perspectives in these regards we also in fact, contextually notice that concerning these outer-space related and also, technologically so complex as well as the cutting-edge scientific and technological challenges, the crucial 'health-monitoring' of various systems and subsystems too, especially regarding the aero-gas turbine engines, becomes extra-ordinarily important. And precisely in these contexts, noted scientists, **Vijay Petley, R Anbazhagan, M Tamilselvi, U Chandrasekhar** of the Gas Turbine Research Establishment, Bangalore, and then, **Thulsiram and G Mohan Rao** of the Indian Institute of Science, Bangalore, in their Technical-Paper, titled, 'Health Monitoring of Aero Gas Turbine engines through Conformal Thin Film Sensors', published in *Technorama* (pp. 25, 29 and 30) have collectively mentioned that, "Structural health monitoring of aero gas turbine engines entails extensive use of sensors that can provide accurate in situ data on the performance of critical hot section modules such as turbine and high pressure compressor modules. Among the multitudinous physical parameters that need continuous monitoring, strain and temperature are paramountcy as they directly impact the engine thrust and endurance. Despite the fact that the thermocouples and pyrometers provide experimental temperature data, thin film sensors are increasingly being used due to their high adhesion, superior structural integrity, low mass and minimal intrusion to aerodynamic flows. This paper describes the functionalities and the latest developments associated with the usage of thin film sensors with focus on high speed compressor and turbine modules of aero gas turbines.... The work related to the research and development of the thin film sensor technology has realized a multitude of sensors for use in harsh environments. The technology has been demonstrated enabling measurement of multitude parameters with the use of individual sensors or with sensor arrays. The thin film strain gauge and thin film thermocouple deposition on complex shapes that of an engine blade has been realized. Evaluation of strain sensitivity and the thermal stability of the sensor films is one of the important aspects in the development of the thin film strain sensors. Influence of thermal coefficient of resistance on gauge factor is very important for sensing accurate measurements at high temperatures.... The need of the contemporary gas turbines with reference to health monitoring demands, the usage of sensors that have superior sensitivity and good temporal stability. The turbine entry temperature of approximately 2000 degree C is envisaged under the program on Versatile Affordable Advanced Turbine Engines (VAATE). The thin film sensor technology will have to mature to meet these requirements and the use of diffusion based sensors, embedded sensors and wireless sensor packaged device for use in harsh environments, will be the area of focus in near future. Advanced processes like focused ion-beam machining, laser micro-machining, need to be explored for realizing the thin film sensors on a more reproducible fashion. The major challenges in the domain of thin film sensor technology include repeatability of gauge characteristics and availability of telemetry systems with high temperature capability." (Petley, Anbazhagan, Tamilselvi, and Chandrasekhar, Technorama, pp. 25, 29 & 30)

At this precise point, even though perhaps at the apparent risk of being rather too 'technical', by bringing-in the above recorded selective quotations concerning the multiple fields of the latest high-end sciences and technologies with respect to mankind's intensifying—multi-dimensionally explorative adventures into the outer space—the intended purpose has been just only to render some maximum possible simplistic and also, the broadly panoramic glimpses of indeed, the multiple enormities as well as, the multi-dimensional complexities of the so fast flowering: the latest cutting-edge sciences and technologies in these regards. In fact, anything more than these would professionally be meant only for the concerned scientists and technological specialists and hence, perhaps totally outside the range of intellectual-acumen of the most of us and therefore indeed, including completely that of this ordinary and humble author too. Hence thereby, pausing a little and caring to even simplistically as well as, somewhat imaginatively to reflect upon the above mentioned enormities and complexities in these regards, and then perhaps we may mentally permit ourselves to even quite enchantingly get overwhelmed for some time, by the **captivating marvels of the so vast fraternity of scientists and engineers since relentlessly engaged in the multitude of these multi-dimensional and indeed, the highly challenging assignments and research-activities across the world. And precisely in these multiple path-breaking and history-making contexts, the USA decisively remains the world-leader and the fountain-head as well as, substantially maintaining her command over the scientific-cum-technological 'knowledge-pool'**. And thus at the same time, and quite encouragingly thereby; permitting considerably but no less closely monitored access to some of the friendly democratic nations towards this so momentously flowering and expanding know-how, and wherewithal of the cutting-edge scientific-cum-technological expertise; and also in addition, opening the doors of the numerous US Institutional facilities, and Training Establishments. Accordingly India too, has substantially been benefiting in these latest scientific areas in many ways especially after the highly relieving demise of erstwhile extremely menacing cold-war era. But at the same time, it is also true that Russia too, in her own ways has been very cooperative to India in these regards.

Now in fact, almost the instant transmission of the multiply vast amount of scientific as well as, the many other kinds of data, through 'LiFi technologies' may also lead to the possibilities in a foreseeable future that these too, may have substantial potential of applications in the vast, multi-dimensional, and always indeed, the highly complex domain of the presently flowering 'Space-Age'. In these captivatingly engaging contexts, the 'Tech-News' in the 'Technorama'; has informed us that, "**LiFi or Light Fidelity is a technology that uses light emitting diodes to transmit data wirelessly**. It was first demonstrated at a TED talk in 2011 by German Physicist **Harald Hass**.... The technology is capable of transferring thousands of streams of data simultaneously, in parallel, at higher speeds, with the help of special modulation, using a special signal processing technology....10000 times the visible Light-Spectrum is larger than the Radio-Spectrum....(i) LiFi can be used in places where it is difficult to lay optical fibres. (ii) In Operation-Theatres. (iii) At traffic signals, LiFi can be used to communicate with LED lights of cars, thus controlling traffic. (iv) In aircraft, the overhead lights can be used for data transmission. (v) It can be used in petroleum or chemical plants.... Present-Status: The LiFi consortium, a non-profit tech Cos and Research Institutions have prepared a roadmap. PUREVLC and Outstanding Technology has already released products based on this tech." (*So far now scientifically, the core-details, designs, and the technical-data in these still fructifying and, the latest kinds of the cutting-edge technologies are yet to come onto the open domain. The related 'R&D' activities in many dimensions are already being pursued and hence, there is still time to get them flowered as well as, proactively matured in these so momentously paths-breaking endeavour.*).

(g) Population Explosion and Environmental Degradation

While being in this world, man has to live, work, and realise his many aspirations and ambitions, within this all-encompassing greater environment, and obviously therefore, the man has to keep sincerely obeying and completely adhering to the commanding dictates of this environment since these environmental dictates are in a way, nothing but the commanding manifestations of the supreme-laws of Mother-Nature Herself. At this precise point, we may also contextually recall that these very specific points, concerning the overall, all encompassing, as well as since also, the all-powerful supreme laws of the Mother-Nature—have since already been discussed and always quite relevantly emphasised from time to time; while critically delving upon the different issues, events, and developments in the previous chapters, and also, earlier in the present eighth chapter too. But no less ironically, we do notice that in spite of fully knowing, and being fully aware of the many kinds of extremely tragic, as well as the absolutely inevitable consequences, man is not at all hesitant to cut the very 'branches' onto which, he is always required to live, survive, and also multiply thrive in this world.

What else one has to so sadly term these phenomena, but as precisely indeed, the greatly unbecoming negations of the human-civilisations themselves. And obviously at the same time and, at this precise point of ours—of the contextually ongoing discussions—**what else could be identified as the mankind's greatest civilisational negations; but the presently so daunting menaces of both, since the manmade 'human-population-explosion' and the 'environmental-degradation'. In fact, it won't be any exaggeration to say that, if both these twin-menaces are not confronted effectively and conclusively, to solve and neutralise them while there is still time, they could be more devastating than even the cumulative destructions and losses, caused by all the previous wars since fought by the mankind onto this planet Earth so far. And consequently in these regards, otherwise, what needs to be urgently understood globally that the mankind won't tragically be left then, with any viable kind of an 'escape-route' too, to save itself, its home and hearth; and thereby also, the survival-cum-well-being of its future generations**. Top grade scientists, environmentalists, and other visionary experts, since the last few decades of the recently bygone twentieth century itself, have been very scientifically emphasising these paramountcy issues and warning the highest-levels policy-framers and the decision-makers all across the world, precisely regarding the above mentioned indeed, the unprecedented twin-menaces, now all the more so dangerously confronting the mankind. In these and other allied broader perspectives, we may recall and since already been noted earlier **one very famous saying of Mahatma Gandhi that, 'the Mother-Nature in this world, has enough to cater for everyone's need, but certainly not to fulfill anyone's greed'**. But the 'million or more accurately, even much more than a trillion-dollar question' remains that, what more ultimately be needed to vigorously shake and wake-up the mankind in these regards, before it is too late? Or is it so perhaps that, *'you can take a horse to the water but, can't make him drink'*.

Multiple-aspects of the 'environmental-degradation in fact, have many dimensions, such as, deforestation, callously irresponsible exploitation and abuse of the natural resources and, ransacking of the Nature's vital-cum-huge: the sweet water-reservoirs, and the numerous river-systems on this planet-Earth; the global-warming, greenhouse effect, climate-change or climate-crisis, reckless and increasingly irresponsible burning of the colossal quantities of the different kinds of fossil-fuels in fact, so self-destructively causing holes through the vital and very fragile Ozone-layer, since situated in the upper segment of the Earth's atmosphere, and so on. And needless to say that, if these are not prevented while there is still time the consequences could be in fact, irreversibly devastating for the mankind. Lot many articles, proceedings of the numerous national-cum-international seminars, and many books too have already been published in many languages all across the world, for indeed contextually educating, as well as coaxing the all concerned in these regards.

Al Gore, a former US Vice President during the **Clinton-era**, has mentioned that, "The Sun's energy enters the atmosphere in the form of light-waves and heats up the Earth. Some of that energy warms the Earth and then is re-radiated back into space in the form of infrared waves.... Under normal conditions, a portion of this outgoing infrared radiation is naturally trapped by the atmosphere—and that's a good thing, because it keeps the temperature on Earth within comfortable bounds. **The greenhouse gases on Venus are so thick that, its temperatures are far too hot for humans. The greenhouse gases surrounding Mars are almost non-existent, so the temperature there is far too cold. That's why the Earth is sometimes referred as the 'Goldilocks planet'—the temperatures here been just right.... The problem we now face is that this thin layer of atmosphere is being thickened by huge quantities of human-caused carbon dioxide and other greenhouse gases. And as it thickens, it traps a lot of the infrared radiation that would otherwise escape the atmosphere and continue out to the universe. As a result, the temperature of the Earth's atmosphere—and oceans—is getting dangerously warmer.... That's what the climate-crisis is all about.**" ...*Well, what exactly are greenhouse gases*? When we talk about greenhouse gases and climate change, carbon dioxide usually gets the most attention. But there are also some others, although CO2 is the most important so far.... What all greenhouse gases have in common is that, they allow light from the sun to come into the atmosphere, but trap a portion of the outward-bound infrared radiation and warm up the air.... **Having some amount of greenhouse gases is beneficial. Without them, the average temperature of the Earth's surface would be right around zero deg. F—not a very nice place to live. Greenhouse gases also help keep the Earth's surface at a much more hospitable average temperature—almost 59 Deg. F. But due to increasing concentrations of human-caused greenhouse gases in modern times, we are raising the planet's average temperature and creating the dangerous changes in climate we all see around us**. CO_2 usually gets top billing in this because it accounts for 80% of total greenhouse gas emissions. When we burn fossil fuels (oil, natural gas, and coal) in our homes, cars, factories, and power-plants, or when we cut or burn down forests, or when we produce cement, we release CO_2 into atmosphere.... **Like CO_2, methane and nitrous oxide both predate our presence on the Earth, but have gotten huge boosts from us**. Sixty percent of the methane currently in the atmosphere is produced by humans. It comes from landfills, livestock farming, fossil-fuel burning, waste-water treatment, and other industry. In large-scale livestock farming, liquid methane is stored in massive tanks that emit methane. Dry manure, left on fields, by contrast, does not. **Nitrous oxide (N_2O)—another greenhouse culprit**—also occurs naturally, though we have added 17% more into the atmosphere just in the course of our industrial age, from fertilizers, fossil-fuels, and the burning of forests and crop residues... **Sulfur hexafluoride (SF6), PFCs, and SFCs are all greenhouse gases that are produced exclusively by human activity. Not surprisingly, emissions of these gases are on the rise too**. HFCs are used as substitute for CFCs—which were banned because their emissions in refrigeration systems and elsewhere, were destroying the ozone layer. CFCs were also very potent greenhouse gases. PFCs and SF6 are released into the atmosphere by industrial activities like aluminum smelting and semiconductor manufacturing, as well as the electricity grid, which lights up our cities.... **And finally, water vapor is a natural greenhouse gas that increases in volume with warmer temperatures, thereby magnifying the impact of all artificial greenhouse gases**." (Gore, 2006, pp. 26-28)

While critically looking little deeper into the above quoted paragraphs in our these specific contexts and hence, at the first glance itself we find the already scientifically well-established perceptions are in fact, not only quite disturbingly queer but also unprecedentedly so extremely dangerous that—during the forthcoming times now being not so very distant—that even the very so far continuing survival of the mankind, as well as the other kinds of living-beings too would soon be getting precariously tied to a very 'slender thread', which may get snapped anyway; if multiply scientific-cum-effective measures are not sincerely implemented soon. **Here it appears to be a great irony so far perhaps, that with our all these ongoing 'gung-ho' reviews, within our this since already continuing—the 'strategically critical-journey in specifically spelled-out broader perspectives'—we have been multiply deliberating upon indeed, the numerously captivating aspects of the mankind's, both evolutionary-cum-revolutionary, the multi-dimensional enhancements. And also thereby, the multi-dimensionally continuing advancements of the human-civilisations onto obviously the mankind's all-encompassing, as well as the expanding civilisational-canvas, and therein meanwhile we do also observe that, due to nothing else but always the singularly manmade—so many kinds of the irrationally greed-oriented and reckless excesses—the man's truly enchanting civilisational accomplishments, have now the already no less menacingly, begun to turn a 'full-circle'.**

In course of these indeed multi-dimensionally long journey and then, especially after the unprecedentedly momentous advent of the **'Industrial-Revolution'**, as well as, consequent to his multiple-entry into the **present times' 'Super high-tech Age'**, man appears to have quite arrogantly discarded **the most fundamental scientific truth that, this planet-Earth on which man himself has been living and also, multiply thriving; is not only an integrative part of its much larger Solar-System but in addition also, remains endowed with an overall, a very fragile eco-system as well**. And thereby both evolutionarily and scientifically, the mankind is not supposed to fall too much 'out of line' and then so unmindfully begin to even increasingly take all sorts of rather self-destructive liberties, with though at times, quite benevolent but also the no less harsh, whenever she thinks so—the supremely powerful mother-Nature. **And hence, if all these excesses and the mindless indiscretions, after crossing certain kinds of the clearly laid-down mother-Nature's 'limiting-lines' begin to boomerang onto the mankind itself in many ways and then, who else man could ever blame but to only himself**. In these multiple-contexts at the same time, it remains also equally true that, since the very beginning of the mother-Nature's always supremely ongoing evolutionary phenomena of the indeed, her incomprehensively vast and complex creations, transformations and the seeming 'destructions', it is the mother-Nature which always unassailably remains supreme and the most powerful and hence, none of its any particular creation; howsoever exclusive, as well as, whether living or otherwise, within her all-inclusively ongoing 'scheme of things', has any unquestioned powers to take any unchecked liberties, and then within the infinite Universe itself, which being perhaps, with no beginning or any end; as per its any dimensions of the space and time whatsoever.

With our all these since critically continuing panoramic reviews and the multiple-discussions, always inclusive of the human civilisations, progresses and also the mankind's multi-dimensionally enchanting accomplishments and marvels, if today the man finds himself having landed in the above so sharply-focussed and precise—the so extremely unbecoming state of affairs—wherein his own continuing survival, comes to indeed the so much menacing risks and then how can the man rationally claim to be really 'civilised', as per the any really worthwhile civilisational-connotations? We may interestingly recall, and since already noted by us during our earlier discussions in some other contexts too, that Bertrand Russell having said that, *'Power corrupts, and the absolute power corrupts absolutely'*; but that was observed by the learned thinker and philosopher in some totally different as well as, relatively in a very limited contexts of the imperialism-fixated or some 'clashes of arms' oriented—the political and military powers towards control and governance—over any targeted nation, peoples or even onto one's own nation. These aspects seem to be perhaps, not enough to provide some analytically perceptive answers, as to in spite of the mankind's such indeed multi-dimensionally marvellous advancements-cum-accomplishments and enrichments, onto its simultaneously expanding civilisational-canvas in this world; then why and how the mankind has found itself to have landed in such a magnifying as well as, the no less increasingly dangerous 'mess' in fact, wholly consequent to its so mindlessly callous-cum-arrogant continuation of the all sorts of ransacking of the planet-Earth's overall such fragile eco-system in all the more accelerating manners, especially during the last sixty to seventy years or so with the so momentous commencement of World-War-II and thus, approximately beginning from the 1940s onward? If phenomenally the so fast-paced and also, the multi-dimensionally continuing enrichments of the human-knowledge along with the simultaneously ongoing advancements in multiple domains of the sciences and technologies and then, the no less greed-based recklessly unmindful applications have been responsible for all these ecological-cum-environmental 'mess-up', which indeed seem to be rather irrefutably true then where is the mankind's civilisational conscience, which is expected to always proactively occupy a rather highest station, in the mankind's individual, economic, cultural, socio-political, and also all the more geopolitical dimensions of his life on this planet-Earth itself?

Precisely in these multiple-contexts here, we may prefer to quote some very perceptive thoughts of India's **Dr. M.S. Swaminathan** since picked up from his highly informative book, *From Rio de Janeiro to Johannesburg: ACTION TODAY and not just promises for tomorrow* (page xvi) in the 'introduction' chapter itself this learned scientist has said that, "Food and drinking water occupy the first position in the hierarchical needs of human beings. I have, therefore, chosen sustainable food and water security as the index of our progress towards achieving the humanistic goals of Stockholm, Rio de Janeiro, and Johannesburg. **Appropriate in this context is George Herbert's verse:**

> Sweet Peace, where does thou dwell? I humbly crave, Let me once know. I sought thee in a secret cave, and asked if peace was there, a hollow wind did seem to answer, No: Go seek elsewhere. At length I met a reverend good old man: Whom when for Peace I did demand, he thus began: 'Take of this grain, which in my garden grows, And grows for you; Make bread of it: and that repose And Peace, which everywhere With so much earnestness you do pursue, Is only there."

So whatever may be the various kinds of priorities and the compulsions for the mankind in this world in fact regardless of all these, the mother-Nature has been having her own kinds of the supremely commanding evolutionary phenomena, since unendingly continuing within this infinite Universe and our planet-Earth remains a tiny segment of the same Universe and thereby, the man's so visible world in the first place, has come into being onto nowhere but this planet-Earth only and hence logically in these regards, one may easily visualise as to where and how precisely, the mankind continues to exist and flourish in his so seemingly enchanting and phenomenal world. So how can the man come to any position whatsoever wherein, he can even dream to control and command in any manner, the supremely powerful mother-Nature herself? Hence, earlier the mankind sincerely understands—this in fact, the perpetually vibrant, living as well as, the irrefutable truth—and then wholeheartedly abide by the same; better for its continued survival and well-being, onto the planet-Earth itself. **At this precise point of our ongoing discussions, the issues involved are since also concerning—the unchecked growth of human-populations in the world, as well as, the extremely callous and mindless ransacking of the Earth's fragile eco-system—which in any case, always unalterably remains very much an integrative segment of the supreme mother-Nature's unendingly continuing evolutionary phenomena onto indeed the incomprehensively infinite-canvas of the Universe itself.** This is now indeed the basic global question, which is so powerfully confronting the mankind and ironically not only naively but also in a rather foolhardy manner, the man is still quite self-destructively busy with his typical 'culture of fooling around' in these regards. It appears that perhaps the entire humanity, except with few exceptions here and there; **has begun to suffer from an ancient-era *'Nero-syndrome'*, i.e. as the saying goes that, *'when the Rome was burning, Emperor Nero was fiddling with his flute or harp, or whatever.'*** The question is that if not in all these then, in what else manners one may endeavour to picturise now, the so precariously prevailing perhaps the most dangerous overall environmental-cum-ecological scenario in these regards in which the mankind has so much already, both naively, as well as arrogantly now seems to be blindly busy in getting itself trapped regardless, with no likely 'escape-route' available so far?

Hence contextually, at this precise point of our ongoing discussions; it seems very appropriately logical to ask another, **the no less important and fundamental question, as to what's the so big a deal and any worthwhile rationale for the mankind, to be so much proud of indeed, the enchantingly multi-dimensional civilisational advancements and the multiply ongoing progresses, over thousands of years of its 'living-history' since continuingly in the making if ironically in the end, the mankind finds itself perhaps, uncontrollably slipping fast, towards an 'abyss' of darkness and possibly, the inevitable annihilation**? In these specifically larger perspectives and also, as the ransacking of our overall environment is going-on regardless, we need to also take note that the **still continuing unchecked population-growth** since all the more menacingly happening in the continents of Asia, Africa, and even in the South America (i.e. Latin America) is no less a menacing culprit, as this singular aspect too, happens to be equally responsible for playing havoc with the Earth' fragile eco-system. As per the contents of above paragraphs, anybody anywhere and in whatever contexts and in multiply logical, as well as in the no less scientifically analytical manners makes efforts to emphasise these now indeed, supremely paramountcy aspects; in fact, so momentously linked to the very survival-cum-well-being of the further human-generations, howsoever and irrespective of the anyone having any contextual reservations or differences in these regards, he or she just cannot justifiably ever, be in a position to accuse the other one of unnecessarily raising any kind of the 'false-alarms' and thus, trying to scare the peoples all across the world. But the point is that, is anybody in any 'corridors of powers' since anywhere in today's world ever taking a pause and thereby, trying to both sincerely-cum-proactively listen to these indeed most worrisome 'alarm-bells' in his/her utmost meaningful manners? And here precisely, lies the 'rub' in these regards.

As since already highlighted in fact, in their unprecedentedly far-reaching multiple-contexts; we have now been quite interestingly noticing a typical culture of the periodically holding some international seminars and conventions, to contextually discuss **the numerous kinds of problems related to the various global consequences so lethal—of the 'population-explosion and environmental degradation'**—and so on; with rather seemingly intended purpose of identifying their viable solutions. These international efforts are meaningful only, if there are openly visible and the sincerely worked out solutions coming out of these intellectual exercises, wherein the leading nations concerned are both sincerely-cum-proactively committed to abide by the resolutions passed, mostly in the best long-term interests of the entire humanity in this world. And if that be not so, then all these would turn out to be nothing but rather, **a 'naked' display of the man's age-old 'culture of fooling-around'**. Now, in all these larger perspectives; where these periodical international efforts, resulting in some kinds of the off and on throwing-up of the any **'Kyoto-Protocol'** or any system of awarding some **'carbon-credit'** etc. to any concerned nations and so on in fact and in what precise manners, all these may contextually be considered and valued are best left to the judgmental acumen of any interested person in these regards. **Such menacing problems are in fact, so overwhelmingly sprouting, as well as magnifying that any kind of just fiddling-around or some cosmetic 'window-dressing', are not going to lead us anywhere; and thus needless to say that any belief towards achieving any kind of the desirably effective solutions in these regards, by resorting to just keep-on conducting an international culture of some politico-diplomatic 'orchestra' in any case, would turn out to be nothing but a plain 'wishful-thinking' interestingly like an ostrich, who displays a tendency to bury its head in sand, thinking that any lurking danger would only bypass it**. Hence, mankind all over the world, individually, collectively, and sincerely too, immediately needs to get geared-up and begin striving towards consistently adopting the scientifically well-planned and truly comprehensive remedies, before it becomes too late. As it is said that, 'time and tide waits for no man' and hence similarly the mother-Nature, under no circumstances, is going to give even little margin in any of her—the unendingly ongoing evolutionary schemes of things—within the infinite Universe, regardless of what the man thinks, plans, or does in these regards. **Hence, let us not dare to confront the supremely mightiest mother-Nature; but to sincerely listen to her commands and thereby, be always proactively innovative to fulfill our only agreeable aspirations, always well within the vast parameters of her indeed tremendous patience, as well as, her so considerate benevolence too**.

Precisely in the above broader perspectives, **Dr M.S. Swaminathan** has mentioned that, "**Managing the consequences of human-induced changes in climate will undoubtedly be one of the biggest challenges of the twenty-first century**. The Inter-Governmental Panel on Climate Change has concluded that, 'the balance of evidence suggests that there is discernible human influence on global climate.' The U.N. Framework Convention on Climate Change adopted at Rio de Janeiro in 1992 was designed to stabilise the concentrations of greenhouse gases (GHG) in the atmosphere at a level that would prevent dangerous anthropogenic interference with the climate system. Unfortunately the necessary political will and action to implement in both letter and spirit the Kyoto Protocol of the Climate Convention are still lacking in many industrialised countries, which are responsible for most of the GHG emissions. It is not yet widely realised that the time scales of reversing environmental change often range from decades to centuries. Also, certain effects, like the loss of genes and species are irreversible.... **Adverse changes in climate as a result of a rise in GHGs in the atmosphere, could induce regional increases in the intensity and frequency of droughts and floods, inundation of coastal areas, outbreaks of fires, pests, and diseases, loss of agricultural production, and threats to human health and safety. Such changes will affect the poor countries and the poor in all nations the most, since they lack the necessary coping mechanisms.... As our power over nature grows, only a sense of humility and a commitment to the principles of ethics and equity can save us from the adverse impact of human innovation and ingenuity**. Also, knowledge that can confer the greatest public good should remain in the public domain and should not come under propriety control. Social exclusion in science is both unethical and dangerous. It is in this context that the recent announcement by **President**

Clinton and **British Prime Minister Tony Blair** that, 'the data from genome mapping should be freely available to the entire world,' marks an important step in mainstreaming ethical principles in the use of outputs of the human intellect. **Let the principles of humility, ethics, and equity become the guiding principles in our daily life**." (Swaminathan, 2002, pp. 152, 153 and 154)

The last line of the above quoted paragraphs, says that, 'let the principles of humility, ethics, and equity become the guiding principles in our daily life'; but the most weighty question in these regards, now obviously arises so as to that have these words with their intended meanings, been both sincerely-cum-proactively sunk even now, in the very psyche of the mankind so proudly living in the presently ongoing modern-era's, the super high-tech age in the today's world? Most depressingly, the answer is obvious for the any objectively discerning citizen, since living in any corner of the today's world. In fact, the irony of ironies now happens to be that, instead of sincerely listening to, as well as desirably understanding these paramountcy aspects the concerned peoples since occupying the various kinds of the decisions-making positions in the numerous corridors of power, in the various leading nations of the world and thus, presiding over the further destiny of the entire global humanity, are just posing around with all sorts of the artificial gestures and concerns, to only be pretending to protect the humanity from indeed, the unprecedentedly momentous lethality of nothing, but the impending fallouts of the menacingly fast manifestation of the—population-explosion and environmental degradation—on an undeniable global-scale. But what is indeed a great paradox that all the concerned peoples since, occupying the seats of power all over the world, as well as, so arrogantly presiding over the future destiny of the whole humanity little do they realise that they too, are also in the same 'global-boat', as rest of the ordinary masses all over the world. But sadly, that's precisely the way, that the man's 'living-history' since continuingly in the making, onto his expanding civilisational-canvas; finds itself, i.e. in a 'logjam' of the fast slipping towards nothing but the already earlier mentioned: an 'abyss' of darkness and hence, the consequent annihilation. On top of all these, as if these are not enough; and so those very peoples since, occupying the seats of power in the various leading nations of the world are even refusing to counter the various kinds of the indeed, amazing myths floating all around on even perhaps a global-scale intended to just so harmfully to refute the globally magnifying threats of the 'population-explosion and environmental degradation', for misinforming and misguiding the common masses across the world. What else could be any more dismal, as well as precisely, a most lethal scenario in these regards?

And hence, precisely in the above multiple-contexts, **Al Gore**, a former US Vice President (pages 308, 309, 312, 313, 315, 316, 317, 318, 320 and 321) has listed the ten misconceptions in these regards:

"**Misconception – 1**: Scientists disagree about whether humans are causing the Earth's climate to change.

Misconception – 2: Lots of things can impact climate—so there are no reason should single out CO_2 to worry about.

Misconception – 3: Climate naturally varies over time, so any change we are seeing now is just part of a natural cycle.

Misconception – 4: The hole in the ozone layer causes global warming.

Misconception – 5: There is nothing we can do about climate change. It's already too late.

Misconception – 6: Antarctica's ice sheets are growing, so it must not be true that global warming is causing glaciers and sea ice to melt.

Misconception – 7: Global warming is a good thing, because it will rid us of frigid winters and make plants grow more quickly.

Misconception – 8: The warming scientists are recording is just the effect of cities trapping heat, rather than anything to do with greenhouse gases.

Misconception – 9: Global warming is the result of a meteor that crashed in Siberia in the early 20th century.

Misconception – 10: Temperatures in some areas aren't increasing, so global warming is a myth.

Mankind can so audaciously ignore these extremely serious alarms to nothing else but, indeed its own inevitable 'peril' only. **Well, interestingly in the above multiple-contexts, concerning the various most worrisome fallouts of the numerous menaces of the 'population explosion and environmental degradation' could there be anything more bizarre, as well as tragic, than the above quoted ten misconceptions and that too, during indeed the present 21st century, of the so fast advancing super-age of the ultra-modern sciences and technologies?** In spite of the all kinds of exclusively special attributes, as well as, also having been gifted with always the 'analytically thinking and creative mind' by the supremely evolving mother-Nature man indeed happens to be very 'queer' mammal, with the so many stunningly strange and intriguing contrasts in this world. And therefore precisely here, lies the proverbial 'rub' in these regards. Especially now in these paramountcy perspectives, the mankind urgently needs to overcome its above mentioned limitations and contrasts, if it desires to continue with its so aspiring and healthy existence and also, ensuring the multiple kinds of tranquility for itself and indeed, for the Earth's so fragile eco-system as well. Must the mankind even in the presently ongoing 21st century, with all sorts of indeed the multi-dimensionally outstanding civilisational accomplishments, still be so blindly greedy and continue to indulge-in the so regressively-fixated, as well as, the self-destructive foolish-acts regardless?

(h) Threats of 'Future-Shock' and Ironical Military Lethality

The phenomena of the different kinds of situations based and, any localized or otherwise, any impending threats of 'future-shock', likely to be triggered by possibly any violently manifesting 'clashes of arms' have been continuing, since the very advent of human-civilisations themselves. Over a period of time, and also as the human-civilisations continued to multi-dimensionally advance we have also critically observed that, almost all the aspects concerning the 'matters-military' too have simultaneously been advancing in parallel, as well as the no less multiply, in tune with the mankind's 'history-making' phenomena as well. Hence, precisely in these multiply complex and also, the vast historic perspectives concerning the mankind itself these aspects have indeed been the basic, as well as, the intellectually very captivating features of the man's 'living-history', since continuingly in the making in this world. And accordingly thereby in parallel manners, the consequent aspects of the fructifying military-lethality too, have been enhancing, and expanding as well. But to cut this long story short, and with no less including the multiple-aspects of their historic perspectives, we may now also say that the multi-dimensionally momentous advent of the **'Industrial-Revolution'** and its so many fallouts too, have indeed been unprecedentedly instrumental in rendering again the multiply complex kinds of the far-reaching boosts since, to the already continuing processes of enhancements-cum-expansions of historically the timelines-based—ongoing manifestations of the different kinds of periodical phenomena of 'future-shock', as well as, the ironical military lethality—onto the mankind's expanding civilisational-canvas itself, in this world.

Precisely in the above multiple-contexts, we may now recall that we have since already discussed in some details as to how, consequent to the historically so momentous advent of the **'Industrial-Revolution'**, as well as its so many far-reaching fallouts, the multi-dimensionally gigantic and indeed the militarily revolutionary advancements were achieved in the domain of weapon technologies, designs and developments of the weapon-platforms and thereby the entire gamut of the further planning-cum-conduct of the mankind's warfare activities too, got totally revolutionized. And consequent to all these inevitably, sooner than later; the mankind came face to face with indeed the numerous kinds of the totally new and much deadlier aspects of the previously unknown kinds of 'military-lethality' during the (1914-18) First World-War itself, and then so tragically just after twenty-one years or so; the man had to suffer all the more newer and extensive kinds the 'military-lethality', during the (1939-45) Second World-War with more far-reaching global-scales. Whenever any person anywhere and also with some discerning mind, is able to find some time to browse through even the systemically comprehensive pictorial-histories of the above mentioned two World-Wars one may be able to have the very detailed glimpses of both the applications, as well as the consequences of the then latest kinds of the 'military-lethality' involved and thereby; the unprecedented sufferings visiting the mankind all over, indeed perforce with a lump in his or her throat. **Whenever a naively innocent and also contextually ignorant person is able to lay hands onto the bulky volumes of those comprehensive pictorial-histories of the erstwhile two World-Wars, he or she may quite ironically accuse soldiers of the opposing alliances, for all these so extensive destructions and deaths and thereby, the consequent human tragedies-cum-sufferings, when the larger realities are mostly different. In fact, if we may care to critically delve little deeper into the larger realities in these regards, we find those additional dimensions to them, than meets the eyes at the first glances. The universally common fact is that the soldiers largely anywhere, don't make the war, but they are just made to fight the war at somebody else's bidding.** And systemically, that 'somebody else', includes many things and hence understandably, has many differently emerging situations-based dimensions, and thereby as per the historically moving timelines-centric, a number of nodal triggering-points in these regards.

The legendary US General Douglas MacArthur had once said that, very few persons in the world would have seen more of war than he himself; and he happened to be indeed right and hence, he knew what he was saying when he talked of war and its multi-faceted horrors. In his already earlier referred autobiography, while talking to the 'Corps of Cadets' of the United States Military Academy, the 'West Point' he had said that, "**The soldier, above all other men, is required to practice the greatest act of religious training—sacrifice. In battle and in the face of danger and death, he discloses those attributes which his Maker gave when He created man in His own image. No physical courage and no brute instinct can take the place of the Divine help which alone can sustain him. However horrible the incidents of war may be, the soldier who is called upon to offer and give his life for his country, is the noblest development of mankind....You are the leaven which binds together the entire fabric of our national defense. From your ranks come the great captains who hold the nation's destiny in their hands the moment the war tocsin sounds. The long Gray Line has never failed us. Were you to do so, a million ghosts in olive drab, in brown khaki, in blue and gray, would rise from their white crosses, thundering those**

magic words—Duty-Honor-Country.... This does not mean that you are war mongers. On the contrary, the soldier above all other people; prays for peace, for he must suffer and bear the deepest wounds and scars of war.... The shadows are lengthening for me. The twilight is here. My days of old have vanished tone and tint; they have gone glimmering through the dream of things that were. Their memory is one of wondrous beauty, watered with tears, and coaxed and caressed by the smiles of yesterday. I listen vainly, with thirsty ear, for the witching melody of faint bugles blowing reveille, of far drums beating the long roll. In my dreams I hear again the crash of guns, the rattle of musketry, and the strange mournful cry, always I come back to West Point. Always there echoes in my ears—Duty, honor, Country.... Today, marks my final roll call with you. But I want you to know that when I cross the river my last conscious thoughts will be of the Corps—and the Corps—and the Corps.... I bid you farewell." (MacArthur, 1965, pp. 482, 483 & 484)

In the highly consequential and the no less complex areas of since the always universally enhancing multiple-aspects of the 'military-lethality'; and thereby during the ending phases of the so far most extensively violent and destructive—the 1939-45 Second World-War—the mankind was forced to very shockingly confront an altogether the then completely unknown and unexpected kind of **an unprecedentedly the very far-reaching instruments of 'military-lethality'; and that precisely being the 'atom-bomb' frighteningly indeed, capable of an unparalleled kind of the nuclear-holocaust and, to threaten even the very survival of the mankind, as well as many other kinds of lives on this planet-Earth**. This came historically indeed, as a zenith of the 'military-lethality', the mankind has been confronting so far. And obviously in the all possible geopolitical, as well as in politico-military terms, the further fallouts and consequences have come out to be unprecedentedly phenomenal, having both intra-national and transnational dimensions, for any concerned Nation-State in these regards. But interestingly, there could be two distinct ways of both the—geostrategically-cum-politico-militarily—of critically looking into perhaps the all aspects, quite discerningly and now also, including the various kinds of the nuclear-weapons and their land-based, airborne, as well as, the underwater delivery-systems. Of the two, one with geopolitically—having considerably 'positive connotations'—which understandably, may appear to be quite controversial and so even bizarre, and then another: quite menacingly with negative fallouts. In any case, for more than past half a century or so, lot of extensive debates have already been going-on keeping in minds the various kinds of contemporarily, the very far reaching perspectives the world over; and thereby, voluminous literature, books as well as, the very many kinds of the seminar proceedings too in different languages, have come out in these regards. Surprisingly, as per the above and so for the many, the first method of geopolitically looking at the entire gamut of nuclear-weapon systems, having perhaps considerably positive connotations could be that it has ensured the globally enduring 'nuclear-balance' in the very vast, as well as, the highly complex domains of the matters-military in any mutually rival camps; and in addition, these have perforce turned out to be quite consistently preventing any all-out major war and then, even effectively countering some kinds of the any likely 'Third World-War', between any two major power-groups in today's world. The weighty point in these regards, happens to be that if we may critically care to geopolitically delve into all these aspects, with their desired strategic comprehensiveness and also, with their contextual objectivity; then we may quite justifiably contemplate that perhaps, during the height of now the bygone Cold-War era rivalries between the erstwhile two Superpowers, and if they won't have been equipped with nuclear-arsenals, they could have even triggered the then most feared the 'Third World-War' sooner or later, during those highly volatile times. Now in retrospect perhaps, all these may sound nothing but, the matters of some academically theoretical conjectures only. But if we may care to critically relive the recently passed historic timelines, with their highly turbulent episodes and the politico-diplomatic clashes in any global forum; then we may be able to have glimpses of some apparently hidden truths in these very so-called conjectures. After all in numerously far-reaching terms and, since the thousands of years of the human-civilisations, what else have been the records contained so far in the voluminous pages of man's 'living-history', than largely the 'history of the all sorts of conflicts and also, the warfare' itself? But at the same time and, the no less logical **prevention of any possible all out or any substantially major 'clashes of arms', in any continent in the today's world may even quite ironically be taken as a singular big contribution of the presence of 'nuclear-arsenals' in the modern world**. And equally at the same time, these specific contentions happen to be quite debatable as well.

But precisely in these multiple-contexts, **what happens to be the most dangerous scenario is that in today's world some kinds of the already 'failed' or fast 'failing' nations—being socio-culturally, as well as socio-politically; heavily soaked in religious fanaticism and then on top of it also, violently infested with irrationally fixated cult of the extremism and terrorism—are in fact most menacingly being seen, quite irresponsibly—in the possession of nuclear-weapons and some kinds of their delivery-systems too—which is indeed an extreme form of nightmare for their some immediate neighbours and, even to the larger world as well**. The world bodies too, appear to be quite at a loss so as to how to effectively tackle this menace, and consequently; the democratic, as well as the reassuringly responsible nations in today's modern-world, with its global-systems of super-connectivity are losing their sleep. This precisely is the second method, of contextually looking into these specific aspects. And accordingly perhaps, **the biggest dangers in these regards happens to be that, if any of these terrorists-cum-extremists' groups (i.e. in other words, the non-state actors.) since, quite actively stationed in any of their 'failed or failing' nations, could succeed in laying their hands on any such nuclear-weapons, or even onto any weapon-grade nuclear materials; and then needless to say that, the already many kinds of lurking dangers to the larger humanity do in turn, magnify.** In fact in these specific contexts; we do find that in the overall geopolitical terms, as well as simultaneously having all the more far-reaching and long-term international repercussions the undisputable ground-realties have to be that, the internal systems-cum-mechanisms of governance, and at their various hierarchical-levels of the official operations in these regards, **in the most of such 'failing' or the already 'failed' nations** are not at all in their essentially desired functioning health, and thus in their effectiveness too. Hence, they are found to be politico-bureaucratically stuck in their self-created 'mess' which in turn, perforce make them totally unable to deliver. So it is needless to say that, the larger international community is quite justifiably worried and also very strongly doubting; that **their highly corrupt governments, as well as their highly politicized, and at times also, the no less religiously fanaticized armed-forces are largely unable to measure up to the mark, in protecting their unsafe 'nuclear-arsenals', from the extremely dangerous evil designs of their all kinds of the incurably fixated 'non-state actors'**. These are indeed, extremely dangerous and complex issues of the present times; which the international community can only ignore to their own perhaps nothing else, but inevitable perils.

And hence, this precisely is another, as well as dangerously, the most risky possibility staring into the eyes of entire humanity today. In terms of contemporarily the multiple-aspects of 'military-lethality too, this particularly singular and, the persisting danger in the today's world is a matter of perhaps, the greatest concern for the international community; where undeniably, there are numerous kinds of the readily visible 'soft-targets' as since likely to be perceived by these enemies of the world-peace as well as even, those of the human-civilisations themselves. And hence, the larger international community needs to display much greater geopolitical prudence and thereby get both geostrategic-cum-politico-militarily united, in the futuristically focussed manners to catch-hold of the 'horns of these bulls of the enemies of world-peace and progress', for breaking their 'necks' once for all. Now as we continue to critically delve deeper and deeper into these multiple-aspects, concerning 'future-shock' and ironical military-lethality, and then in addition to all the above, **we quite disturbingly notice another no-less menacing developments consisting of the all sorts of regressive culture of the irrational 'arms-race'; quite recklessly being practiced principally, by those of the already above mentioned 'failed' and 'failing' nations, since especially found to be located in some parts of the continents of Asia and Africa. In these respects, what indeed, is of the much greater worries that, during the present times, these highly volatile regions of South-Asia and the Middle-East happen to be perhaps, the most dangerously unstable 'trouble-spots' in the world**. In these perhaps now, the world's most volatile—Afro-Asian regions—what we find are the two historically glaring socio-cultural-cum-political aspects of the lingering societal failures on the part of peoples inhabiting these regions, barring few exceptions; and those being that, **the peoples are fanatically fixated with an extremely distorted form of a beautiful religion, i.e. Islam**; which so overwhelmingly in a maximum possible manners, get proactive towards most powerfully influencing and controlling all dimensions of their lives; and then also in addition to all these what is adding 'fuel to the fire', is almost the total absence of the any meaningfully vibrant and truly functional form of the some kinds of liberal-democracies in those in fact, the 'failed' or 'failing', and also the so-called 'Islamic' nations. So the recipe for the all sorts of lingering turmoil, flash-points and violence and thereby any sudden surfacing of the perhaps, quite uncontrollable instability happens to be always proactively present in their entire socio-cultural-cum-political canvas, and also in the very psyche of the unfortunate peoples inhabiting these regions. The more one tries to decipher and analyse indeed, the dauntingly complex problems of these highly volatile Afro-Asian regions, more their complexities continue to magnify and thus at times, scaring-away any sincere multi-faceted, multi-national or even some U.N. sponsored international-efforts, to try and develop some 'grips' over them, thereby proceed to help the concerned 'stuck-up' peoples over there somewhat imaginatively, to solve their such menacing problems.

In specifically these multiple-contexts, while critically delving deeper into the mankind's very recent 'living-history' since still multi-dimensionally continuing to be in the making only; we find that among some of those Afro-Asian nations being quite menacingly seen to be so much regressively fixated with the above talked about—**the extremely distorted-cum-subverted forms of 'Islam as a religion'**—the already 'failed' or 'failing' Nation-State of Afghanistan has already set an example as to how to get '***Talibanised***' and thereby, quite irrationally and multiply begin to undo and also, ruthlessly destroy almost all the so-far well-known, and well-accepted tenets-cum-traditions of the otherwise perpetually advancing human-civilisations across the world in that 'Talibanised' territories. Contextually here, it appears perhaps essential to look into as to **what is 'Taliban'** and also; what could be the unending tragic consequences, if '***Talibanism***' is enforced onto the any unfortunate society. In precisely these multiple-contexts, **Neamatollah Nojumi**, who was raised in Afghanistan, was participant in the Mujahideens fight against the Soviet invasion and was a contributor to humanitarian efforts providing medical and other services to those Afghans who have suffered most from the external and internal fighting: the refugees and internally displaced. Hence Nojumi, with his intimate and the first-hand, and indeed with substantially detailed knowledge about all these so much turmoil-cum-violence-soaked land and the unfortunate people of the present-day Afghanistan, in his book, titled *'The Rise of the Taliban in Afghanistan: Mass*

Mobilization, Civil War, and the future of the Region' has given a geopolitically very analytical, and detailed account of all these extremely tragic and also, very far-reaching—a series of the developments and events—which in turn, have caused a ruthless ransacking of the entire socio-cultural-cum-political fabric of the so tradition-bound, as well as the no less naively innocent, peoples of Afghanistan. In fact, this book should be objectively, as well as properly, edited and then translated in the Arabic, Urdu, Persian, Pushtu, and also the Bengali languages; and (*which, no doubt, can indeed be made possible under the U.N. sponsorship*), and distributed free among the peoples of these affected regions, for ensuring the multiply long-terms good of the entire global humanity. In the above referred book (pages 154 and 155) **Neamatollah Nojumi** has indeed, very perceptively written that, "The Taliban enforced numerous law decrees according to their interpretation of *Shari'ah*. To make sure that these decrees were implemented in Afghanistan, they established a religious police. The main objective of the religious police is to carry out what is decided in the court of law of the country and conduct punishment. **In accordance with the Taliban, the *ameer-ul-mumineen* is a political leader who has the legitimate authority over the people living in a territory controlled by his followers. Obeying *Ameer-ul-Mu'mineer* is *fardh* (God's demand and must be practiced), and it is based on the law of God.... Moreover, if the *Ameer* called the people to jihad, it is also *fardh* for all qualified Muslims to follow his order and bear arms against those who are enemies of the *Ameer*....** On September 1996, Wakil Ahmed, the Taliban spokesperson, stated to reporters in Kabul that 'women should not report to work, they should stay home'. In a similar line of order, a source of the Taliban stated:

> **'In the areas under the Taliban government every kind of wickedness and immorality, cruelty, murder, robbery, songs and music, TV, VCR, satellite dish, immodesty [be purdagi], travelling [women] without a mehraum [immediate blood related person], shaving-of or trimming the beard [among male adult], pictures and photographs, interest, have all been totally banned'.**

To implement these decrees, the Taliban formed an organization called *Amre-e-Bil M'arouf Wa Nabi Anil Munkar*, the General Department for the Preservation of Virtue and the elimination of Vice. This organization, headed by Maulwi Qalamuddin, 'has thousands of informers in the army, government ministries and hospitals who monitor foreigners and Western aid agencies.' Most of these informers are teenagers and recent graduates of the Pakistani *madresahs*. The young members of this organization patrol the streets with long sticks, making sure that the people go to mosque at the time of daily prayers, women are covered from head-to-toe with a garment (*bughrah*), and men have not shaved their beards.... **Capital punishment is widely enforced for all cases dealing with manslaughter, and this type of punishment is usually in the form of *qisaas*, through which a male member of the victims' family carries out the execution. Under the Taliban rule, this type of punishment takes place in public, in the presence of the victim's family**. All Afghan families have to pay *zakat* (similar to income tax), and the landowners and farmers have to pay *ushr*; 10 percent of their agricultural products. Under the law, 'gambling, betting, pigeon-flying, dog-racing, sodomy are strictly forbidden. Thieves are subjected to surgical amputation of their hands and arms and adultery is punished by stoning to death in the eyes of the public."

The facts contained in the above quoted paragraphs here, have very precisely and also contextually in a sharply focussed manner been able to highlight **Taliban's extremely demonic socio-cultural-cum-political ideologies**; and if allowed to have their sway over any Islamic-society anywhere unchallenged, one can imagine as to how multiply devastating these could be for the peoples concerned. Further, we had discussed that in any alien territory anywhere and then all the more less so, in any regressively fixated with an extreme form of religious fanaticism; it is nothing but largely a pipe-dream to somehow transplant some modern form of democracy and make the same viably functional, for the any otherwise greater, long-term good of the people concerned. The mankind's very recent episodes in its 'living-history' clearly show us that, as and when any such efforts are made in any particular 'nation', since so irrationally fixated with some extremely distorted forms of 'Islam' and also, especially located in these highly troubled Afro-Asian regions, have indeed been frustratingly unsuccessful as well as, often embarrassingly, the results have been counter-productive too; and then also in addition, have made many of its already existing problems ironically, even worse than before. As we continue to critically delve deeper into the highly bothersome, but also not at all ignorable—these certain so much menacing 'trouble-spots' in today's Afro-Asian Region since perhaps, incurably fixated some extremely distorted forms of 'Islam'—what we clearly notice that, these static socio-cultural-cum-political entities and, especially their largely illiterate rural-masses are found to be psychologically living in a kind of the medieval-age oriented 'time-warp', and thereby pathetically stuck with an overwhelming 'cult of senseless-violence and religious-fanaticism', along with some kinds of the socio-psychological regressions too.

Hence most of the times, any purely in-house, as well as, the no less well-intentioned saner voices etc, very tragically keep falling onto the uncaring 'deaf-ears' of the peoples concerned. As if these are not enough, and so at times; some of these well-meaning and learned internal counselors are ruthlessly done away with, by sinister conspiracies and thereby; if the need be, through 'fatwa' of any readily available **'Talibanayzed' Mullah** too. And so their stories go on and on regardless; in fact ironically, as well as very painfully indeed, staining the mankind's enlarging civilisational-canvas with the human-blood itself. Well, actually all these are certainly not all and hence in addition to the above, there is another no less menacing dimension since, pushing these so much regressively-fixated Islamic-entities into a dark and self-destructive 'abyss'; and that being the always no less demonically lurking dangers of the forcefully thirsting-upon a ruthless military-dictatorship, onto the any concerned helpless and unfortunate peoples; by no one else but, their own already **highly politicized-cum-Talibanayzed military-forces** in turn, duly aided and abated by their extremely callous and corrupt politicians, as well as, their 'Talibanayzed Mullahs' too. Hence these precisely are the kinds of 'Catch-22' situations, into which those Afro-Asian—so much distorting-cum-fanatically Islamic-fixated political entities and their unfortunate Muslim-populace—find themselves helplessly held in their 'vice-like grips'. So-far with no visible avenues to escape to otherwise, the globally much larger world of the freedom and tranquility; and thereby be able to breathe in some truly vibrant and humane environment, to which quite justifiably in any case in the world any human-being always remains entitled in unrestricted manners.

But even if all the above discussed subject-matters, as well as their multiple dimensions taken together still do not constitute the complete scenario in these regards, as there are more additional dimensions to them and so, one of them specifically being the multiple-aspects of the both legal and illegal, all sorts of the **'arms-bazaars'** flourishing both within and outside these highly turmoil-ridden Afro-Asian Islamic regions and hence always being busy in feeding and serving their **'Talibanayzed militias'**, as well as the other **fanatical 'Jihadi elements'**; as per the dictates of their rather incurably 'Talibanayzed' ruling-classes, Mullahs, and their Armed Forces too. These extreme and irrationally violence-loving elements since, active in these regions, and if and whenever possible then, striking in their neighbourhood and also in the outer world too are getting not only their firearms and some kinds of the weapon-platforms, from those 'arms-bazaars' but also perhaps if feasible, some crude varieties of **'chemical and biological agents'** as well, and thus always projecting some extremely dreadful dangers to the entire humanity.

Now how to go about these indeed, unprecedented menaces in today's world; we may recall that, we have since already discussed them in the earlier 'Sub-Chapter' dealing with the multiple-aspects of the always ongoing 'Great-Games', onto the vast, highly dicey, and the multiply-complex arena of the 'international power-politics'. If the multiply-feasible solutions lie there and then at the same time, no less 'rubs' as some kinds of the obstacles lie therein only. But apart from these indeed, very important aspects if we may pause a little and thereby, may freshly delve still deeper with even more geopolitical criticality and then the larger and all the more important questions arise, as to why and in what all manners, these now so dangerously menacing situations began to so demonically fructify in the first place? And precisely in these specific perspectives, as we begin to critically explore into these aspects, we do find that principally nothing else, but shortsighted, as well as naively selfish kinds of motives—aiming at nothing but largely very short-term gains—and today's modern leading powers having been found to be rather hastily indulging in some recently past and the ill-planned 'great-games'; with respect to the some nations located in these highly turmoil-ridden Afro-Asian region. **And at the same time, to perhaps the utter dismay of those ill-conceived initiatives taken by Western-Powers, some of the targeted 'nations' thereby, were also found to be the no less fanatically-fixated with, largely the already earlier mentioned some highly distorted forms of 'Islam'. It is needless to say that, and may even be quite unintentionally on the part of those leading Western-Powers that now the unchecked manifestations of the so much unprecedented, as well as the extremely menacing demons of the Jihadi terrorism and extremism happen to be nothing else, but few of the numerously negative fallouts, consequent to those of the above mentioned—quite naively thought-out, ill-planned, and also the no-less clumsily-executed, some of those very 'great-games'—since launched by them and also, principally led by the today's sole Superpower the USA over there, in the recent past**. In these specifically very far-reaching contexts, and simultaneously as per our the already broadly well-defined perspectives in these regards, we may even contextually opine that, no doubt geopolitically in the fitness of things, it was imperative on the part of sole Superpower: the USA, to lead any such in fact, some 'great-games-oriented' ventures; but at the same time it was no less imperative on the part of USA, both geostrategic-cum-politico-militarily, as well as, while keeping many of the still highly relevant historic-perspectives in the minds to prudently-cum-intimately consult her more experienced close ally, the UK. In fact interestingly in any case, nothing stopped the USA from doing so (*but what else....?*). During the present times, it is understandably true that, now the UK, is no longer any greatly unrivaled 'Imperial-Power of the truly global-reach and consequences' like since, in the already bygone colonial-era; but the intellectually well-gelled historic perceptions and knowledge, extensively based-on the multiple-kinds of actual experiences does always matter and thus, remains valuable in many ways.

(j) The Indian Sovereign Democratic Nation-State: A Vast and Diverse Landmass of Many Contrasts

We may recall that in our earlier fourth chapter itself, we have since critically discussed some of the numerously relevant and, the multi-dimensional aspects of the various human-civilisations and thus studying and also evaluating them specifically, in their overall global geostrategic and politico-military perspectives. And as we have already seen that contextually, we have also covered ancient Harappa-cum-Mohenjodaro, as well as the Indian civilisations; in fact, both are quite integrative as one larger magnificently unique civilisational entity since with its still the ongoing continuity and thereby, continuing to multi-dimensionally flourish in the huge and diverse landmass of the Indian subcontinent.

This is the land of the world's oldest books/treatises like: the Vedas, Puranas, Srimad-Bhagavad, Geeta, Upanishads, Ramayana and Mahabharata, as well as, the so many other remotely ancestral-cum-magnificent literary creations in the uniquely perfect and the world's most ancient language: the Sanskrit on whom indeed, the unendingly devastating and powerful impacts of the dusts of time and the history as well, have not been able to make even any slightest dent in their shining-glory, and their timeless relevance indeed. This is also the land of so many greatly brilliant, learned and outstandingly visionary Rishis, Sages, literary giants, mathematicians, medical (*Ayurvedic* system) experts and, many ancient scientists-cum-thinkers too; like: **Vishwamitra, Vashistha, Agustya Muni, Dronacharya, Aryabhata, Patanjali, Charaka Rishi, Ved-Vyas, Kalidasa, Chanakya, and then equally glorious so many others as well**, who have continued to endeavour till their last breadths and thus indeed, the multi-dimensionally, as well as the timelessly and no less enduringly, enriched the entire humanity in the world.

Again, this is also the land of great warriors, Military-leaders par-excellence, and strategists too since the very antiquity of times and, then no less continuingly to the presently ongoing modern-era periods. These as per their historically passing timelines, have been able to leave their indeed, indelibly magnificent-cum-historic footprints onto the unendingly accumulating dusts of time. Just to contextually cite a few, we may name **Lord Rama and His ancestors, Lord Krishna, Bhism Pitamah, Bheem, Arjuna, Karna, Vikramaditya, Chandragupta Maurya, Emperor Ashoka, Kalinga Kings, Chola and Pandyan Kings and their military-commanders, and so many others all around the subcontinent**. Astonishingly rather, in this subcontinent even the so-called 'mythologies' become rather inseparably integrative with, since continuingly in the making, its 'living-history' itself. This is India and her indeed, enchantingly magnificent: the 'Indian-Civilisation' with its unbroken continuity and, as the numerous centuries have been rolling-by, since the remotely ancient of times; continuing to flourish, undergoing and enduring all sorts of folds, twists and turns and then from time to time, even experiencing so many kinds of the history-making 'clashes of arms'; onto the Indian Subcontinent's expanding civilisational-Canvas itself. As further centuries passed and more and more interactions between the mankind's different civilisations began to mutually occur and thus, their consequent civilisational enrichments too happened. Since the very beginning, the multi-dimensional civilisational-flexibility of the diverse populace of the Indian subcontinent has been such that they have seldom found any difficulty or any kind of the socio-cultural-cum-psychological limitations in welcoming, accepting and then integrating others, as they continued to enter this indeed, quite mesmerizingly magnetic landmass of the Indian subcontinent. **World's oldest and voluminous treatise, *'The Rig-Veda'* is known to say that let all the socio-cultural-cum-mental doors and windows be always kept wide-open and thereby, let the freshly enriching breezes continue to flow-in from all directions, for the larger good of the entire humanity**. Do we need to say anything more in these regards? And as the more and more centuries have passed, besides Hinduism, some the then newly manifested other religions also, took their interestingly momentous, the historic births in this subcontinent itself and then not only flourished but, even peacefully, as well as quite overwhelmingly; moved to the foreign lands, took roots and hence, flourished and got further enriched over there too.

But as we may contextually begin to again peruse them with the same criticality and delve still deeper, we may also paradoxically observe that indeed, all these are not reflecting the complete historic narratives in these regards. It is true that in fact, the very systemic nature of since the perpetually ongoing making of the mankind's 'living-history' happens to be such, that as the dusts of time gradually keep-on accumulating onto the man's expanding civilisational-canvas, sometimes along with history's timelines, many kinds of the 'myths' and fictions' too do continue to get rather unidentifiably intermingled with some relevant historic-facts as well. Hence with respect to the any precise kind of 'the strategically critical-journey' in these regards and certainly all the more so, the same being undertaken in the 'global geostrategic and politico-military perspectives', it always remains no less essential to be rather much more thoughtfully watchful and critically alert so that desirably at any time, the ongoing explorative discussions all the way in any manner; are analytically prevented as much as possible, from going tangential to the contextually relevant historic-facts in these regards. And in these broader perspectives, and then precisely concerning the Indian subcontinent and the so colourful peoples inhabiting therein down the ages as we may delve critically deeper, into the historic-cum-civilisational messages contained in the above paragraphs to our perceptive dismay we find that down the ongoing movement of their historic timelines, that both historically, as well as geopolitically, something of always the multiply paramountcy values; has gone almost totally **'amiss'**, so far as the peoples of this huge landmass and their destiny-makers are concerned. This has been perhaps ironically, the greatest 'negation' on the part of various kinds of the Indian Monarchs/rulers down the ages since the antiquity of times, as well as also unfortunately; for the otherwise so magnificent, the Indian-civilisation too.

And as if, all these are not enough for India and the Indians even during the present high-tech era modern times that very pathetically, as well as paradoxically the same in fact age-old indeed, the multi-dimensional negations in these regards are still persisting regardless, in such a 'foolhardy' manner that a kind of the so arrogantly 'nexus-fixated'—the extremely callous-cum-corrupt political-class and the entrenched bureaucracy—even in today's sovereign democratic India and as expected are refusing to fruitfully learn from the invaluable lessons of the mankind's 'living-history', since always in the making onto its expanding civilisational-canvas in this world. We would contextually be discussing these aspects little more in detail, shortly later in this sub-chapter itself. But here for the time being, we also may contend that the irony or ironies remains that even today in the presently ongoing super high-tech age of the modern-era too, no consistent efforts are being made to indeed achieve the always quite feasible and also, timelessly so valuable—the course-correction-oriented improvements—so that many kinds of the historic turmoil, as well geopolitical embarrassments down the pages of the mankind's 'living-history'; which the peoples inhabiting this subcontinent were subjected to from time to time, by the always unsparing heat and dusts of the onward movements of historic-timelines are prevented further in future quite reassuringly, onto the mankind's expanding civilisational-canvas itself. And what precisely remains **'amiss' in the first place**, which has in fact, quite singularly caused all these, down the thousands of years of since the ongoing making of this Indian subcontinent's highly chequered 'living-history'?

At this precise point, we may again recall that all along our 'strategically critical-journey' and therein, our multiply-ongoing discussions too we in the numerous contexts, have been discussing, as well as emphasising the timeless-cum-universally paramountcy importance of the **multiple-aspects of the 'essential strategic-culture and thoughts'**, for always the larger good of any viable and historically enduring socio-cultural-cum-political entity in this world. Then to our utter dismay we are indeed made to perforce notice that, India since the very ancient of times, has not been able to successfully, as well as quite enduringly-cum-robustly; the multi-dimensionally cultivate, develop, groom, imbibe and then; flexibly-cum-fruitfully place those since always paramountcy: **the multiple-multiple-aspects of 'essential strategic-culture and thoughts'**. And needless to say now, that this precisely is the **'amiss'** in these regards.

Shri Jaswant Singh, himself an old-time Indian Army Officer and then India's veteran Parliamentarian and well-known, and also well-accomplished diplomat in the above multiple-contexts and thereby, while quoting **Stephen Peter Rosen**, from his book, *India and its Armies, Societies and Military Power* and in his earlier referred book *Defending India* (pages 6, 7 and 8) has mentioned that, "Intellectually, strategic culture is related to the specific idea of a Bolshevik 'operational code', as advanced by Lucian Pye, Sidney Verba, and others. Leites' concept of the operational code was subsequently elaborated during the cold war explain different approaches by the Soviet leadership toward nuclear weapons.... The other source of the concept of strategic culture is the more general concept of 'political culture'; which, in turn is the product of 'non-political beliefs'—such as feelings of basic trust in human relations, orientation towards time and the possibility of progress, and the like—[which] can be of overriding importance' in shaping political institutions. These sets of beliefs constitute, in each country, that nation's political culture. Political culture is not uniform across cultural boundaries, [and] even if countries adopt common forms of economic organisation, because [they] would be overlaid on older systems of beliefs; and would not simply replace them. **Particularly as Western colonial patterns of government receded in time and prestige, local 'subjective attitudes, beliefs, and values prevalent among the dominant groups in the society' would increasingly shape the political development.... This Stephen Rosen calls 'a short-hand expression for a mind-set'.... And this 'mind-set' as...societies organized themselves differently...might lead to differences in the amount of military power they could generate. That is why 'whether or not culture, defined as the dominant set of beliefs, matters, for strategy, domestic social structures could still affect the generation of military power.' But whereas an entire society will obviously have distinctive attributes, not all constituents [of it] are expected to contribute, to a formulation of its strategic thought, not even the 'entire political class, only that 'subset' who are 'political-military decision makers'**. (*The meanings contained in the above highlighted lines here; are very important and indeed, especially very relevant now to the modern, sovereign, and democratic India; and her intelligentsia and above all, to her political-class as well.*). [It is] their beliefs and assumptions, which determine 'military behaviour, in particular, the choices concerning decisions to go to war; preferences for offensive, expansionist, or defensive modes of warfare.' Alastair Iain Johnston refers to the assumptions, symbols, myths, and beliefs held by national leaders that affect their perception of available acceptable strategic options, 'a nebulous ideational milieu which limits behavioral choices' and 'which acts to establish long-lasting strategic preferences by formulating concepts of the role and efficacy of the military force in interstate political affairs...the role of domestic politics in shaping the attitudes of a society toward its military and the political views of the military with regard to its host society, all while acknowledging the importance of the internal culture of the professional officer corps, a culture that was not determined by the distribution of domestic political power. The problem [however] with strategic culture tends to lie in its application. In a proper understanding of power, in the ability of a people and society to generate power; thereafter, to have the necessary social will and ability for a full and effective employment of that power.... **How does India rate when we apply these broad criteria?** To start with, the essence of Indian civilisational thought engages itself more with the other-worldly than this. As a great Indian philosopher and sage (**Sri Aurobindo**) expressed early in this century: 'India's central conception is that the External, the spirit...involved and imminent in it and evolving in the material plane by rebirth of the individual...till in mental man it enters the world of ideas and realm of conscious morality, Dharma. India's social system is built upon this conception; her philosophy formulates upon it; her religion is an aspiration of the spiritual consciousness and its fruits; her art and literature have the same upward look; her whole Dharma or law of being is founded upon it. It is her founding of life upon this exalted conception and her urge towards the spiritual and the eternal that constitute the distinct value of her civilisation."

But since the very ancient of times, nowhere the legendary scholars, thinkers, philosophers, as well as the sages and the spiritual Gurus in India have ever discarded, or even marginally minimized—the universally timeless and paramountcy importance of *the multiple-aspects of 'essential strategic-culture and thoughts'* for the India's various ruling-classes, history-makers and, the guardians of her destiny, in any manner whatsoever. When contextually, we closely, as well as the no less critically, go through India's Vedic-Era immortal scriptures and treatises and, the voluminous history/military-history of the ancient India, we find that learned sages and spiritual gurus of the various Monarchs and their royal households, then ruling over their various empires/ Kingdoms, since spread all over the huge Indian Sub-continental landmass were always accorded very high respect and the most exalted positions. But at the same time those learned peers and saints, since always proactively aware and also, knowledgeable of the universal importance of arts and crafts of the warfare and the matters-military; as well as also, mastering the expertise in handling the all kinds of arms and weapon-platforms and thereby; always expected and insisted-upon their ruling Monarchs and the royal princes and princesses, to be knowledgeable and experts in these paramountcy subject-matters. Those peer and sages since centuries during the ages in ancient India, always maintained that if the Monarchs and their military-commanders were first-class in these matters; then their armies would always remain capable of ensuring the overall security of their Kingdoms and thus, the people too would always remain safe and peaceful and, safely go-around with daily chores of their lives. **So interestingly perhaps, the *'historic-rub'* in these regards, indeed lies somewhere else**. Then contextually as the further centuries continued to roll-by, the overwhelmingly far-reaching and thus, the most important point precisely here arises in a very focussed manner as to why, how and what all have multiply been going wrong in India and for her peoples down the line of her 'living-history', since so many centuries? At the first sight itself, all these indeed, may appear to be matters of her great 'civilisational-paradox', and also hence the so magnificent saga of her historicity as well, won't have been rather so embarrassingly 'peppered' with such indeed, very unbecoming civilisational-negations in these regards.

We may recall that in the earlier fourth chapter and, in our broader perspectives when we were critically dealing quite in detail, with the multiple-aspects of the very ancient Indian-civilisation, spanning thousands of years of its unbroken continuity and in differently very relevant contexts therein, we have since also talked about **the very self-centered, all kinds of the greed-driven, callously regressive and indeed, very powerfully-entrenched *Brahmin 'priest-class'* who besides in so many ways, were holding their sinister sway over even the 'core-points' of the psyche of both the rulers, as well as the masses and thereby very selfishly, not allowing them to fruitfully cultivate, multiply-develop, and also flexibly imbibe, the timelessly paramountcy—the multiple-aspects of 'essential strategic-culture and thoughts'—since the very beginning in ancient India. And then quite damagingly, this state of affairs continued later all through the medieval-ages and then lo and behold, and in spite of now our modern democratic Constitution in force with its unquestioned supremacy too; in some manners or other, the modern Indian Nation-State is yet to fully succeed in loosening this selfish and highly corrupt Brahmin priest-class's tentacles over almost all segment of the Indian societies, including the masses in otherwise now, even in the progressive and democratic Indian Nation-State.** ***Now even for a moment here, let us not get confused and thereby quite naively; begin to bracket these pedestrian kind of sinister and—regressively entrenched—'Brahmin priest-class'—with the Indian ancient, as well as later***

eras' so glorious, learned and also multiply-talented: sages and seers—who all since the very advent of human-civilisations on this planet-Earth remained so magnificently, occupied in always multiply-enriching the unique Indian-civilisation; which even now, is so healthily maintaining and nursing its unbroken continuity, onto the mankind's expanding civilisational-canvas. No wonder that since centuries; volumes have been written on the multiply enlightening and learned messages, and the timelessly magnificent teachings of these great Indian seers, saints and peers by the scholars in India and abroad and hence all these indeed, the inexhaustible fountainheads of both the intellectual and also, the spiritual knowledge-pool. Well at the same time, it is also true that the many kinds of very enchanting contributions made by the great scholars, thinkers, philosophers, and the writers from the very ancient of times and also, belonging to the different human-civilisations flourishing in many foreign lands too are equally valuable, as well as magnificently enriching even now, for the mankind all across the world. At this precise point, it would obviously be so appropriate to say that all these unique and magnificent store-houses of human-knowledge and the timeless wisdom—are indeed very vast, as well as multi-dimensionally so complex that—it is impossible for any single individual anywhere in his or her entire lifetime, to even little bit comprehensively delve through all of them. Hence just somewhat attentive glimpses, through even any part of them; then quite overwhelmingly, may make us aware of the colossal enormity of these timeless store-houses of the human knowledge and wisdom all across the world, since the very ancient of times.

In the above multiple-contexts and also, precisely concerning the Indian subcontinent, just to have a feel of the above mentioned colossal enormity in these regards, **Subhendu Bhattacharya**, in his article titled, ***'Vivekananda and Vedanta'*** published in *The Statesman* (Kolkata, dated 07 November 2013), mentions that, "**In norms of concepts and dictionary, the implications of the word 'visionary' are diametrically opposite to that of the expression, 'realist'. The former denotes divine or highly imaginative thoughts, whereas the later is concerned with facts of the mundane world. But this schism between the two did not hold water in Swami Vivekananda's reckoning because his unflinching belief in Vedanta led him to come close to the arena of reality**.... **Sankaracharya**, one of the greatest interpreters of the Vedas and Vedanta, advocated the concept of monism, according to which Brahma, inherent in the soul, is the only reality and the phenomenal world in which we breathe in and breathe out is illusory. The façade is *Maya*. Hence the mundane world is ephemeral. But **Swami Vivekananda**, a modern interpreter of Vedanta, went a step forward to uphold that Brahma is not only inherent, but also transcendent. It is omnipresent and as such all-pervading. According to **Ramakrishna**, Brahma is realised not only in the inner-self of man, but also in natural objects surrounding him. In this respect, Swamiji followed the path of his guru.... Vivekananda is a practical Vedantist. Vedanta is the philosophy of the Upanishad and the quintessence of this thought is to 'know thyself'. In other words, Vedanta teaches us about the existence of Brahma, inherent in the soul, and to reach him is to derive supreme bliss or *ananda*. Brahma is the Supreme Being and is absolute by Nature. He has neither a beginning nor an end. He is both *Advaita* (non-dualist) and *Dvaita* (dualist). He is both *Saguna* (qualified) and *Nirguna* (non-qualified).... He (i.e. Vivekananda) was a realist par excellence. He was not only a visionary or a thinker, but also a practical man, whose philosophy was rooted within the country. The spiritual, cultural and social activities, performed by **Ramakrishna Mission and Math**, bear testimony to his realistic approach to life. Vivekananda's heart was as broad as it was passionate. Thus was he able to expose the stern realities of human life and express his concern for the misery of the downtrodden.... The observation of **Dr. S. Radhakrishnan**, former President of India (Swami Vivekananda Centenary Memorial Volume, Calcutta, 1963) is particularly relevant—'He looked at the millions of who died of poverty and hunger and he said, 'I am the worshipper of *Daridra Narayana*, of the *Narayana* (*i.e. the supreme Lord*) who is located in all the poor people of this world. So long as they are there, how can I content myself with my salvation or with my own attitude? It is my duty to look after them all, the best way to reach God is by the service of man.' (*The same thoughts were the 'spiritual-anchor' to the life-long mission of Mother Teresa as well, recently in India*). The expression, *Daridra Narayana*, coined by his Guru, reflects the sincerity of his service to suffering humanity.... Like **Gautama Buddha, Vivekananda** was passionately touched by the lot of the destitute and the downtrodden. His clarion call to his countrymen for the uplift of the poorest of the poor and the lowest of the low testifies to his love for this segment in unequivocal terms. As he once said: 'forget not the lower classes, the ignorant, the poor, the illiterate, the cobbler, the sweeper are thy flesh and blood, thy brothers. Thou brave one, be bold, take courage, be proud that thou art an Indian, and proudly proclaim I am an Indian, every Indian is my brother.'.... According to Vivekananda, India's development could not be accomplished unless the socio-economic condition of the poor was improved. **Actually the unprivileged and under-privileged were the victims of the Brahminical casteism, a legacy of the past.** (*As we have already observed earlier, this indeed greatly unbecoming civilisational-negation of the Indian societies, since the very ancient of times were in fact, very clandestinely, as well as sinisterly, perpetuated by the age-old and also, very selfishly vested-interests-oriented and entrenched—the* ***'Brahmin Priest-class'****—which is sadly, yet to be completely, as well as effectively, discarded by the Hindus within the Indian societies.*). They could not come close to the mainstream of social life. They were denied the right to worship in a temple. Economically, they were subject to exploitation and torture, inflicted by the British rulers and their agents—the so-called zamindars.... The ideas of Swami Vivekananda need to be followed to make our country a better place to live in."

We have also discussed that during the very early period of the medieval-era itself, the **historically momentous birth of 'Islam'** dawned in the world. Subsequently, in very far-reaching and 'history-making' manners, its advent, propagation, and also territorial expansion too, to begin with, took place in a highly chequered manner, as well as indeed, as a highly momentous-cum-history-making—the religious-cum-socio-cultural revolution—in the then pre-Islamic Arabia and the adjacent lands. And very soon, Islam came into the Indian societies too and as usual, according to the typically unique socio-psychological-cum-cultural features prevailing in the then Indian subcontinent, Islam along with its own many unique attributes, indeed got assimilated and integrated into the vast Indian landmass. **It is also a history now, that in the process different kinds of the 'clashes of arms' too, took place.** But unique to India and her legacies, as the socio-cultural-cum-political assimilation became truly substantial in these regards, Muslim peoples never remained so much a separate entity and hence all across the Indian subcontinent, societies became a homogenous one, and thereby as the timelines kept-on moving ahead both Hindu and Muslim feudal lords-cum-kingdoms had surfaced, along with masses of harmoniously mixed populaces. **And then all the more interestingly, in such a multi-featured and diverse socio-cultural-cum-political scenario in the medieval-India there happened to be even both Hindu and Muslim feudal lords, having some vast majority of the Muslim or Hindu population respectively under their rule. In these historically magnificent socio-cultural-cum-spiritually awakening and enriching phenomena in the Indian subcontinent not only the Hindus, but also the Christian, Muslim, Jain, Buddhist, Sikhs and other seers, saints and the spiritually enlightened souls too began to contribute in these indeed multiply diversifying, and all the more flowering enrichments of the Indian societies, onto the expanding civilisational-canvas of the mankind in this world. All these and even all the many more, in their befitting comprehensiveness, since multiply concerning the**

human-civilisational connotations, as well as in their globally vast perspectives, have indeed been perhaps exclusively unique to the Indian societies and also, to the very ancient antiquity of their magnificently multi-coloured Indian civilisation. In the above multiple-contexts, **Professor Subrata Mukherjee** has mentioned that. "The period between **Rammohan Roy** (1774-1833) and **Swami Vivekananda** (1863-1902) can be described as a phase of cultural churning that led to the Vedantic revival. Except for **Ishwar Chandra Vidyasagar** (1820-91), all the leading intellectuals were attracted to the phenomenon. In the context of the achievements of **Sir William Jones** (1746-94) and his associates in the Asiatic Society, there was great enthusiasm to get to the roots of India's ancient civilisation. Rammohan and Vivekananda revitalized religion and through this medium, created new national consciousness. For Rammohan, their framework was provided by a reinterpretation of the ancient texts in an effort to end such inhuman social customs like *sati*. His work was followed up by Prarthana Samaj, established by **Mahadev Ranade** (1842-1901) in Maharashtra. However, Vivekananda's mission was rather different; he wanted to portray Vedantic principles to oppose Western philosophy and materialism.... Rammohan took up legal issues, deriving his strength from Vedantic literature with interpretations. He accepted the fact that Western religious beliefs were superior to that of the Indian. He believed that it was the lack of scientific knowledge, rather than institutional weakness that led to India's backwardness. (*At the same time it also appears that, this was further aggravated by the no less historic fact that when the time came; due to the peculiarity of her geographical location, India was perforce made to miss the momentous advent of the 'Industrial-Revolution'. We may contextually recall that, we have since already discussed the various aspects of the 'Industrial-Revolution'; as well as its numerous fallouts, in the earlier chapters.*). He was an ardent admirer of the Western liberal, rational and secular tradition. He realised that this needed to be assimilated with the values and cultural traditions of India, which had been distorted by the superstitious beliefs and idolatry. It was this confidence in Indian tradition that made it possible for Rammohan to think of a modern Indian State, based on democratic values and cultural plurality.... His philosophy became more pronounced in the writings of **Keshab Chandra Sen** (1838-84). **Vivekananda** and **Dayananda Saraswati** (1824-83), Keshab argued that while the West could teach us technology, science and practical arts, it could learn religion and spirituality from India. The assimilation of the East with the West became the cornerstone of Keshab's argument, and was further developed by Vivekananda, **Sri Aurobindo** (1872-1950) and **Rabindranath Tagore** (1861-1941). Dayananda and the Arya Samaj, in contrast to the Brahmos and Rammohan, were conservative and aggressive in upholding India's past. Dayananda was not influenced by Western thought. In his reckoning, India's salvation could be attained by the exaltation of the Vedas. The importance accorded to the Vedas appealed to a section of educated Hindus who resented the intellectual and cultural domination of the West.... Unlike Vivekananda, Dayananda appreciated the good qualities of the British, notably the absence of the child marriage and the fact that they chose their partners in marriage. **Furthermore, conscious of the dynamics of democracy, he could easily perceive that one of the important reasons for British ascendancy was the fact that every important policy was discussed in a representative assembly**. Placing his argument in the Indian context, Dayananda put forward the idea of a golden past and also stressed the importance of the values like devotion, purity, patriotism and self-rule. In a sense, his elaboration of the essential principle of a nation followed the outline laid down by **Bankim Chandra Chatterjee** (1838-94) in his Anand math, published in 1882.... Vivekananda's primary focus was on the development of the human character. Rammohan constructed the past to initiate social reforms, but Vivekananda like Dayananda, revived the past to meet the Western challenge. **J.S. Mill**'s *Three Essays on Religion* (1874) that had incorporated **David Hume**'s scepticism and **Herbert Spencer**'s agnosticism influenced him profoundly....

Vivekananda lost his faith in the Indian princes as they were virtually incapable of leading any reform movement. **Even his famous trip to Chicago was more political and economic than religious; his primary motivation was to deal with poverty in India. For him, religion had an instrumentalist value of reform though, like Keshab, he consciously avoided politics**. This is clearly manifest with the establishment of the Ramakrishna Mission which was a spiritual and humanitarian institution, with no links with politics. As the chosen successor of Sri Ramakrishna, he laid stress on vocational education, Vedantic studies and social services, as the three most important aspects of RKM's activities." (Mukherjee, *The Statesman*, 04 December 2013)

And now coming back to some of the politico-military aspects of the historically engaging medieval-era Indian subcontinent and accordingly, as we continue to delve deeper into the India's highly chequered 'living-history' we do also notice that whenever any kind of conflicts took place among those feudal chieftains/ruling Monarchs, for any political, socio-economic-cum-territorial or feudal reasons their clashing feudal armies, consisted of soldiers of the all sorts of mixed castes, creed and religions. Now in these multiple-contexts, we may also need to keep in our minds that even much before the advent of Islam in this world; Christianity had already arrived onto the southern shores of the western side of Indian Peninsula, i.e. on the coast of today's Kerala. **In fact the then, one of the direct disciples of Jesus Christ; St. Thomas had already landed onto the coast of Kerala**. Hence it is true that arrival of the Christianity in Kerala magnificently predates by even few centuries compared to the arrival of the Europeans as traders to begin with, onto the coasts of Gujarat in the western India, and the rest is already a no less highly chequered, as well as the so well-known, India's 'living-history' of those indeed, quite extra-ordinary times. We may recall that, we have since already discussed these historic aspects in some details in their numerously different contexts in the earlier chapters. Hence as since already noted in detail earlier, in a number of differently relevant contexts; when the then medieval-era leading seafaring European colonial-cum-imperial Powers began to arrive one by one, with their numerous kinds of the imperial ambitions and the related greed; they obviously brought along with them, their many kinds of the quite irreconcilable feuds, rivalries, and also their age-old legacies of the extremely violent conflicts and wars, geostrategic-cum-politico-militarily indeed, as their rather inseparable baggage in these regards.

Mankind's 'living-history', since continuously in the making onto its expanding civilisational-canvas in the world has been showing us from the beginning itself that, very many different kinds of the rivalries and thereby various types of the conflicting interests of the intensely competing Powers anywhere, do invariably lead to the violent 'clashes of arms'. Medieval-era European history shows that, the above mentioned the then leading European seafaring imperial-cum-colonial Powers in their home continent in fact, had relentlessly been engaging themselves in the all sorts of rivalry-ridden violent clashes and hence, living with anything but any kind of the 'peaceful coexistence' at all. Past is past and that cannot be wished away, and in any case then subsequently the much later and so momentous—**the 20th century dawn of their 'European-Union'**—as per its perhaps destined timeline, was still quite far away. And obviously therefore, when they began to arrive onto the coastlines of Indian subcontinent; they were not expected to be cooperative or be peaceful among themselves. During those times, **India had a typically lingering reputation of being a kind of the highly attractive 'golden-bird', which happened to be rather readily available for the grab by any adventurous foreign Power,** due to its numerous kinds of the inherent socio-cultural reasons, as well as also, rather quite 'strange' kinds of its geopolitical-cum-strategic negativities and naivety and thus always being unable to act—either economically through the 'crafty' avenues of international trade and commerce or imperially through any situation-based military wherewithal—for the various kinds of the possible benefits and fallouts for own its peoples and also, safeguard its territories.

In whatever perspectives, one may view all these, but these have been the highly chequered historical truths concerning India and the Indians, since the very ancient of times; and hence obviously, always encouraging the age-old practices of the numerous kinds of the alien imperialism-cum-colonialism, getting rather irresistibly attracted towards the Indian subcontinent. Well historically, we have noted that such periodical phenomena in fact, may have begun with the **invasion of Alexander** even before the momentously history-making advent of the Christ era in the world. We may recall that all these historic-aspects, in their numerous contexts, and according to our broader geostrategic-cum-politico-military perspectives and also, as per their civilisational historicity too, have already been discussed in some detail, in the earlier sixth chapter. **Contextually in their historic retrospect, we may not now hesitate to pragmatically accept that during those highly chequered times; and much later consequent to the irreversible decline and passing away of the earlier, once mighty Mughal Empire, the Indian subcontinent became rather geopolitically wide open again for the 'grab', by either of the rival seafaring European Imperial-Powers; as the native Indians then, geostrategic and politico-militarily, were fast sliding towards their rather inevitable fait-accompli. These are also some segments of the India's 'living-history', and so history whether sweet or sour, cannot be denied or wished-away in any manner. But contextually in the same larger perspectives, we may also pragmatically admit that to India's good-luck comparatively; it was much better that finally the British succeeded in colonizing the Indian subcontinent, rather than any other the then rival European Imperial-powers; since having already registered their highly ambitious presence in the subcontinent, with their different kinds of colonial-agendas, greedy ambitions, and also their not so humane imperial legacies**. In these precise contexts and, in order to historically substantiate the mankind's plain civilisational reality we may in fact, just need to have some passing glimpses into the not so palatable—the very recent 'living-history'—of a number of the erstwhile French, Dutch, Spanish, Portuguese, Japanese and any other similar imperial colonies; once located all across the world, in the vast and so varied continents of Asia, Africa, and even the South America (*which is, even now, so commonly known as the Latin America.*). Having said that, we may also historically, must not hesitate to simultaneously add that in fact, it remains equally true that in broader human-civilisational perspectives; all were far from being substantially good and laudable, within the so much imperialism-driven and arrogance-soaked vast and global British-Empire; where even many of the basic tenets of humaneness and civility, were quite frequently being violated at various levels; by the concerned British colonial-masters, in course of their processes of the imperial governance, all across their subjugated territories. Hence, the above highlighted perceptions have only been put here contextually, in some important comparative perspectives.

We know that whenever anybody dares to say all these rather quite irrefutable historic-facts concerning India's past and also, not so palatable some of her geostrategic-cum-politico-military legacies, in such plain and frank manners; it remains obvious that he or she is bound to disturb and also, irreconcilably provoke all sorts of the typically fixated ideas, as well as 'schools of thoughts' since in any case having their different kinds of the intellectually half-baked free-runs, in their modern, sovereign, democratic, Indian nation-State and also no less in addition at the same time, they even refuse to have somewhat historically pragmatic, as well as geopolitically prudent debates in these regards. If for nothing else then, at least for even now in today's—the overall highly challenging geopolitical scenario, as per its multiply numerous geostrategic-cum-politico-military dimensions—essentially towards more proactive and strategically invaluable enlightenments of India's presently upcoming and also, the future generations. **In fact in these specific geopolitical contexts here, we may also say that anything which is finally dead or become totally inert, can never contemporarily teach anything fruitful to any societies or the Nation-State and that's precisely the reason, as to why we have called the mankind's history—the vibrantly 'living-history'—which remains continuously in the making onto its expanding civilisational-canvas in the world and thereby, it is able to always contain many kinds of the invaluable lessons for the mankind's present, as well as, for posterity too.**

So far as the mankind's socio-cultural-cum-political aspects in the world, are concerned, it is true that, man's incredibly vast and the multiply complex 'living-history' is always containing many things of indeed great values; but it all depends upon the well-cultivated, well-groomed, already duly imbibed, as well as the flexibly well-placed—**the multiple-aspects of 'essential strategic-culture and thoughts'**—with respect to any particular sovereign Nation-State, for fruitfully learning the requisite lessons in these regards. And hence precisely here, and onto the mankind's expanding civilisational-canvas, with its simultaneously progressing history-making timelines, in fact lies the universally timeless 'rub', so far as any society or Nation-State is concerned and also, at the specifically given point of its 'living-history' since in any case, continuously in the making too. In precisely these multiple-contexts and so in other words, we may also say that in any case as and when we may critically review and delve deeper into the 'living-history' of any particular Nation-State, we may even be able to multiply gauge and also geopolitically evaluate the prevailing state of its ***multiple-aspects of the 'essential strategic-culture and thoughts'***, at any specifically considered historic timeline and which all in turn, may even convey many things palatable or otherwise, in these regards. Hence the timeless essential strategic prudence, as well as the related imperatives of the present and future well-being of one's nation-State; demand that the leaders, decision-makers, and also the peoples of that Nation-State, do always be visionary and courageous enough, to be able to display their desired objectivity for positively striving with their multiple-synergy towards suitably equipping themselves, for always being capable of fruitfully learning the requisite important lessons, as since elucidated above.

Well, it may be expected that some may ignore these thought-processes as matters of some conjectures only and then, some even may not perhaps hesitate to dismiss them as 'wild and thus untenable'; but all these are matters of not so reasonably perceived opinions, but they too are equally entitled to their opinions. But in any case, if one may pause a little and pay some analytically perceptive attention to them then, one may contextually find some logic therein, for further critical examination of the larger historical-canvas in these regards. And hence thereby precisely, as we may again contextually proceed further with our ongoing deliberations concerning India and the Indians, we may quite interestingly notice that due to the above noted some socio-cultural-cum-politically lingering lacuna and also, socio-psychological limitations, some additional not so desirable dimensions too begin to surface and, which in their typically regressive manners since the very ancient of times have been pulling back the Indian societies and thus preventing them to fruitfully cultivate and proactively imbibe, some very essentially needed paramountcy reforms, as well as socio-cultural uplift and the multi-dimensional progresses. **In precisely these multiple contexts, what we may clearly notice remains the numerously damaging negative-aspects of the rather malignantly lingering connotations of the indeed age-old *'feudalism'*. We may recall that in earlier chapters and in different contexts, we have since already discussed many dimensions-cum-aspects of this, even now lingering 'feudalism' and its fallouts.**

And when the British arrived and achieved their firm footholds onto the Indian-soils, this feudalism could surface in even some differently new forms and textures. In fact with their already well-gelled, prudently proactive, and also, the flexibly well-placed—**the multiple-aspects of 'essential strategic-culture and thoughts'**— the British could in no time, very shrewdly see through even the minutest details of the many dimensions of India's age-old feudalism and also, the highly chequered, as well as the no less complex: her socio-cultural-cum-socio-psychological milieu and thus very craftily proceeded further with the highly advantageous fine-tuning of their quite long-term imperial strategies in the subcontinent. It is also historically true that the British, as per the not so 'visible-cum-audible' dictates of the then prevailing situations and their historic timelines excelled in listening and geopolitically understanding them for their far-reaching imperial benefits and thus indeed, reaping their colossal colonial-harvests in the Indian subcontinent. Viewing all these in their historically broader perspectives, it also remains equally true that these aspects were not limited to just the Indian subcontinent; but they practiced the same in their typical geostrategic-cum-politico-military prudence in their various other far-off imperial colonies too suitably, with indeed their differently well-crafted geopolitical strategies and long-term plans. And hence in all these historic aspects and colonial-era dimensions, lie the 'keys' to their once so enduring imperial success.

(i) Independent India: Democracy and Nehru's Geopolitical Naivety

While contextually highlighting some of the most regressive and lingering aspects of the Indians' 'feudalism', culture of individual-cum-social jealousies, and the corruption etc, **Mr. Pavan K. Varma**, India's Foreign Service officer has indeed very perceptively written that, "**In the Indian tradition the powerful are not expected to be reticent or modest in the projection of their power.** (*Sadly could there be anything more ever, as a kind of rather nauseating negation of the higher-values of socio-cultural civility; in any otherwise indeed, a very great ancient civilisation? And here lies perhaps, the original-roots of the age-old and also, so malignantly lingering 'feudalism' in the Indian societies.*). **In the eighteenth century Lord Wellesley told the East India Company in London that to rule the 'natives', it was essential to build palaces to *awe* them into submission. (*This is irrefutably precise and historically, so visionary too in these regards; and the same remains no less valid even today.*). The Company was persuaded to accept his logic, and the massive residence for the governor came into being in Calcutta**. (*These very interesting historic-aspects, concerning the British colonial-era India; have since been also contextually, very vividly discussed by* ***William Dalrymple*** *too, in his already earlier referred book: the 'White Mughals'.*). More than fifty years after the British left, the imposing palace fulfills more or less the same purpose. The governor of West Bengal lives in isolated splendour in the sprawling estate, maintained by 168 underlings.... **The departure of the British effected a transfer of power. A transfer of the *paraphernalia* of power accompanied the transfer of power**. (***And hence in the Indian subcontinent, the erstwhile 'white sahibs' were so naively-cum-comically replaced by the native 'brown sahibs'; as a great betrayal of the impoverished and oppressed indigenous population. And on top of it, the political leadership led by Nehru, thought that nothing had gone wrong.***). This was not an uncommon occurrence. Colonized elites all over the world, emulate the pomp and ceremony of their oppressors, once rid of them. But the Indian elite could have been different. The man who got them independence, and whom they chose to call the **Father of the Nation**, shunned the trappings of power. He did not live in government palaces. He travelled in the lowest class in trains. He celebrated his austerity, wearing little more than a loincloth. He wanted the massive Viceregal Palace in New Delhi to be converted into hospital. (*Well now in retrospect, we may even say that this too, happened* *to be the Mahatma's one of some quite 'kinked' ideas. No wonder that on learning this, the outgoing last British Governor General, Admiral Mountbatten had commented that, 'the mahatma should know the design and layout of a large hospital. Anyway, this is nothing but a very trivial matter; and what the Mahatma really meant was indeed quite well-intentioned and also very far-reaching, for the then already extremely poverty-ridden newly independent Indian-Nation; and hence in India's highest national interests as a sovereign, modern, and fast-developing democratic Republic; very painfully now, one may not hesitate to say that—**the extremely corruption-ridden, selfish and arrogant: the politicians-bureaucrats nexus**—have quite shamelessly betrayed the Indians and their Motherland.).* **He exhorted the members of the new Indian government to live, and rule, with humility. Undoubtedly, his examples were very difficult to follow. Even those who believed in him could hardly be expected to be as austere, or as utopian, or as self-denying as him. But, considering the massive following he commanded, why were his values betrayed so easily and almost immediately?** (*Well very interestingly, Pavan Varma himself, has very perceptively indeed, tried to unravel this 'mystery' in his following quoted paragraphs.*).... **Kapila Vatsyayana**, who has spent a lifetime studying India's cultural traditions, told me once in all seriousness that the only enduring emotion (*sathayi bhava*) among Indians is envy (*irshya*). Even educated families routinely succumb to the need to *'utaro nazar'*. A handful of chillies sprinkled with salt are waved around the victim of the ubiquitous eye. The chillies are then thrown into a fire. If the smoke does not hit the eyes or cause one to cough, it is proof if proof were needed of the need to ward off the evil eye.... An instinctive suspicion pervades the Indian social milieu. The Mahabharata is a labyrinthine epic about human deceit and betrayal, but even the other great epic, the Ramayana, which narrates a more simple and straightforward theme of the triumph of good over evil, 'is full of suspicion and doubts—every character and virtue, even the chastity of Sita and the fidelity of Laxman, are tested in the crucible of doubt. Altruism, the ability of an individual to act in the public good without a self-serving ulterior motive, is deeply suspect. This ambience of distrust creates two opposed but coexisting worlds, one of deference, the other of hostility. Deference exhibits itself when the gravitational pull of the centre of power is strong; hostility surfaces when this pull weakens; both coexist when the situation is in balance.... The dialectic of deference and hostility explains why a person whose power is ascendant will always find enough Indians to salute him. Conversely, a person who falls on bad times will be deserted without much ado by those who happily fawned on him. This is not an uncommon trait in other societies as well, but in Indians such behaviour is vivid and extreme. **A rising star is greeted with disproportionate adulation, a fallen hero is condemned with unjustified vehemence.... Corruption, is of course, not unique to India. What is unique is its acceptance, and creative ways in which it is sustained**. Indians do not subscribe to antiseptic definitions of rectitude, as are common in the Scandinavian countries. Their understanding of right and wrong is far more related to efficacy that to absolutist notions of morality. An act is right if it yields the desired end; it is wrong if it does not. **To pay *'chai pani'* (*i.e. for 'proverbial costs for some tea and snacks.*) costs to an official flunkey is right if it smoothens the way to the desired goal. The payment of baksheesh is matter of the right investment, not morality**. No invocation to **Lakshmi** (***i.e. Goddess of wealth and money***) emphasizes the importance of making money only by conventionally legitimate ways.... **And here there is little doubt the areas of excellence that have confidently assimilated the influence of the West are the exceptions, not the rule. The rule is a vast imitativeness that has dulled the pursuit of excellence and creativity, and made common a debilitating acceptance of lower standards**. When people think they cannot be as good as someone else, or spend their energies trying to be like someone else, they usually end up by being much less that what they can be. A nation that *internalizes* a sense of inferiority begins to accept inferior standards for itself. If, as Murthy (***i.e. Narayana Murthy***) says, there not a single field where India can say it is best in the world, it is because Indians do not believe they are capable of being best in the world. In the barren stretch between ability and expectation, between what Indians can do and what they have reconciled themselves to doing, there is a mutilating mimicry, which would be comical if it were there not a symptom of something far more pernicious eating away at the identity and genius of a people.... **The reasons why very few Indians working in India have produced world-class products is because Indian society encourages status quoism and tolerates mediocrity. When caste and regional loyalties are more important than merit, when the use of money can determine recognition, and when networking and nepotism overshadow real achievement, there is no incentive to excel.** Over time, the habit *not to excel* becomes a Pavlovian reflex; only those are rewarded who have extraneous factors in their favour. Mediocrity becomes the means to preserve the status quo. Radical departures that introduce the new coordinates to measure achievement are anathema to the sanctified inertia of such a milieu. Old ways of thinking continue to influence modern aspects of behaviour. **If the dominant goal remains not the pursuit of excellence but the pursuit of status and power, more and more engineers and doctors will—as has been happening for the last several years—junk years of specialization and join the Indian administrative Service. (*What could be more glaring wastage of the so much valuable young technical-cum-scientific specialists, in whose education and training, the nation has spent colossal sums of money. And the reasons behind all these are also commonly well-known; so much for the modern India's, much talked-about, 'national-character.*). If material wealth motivates in exclusion of the innovation that should sustain it, entrepreneurs will not invest in Research and Development. Not surprisingly, private industry accounts for only 15 percent of the nation's total spending on R&D**, prompting **P.V. Indresan** to contemptuously say: 'Indian industrialists are nothing but traders'. A nation will glorify the short cuts if its vision is limited to the system as constituted. Only a vision that has the incentive to transcend it can motivate a different kind of behaviour. When people live only for immediate returns on traditional goals, society cannot nurture the excellence that creates innovators... Among the major civilisations, India is uniquely unassertive towards others, and as uniquely forgiving. In 1948, while repulsing Pakistani aggression in the Valley (*i.e. Kashmir Valley*), the Indian army could have taken all of Kashmir, but **Nehru** stopped **General Thimayya** in his tracks, at the urging of Mountbatten. When China attacked India in 1962, the armed forces—and Nehru—were taken by surprise. During the 1965 war against Pakistan, the Indian army victorious, but Prime Minister Lal Bahadur Shastri generously returned the strategic Haji Pir to the enemy. In the 1971 war, a magnanimous Mrs. Gandhi handed back almost all the military gains to Pakistan. India could have become a nuclear weapons power in 1974, but chose not to. There is no other example of a nation that was that has voluntarily renounced its proved nuclear capability. India opted to become a nuclear power in 1998, almost quarter of a century after it had demonstrated that it could become one.... **In no other recently independent country does the army have such a small role in the power structure**. (*Well, there cannot be any more glaring proof, as to why the so much vested-interests-oriented, and regressively entrenched—the politician-cum-bureaucratic classes—have not even cared to pay some duty-bound sincere attentions to the always paramountcy* ***multiple-aspects of 'essential strategic-culture and thoughts'*** *in fact, in India's highest national-interests. And here lies the keys, as to why and how—the independent India's always naively clueless political leadership— have been so callously squandering away the invaluable military-gains; achieved by the Indian armed-forces, with their gallant bravery, well as their bloods and sweats.*). **If history is a guide and the Indian temperament an indicator, there**

is every reason to believe that modern India will be a responsible nuclear power, in fact more than those who consider themselves to be above scrutiny in such matters (*yes, indeed.*)." (Varma, 2004/5, pp. 18, 36, 37, 76, 134, 135, 167 & 168)

Well historically in retrospect, it won't be fair to 'bracket' **Lal Bahadur Shastri** with respect to precisely the 1965 India-Pakistan conflict, along with the other Indian Prime Ministers concerning their differently historic episodes, in similar manners, pertaining to the independent India's overall recent politico-military history of the last 65 years or so. As Shastri's case was indeed substantially different in terms of many the then prevailing geopolitical realities, so much callously neglected by Nehru earlier. Further, **Shastri** never naively bothered Pakistan; as **Nehru** so unimaginatively did to China, prior to the 1962 conflict and that too, without even caring to ensure any tangible military-cum-logistical preparations of the Indian armed forces in advance, and so the 1962 resulted in an embarrassingly avoidable 'fiasco'. **What to talk of 1962, even the earlier 1947-48 'Kashmir-War' too had not been handled by Nehru, in any desirable strategically prudent manners.** We will discuss more about the indeed multiple-dimensions, as well as numerous connotations concerning the 1947-48 war and its still continuing fallouts little later, in this sub-chapter itself. Well good development was that immediately after 1962 'military-setback', a state of internal emergency was declared and thereby, feverish military preparations were already in force and the same, indeed came very handy for Shastri during the 1965 conflict on India's western borders. And further, unlike 1962, Shastri acted very decisively and with requisite politico-military courage and moreover; he was quite open and transparently sincere to his military commanders and, had no bias or prejudice against India's military-men, much like **Sardar Patel** during 1947-48 Kashmir war or otherwise, during his sadly very short spell as India's towering Deputy Prime Minister. But historically those years, being the times of intense cold-war between the then world's twin Superpowers, the USA and USSR and also Pakistan was already a member of the SEATO and SENTO, so it was getting enough favours from the USA in many ways. Those were also the years of relentlessly ruthless, as well as intensely competitive—saga of the all kinds of 'great-games' between the world's these two Superpowers— naturally being regardless of any consequences to the any so called 'third-world nations. And Shastri helplessly got trapped in the cobweb of these twin Superpowers' some 'great-games' in South Asia, as the USSR was then very keen to 'broker peace' between the two South Asian warring nations, with its own larger Soviet geopolitical intentions. **Shastri never disregarded his top-level crucial military advises, but he was cornered in Moscow since those were the times typically not in India's favour geopolitically and all these indeed came very heavy on the then Indian Prime Minister, as he was acutely aware of all these adversities resulting in a great tragedy for India. And needless to say that finally he didn't return from Moscow, but only his body duly wrapped in the Indian tri-colour came to New Delhi; to be received by the then deeply shocked and shattered, the entire Indian-Nation**.

Any nation to begin with, if she is economically, industrially, as well as geostrategic-cum-politico-militarily, considerably on a very 'weak-wicket' then her top-level political leadership is never supposed to 'play any big' and hence, not expected to project itself as a some sort of important 'world-leader'; since all these imperatives get always drawn from the very fundamentals of the multiple aspects of international geopolitics and also the otherwise, relentlessly ongoing 'great-games' onto the 'highly complex and dicey arena of international power-politics'. **Hence with respect to the quite differently multiple-aspects of the then newly independent and also deeply crisis-ridden, as well as fledgling democratic Indian-Nation in these regards it appears reasonably clear that Nehru, then as the first Indian Prime Minister, perhaps never cared for**

these indeed ruthless but also, timelessly paramountcy geopolitical realities with due statesmanlike seriousness and geostrategic-cum-politico-military pragmatism, while presiding over destiny of the sovereignty of India. It was a very crucial 'dark-side' of his, otherwise a very charismatic personality as well as, indeed his history-making top-level national leadership in India. It remains true that man, who wears the 'crown', must always be capable and prepared to endure and also successfully 'handle the inherent thorns in the crown'; with due multiply visionary prudence, so that the vital national-interests are always taken care of successfully. **There can't be any two-ways about it, as in international geopolitical-arena; there are no permanent friends or enemies, but only one's paramountcy national-interests. Hence it has been correctly said that, anyone who refuses to learn from history, is condemned to repeat the same and that too if possible, again and again**. It was not that Nehru lacked diplomatic acumen, since it was he who could succeed in concluding the **'The Indus Waters Treaty'**, with otherwise largely 'unpredictable' Pakistan in the year 1960 itself. But what he very sadly lacked was his rather inescapably mandatory interest and seriousness in the various crucial aspects of the 'matters-military', as sovereign democratic India's first Prime Minister, and hence his limitations in these regards, did bring some avoidable losses and also geopolitical and military embarrassments to India; and now in historic perspectives, for all these he had no one but himself to blame. And as since we have been repeatedly emphasising, here too comes indeed the unassailable paramountcy of the **multiple-aspects of 'essential strategic-culture and thoughts'**; which sadly in India since centuries, have never been multi-dimensionally attended-to and taken care of adequately.

Field Marshal Ayub Khan too of Pakistan like almost all the Pakistani leaders, diplomats, intellectuals, as well as military men; remained typically fixated with already absurd and defunct 'Two-Nation theory based Hindu-Muslim divide' and in addition also, with geopolitically untenable Pakistani stand on Jammu and Kashmir and so on; and accordingly, he has not hesitated to harp on them in his political autobiography 'Friends Not Masters' as well. Anyway, he too was entitled to his opinion. But certain matters have indeed been brought out by him in his book interestingly, in somewhat reasonably objective manners. Hence needless to say that strategic prudence always demands, that the writings of one's adversaries too, should never be ignored.

Ayub Khan has written that, "The Indus Waters Treaty was based on the division of the rivers, according to which, after a transitional period of ten years, extendable at Pakistan's request up to thirteen years, the three eastern rivers—Ravi, Beas, and Sutlej—will be allocated exclusively to India, while the waters of three western rivers Indus, Jhelum, and Chenab—will be available exclusively for Pakistan, except for limited uses by India on upstream areas in Indian—occupied Kashmir, eastern Punjab, and Himachal Pradesh.... The Indus works programme will be the largest of its kind to be undertaken anywhere in the world and will cost about 1,070 million dollars, of which about 870 million dollars will be spent on the works in Pakistan.... During the course of protracted negotiations, it had become apparent that the cost of financing the system of work in Pakistan and India, to which the two governments had agreed as one of the features of an acceptable settlement, was far beyond the capacity of these two countries. The World Bank, therefore, set up an Indus Basin Development Fund to finance the whole programme. India undertook to contribute to the Fund about 174 million dollars, and the cost of works in Pakistan will be financed out by the Indus Basin Development Fund.... Details of agreement show that the division of the total waters of the Indus system under the Treaty be in the proportion of 80 percent for Pakistan and 20 percent for India. The Treaty was signed in Karachi on 19 September 1960 by me, Mr. Nehru, and the World Bank Vice President, Mr. Illif.... Later, while speaking to the Press, I suggested that there was 'need for reappraisal, for forgetting and forgiving, and for a more realistic, rational, and sensible relationship with each other.... I did not get the impression that Pundit Nehru was extraordinarily pleased to see me, but he was quit moved by some of the suggestions I made to him. He struck me as a tired man, though he still had a reserve to fight and political acumen. I missed the idealism or starry-eyed thinking often attributed to him. I told him I considered that relations between India and Pakistan had been dictated by drift rather than any rational design. The reason, I thought, was that neither side had drawn up any plan for neighbourliness. People on both sides thought that all social and economic ill were due to the British and once they left everything in the garden would be lovely. Event proved that they were not entirely right, though a good deal of our misfortunes were due to British occupation. Much distress and hatred was generated and people on both countries began to believe that they would have nothing further to do with each other. I felt it was not too late to formulate a plan to put our relationship on a rational and sensible basis.... **In military thinking the crucial factor was 'capability'. If a country had decisive military capability it could always change its intensions and act aggressively**. I reminded him of the concentration of the Indian Army against us in 1950 and again in 1951 when there was hardly any military cause to do so. (*What could be any more myopic views and opinions in these regards? Ayub Khan, while talking these so called 'tall perceptive', very conveniently chose to forget or ignore the hard historic fact that, even immediately after the Pakistan's so turmoil-ridden birth on 14th August 1947 and, very hurriedly in October 1947 only; it so brazenly decided to invade Jammu and Kashmir, with many nefarious and clandestine designs—and the rest is history. So how could India rely on the Pakistani leadership, and thereby India always has every conceivable and sound geostrategic-cum-politico-military reasons to deploy reassuringly requisite military wherewithal in her strategic State of Jammu and Kashmir. Hence, the Pakistani Field Marshal had no rationale for any grudge on this point.*).... On 12 October 1962 Mr. Nehru announced that he had instructed the Indian Army to 'throw the Chinese' out of disputed areas. This announcement was described by the New York *Herald Tribune* (15 October) as 'tantamount to a formal declaration of war', while the British newspaper, the *Guardian*, called it an 'ultimatum'. That the Indians fired the first shot was suggested by a subsequent report in the *New York Times* of 19 April 1963, which mentioned that 'General Maxwell D. Taylor, Chairman of the Joint Chiefs of Staff, indicated in secret Congressional testimony made public here today that India might have started the border fight with Communist China'. On 20 October 1962, fighting flared up between China and India at a number of points along their disputed border in Ladakh (Kashmir) and in the North-East Frontier Agency (NEFA) area lying east of the state of Bhutan.... What happened subsequently is, however, fairly generally known." (Khan, 1967, pp. 111, 112, 123, 124, 133 & 134)

And hence the much larger question concerning the India's overall well-being arises as to why the **Indian 'Politicians-cum-bureaucrats nexus'** are still stubbornly refusing to heed to the crucial 'wake-up calls' and thereby, pay their duty-bound fruitful and proactive attentions to the always **paramountcy multiple-aspects of the 'essential strategic-culture and thoughts'**? They must realise without any further delay and also, by discarding their typically fixated arrogant callousness that, this is in no way any kind of a trivial or small matter, not worth bothering about so much for giving them any serious thoughts; since these would prevent India's all sorts of periodical 'fiasco' in the **ruthless arena of the 'international power-politics-oriented great-games' further**; since India can in no way remain unconcerned or disconnected in any manner, as these would always perforce keep affecting India's supreme national-interests in many uncontrollable ways. **At the same time, all these would also ensure that in India's unquestioned democratic polity, her military too always needs be given its proper station within her overall gamut of the power-structure and policy formulations. If and also as it has been happening regardless, these**

crucial aspects are not sincerely attended-to in the overall larger paramountcy national-interests and the entire top-level Indian leadership still so regressively keeps on refusing to learn from history and then tragically, it would remain condemned to keep repeating the same and that too, at the greater and greater otherwise avoidable costs to the entire Indian-Nation itself. We would contextually be coming to all these since no less pressing issues, in greater details little later.

Needless to say that all these factors are always crucial to India's long-term well-being, and also in her multiply supreme national-interests, and therefore it remains unquestionably mandatory on the part of her top-level leadership to remain proactively, as well as ceaselessly focussed onto them, in India's overall best national-interests. At this precise juncture of our ongoing discussions in these regards, if we pause a little and again may care to closely read in between the lines in the little earlier quoted paragraphs from Pavan Varma's book, then we do notice indeed with a great concern, that since even remotely ancient of times India is suffering from a number of rather quite unbecoming *'kinks and twists'*, in the socio-cultural-cum-psychological make-up of the Indian peoples, which have been so damagingly causing many holes into her expanding civilisational-canvas, her security concepts, as well as her military-armour over the ages, and still even today, all these are so irrationally being accepted by India's leaderships and the peoples. We may also contextually recall that, all these subject-matters have since been discussed in the earlier chapters as well. In the overall human-civilisational terms, so far as India is concerned, it is an amazingly strange phenomenon, which in any case perhaps, defies any possible rationale and all the more so when viewed in terms of the mankind's 'living-history', when one visualizes India as a great ancient civilisation with its numerously enchanting facets as well as, with its no less amazing continuity too in the world. In global perspectives and onto the mankind's much larger expanding civilisational-canvas and also in terms of their overall historically timelines-based, since always altering geopolitical considerations, it appears to be rather a paradox of paradoxes, that even today the modern, sovereign, and democratic India's highest-level leaderships and decision-makers are not caring to fruitfully look into these lingering lacunae for initiating corrective-actions in comprehensive manners, for the larger good and future, of the entire Indian-Nation.

Precisely in the above multiple-contexts, as well as in their comprehensively broader perspectives in these regards; it is indeed universally true that always the well-cultivated, well-developed, well-groomed, well-imbibed, as well as the no less flexibly well-placed—**the multiple-aspects of 'essential strategic-culture and thoughts'—do remain the sole and decisively supreme 'keys and also the multi-dimensional instruments' for developing and fine-tuning of the any otherwise 'worthwhile' nation's multi-dimensional-cum-proactive—the geopolitical skills and acumen always being paramountcy—in looking after its overall both short and long-terms highest national-interests**; in the highly challenging, so much 'slippery' and also, the almost perpetually turmoil-ridden—'arena of the all sorts of great-games of international power-politics—since perpetually being played so ironically, and also regardless of any otherwise larger humane considerations, in the world. These are the plain and global, geopolitical 'ground-realities' in which the comity of nations are always perforce made to survive, as well as, strive to fulfill their aspirations, in the world. In other words, we may also say that in any case no nation can totally get out of the above mentioned 'arena of the otherwise continuingly ongoing the multiple-kinds of the ruthless great-games of international power-politics' and thus, can never remain isolated, unaffected or uninfluenced in the world. Hence in these multiple perspectives, as well as in the overall geostrategic-cum-politico-military considerations, we may also say that precisely the above discussed, and the apparently unseen, but always powerfully proactive, the multiple-aspects and dimensions of the 'great-games', since relentlessly being played in that very 'arena of international power-politics', being found to be the 'fulcrum', around which the so vast and highly complex, the mankind's 'living-history' revolves in its unending phenomena of the making, onto the man's expanding civilisational-canvas in the world. We may also recall that in one of the earlier sub-chapters here and hence, in the variously different but also the broadly allied contexts; we have since already very critically delved in much greater details, upon the numerous aspects-cum-dimensions of the 'great-games' in these regards.

So in these multiply broader perspectives, what we may be able to visualise that precisely since the already earlier discussed in numerously different contexts that those very **'multiple-aspects of the essential strategic-culture and thoughts'** only, can provide the requisite vision to any concerned nation's highest-level political, military, and bureaucratic leaderships for being able to successfully sail-through the murky, as well as the dangerously dicey geopolitical 'waters' in the above mentioned 'international arena' while very prudently looking after that nation's supreme national-interests and hence, be able to also successfully face the any challenges in the world. And as we may contextually review with a desired criticality in these regards—the India's 'living-history' since the ancient of times—we do observe that, as per the above discussed 'conceptual-framework' of these very relevant and important parameters; we are indeed rather compelled to notice that precisely in these respects, India has always sadly been found to be faltering and thus landing into the various kinds of the geopolitically quite disadvantageous and also, embarrassing logjam from time to time, in the 'very dicey, as well as the no less always extremely turmoil-ridden arena of the international power-politics'. We may recall that in the earlier chapters and then also in the different contexts as well, we have since perceptively observed that ***concerning any nation in this world, 'there are no permanent friends or enemies, but only one's permanent national interests'***. And hence any nation-State and also, at any given point of its 'living-history'; if it is geopolitically callous and unimaginative enough in these regards then, it may do so to its own peril or at least, perforce leading it towards landing into some fairly serious disadvantageous situations.

In these multiply paramountcy perspectives and so far as the modern, sovereign and democratic Indian Nation-State is concerned just to conceptually substantiate the above discussed precise perceptions, we may consider the India's just around half a century old and very embarrassing 'fiasco', with respect to 1962 India-China border conflict; which are still at many a times, being discussed and analysed in the Indian academic and strategic circles in many geopolitically explorative ways. Hence, particularly with respect to the multiple-aspects of larger and since always the pressing overall gamut of the otherwise no less paramountcy 'matters-military' it would rather be very relevant, as well as perceptively no less important, to also contextually quote some paragraphs from the writings of at least few of the independent India's well-known and learned erstwhile military personalities; for putting these crucial subject-matters in their proper perspectives, for the intended larger wellbeing of the modern, sovereign and democratic Indian-Nation. At the same time and also particularly at this precise juncture of our ongoing discussions; and hence unlike the more developed, experienced, and matured Western democracies in the presently ongoing super high-tech and fast-changing modern-era, so far as specifically the modern, sovereign and democratic Indian nation-State is concerned we may not hesitate to even admit with due objectivity, that barring indeed not very many experts, authors and thinkers among both the civil, as well as military communities, people in general are ironically, not found to be equally and seriously concerned and interested in the thoughtful writings and also no less valuably critical-analyses always aimed at serving the India's best national-interests, and also for both her present-cum-future wellbeing since coming from mutually the other sides of their respective domains, in course of their intellectual pursuits in these regards. Needless to say that all these otherwise largely naïve 'fixations', do create many kinds of rather the easily avoidable limitations in their above mentioned intellectual pursuits; and at the same time in contextually larger considerations, such limitations do also create some serious hurdles in the multiple-paths towards cultivating, developing, grooming, as well as flexibly imbibing the indeed, universally paramountcy—**the multiple-aspects of 'essential strategic-culture and thoughts'**—among almost the all strata of the modern Indian societies.

Concerning the 1962 India-China border-war; **Shastri Ramachandran**, in his a 'book-review' article he mentions that, "Prime Minister Jawaharlal Nehru 'shut the door to negotiations on the (India-China) boundary on July 1, 1954', according to just released book by **AG Noorani** based on archival research and hitherto unpublished material.... And this—Nehru's refusal to negotiate and the 1960 rebuff to Chou En-Lai when he was visiting and appeared ready to settle the issue—may well have sowed the seeds of the 1962 India-China war.... The important and explicit directive, from Nehru, in the 17-para memorandum, cited by Noorani in his book, says, 'Both as flowing from our policy and as a consequence of our Agreement with China, this frontier should be considered firm and definite one which is not open to discussion with anybody (*Wah.*). There may be very minor points of discussion. Even these should not be raised by us. (*Look at this, what could be more geopolitically naïve, as well as so undiplomatic and that too; on the part of a nation, which the then militarily happened to hardly anything more than largely a 'pushover one', especially at those remote and the extremely harsh high altitude border areas. Again courtesy: the then Nehru government only.*).... Noorani, an expert on legal and constitutional issues, known for his study of the boundary issue, writes in the book, 'Indi-China Boundary Problem, 1946-1947: History and Diplomacy', that 'India unilaterally revised its official map. The legend 'boundary undefined' in the western (Kashmir) and middle sectors (Uttar Pradesh) in the official maps of 1948 and 1950 were dropped in the new map of 1954. A firm clear line was shown instead'.... The author says that Nehru's directive of July 1, 1954 was apparently in pursuance of a decision taken on March 24, 1953 to formulate a new line for the boundary. 'It was a fateful decision, old maps were burnt. On former Foreign Secretary told this writer how, as a junior official, he himself was obliged to participate in this famous 'exercise'.... Describing as 'historically untrue' every one of the statements of Nehru in his letter to Chinese Premier Zhou En-Lai of March 22, 1959, Noorani states that as late as 1950, Indian maps showed the boundary as undefined.... The unpublished archival material in Chapter 12 titled 'Aftermath', quotes extensively from the 17-page memo and says a 'divided Cabinet, an irresistible opposition, an uninformed press and a restive Parliament, all fed on, bad history held Nehru a Hostage.' (*Well in retrospect and indeed historically, it is clear that his all these naivety, as well as his diplomatic arrogance too, were entirely making of the Prime Minister Nehru himself. In fact, he even quite arrogantly refused to sincerely listen to the saner and professionally sound advices of the then few competent and selflessly patriotic generals and other thoughtful geopolitical experts. So much and much more historically shows the dangerous 'cracks' in his visionary 'grip', over the many larger geostrategic-cum-politico-military compulsions of the then newly independent Indian Nation-State.*).... **The conclusion is hard to resist that there was a total disconnect between the facts of history and India's policy on the boundary problem and later boundary dispute. Its diplomacy became inflexible because it espoused a policy which barred a give and take**. Each one of the propositions stated earlier in Chapter XI was flouted the 1842 Treaty; and undefined boundary; and, worst of all, an impressible recourse to unilateral change of frontiers'.... This, is a dispute pre-eminently susceptible to a fair solution; for, each had its vital nonnegotiable interest securely under its control. India had the McMahon Line while China had the Xinjiang-Tibet road across the Aksai Chin in Ladakh'.... 'Zhou En-Lai was all too ready to accept such a solution during his visit to New Delhi in April 1960. He was rebuffed. **China proceeded to practice his own brand of unilateralism, sanctifying territorial gains won by armed force'**.... There was nothing inevitable about this impasse. A settlement was possible at the summit in New Delhi in April 1960, despite the fact that public opinion had been ignited over the armed clashes in Longju and the Kongka pass in 1959.... **"But history was cornered and it took its revenge; paving the way to a wild, territorial play of military might and the politics of power to shape a border dispute inherently and pre-eminently susceptible to compromise. The diplomatic consequences of the deepening rift between India and China are incalculable; especially in India's relations with its other neighbours, particularly Pakistan."** (Ramachandran, *Times of India*, 20 December 2010)

In historic, as well as the no less contemporary perspectives concerning the modern, sovereign and democratic Indian Nation-State the above quoted paragraphs, quite perceptively and also no less irrefutably, convey many seriously disturbing facts, with respects to her overall governance, not so laudable conduct of her foreign policies; and at the same time, so callous, naïve, arrogant, and also acutely self-damaging disregard and geopolitically persistent neglect of the India's otherwise unassailably paramountcy multiple-aspects of the **'matters-military'** in all sorts of manners, prior to the 1962 India-China border clashes. Concerning particularly the modern India, it is a very sad historic saga, which still keeps bringing some 'lumps in the throat' of any right thinking patriotic Indian today and thereby, it is needless to say that all these historically unsettling and not at all any palatable episodes, do historically, as well as undeniably put Nehru in the 'dock' in the annals of the modern independent India's contemporary history since continuously in the making, onto her expanding civilisational-canvas in the modern world. **Though it remains a plain fact that after the Indian independence in 1947 and also, prior to the so unfortunate India-China border-clashes in the year 1962 the overall national political-level management of India's 'matters-military' indeed happened to be almost in total neglect and thereby, all these not only resulted in very tragic kinds of the policy-paralyses in the Indian national-defence spheres but also the otherwise always mandatorily ongoing nation's strategic-level military-cum-logistically multi-dimensional preparations too, remained almost totally derailed.** The recent histories regarding all these otherwise indeed, unpardonable neglect of the then fledgling Indian democracy's overall gamut of the 'matters-military', are now fairly well-known and also lot many analytical, as well as historically highly valuable writings by the noted experts from different disciplines of learning have since already appeared during the last half a century or so, for the larger good of India's subsequent generations. And hence it is needless to say that quite a number of them would quite relevantly be referred too, during our ongoing discussions in the subsequent paragraphs below, to contextually substantiate our further derivative perceptions in these regards, we have been critically studying the above mentioned writings by some of the learned experts we are almost compelled in many ways to perceive that, it won't perhaps be too far-fetched to historically even comment that indeed the 1962 India-China border-skirmishes—not only gave a big jolt, as well as rather a 'well-deserved shock-therapy' to the Indian Establishment, but also a kind of the much needed 'wake-up call' to the entire sovereign, democratic, Indian-Republic—for the future overall geostrategic and politico-military wellbeing of the peoples of India. Historic facts are indeed always irrefutable facts, though they do also remain otherwise debatable as well.

We may recall that in the earlier chapters, and in differently no less very relevant contexts; whenever we have been taking-up the multiple aspects of the Indian societies, her peoples, and also her intriguing histories since the very ancient of times, in course of our overall ongoing critical discussions with their numerous dimensions, we have also been noticing that along with the unbroken saga of the magnificence of India's many civilisational marvels, there have also been some periodical geostrategic-cum-politico-militarily not so minor 'hic-ups' too; and needless to say that all these have in turn, been pointing towards the surfacing of some worrisome and also typically lingering the—India's socio-cultural-cum-geopolitical naivety, and the many kinds of 'fault-lines'—along with the Indian history's advancing timelines onto the India's expanding civilisational-canvas itself. In overall global geostrategic-cum-politico-military perspectives, these glaring and also the no less geopolitically malignant 'chinks in India's civilisational-armour' indeed, do not allow her multiply historic-saga, to be all that reassuring and motivating in these regards. We may also contextually recall that we have since been critically analyzing and comparing them with other civilisational-cum-geopolitical 'entities' and then at the same time, rather perceptively trying to identify the pivotal reasons therein, which have been centrally instrumental in creating any such not so laudable developments and hold-ups. And hence as we may also recollect that unfortunately, one so much lingering lacuna in these crucial respects; has been the almost total disregard for always universally, the paramountcy **'multiple-aspects of the essential strategic-culture and thoughts'**. And so historically, this specific 'void' has been perhaps singularly responsible for the occurrences of not at all palatable episodes concerning—the many geostrategic and politico-militarily very embarrassing set-backs—getting repeated from time to time since centuries as the timelines of India's living-history, have been advancing onto her expanding civilisational-canvas.

At this precise point what we do very crucially need is, to all the more critically concentrate onto the presently ongoing post-1947 era of the modern, democratic, and sovereign Indian Nation-State. Needless to say that these are indeed now, of paramountcy importance and at the same time, these current aspects do very critically concern modern India's, both short-cum-long terms future as well. As we begin to critically delve little deeper into these aspects of the contemporary sovereign and democratic India, we do notice some totally new dimensions therein, which couldn't have been imagined even, during the so prolonged times of her ancient, feudal, medieval and then, colonial eras. **It is indeed historically undeniable that during the British colonial-era itself, India had ceased to geographically remain socio-politically so much divided—just as a civilisational-entity only—since, she had been already quite firmly recast into geopolitically, the single and huge centrally governed imperial-entity, encompassing the entire Indian subcontinent, so geostrategic-cum-politico-militarily covering a largest geographical chunk of the South Asian region**. This was historically so phenomenal, strategically invaluable, and also so multiply enriching for the then British Empire that, the British colonial masters in London had named India as the **'Jewel in the British Crown'**. So far as the multiple-aspects of the entire gamut of governance of the colonial Indian subcontinent was concerned the British had quite imaginatively and along with their far-sighted strategic prudence-cum-vision, had setup a number of the different kinds of very valuable institutions and systems. And these indeed turned out to be very handy and also, as the no less greatly valuable assets, when India finally gained her political independence in the year 1947 to be further built and groomed upon suitably with due alterations, changes in their professional-cum-socio-cultural ethos, and also encouraging the paramountcy needs for cultivating their unwavering commitment to the Indian Constitution in the highest national-interests, by the subsequent democratically elected constitutional Governments of the sovereign Indian-Republic.

We may just say here that in many ways these historic aspects too, speak volumes regarding the indeed very laudable—**the so enriched 'multiple-aspects of the essential strategic-culture and thoughts' of the British**—which would even now; continue to remain the very valuable sources of learning and the multiple-enrichment, for the modern and fast developing Indians and their Nation-State. The multiple-aspects of the geostrategic-cum-politico-military pragmatism, as well as socio-cultural and also, the democratically political large-heartedness-cum-maturity do always demand that, one must not hesitate to bestow the any due appreciation and praise onto others too, wherever it clearly appears to be befitting in the larger civilisational considerations and also, enhancement of the overall human-values in the world. But that is altogether a different matter and hence, do remain fairly valid in its due place, in our contextually continuing arguments and discussions. **But at this specific juncture and also in the above multiple-contexts concerning the post-1947 sovereign, democratic Indian Nation-State two precisely, very pressing questions very glaring arise for the Indian peoples, their leaders and decision-makers and also, the presently ongoing the Indian establishments and the same just cannot be wished-away in any manner whatsoever, in the always critically discerning 'eyes' of the annals of the modern India's contemporary living-history. Therefore at this precise point, these two very specific questions are that:** ***(a) Have the above mentioned 'Institutions and Systems' as left in India in 1947; been duly expanded/altered, redesigned and desirably groomed, since as per the paramountcy dictates of the so greatly changed and revolutionized times? And, (b) have the above mentioned so 'enriched the multiple-aspects of essential strategic-culture and thoughts' of the British, conceptually been 'taken-over' in India after the British departed; in the Indian Nation's highest national-interests, since as enunciated above?***

Needless to say that in the pragmatically precise and also, the no less comprehensive answers to the above derived two paramountcy questions in these regards here; lie the 'keys' to decipher the presently prevailing, as well as so menacingly enlarging—the numerous ills, malfunctioning, derailments, and confusions in the present Indian 'Institutional-Systems'; along with the so overwhelming crisis in the entire gamut of governance—in today's sovereign and democratic India's overall polity. In these multiply paramountcy perspectives concerning the fast 'developing', sovereign, and democratic Indian Nation-State; and accordingly at this precise point here, we may like to take-up the above mentioned first question for our further analytically critical discussions. And hence during the now bygone British colonial-era, concerning the India's imperial-governance (or miss-governance!); the highest-level seat of political-powers, as well as the decision-making mechanisms, were stationed in the far-off London and under the direct political-executive control of the 'Secretary of State, India', who happened to be member of the British Cabinet, with his direct subordinate administrative 'Chief Executive'—the Viceroy-cum-Governor-General of the British-India—since located earlier in Calcutta and then, subsequently in New Delhi. In some ways it may appear to be rather a 'queer' arrangement now, but that was precisely the way, the British colonial-rule in the Indian subcontinent functioned; and that too, functioned well, so far as exclusively the then British long-term imperial-interests in India and also, those of the once truly transcontinental British-Empire were concerned. **As we may delve little deeper into these historic subject-matters, then even now a very relevant moot point emerges here that, all these aspects in many ways speak volumes about the geostrategic-cum-politico-military vision, the far-reaching geopolitical-prudence, and at the same time no less; the so much enriched, as well as the quite flexibly well-stationed, the 'multiple-aspects of essential strategic-culture and thoughts', on the part of British as a far-off and an alien European Nation-State in these regards.** Well historically, that is altogether a different story and in any case, these just don't compensate in any manner for now the numerously colossal post-1947 'failings and callousness' of the Indian peoples, their leaders, and the establishments. We are very consciously aware that these perhaps some 'sharp' comments, may even stir some 'heated debates' specially among the political-cum-bureaucratic, the extremely nexus-oriented or fixated establishments, in now the independent India's various 'corridors of power', largely remaining busy in nursing the extremely corruption-infested: their 'vested-interest-nestles'. Volumes have already been written and also being written by different scholars, journalists and analysts, on these not so laudable dimensions concerning the present Indian-Nation's 'Institutions and Systems', within India itself; and even being critically commented-upon abroad too, from various schools of thoughts for anybody's ready reference and research, but that's a different matter.

Now in these multiply important contexts here, we may like to take-up the pivotal aspects of the independent India's political-governance first and precisely here what many among the Indian intelligentsia, do observe with perhaps 'lumps in their throats' that, immediately after 15th of August 1947; it was Mahatma Gandhi alone, who indeed in a very visionary manner; could so futuristically sense then, the real dangers and consequent derailment of the entire constitutional gamut of the 'political governance' of this newly independent country, if certain most fundamental corrective steps were not immediately adhered to by the then concerned top political-leaders of the Congress Party to begin with. **Mahatma Gandhi was indeed even irrefutably and, so prophetically forthright that he emphatically advised them to disband the 'Congress Party' first before anything else since the main purpose, for which it was formed, had already been fulfilled and hence there could not be any further rationale and relevance in its continued existence, in the sovereign democratic India. His indeed far-reaching, visionary, and so prophetic argument was that if this 'Congress Party' was going to be used for participating in the independent and democratic India's electoral-politics for gaining power for forming the new democratic governments by making use of and also, encashing its goodwill among the teeming millions of the Indian masses, then they or their successors in the Congress or any other Political Party, would go corrupt, and thereby the entire systems of India's political-governance very sadly, would consequently be thoroughly corrupted with menacingly multiple-fallouts for the future well-being of the entire Indian Nation-State**. Gandhiji was then alone happened to be, with great sense of history, as well as with indeed a very far-reaching futuristic vision too, so prophetic and precisely right that **the real roots of all the mess-up, and the extremely unbecoming turmoil that now we find in the entire political governance and also, in the otherwise systemically so necessary, the bureaucratic constitutional-functioning of both Central and the States-levels' civil-administration in fact, lay in the phenomenon of the so callously non-listening of the above mentioned, the historically prophetic words of wisdom and suggestions of the Father of the nation, in the year 1947 itself**. In other words we may not even hesitate to say here that all this subsequently turned out to be indeed, an amazingly great man-made tragedy with adverse consequences for the entire democratic-polity and thereby, for the future of independent India from the very beginning of the further making of her 'living-history' itself in these regards. In fact, 'the faults didn't lie in the sovereign and democratic India's fate-lines but squarely and that too immediately after the 15th of August 1947, in the otherwise indeed the then so brilliant, scholarly, visionary, and also charismatic—the great top-level Indian political leaders—almost all of whom also happened to be the great and truly selfless and non-violent freedom-fighters for the teeming Indian masses.

Immediately after successful conclusion of the Indian peoples' globally unprecedented and non-violent freedom movement—under his uniquely charismatic and history-making saintly leadership, what Mahatma Gandhi then wanted from those Congress stalwarts was only, to liquidate the old 'Congress Party' and form their altogether a new Political-Party and from then on, to continue with their indeed necessary and well-deserved electoral-politics in the sovereign, and the then fledgling Indian democracy. In fact, Gandhiji never ever wanted them to refrain from their own totally justified, and also desirable political-ambitions, and hence for the sake of these, he was never against their active political participation in the then newly sovereign India's democratic-polity. All he desired was that, in the India's best long-term national-interests; it was essential to discontinue with the anymore existence of the old 'Congress Party' and hence, they should form their own new Political Party or Parties, for their further political activities. In the year 1947, the Congress stalwarts were no 'pigmies' in any manner; like so many of the present times' political-minions of the Congress and also any other Political Parties; now since being seen to be going around all over, with all sorts of their highly nauseating corrupt-practices and on top of all these, so shamelessly masquerading as the modern India's 'great political-leaders' too.

It is also equally true that the present times' all political-leaders in India are not corrupt and some of them are indeed honest, very hard-working, and truly patriotic too, who are always sincerely eager to selflessly serve their nation, with the best of their abilities. And at the same time, it also remains a matter of great historic conjecture that, if the then old 'Congress-Party' would have been dissolved as desired by Gandhiji then, no political-level corruption would have sprouted subsequently, in the India's democratic polity; but one thing appears to be fairly certain that the same would in any case, have remained much more limited in its scope, range, and the contents. Hence contextually in a historic retrospective, we may not perhaps hesitate to conclude that when Gandhiji was so sincerely advising those truly great and highly patriotic Congress stalwarts immediately after India gained her political freedom, it was for all of them indeed, great moments of the **'tryst with destiny'**; for further shaping-up of the Indian masses' future-wellbeing, and so at these moments to India's great misfortune, they so inadvertently slipped-up and the rest is now history not so palatable in many ways, with hardly any possible end in sight for the present in these regards.

Mr. K. Natwar Singh, in his autobiographical book, has written that "In this over quoted **'tryst with destiny'** speech, he conceded that the freedom movement had ended in partial failure. All through the movement, the aim and destination was the achievement of an independent, undivided India. Few remember that Pandit Nehru's first cabinet of fifteen ministers included six non- congressmen: S.P. Mookherjee, B.R. Ambedkar, Sardar Baldev Singh, John Mathai C.H. Bhabha and R.K. Shanmukham Chetty. It is futile to debate the pros and cons of Partition. As Pandit Nehru conceded, Partition was the lesser evil.

M.A. Jinnah never went to jail, never participated in any street protests, never faced a lathi charge. He was a single-minded person clear about his goal. He outwitted the entire Congress leadership with the exception of Sardar Patel. It is often said that Sardar Patel would have made a better Prime Minister. He died on 15 December 1950, three years after Independence. Who would have succeeded him? Jawaharlal Nehru without a doubt." *(These points are well—taken, but here Mr. Natwar Singh has very 'craftily' chosen to ignore the obvious hidden reality of great national import. And that point is that as the above mentioned option, Sardar Patel would have been India's P.M. for precisely three years and four months, i.e. from 15 August 1947 to 15 December 1950, and during that brief period, Sardar Patel would have constitutionally involved and built-up of the basic tenets and structuring of the culture and systems of India's governance, in an entirely differently manner and thereby, Nehru won't have* *been able to indulge in his the then usual, all kinds of* ***'Kite-flying'****, since, not desirable for the Nation at all. Further Sardar Patel won't have handled the then emerging* ***'Kashmir Problem'****, in Nehru's methods. And rest would have been India's post-1947 history altogether, in different ways—so much more good for the whole Nation!).* (K. Natwar Singh, 2014, p. 109)

But all said and done, it is also the no less true that all has certainly not been lost; and so the future of now the fast developing democratic India quite reassuringly remains fairly bright with the indeed, amazingly versatile colourful and uniquely resilient-cum-talented Indian peoples' hopes and aspirations now along with the already momentous dawn of the present 21st century in the world. And to the India's good fortune for the present that, the few new kinds of the very dynamic, focussed, so popular and very charismatic leaders too are being thrown-up onto the Indian political-arena, for charting new and revolutionary paths for development, as well as, the much awaited course-corrections. It is also true that the truly robust, very far-reaching also the very long-term multi-dimensional foundations for India's these multiple-advancements to begin with, during the present-era's fast changing now the super high-tech modern world were indeed laid by now the yesteryears', some those above discussed, the very towering political stalwarts of the old Congress-Party, as well as, of the no less some other Political Parties too; along with many leading Indian personalities belonging to both the civil and military areas of leadership and human-endeavour. And hence, their contributions too; just cannot be minimized in any manner whatsoever, during any objectively apolitical and truly analytical discussions, otherwise in these regards. In fact in these presently prevailing the Indian multiple-scenario, as well as in the critically broader perspectives one singular failing has now become an extremely menacing drag onto the multiply developmental-paths of the Indian Nation-State, and that happens to be the not at all any effective-check on now the monstrously growing population. And ironically, and also rather so unfortunately, any serious and sincerely proactive thoughts are still not being given in the India's various 'corridors of power' for any effective check on this indeed, most serious menace. All the more sadly even now any leading Political Party and their top-level leaders, especially during the times of general-elections in India are perhaps deliberately shying away from this most menacing national-problem, obviously compelled by their only short-term electoral-gains, with no concern for this great socio-economic and ecological perils of Indian peoples. **On this specific issue of India's largely unchecked population-growth (or explosion perhaps.), one detailed 'article' with some suggestions, has been attached as 'Appendix-A' of this sub-chapter itself, for contextually a ready reference in these regards**. One point is rather quite irrefutably clear that the crippling menace of this 'population-growth' problem in the presently fast developing modern, sovereign and democratic India, cannot be wished-away in any manner, and hence this urgently needs to be proactively confronted with all the necessary sincerity and, with clear-cut strategies and comprehensive planning, to effectively preventing the Indian Nation-State from ultimately falling into an 'abyss' of a preventable kind of failure and darkness.

In fact, these 'alarm-bells', are but calculated and strategically prudent words of caution, which are indeed too serious to be ignored. We may recall that this specific issue has also been discussed in its globally broader perspectives, in the earlier sub-chapter on the 'Population-Explosion and Environmental Degradation'. But for India this rather most dangerous menace; indeed assumes all the more extra dimensions and, all these would contextually be clear in the above mentioned 'article', since attached at ***'Appendix-A'*** **in the end of this sub-chapter**. Unfortunately even during the times of independent India's the first Prime Minister, **Shri Jawaharlal Nehru**, this problem was not even seriously considered, in spite of very forceful persuasions of **Shri J.R.D. Tata** to begin with, in the India's highest long-term national interests, as would be clear from this attached 'article'; and the rest is a not so encouraging saga in these regards. This too may be termed as, an unfortunate fallout of the continued absence of the universally always paramountcy—**the multiple-aspects of 'essential strategic-culture and thoughts'**—from the very fundamentals of India's national-being in many ways. But otherwise, Nehru's contributions in the still continuing phenomena of the India's multi-dimensional nation-building are truly great and far-reaching and volumes have already been written on this important subject-matter. In fact, it was Nehru who laid the foundation for the modernisation and industrialization of the modern, sovereign and democratic India and also so valuably brought-in the much needed 'scientific-temper and outlook' into the very psyche of the Indian peoples and thereby, the so encouraging and valuable results are visible everywhere, in the vast and multiple-arena of the modern India's fast progressing scientific and technological endeavour. If there are still some deliberate hiccups and obstacles on the way, then they are principally due to the politico-bureaucratic mess-up, arrogant callousness and also, the unashamed corruption; and the seeds for these were unfortunately sown-in to begin with, largely during the Nehru-era itself.

But even before that, even preceding Nehru's momentous efforts in these regards, **Sardar Ballavbhai Patel's contributions, as India's first Deputy Prime Minister in fact, unprecedentedly prepared the requisite ground-ground by politically uniting the whole of India's vast and so diverse and huge, the geographical landmass indeed so astonishingly, within a span of two to three years only by charismatically integrating the then so regressive, the more than all sorts of the five hundred very big, medium, and relatively small, 'Princely States' then continuing to exist since centuries, before India gained her political freedom. We don't think there could ever be found any parallel to this historic episode, since so brilliantly achieved by the revered 'Iron-man of India', during those extremely trying times after 15th August 1947; especially when 1947-48 India-Pakistan War in the State of J&K had already started. But the strong-willed India's great 'Sardar' never wavered even for a moment, in spite of his some lingering health problems; and so before he breathed his last, so greatly tragic for India then, he had bravely achieved the indeed unique victories on both the above cited political and military fronts, for the future of India's security, unity and wellbeing.** In order to at least know some worthwhile aspects and dimensions concerning the amazing greatness, character, patriotism, selflessness, brilliant vision, strong leadership qualities, as well as his unique contributions to his nation, i.e. *India*; one needs to at least attentively read the biography, written by **Shri B. Krishna**, of this one and only one—the great 'Iron-man of India—so ironically about whom, the Indian peoples are yet to substantially know so far. It cannot be overemphasized that, the passing-away of Sardar Patel in the end of year 1950 itself; came out to be the not only a colossal tragedy for the then so much crisis-ridden—the fledgling Indian democracy—but a very depressing truth remains that even now, the many of the incomplete tasks as left by Sardar Patel and, as the destiny would so unfortunately have it for India; especially in the highly challenging and complex domain of the India's 'matters-military', are yet to be rather 'sensibly completed' by the subsequent top-level Indian political-leaderships and the governments so far. It so regressively, as well as the no less malignantly so far, remains an embarrassingly great paradox for the entire Indian nation-State; even as the Indian peoples too, have already so enthusiastically entered the present 21st century. But even more astonishingly remains a much greater irony that all along these decades after the year 1950 or so, barring some few exceptions here and there; the independent democratic India's so damagingly entrenched—the political-cum-bureaucratic, the all sorts of 'vested-interests—so callously-cum-arrogantly, always remain ready to further damage the multiply important causes and dimensions of the India's 'matters-military' regardless of her national-security related, the many paramountcy national-interests, in both short-cum-long terms' perspectives. And hence even a much bigger irony in these regards happens to be today that, what to talk of India's friends thereby even her diehard adversaries too, are no less astonished to witness the same.

The India's second Prime Minister, Shri Lal Bahadur Shastri, unfortunately did not survive for long but within a short spell of nearly two and half years only of his Prime Ministership, he made history and could undo up to a great extent, the 'military mess-up' in fact, so naively created by otherwise truly so great and charismatic Nehru himself; and we may recall that we have since already briefly touched upon these historic aspects in the earlier chapters too, in differently relevant contexts.

At this point we may prefer to take note, as to what the Indian Navy's **Vice Admiral Vijay Shankar** has to say in these regards, he has written that, "India has faced many traumatic events since 1947 that have had critical impact on security of the nation including five wars, insurgencies, gory terrorist acts, periodic crumbling of the law and order mechanism and even a period of authoritarian rule (*i.e. the 'internal-emergency since, both so sinisterly, as well as selfishly; imposed by **Mrs. Indira Gandhi**, during 1975-77.*); but what stands out is a lack of strategic vision instigated by a distorted civil-military correlation. India's military is characterized by an apolitical nature and exults in its professionalism. It responds best to objectivity in purpose and control as can only be provided by a synergetic relationship between the civil authority (*i.e. political authority*) that defines policy and the military that gives to policy the necessary energy. **Prime Minister Nehru had an abiding undertone of aversion to matters military**. Resolute action and control of the military in the more successful initial operations in Jammu and Kashmir and Hyderabad was excised in both instances by **Vallabhbhai Patel**, the then Home Minister and Deputy Prime Minister of India. Post Patel's demise, the Defence committee of the Cabinet system bequeathed by **Lord Mountbatten**, broke down and the military was totally marginalized. Politico-bureaucratic unilateralism, in the formulation of military strategy botched calamitously, in the 1962 war against China. It brought into stark contrast the relative efficiency with which the First Kashmir War and Hyderabad operation was planned, coordinated and waged. Therein lies the roots of the recurrent crisis in civil-military relations in India.... In 2008 India's Defence Minister **AK Anthony** told Parliament **that the *Henderson Brooke-Bhagat Report* could not be declassified because its contents 'are not only extremely sensitive but also of current operational value'. Half a century on it is challenging to comprehend what the report could contain that would warrant such sarcophagal silence**.... In the causation of a newly independent nation confronted by a variety of mortal security challenges, the latter lack of impulse posed an awkward dilemma which as events unfolded, only served to elbow the military establishment to the status of an 'attached' office, to be heard only when consulted shorn of any part in strategic decision-making. Why this came to pass is a question that is not easily answered, but clearly two dynamics were at play, the first was the **misguided fear of a 'Kitchner redux'** and therefore the misplaced trepidation of military control of the state, and the second was flawed belief that civilian control of the military not only implied superior dual control by the politico-bureaucratic alliance but also a self-fashioned conviction that military matters were essentially of execution and had little to do with policy making or strategic planning.... What was conspicuous in its omission was a statement of **the blemishes in higher defence management, the failings in the political direction of the war and 'courtiership' being promoted in the military**.... The woeful neglect of defence preparedness, an unreal approach to international relations, particularly with China and the dangerous inadequacies of higher defence management significantly the intrusive, extemporized and incompetent manner in which civil-military relations had evolved.... Clearly it was dominant persona of Nehru, his anti-militaristic disposition, his flowed obsession with internationalism and allure of non-alignment (in a milieu that was driven more by opportunism than idealism) that must have had much to do.... The larger-consequence of the absence of controlling stimulus in civil-military relations was the inability to articulate a strategic vision that linked with policy and to promote long-term planning and force development that serve to implement and attain a strategic posture and in turn influence the future.... If we were to analyse the civil-military correlation in India and in China against two attributes of efficiency and coherence of response, we would find that in India's case, the nation is often swayed by its democratic mores, cultural traditions, historical and pluralistic ideals to fully realise the significance of its military as a direct consequence of which there is a persistent undertone of friction and unease in the relationship." (Shankar, Defence and the Security Alert, November 2012)

If we may care to closely read an Indian learned Admiral's above quoted paragraphs and also in between the lines then, we may clearly visualise that Admiral Vijay Shankar has indeed succeeded in conveying many kinds of persistently malignant and the no less depressing facts about the many chinks in India's civil-military constitutional relationship's armour, with far-reaching adverse consequences for her democratic-polity and also, the always supreme national-security aspects. And precisely in these multiple-contexts, what the today's otherwise fast developing democratic India's top-level political leaders and the so regressively entrenched bureaucrats in the corridors of power in India's national capital New Delhi, have been doing since last more than six decades or so; happen to be doing nothing much substantially—but mostly brushing the so forthrightly mentioned above—the paramountcy concerns in these regards—under the 'thick and heavy carpets' of theirs: the all sorts of relentless politico-bureaucratic corrupt practices, as well as, the highly damaging geostrategic-cum-politico-military 'mess-up' as well. Consequently, not only the India's vital national-security essentials continue to get damaged and compromised; but also the no less important—the morale and motivation of the gallant military officers and men too—are getting dangerously 'smashed'. As if all these are not enough to shake-up the India's corruptly nexus-oriented politico-bureaucratic class, that on top of all these, this callous-cum-arrogant and only power-hungry 'class', has even gone up to the extent of quite sinisterly ensuring that for them under the existing—extremely corrupt and confusing systems of the decision-making and governance (*or largely the non-governance?*)—No threats of the any kind of lawful accountability are even remotely able to visit their 'stinking' doors and premises; hence, they continue to remain safe and under their 'sinful covers'.

(ii) Matters Military: Some Crucial Points

In the above multiply crucial contexts, as well as precisely due to the above discussed politico-bureaucratic arrogance, corruption and their irrational culture of 'mess-up', of the democratic India's both internal and external security-aspects and also the larger national well-being and hence so far as India's always totally apolitical Armed Forces are concerned, in the broader terms—two-pronged damages have been seriously inflicted onto the military command and control systems and also, onto the always ongoing 'multi-dimensional soldiering aspects' of the rank and files—**(a) the morale and professional motivation** of the always very crucial young-cum-junior leadership cadres have been receiving very serious beatings, and **(b) the erstwhile indeed so marvellously impeccable integrity, credibility and also the professional competence** of the military's middle and higher levels of the no less crucial command and staff functional-systems too; have developed serious 'cracks' and thus, stand terribly vitiated and compromised as well. **And alas no wonder that in the process, the military's higher and at times even the very top-level command-cum-staff levels functionaries too have so sadly been losing much of the sheen of their professional-armour. And hence as matters of a great tragedy the above identified two factors in totality and, not only for the Indian Armed Forces but also for the entire Indian-Nation too, the aura and unique respects for the uniforms, in the eyes of the Indian citizenry are fast getting corroded. Hence historically from the very ancient of times, history has been clearly showing that if one wants to 'crack and damage' any nation's security-systems and also its Armed Forces from within then, the easiest and sure way is to subvert the morale, professional honesty, and the financial integrity of that nation's military-men, and thereby the rest would rather quite smoothly follow for the benefits of its adversaries.**

Air Vice Marshal in his well received and highly scholarly book, "India's Wars (A Military History)—1947-1971, has very informatively mentioned is the pages 197, 198 & 199 that: "The India-China conflict of 1962 did not own its origin merely to the jostling for Himalayan space between two emerging Asian powers. Nor was it only a result of the fallout of the 'great game'-orchestrated by Britain to protect its northern frontiers from perceived Russian ambitions of territorial expansion towards the warmer waters of the **Arabian Sea**. Control and influence over Tibet—considered by many strategic commentators to be the main bone of contention between India and China—did act as trigger for the flare-up, particularly after the Dalai Lama sought refuge in India in 1959. However, a closer look at the history of the region encompassing Ladakh and Tibet reveals a complex and myriad interplay of geostrategic events. These mainly comprised attempts by various players over almost two centuries (early nineteenth to mid- twentieth) to dominate a high-altitude wasteland that provided strategic depth to a declining power like Britain, and offered the same pay-offs to the two emerging Asian powers of the twentieth century, China and India.

Additionally, it also appeared that the 1950s saw a concerted attempt by a resurgent China to identify with the non-Han people of all the contested territories, comprising communities that resided in the regions around the Tibetan Plateau. The Chinese shrewdly offered a racial argument to what was primarily geopolitical problem. India, on the other hand, had embraced a multicultural and multi-ethnic way of life, aspiring to amalgamate many tribes and communities in ladakh and the north-eastern parts of the country. It felt that china's argument of racial affinity was not tenable when it came to coveting Indian territories.

A number of British geographers, explorers, diplomats and soldiers like **Forsyth, Agnew Cunningham, Johnson, Ardgah. Mc Cartney, Mc Donald and Younghusband** have, at various times between 1850 and 1914, offered compelling arguments to support their view of where the boundaries between **Kashmir, Tibet** and **Sinkiang province of China** ought to have been. Despite forays by Indian Explorers like Pandit Nain Singh, who first accompanied Forsyth on his expedition to Kashgar and Yarkand in 1873 and then forged out on his own eastwards from Leh to Tawang via Lhasa, the region did not attract the attention of independent India's military or political leadership. For his exploration of nearly 2,250 km, which he measured by the length of his fixed strides and recorded secretly on prayer wheels; and contribution to unraveling the mysteries of Kashgar, Yarkand and Tibet, **Pandit Nain Singh** was awarded the Patron's Gold Medal by the Royal geographical society. No other Indian would receive this award for over 125 years. By the late-nineteenth century, so fascinated were the British with Tibet that they encouraged feisty Indian explorer to venture into Tibet from Sikkim. One such explorer was Sarat Chandra Das, who ran a school in Darjeeling and whose covert role was to **'train interpreters, geographers and explorers who may be useful if at any future time Tibet is opened to the British'**. Between 1877 and 1882, Das made two trips into Southern Tibet through Sikkim and finally reached Lhasa in mid-1882, purportedly to study the Tibetan scriptures. He returned to Darjeeling in December that year and wrote about his journey in a book titled a journey to Lhasa and central Tibet, Das was probably among the last of the adventurous Indians to venture into Tibet in the nineteenth century.

After they became nation states, few cartographers or explorers of eminence from India and China ventured into the icy wasteland in the 1950s even after it emerged as a geopolitically combustible region. An exception though was a 400 km exploratory trek through Tibet in 1949 by **Major Zorawar Chand Bakshi** in the guise of a Tibetan monk. He was awarded the Mc Gregor Memorial Medal for military reconnaissance, the first Indian to be so awarded, Bakshi would go on to become one of the Indian Army's most accomplished and decorated commanders in the years ahead.

For want of a better argument, the Indian government accepted the British Position as a fait accompli, sticking to it rather adamantly during almost a decade of negotiations with China till it entered the psyche of the Indian people as a matter of national pride. This more than anything else galvanized public opinion with India to contest any use of force by China to redraw these boundaries. China on the other hand relied on scanty historical data for its claim on Aksai Chin taken from two narratives ordered by the Qing dynasty in 1782 and 1820. Except for travelers' reports provided by Chinese caravans traversing from Turkestan and Sinkiang to Tibet, no cartographic exploration of the region was carried out by the Manchu's of the Qing dynasty as their power waned in the region in the mid- to late-nineteenth century. Despite the plethora of accurate maps drawn by the British, communist China followed a different approach to accepting colonial boundaries.

Mao's China expected Nehru to enter negotiations considering India's own bitter experiences with the British and understand some of its geopolitical aspirations and apprehensions in the region. When India did not respond, China reacted like any powerful authoritarian state with a clear understanding of the benefits of using military force to prove a point and teach a weaker adversary 'a lesson', as Mao put it. Was it as simple as that? Obviously not! That is why a detailed historical-geopolitical-military-study is an absolute imperative in any serious analysis of the conflict". (Air Vice Marshal Arjun Subramaniam, 2016).

These are too worrisome matters to be ignored in any manners, for all segments of the Indian Nation-State. **It also stands irrefutable and the no less crystal-clear that, much of the extremely 'polluting-muck' in these regards, have principally been flowing into always the so important—multiple-aspects of the matters-military, as well as among the all levels of India's Armed Forces—from the above noted unmindful rampant corrupt-practices and the relentlessly ongoing 'culture of fooling around'—in the politico-bureaucratic 'corridors of power' in the country**. It is indeed very painful for any upright and patriotic Indian-citizen to watch, read and hear from all sorts of the substantially ill-informed, people in the media, the political and bureaucratic classes; to rather irresponsibly, as well as shamelessly, pass all sorts of the quite unbecoming comments, onto the Indian Armed Forces as a whole. In fact many of them among themselves are largely so arrogant, callous, corrupt and unpatriotic that while doing so, little do they pause and sincerely search for, in their very desirable depths and dimensions, the real reasons for such 'tragic-fall' of the India's Armed forces and for which they themselves, have mostly been indeed, unpardonably guilty for the so much serous and also, the no less momentously dangerous betrayals of India's always, the paramountcy multiple-aspects of the matters-military. Thanks to the multiply geostrategic-cum-geopolitical far-sightedness on the part of even the erstwhile British imperialists that even during the zenith of their prolonged colonial-rule in the Indian subcontinent, they were very prudently be able to always keep their British Indian military in the subcontinent—healthily isolated in theirs so strategically located numerous military cantonments, and hence, away from even the routine 'heat and dust' of the vast civil areas and masses—and thus the Indian military continued to remain totally professional and impeccably apolitical too. But it is also historically true that during those times, they were never able to totally trust their native soldiers, especially after the '1857', but for all these they themselves and their irreconcilably arrogant racial-cum-skin-colour prejudices, were principally responsible for any such 'trust-deficits'. Moreover, in spite of their numerous qualities and high accomplishments they were nothing but intruders and invaders, and hence as per the broader geopolitically ethical-cum-moral perspectives, their intrusive and the so much greed-driven and exploitative colonial-rule in the Indian subcontinent, could never have been justifiable.

Post-1947 in the politically independent India the then military-men too, since continuing to serve since the pre-independence days were unquestionably patriotic enough to administer their fresh oath, in the names of Indian tri-colour and also the new Indian Constitution to remain totally apolitical, professional and loyal to their democratic Indian Nation-State. Nobody could ever doubt their moral uprightness and loyalty to the Indian Constitution; irrespective of the whatever unfortunate politico-military 'twists and turns', were periodically seen to be happening anywhere in the India's neighbourhood, as they never gave any ground to anyone to harbour any 'hiccup' in these regards. **But even then, India's, otherwise so great, learned and visionary—the first Prime Minister Shri Jawaharlal Nehru—ironically in spite of his fairly prolonged and the 'multiply-charismatic' leadership and, so high a political-stature, never felt at ease unlike Sardar Patel, and at the same time no less the constitutionally mandatory, his politically fair-cum-trustworthy accommodativeness towards India's indeed impeccably apolitical Armed Forces—in the highest national-interests of the Indian peoples.** Thus, for his any such myopic-outlook and also the geopolitically erroneous misconceptions, historically he remains himself to blame and that too, for causing the great 'damages' to the India's gallant military, her paramountcy geostrategic-cum-politico-military imperatives, and thereby to the democratic Indian-nation's entire gamut of the 'matters-military' for years to come. And even now with respect to the nation's these always so crucial national-defence imperatives, only insincere 'lip-services' are being noticed at the highest political levels and thus, these so dangerous political and strategic-derailments are yet to be even tangibly rectified.

So specifically in the Indian contexts, the totally baseless, unfounded and also, irrational fear of the proverbially fictitious—the **'Kitchener-redux'**—still so damagingly lingers-on, into the sovereign and democratic India's so regressive-cum-corrupt—the every politico-bureaucratic nook and corner in the Indian Government's 'corridors of power'—to the great detriment of the entire Indian Nation-State. In these specifically multiple perspectives, so far as the entire gamut of now the modern and democratic India's national-governance is concerned, still in the presently ongoing 21st century these indeed continue to remain a very sad and frustrating saga, for the India's democratic polity so far. **These are no simple and any passing matters for the Indian-nation's national security, as well as the multiply paramountcy national-well-being, as the nation's political-class would like the peoples of India to believe, while talking with all sorts of their superficial political-jingoism in the Parliament and thereby so insincerely playing around with**

the periodical national defence-budgets, principally for their electoral propaganda purposes. These always happen to be the too serious and paramountcy subject-matters concerning the nation's highly complex and challenging—the multi-dimensional 'matters-military'—and hence in no way, the political-class whether in Parliament or outside is supposed to 'beat round the bush', by so callously indulging in their typical 'culture of the fooling around'. In the multiply crucial India's matters-military aspects, always unmindful to any extremely dangerous consequences to the entire Indian Nation-State the above discussed highly corrupt, vested interests-oriented, as well as the so deeply entrenched politico-bureaucratic combine's largely principal programmes appear to be so regressively always ensuring—the continuing marginalization of the Indian military—in the overall gamut of national governance and, the national-level policy-formulations, and the decision-making processes. **In the Indian democratic polity, it amazingly appears to be a very queer scenario wherein instead of comprehensively attending to the national security, as well as the matters-military related issues and problems, the political-cum-bureaucratic 'clueless' functionaries, since sitting in the nation's 'corridors of powers', are always keen to display their highly irrational aversions to the indeed irrefutably apolitical Indian citizens in their military-uniforms, being so ready to serve the nation with their unquestioned readiness to make even supreme sacrifices and also, always with their indeed unshakable oaths of loyalty to the Indian Constitution.**

Admiral Vijay Shankar, in his already above quoted article, has identified the pivotal reasons for the beginning of India's 'mess-up' at the top-level political directions of war and the management of matters-military; principally to the multiply-dominating persona of Nehru to begin with and, especially after the demise of Sardar Patel, and his recklessly regressive aversion to the matters-military. And these historically factual points are indeed well-taken; but how Nehru being so learned and wonderful writer, got into geopolitically such not at all any laudable mould, and the reasons can be traced back to the then British India's colonial times of 1920s, 30s, and even 40s. **But whatever and in spite of all these of India's colonial past and so on, Nehru was himself alone responsible for all these so much unbecoming 'kinks and twists'—unlike the much more matured personality and vision of Sardar Patel and also, his brilliant acumen in the domain of statecraft—in Nehru's otherwise no less charismatic persona and thereby, for all these geostrategic and politico-military 'hiccups', during the years of his Prime Ministership of the country.** As anyone rather attentively goes through the pages of Nehru's book, like the 'Glimpses of World-History' and also the 'Discovery of India'; one may get very thoughtfully overwhelmed by his range of learning, hard-work, powers of his pen, and also his intellectually analytical thought-processes, especially when one goes through his captivating descriptions of Emperor Chandragupta Maurya's military-might, generalship, systems of governance of his empire; and then the no less, his writings on Chanakya's vast knowledge of the entire gamut of the ancient India's statecraft, along with those of the other Indian and non-Indian historic thinkers, rulers and warriors; down the ages, onto the mankind's expanding civilisational-canvas in the world. Hence it remains all the more very paradoxically astonishing as to what happened to his all these indeed extraordinary attributes, when his times with his 'tryst with destiny' after the 15th August 1947 came while presiding-over the newly independent and democratic India's destiny so historically, as her first Prime Minister? It is precisely in these multiple-contexts; that the otherwise great-man Nehru is found to be standing on quite 'loose and shaky sands', in the eyes of the mankind's discerningly judging 'living-history'. **Major General AA Rudra**, one of the earliest so few Indian officers in the British India and that too during the last years of British rule; in his biography and since narrated and hence penned by another well-known Indian officer scholar, **Major General D.K. Palit, Vr.C**, who has been able to very clearly and also quite interestingly, describe these, as well as many other historic aspects concerning India.

Hence in the above noted book and as per General Rudra's own narratives, General Palit with his fairly well-known scholarship, sense of history, and intellectual precision has written that, "Shortly after the incident of abortive (*times: almost immediately after the 15th August 1947*) move to the Red Fort, the C-in-C Army convened a high powered meeting to discuss defence strategy and to prepare a policy paper for the Government. Present at it were the two other Chiefs, **Vice Admiral Parry** and **Air Marshal Elmhirst**, the Defence Secretary **Baliga** (*must have been an ICS officer.*), **Lieutenant General Sir Dudley Russell** (Adviser, General Staff) and myself (no Indian officer had yet been appointed to replace the C.G.S., **Sir Arthur Smith**, who had moved to **Auchinleck**'s Supreme Command H.Q.). We discussed strategic matters such as the military strengths of neighbouring countries and the threats they could pose against India; the types of terrain across our borders—the high mountain ranges to the north; plains and deserts in the west; jungles and mountains to the east; and the requirements for coastal defence of the peninsular region of South India.... **The Chief drafted a paper containing all this information, together with a gist of our recommendations—and he finished up with a minute to the Prime Minister asking for a Government directive on Defence Policy. In the afternoon he went to see Nehru.... I received a call from him at tea time. He said: 'Jick, if you have a *burra peg* in your office. I am coming down. I need a drink'.... I told him that of course I did not have a drink in my office, but that I would walk over to his office at once.... When I entered the Chief's office I found him sitting in his desk, his head in his hands—gazing at the table-top. 'What happened, Sir?' I asked.... 'What happened? What happened? You may well ask. I was thrown out of the P.M.'s office, that's what happened.The P.M. took one look at my paper and blew his top. 'Rubbish. Total rubbish. He shouted. We don't need a defence plan. Our policy is *ahimsa* [non-violence]. We foresee no military threats. Scrap the Army. The police are good enough to meet our security needs'.... Rob and I looked at each other in consternation—the mutual commiseration. No defence policy was issued to Army HQ. And that put a stop to strategic planning. (It was the Kashmir war that saved the Indian Army.)....** I was at last able to concentrate on my main task—that is, to find Indian officers for the various senior appointments fallen or falling vacant." (Palit, 1997, pp. 320 & 321)

Whatever, but the so dedicated and always a very passionate sense of patriotism of Nehru, as the then newly independent and democratic India's first Prime Minister, cannot be minimized even marginally in spite of his lingering geopolitical naiveté concerning the nation's geostrategic-cum-politico-military imperatives, the mindless prejudices against the always important aspects of the matters-military and also, his no less unjustifiably arrogant aversion towards India's military men. And these were indeed not only, going to cost the nation's honour and strategic reputation very dearly but also, ultimately turned out to be the no less devastating to his own persona; and consequently Nehru stood rather a completely broken and disillusioned man after 1962, during nearly the last one and half years of his lonely life.

Now again coming back again to the overall post-1947 aspects of the India's 'matters-military', what we very depressingly notice that not only at the apex political-level, but also at the levels of the senior most bureaucratic functionaries too, in some crucial departments of the central government of the then just newly independent nation; were found to be no less clueless with all sorts of their half-baked ideas, to even harm the national-defence imperatives, by quite sinisterly conspiring against the Indian military for serving their selfishly arrogant vested-interests only. These so worrisome and the no less sinisterly manipulative developments, regardless of the adverse consequences to the nation's matters-military-aspects were noticed amazingly among the 'native

Indian-ICS fraternity' and that too, even before the some top-level British persons were still functioning in India on request, in the then totally changed circumstances. If that being so historically and that too, immediately after India gained her much cherished freedom then, what to talk of shortly the later period's much less reputed, trustworthy, and the present much more callous and regressive IAS fraternity of the nation's entrenched bureaucracy. To cut a long story short and also to substantiate the above mentioned not so palatable observations regarding the Indian bureaucracy now, a reference to the quoted paragraphs just below; would rather adequately serve the specific purpose in these regards. **And if the so sinisterly intended damages were prevented, it was not because of Nehru or anybody else Indian; but solely due to Lord Mountbatten, who was still then, occupying the India's soon to be the 'Rashtrapati Bhawan' as the independent India's last British Governor-General, minus his pre-15th August 1947-era's Viceroyalty.**

Lt Gen S.K. Sinha, independent and democratic India's a former Vice-Chief of the Army Staff, and an even internationally well-known military-thinker, Statesman, and a writer, on the various subjects concerning the matters-military, while delivering one of the USI's 'National Security Lectures (1990)', has very frankly and also very knowledgably talked about the multiply paramountcy aspects of the India's 'Higher Defence Organisation'. In the above referred USI publication, the so distinguished and learned General; had very perceptively spoken before a very distinguished audience that, "At the time of independence, control over Ministry of Defence had passed from the Commander-in-Chief to the Defence Minister. This had happened when the Interim Government had come to power, a few months earlier. However the role of the ministry was limited. It was geared only to issue government orders as agreed upon by Army Headquarters and military finance Department. The limited role of the Ministry was reflected in the protocol status of the Defence Secretary who had been a subordinate to the Chief and still ranked junior to all the Principal Staff Officers at Army Headquarters. There was a lobby among civil servants which wanted the Ministry to become all powerful. A Committee consisting of **Mr. R.N. Banerjee**, the then Home secretary with **Mr. Vishnu Sahay**, Secretary Kashmir Affairs and **Mr. H.M. Patel**, Defence Secretary, was pressing that Ministry of Defence in relation to the Defence Services, should function on the same line as the Ministry of Home Affairs in relation to the police. (*Look at this, could there be anything more naively arrogant, and laughable so far as the Indian bureaucrats' rather incurably regressive mindsets were concerned?*). Starting from the lowest level, Police officers have to function under civil servants at all levels, who even write their Annual Confidential Reports. The Home Secretary is virtually the boss of all Police Officers and the top Police Officers have a subordinate status to him. Luckily at that time **Mountbatten** was the Governor General and **Ismay** was his Chief of staff. **The Service Chiefs could retain their higher status over the Defence Secretary and Ismay was asked to work out a new higher defence set up for India.... During the Second World War, Lord Ismay had been Secretary to the Chiefs of Staff Committee in UK, and also had been the Principal Staff Officer of Churchill. He was an expert on Defence functioning at national level. After the war he had been to USA to help the Americans in reorganising their higher defence set up. We were lucky to have him in Delhi at that time....** The Indian Defence Services had been through a surgical operation in 1947 when various units of had been divided between India and Pakistan. Senior Officers who were mostly British, had left causing a vacuum. The Army had been fully committed in the Partition bloodbath followed soon after by the war in Kashmir. In those circumstances Ismay rightly appreciated that any radical reorganisation of the Defence set up was not feasible. At the same time, a system had to be devised which would provide for the supremacy of the civil (*i.e. at the political-level only.*), proper coordination between the three services and minimum red tape or bureaucratic control. However, if an integrated organisation could not be evolved, there was certainly a case for ensuring integrated functioning. Accordingly, he suggested a series of committees. At the highest level, there was the Defence Committee of the Cabinet comprising the Prime minister, the Defence minister and some selected ministers. The Service Chiefs and Defence Secretary were required to be in attendance at all meetings of this Committee. Then there was the Defence minister's Committee, with the Defence Minister in the Chair and the Service Chiefs, the Defence Secretary and the Financial Adviser as members. The Service Chiefs were give equal status even though the Army Chief was in four star rank and the Chiefs of other two Services were in three-star rank. The chairmanship of the Chiefs of Staff Committee was made rotational with the Service Chief longest in the Committee becoming the Chairman." (Sinha, National Security Lecture – 1990, pp. 19-22) (*All these have naturally been well intentioned, multiply well-balanced, very far-reaching and futuristic in the politico-military-thoughts and strategic-concepts, aimed to be healthily functioning and also, fully democratic to the core and thereby always quite unquestionably ensuring the constitutional supremacy of the civil over the military and that too, exclusively at the political-level only, and hence from then on very befittingly to be always, as per the overall gamut of the sovereign India's national-governance*).

And hence both historically, as well as all the more amazingly shocking, happens to be that—the so much talked about, Nehru's geopolitical, diplomatic, and strategic-vision, his political seriousness towards the crucial understandings of the matters-military, and also about the **'multiple-aspects of essential strategic-culture and thoughts'** as the first Prime Minister of the newly independent India, especially during the very first years of the India's fledgling democracy, and thus in his own words—India's 'tryst with destiny'—so far as the Indian-Nation's geostrategic and politico-military perspectives were concerned, in India's already the then greatly changed overall security environment, and then the always quite fast changing aspects of her threat-perceptions; **Nehru remained quite prone to always 'trip and slip', while dealing with the nation's matters-military, and the rest is not so palatable history for the sovereign and democratic India and the Indian peoples**. And hence precisely in these multiple-contexts the national-defence-related adverse fallouts have been far too many and also dangerously very far-reaching, and their ill-effects are still quite malignantly lingering and thus no less adversely continuing to affect the India's overall security environment, as well as, her defence preparedness. We have since already looked into the very crafty and indeed, the equally undesirable 'machinations' of the earlier ICS-bureaucrats.

And then subsequently in addition to all these and, with not even slightest concerns for the vital aspects of India's matters-military, as well as, with their total disregard for the highest national-interests soon to follow the IAS-bureaucrats too, were also no less unsparing, and always ready to conspire in all sorts of their sinister ways to bring down the status and importance of the democratic India's unquestionably apolitical military-brass to only satisfy their arrogantly inflated-ego and also, false superiority-fixations. And consequently, the otherwise always paramountcy national-security aspects have repeatedly been damaged by such kinds of the extreme bureaucratic callousness. We would contextually be coming onto all these seriously worrisome aspects later from time to time, as we proceed further with our ongoing discussions. Nevertheless for the time being, in order to further substantiate now the historically and even contemporarily irrefutable—the above discussed observations in these regards—it no less perceptively appears to be very important and contextually quite relevant to quote certain factually very revealing paragraphs; from the book of again, one of the independent India's famous, patriotic, and the professionally brilliant Generals; **Lt. Gen. SSP Thorat** of the 1950-60s era. **The upright and honestly thoughtful General, in his well-known book on those bygone times, *Reveille to Retreat* was also able to very frankly picturise the independent India's clueless and arrogant politico-bureaucratic combine in a dangerously very poor-light.**

And accordingly in the above referred book, **General Thorat** has very revealingly, as well as with a great sense of history, professionally with a befitting apolitical courage, and hence for posterity has written to straighten the records that, **"Gen. Cariappa's PSOs made a happy and united team which was a fortunate combination**, for the relations between the Army HQ and the Min. of Defence were, to say the least only lukewarm. From our side we blamed the Defence Ministry and the Financial Adviser; who in our opinion, lacked a proper understanding of the Army's problems and needs, and had an unsympathetic attitude towards them. We also had a cause to believe that they and most of the senior officers of the Defence and Finance (Defence) Ministries had an exaggerated opinion of their own ability and status, and rather looked down upon the senior Army officers. (*Nehru's own mindset was similar and hopelessly 'out of tune' of the independent India's 'stark realities'.*). Some of them believed that they knew more about Army affairs than we professional soldiers did. (*It appears that right from the beginning after India achieved her freedom; among the bureaucrats in the country, there has been no limit to the totally baseless, highly inflated and, having an arrogant culture of always being false and blindly self-opinionated, on the vast and highly complex domain of the 'matters-military'.*). They were in a position to rough ride us because of the tremendous influence they wielded with the Defence Minister. **Sardar Baldev Singh** was a perfect gentleman but did not have the courage to control the officers of his ministry, even when he was satisfied that they were being unreasonable with the sole object of 'establishing the superiority of the civil (*i.e. so sinisterly, the civil servants.*) over the Army.... Even while (*i.e. as GOC-in-C Eastern command*) I was attending to the many civil and military problems arising out of the Naga unrest, I was seriously worried about what the Chinese were likely to do on our northern borders. They had full seven years to consolidate themselves in Tibet after annexation in 1950 and had made full use of this period. Judging by reports from the I.B., the Chinese were building a road link in China with Tibet. They were also constructing a net-work of roads within Tibet, many of which were undoubtedly intended to facilitate rapid troop movements." (Thorat, 1986, pp. 115 & 189)

But all the above mentioned highly worrisome problems then, so dangerously threatening India's national-security aspects and even, the Indian peoples' larger well-being amazingly were perhaps, looked like matters of no concern for Prime Minister Nehru and his clueless teams the so callous 'politico-bureaucratic combine', as they remained busy in altogether singing the foolishly make-believe song of the fictitious 'Hindi-Chini bhai-bhai', i.e. 'Indians and the Chinese are brothers'. **Well, so much for Nehru's, as well as his political-cum-bureaucratic minions' concerns for the multiple-aspects of 'essential strategic-culture and thoughts'! And the regressive 'cobweb' of the clueless Indian establishment led by Nehru at the nation's indeed apex-level, kept on living in their all sorts of the geopolitically so naïve and imaginary worlds, indeed totally devoid of the any serious concerns for the harsh ground realities till the Chinese military's 'kicks at their backs' in October 1962.** Nehru's otherwise damaging and the absolutely unjustified aversions for the matters-military and, the always unquestionably patriotic officers and men besides playing havoc with the nation's essential military preparedness in turn also, quite untenably encouraged Defence Minister **Krishna Menon**, who seldom spared the outstanding Army Chief, **General K.S. Thimayya**, from misbehaving with him. **Those were indeed the darkest hours for the Indian military courtesy—Nehru, Krishna Menon and also the no less clueless, selfishly egoistic, and**

relentlessly conspiring bureaucrats, since seated in various Ministries of the Government of India. And hence due to all these historically unbecoming developments indeed the 'recipe' for the greatly embarrassing 1962-fiasco was fast getting completed, by the so much arrogantly-cum-geopolitically 'blind', the Indian politico-bureaucratic combine. **These extremely sorry states of affairs, had even once forced General Thimayya to resign from the Army.** But indeed objectively speaking in the larger historic perspectives in these regards one would not hesitate to say that, if Nehru would have been able to manage India's these paramountcy geopolitical-cum-military challenges, with requisitely sound strategic prudence and vision, the lion's share of glory would have rested exclusively at the doors of his Prime Ministership only. Precisely with respect to the sad commentary on indeed the so unfortunate, 'General Thimayya-Krishna Menon episode'; we may like to refer to the historically well-crafted biography of General Thimayya, by the author: **Brigadier C.B. Khanduri** (Retd), of the Indian Army very informatively, as well as for putting the records straight has written that, "**Thimayya's popularity, despite the admonition by Nehru in the Parliament, continued to grow**. (*Well in any case, it was most uncalled-for and ungentlemanly on the part of Prime minister himself, on his 'resignation', since Nehru was hitting him so irresponsibly and unfairly, 'below the belt'; as the General had already withdrawn his resignation like a disciplined and patriotic soldier, on Nehru's own persuasion only, and in addition nevertheless Nehru should have at least remembered that Thimayya was in no position to defend himself in the Parliament. In fact Nehru, only in a very selfish and short-sighted manner, was trying to 'save his face and image' before the nation.* ***But in any case, it was all very short-lived for him, since the Chinese 'military-kick' soon, exposed him thoroughly and, he and Krishna Menon, so unfortunately stood discredited and at the same time Thimayya and his strategic-thoughts, stood all the more vindicated before the Indian peoples.***). In fact, it grew beyond description. One such example was his visit with Menon to Secunderabad a day after the inauguration of the School of Land-Air Warfare.... **The School of Land-Air Warfare, Secunderabad, had been raised by the concerted efforts of Thimayya and other Chiefs as a small example of integration of the Services**. Here it took shape when Menon inaugurated it on 03 September (1959)—and when the 'episode' was on everybody's lips. Thimayya had arrived here on 2 September while Menon flew in on 3 September morning.... Opening this **'first of its kind in Asia' Institute**, Menon took the opportunity to play up the issue of 'friendly rivalry among the Chiefs of Staff'. 'Our Chiefs of Staff are not given to any sense of separatism but are bound by loyalty to the service of our land,' said he. Then he smiled quizzically as he remarked, 'Friendly rivalry however, would be helpful to the development of forces'. (*Look at this, limit. No scruples perhaps, even as a Defence Minister.*).... Then Thimayya rose to speak and the whole gathering, including the VIPs, were on their feet, giving him a standing ovation. A tremendous ovation for a hero. So jubilant was the audience that a fly-past by a few Otters aircraft did not distract them, until the airborne troops began to 'land from the sky'. Menon's deep-rooted envy had already turned into malice; now it was pure and simple hatred, turning eventually into a mad vindictiveness, which he would show on his return from New York. .... The resignation episode had to be erased from public memory, especially of the officers and men. By October beginning, as Nehru conveyed the correct impression, he wrote to the army commanders and commandants of the schools of instruction explaining his position. Belatedly there followed the deep and well-founded anguish of **General Cariappa** whose letter of 7 November made a deep impact on Nehru. For he wrote scathingly:

> 'Thimayya put in his papers to leave the army for reasons, which he, after long and deep thought, must have thought, were strong enough to do what he did. It was a major decision in his life and could not have been taken just as a school boy 'joke. Apparently, it was

> said in the Press, the other two Chiefs of Staff were likely to follow suit. However, you persuaded Thimayya to withdraw his resignation, which he, like a good soldier did, with his sense of un-bounding loyalty to you and to our country. You then went to the Lok Sabha and told a packed house that Thimayya on your advice had withdrawn his resignation, but you thought his reasons for sending in his resignation were very trifling, and you could not congratulate him on what he had done, and so on. **You emphasised that civil is supreme, although this fact had never been even once raised by anyone in the army and other Services. Panditji do you realise how humiliated Thimayya and the army are about your statement in the Lok Sabha?'**

Nehru replied post-haste to General Cariappa's letter on 19 November: 'I do not understand why you should consider what I said about Thimayya, taken as a whole, as unfair to him. (*In fact in retrospect, one is able to clearly see the plain-truth that, Nehru was too arrogant and awfully occupied promoting his all sorts of—the fictitious geopolitical-ideas across the world—to be able to understand the fine 'nitty-gritty' of India's so compelling matters-military; as the nation's first Prime Minister, with any long term geostrategic-cum-politico-military vision in these regards.*). You must remember that I was answering supplementary questions put to me after I had made a statement. I paid a handsome tribute to Thimayya. I certainly did say also that I did not congratulate him for his offering his resignation. I thought, then and I still think, that it was not right for him to offer his resignation for the reasons he gave me and, more particularly at that juncture when we were facing a rather difficult situation on the frontier. I told Thimayya so when I asked him to withdraw his resignation and he agreed to do so. He had mentioned some incidents to me which seemed to me rather trivial, even though they might have been irritating. (*Sick. No limit to Nehru's 'double standards' in his so much publicized but 'dubious matters of principles'.*). It is to these incidents that I referred in the Lok Sabha and not to other causes that might have influenced him. It was, indeed, very far from my intention to 'humiliate', as you say, Thimayya. It would have been absurd for me to do so and yet ask him to continue in the high and responsible office that he holds. In fact, the whole trend of what I said in the Lok Sabha was different and was more of a compliment to him. When I am asked question in the Lok Sabha, I cannot refuse to answer.... (***Well, whatever might have been Nehru's intentions behind his apparently some very 'pedestrian' kinds of logic to justify what he so callously did, one historic fact is very sadly clear that Nehru had no moral or any strategic grounds then, to humiliate Thimayya in any manner. In fact, what indeed the very harsh history shortly unfolded then that, it was Nehru only, as India's the then first Prime Minister who was so squarely humiliated by the Chinese, onto always the ruthless and very dicey 'arenas of international power-politics and also, the no less very crafty international-diplomacy' too; and thus at the same time obviously in turn, bringing-in some very frustrating embarrassments to India and her innocent peoples. And sadly, Nehru stood alone and dejected; and he had to blame himself and none else.***).'

To tide over the problems created by the episode, Nehru diplomatically smothered the rough edges of the matter. He pacified the aggrieved and deeply hurt army chief. Thimayya began to trust him again. Then he informed his chief ministers about the episode through his newsletters. He would pacify a deeply annoyed President. And finally, he told Menon to quit the 'battlefield' for two to three months by going into the UN, paradoxically, for preparing India's stand on the Chinese entry into the UN (besides taking a non-communist stand on the admission of Tibet's case in the UN, at the behest of Dalai Lama. By these overtures to the Chinese in the world forum, Nehru was hoping to diplomatically win them over.).... In an overt display of full trust in Thimayya, Nehru visited Rawalpindi, Tehran and Kabul, during the second half of September. (*All these futile, as well as certainly not in India's strategic interests, initiatives on the part of Prime Minister Nehru, speak volumes about his geopolitical-cum-diplomatic naivety and fictitious thoughts. Instead of all these 'exercises in futility', since so 'ambitiously' undertaken by Nehru then, it would have been much more far-reaching and also, would have greatly served India's vital geopolitical interests; if he would have brought his top military commanders proactively closer and more involved into his policy-level loops and thereby, paid due attention to India's already almost 'criminally' neglected matters-military, and due attention to the preparedness of her the no less arrogantly marginalized Armed Forces then, the Indian peoples would have been saved from many anguish and humiliations, in the eyes of world*). With Menon in New York, the MoD (*i.e. Ministry of Defence*) virtually had Thimayya as the senior most, besides Majithia who officiated as defence minister. Thimayya and Majithia utilised this opportunity by visiting ordnance factories as also the Chinese border in NEFA.... 'This is the time of productivity and consolidation', he told Majithia 'and I hope we can get a few urgent cases through'. Majithia helped. He began to pursue the old cases of Army Headquarters gathering dust over the past 30 months or so." (Khanduri, 2006, pp. 268-270) (*But, it was all too late anyway; as per the India's, both short and long terms, much larger 'geostrategic-cum-politico-military' compulsions.*).

Now coming back to some important aspects of India's well-known and also very embarrassing fiasco, during the 1962 border-clashes with China in the north, another military legend **Arjan Singh, DFC, Marshal of the Indian Air Force (i.e. India's first five-star rank Air Force officer)** too, has made some professionally very revealing comments. Precisely specific to the domain of matters-military in these regards, and then also covering the glorious role of the Indian Air Force during 1965 Indo-Pak War, Arjan Singh's very perceptive comments cannot be taken lightly in any manner and at the same time, it contains some very valuable lessons for future too, of the Indian Nation-State. In a very brief and pictorial biographical-sketch of the legendary Indian Marshal of the Air Force, the author **Roopinder Singh**, has brought some historically interesting and also the no less militarily revealing sides of India's this so famous Air force officer and his times.

In the above referred book the author writes that, "In 1960, Air Vice-Marshal Arjan Singh went to England for the one-year course at Imperial Defence College (now called Royal College of Defence Studies) in London. During the break at IDC in August/September, he went to the USA and Canada, where he got good exposure to the operational procedures of their air forces and also met a number of officers, some of whom became good friends. He says that he was surrounded by highly decorated officers and 'felt like a mini' out there. After this course, which prepares officers for higher postings, he held various senior appointments at Air Headquarters. He was Air Officer, Administration (1961), Deputy Chief of the Air Staff (1963) and subsequently Vice Chief of the Air Staff.... He was in the Air headquarters when the Chinese invasion in 1962 took place. As Arjan Singh recalls; 'It was not because of the fighting ability of our soldiers that we suffered; it was because of bad leadership in the Army and the political interference'. It was decided that the Air force should not play a fighting role in this war, because once the air forces get involved, it ceases to be a limited engagement. As it turned out the Air Force did play a 'non-fighting' role in logistic support and in evacuating wounded soldiers. Even though he was not in command, Arjan Singh says the Air Force was prepared to fight and the Chinese advance would have been much restricted had the Air force also participated in that operation.... Arjan Singh took over as **Chief of the Air Staff (CAS)** on July 15, 1964, and within a year came the most testing time of his life.... Pakistani forces infiltrated into Kashmir in August, 1965, and later attacked the Chhamb-Jaurian sector with regular army formations on September 1, 1965. It was a fierce attack in which a Pakistani force of two infantry brigades and two armoured regiments were involved. The 3rd Mahar battalion bore the brunt of the attack. They were supported by a squadron of tanks of the 20th Lancers, but there was no artillery support since the enemy shelling had damaged and dislocated some of our artillery.... The Army in Kashmir sent a request for an air support at 11 am. **General J.N. Chaudhury**, Army Chief at the time, asked the Air Chief if the Air Force could help. Both the service chiefs went to the Defence Minister, **Y.B. Chavan**, who gave the permission within few minutes. He asked the CAS: 'Can you do it?' As he recalls, the CAS replied: 'We just can. If you give orders now, we may be able to attack before night.' The urgent note was justified because it was 4 pm when the order was issued and the Indian Air Force lacked night vision equipment and the attack would be launched in twilight, which would be risky.... Chavan gave orders though, as he recorded in his diary: 'Had no time to consult ECC or Prime Minister. Time was a vital factor. Took decision on their (the Chiefs) advice and asked them to go ahead'. Chavan maintained a diary and his records have been printed in *Debacle to Revival: YB Chavan as Defence Minister 1962-65*, by **RD Pradhan**, the Defence Minister's aide.... The air attack was launched from Pathankot, near the border of Jammu and Kashmir that day. Vampires of World War II vintage were used for the sorties in addition to Mysteres and Gnats. The first wave of four Vampires took off at 1719 hrs, as soon as they got orders. They also attacked, some of our own Army positions and troops. One Vampire was lost to ground fire in this attack. The second wave of four Vampires was attacked by Sabre jets, which were superior in performance and armament, and three Vampires were lost in that attack, with one managing to escape. The fourth wave did not encounter any aircraft and attacked enemy troops, armour and vehicles.... Arjan Singh recalls: 'We were able to attack Pakistani troops but, unfortunately, hit some of our own troops as well. We lost four Vampire aircraft, but the whole operation was a major success. Though planning and executions could have been better, the objective of stopping Pakistan's advance was achieved.... This was the beginning. The war escalated and the Indian Army and Indian Air Force performed well as judged from some of the headlines in the various newspapers during the period as well as from Chavan's diary.... During the war, the Indian air Force penetrated deep into Pakistani territory and attacked military targets around Peshawar, Quetta, Rawalpindi, Sargodha and other Pakistani cities. As many as 72 trains and many vehicles were attacked, disrupting the Pakistani army's supply lines. They scrupulously followed Prime Minister **Lal Bahadur Shastri**'s instructions to attack military targets only.... As the war progressed, the pilots' confidence increased. Initially, there had been some doubts about the ability of the indigenous Gnats against the highly sophisticated American-built Sabres, but even those flying the Gnats showed a lot of élan, particularly after they shot down two Sabres of the Pakistani Air Force in the initial days of war.... It was a short war. Mainly under international pressure, a UN brokered cease-fire was announced on September 23, 1965. The Army had conquered considerable Pakistani territory, and the war had seen the biggest tank battles since World War II. Yet the fighting men were itching for more." (Singh, 2002, pp. 48-57)

As per broader dimensions regarding the overall geostrategic and politico-military perspectives concerning the Indian Nation-State, as well as its style-cum-ethos of governance the above quoted paragraphs contextually, conveys many things to the critically discerning minds, whether Indian or from abroad, who may care to historically study, analyse and evaluate the overall conduct of 1962 Indo-Chinese and the 1965 Indo-Pak short wars and that too, mainly in India's so much politico-militarily always highly volatile northern and western sectors. **As the numerous lessons of military-history, since the every advent of human-civilisations, as well as at the same time always inseparably remaining organically integrative segments of the much larger and multi-dimensional the mankind's overall living-history, being continuingly in the making onto its expanding civilisational-canvas have been conveying that—though the military ranks and files and the various kinds of weapon-platforms, as and when deployed and activated in the**

battlefields, do the actual fighting against the opposing forces—another kinds of the no less intense and challenging battles/wars are mentally fought in the minds of the higher military commanders and the concerned nation's political leaders at strategic, and the politico-military-cum-diplomatic levels, for ensuring the requisite success in these regards. The imperatively multi-dimensional as well as the visionary awareness, as being always proactive segments of the inescapable thought-processes, in the crucial domains of the any concerned nation's just above identified much higher-levels of the politico-military leaderships and their potentials in fact do principally-cum-fruitfully get triggered as per the systemic patterns of the proactive presence of since, that nation's, 'multiple-aspects of the essential strategic-culture and thoughts', in their entire gamut of governance, as well as their prevailing culture of the decision-making processes. And precisely in these areas, lie the proverbial 'keys', to desired success or the reasons for any failures. In precisely the above multiple-contexts, as well as particularly concerning the sovereign and also vibrantly democratic Indian Nation-State, and the very dawn of her political independence in 1947, what we have since been very ironically noticing the one specifically crippling lacuna and that being that, there has not been any constitutionally well-crafted, healthy, and really effective institutionalised arrangements in the always so crucial areas of the highest-level politico-military decision-making processes. And obviously due to this so apparent and specific lacuna, not only the multiply paramountcy synergy in these regards are prevented from being achieved, but also when any really challenging situation crops-up in the highly complex and dicey areas of the national-security aspects, much of the politico-diplomatic as well as at times, the nation's some decisively advantageous military-assets too are even very embarrassingly prevented from being timely put to their full applications, as per the dictates of India's highest national-interests. And hence, these always so pressing points and thereby the so well-known lacunae therein, concerning the Indian-Nation's geopolitical, as well as the geostrategic-cum-politico-military domains, would contextually be discussed in some more, and also all the more important very contemporary details, little later in this sub-chapter itself.

For the time being and precisely in these contexts, we may very appropriately prefer to take a note of the some very thoughtful perceptions of another very famous and truly legendary officer, **Lt Gen PS Bhagat, V.C.**, to further drive the points home. General Bhagat, during his times and especially after the Indian debacle during 1962 border-clashes with China, as well as quite justifiably upbeat (*i.e. considerably unlike the 1962 Indian debacle.*) in the 1965 Indo-Pak War, had recorded his thoughts with great professional precision, and the same have since been published as a valuable handbook in the year 1967. General Bhagat has written that, "It is little wonder therefore that even responsible persons in India have little concept of war and its concomitants. If we are not to grope in the dark, if we are not to be intimidated by threats that we cannot militarily evaluate, and above all—if we are to understand war, it is vital to have knowledge of it. This does not necessarily mean that each one of us has to be a military expert, but it does require that the basic essentials of war should be much more widely understood.... In Western countries, where succeeding generations have experienced war, knowledge of its general principles and of its effects has become almost second nature. This is not so in India nor in countries similarly placed.... **Prosecution of war is the responsibility of the government but preparedness for war involves the entire nation. Without a general understanding of what war means, a nation will be gravely handicapped.... It has been truly said by an eminent British General, that unpreparedness in a nation is of three kinds—moral, material and mental. The last is the most dangerous. It is from mental unpreparedness for war that conditions for the other two generate. Fear of the unknown is proverbial. Without an understanding of war, dangers and threats are magnified or else people**

are lulled into a false sense of security; both are equally harmful. War and civilisation seem to have been inseparable. As civilisation progressed over the centuries, so also have wars increased in magnitude, scope and power of destruction. Where once a tribe was wiped out now a race or a nation is involved. The next war could destroy civilisation. Yet while there are strong nations and weak, rich and poor, warlike and peaceful, the possibility of war between nations cannot be ruled out. It is possible that civilisation will learn from history and outlaw war.... At present there is little sign of it and wars can be on many scales.... That being so there should be no inhibitions against fighting for survival. There is no gainsaying the fact that war is bestial, but this very fact means that a nation should be conditioned to face it rather than avoid it.... Diplomacy must first ensure the integrity of the country and no other considerations must be allowed to jeopardise it. Diplomacy must be aimed at cutting off the supplies to the potential enemy or, conversely, at securing aid to balance it. If neither is achieved diplomacy has failed, and the nation endangered. If both succeed, security is ensured.... It stands to reason that unified coordination of the three services under a military expert would be more effective than a loose arrangement under the Defence Minister. Admittedly the Chiefs of the three services will have a major say in military policy-making in either case. In both, there has to be a Chief of services Committee to consider inter-service and military problems. The difference is that where there is a Chief of Defence Staff, he heads the Armed Services and is the permanent Chairman of the Chiefs of Services Committee. Where there is no such appointment, the Service Chief longest in the appointment takes over as Chairman.... Where there is an appointed Chief of Defence Staff, as head of the Armed Forces, he has the overall statutory responsibility for their coordination. Admittedly at that level there can be no arbitrary decisions and there must be a degree of acceptance of the service Chiefs before a policy can be finalized. On the other hand, being above and not part of any of the Services, albeit that he originally belonged to one, he is not to that extent influenced by any particular service requirements.... **Final decision on major policy matters will of course be taken by the Defence Minister or in some cases even the War Cabinet. The Chief of Defence Staff, however, is in a better position to advise the Defence Minister than a Service Chief, who is incidentally the Chairman of the Chiefs of services Committee. This is not to say that the Services Chiefs are not consulted by the Government. They are, but only for advice regarding their own service, within the overall accepted inter-service plan**." (Bhagat, 1967, pp. 9, 10, 15, 20, 59 & 60)

At this precise point, we may also recall that in differently very relevant contexts we have since been emphasising the universally timeless **'multiple-aspects of essential strategic-culture and thoughts'**, and which India has since traditionally been very sadly lacking, barring in case of the some historically noteworthy and very outstanding personalities coming onto the Indian scene off and on, from the very beginning itself. Now further at this precise point of the sovereign and democratic India's 'living-history', we may very logically and also justifiably expect that if the above discussed institutional mechanism is proactively ensured in her since multi-dimensional domain of the 'matters-military'; then the **'multiple-aspects of essential strategic-culture and thoughts'** too, would remarkably begin to get fruitfully cultivated, evolved, developed, groomed and thereby also, flexibly-cum-proactively would get imbibed across the nation.

In these precise and very crucial perspectives, as well as their action-areas, the main hurdle remains the deeply entrenched, sinisterly seeking self-importance—the almost all pervasive bureaucracy and the no less regressive bureaucratic mindset—in the entire gamut of the India's national governance. We may also contextually recall that, we have since already discussed these crucially important aspects of the India's civil-services in their differently historic, as well as contemporary perspectives in the earlier chapters too, in somewhat their desired analytical criticality. **In these multiply important perspectives, we may need to realise that the essentiality of civil-services in the actual functioning of the entire governmental administrative systems at different levels of the civil administration across the nation cannot be minimized and at the same time thereby, their contributions too cannot be overlooked. And then so far as the multiply vast and also the no less complex aspects of the soldiering, the always multi-dimensional-cum-multi-layered functioning of the nation's Armed Forces and, the Defence Ministry too are concerned the valuably important place and contributions of the civil-services personnel cannot be sidelined. Further on, the coordination, as well as ensuring the differently vital cooperation of the so many different kinds of civil administrative departments as the military's facilitators too, cannot be overlooked in anyway.**

But the centrally most important point in these regards remains that, while a positive and broadminded, patriotic, truly dynamic and visionary bureaucracy, remains not only the most desirable but all the more essential too; at the same time any regressive, with a very damaging and egoistic-culture of mostly the sinisterly scheming and conspiring bureaucracy so much entrenched in the Defence Ministry, can turn out to be nevertheless an extremely problematic-drag onto the entire preparedness-cum-functioning of nation's Armed Forces, thereby even dangerously affecting the paramountcy national security-aspects. And hence so far as these highest national-security aspects and the nation's overall wellbeing are concerned specifically the numerously vital, as well as always the multiply desirable reforms and also, the complete overhaul of the whole systems of India's civil-services, are so much called-for, especially if the Indian-Nation's entire gamut of governance and thereby, the civil administrative-systems are to be made more healthy, dynamic, effective, and always positively results-oriented, down the line at different levels. And needless to say that, precisely here lies the 'rub' since the independence of India, as well as the formation of the so much cherished Indian-Republic; and so far as the many hopes and aspirations of the Indian peoples are concerned.

Now we may not hesitate to say that the long years of India's Prime Ministership of **Mrs. Indira Gandhi**, happen to be indeed a 'water-shed' development in the independent India's history so far. In fact, the whole socio-political-cum-administrative culture of the entire gamut of the governance of Indian Nation-State underwent a 'sea-change'; with Mrs. Gandhi coming to power, after the unfortunate death of the Prime Minister **Lal Bahadur Shastri**, in the then USSR's city of Tashkent, in the year 1966. All these were so colossal, as well as far-reaching; that in the process, the entire Indian-masses' age-old and the largely indigenous culture-based value-systems—concerning their various aspects of human morality, and the corruption too—were mercilessly turned upside-down. And needless to say that the momentous consequences in terms of the socio-cultural-cum-economic and indeed politically too, have been almost 'devastating' for the nation. These numerously all pervasive and also, the extremely sad-aspects concerning the overall Indian polity; are clearly as well as massively so visible that, all these in any case don't now require, any kind of the even remotely laboured 'proof' for any substantiation whatsoever. Though Mrs. Gandhi's strong political leadership during the 1971 Indo-Pakistan War; resulting in the creation of a new nation of Bangladesh out of erstwhile East Pakistan, had indeed been India's finest hour, but ironically she couldn't prevent the squandering away of the unique geopolitical-cum-military gains so brilliantly achieved by the Indian Armed Forces, principally due to her own geostrategic-cum-politico-military naivety and, the short-sightedness of her clueless advisers. Further the non-inclusion of the top-level military professionals as experts into the team of her advisers, prior to and, during her summit-meeting with the then Pakistani President **Mr. Z.A. Bhutto** at Shimla in the year 1972 resulted in again rather a well-known and also considerably embarrassing historic-fiasco for the Indian-Nation. A number of the analytically well-informed and scholarly books by the various civil and military experts have since already appeared on these historic subjects, for the benefits of readers across the societies and, for posterity too. **But again further-on, the no less quite paradoxically, the 'unlawful' imposition of the internal-emergency in the year 1975 by Mrs. Gandhi; to politically silence and harass her political opponents was in fact, a great constitutional subversion of the entire Indian-Nation, as well as the systems of its political governance. And further in process, she no less tragically didn't even hesitate to subvert and somewhat even 'ransack' many other constitutional and important Institutions, including the Judiciary and the Armed Forces and thereby; many damages as inflicted onto the India's body-politic, are so serious and far-reaching that these have not yet been fully repaired and constitutionally restored.**

In precisely the above multiple-contexts, **Simantini Krishnan**, in her article very perceptively, as well as with a great sense of history has written that, "The restoration of democracy in 1977 notwithstanding, the emergency did not penetrate public consciousness. The democratic upsurge that gripped Indian politics in the post emergency period did not sustain long enough to loosen the Congress Party's stranglehold on the telling of history. Rather, the Congress' return to power in 1980 ensured that very few accounts of the emergency remained available. The Shah Commission appointed by the Janata government had submitted a detailed report of the atrocities committed during the emergency years. But Indira Gandhi rejected and withdrew the report on her return to power. It was only in 2010 the sole surviving copy of the Shah Commission was retrieved and published by veteran writer and DMK politician **Era Sezhiyan**. The odd book such as John Dayal and Ajay Bose's 'Delhi under Emergency' remains out of print, while text books for school children mostly gloss over the chapter of India's past. Not surprisingly, the emergency years have little recall value or relevance for Indians aged less than 45 years.... **Contrast this with a country like Germany that was left broken and shattered after the Second World War, children were taught about the country's tryst with a fascist dictatorship with the specific intent of inculcating responsibility for burying racism, anti-Semitism and xenophobic nationalism. It was a challenging proposition, for ordinary citizens had been complicit in a regime that promised careers and professional security at a time of great economic adversity. Even after the fall of the Berlin Wall, the federal government in Germany pursued the policy of acquainting young Germans with grim details of the holocaust. The sense of history imparted by such an approach may initially have come at the cost of pride in Germany's rich culture and heritage. But gradually, the new generation of Germans was able to emerge with a deep understanding of the past minus the burden of guilt for the occurrences in Hitler's Germany**.... Much has changed in Indian politics since the 1970s. Most significantly, the rise of regional forces has shifted the centre of gravity away from New Delhi to the states, checking the centre's tendency towards authoritarianism. In the 70s and 80s, the Congress Party did not win more than fifty percent of the elections in the state assemblies across the country. The government in New Delhi responded by rolling back the fiscal and political powers of state governments. Additionally, Article 356 was invoked about seventy five times in these twenty years. The incidence of Article 356 has reduced drastically over the last two decades." (Krishnan, *The Sunday Statesman*, 12 January 2014)

As we may care to read closely and 'in between the lines' of the above quoted paragraphs and specially concerning the multiple-aspects of the socio-cultural-cum-political value-systems, as well as the long-term strategic-vision of the German people and their leadership, and that too even after their so much devastating defeat in the 1939-45 Second World-War, we would indeed very enchantingly get mesmerized by their so pragmatic, matured and visionary 'sense of history' and thereby their multi-dimensional keenness to learn from their own past mistakes, and some historically cruel blunders. **These indeed invaluable civilisational-cum-socio-political attributes, concerning the Nation-State in any case do also sprout, multiply flourish, as well as the no less in their all-encompassing imbibing-manners; then flows-down from that nation-specific—the 'multiple-aspects of essential strategic-culture and thoughts'—and thereby consequently, their imprints too do get implanted onto the pages of that nation's 'living-history', since continuously in the making onto its own distinctly expanding civilisational-canvas in the world**. In other words we may also contextually say, that these aspects do always form quite importantly, the integrative segments of that specific nation's dynamically fruitful 'sense of history'; and needless to say that all these in turn, go a long way in further positively shaping-up its living-history for the overall well-being of the peoples of that very nation.

Now contextually coming back to the highly chequered times of the India's Prime Ministership of **Mrs. Indira Gandhi** the very shocking controversy related to the operation 'Blue Star', since then so recklessly launched into the Golden Temple at Amritsar in 1982, indeed happens to be a greatly nauseating recent blot onto the pages of India's 'living-history'. Lot of writings have since already appeared, covering this extremely tragic episode, for the Indians to learn from their own blunders. India's top political-journalist, **Mr. Kuldip Nayar**, in his article has written that, "Mrs. Gandhi sanctioned the operation, provoked by the fear she felt of the militants who used the Golden Temple as their shelter.... The then British Prime Minister, **Mrs. Margret Thatcher**, was Mrs. Gandhi's close friend, and I believe that Mrs. Thatcher was privy to the Operation Blue Star, because of the relationship she and Mrs. Gandhi enjoyed. They talked daily on the telephone. Mrs. Thatcher herself told me, when I was India's High Commissioner to London, that they would converse on affairs related to India and the UK. They could not be discussing the weather. Sikh militants would have almost certainly figured in their talks. (Once I asked Mrs. Thatcher whether she was in touch with Prime Minister Rajiv Gandhi, she peevishly replied that 'he was a different person'.).... Indeed, before sending the army into the Golden Temple, Mrs. Gandhi was frantically seeking opinions from different people on whether or not to undertake the operation. **RK Dhawan**, her aide, came to my residence to find out what my reaction would be, should the troops enter the temple. He indicated specifically that Mrs. Gandhi had sent him. I told him that he would not have come without her permission. Knowing well how annoyed she was with me because of my writings.... 'Don't you ever think of sending the army because the Golden Temple is the Sikh Vatican,' I warned him. 'If you ever did that, the Sikhs would never forget or forgive this sacrilegious act.... Whatever the government's compulsion, the onslaught on the Golden Temple was not justified. In fact, they should have found some other way to tackle the **Bhindranwale** challenge." (Nayar, *The Statesman*, 30 January, 2014)

What Mr. Kuldip Nayar says in his above quoted lines, in fact appears to be very logical and correct. It was the prime duty of Indian Government of the day, to make serious, as well as politico-strategically pragmatic efforts to solve this indeed very dangerous problem; while there was still time in the beginning itself and also, when Bhindranwale was yet to grow in some 'stature' as a menace; i.e. when he was still sitting in Amritsar's 'Mehta Chowk' (i.e. a road crossing with a pedestal). But when the concerned top political-executive is so focussed on his/her domestically sinister designs and un-statesmanlike 'plans', then the leader gets confused and develops a totally blurred-vision and hence, it is only a matter of time when all possible kinds of blunders get triggered, with some extremely adverse consequences. Almost all essential aspects and parameters of governance were twisted and done away with, all geopolitical compulsions being then so mandatory, were also callously ignored and thereby the no less paramountcy imperatives concerning the Indian-State's geostrategic-cum-politico-military dimensions with respect to highest-level of the then Indian Government's decision-making processes too, very arrogantly flouted and no wonder that the rest still remains indeed, a very deeply tragic and unpalatable saga of the modern and democratic India's very recent 'living-history'.

It is now historically true and rightly said that a 'job' which, an ordinary Police Sub-Inspector would have been able to easily do by lifting Mr. Bhindranwale and arresting him in the beginning itself, while he was still sitting onto the Mehta-Chowk pedestal with his few ill-informed and fanatic followers became a 'monster' when the Indian Army had to be called-in to 'liberate' the Golden Temple from the 'Khalistani' terrorists. This is a classic example of the very recent history of the mankind anywhere, depicting indeed a very tragic and violent saga as consequences of the so callous blunders since committed at the highest politico-military levels of any independent, modern and democratic Nation-State.

Precisely with respect to the above, **Mr. Khushwant Singh** a very popular and celebrated scholar and author has as usual very frankly written that, "I was opposed to the **Anandpur Sahib Resolution** which described the Sikhs as a nation apart from other Indians and thus germinating the seeds of a demand for a separate Sikh State—Khalistan. I felt passionately that Khalistan would be fatal to the interests of the Sikhs and to India.... I had an exchange of views with retired **General Shahbeg Singh**. He was another tall, lean man with a long, grey beard. He had been clean shaven when he trained and led the **Mukti Bahini** in the liberation of Bangladesh. He had been cashiered on charges of corruption and was very bitter with the government. I discovered that, without a single exception, all the army officers and men who joined the Akalis had been dismissed or superseded in service and had personal grievances for the way they had been treated. At one time there were over 10,000 with the Akalis.... 'You don't know the Sikhs,' I once told the Home Minister P.C. Sethi, a peace-loving Jain. 'They can be like a swarm of hornets. You put your head in their nest and you will be stung all over your face.' He assured me that the government had no intention of sending the army into the Golden Temple. So did Mrs. Gandhi, more than once.... it is not known when Mrs. Gandhi came round to the view that she had no option but to order the army into the Golden Temple, and who her advisers were at that time. The names of **Rajiv Gandhi, Arun Nehru, Arun Singh and Digvijay Singh** were mentioned. It was also unknown who chose the date when operations should commence. There is no doubt that President Zail Singh was kept in the dark. When Mrs. Gandhi persuaded him to put Punjab under military rule, she did not tell him that she had decided to order to clear the Temple of Bhindranwale and his armed followers. When it came to Punjab or Sikh affairs, she did not trust Gianiji. And none of her advisers had the foggiest notion of Sikh traditions. They chose 5 June 1984 as the day to launch the operation. It was the death anniversary of Guru Arjun, the founder of the Har Mandir, a day when hundreds of thousands of Sikhs were expected to come on pilgrimage from remote areas. Nor were alternative methods of getting at Bhindranwale considered seriously.... However the army stormed the Golden temple with tanks, armoured cars and frogmen, with helicopters hovering overhead to give directions. The battle that ensued lasted two days and nights. In the cross-fire almost 5,000 men, women and children perished. The Akal Takht was reduced to rubble by heavy guns fired from tanks; the central shrine which both parties had declared *hors de combat* was hit by over 70 bullets.

The entrance (*deohri*) had a large portion blasted off; archives containing hundreds of hand-written copies of the Granth Sahib and *hukumnamas* (edicts) issued under the signatures of the Gurus were reduced to ashes. Even Mrs. Gandhi, who had been assured that, the operation would not last more than two hours was horrified, at the extent of damage caused to sacred property and the horrendous loss of lives. Instead of admitting that she had blundered, she decided to cover-up the whole thing with a barrage of lies." (Singh, 2002, pp. 312, 315, 324 & 325)

What could be more and so much irresponsibly bizarre in thoughts and action, callous short-sightedness and dangerously fixated arrogance, on the part of any top-most political chief-executive in any democratic nation anywhere? At times during the hours of extreme stresses in any highly complex crisis situation anywhere and that too in a multi-party parliamentary democracy, if the political chief-executive, whether somewhat naively or unknowingly commits a blunder then, situation could be different and also at times be managed. But if that concerned political executive-head of a democratic nation, rather with stubborn arrogance and for any unquestioned greed for political-power, somewhat knowingly and also so much 'foolishly' gets bent-upon to commit serious blunders with almost total disregard to the any possible very dangerous consequences and at the same time, even by ignoring many saner and sincere advises then, it indeed becomes extremely sinful and thus, never to be forgiven by that unfortunate nation's 'living-history' itself, since continuingly in the making onto its expanding civilisational-canvas. And obviously these multiple aspects in turn historically, gets gelled onto the pages of that very 'living-history', as an extremely unfortunate saga of pains, tears and human-tragedies, about which we have since already discussed little earlier. But even any such so much unfortunate saga too, does also contain very valuable lessons for posterity and thereby the future generations and their leaders are always fully and mandatorily duty-bound to sincerely learn those lessons, in their highest national interests and all round well-being.

And again for these to be truly meaningful the already earlier briefly discussed—the multiple-aspects of the **'sense of history'** and thereby also, the **'fruitful traditions of learning the valuable lessons from history'**—remain always proactively essential under all circumstances. Now in the Indian-Nation's contexts and thus specifically, while coming back to the 'Bhindranwale-episode' and that of the indeed extremely unfortunate 'Operation Blue-Star', what we do find that the then Indian establishment at almost all levels, had in fact probably totally done-away with the above mentioned and also timelessly paramountcy multiple-aspects, concerning the man's living-history, as well as those of the essential strategic-culture and thoughts in these regards, so callously with no sincere concern for the indeed some inevitable consequences. And all the more paradoxically, the same so much unbecoming and tragic scenario appeared to have prevailed in the then India's military-establishment, as well as in the always highly complex domain of the 'matters-military' as well.

Historically in retrospect, it appears to be extremely bizarre and hence, the no less irreconcilable too that now, these are also substantially corroborated by the so much objective and knowledgeable 'quotes' from the writings of a number of experts and scholars, since contextually put-in here in these regards. So far as the conduct of the then likely 'Operation Bluestar' was concerned and also while there was still time during its actual planning stage itself, some other professional military leaders' suggestions and expert-thoughts, as since already officially put-in earlier before their retirements as guidelines in the readily available records, and which could have been of great values for the subsequent military-commanders, on whose shoulders the real responsibilities came to proceed further. It appears to be very strange as to what prevented them from to make use of these as valuable inputs, when their real need had arrived at the time of its operational reckoning? Indeed always true and also, the so much talked about—**the irrationally entrenched and powerfully almost all-pervasive culture of the 'unholy politico-bureaucratic nexus'**—apart at the time of reckoning, the nation had justifiably expected its active military-leaders as usual, to show themselves in altogether in their differently glorious professional light, within their exclusive domain of authority and the systems of functioning. **In Indian contexts, it is generally accepted that the 'iron would in any case rust as usual, but what to say and where to look for if the so-called gold too at times, begins to rust'? The people in military-uniforms in India and their retired fraternity too; now need to sincerely ponder over any such disturbing scenario, both individually and collectively and with focussed intentions, for improving upon their exclusive domain of professionalism and also, their soldierly uprightness**.

Now specifically in the above multiple-contexts, as well as precisely related to the no less matters-military aspects in these regards; a well-known visionary author and thinker, **Lt. Gen. S.K. Sinha**, former Vice-Chief of the Indian Army, has very informatively and also with indeed thought-provoking historic perspectives, has recorded that. "Only a few days earlier the Home minister had announced in the Parliament that troops would not be sent to the Golden Temple and the President of India had also issued a similar statement. There was no need to issue such statements, when the Government was to act in a manner contrary to what had been stated. The Saudi Arabian commandos had gone inside the Holy Kaaba at Mecca to deal with the terrorists who had lodged themselves there. The police had entered Vishwanath Temple at Varanasi and had arrested the priest for theft. Therefore the troops could certainly enter the Golden Temple where circumstances justified their doing so. However, it should have been ensured that we took the due cognizance of the psychological aspect and not only acted correctly but should have been seen to be acting correctly. I wish the operation had been conducted on the lines of what I had laid down, when I was the Army commander in the West. That may have prevented a sense of hurt and alienation amongst the Sikhs. **After Operation Bluestar, I received a number of letters from my Sikh friends saying that if I had been the chief, Operation Bluestar would not have taken place. All that I can say is that if I had been in the Army, I would have conducted this operation in an entirely different manner**." (Sinha, 1992, back jacket)

In fact, what all have contextually been visualised, identified, and since analytically discussed as an overall geostrategic-cum-politico-militarily prevailing scenario within the entire gamut of governance of the today's sovereign and democratic Indian Nation-State, is indeed far from being socio-politically reassuring for the citizens of this principal South-Asian nation. And the reasons are also numerous but the main 'culprit' in these regards, largely happens to be the already earlier talked about—the so much regressively entrenched, so far unassailable as well as, being powerfully almost all-pervasive the **'unholy politico-bureaucratic nexus'**—since always very sinisterly lingering in India's various 'corridors of power' spread all over the nation.

The pivotal point in these regards is that, why and how it is so and at the same time, how the above mentioned 'unholy politico-bureaucratic nexus', always so callously as well as sinisterly, is able to perpetuate these so obnoxious 'state of affairs', with almost total disregard to the all sorts of extremely adverse and dangerous consequences to the larger well-being of the Indian nation-State? Well that is a different matter that, whenever and in any differently important contexts; any senior bureaucrat, whether serving or retired, or any so-called 'heavy-weight' politician belonging to any political affiliation, is confronted with these extremely important questions related to the nation's far-reaching matters-military, as well as those of the highly complex and paramountcy national-security affairs, they tend to quite disturbingly look the other way and thus, try to brush-aside and confuse these pressing questions, since always unfailingly staring and casting many kinds of the adverse shadows on the future of sovereign and democratic India itself. What could be more bizarre and hence it is rightly said that, sometime hard facts happen to be stranger than even fictions. But whatever, the Indian-Nation by itself is now so pressingly demanding the comprehensively pointed explanations from all sorts of such bureaucrats and politicians and, asking them to mend their ways and also bring-in multi-dimensionally systemic-cum-cultural changes in their functioning without any further delay; for the well-being of future Indian generations, before it is too late. So far as India's multiple-aspects of the highly complex matters-military are concerned, we may contextually recall that we have since very relevantly discussed the regressive fixations, concerning the **'Kitchner-redux'**, since even now irrationally prevailing in the India's concerned 'corridors of power'.

It remains to be seen that when the always mindlessly obsessed tribes of the Indian bureaucrats and politicians would be able to come out of their fictitious and self-created dark 'cocoon' in these regards and thus, be able to breathe in the open fresh air of modern times? **In the presently fast moving super high-tech modern-era, it is indeed a great paradox for the modern sovereign India's leaders and the Indian-citizens, that on one side India remains a thriving parliamentary democracy, as well as our so dynamic scientists and engineers are now able to successfully launch their space equipment-systems to travel all the way up to the Mars and land there for their multiple scientific explorations and research; and on the other side the naively superstitious Indian establishment gets busy in even so 'foolishly digging for the factiously buried gold treasures' and then in addition to all these—the so much regressively sinister and also the powerfully entrenched politico-bureaucratic nexusis even now always so baselessly doubting the unquestioned patriotism, impeccable loyalty to the Constitution of India, and also the superbly apolitical ethos of the Indian Armed Forces. In the larger modern Indian national perspectives and for the any rationally thinking Indian citizen today, nothing could be more socio-politically nauseating as well as bizarre, than these strangely negating fault-lines across the Indian societies and their establishment**.

In the above precise contexts, and also concerning the overall presently prevailing state of the nation as well, the otherwise highly praiseworthy multi-dimensional potential of the democratic Indian-State, **General V.K. Singh** has very frankly as well as quite perceptively written that, "Since independence, the top political leadership in the country has been haunted by the possibility of a military takeover. **It is no secret that people around Nehru exploited his paranoia of a military coup and started chipping away at the army in an evolving civil-military relationship**. The appointment of **Baldev Singh** as India's first Defence minister—a man known for his political 'fix-it' ability rather than any military acumen—set the tone for the future. Had Nehru had his way, **Field Marshal Cariappa** would never have been the Chief. Subsequently, **Thimayya**'s popularity as an individual would give Prime Minister Nehru more nightmare than the Chinese on the border (*what could have been more astonishingly shocking and bizarre in these regards?*). When the attack did come in October 1962, the Indian Army was engaged in 'Op Amar' where they were building houses while our ordnance factories were making coffee percolators.... **Thorat, Prem Bhagat, S.K. Sinha**—they were all dumped by the wayside. **Indira Gandhi** not only inherited the leadership of the country from her father, she too never stopped looking over her shoulder. **The examples are endless, the lesson the same. If you want to spook the system, raise the spectre of a military takeover. It always works. (*What a pity and the no less depressingly sad scenario indeed.*)**.... (*And* *on the highly complex and challenging—the overall emerging security scenario around the Indian Nation-State now—Gen. V.K. Singh has given his well-perceived observation and suggestions in the following lines*): What are the solutions? We have the institutions we just need to let them function without interference. Define the road-map for every Indian, bring him into the loop of national security and leave it to those who know the subject to deliver. Whether we are willing to admit it or not, today we face greater security challenges than ever before. The US is withdrawing from Afghanistan, creating a void that this time round is likely to be filled by the Chinese. **History has taught us valuable lessons, but we can only see something if we are willing to look**.... The Transformation Study recognized the threats and analysed various likely scenarios. Whatever possible, we pushed hard and got things done, especially in areas where the ministry's direct intervention was not required. The first of our accretions had started coming in while I was still in Kolkata and a detailed roadmap to absorb new subsequent additions was made. At Army HQ we had worked out details for the required accretions. The locations and infrastructure needed were worked out in great detail. Yet, it is a shame that the proposed corps is still stuck in a bureaucratic maze. **The former defence secretary wryly commented that since this was a V.K. Singh proposal, it will not get through. How much more myopic can we be? It is a pity that none of our bureaucrats are ever held responsible for any shortages and failures. There is an urgent need to bring about accountability if the nation wants to protect itself**.... And after having spoken to a lot of people in the country, I find there are enough experts available who have solution to these problems (*i.e. the numerous national-cum-social problems*). Whether it is the judicial system, the education system, or the power sector, people have ideas that they have converted into workable solutions. What holds back India today is the personal greed of the political and the bureaucratic class that has merged into one common system (*that precisely is the extremely sinister 'nexus', that we have since been talking about.*), that does not let anything happen in this country unless there is something in it for them. Until this changes, nothing will happen and that is why this *parivartan*/transformation/reclamation is required across the board.... **I have interacted with the young officers and men of the Indian Army all my life and now as I speak to university students, farmers, people from all walks of life, I have no doubt that we as a nation can achieve whatever we want for our land and for our people. We can only achieve that if we are willing to walk the selfless, moral path that puts the country ahead of everything else. Every soldier in the Indian army has always done just that. We now need to borrow again from the British who influenced all our institutions and even our Constitution to a great degree. The famous words of Lord Nelson at Trafalgar have to be adapted and be our guiding light forever: India expects every man and woman to their duty**." (Singh, 2013, pp. 335, 341 & 347)

Perhaps the biggest tragedy for the modern, sovereign and democratic Indian Nation-State has been that even the modern Indians today as free citizens of the nation, have yet not succeeded in fruitfully getting rid of the age-old, highly regressive, and even now so malignantly lingering: the bygone medieval-era traits of the mindless feudalism. These so much counterproductive and socio-psychological-cum-culturally irrational feudalistic fault-lines, besides socially elsewhere; are even allowed to run through the corridors of power and thus, even though the various governmental establishments and thereby typically facilitating the colonial-era mindsets of the official functionaries. All these negations in their totality thereby are preventing even now the present democratic Indian Nation-State from being able to successfully cultivate, develop, groom, imbibe and then station always, the proactively-cum-flexibly well-placed the, **'multiple-aspects of essential strategic-culture and thoughts'**. In fact after the British left, the all-round highly adverse fallouts of these lacunae began to be visible; because the Indian citizens had fallen short of indeed, to so valuably learn some crucial lessons from the departing British. And needless to say that the Indian political leadership since now constitutionally bestowed with the responsibilities to preside over the destiny of India, have proved themselves to be shortsighted and naively selfish enough and thus, not been able to correct these lacunae in the peoples' highest national-interests, in the modern-era comity of nations.

And in the above multiply important contexts, so far as the modern and democratic India's impeccably apolitical Armed Forces are concerned and thereby, in the complex domain of civil-military relations, the consequences have been very paradoxical and unbecoming. In fact, all these have principally been due to the essentially certain additional dimensions and compulsions in the highly complex domain of the multiple-aspects of nation's matters-military. In India's highest national-interests, as well as for her assured future wellbeing; it remains inescapably paramountcy that the earlier mentioned, the arrogantly sinister and snobbish, regressively fixated, and also the so much powerfully entrenched—**the politico-bureaucratic combine in the nation's corridors of power—that this combine is somehow effectively persuaded and democratically forced to reform, refine and also made to realise that the nation's military has to be given its proper constitutional place in its overall gamut of national-governance, by fruitfully bringing the military within the decision-making loop in these regards. In fact and specifically in these multiple contexts; there is never any scope for the any kind of superficial 'window-dressing' and thus, no space available for the typically regressive 'culture of fooling-around', on the part of so much arrogantly self-centred 'politico-bureaucratic combine'; and hence they would continue to be only doing so possibly, to the great peril to modern and democratic Indian Nation-State; especially in today's highly ruthless and dicey 'arena of the international power-politics'**. Hence, they must always effectively as well as constitutionally, continue to be countered and be compelled to mend their ways in India's highest national-interests.

Well, all these in any case won't be any easy tasks while looking into and analyzing the so much short-sighted and highly regressive psyche of the Indian politico-bureaucratic combine in spite of the super high-tech modern era's instant global connectivity, as well as the all sorts of I.T. triggered 'information-bombardments' and thereby, even due to ready availability of the relevant high-tech wherewithal and its multiple-potentials for the needful. Ironically a very depressingly visible sad scenario in these regards, in fact, appears to be that the modern and democratic India's so much powerfully entrenched—this politico-bureaucratic nexus—remains mostly busy in making use of these so far-reaching IT-related wherewithal, for obsessively feathering their own 'selfish nests' with no concern whatsoever, for the all kinds of highly adverse consequences for the nation. All these highly regressive factors collectively, indeed lead to an extremely bizarre and also dangerously unbecoming scenario since, so all-pervasively prevailing all across the today's democratic Indian Nation-State. But that does never mean that persistently forceful efforts should not relentlessly be made by the patriotic Indian military-leaders, some truly honest, strategically prudent and far-sighted bureaucrats, professionally competent diplomats, leaders of the India's multi-dimensional academia as well as the media, thinking citizenry and also some honest, visionary and patriotic political leaders and thinkers, and even the leading corporate-sector personalities with their well-coordinated and desired geostrategic-cum-politico-militarily synergy-making, and thus collectively; to correct the presently so much faulty-systems and thus, achieve the fastest results in India's larger and the multiply long-term national-interests. And hence, in India's highest national-interests and, for the above mentioned so much knowledgeable leaders and patriotic citizens of the modern, fast-developing and democratic Indian Nation-State, it becomes the no less urgently mandatory to collectively proceed with an unwavering determination and their missionary zeal before it gets too late, in today's so much geopolitically complex and fast moving world. On the one hand, the democratic and fast developing Indian nation-State aspires to be some kind of a credible 'Superpower' or so and simultaneously on the other hand, the so mindlessly shortsighted and hence, both mentally as well as functionally static—the fossilized Indian-establishment—no less persistently continues to be bogged down in its self-created muck of all kinds of the above discussed lacunae in the nation's corridors of power.

If that being so then, how can the modern Indian-citizens ever successfully fulfill theirs befittingly well-deserved numerous aspirations, in today's indeed unimaginably super high-tech era's fast moving world? This perhaps is the most urgent, weighty, and crucial question, which the today's conscious and thinking Indian-citizens from all walks of life, need to proactively ponder over, in their highest national interests and the well-being of the teeming Indian masses. Contextually at this precise and the no less weighty juncture, an interesting point arises as to who happens to be the bigger sinner—the politicians belonging to any political affiliation, or the deeply entrenched, mostly self-serving and also always very craftily scheming-cum-manipulating bureaucrats in India's corridors of power, as well as the decision-making systems? Well it is rather a quite 'slippery-scenario' to make any kind of judgement or grading but whatever, the entrenched bureaucratic-system is mostly seen to be manipulating behind the 'jungle of official files' and thus, quite justifiably perceived to be 'shooting from the dark' with no responsibility and accountability whatsoever and thereby, mostly with almost total disregard to the any adverse consequences to the nation. And to serve its vested interests, the Indian bureaucracy remains always busy in the all sorts of highly unhealthy manipulations in marginalising the military and downgrading their protocol. And precisely that's what a former Army Chief Gen. V.K. Singh, has called the 'chipping away at the Army in an evolving civil-military relationship' in today's modern and democratic India'. Many strategic-thinkers and scholars, in fact coming from the both civil and military streams of thought-processes, have voiced their intellectually well-gelled similar opinions in these regards.

In precisely the above discussed multiple-contexts; **Lt. Gen. S.K. Sinha** has very perceptively revealed that, "I once had an interesting exchange of notes with **Sushital Banerjee (*the then Defence Secretary*)** on the file. **I had written in a note that before 1947 in terms of emoluments and status, the Army officer and the ICS officer were equated and the evolved a higher position than officers of all other Services, that equation had been disturbed after independence by providing an edge to the IAS (successor of ICS) and equating the Army officers with the Police and other Class I officers**. Sushital had recorded a contrary view that there was no basis to assert that there was any equation between the ICS and the Army before independence. This gave me the required opening. I demolished his contention on the basis of facts ascertained by me from Civil Lists of 1903 onwards. He was gracious enough to tell me that much of what I had written about the Civil Services was not known to him. After these two incidents mutual respect and understanding developed between us. This helped me not only in my day to day functioning with the civil officials of the ministry but also made my officers appreciate that what I preached, in terms of trying to maintain the dignity of the Service. Unfortunately a few months later Sushital suffered a heart attack and died suddenly. I had met him in the hospital only a few hours before he passed away and as usual he was very friendly and courteous. He insisted that his son went out of the hospital building to see me off to my car. I his premature death, the nation lost a brilliant civil servant and the Army a good friend.... I made a point to address gatherings of Army officers whenever I went on tours. I appraised them of we were trying to do both for the officers and the soldiers. I used to emphasise what I called the three Ps—Pay, Prospects and Prestige. This had a healthy effect on morale. We had won the first round in the battle for pay with the Third Pay Commission. We were now in the process of making further improvements in career prospects of the officers and soldiers.... **As for prestige, I again took up the issue of Table of Precedence, which I had done earlier as Deputy Adjutant General. I brought out how Army ranks had been systematically downgraded vis-à-vis Civil Services in every revision of Table of Precedence that had taken place after independence. Further, the State Governments had issued Tables of Precedence for ranks below Major General which were not covered by the Central Tables of Precedence and some of the States had virtually gone berserk, ignoring the policy guidelines issued by the Central Government, they had downgraded Army ranks most unjustifiably**.... I advised local Commanders to take up the matter with State governments and have the anomalies removed. If State Governments did not do so, the local Commanders were authorised to direct Army Officers not to attend any ceremonial functions being organised by the State Governments. At all Army functions in Cantonments, the pre-1947 protocol for relative Army and Civil ranks for officers of the rank of brigadier and below was to be observed. The Ministry of Defence was under pressure from the Ministry of Home Affairs to get me to withdraw the letter. I dug my heels and said that the letter was in full conformity with the existing policy of Government of India. I asked them to point out the deviation if any. If the Government now wanted to change its instructions, it may do so in writing and based on that I would issue an amendment to my letter. They did not issue a revised policy letter and my letter was never withdrawn." (Sinha, 1992, pp. 264-266)

Further on **General Sinha**, in his book, has systematically proceeded in fact to very informatively explain the 'nitty-gritty of Warrant of Precedence' quite in detail as to how so much healthily in a balanced manner, the things stood in our country before 1947. In these contexts, the President's Secretariat (i.e. in Rashtrapati Bhawan, New Delhi) Notification No. 68-Pres/63 dated 03 Sept 63 (as amended up to 22 Nov 66), and as also reproduced as Appendix 'A' to the Army Order No. AO 235/67 may be referred for any clarification. Quite interestingly, it may be obvious that the so much mindlessly entrenched Indian bureaucracy as per their all sorts of fixated arrogance, may ever be ready to raise their distorted 'boggy' of always the unquestioned supremacy of the civil over the military, in any worthwhile democratic-system of governance anywhere and hence in any case, that remains absolutely unquestioned in the today's modern, sovereign and democratic India too. And this point is always well-taken and remains constitutionally unassailable as well and hence, nobody is questioning this. But also the always well-settled meaning of the 'civilian supremacy over the military', clearly and the no less unquestioningly means civil supremacy at the political-level only and certainly never at the bureaucratic levels. And precisely here lies the 'rub' for the bureaucracy and hence thereby, the arrogant Indian bureaucracy with such nakedly inflated-ego, is always fixated with its culture of dubiously not wanting to lawfully reconcile with this constitutional truth and hence, not willing to rather magnanimously accept and remove the anomalies in the larger good of the country.

Moreover, concerning the all-important and also well-accepted, as well as well-practiced culture of always constitutionally maintaining exclusively the inviolable culture of political-level supremacy of the civil over the military in any truly and well-rooted democratic-system of governance; it has been the British who have historically been pioneers in systematically conceptualising, healthily practicing and also mastering the same as well-accepted norms and thus, being rather a shining example for the whole world to copy wherever possible. It also remains no less historically true that as timelines of the mankind's living-history, are continuingly advancing onto its expanding civilisational-canvas, the British have not only been quite vibrantly maintaining the same but also during now their already bygone colonial-era, when even the 'Sun once never used to set' over their globally vast British-Empire; they geopolitically did always bring and scrupulously maintained the similar protocol-related norms of precedence in their numerous colonial empires in every continent of the world. And naturally, the then colonial Indian subcontinent, then being 'Jewel in the British-Crown' happened to be no exception in these regards. In addition it is also historically true that they left behind the same system intact and functioning in their respective colonies, when subsequently these colonies gradually began to gain their indeed well-deserved independence and the rest is history. Hence in these precise respects, any problem now still being persisting with the today's Indian bureaucracy shows their inflated-ego and the totally unacceptable bureaucratic-arrogance only.

We may recall that General V.K. Singh has clearly stated as to how the ill-informed bureaucrats sitting in the Ministry of Defence, is always quite arrogantly bent upon to 'shoot' any well-intentioned matters-military related proposals emanating from the Services headquarters. In fact, Gen V.K. Singh has clearly put-on record as to, how once one Defence Secretary could even say that just because the 'proposal had come from Gen. V.K. Singh', it would not get through. What could be more bizarre and even amounting to a greatly unbecoming disservice to India's national defence aspects? Certainly no Indian Army Chief, or for that matter any Navy or Air force Chief, ever sends any kind of the vital matters-military related proposal to the Defence Ministry for his any self-interest or for any kind of a personal-gain whatsoever. Proposals are always well-considered, well-analysed, as well as thoroughly vetted through systematic military staff-channels from below to the top, since always by the well-organised and also proactively well-stationed and expertly well-specialised systems in the Services Headquarters before sending to the Ministry and thereby, in the best possible military-cum-strategic interests of the nation. And if any bureaucrat, with his/her whatsoever misguided or baseless prejudices against the military, since sitting in any ministry at any level ever tries to 'shoot-down' the proposal or keep sitting over it, the he/she would only be betraying the oath taken while entering the government service and thus thereby be adversely affecting some vital aspects of the nation's military preparedness and also, geopolitically harming the national security affairs. Which all in any case, remain totally unacceptable for the nation and thus even obviously amounts to being quite willful and serious acts of treason and hence, totally unpardonable too.

This strange and the no less foolishly harmful bureaucratic mind-set must politically be done away with without any delay; if the modern, Sovereign and democratic Indian Nation-State wants to keep itself safe and also the citizens assured from the dangers of the possibly any kind of internal, as well as external security-related threats. As discussed in some considerable detail in the earlier paragraphs, all kinds of sinisterly bureaucratic scheming and dirty manipulations both at the Centre, as well as at the State Government levels; to quite trickily downgrade the otherwise always well-deserved status and prestige of the military officers; the bureaucrats may have only superficially gained to massage their inflated-ego, but they too have harmed their reputation and also quite irrecoverably damaged their standing and status before the peoples across the nation, obviously in many ways. We may also recall while referring to the already earlier quoted few paragraphs from **Lt. Gen. S.K. Sinha's historic 'USI National Security Lecture—1990'** and wherein, the learned General has very perceptively narrated as to how even soon after 15 August 1947, the then highly placed few senior Indian ICS officers tried to play havoc in these regards but could only be prevented by **Lord Mountbatten and his Chief of Staff, Lord Ismay**, who happened to be still in India and serving the Independent India's democratic government.

But further on and in spite of the above, all possible kinds of the dirty manipulations of Indian bureaucrats did not cease at all. In fact, after the final departure of Lord Mountbatten and then also with no less encouraged by the nationally so much damaging—the first Prime Minister Nehru's now so well-known irrational aversion for the independent India's highly disciplined and patriotic military officers, as well as the nation's always so crucial multiple-aspects of the matters-military—the powerfully entrenched bureaucracy since perpetually busy in mostly massaging their own inflated-ego above anything else could get 'wide-open fields' to perpetuate all sorts of their evil-designs against the democratic India's neglected military, with total disregard to the any adversely humiliating consequences to the hapless Indian-nation. So far as the India's multiply highly complex aspects of the matters-military are concerned, most paradoxically even in the presently ongoing 21st century, what else but only these have continuously been the highly frustrating and unfortunate saga, and even with no end in sight in these regards. And hence at times contextually, if any military chief happens to be politico-bureaucratically not being desirably shrewd, critically discerning and also, pragmatically visionary enough to see through and decipher different kinds of the bureaucratic manipulations—to safeguard the various long-term interests of the defence services—then the crafty bureaucrats have always been ever ready to manipulate and, get any 'official-paper' signed as a matter of official concurrence from the military. This happened with the first Indian Commander-in-Chief, the then **General K.M. Cariappa** only, with highly damaging consequences for the Indian military, as would clearly be seen in the following quoted paragraphs. What it in fact boils down to is that, the professionally active officers while sitting in the high places on the nation's military hierarchy essentially always need to unwaveringly rise above the self in many ways, if they ever prove themselves to be really deserving, in the eyes of his officers and men serving within his command, to be in the high positions, which they happen to be really occupying. Otherwise, things would never be crucially in place which they must remain, irrespective of the any kind of adverse situations developing around whether in times of peace or war. **One rather quite unbecoming 'drag', now even gradually getting more and more professionally vitiating, happens to be the irrationally negating-aspects of the always undesirable culture of even being socio-culturally 'rank-fixated' attitudes among officer-corps of the Indian military**. We would be coming to this always very damaging issue, in more detail little later.

And precisely with respect to the above, **Lt Gen. Harbakhsh Singh, Vr.C, who so successfully commanded the India's Western Army in the historic 1965 war with Pakistan, in his autobiographical book** has very revealingly written that, "His tiff (***i.e. General Cariappa's***.), with the first Defence Secretary of the country, **Mr. HM Patel**, did the Army more harm than good. Army officers lost out on prestige and protocol. When it came to fixing the pay of officers, immediately after independence, he sided with the bureaucrats (mostly ICS), and gave the king's Commissioned Indian officers (KCIOs), a higher pay packet than Indian Commissioned Officers (ICOs), who had passed from Indian institutions, although both categories did the same, if not similar jobs. This was nothing but discrimination, and it is a pity that our first Commander-in-Chief was a party to it. It built resentment between the Indian Commissioned Officers (ICOs) and the KCIOs, who were nicknamed 'kassais' (*i.e. butchers.*). The discrimination did not end here. I have already mentioned my personal case, when for my conduct of the Tithwal Operations, my General Officer Commanding, **General Thimayya**, had recommended me for Mahavir Chakra (MVC) and not Vir Chakra (which is meant for personal bravery). But General Cariappa, as Chief of Army Staff, sitting in his office, decided for some reason or another, which I need not go into here that it would be Vir Chakra (Vr.C). General Thimayya told me about this himself, and felt sorry for it, but there was nothing he could do. **The trouble with General Cariappa was that he had never commanded troops in battle; nor even heard a shot 'fired in anger' so to speak.**" (Harbakhsh, 2000, p. 386)

Concerning some universally crucial aspects of **successful generalship in war** Lt Gen Harbakhsh Singh also has quite valuably reflected upon and that too, based on his some personal experiences. **In his already above referred autobiography; the General has mentioned that, "It has been my experience that he, who plays 'safe', lies low and awaits developments, usually comes off second best. The Commander who always tries to dominate a situation, is eager to seize the fleeting opportunities of war and is prepared to pounce upon the enemy with inspired 'audacity', usually carries the day, however adverse the circumstances of the battle. There are always more ways than one of fighting a battle successfully, but few of winning it cheaply.** For while the rigid, slow plodder, the stickler for textbook conventions, too, reaches the goal, his route of success is often littered with prohibitive costs in men and material. Quick, spectacular achievements, at comparatively low cost usually come to the audacious, the bold and the enterprising in battle." (Harbakhsh, 2000, p. 391)

The nitty-gritty of successful generalship—as narrated by **Lt. Gen. Harbakhas Singh** in the above paragraph—has been very successfully and also in action-mode, has been practiced and displayed by **Lt. Gen. Sagat Singh** in 1971 war in the erstwhile east Pakistan (now sovereign Bangladesh). Please refer to the book, 'Talent for War: Military Biography of Lt. Gen. Sagat Singh; and authored by **Maj Gen. (Retd.) Randhir Singh UYSM, AUSM, SM**; with a 'Foreword by **Lt. Gen. S.K. Sinha (One of the Legends of Indian Army)**—and published by 'Vij Books', New Delhi and through USI New Delhi. Here it may be interesting to note that, **Maj. Gen. (Retd.) V.K. Singh (Corps of Signals)**, in his book title as, 'Independent India's Twelve Outstanding Generals'—has mentioned about **Lt. Gen. Sagat Singh**—as India's best fighting General!

Precisely in these multiple perspectives, since in any case always having conceptually—an overwhelmingly regressive and also a 'blanket-kind' of any sweeping policy-approach across the board—for equating the entirely different and dissimilar kinds of the nitty-gritty of both in fact, the poles-apart—civil and defence services—are bound to be adversely counter-productive so far as the all-encompassing and overall gamut of the nation's governance is concerned. Moreover ironically, any such approach may lead to a very damaging situation of 'being penny-wise and pound-foolish' as well. And that's what precisely has happened after 1947, in the mostly so much shortsighted, callously naïve, and also geopolitically clueless Indian Establishment. And consequently then, these have all along been very adversely affecting India's Armed Forces and thereby, these negative developments in the sovereign and democratic India's overall and the multiply-complex security-domains are indeed matters of great concerns for the patriotic and also informatively thinking modern Indian citizenry, in today's so much super high-tech-oriented, fast-moving, dicey and also unpredictably challenging world. All these too again, prove that India has been in fact, quite awfully remaining clueless about the always paramountcy: **'the multiple-aspects of essential strategic-culture and thoughts'**. We may recall that we have since been discussing these always crucially important aspects in their differently multiple perspectives and at the same time little later, we would again contextually be taking-on these points for further discussions as well. The pivotal point remains that—the highest national-interests—must always continue to irrefutably take precedence over any kind of the partisan considerations, since emanating from any concerned quarters in the Indian Establishment. And at the same time, there must remain a judiciously pragmatic blend of the 'qualitative-aspects-vis-à-vis-quantity; while getting engaged with the multiple-aspects of the personnel/manpower related organizational planning, concerning any national institutions/establishments, and so on to serve the national-interests in their best possible manners. And therefore, needless to say that concerning the Armed Forces precisely these have always being important aspects, and become all the more overridingly crucial here.

Contextually in these multiple respects, a former Indian Army Chief, **General K. Sundarji** in his unfinished book has mentioned that, "The pressure for unpaid 'perks' started quite early, in 1947 just before independence when the 'new pay code' with highly reduced scales, was promulgated for all central government servants including the armed forces. This was still under the British aegis, but the Indian interim government of Nehru was making its presence already felt. In the drawing up of these scales, the old Indian Civil Service (ICS) had a major hand and the politicians who were mostly idealistic and dedicated (*well, now what about today's politicians: how indeed so much depressingly, things have declined and also got rotten in the almost all possible manners?*) were wanting no more than five hundred rupees per month as pay for minister. So, it was easy enough to please them by an across-the-board reduction from what was considered bloated colonial salaries.... The ICS lot, of course it goes without saying, covered their own derrieres by stipulating that they were to retain their old scales of pay and the new dispensation would only affect the postwar 'Indian Administrative Service'. They offered the same benefit to armed forces officers, but cleverly only to those who were referred to as KCIOs (King's Commissioned Indian Officers) who were trained in Britain and which there were only about a hundred, as the breed was no longer produced after 1934. All those regulars who were commissioned from the Indian military Academy after 1934, who were a few hundred strong, were to come under the new pay code along with the war time emergency commissioned officers, who might be granted regular commissions and these would run into many thousands. For example, a lieutenant general who used to start at four thousand five hundred rupees per month, *finished* under the new scales at about two thousand five hundred rupees. The four-star general finished at three thousand rupees which was the maximum for all services, which appeared to the idealistic politicians to be very fair considering the very low scales that were being accepted gladly by patriotic MPs and ministers. With true socialistic urges, the ratio between the lowest paid and the highest paid, whether it was the military or the civil, was reduced drastically. (Sundarji, 2000, pp. 116 & 117) (*Well, those were the days—understandably based on India's freedom-fighting times and spirits—where idealism soaked in patriotism was the call of the day, but in fact, all these survived during the only very few beginning years immediately after 1947. The overriding reasons were indeed many, and which have since been somewhat critically looked into contextually, during our earlier discussions and also, we would be doing the same still more intensely little later soon*).

In any case precisely as per the above, **General Sundarji**, again in his above referred book and therein, he has quite interestingly said that, "Around 1940 or so, war time Delhi as an Allied Base felt the need for a lot of new buildings to prosecute the war against the Germans and Italians in the Middle East and the Japanese in the east. Lutyen's well laid out Delhi had plenty of wide open space to accommodate the lawns and gardens which were liberally interspersed between the imposing buildings as well as between them and the majestic roads. These became the inevitable casualties in this construction boom, and in addition to the unaesthetic locations of these new buildings, they were the eyesores being constructed to wartime 'hutment specifications' that were supposed to last for just three or four years, from 1942 onwards. The fact that some of these have lasted for very nearly four or five decades (*well even now, many of them are still very usefully continuing to exist and serve.*), whilst Indian 'permanent' postwar constructions crack and crumble much sooner is **a sad commentary on independent India's galloping propensity for graft.**" (Sundarji, 2000, p. 115)

Now coming back to the highly damaging aspects of imaginably the all possible kinds of always extremely sinister politico-bureaucratic manipulations and thereby, the no less creepy disservices being affected onto the India's highly professional, as well as impeccably apolitical Armed Forces since started right from the highly checkered beginning of the Nehru-era itself and then still continuing regardless, in the all sorts of manners—even in the presently ongoing and all the more significant 21st century too—do have numerous kinds of the ill-effects onto the overall national security situations. We may recall that in India's these very adversely far-reaching multiple-contexts and thereby, in order to rather irrefutably substantiate these indeed extremely worrisome perceptions we have since already quoted earlier some of the very relevant paragraphs from the writings of a number of the well-known civil/military personalities. But irrespective of all these, India's arrogantly callous—the politico-bureaucratic combine—is just busy in hiding their anything but innocent faces inside heaps of dust-gathering files stacked in the corridors of North and South Blocks' corridors of power in the Raisina Hill, New Delhi. And hence, regardless of the all sorts of patriotically well-intentioned alarm-bells being always loudly sounded from the considerably knowledgeable segments, since belonging to the nation's military, academia and the vocal media circles and also, to the so much conscious citizens too, nothing is being purposefully effective in waking-up the always powerfully entrenched—political and bureaucratic classes—from their irresponsibly deep and callous slumber. What could be more bizarre for the modern, sovereign, democratic, and fast developing Indian Nation-State even after more than half a century of its independence? My God, even the other nations are amazed; but the Indian politico-bureaucratic combine remains almost totally unconcerned, irrespective of the Indian Nation-State being turned into some sort of a 'laughing-stock' before the whole world.

India boasts of possessing a third or fourth largest Army, fourth largest Air Force and also, the fourth or fifth largest Navy in the world, but irrespective of all these military wherewithal, the Indian-Nation is very disgustingly treated largely as a 'push-over', even in its immediate neighbourhood too. What to talk of indeed geopolitical-cum-militarily—the extremely challenging situations all along its northern and western international borders—even the overall internal security scenario across the Indian heartlands are far from being peaceful and thereby, not even being under any assuring control of the Indian State-machineries and their any political-cum-administrative reaches over there, in any tangible manners whatsoever. But sadly even this is not any complete story of the modern democratic India's overall security-related scenario since in addition consequently, the no less a highly unpalatable truth still persists that, in a number of the substantially expansive pockets within the sovereign Indian territories, the writs of the State don't prevail at all, whether due to the many kinds of nefarious activities of the Islamic Jihadi terrorists, or the entrenched Maoist insurgents being always hyperactive in the so-called 'Red-corridors' across the Indian heartlands. It is true that after 1947, due to their own typically-fixated psyche, the mostly arrogant bureaucrats began to sinisterly connive in all sorts of manners for their ulterior motives and greed and thereby, lost most of their erstwhile civil-services-related sheen; and consequently along with the political-class, they too got relegated to become one of the two sides of the hapless Indian-Nation's highly damaging, so stinking and now the well-known 'corruption-coin'. But India's highly regressive—the politico-bureaucratic class—did not stop there and on top of it, it began to very seriously harm the prestige, socio-official status and protocol, and even the pay and perks of the nation's highly disciplined and totally apolitical military-men and their officers; in most unfair and arrogantly callous manners, and the rest is history.

And while doing so in such socio-politically unacceptable, as well as officially in very secretive-cum-sinister manners; India's arrogantly entrenched 'politico-bureaucratic combine' has never cared to pause a little even once, and sincerely think as to how much serious damages and injuries, it has been inflicting onto the always vitally paramountcy—the nation's matters-military and morale of the people in uniforms—with even extremely adverse long-term consequences. And therefore, even in today's modern and democratic India in fact what these arrogantly callous political-classes and theirs rather so sinisterly 'obedient' bureaucracy are doing, but only 'replaying the ill-famed Nero-syndrome' and that too in today's so fast-moving and super high-tech era of the 21st century; i.e. as the saying goes, 'when the Rome was burning, its naive emperor Nero was fiddling with his harp or any other musical instrument in his palace'. As the centuries old high-valued traditions in India have been that, young people never used to join the nation's military-services only for bread and butter or to earn any fat wealth. India's fit and highly self-respecting young men have traditionally been responding to the so much soul-stirring call to arms, to always magnificently uphold the security, honour and well-being of their nation and thereby; maintain their own since, always well-deserved honour, prestige, and also earn a respectable means of living in their societies. And needless to say that so far as the Indian-Nation's military officer-corps are concerned these aspects and much more in these regards, obviously as well as since very compulsively assume a number of the additional and important dimensions; therefore the nation can only ignore or be callous towards these always multiply geostrategic-cum-politico-militarily vital aspects and their numerously complex dimensions, at indeed its own great peril. These timelessly paramountcy points are certainly not of academic or the any so-called some seminar-related importance only but always carry the multi-dimensionally far-reaching importance, concerning the nation's defence forces, their multiply strategic threat-perceptions and thereby its overall both short and long terms security situations as well as, the overall national well-being.

Hence in a democratic nation like India, if her since already earlier analysed—the so crafty politico-bureaucratic combine—remains hell-bent to play havoc with the morale, prestige and honestly indeed the fair entitlements of the nation's military, then the fallouts are bound to be dangerously very grave, and precisely that's what has happened in the modern and democratic India. **If the politicians and bureaucrats are allowed to go berserk, with almost total disregard to the military-morale and the fighting acumen-cum-potential of the nation's defence forces, the many gaping holes and chinks begin to very alarmingly appear in its overall the multi-dimensional and highly complex military armour-systems and thus the whole nation becomes dangerously vulnerable, in the ruthless and the extremely dicey 'arena of the international power-politics' and therein, the always continuingly ongoing, the all sorts of international 'great-games'**. The universally timeless truth concerning—the mankind's living-history since continuingly in the making onto its expanding civilisational-canvas—maintains that a nation has to remain both geopolitically, as well as strategic-cum-militarily, always dynamically well-prepared, if it wishes to reassuringly prevent its any security-related peril and humiliations among the comity of nations in the world. If these vital aspects are not always taken care-of properly and also since the no less important and also non-detachable—the multiple-aspects of morale, prestige, status and also the comparable well-being of the defence forces—are not sincerely and adequately maintained within the social-fabric across the nation then all these callous failures collectively, may at any time result as sure invitations to the nation's—any possible humiliating disasters and defeats. And needless to say that since 1947 itself, the Indian establishments have been continuously failing on these multiply vital counts and thereby, becoming so callous and even thick-skinned; that no amount of any humiliating 'kicks at their back' by any neighbouring power from time to time, are making any dent to their corrupt attitudes and practices. All these indeed constitute a very depressing scenario and thereby rather, compel any self-respecting and patriotic-minded Indian-citizen to feel betrayed and also extremely frustrated.

(iii) Why Chinks in the Indian Military's Armour?

In the nation's military-domain, as the otherwise always never to be compromised—the qualitative parameters and character-aspects—for the men in uniforms, and all the more especially for their officer-corps too, are very callously being pushed towards their serious deteriorations and thereby causing many dangerous chinks in the nation's armour; the always time-honoured call to arms and thus, leading towards the profession of arms have fast been losing their high esteem, honour and socio-economic attractions before the nation's eligible men. **And in addition also, the otherwise always so nauseating, a 'culture of corruption' and the any possible kinds of 'corrupt practices' too have since very sadly begun to seep into the defence services. And very horribly that's what is now increasingly being seen in India, since the last few decades. Officers are becoming somewhat more self-centred, not remaining above any kind of doubt concerning their integrity, and being only careerists, and thus not being adequately careful, both individually as well as collectively as an honourable institution towards their socio-official prestige and reputation in the larger Indian civil-societies.** It is indeed a very sad scenario, and that precisely is being done so sinisterly by the entrenched—Indian 'politico-bureaucratic combine'—as since being the so much multiply-injurious, as well as a great disservice, to the nation's once so upright and honourable defence forces, being a great envy to any hostile power.

In the context of current larger Indian scenario, it appears that the Indian political-classes and the bureaucracy are perhaps seen and heard to be forcefully saying to the nation's officers and men in uniforms that, "look in any case, we have since already fallen from our earlier respectable pedestals and forced our nation to be an extremely unfortunate and hapless victim of our increasing greed and corruption. Hence we are not going to spare you anymore and allow you to continue with your untainted existence and thus, honourably pursue your patriotic profession of arms in the always highest and dedicated service to the nation. In fact, we are hell-bent to ensure that our 'stinking-muck' and also our 'shameless culture of corruption' are pushed to seep into your exclusive domain too, to bring you down in the so-called starry-eyes of our simple populace." In fact, the extremely 'corrupt centres of power' in the Indian politico-bureaucratic establishments everywhere and almost at all levels, are allowed to thrive with no worthwhile hurdles, then does India need any powerful external enemy as any threats to her security and wellbeing? Well, these are not just any wishful 'figments of imagination'; but hard and the indeed irrefutable ground-realities in fact all across the presently modern, sovereign and democratic Indian Nation-State, as well as also, at almost every level of its stinking and corrupt systems of governance. What could be the anymore so frustratingly unbecoming, the overall 'state of the nation', for the largely innocent and hapless Indian masses today? Most dangerously now, what we see that even the top-hierarchy of the defence forces, both individually as well as collectively, are not seriously careful enough to always remain impeccably above board and hence, all the more above their petty selfishness and thus not come under the shadow of any kind of doubts among the civil social-segments; and which in any case would always quite unsparingly hurt the professional esteem and social-prestige of indeed, still the no less trustworthy Indian Armed Forces. It is true that 'water flows down the gravity'; and so far as the always paramountcy multiple-aspects of the cultivation, development and grooming of the nitty-gritty of professional military leadership are concerned, the spirits of this 'law of nature' with

its numerously complex dimensions get all the more important and pronounced in the vast and highly complex, modern domain of the 'matters-military', as well as the profession of arms.

Precisely therefore in the above multiple-contexts, immediately after 1947 onward, giving a historically very disturbing glimpse of the politically-triggered—the gradual moral-cum-militarily professional fall of the 'soldierly-fibre'—at India's top levels of the military-command system, **Lt Gen S.K. Sinha**, in his already earlier referred autobiography has very revealingly written that, "My Division also had a likely operational role in Jammu and Kashmir. I had to attend a war game at Pathankot conducted by **Lieutenant General PS Bhagat**, who was then the Northern Army Commander. I had served under Bhagat when he was my divisional Commander and I was fully aware of his professional competence. I found attending the war game conducted by him very educative. Bhagat was an officer with vision who could think big and provide interesting solutions to intricate problems. He was supposed to take-over as Army Chief shortly. We were confident that under his brilliant leadership, the Army would give an excellent account of itself. Unfortunately our hopes were belied. **General Bewoor**, who was then the Chief, got a year's extension and this meant that he would outlive Bhagat in the Army; the retirement age for a Lieutenant General then being 56 and for a General 58. Bhagat would now retire a few months before Bewoor and the Army would be deprived of the opportunity of having him as its Chief. Denying the top appointment to such an outstanding General like Bhagat, was not only very unfair to him but also very unfair to the entire Army. We were shocked that Bewoor had allowed himself to become a party to the sordid manipulation. He and Bhagat had studied together at the Royal Indian military College at Dehradun before joining the Army. It is a pity that he did not show any regard either for his old school tie or for the interests of the Army. He should have refused extension to ensure that such a gross injustice was not done to such an outstanding colleague. Unfortunately he was not made that way. After retirement he did not bring glory to the Army by accepting a junior diplomatic assignment as Ambassador to Denmark. I recall how **Rajinder Sinhji** had refused to become the first Chief and offered to resign if the Government superseded **Cariappa** and gave him that appointment. How great was the difference between Rajinder Sinhji and Bewoor." (Sinha, 1992, p. 243)

In his above quoted paragraph, the learned General has indeed frankly, as well as very befittingly, commented that 'how great was the difference between Rajinder Sinhji and Bewoor'. Well contextually at hindsight, we may now say that it was not just a matter of difference between the persona and character of the above mentioned two erstwhile top-level Indian Generals, but it also conveys a very sad and even painful saga of differences between two 'eras' and thereby that is, indeed a highly undesirable, as well as very non-soldierly phenomenon of the gradual fall of the once so glorious traditions, professional value-systems, 'moral-fibre', and also the character-aspects concerning the entire officer-corps of India's Armed Forces. Any patriotic and thinking citizens, whether coming from the civil or military streams of the modern and democratic Indian nation-State, while now quite haplessly witnessing these depressingly unbecoming scenario of their nation's matters-military-related decline and fall from grace of the defence-services officers; are bound to be very sad and feel frustrated. There is no point again in contextually raking-up as to where precisely the roots of this malaise belong to, now since very dangerously spreading like an 'epidemic' not only into the always crucial aspects of the persona and character of the defence-services officer-corps, but also in the multiply vast and highly complex areas of the ongoing processes of always the multi-dimensional modernisation of India's Armed Forces.

Where else the irrefutable blames must rest, but at the doors of so much always conspiring, sinisterly conniving, and the recklessly corrupt, and also powerfully entrenched nexus between the Indian political-classes and bureaucrats in the New Delhi's corridors of power? Now coming back to the personal case of Gen. Sinha himself, what happened to him and in what an extremely unfair and dirty manners he too was deprived by the them Prime minister Indira Gandhi, from so befittingly becoming the Indian Army's Chief, when the time for his indeed well-deserved 'tryst with destiny' had arrived? This not at all any distant historic details in these regards are very well-known and also, many articles and even a book have already been published by the many well-known leading personalities, in fact, clearly exposing the concerned stinking truths pertaining to all sorts of politico-bureaucratic manipulations for posterity. **It has become very clearly visible that far from being upright and farsighted ever—the Indian political-classes—since immediately after 1947, have never wanted any upright and professionally very competent officers, and since also having a befittingly strong 'spine', as Chiefs in the Indian Armed Forces. We may recall and since already noted earlier, that General V.K. Singh has very franking stated that, Generals Thorat, Bhagat and Sinha were so unscrupulously dumped by the wayside, in fact by none else, but by the then Late Prime Ministers Nehru and Mrs. Gandhi respectively.** What great betrayals of the trusts of Indian peoples as repeatedly being inflicted onto their hapless Nation-State, by its extremely short-sighted, irresponsible, and also highly corrupt 'politico-bureaucratic combine'; and in addition, they are also never concerned about any serious damages being caused to India's military wherewithal, by theirs numerously shameless omissions and commissions and thus obviously jeopardizing her many paramountcy geopolitical, as well as geostrategic-cum-politico-militarily related national security dimensions. In the process, they have now become so arrogantly callous and even 'thick-skinned' that, they appear to be ever inclined not to pay any sincere attention to the peoples' well-meaning disgusts; as since also regularly being expressed by the nation's many highly worried and well-known personalities belonging to the different walks of life both civil and military too. All these in turn obviously, and very understandably, increase their frustrations and anger.

In precisely the above multiple-contexts and also, especially concerning the so much sinisterly unfair supersession of **Lt Gen S.K. Sinha** as per the above, we may prefer to at least reproduce the five comments out of the many as below:

- **(a) Shri SS Khera, ICS**, former Cabinet Secretary and former Principal Defence Secretary: 'If the Government's future (not yours) has deprived you of the top Army post, it seems to be due in part , if Press reports are true, to the machinations of the civilian bureaucracy. **I know what everyone with new ideas comes up against the indescribable arrogance of an entrenched bureaucracy**. By putting in your papers, you have earned the respects of many like me'.
- **(b) Justice H Mahapatra**, former judge of Orissa High Court: 'I am as happy and proud of your achievement as your father must have been and you must have noticed that the whole country is with you and voiced what should have been done in your case'.
- **(c) Shri CS Jha, ICS**, and former Foreign Secretary: 'I need hardly tell you of our profound disappointment and shock of the news.... **Strange things happen in our country these days.... I am proud to read the report of the interview you gave to the Press and your impeccable dignity, deportment and reactions, which are in the highest traditions of the Indian Army'**.
- **(d) Shri Vijay Singh, MLA**, from Himachal Pradesh: 'I have been deeply pained by the recent happenings. **Such decisions only reinforce my conviction that there is something**

rotten inside the State.... This country does not require a mere face-lift it needs to be completely purged. God willing we shall achieve this one day. By your not being there at the top, we shall be poorer. But I know that is not the journey's end'.

(e) **Shri SV Neetu, Advocate** Nagpur: 'If a humble soul can express, he only sees a great General in you. **Great Generals like you will always save the country from without that is outside. I do not know as to who will save the country from within'**. (Sinha, 1992, pp. 304 & 305)

If all these still remain as frustratingly continuing 'states of affairs', concerning the very vitals of the nation's always paramountcy and multiply complex aspects of the matters-military, and hence thereby the nation's national security-related fortifications-cum-wellbeing getting severely derailed and damaged then whom shall the people, since occupying the seats of power with their crucial decision-making authorities too in the Indian Establishments, should in any way blame—any Foreign Powers, or some 'astrological-stars' of the nation itself—or none else but themselves, who singularly happen to be the offenders and indefensibly guilty too? The whole nation is asking this question and also, insisting upon precise and total accountability as well since it is always the nation's colossal amount of hard-earned money, that remains involved, and none else's any private wealth or any assets. Moreover, these are not only related to the top policy-level management-matters concerning the nation's public-money only, but also do pertain all the more, to crucial aspects of the continuing processes of modernisation of India's always so praiseworthy and impeccably apolitical Armed Forces, as well as their perpetual readiness to defend the country from any kind of the security-threats from any hostile power. And thus, the politico-bureaucratic sinners, shamelessly guilty of craftily shirking their irrefutable responsibilities and accountabilities, must inescapably be brought to book as per the laws of the land. But alas, the various stinking-cum-systemic rots, since already having taken deep roots all across the multi-dimensional gamut of governance in the modern, sovereign and democratic India are very frustratingly preventing the requisite legal-actions in these regards.

We may recall that interestingly earlier in some different contexts, we have since already noted the meaning of a couplet of a well-known Urdu poet, Faraz, conveying that 'only a single owl is sufficient to ruin a garden; but what would be its fate, if almost onto every branch of trees in the garden is sitting an owl?' God knows as to why and how India's this 'politico-bureaucratic nexus' is bent upon doing, of what all they still remain so corruptly busy in doing, with indeed total disregard to any severely adverse consequences to the Nation. Who will forget India's so much humiliating '1962 fiasco' during the then Nehru-era and the subsequent—**the still so naively called 'top-secret' Henderson brooks/PS Bhagat Enquiry-Commission Report**—which India's so fake and disreputably 'tainted' Political-Class is yet to cultivate any moral character and courage, to officially and fully unfold the contents of this historic report before the Indian peoples, which they are indeed lawfully entitled to. And on top of it in these specific contexts, all the more damaging and irrepressible fact remains that, the so much powerfully entrenched, and always occupied in the many kinds of sinister manipulations—this 'politico-bureaucratic combine' sitting in India's corridors of power—are never inclined to most dutifully learn and sincerely abide by the essential lessons therein, in the highest national-interests.

Irrespective of the all sorts of politico-bureaucratic as well as quite ill-intentioned manipulations in the Indian 'corridors of power' in New Delhi, it is now an open fact that the full-copy of the above mentioned 'Henderson Brooke/PS Bhagat Enquiry commission Report' on the 1962-fiasco, couldn't be kept under a very dicey official-cover any more. And hence though belatedly but nevertheless as per the Press-reports now in March 2014, the concerned and well-known but also somewhat controversial Australian Journalist, Neville Maxwell, who wrote the book, 'India China War', since published way back in the year 1970 (the no less naively and amusingly then banned in India.) has confirmed that most of the contents of the then so much sensationally controversial book, was based on the findings of the above mentioned enquiry-report. In fact he has now also confirmed that indeed, he had somehow managed to get hold a copy of this so-called 'top-secret' report, during his stay as an accredited foreign Journalist in India from the year 1963 onward, for his research-cum-study and reporting on India's this 1962 border-war debacle itself. He has also confirmed through the international media that after even abnormally more than 50 years' of so much tardily waiting for the Indian Government's any tangibly formal official-clearance; understandably indeed in the evening of his life, he has now posted the full version of this 'enquiry-report' on his official website, before it becomes perhaps too late for him. So much for any worthwhile Politico-bureaucratic strategic-acumen, as well as its geopolitical-prudence in the Indian Defence Ministry, and what more to say....

And hence precisely with respect to the above, now **Neville Maxwell** in his article very perceptively and also frankly writes that, "Those who gave me access to the Henderson Brooks Report when I was researching my study of the Sino-Indian border dispute laid down no conditions as to how I should use it. That they would remain anonymous went without saying, an implicit condition I will always observe, otherwise how the material was used was left to my judgement. I decided that I would quote freely from the Report, thus revealing that I had had access to it (and indeed I had a copy), I would neither proclaim nor deny that fact; and my assumption was that the gist of the report having been published in 1970 in the detailed account of the Army's debacle given in my (*book*) India's China War; the Indian government would release it after a decent interval.... The passing of years showed that assumption to have been mistaken and left me in a quandary. I did not have to rely on memory to tell the falsity of the government's assertion that keeping the Report secret was necessary for reasons of national security (*What an extremely irrational and corrupt thinking in the India's top-level politico-bureaucratic corridors of power in the South Block—as to when and where all they could in fact, ever be patriotically sincere and honestly trustworthy for the hapless Indian peoples and their apolitical Armed-forces, particularly towards their extremely slow-paced and also the topsy-turvy military modernisation and up-gradation? Courtesy: India's so much nefarious, corrupt, and also powerfully entrenched politico-bureaucratic nexus only; and well, look at its concern for national-security.... limit.*). I had taken a copy and the texts, nowhere touches on issues that could have current strategic or tactical relevance. **The reasons for the long-term withholding of the report must be political, indeed probably partisan perhaps, even familial (*What more to reveal now.... and what more to say.*)**. While I kept Report to myself I was therefore complicit in a continuing cover-up.... I marked the new century by publishing as an 'Introduction to the Henderson brooks Report' a detailed description, and account of the circumstances in which it was written, explain its political and military context and summarising its findings (EPW, April 14, 2001): there was no public reaction in the Indian Press or even among the chauvinist ranks of the academic security establishment (*Wah.*).... Although surprised by this reaction (*i.e. while putting the Report on Website.*), unusual in the age of Wikileaks, I could not argue with their reasoning.... So my dilemma continued—although with the albatross hung, so to speak on Indian necks as well as my own.... As I see it now I have no option but rather than leave the dilemma to my heirs, to put the Report on the internet myself. So here is the text (there is two lacunae, accidental in the copying process)." (Maxwell, *The Statesman*, 19 March 2014)

In the same article he further says that, "Those who had read Maxwell's book (never allowed to be sold in India) would claim that what is now in the public domain is no more explosive that what the journalist had revealed long ago. **And that is widely accepted that the blood-soaked disgrace was the result of the political blunders of Nehru and Krishna Menon, and their 'chosen' military sycophant Lt. Gen. BM Kaul**. From *Hindi Chini Bhai-Bhai* to 'throw the Chinese out of NEFA' was the undoing of the man credited with the building of modern India—the failure that 'broke' him. All this now stands re-authenticated. And the slanging match will long continue.... What is now required is detailed unemotional professional military analysis of the manner in which no war should be fought. Scanty 'intelligence' (*again, courtesy: another Nehru crony: BN Malik, the then Director, Intelligence Bureau.*) tailored to suit political interests, pathetically unprepared troops, communication gaps between the men on the ground and Army Headquarters.... **Discard the report of the political embellishment for which its authors are not responsible and it emerges (from needless clouds) as an exemplary document: detailed, honest—with pregnant 'quotable quotes'—and indeed a tribute to Gen. JN Chaudhuri for having selected the men who conducted a difficult assignment with a most 'soldier-like' thoroughness. It must be no academic analyses, for in so many way the situation on the Indo-China front reflects many of the shortcomings of 1962—asymmetry on the ground in terms of infrastructure and military muscle, and an inability to 'read' the Chinese mind.... The Indian establishment's inability to 'ride a punch', and the military's living in a state of denial of its own inadequacies are not side-stories**. They perhaps throw light on Maxwell's observation that should answer many a query: The reasons for the long-term withholding of the report must be political, indeed probably partisan perhaps, even familial. While I kept the report to myself, I was therefore complicit in a continuing cover-up." (Maxwell, *The Statesman*, 19 March 2014)

Hence what one is compelled to notice is, that perhaps the all sorts of 'fault-lines' are running all over the places and thus almost in every segment of the modern Indian Nation-State; and the nation's 'power-centres' instead of multiply taking the requisite corrective-steps, are only persistently bent upon clandestinely to keep looking the other way always. We would be coming to these aspects in some more details and contexts, little later. **Why is it that after so many years of India's political independence and that too now in the present 21st century the Indian peoples, their political leaders, the media, intelligentsia and even the many professionals are still regressively suffering from the numerous kinds of selective amnesia, towards the always multiply important spheres of objectively-cum-discerningly studying and analyzing the India's living-history and historic personalities, solely for the future well-being of the nation**?

Must India always and that too even today in the age of I.T. and instant connectivity; 'remain condemned to keep repeating the blunders of the past and thus, keep-on reliving any humiliating saga of her own past living-history, with no concern or any awareness of rather the no less amusing amazements of almost the whole world? These still continuing states of affairs, concerning the modern and democratic Indian Nation-State are indeed bizarre and totally unpardonable, in the present-era's fast altering international geopolitical scenario where the Indian leaderships are always inescapably expected to be geostrategic-cum-politico-militarily alert and prepared to safeguard the nation's honour and well-being, with due strategic vision and cool pragmatism. But very sadly, they are bent upon to be rather criminally myopic, as well as remaining mostly busy in petty politicking and also in the all sorts of rampant corruption. And the all the more frustrating is that, after the Indian military-debacle at the nation's northern Himalayan frontiers in 1962, and the especially concerning India's overall national security affairs and the matters-military much forceful writings and debates have regularly been taking place both within the country and outside but very sadly, even these are largely failing to make any worthwhile effective ripples in the nation's corridors of power.

Hence, even the mankind's much larger living-history, since continuingly in the making onto its globally expanding human civilisational-canvas, is perhaps also demanding as to why the Indian leaders and the 'politico-bureaucratic combine' decision-makers are still refusing to have national-pride and remain incapable of learning the valuable lessons of history and thus, keep compelling the innocent Indian peoples to somewhat repeatedly relive and suffer the different kinds of indeed, humiliating saga of their highly chequered past? The possible answer could be that even after more than 65 years of India's political independence, the populace and its leaders are yet to reach some effective and substantial levels of politico-democratic maturity and thus, be able to successfully take care of their these lingering, as well as so humiliatingly unbecoming lacunae. And thereby to come to the roots of even the modern India's these rather quite irreconcilable geopolitical limitations, we have to even discerningly revisit certain very enlightening and yet largely hidden historic factual aspects of the last few decades of the British-Raj in the Indian subcontinent, and hopefully the same would be able to throw some fresh and valuable lights onto these still persisting and far-reaching 'dark-spots' concerning the modern, sovereign and democratic Indian nation-State. And needless to say that collectively all these too, have also been preventing India from developing her proactively fruitful grips over the always paramountcy**'multiple-aspects of the essential strategic-culture and thoughts'**.

It is quite justifiably said that the very decisively sharp and swift 14-days War with Pakistan in 1971 indeed turned out to be the most glorious hour for the Indian military and both geopolitical-cum-politico-militarily at that time, there were understandable reasons for it, and understandably these historic aspects have since already been contextually debated and written about extensively for posterity. Hence, how historically far-reaching and indeed the so much glorious times were for the Indian military during the Indo-Pak War of 1971, can contextually be gauged by going through **a book written by Lt. Gen. AAK Niazi, the former commander of the Pakistan Eastern Command, then stationed in Dacca.** General Niazi has very interestingly and also frankly written that, "The delay in publication of this book has been a blessing in disguise. It is true that time is great healer, but it is also a great exposer of lies. Had I written the book earlier, many questions would have remained unanswered? **With the passage of time quite a few facts have come to light. It has been proved beyond any shadow of doubt that the disintegration of Pakistan was an arranged capitulation and not a military defeat (*sic.*). People at the helm of affairs at that time had decided to quit East Pakistan much earlier than December 1971**.... According to the Larkana Plan, finalized between **General Yahya and Mr. Bhutto** East Pakistan was to be left without a successor Government. **Mrs. Farida Ahmad**, author of *Pakistan's National Security Problems*, told me in writing that Mr. Agha Shahi had told her during an interview that orders were sent by President Yahya Khan to the Pakistan Representative at the United Nations on 09 December 1971 to proceed with the preparations for the surrender of troops in East Pakistan. At that time we were still holding the enemy on or near the borders, and having repulsed continuous attacks the troops were physically and mentally ready to continue the fight (*Really?*).... The tragedy of 1971 still resounds with bitter memories of strife, bloodshed, and destruction. It is a story of physical ambition backed by military might, aided and encouraged by our enemies from outside which tore the country asunder, culminating in the birth of Bangladesh.... **The breakup of Pakistan and creation of Bangladesh is the most important event in the recent history of South Asia. The succession of East Pakistan has altered the political**

balance in the area. The magnitude of the event attracted the attention of innumerable writers, but only a few have been able to view the crisis in its true perspective, whilst the majority have presented a version far removed from the factual happenings, based on hearsay, conjectures, and unsubstantiated records. Most of the accounts are still written from a safe angle for various reasons, mostly for personal gain. The writers have conveniently avoided mentioning the names and misdeeds of the people who mattered most.... **Granted, the commanding general and his troops play the leading role in a battle, but there are many factors beyond their control which have far-reaching consequences on its outcome, and eventually on the war.... The military action is a means to an end and not an end in itself; it is wheel within a wheel. Therefore, it is important for a writer to understand the national strategy and its implications during a war.... The most annoying thing is that none of the writers of the books on the 1971 War kept in view the tasks allotted to me, or the achievements of my small, tired, ill-equipped force under the worst possible circumstances against a far superior enemy**." (Niazi, 1998, pp. xxvii, xxiii, xxx & xxxi)

Hence it remains always important for the highest level political leadership of any nation anywhere and also under any given situation, along with numerous other external and internal factors, to always be able to command proper and beneficial grips over even the finer nitty-gritty of any particular situation-specific broader geopolitical developments and their implications all around the vicinity of one's immediate neighbourhood too. And thereby if the leadership is also able to prudently keep its nation's 'powder dry' and also, maintain its geopolitical compass in order; then it invariably is able to advantageously siege any opportunity coming on the way, indeed in a very far-sighted manner. Principally due to the India's multiply fast-paced advance diplomatic-initiatives-cum-preparations and also very timely entering into a well-conceived 'military-pact' with the then USSR, and then simultaneously also allowing the nation's military-chiefs and their top field commanders to strategically siege their valuable times in-between, to make almost all possible strategic-cum-logistical preparations, and thereby they were able to decide the date, time and places of their own professional choosing to launch their multi-directional invasions into the enemy's targeted heartlands, to finally achieve their brilliantly swift and decisive military-victory of the war. Hence, at the end of day, the Indian Armed Forces were able to display their excellently integrative tri-services' synergy and thereby finally, fructifying into the unconditional military-surrender of the entire Pakistan Army's Eastern command, along with the components of civil administration therein, in presence of the Indian-Nation's the then Eastern Army, Air force and naval Commanders-in-Chief, and the rest is now all so well-known history.

In fact the Indian military's historically so fast-paced and decisive victory in this war against Pakistan was so momentous and far-reaching that, it did trigger almost very significant changes in the further course of the continuing living-history of whole of South Asia. The momentously further history-changing reach, as well as the multiply far-reaching fallouts of these developments, as since affected by the outcome of 1971 Indo-Pak War; can indeed be gauged, if we may care to closely read 'in-between the lines' in the contextually above quoted paragraphs, as taken from the above referred book of the Pakistani General Niazi and also then, by visualizing his mental-state so clearly while reading his this book.

Hence logically in any given historic situation anywhere, it may be quite appropriate to say that the victor's 'finest hour' is indeed the 'worst hour' for the vanquished, in almost equal and opposite proportions. So as the war ended and the so much momentously history-making, massive, and the totally unconditional military surrender took place, being largest after the Second World-War times; the Indian military was able to bring-in before its political-masters—the many historically unprecedented, as well as geopolitically, the invaluable and also, geostrategic-cum-politico-militarily far-reaching gifts on a platter—meant to be very prudently used and also quite expectedly, in their mandatorily befitting and visionary manners to essentially serve the India's highest and long-term national-interests. But historically indeed as a great anticlimax, the independent and democratic India's always so rigid and regressively fixated, as well as the no less powerfully entrenched 'politico-bureaucratic combine' then—to the entire nation's greatly traumatic-shock—the political and bureaucratically so naive personas sitting in the South and North Blocks in New Delhi, and also to the great amazements of many other nations too, didn't at all prove themselves equal to their these paramountcy strategic duties when the time for theirs, another 'tryst with destiny' had so advantageously arrived and hence thereby, those invaluable 'gifts' were just very 'foolishly' squandered-away as a great loss and disservice to the nation. And even now, this 'politico-bureaucratic combine', since always being a great drag onto the nation, is very arrogantly refusing to have any remorse and thus may care to learn many valuable lessons of history and thereby, be sincerely visionary enough for not to compelling the nation to suffer the agony of their callously repeating blunders anymore.

Now with a hindsight and also while historically looking-back and objectively reevaluating the then prevailing overall geopolitical scenario in the South Asian region, after the 1971 Indo-Pak War, we find that when India's Mrs. Gandhi, along with her team of close-advisers (*i.e. mainly the then so-called powerfully entrenched 'Kashmir-mafia'!*) was proceeding to Shimla to attend her historic summit with Pakistan's Mr. Z.A. Bhutto she was comparatively indeed, in a terribly stronger position with around 93,000 prisoners-of-war in her bag, to serve India's some most vital long-term strategic-cum-geopolitical interests very conclusively. In fact geopolitically, she was in such an unassailably commanding position then, that very firmly and indeed with requisite diplomatic prudence and due politico-strategic vision she could have dictated certain very well-crafted and sugar-coated terms and conditions, for lasting solutions to the vexed 'Kashmir-problem' and thereby, ensuring some desirably tangible peace at India's now only remaining western-borders with Pakistan. But the golden opportunities were lost with such foolish naivety that the Indian apex leadership looked quite pathetically cheated and, it had to blame itself only. Any successful diplomacy and that too in any crucially far-reaching situation, always needs to be conducted from a position of strength. Hence it is rightly said that any successful diplomacy with any long-term endurance and effects, and that too with any dicey and highly unreliable adversary in one's neighbourhood, always needs to be conducted with a 'fist of steel in velvet gloves'. Astonishingly and also as a great tragedy for India then at Shimla; the then highly manipulative, as well as very craftily entrenched—the so-called *'Kashmir-mafia'*—had already bureaucratically encircled Mrs. Gandhi that, they quite clandestinely ensured that no very senior-level, professionally knowledgeable, and also the then freshly experienced military-adviser was included in the top-level delegation to Shimla and the rest is rather a very unpalatable history.

In fact a similar kind of very foolish naivety even little earlier, was so callously adhered to by Mrs. Gandhi's same team of the very dicey cronies when it came to the many geopolitical-cum-politico-militarily crucial discussions with Mr. Mijubur Rahman, regarding the assistance to be given to him in setting-up indeed the multi-layered civil administration and also maintenance of peace in his the then newly born nation of Bangladesh. Nevertheless, the Indian military-brass too committed some kind of an embarrassing 'fiasco'; when they failed to catch hold of any Mukti-Bahini commander as available in the vicinity, to be present as a witness to the signing of historic surrender ceremony at Dacca, by the two opposing Army commanders and thus, get included in its historically unique photograph. This very prudent and courteous gesture on India's part, would have had very long term and positively lasting and great positive effect with unique perpetuity on Indo-Bangladesh multiply important relationships and neighbourhood. But quite ironically, a golden opportunity was so naively lost in these regards. Such opportunities if gone can never be availed again, and that's what mankind's living-history has been showing us since the very ancient of times. During top-level politico-diplomatically close and far-reaching discussions with Mijubur Rahman and his associates, during those history-making days both in Delhi and Calcutta; many strategically very important matters to India's great advantages, could have been successfully concluded, if the then Indian 'politico-bureaucratic combine' happened to be geopolitically prudent and strategically visionary enough; as per India's the then emerging and situation-specific certain vital national-interests. If the Indian leadership would have conducted these, the then historically so important and also crucially far-reaching geopolitical and geostrategic-cum-politico-military tasks with due strategic prudence and diplomatic-vision; vis-à-vis Pakistan, and also the newly born Bangladesh, these would have indeed turned out into a greatly successful saga of India's highly advantageous 'great-games' in South Asia. **Where precisely in fact, lies the blame for these indeed very 'foolishly-triggered fiascos' but in fact totally and squarely at the doors of India's so much regressively short-sighted, arrogant and powerfully entrenched—the 'politico-bureaucratic combine'—in the Indian Government's corridors of power in New Delhi, with no accountability whatsoever**.

Precisely in the above specific contexts, **Lt. Gen. JFR Jacob,** the General who was then Chief of Staff, as a Major General, of the Indian Eastern Army Command in its headquarters at the Fort William, Calcutta; has indeed very informatively written that, "Meanwhile, the External Affairs Ministry in Delhi stepped up the degree of liaison with the Bangladesh Government in 'Mujibnagar' in Calcutta. **AK Ray**, a joint secretary in the Ministry, was sent to Calcutta for this purpose. Ray was a capable officer who was well-acquainted with the current situation. He was practical in his approach and established a good rapport with the Bangladesh leadership.... **DP Dhar** was the head of the policy planning division in the Ministry of External Affairs. The External Affairs Ministry was largely ignored and foreign policy was formulated by a very competent and influential group of Kashmiris, pro-Soviet in their views: **DP Dhar**, Foreign secretary **TN Kaul**, **PN Haksar**, and **PN Dhar**. This powerful group was unkindly often referred as the **'Kashmir Mafia'**. DP was sympathetic to the pro-Soviet elements in Mujibnagar. From the end of August, whenever DP Dhar visited Calcutta he operated out of the owner's suite in the Grand Hotel. I had known DP Dhar earlier in Kashmir and he invited me for discussions several times in his suite. Our meetings were spread out over several hours and interspersed periodically by shorter meetings with many others. His views on communists were revealing. According to him pro-Soviets communists were 'good guys' and the Marxists, in East Pakistan were not. I tried to argue that these groups should not be supported either physically or militarily. His answer was always 'Jake, I know what I am doing'. He bypassed AK Ray, which was indeed unfortunate. **I last met with Dhar a few days after the surrender, at the VIP lounge of Calcutta airport. I suggested to Dhar that it was important to get from the Bangladesh Government an agreement on what I considered to be the three essentials: guarantees for the Hindu minority, rationalization of the enclaves, and transit rights by rail and inland waterways through Bangladesh with use of facilities at Chittagong port. DP, with his charmingly disarming smile, turned around and said, 'Jake, you are a soldier. These problems are political. They can be sorted out at the appropriate time'. I replied that the appropriate time was now and it would be very difficult later to obtain any agreements. He smiled again. The Chief Minister of West Bengal was present at this meeting but did not respond. (*Naturally, who***

cares in India about these geostrategic-cum-politico-militarily indeed, very long-term and also crucially important matters; among the India's always so much shortsighted and self-centred politicians, since nothing else but only fixatedly occupied with their electoral-politics-oriented-thoughts and activities; and also regardless of the Great-Powers' many kinds of the international 'great-games', as since always being played in the region.).... **Ambassador Keating** visited Fort William during this period. The Army commander being out on tour, I apprised him of the current situation, particularly the refugee problem and the large-scale atrocities being perpetrated by the Pakistanis. I told him that I was at a loss to understand why the Government of the world's most powerful democracy should support a brutal, repressive, military regime which had completely disregarded the results of the elections in East Pakistan. Keating was a polished diplomat. His sensitive face coloured slightly but he did not resent what I said and did not defend his Government's stance either. Perhaps his relations with the State Department and **Dr Kissinger** were not sufficiently close to counter **Nixon's tilt towards Pakistan**." (Jacob, 1997, pp. 98-100)

India's regressively fixated, arrogant, largely clueless, and powerfully entrenched 'politico-bureaucratic combine' about the highly complex, but-extremely crucial nitty-gritty of the multiple-aspects of the geostrategic-cum-politico-military related subject-matters, even today refuses to understand that at the top-military levels of command, these civil-cum-political issues and also, just superficially appearing to be largely the military-issues, just cannot be separated and locked-up in two distinct water-tights compartments. In fact at the apex military-levels, all these become so much mutually overlapping and integrative that, they can't be dealt with separately or in isolated manners by the respective functionaries both at the top political and military levels since both become inseparable, as well as mutually influencing-cum-complementary too. From the points of any worthwhile and thus effectively seeing through some kinds of this conceptually-misguided, misconceived and separating mindset we find them to be indeed totally untenable, as well as counter-productive too; and hence at any crucial stage, could even be dangerous to India's vital security related national-interests. But concerning the modern democratic Indian Nation-State, these so depressing apex political-level governance realities are still so callously continuing regardless that in spite of now being impeccably apolitical, trustworthy, and highly professional top military-brass in fact, so relentlessly banging their heads against the no less sinisterly-cum-clandestinely erected—the politico-bureaucratic 'stonewall' all across their exclusive corridors of power nothing seems to be changing at all. And moreover at the same time and irrespective of any seriously adverse consequences to the nation this arrogantly fixated 'politico-bureaucratic combine', does always remain bent upon to resent the military's any intrusion through its above noted so-called 'stonewall' even if it is indeed consistently meant to nothing, but always to wake them up towards unfailingly serving the nation's paramountcy national-interests.

It is now historically an open secret that the independent India's even the first Prime Minister, **Mr. Jawaharlal Nehru**, being a maker of modern India and with all his colossal learning, attributes of his charismatic leadership, his some outstanding vision for the modern, dynamic and democratic India; quite amazingly fell short of seeing through the already above discussed and since always being paramountcy tenets concerning the nation's vital aspects of the geostrategic-cum-politico-military imperatives and challenges and thereby, couldn't enable the nation to successfully sail through the always dicey and choppy 'geopolitical-waters' around India's neighbourhood. **Historically on the other hand now, it is also very befittingly said that Nehru took the lead with great vigour to firmly lay the vibrant foundations for triggering India's economic, industrial and technological developments and simultaneously also, provided valuable visions and**

conceptual-inputs towards building a modern, democratic India with socio-culturally having a dynamically fructifying, the peoples' multi-dimensional scientific-temper too. This indeed remains a great irony of the independent India's comparatively now the newly manifesting, but the no less engaging, living-history in today's modern world. But even the subsequent leaders in India in course of their reign of power, have not cared to somewhat meaningfully pause and look back to learn lessons in the highest national interests. Concerning historically now the bygone socio-political crises in the Indian State of Punjab during 1980s, since the then culminating into the highly tragic 'Operation Bluestar', we paradoxically find the same very embarrassing history being repeated again.

And precisely with respect to the above, India's former Vice-Chief of Army, **Lt. Gen. SK Sinha** has revealed that, "**Mr. R Venkataraman**, the Defence Minister, visited Bhatinda (Punjab) in my Command (*i.e. when General Sinha happened to be India's Western Army Commander*). The Defence Secretary, **Mr. PK Kaul** accompanied him, I received them at Bhatinda. They were with me for two days.... Mr. Venkataraman was member of the Cabinet Negotiating Committee for talks with Akalis. I felt that I must take advantage of his visit and put across my view to him. The situation had begun to deteriorate alarmingly in Punjab and a Hindi-Sikh divide had started taking place. The old tradition in Punjab had been that one brother would be a Sikh and the other would remain a Hindu. Inter-marriage between the two communities was very common and they used to freely participate in functions of each other's places of worship. All the closeness between them had begun to evaporate for political reasons. Religious fundamentalism was being exploited for political purposes.... **I spoke to Mr. Venkataraman that I would like to discuss the Akali problem with him. He was a little surprised at my raising this issue. He told me that it was a political matter and I as a General should not get involved in politics. I agreed with him that General should keep away from politics. However, sometimes the line dividing strategy and politics often gets blurred. The Akali problem was a case in point. As Western Army Commander, I was deeply interested in the issue being resolved amicably soon because of its serious military implications. There were three specific reasons for my interest in the matter**. First, I was responsible for the defence of Punjab. If the controversy continued or got aggravated, it would have direct bearing on my operational plans and I might even have to consider changes in deployment of troops, in the event of war. Second, if the situation went beyond the control of the civil administration, I would be called to render aid and restore order. Thirdly, out of the nearly three lakhs troops under my command, some 80,000 were Sikhs. This controversy in the long run might affect their morale and discipline. For all these reasons, I was deeply interested in an early political solution in Punjab. Mr. Venkataraman saw the logic in my argument and enquired if I had any solution to offer. I replied that my interest lay in the problem getting resolved quickly and I was not qualified to offer any solution as such. However, since he wanted to know my views I would put them across, for whatever they were worth. The Akalis had put forward both religious and political demands. So far as the former were concerned, they wanted recital of Gurubani at the Golden Temple relayed on the AIR, removal of cigarette shops from the vicinity of the Golden Temple and renaming of the Flying Mail as the Golden Temple Express. It should not have taken the Government more than a minute to concede these innocuous demands. Mr. Venkataraman said that it was all politics. You give an inch you will have to give a yard. Today they were making these demands but tomorrow they will ask for Amritsar being given the status of Vatican City. I replied that in the Army we have a saying that you cross a bridge only when you get at it. The demand for Vatican City status has not been raised and when raised could be suitably dealt with. For the present, if their reasonable demands were not conceded quickly, they would exploit these issues for political purposes promoting religious fundamentalism. The Minister then asked to suggest a solution for political demands. I said that the transfer of Chandigarh to Punjab had already been agreed to. As regards of river waters and adjustments of territories on linguistic basis, it should not be difficult to find a mutually acceptable solution on a give and take basis. **The impression I got after my discussion with Mr. Venkataraman that the Government lacked the political-will to resolve the problem quickly. The trauma of Punjab would continue and would get much worse**." (Sinha, 1992, pp. 287-289) (*And contextually, what we do find here with our heavy hearts that with the continuing passages of time since 1947, nevertheless, a relentless saga of the similarly depressing stories are being repeated regardless.*).

Now historically in retrospect, we indeed very unambiguously find that both geostrategically, as well as politico-militarily, how professionally visionary and also prophetic were the submissions of the one of India's most outstanding Generals: S.K. Sinha, to the then Defence Minister; specifically concerning the highly complex and menacingly volatile situation in Punjab, during the 1980s. But then again to the great misfortune for the peoples of India; as usual, the arrogant and also extremely shortsighted the apex-level Indian political leadership and theirs, the no less arrogant and clueless cronies—the bureaucratic class, since sitting in the corridors of power—didn't care to sincerely listen to the any timely valuable advices from the top-level expert military-professionals shouldering heavily sensitive responsibilities, and thus failed to mend their thoughts and make multiply crucial efforts to meaningfully live up to their always mandatory commitments to the nation and sadly, the rest is again a very well-known but equally shocking history in these regards. And needless to say that specifically concerning the problems concerning the Indian State of Punjab in the 1980s very unfortunately all these cumulatively in turn, finally culminated into the very violent and tragic, socio-cultural as well as politico-military fiasco: now recorded on the pages of India's post-1947 era living-history, as a greatly embarrassing 'blot' since equally infamously known as the 'Operation Bluestar'.

After the very tragic assassination of Prime Minister Mrs. Indira Gandhi in 1984, her elder son **Mr. Rajiv Gandhi**, in a very strange and democratically in the no less bizarre manner, took charge as the next Indian Prime Minister, with no administrative or political experience whatsoever. And moreover, by this time once being so glorious and indeed a history-making political outfit—**the Congress Party**—had already got reduced to not even any remote-shadow of its once shining past and thus appeared to be totally deprived of its earlier socio-political-cum-moral fibre and thereby; being full of sycophants, its many senior leaders' cronies and also the all sorts of hangers-on in the Party's rank and files, almost everywhere in the nation. He being politically naïve and thus in no time, he got surrounded by all sorts of rather quite selfish and manipulative party-men, to the great detriment of his official and the numerous geopolitical tasks and so to his paramountcy responsibilities to the nation. His some nefarious advisers and cronies were having a 'field-day' and hence, while pursuing their many undesirable designs, greed, and the games of 'vote-bank politics' they began to disturb the then prevailing peace in the border State of Jammu and Kashmir, wherein even a J&K political party, the 'National Conference' i.e. NC, as a partner of the Congress Party, and being mostly obsessed with its own vote-bank politics under **Mr. Farouqqe Abdullah**. They both became so much short-sighted to obtain any political mileage electorally in any manner and thus together creating dangerous situations, consequently being made available for Pakistan to fish in the geopolitical-cum-politico-militarily troubled-water in J&K, and the rest is now so problematic and cumbersome history, currently being made a burdensome saga, like a malignant 'pain in the neck' for the Indian Nation in this border State. But in these multiply serious contexts for the Indian-Nation very sadly; all kinds of the anti-national and highly nefarious designs and the many acts of treason too in the State of J&K then, began to be conceptualised, planned and thereby, the related 'blueprints' too were being affected on a very large-scale, since the times of Mrs. Indira Gandhi as India's highly controversial Prime Minister. Cumulatively all these extremely adverse developments in the Kashmir valley in turn, also began to facilitate the numerous die-hard Pakistani ISI agents to make very dangerously deep inroads almost all over in the J&K.

In fact, since always continuing to massively indulge in all sorts of their shameless corruption and nepotism these political parties and their leaders also have no time to sincerely think for their nation's geostrategic-cum-politico-military security-aspects, both internal and external, in India's this so much strategic border State of J&K. Hence, whom the Indian peoples would blame for these massively complex mess-up, as well as the extremely nefarious birth-cum-expansion of the all sorts of Islamic Jihadi terrorism and extremism onto the soils of this unfortunate State, but these unscrupulous, corrupt and irresponsible politicians, since both in Delhi and in the J&K? How and why this culture of Islamic Jihadi terrorism could take roots and thus flourish in the State of J&K; is indeed a very complex story and lot of fairly in-depth writings in these regards, have since already been published, helping the interested people to decipher and understand, the many 'cobweb' kinds of the nitty-gritty of these nationally dangerous problems. **Similarly there is one book, titled *Terrorism and Insurgency in India: A Study of the Human Element* authored by Captain Ashish Sonal, VrC, of the Indian Army. Contextually, it is indeed important to note that this officer was awarded the Times Fellowship for the year 1991 and this work is the result of the research undertaken in that capacity.** The support of the Times of India Group has gratefully been acknowledged by the author. Besides J&K, Capt. Ashish Sonal, in his in-depth studies and research, has also covered the North Eastern Indian States too while studying these complex problems and in the process, has been able to throw much light towards understanding these problems in their objectively proper perspectives and thus, been also able to considerably remove much of the related confusions. In fact deciphering and analyzing the roots of the highly complex phenomena of these numerous kinds of the terrorism and insurgency related problems and their multiply menacing growth are no easy tasks as they happen to be unpredictably fluid and dicey in nature, like 'quicksands' and hence being always very problematic to really understand, control, and thus effectively neutralise them, in the larger national socio-cultural-cum-political peace and tranquility. But even then, Ashish has indeed succeeded greatly in achieving his aims, while pursuing his studies and research in these regards.

And hence precisely in the above multiple-contexts, the then Indian Army's a young **Captain: Ashish Sonal** has also very thought-provokingly written that, "The main contributory factors to the spread of terrorism/insurgency in India have been widespread economic frustration, communalism, poor moral health of Indian politics and its democracy, administrative failures and to a lesser extent secessionist aspirations. Further impetus to the above causes has been provided by some of our neighbouring countries by their willingness to extend material, ideological and other help to terrorist/insurgent organizations. Outside help has played a crucial role in the emergence and growth (as well as decline) of most of the terrorist/insurgent movements that India has experienced.... **In Kashmir Islam has traditionally been one of most secular cultures. The Kashmiris can be termed as nationalists—but certainly not Muslim fundamentalists. However, since mid-1970s there has been a conscious effort to 'fundamentalise' the religious scenario there, by highly articulate elements propagating the closeness of Kashmir to 'Islamic' Pakistan. The Kashmiris, as part of India, have never felt any restrictions on their freedom to practice their religion. Yet the militant organisations in the valley call the young men for *Jihad***

against Indian oppression threatening Islam in Kashmir. The unique cultural identity and the affinity of Kashmiri Muslims to Pakistan because of the Islamic brotherhood factor, is stressed. Starting from mid-1970s, with a surfeit of money from the Islamic Middle East countries (*principally Saudi Arabia.*) coming into Kashmir for religious purposes, a large percentage of villages in Kashmir came to have about 5 or more mosques instead of the earlier one or two. The *maulvis* of these mosques were not Kashmiris brought up in the secular Kashmiri culture but fundamentalist Muslims hailing from Uttar Pradesh and Bihar. They helped to attract the youth towards the idea of Islamic nationhood for Kashmir by sensitizing their minds with the idealism of fundamentalist Islam. Since the mid-1970s, a generation has grown up (*now naturally, even two successive generations.*) in Kashmir which has been brought up listening to the articulation of separate cultural and religious identity and its affinity to Pakistan (*duly acknowledged and quoted from Rajesh Kadian in the 'The Kashmir Tangle—issues and Options', 1992, Vision Books, page – 13, since also describes this phenomenon*).... In the Naga insurgent movement one does not find revivalist efforts because the requirement was never felt. Since the various **Naga tribes always felt that they were never part of India** – culturally and historically. ULFA (*i.e. United Liberation Front of Assam*) has attempted to give the Assamese community the status of an ethnic nationality by emphasising the glorious past of the **Ahom Empire** and the distinctiveness of the Assamese culture. 'Assam was never a part of India, we had our own empire, it was a glorious period till the British came and subjugated us (the Treaty of Yantabu in 1826) and finally made us a part of India – so it was by an accident of history that we became a part of India.' This is how the ULFA projects its views to the people negating the history of cultural, historical and racial interaction with other parts of the country. In **Manipur** there is a strong revivalist movement emphasizing reversion to the **original Meitei script, culture and nature worship practiced by the Meiteis before their conversion to the Vaishnav sect of Hinduism**. (The origin of Meitei insurgent movement lies in the revivalist efforts which were started in 1930 by 'Meite Marup' – and organisation which was formed to revive tribal Meitei religion i.e. 'nature worship'. By 1960s the feeling that Meiteis were under domination of outsiders, both culturally 'their Hindu religion', and economically had taken firm roots among the Meitei society). The revivalist movement, here as in other areas, has been used by the extremist elements to emphasize the distinct identity of the Meiteis. One can detect similar efforts even in the nascent Hindu extremist movement.... **The extremist elements supporting the terrorist/insurgent movements attempt to involve *religious and student organisations* in their movements and such bodies have often played an important role in the revivalist and the de-Indianisation efforts as also in the propagation of their political doctrine**. The religious and student organizations provide a public platform for voicing extremist beliefs and are prone to easy manipulation by elements sympathetic to the terrorist/insurgent movement. Students and student organizations, due to the innate tendency of the youth to rebel against the established social system, form a ready source of active supporters and are an important support base for the terrorist/insurgent movements. However, these educated or partly educated youth belonging to student organizations are *not a major source of cadres* for the terrorist/insurgent organizations. The educated youth prefer to give their support from the periphery without fully joining the movement and sharing hardships. The leadership of students often falls into the hands of active supporters of militancy.... **Religious issues become one of the main means of revivalist efforts because of the strong emotional hold of religion on people. Religious heads, functionaries and institutions exercise tremendous power in terms of their ability to influence public opinion therefore they are often extensively incorporated in the terrorist/insurgent movement. In Punjab, gurudwaras were the sanctuary as well as the focal point of all militant propaganda. The Kashmir**

extremist movement aims to create the Nizam-e-Mustafa for the people and the NSCN in Nagaland the 'land of Christ'. The church over the years played a pivotal role in the development of insurgency in Nagaland, so much so that the government peace process and negotiations have also been affected through the church. Religion is an important factor that gives a community its distinct identity. The terrorist/insurgent movement therefore endeavours to reinforce the distinctiveness by religious fundamentalism – it is in this role that the militant movements in Punjab and Kashmir attempted to use Sikh and Muslim fundamentalists.... **Violence as a means of quick solutions holds a special charisma for frustrated youths – millions of whom are available in India for exploitation by the leadership of the present and future extremist movements (*courtesy: India's deliberately unchecked population-growth.*). In the face of unequal opportunities, grossly deficient means of development, rampant corruption, criminalization of politics, unresponsive executive and judiciary, but free recourse to knowledge and increasingly greater availability of the means of terror (the weapons), in times to come to India is likely to see an increase in militancy**. Not the kind of high intensity, well organized militancy which Punjab, Kashmir and Nagaland have experienced, but the kind which is prevalent in the urban areas of Manipur—deriving its substance solely from the frustrations of the people, and organized in the form of politically conscious small sized extremist groups. **The role of the future militant leadership need only be to project violence as a viable option, and both criminal and frustrated elements would respond in adequate numbers. The connivance of political elements with terrorist/ insurgents which is blatantly visible in Punjab, Assam, Nagaland and Manipur, or for that matter also in Bihar, can only lead to increased percolation of violence into the country's political process, depriving the State of the moral justification and determination required to deal with terrorism/insurgency**." (Sonal, 1994, pp. 8, 9, 23, 24, 25, 26, 120 & 121)

In the above quoted paragraphs, and specifically so far as the very sinisterly anti-national plantation of 'seeds', the demonic sprouting, as well as their highly menacing fructifications inevitably leading to the phenomenon of extremely violent and also, unprecedented forms of the 'Islamic Jihadi Terrorism' in India's North-western border State of J&K being adjacent to Pakistan, and hence duly aided and abated by the Pakistani ISI, are concerned even the sensational disclosures, since already made way back during 1994 itself as per above have not made any dent in the consistently callous, arrogant and also the no less regressive thought-lines and attitudes of the so much powerfully entrenched 'politico-bureaucratic combine', in the Government of India's apex-level corridors of power in the North and South Blocks of New Delhi. The overall scenario in these regards, are not only very frustrating but also being so much dangerously anti-national that all these can even lead the present super high-tech modern-era's, any reasonably educated, patriotic, as well as also consciously sensitive and thinking Indian-citizen, to some sort of restlessness and 'intellectual-nightmares'. What more to say and, how to wake-up the Indian Political and bureaucratic classes, to fruitfully compel them to see reasons and thus, prevent them from taking all possible sorts of the highly short-sighted and also the no less dangerously irresponsible liberties with India's paramountcy geopolitical interests, national-security aspects, and also the overall well-being of the nation, where any kind of the whether naïve or otherwise 'slip-ups', may even turn out to be seriously harmful too?

When the extremely fundamental, religiously blind and fanatic Mullahs from the Indian States of U.P. and Bihar, had initially begun to make inroads into almost all across the Indian State of J&K, with very sinisterly planned intentions to make them 'settle-down' in the newly constructed numerous mosques over there; and also duly facilitated by some highly fanatic and anti-national

Islamic extremist outfits, whether international or intra-national; during the Premiership of **Mrs. Indira Gandhi**, what were the nation's administrative, intelligence, and internal-security organisations doing and thereby what precisely, the multiply preventive actions were being taken then, while there was still sufficiently actionable time available to begin with and then, to what effectiveness if any? The nation since now already being an established multi-party parliamentary democracy, has every right to get the full and also the no less compulsively unambiguous answers to these questions. But sadly no answers are still coming, related to any dangerously serious 'fiasco' in these regards; where the bucks are always unfailingly stopping at the doors of India's powerfully entrenched 'politico-bureaucratic combine', and then to all the more dismay of many Indian citizens, even no signs of the any meaningfully corrective measures are being visible in any manner in the nation's corridors of power. And thereby to the great misfortune of the peoples of India, their nation is being made to repeatedly suffer. The larger point in these regards, happens to be that the overall geopolitical-cum-military scenario around the Indian-Nation in Asia remains hardly friendly and at times even highly hostile due to various reasons, some as fallouts of the various kinds of 'omissions and commissions', since periodically being committed at the Indian apex-level political leadership and at the same time others as largely the different kinds of the 'freaks of history' and thus obviously, being beyond India's control.

So what all these boil-down to be in fact is that the continuation of the numerous kinds of sad recurrences of the national setbacks and embarrassments do indicate that periodically from time to time—only the Prime Ministers in India have been changing—and the overall geopolitical realities around the modern Indian-nation today continue to be always dicey and highly worrisome and thus, largely remaining beyond the reach, as well as effective influence of the apex-level Indian leadership and its any so-called multi-dimensional visions. But to India's great ill-luck, the malignantly entrenched, as well as the highly regressive and corrupt bureaucratic culture remains fixatedly unchanged and thereby, being a hugely heavy drag onto the multiple-aspects of national-governance, multi-dimensional socio-cultural-cum-political and economic progress and in addition, also equally onto the crucial imperatives of the upkeep of nation's geostrategic-cum-military wherewithal, as well as the politico-diplomatic strengths and assets. Now coming to the highly vexed problems being faced by India in J&K; first and foremost and also as per the findings of **Captain Ashish Sonal** through his extensive research and studies, we have clearly seen the pivotal reasons, as to how the Muslims in the Kashmir valley and adjacent areas were radicalized and thus, many of them especially the youth, got converted into the irrationally fanatic Islamic Jihadists with extreme hatred towards their own nation, i.e. India, and in spite of being treated as equal Indian citizens. It is a very sad story and, all these won't have been possible if the India's so much clueless, egoistic and powerfully entrenched 'politico-bureaucratic combine', would have been sincere and competent in doing their jobs with requisite strategic-vision and also with a flexibly open-minded coordination, as well as proactive cooperation with the nation's concerned military professionals. In any case, only India's these military-men have had to be finally asked by the Government to undertake the heavy and indeed multiply-demanding responsibilities of ultimately crushing the anti-India Jihadi-militancy in J&K and thus successfully safeguard the integrity and security of the modern, sovereign, and democratic Indian-Republic; by effectively countering the all possible kinds of nefarious designs of Pakistan and its ISI.

We may recall that interestingly the wisdom-oriented writings and teachings of the ancient-era's two greatest Chinese and Indian strategic-thinkers, **Sun Tzu and Chanakya** respectively. But ironically to the no less tragic disadvantages to the modern Indian-Nation, her clueless political and bureaucratic classes since always egoistically fixated with their highly inflated and irrationally false sense of 'superiority-complex', have since consistently been displaying no patriotic interests in these invaluable ancient treatises containing the multiply-outstanding vision and knowledge; then what to talk of the super high-tech modern-era's already multiply-expanding storehouses of the mankind's knowledge-world. Hence alas, no wonder that now the currently fructifying independent India's modern 'living-history' continues to be so much embarrassingly clumsy and unbecoming regardless wherein, the Indian leadership is mostly found to be regressively reactive, rather than being strategically visionary, as well as preemptively proactive and also geopolitically confident, in the highest national-interests.

Precisely in these multiple-contexts, concerning the State of J&K since now constitutionally and legally being an integral part of India, and also regarding the other insurgency-infested States in the North-East, **Mr. Manoj Joshi** with his deep insight has written that, "**To say that the sudden rise of the gun-culture altered the dynamics of Kashmiri politics would probably be an understatement. But the systematic manner in which the practitioners of 'normal' politics were attacked and eliminated points to a design, ruthless calculation, if you will. By physically eliminating those who were deemed to be structures of Indian rule, the militants wanted to create a vacuum that they alone could fill.... All political formations, be they anti-Indian, pro-plebiscite or pro-Indian, had to adjust to a reality where the AK-47 was the arbiter**. The manner in which existing political equations were altered can be seen from the way in which the Jamaat, never particularly popular in the valley, suddenly became a major force, courtesy the gunmen of the Hizbul Mujahideen. Other formations scrambled to deal with the situation.... The multiple groups came up because of the pre-existing fault-lines in the valley's politics that divided the People's League or People's Conference from the Jamaat-e-Islami, in addition to personal biases. The ISI had its own reasons for encouraging them. One was the need for security. They realized the brutal truth that if one member of the group was captured, it was almost certain that he would be tortured to reveals the names of others in the Group. So to prevent a domino effect, several groups were created to ensure security.... The second, perhaps more important, reason was to have levers to control the movement of the members and ensure that no single group or leader became known or well-established since they could pave the way for either a political force capable of dealing directly with India, or establishing its own authority over Kashmir, leaving Pakistan out in the cold.... But when the die was finally cast, and no one is clear by whom—the underground leadership in the valley, their controllers in Pakistan, **the incompetent Indian authorities**, or by any sheer force of circumstance—it took the conflict to a nasty, brutish action-reaction cycle which has since characterized it.... **The loss, though heavy, did not end the ISI operations. The Indian authorities soon realized that the ISI was anywhere and everywhere, where there was a chance of sparking off a revolt**. There were even charges that Pakistan was aiding the **LITTE**, in the 1987-90 period.... The arrest and interrogation of **Khayao Hurey**, the finance minister of the 'Sovereign Government of Nagaland', set up by the National Socialist Council of Nagaland (*i.e. NSCN*), revealed extent of this support. The ISI, Hurey revealed, had given his outfit US $ 1.7 million, spread over the previous two years in three installments, which were used to buy weapons from Thai arms bazaar. Hurey's contact was an ISI official operating under the cover of an education counsellor in the Pakistan High Commission in Dacca. Writing in *Frontline*, **M.S. Prabhakara** wrote in February 1994, that the **ULFA-ISI** link had come up in the late 1980s, though the first batch of ten ULFA militants were taken **to Pakistan for training** in mid-1991. In all, some seventy militants were trained, all of them going **via Bangladesh**. ULFA militants told Prabhakara that the ISI wanted them to initiate actions which they knew would harm not just Indian interests, but also some of Assam, such as blowing up oil installations.... Shortly after **Lal Singh**'s arrest on 16 July 1992, the police wrapped up his ring that consisted of a number of arms smugglers in Gujarat, a top motivator, Tahir Jamal, and thirteen young men: four from Delhi, five from Mumbai, three from Ahmedabad and one from Modasa, also in Gujarat, who had actually been sent to Pakistan and trained in guerrilla warfare and sabotage. The Bombay boys had been sent on to Afghanistan for 'advanced training'. These boys were told to target Indian economic sites. On 4 July, Singh had planned to blow up the Madras Stock Exchange with the aid of **Student Islamic Movement of India (SIMI)** activists. He had the explosives and equipment for triggering the blasts, but the SIMI activists failed to show up and the plan was aborted. Another scheme was to blow up the main buildings of the Bhabha Atomic Research Centre and two oil refineries in the Bombay area.... Not surprisingly, eight months later in March 1993, when the Bombay blasts occurred, one of the buildings targeted was the Bombay Stock Exchange and another, the Air India headquarters. This time, the ISI found the Bombay underworld a convenient and daring ally.... The arrest of **Yakub Memon** and the subsequent investigations, have only strengthened the situation of Pakistani involvement in the Bombay blasts. Authorities now say that Yakub was given the alias of **Yousuf Ahmad Mohammed**; with a Pakistani national identity card (No. AJ 763587); and a Pakistani passport, (No. AA 763242). While the Pakistanis maintained that these were forged, subsequent investigations confirm their authenticity, pointing to Pakistani complicity.... India has often been compared to an elephant. The lumbering vegetarian pachyderm is not a natural predator yet in fear or anger, it can and does kill, often with the greatest brutality. For four years since the onset of the Kashmir rebellion, the Indian security forces had responded with a tough and sometimes brutal campaign of suppression. However, their actions were more a primeval reaction than a systematic campaign to root out an insurgency or, as the Army saw it, a 'low-intensity conflict'.... In these circumstances, the bureaucracy, the intelligence apparatus, the paramilitary organizations and the Army, began to see Kashmir as the raison d'etre for their bureaucratic empires. (*All these negative developments in such challenging situations, do occur principally due to the mostly always shortsighted, irresponsible and corrupt Indian political-leadership and also, quite wasteful-cum-duplicating number of the ill-founded and multifarious Security-Agencies.*). Beyond stating that the conflict in valley was low intensity one, the Army had done little in 1991-1993 to work out a specific tactics to deal with it. But its experiences in Nagaland, Punjab and Sri Lanka, had made it sufficiently wary of trying to replicate strategy and tactics used elsewhere. By 1993, a doctrine did begin to shape up, based on its institutional experience and history.... Is there, then, no solution to the Kashmir problem? In my view, there is none in the near future. The only chance of a settlement that would satisfy all parties would come when India and Pakistan are more closely aligned to each other than they are today. In other words, mature liberal democracies where the rights of individuals, regardless of nationality, origin, religion and ethnicity are upheld in practical terms, not merely on paper. This may sound like asking for the moon. Perhaps it is. But there is indeed a lesson in the way another tangled problem—that of Alsace-Lorraine, inviting Germany and France—was resolved. An area that was a battleground for centuries is now sleeping backwater because Germany was forced to abandon its militaristic ways and become another democracy in Western Europe.... India and Pakistan are formal democracies, but it is no secret that they have a long way to go before they can be termed 'mature'. Both are beset by powerful forces of religious fundamentalism, overlaying chronic poverty and illiteracy. To forecast a liberal democratic future for either of them, especially for Pakistan, requires great optimism. But then, that is the only way to look at the future." (Joshi, 1999, pp. 51, 52, 53, 168, 169, 170, 402, 403 & 450)

That's true, but indeed the highly vexed J&K problem in South Asia today is quite different, in the sense that, Pakistan remains an artificially created, unviable and politico-economically untenable State, so far surviving on the US 'geopolitical-cum-military and economic drips' and hence unlike India, with an extremely uncertain future even concerning its any durable integrity, without politically proactive (or hyper-active?) involvement of the Pakistani Army, in its so much turmoil-ridden and also very messy affairs of its leading Political Parties. In fact, almost all the leading political leaders of any consequence and since also belonging to different Parties always look-up to their Army for patronage and support, so that they could feel stronger vis-à-vis their rivals. Historically it is also equally true that Pakistan came into being, when the erstwhile European Imperial-Powers' colonial-empires in Asia, Africa and elsewhere, were fast disintegrating with the end of World-War II, since the then triggering numerously great upheavals across the post-war world. And thereby, Pakistan was created by partitioning of the then 'British-India', principally to be used as a valuable strategic-asset in the north-western corner of the Indian subcontinent and also to be used as a 'spring-board', for maintaining the security of Western world, soon to be led by the USA and hence, to counter expansion of the highly ambitious—the massive communist Soviet-Empire—during the then fast manifesting and the highly menacing: cold-war era. And hence it is also equally true that, this geopolitically so futuristic and politico-militarily brilliant idea of the British, behind the extremely violent and geographically the no less ruthless 'surgical-creation' of Pakistan to be later used by the USA, has been paying rich dividends to the western world; and at the same time it is still continuing to be so now for this sole 'hyper super-power' even in the present 21st century as well. *We may recall that in earlier chapters and also in the differently multiple-contexts, we have since already somewhat critically discussed these aspects in their analytically relevant details too.*

Now, in order to geostrategic-cum-politico-militarily see through the very genesis and fructification of the extremely violent Islamic Jihadi-terrorism in the Indian State of J&K, in its objective historic perspectives, we need to closely read and analyse the contents of the paragraphs, since quoted from **Captain Ashish Sonal**'s earlier referred book; in conjunction with the little later quoted paragraphs from the above referred **Mr. Manoj Joshi**'s book. And thereby all these in turn, would clearly show us: as to how and what all these naïve and also so arrogantly clueless Indian political and bureaucratic classes, since occupying the various top-level seats of power both in New Delhi and Srinagar; have been doing which in any case, they won't have been doing; if they were knowledgably visionary enough to understand the history-making burdens of their responsibilities and also, crucially far-reaching implications of their decisions and actions. India's this 'politico-bureaucratic combine', since entrenched all across the corridors of power; have mostly been capable of 'shooting on their feet' only, and thus causing great national-embarrassments for the Indian peoples in these regards. Had these not been so, India won't have been facing the kinds of some security, as well as many other socio-political problems in her north-western and north-eastern regions, which in turn and as since the time has been passing-by, are becoming more and more costly and bothersome in numerous ways for the nation. **But who really cares since the 'insensitivity-levels' in fact, of the Indian 'politico-bureaucratic classes' have now demonically shot-up so high that, nothing seems to affect them and thus be capable of pulling them out of their deeply-intoxicating and irresponsible slumber; regardless of the any kinds of adverse consequences to the 'unfortunate' Indian Nation-State in the present 21st century too**. And from the 1947 itself when India gained her political independence, broadly there has been almost similar kinds of the highly humiliating and depressing chronological-saga in terms of India's apex-level geostrategic-cum-politico-military managements of her national-security related systems and also, to take care of her highly complicated geopolitical-cum-strategic challenges since trying to threaten her from almost all-across her western, northern and also eastern frontiers. And so in other words, we may even say that indeed for the hapless peoples of India, the same identical and very depressing story-line scenarios are getting replayed, and thus repeatedly causing great but highly avoidable disadvantages and losses to the nation.

Specifically in these multiple-perspectives, the earlier discussed phenomenon of the such history-making and the so much forcibly indeed—the then 'artificial-creation' of the 'Islamic-Republic of Pakistan' in South-Asia—may be taken historically, as a very far-reaching and also, an enormously consequential example of a kind of the classic 'great-games'; and the rest is all so much well-known, as well as still so much momentously manifesting living-history in these regards. Further we may also recall that while globally reviewing the ongoing phenomena of these multiply different kinds of the so-called 'great-games' and also keeping in mind theirs even differently manifesting perspectives; we find that the overall results therein, have been quite mixed and at times not at all palatable and hence, culminating into some kinds of even politico-military 'fiasco' as well.

In course of our continuing discussions on India's highly messed up problems in J&K; this author came across two very thought-provoking and also, recently published books of **Lt. Gen. Gautam Banerjee**, on highly complex and multi-dimensional 'matters-military' and that too, being expertly indeed sharp-focused with respect to especially the contemporary Indian contexts. And these books are:

(a) 'The Kashmir Narrative' – A V.I.F., i.e. Vivekananda International Foundation, New Delhi, 2017.
(b) 'India's National Defence: Defining Defence Reforms and Military Modernisation'- Published by Pentagon Press, New Delhi, 2017.

In the first book titled, *'The Kashmir Narratives'*, **Lt. Gen. Banerjee** on the pp. 11, 12, 57, 58 & 59. "Recently, anti-national armed uprising in the vale of Kashmir had escalated to an unprecedented level. It is only a matter of seasons when, bolstered with regular quantum's poison pumped up from Pakistan and strengthening of local religious and money-making vested interests, the tumult could keep escalating for the world to be further misled and Pakistan to hammer its next blow upon the roots of Indian nationhood. Obviously, our policy in assimilating the people of the valley had failed to deliver-seven decades is a long time to accept that. Time has therefore come to devise new approaches to prevent the falsehood of 'Muslim' cause for 'liberation from non-Muslim rule' from blowing away our democratic, secular, all-inclusive way of life. And that purpose is better served by revisiting the problem from its roots upward. Our policy in assimilating the people of the valley has failed to deliver—seven decades is a long time to accept that … time to devise new approaches to prevent the falsehood of 'Muslim' cause for 'liberation from non-Muslim rule' from blowing away our democratic secular, all-inclusive way of life. Pakistan has found it expedient to turn Kashmir into a 'core issue' of her existence as a nation and a plank to secure domestic political power. Enchanted by the favour of the above mentioned two factors, Pakistan is sanguine that its capture of Kashmir, if not the entire State of J&K, is imminent, just a push away. Thus consumed by an innate propensity to use force to have its way while stoking estrangement among the people of the Kashmir Valley, Pakistan remains encouraged to wage its relentless assault to wrest the part of the State of J&K that remains in India's control. Having described Kashmir as its 'jugular vein', Pakistan's existence as a nation-state is inexorably tied to capture of Kashmir and therefore, Pakistan cannot give up its stake. No doubt, the Kashmir issue would remain in contention, unless of course that recalcitrant usurper is left in no position to usurp. Kashmiri's new found fundamentalist Islamic injunctions against non- Muslim governance would not permit them to live satiated in secular India—unless their loyalty is secured with economic progress and infrastructural uplift. Mature democratic governments adapting to extraordinary measures in the interest of national integrity...time to accept that only drastic quantum of soft-and-assuaging civil measures, reinforced with hard-and-disarming military action, over a period of time, would remedy the situation.

And then in his another book on *'India's National Defence, Defining Defence Reforms and Military Modernisation'*, as since mentioned above; **Lt Gen. Banerjee** again, on pages 7,8,9,156 & 157; writes indeed with intellectually professional precision, that, "Many defence policymakers find expedient to continue with status quo in the prevalent system of defence management, while talking recourse to incremental upgrades when unavoidable. Besides containing what is viewed as ambitious and unaffordable demands from the military institution, they find the persisting system to be reliable and time tested. Above all, they prefer to perpetuate the stranglehold of bureaucracy, finance manager and science community over defence policy-making, practically to the exclusion of military professional participation. Liberation from such regressive attitude towards national defence and corrective institution of modern concepts in its management is therefore the matter of discussion in this chapter. To do so, it is intended first to touch upon the contemporary challenges of national defence, followed by a count of the debilities that afflict India's defence management, and finally, answers to the dilemmas over investments that confront the national leadership. The vitiation of the global environment has imposed an unpredictable and uncontrollable burden of massive proportions upon the state in the form of management of natural calamities. Here too, military forces, professionally competent and mentally attuned to standing up to overwhelming odds as these calamities are, must remain the ultimate responder in bringing succor to the people, existence of disaster management organisations notwithstanding.

No doubt, to deal with such consistent impositions of destructive nature, farsighted foreign policies, competent diplomacy and good domestic governance need to be backed up with the language of universal power: a modern military organization that is founded upon a robust defence infrastructure, and guided by a sound process of defence policy-making that leads to the adoption of farsighted defence policies....

As discussed, defence reforms have to be directed at the following aspects of national defence:

a) Modernisation of the military institution, involving force-restructures, command and control hierarchies, joint inter-service force-application, logistics and training.
b) Rationalisation of manpower to fund modernization. This would involve trimming of regular manpower in favour of TA, reserves and a competent system of force-mobilisation.
c) Institution of defence acquisition and procurement process that deliver rather than just promise.

As the American, British and Chinese experiences indicate, an endeavour so fundamental and profound cannot be met by a system so restrained by bureaucratic resistance and military inertia. Defence reforms therefore have to be initiated, implemented and monitored at the strength of political authority that is triggered by a concern for fiscally sustainable national defence."

The point is that in certain specifically given situation, even if the concerned nation is not preparing or not even wanting to take any kind of some geostrategic-cum-politico-militarily 'great-games-related' initiatives, as per its situation-specific, national-interests it is bound to forcibly get sucked into the fast manifesting 'whirlpool' of the concerned 'great-games'. Contextually in these respects, we may befittingly cite the historically very momentous recent example of the Iraqi invasion (*Under its President* ***Saddam Hussain***) of the middle-eastern Nation-State of 'Kuwait', in fact triggering the first Desert-War; and the rest is now well-known history. But the point is that Kuwait was no partner of Saddam's Iraq in his now already 'settled'—the irrationally ambitious 'great-games' in the region, and also to hit at US interests over there—and moreover when, Kuwait being always a substantially rich source of the globally strategic-cum-economically valuable item, i.e. the 'liquid black-gold'. But nevertheless Kuwait got perforce sucked into the momentously manifesting 'great-games' in the Middle-East, and that is now again a living-history. So geopolitically in these regards, at many a times for the concerned nation it seldom in fact remains a matter of choice but mostly the matter of some sorts of its compulsive destiny only.

Hence, these no less hard facts too, need to be understood and grasped by the concerned nation, in their strategically desired perspectives. To substantiate these geopolitically far-reaching realities contextually the mankind's living-history would always be able to provide the numerously different kinds of valuably revealing examples; but the fundamental 'conceptual-framework' in fact to facilitate the requisite understanding in these regards, would perhaps remain principally the same. And at the same time contextually, another no less crucial point that needs to be always taken into account is that the proactive effectiveness and success of the very constitution of this 'conceptual-framework' indeed, would almost squarely depend upon, since already repeatedly emphasised earlier—**the cultivation, development, grooming, fruitful imbibing and then, pragmatically and also proactively flexible placement of the 'multiple-aspects of the essential strategic-culture and thoughts'—in the concerned nation's socio-cultural-cum-political domains as well as, all across the multi-dimensionally highly complex arena of the multiple-aspects of its national-governance too**. And at the same time interestingly, if the above mentioned prerequisites in these regards, are not vibrantly-cum-proactively in place, with its all-encompassing 'geopolitical-antenna' since always remaining duly activated and also multiply alert, in any crucially geostrategic-cum-politico-militarily challenging and given situation, needless to say that, all these may lead to the invitation to probably even quite an 'embarrassing fiasco' too. Hence contextually, if these multiple-aspects perforce always remain so challenging, dicey, mostly changing, and also the no less complex too; and then at the same time in any case, all these multiple-aspects as since already briefly discussed above, would also mandatorily have to manifest, as well as get multiply activated with their quite unpredictably far-reaching fallouts, nowhere but onto 'the always very slippery, complex and also highly fluid kinds of the arena of international power-politics'. Precisely in the above multiple perspectives, we may like to critically study and also overall geopolitically evaluate, her recent geostrategic-cum-politico-militarily botched-up involvements in Sri Lanka; since the overall results of in these regards, cannot be seen adding even any some 'consolation kinds of the feathers in her big-cap' as a 'so called and fast-emerging leading Asian Power'.

When ultimately the Indian forces finally withdrew during the Premiership of **Mr. VP Singh**, as since already very sinisterly let-down by their own powerfully entrenched, and also the always so much negatively all pervasive the Indian 'politico-bureaucratic combine', the Sri Lankan Forces, as being part of their much smaller island-nation located near southern-most tip of the massive Indian Peninsula soon very decisively seized the opportunities in their determined zeal to serve their vital national-interests and at the same time, very imaginatively did make full use of most of the sharp-edged battlefields' related jobs, since already having been largely done then, by the departing and much larger Indian Forces with their sizeable military wherewithal. In any case, as these do understandably happen to be always substantially varying-cum-debatable points, the opinions in different quarters/institutions in India, Sri Lanka and also elsewhere would naturally be varying and at times, even could be quite expectedly controversial too. But that's precisely the way these things do generally manifest themselves anywhere, and also get thrown up for the national/international media.

Historically, now so far as specifically the multiple-aspects of India's recent involvements in Sri Lanka and their fallouts are concerned, contextually the following quoted paragraphs would be able to clear much of the fog, since still encircling India's indeed the very recent aspects of her this 'great-game' related episode in her southern neighbourhood itself. Much before the actual departure of IPKF on politico-diplomatically so badly half-baked mission in Sri Lanka, much of the indeed, quite avoidable socio-political-cum-diplomatic injuries had already been inflicted onto the very national-psyche and pride of the Sri Lankan peoples; by the all kinds of India's politico-diplomatic naivety, short-sightedness and even perhaps real or perceived arrogance to begin with firstly under **Mrs. Indira Gandhi**'s Premiership and then subsequently in continuation, no less under that of **Mr. Rajiv Gandhi** as well. As per the all-encompassing and also the totally unsparing geopolitical-parameters of the so-called 'great-games' and their nitty-gritty anywhere are concerned, and as quite extensively discussed earlier with maximum possible objective criticality too it cannot be denied or wished away that first the above mentioned Indian diplomatic, as well as different kinds of the politico-economic involvements in Sri Lanka and then subsequently leading to the so much violent IPKF-episodes are concerned, all these in fact were some kinds of the Indian-sponsored 'great-games' only. And contextually it is altogether a different matter that, substantial lack of the always paramountcy geopolitical-vision, as well as an absence of the no less crucial multidimensional strategic-prudence, almost all across the Indian Establishment in New Delhi, were indeed able to very damagingly derail India's entire 'game-plan' and for all these 'fiascos', and hence historically from the very beginning itself the entire blame in these regards, irrefutably resides nowhere else but at the doors of India's as usual: the 'multiply-stinking corridors of power'.

The point is not that India in any case, was guilty of triggering these 'great-games' in the first place, but historically indeed the extremely embarrassing point here, does persistently remain that the apex-level Indian political-leadership, along with its mostly clueless, casual and shortsighted bureaucratic-cum-diplomatic advisers too, proved themselves not to be visionary enough and professionally competent to successfully steer the numerously compulsive nitty-gritty of the modern, sovereign and democratic India's 'great-games' in her southernmost neighbourhood. This malignantly lingering lacunae at the highest-levels of India's corridors of power in New Delhi, led to the so much embarrassingly botched-up conduct of her the then some 'great-games-oriented' initiatives in Sri Lanka; and needless to say that irrespective of all these extremely regressive drags onto the overall Indian systems of governance, the deployed components of the Indian Armed Forces in Sri Lanka, did do their best and in spite of the so many hopelessly irrational limitations put on them by the none else, but the Indian 'politico-bureaucratic combine' only. Here we quote one of India's well-read, professionally research-oriented, very objective, as well as competent Journalists **Mr. Apratim Mukarji** who has written that, "While the IPKF left a bitter legacy with the Tamil civilian population behind, **President Premadasa**'s disdainful rejection of the move to erect a monument to the dead Indian soldiers completes the impression of the infinitely thankless job that the Indians were called upon to do in those days of innocence. The remarkable emphasis with which the Indian leadership of today was insisting in May 2000 that India would never take the 'IPKF route' goes to illustrate the deep scar that the IPKF episode has left on the collective psyche of the country.... The situation grew more complicated with the other Tamil groups trying to utilise the peace interregnum offered by the presence of the IPKF to further their interests since this attracted immediate LITTE protests and retribution at times. Jockeying for positions in the proposed interim administration for the North-Eastern Province, as a precursor to the elected autonomous council after the elections; as the TELO, BPRLF, PLOTE and ENDLF; being on the receiving end from the LITTE, sought to utilise the developing situation to their advantage. Thus, the internecine intra-Tamil groups fighting began, complicating the IPKF's agenda and drawing it inexorably into the volatile and highly unpredictable militant politics, carried on more by guns than by rhetorics.... **By early September 1987, the situation had turned so bad that tension was running high and major inter-Tamil group clashes had begun. The LITTE wanted the civilian population in the north-east not to cooperate with the IPKF; the Colombo based media voiced the demand that since the Indian had clearly failed in performing the task of restoring peace and normalcy, they should go back.... All this while, as is well known, the IPKF had succeeded eminently in restoring normalcy to a great extent by reviving the Colombo-Jaffna rail-link, buses were plying all over the north-east; even Tamils who had lost their property to the usurping Sinhalese got them back. Justice was once again visible and accessible, a significant improvement in the situation. It was, therefore, little wonder that the Tamil civilian population was glad that the Indians were there.... This persuaded the LITTE to deploy a strategy which should destroy effectively the trust that had grown between the IPKF and the civilian population** (*In any case, these geostrategic-cum-politico-militarily very bothersome 'hick-ups' over there then in course of time, were inevitable; but fundamentally the most crucial question remains as to why India's 'politico-bureaucratic combine', had not been able to geopolitically visualise all these with essential strategic-prudence and thus, to be able to cater-for any such unforeseen developments with requisite vision; as the most powerful South-Asian democratic Nation-State?*). **M.R. Narayan Swamy** narrates how **Prabhakaran** explained to an Indian journalist his plan to 'play politics' by provoking the IPKF to attack the civilian population, a strategy he would use with deadly effect in the years to come against Sri Lankan government forces. In short, his Machiavellian scheme was to persuade the IPKF to suspect every Tamil at one point, irrespective of his guilt or innocence, get panicky and take reprisal.... The LITTE undertook two major tasks at this stage. On 13 September 1987, it carried out a massacre of its rival militant groups in Batticaloa, killing 70 of them in one sweep and sending survivals running helter skelter into IPKF and Sri Lankan Army camps in the east. This was followed up by a sustained high-pitched propaganda of how the IPKF was acting on behalf of the Sri Lankan government and against the interests of the Tamils. It was clearly building the case for branding the Indians as agents of Colombo and anti-Tamil and thus rendering them eventually ineffective in restoring full normalcy in the north-east.... Irrespective of the truth (*i.e. in any case, the much larger politico-diplomatic-cum-military truth, concerning the entire gamut of India's the then involvement in Sri Lanka; remains that India failed to effectively counter the LITTE's this propaganda-war.*), however, the Indian offer of evacuation could only be seen as an unpardonable diplomatic faux pas in the context of India-Sri Lanka relations. The offer also came on top of New Delhi's steadfast refusal to help Sri Lanka militarily. How the CCS came to decide on an announcement of the offer is yet to be unravelled but enquiries reveal that the defence service chiefs had been pressing hard for extending military assistance to the country. They had a fairly valid argument on their side. They said that if India decided in favour of providing military assistance or even sale of arms to Sri Lankan forces, it would be between two national armed forces while the adversary in war was the LITTE, a terrorist organization known to deploy all unscrupulous means to attend its objectives and which was a major security concern for India.... If the Indian defence service chiefs pressed for military assistance to Sri Lanka, the Indian government walked the other way. Held captive by the fear of antagonizing the NDA partners from Tamil Nadu, the **Vajpayee** government had stretched the 'hands off' policy to such an unwise length that when the denouement came it did not have a cushion to fall back on.... Such a complete lack of options to choose from was surprising since there was essentially one single prism to look at the war in Sri Lanka. The LITTE was a military force and it was its military prowess that was preventing peaceful settlement of the larger ethnic strife.... The conflict could be neatly divided into its military and political aspects. As far as the military aspect was concerned, obviously the only way to deal with it was through military means. Since there was a no question of a military involvement by India, the only option left was to assist Sri Lanka militarily. If India could look benignly at Pakistan (not really, because once India learned of the secret arms purchases it told Sri Lanka plainly enough that this was not the done thing between two close and trusted friends but obviously Colombo wanted to keep its options open), China and now Israel (this was one area where India was quite at ease since the relationship between Sri Lanka and Israel was re-established with New Delhi's blessings) selling arms and weapon systems to Sri Lanka, what was the harm in doing it by itself? The sticky point once again was the impact of its repercussions would be unaffordable in the contexts of the stability in New Delhi. It was government's inability to resolve this contradiction that led to the perpetuation of the kind of faux pas that Mishra (*i.e. Brijesh Mishra, then being a very close aide to the Prime Minister, Mr. Vajpayee*) indulged in May 2000.... **The crisis in April-May 2000 brought to the fore certain lessons for Sri Lanka and India, which should be analysed and the necessary lessons drawn. Sri Lanka realized that the larger world was simply not interested in its civil war even if the country was clearly suffering enormously with serious implications for the present and future generations. President Kumaratunga gave vent to her feelings in a poignant manner when she remarked in a newspaper interview that distant countries would be bothered only when their own interests were affected. They would react strongly only when terrorism would arrive at their doorstep, she said (*how geopolitically precise and prophetic indeed.*).... Her frustration at Western countries not driving out LITTE offices and operatives completely from their territories even while acknowledging the rebel group as dangerous terrorists was all too visible**." (Mukarji, 2000, pp. 51, 52, 53, 168, 169, 170, 402, 403 & 450)

Hence concerning any nation anywhere and in any given situation the crucially far-reaching geopolitical wisdom lies in its geostrategic-cum-politico-military skills in multiply grooming, flowering and sharpening, as well as to fruitfully deploying and activating its strategic-thoughts and wherewithal to protect and further its highest national-interests with requisite focussed-vision and effectiveness. And perhaps to anybody's dismay in these regards and at the same time, in spite of being only a very small island Nation-State and since located very near to the southern-most tip of the huge Indian Peninsula; Sri Lanka has been able to show much better geopolitical understanding, vision with dogged determination to unwaveringly face the many kinds of the crises for a number of years, during the extremely violent ethnic civil-war in its north eastern areas, and come out successfully in the end by finally wiping-out the entire terrorist-outfits once for all. Hence in the multiple-contexts of now the erstwhile and once so indeed overwhelmingly menacing and violent—Sri Lankan ethnic civil-war, and finally after its successful end—the content of a newspaper interview, given by the then **President, Mahinda Rajapaksa** has been able to convey many important messages, both for the present times and for the future as well. In fact presently after the successful end of the war with LITTE, the tallest Sri Lankan leader and this island-nation's jubilant President, Mahinda Rajapaksa, while talking to the Editor-in-Chief, **Shekhar Gupta** (*The Indian Express*, Chandigarh Edition, dated 01 June 2009), has said that, 'The war is over, we should be together. To India's Tamils, I will say that their brothers and sisters will be looked after by Mahinda Rajapaksa.' Some extracts of this interview is given below, and geopolitically in many ways Rajapaksa's statements do convey many valuable messages in these regards, both for the present times, and for the days ahead. "Q. *My guest is President Mahinda Rajapaksa. Welcome to Walk the Talk in a very different circumstance from our last conversation.* A. Yes, when you came here last time, less than one-and-a-half years ago we were not a united, unitary country. At that time, North was north and the East was controlled by the LITTE terrorists. Now, we are free of terrorism. Q. *You said you are a unitary country, but a federal unitary country, I hope, because last time when we had a conversation you said that in one-and-a-half years there will be peace and there will be no LITTE. That situation has come faster that you had imagined.* A. That's right. Q. *Tell us a little bit about the final stages of the battle. How did it progress? When did the breakthrough come?* A. When we walked into Madhu, the North-West of the island, we were stuck there for an eight to nine months. Then it was just a walk over. We knew after we went to Killinochi that we were through. Q. ***What is the biggest mistake that Prabhakaran made***? A. **Killing Rajiv Gandhi was his biggest mistake, because that antagonized the sympathy of India. That is the biggest mistake he did in 30 years. The second was that he didn't gauge the power of the South.** Q. ***There is a third mistake too—a big ego. No guerilla fighter should try to take on a whole country's armed forces in a frontal battle.*** **A. That's right. He wanted not only an area but the whole country**. Q. *Not just the Tamil regions*? A. No. His plan was to take over the country. Q. *How could he have done that? The country is 85 percent Sinhala.* A. He (Prabhakaran) thought that everybody could be controlled with a gun. Q. *I am fascinated by you saying that the biggest mistake he made in his 30 years of fighting was the assassination of Rajiv Gandhi.* A. Yes. **When he started this, everybody knew that they were trained in India**. And finally they killed an Indian leader who was loved by the Indian people—not only the Tamils but entire India. They killed him in Tamil Nadu. A man (Rajiv Gandhi) who gave his own bullet-proof jacket, this is what I have heard, to Prabhakaran and was trying to bring a solution by introducing the 13th Amendment to our Constitution.... Q. *Which, you now honour*? A. We have one. There has to be 13 plus one. We must get together with the Tamil parties and decide on that. I must not force them to have something. It must come from the people, from their leaders. Q. *You were once not happy with the 13th amendment.* A. **Certainly, we were not happy when the IPKF came to Sri Lanka. I was not happy. I was in the Opposition. But now we can see that if we had allowed the IPKF for another one or two months, they would have finished this**. Q. *I came as a reporter here at that point. It looked as if IPKF controlled all the cities and it looked like LTTE was on the run. And then, one of your former colleagues told me that President Ranasighe Premadasa was supplying some arms to the LTTE to fight the IPKF in the TATA trucks given by us to the Sri Lankan Army.* A. That's right. They gave 5,000 T56s when they had only 250 fighting cadres, according to Karuna Amman. Q. *So did you tell the Sri Lankan people that in the past the LTTE was helped by some of their own leaders*? A. During election campaigns, everybody was talking about it, Q. *And yet, they killed Premadasa*? A. Yes. They killed Rajiv Gandhi. They killed Gemini Dissanayake, who tried to give them a political solution and they killed our people. Their own leaders were killed. Q. *One got the impression that there was nothing the LTTE hated more than a Tamil trying to make peace.* A. Yes, that's why they killed Jeyaraj Fernandopulle, a minister. He was a Tamil and instrumental in peace talks. Q. *Prabhakaran had a talent of getting out of every tight spot. He was cornered by the IPKF, he got out of it. He was cornered by the Sri Lankan Army many times and he would bargain for a ceasefire, build-up his strength, and suddenly we'll hear a story, 'LTTE attack, 500 soldiers killed'.* A. That's right. Q. *How did he make such a big blunder as killing Rajiv Gandhi? Have you analysed it*? A. He hated Rajiv Gandhi because he tried to patch up this whole thing. He (Rajiv Gandhi) tried giving a political solution and he (Prabhakaran) had to accept it. And Prabhakaran didn't like it, as he didn't want any solutions to it. Q. *He never realised the consequences it will have in India, particularly in Tamil Nadu?* A. I didn't think he was bothered. There were some leaders in Tamil Nadu who were communicating with him and sympathizing with him. I don't want to mention their names. I think all of them were defeated by the Indian voters. Q. *Do you think these people kept on massaging Prabhakaran's ego?* A. That's right. When you are in an isolated place, in a jungle, without getting much information, you always live in your own world. This happens to some politicians too. Q. *You've arrested many of his followers and interrogated many. Did anybody throw any light on his killing Rajiv Gandhi or did he have genuine regrets after that?* A. He never regretted. Q. *Did they tell you about the discussions over the killing of Rajiv Gandhi?* A. No. I don't think he (Prabhakaran) discussed what he was going to do with his colleagues. He just gave orders without bothering about the repercussions.... Q. *You have got inflation running at 25 percent?* A. At one time, yes. But it has come down now to either two or six percent. Q. *Inflation is out, but interest rates are still there. Is that the cost of this war?* A. Yes. Q. *We are a country of 1.1 billion, i.e. eleven hundred million people and have an army of 1.3 million. You as a country of 20 million people have an army of 200,000 people.* A. We had to increase it. Q. *On a per capita basis your army is hundred times the size of our army. You don't need it now.* A. We can use them for development work. Q. *Will you downsize it over a period of time?* A. I have no such plans. An army of 200,000 is not much for Sri Lanka. We can also send them abroad as UN peacekeeping force.... Q. *You don't worry about the Generals becoming powerful?* A. I am not bothered. I know they are loyal to the cause. I don't think they have any political ambitions. Q. *There are concerns in India regarding your relations with China, Pakistan.* A. India should not have any concerns. India should have a large-hearted approach towards us. We have friendly relations. We bought weapons from China, Israel, Pakistan, Russia and Ukraine. Q. *Why should India not have any concern about the military relationship because we keep hearing about China's interests in Hambantota?* A. They are helping us develop a port. That doesn't mean they are going to stay here. I will not allow Sri Lanka soil to be used by anybody against India. Q. *Can India come and develop a port in Sri Lanka?* A. Why not? They are welcome. Q. *Do you have a hierarchy in your relationships in your neighbourhood, in the world? Do you have a hierarchy or a level of importance you give to your relationship with India, China, and Pakistan?* A. India is our powerful neighbour. We must have a good understanding relationship with India. Q. *Earlier, there were posters calling the IPKF innocent people's killing force. I used to wonder Indian soldiers are sacrificing their lives without a word of gratitude, just curses and abuse.* A. You must remember that when an outside force comes here, there's always a reaction from the society. Q. *You have now built a memorial for the IPKF?* A. Yes. That shows our friendship and gratitude. We are waiting for someone to come from India—the people who fought at that time, IPKF generals and some relatives of the soldiers who died. I want to visit India and discuss about it."

Needless to say that in any case, all these aspects can never be any monopoly, of the so-called any 'elite-club' of the world's more powerful and/or any bigger nations. But at the same time, it remains all the more true that all these aspects do always get multiply-governed by the levels of their in-house and also, their some exclusive kinds of the some synergy-creating manifestations, as well as their multi-dimensionally proactive effectiveness of the any concerned nation's—**the multiple-aspects of essential strategic-culture and thoughts**—at any given point of the always advancing timelines of its 'living-history' in the world. And hence in these specific perspectives, we may even say that, Sri Lanka against its all odds, limitations, and also in spite of the many kinds of geostrategic-cum-politico-militarily trying times has finally come out successfully to maintain its territorial integrity and sovereignty, during its indeed very prolonged and the extremely violent confrontation with LTTE onto its own territory.

Well contextually in some much broader geographical perspectives and also historically keeping in our minds that, since the advent and also continuing advancements of human-civilisations onto this planet Earth, if any particular island-nation happens to be having a fairly powerful Navy capable of effectively displaying any effective deterrent against some menacingly evil intentions of any potential invaders in the vicinity, then obviously its homeland security needs can be much more effectively organised and commandeered. During the erstwhile two World-Wars and even during many wars in Europe, the island territories of the UK do contextually give us excellent examples in these regards. And at the same time, in the overall geographical, economic, techno-scientific, industrial, as well as numerous other geopolitical-cum-diplomatic, and even in the socio-cultural terms, hardly any similarity can be established between the two island-nations of the UK in the Western Europe and, Sri Lanka in South Asia.

But within its own very limited dimensions, and also indeed having considerably very limited, as well as multiply very small military and also a number of other crucial wherewithal in these regards, Sri Lanka's struggles and achievements, have been quite laudable too. Well, it is also the no less true that historically the then erstwhile extremely menacing threats confronting the UK, during the extraordinarily and also fairly long trying times of the (1939-45) Second World-War, were much more complex and colossal, and hence no similarities can be drawn between these two totally different scenarios in the two entirely different and also geographically far-off locations. But interestingly it is also equally true that, in both the cases since having distinctly two different geostrategic-cum-politico-military scenarios and geographically having a commonality of being island-nations, did help considerably to both the nations, during their respective trying times and challenges. And now historically at the same time, it is also very far-reaching and equally true that, before the Sri Lankan Army could later directly take-on the LTTE, earlier the Indian Army was indeed able to very decisively pulverize the LTTE's military prowess and thus, making it militarily possible and much easier for the Sri Lankans to achieve their socio-politically strategic and also militarily paramountcy aims, a fact which the Sri Lankan President has already very gratefully acknowledged without mincing his words, and since already recorded in his earlier quoted interview with an Indian Journalist as well.

It is true that being geographically located in a strategically important, as well as geopolitically extra-sensitive, dicey also highly complex region—like South Asia where even now the various kinds of, as well as the intensely competing interests of the today's sole Superpower and another highly ambitious-cum-biggest Asian-Power are continuously in a state of 'rubbing shoulders', and also clashing diplomatically—for the other big and also equally aspiring to acquire a credible regional-power status at the earliest, i.e. India; the paths do remain very slippery and at the same time, the overall socio-political-cum-diplomatic and also the security related scenario remains not only unpredictably fluid, but also multiply challenging as well. **On top of it, if India also aspires to the otherwise rather befittingly indulge in her own kinds of some distinct 'great-games'; then she has to not only remain multiply-equipped and vibrant, but also be possessing the requisite acumen and competent leadership; to play her 'geopolitical' cards always in tenaciously determined and visionary manners. In fact, there remains naturally, no genuinely well-meaning sympathizers in these regards and thereby, if at all any emerging situations do turn dicey and somewhat out of control; then the concerned nation in the 'arena', must know 'how to bear its own cross' and thus at the same time also needs to remain always prepared to imaginatively retrieve the 'situation' alone and, as per her own highest national-interests.** It remains equally true that as rather in a much larger topsy-turvy manners, the erstwhile Iraqi dictator, **Saddam**

Hussain, who happened be first created, groomed and encouraged by the sole Superpower USA itself and, with whom principally the Americans only had to fight later to finish him off, when he began threatening and also resisting to dance to the tunes of this Superpower. And in broadly in similar manners, now the already slain LTTE chief, **Prabhakaran** in Sri Lanka too, was first 'created', trained and equipped by none else but India only, in this South Asian region, with whom later it was India only, who had to fight to decisively reduce his once so much monstrously growing menacing-power, so that subsequently the Sri Lankan Army itself, could be able to finally finish him and his terror outfits. And needless to say that now the rest is a so commonly well-known Indo-Sri Lankan 'living-history' in these regards. Contextually almost in the similar perspectives but geopolitically, in an entirely different scenario and the very sinister motives of India's western neighbour, i.e. Pakistan, the Pakistani Army and its ISI, very meticulously planned and prepared to very menacingly launch the cross-border—Islamic Jihadi-terrorism into the Indian State of J&K—to perforce make India 'bleed white with thousand cuts' (*as since so recklessly as well as arrogantly, declared earlier by the Pakistani President late* ***Mr. ZA Bhutto***.), so that Pakistan be able to grab Kashmir through its so-called 'dare-devil' military adventure, to be suitably launched against the Indian forces deployed over there. Hence interestingly in these very precise perspectives, these entire systems of the geopolitically reckless thought-processes on the part of Pakistani-Establishment and, with the follow-up actions through extensively exporting the 'Islamic terrorism' into the Indian territories and then subsequently, even indulging in occasional military misadventure against India to throw the Indian military off-balance in J&K so that, it could militarily siege the State from India, do show the **extremely obsessed Pakistani 'great-games' against the Indian Nation's vital national-interests** and thus also, to seriously harm its territorial integrity as well. **It is now a different matter that being senselessly anti-India in all possible manners, and the Pakistani-Establishment's game-plans against her indeed has always very damagingly boomeranged on Pakistan only.**

We may now prefer to critically look into and discuss the **historical 'Kargil-episode'** then occurring in the extremely remote and rugged Himalayan heights in the Indian J&K State's Ladakh-region. Though now in retrospect, it is true that the Indian Establishment during the Kargil-crisis, and geopolitically in many ways, did not cut any embarrassingly 'sorry-figure', but at the same time it remains also no less true that even during it, again many otherwise avoidable, as well as since the always equally manageable in advance awkward 'chinks in the Indian armour' did surface and stared into the eyes of the Indian military-command systems; courtesy: the same powerfully entrenched and also perhaps, the perpetually regressive, callous and arrogant, 'politico-bureaucratic combine', and thus this time too, the typically depressing failures of the mostly as usual and being mentally-static, the Indian Intelligence Agencies as well. Hence broadly speaking in these larger perspectives, it appears that more the various constituents being parts of the friction-infested Indian Establishment make efforts to develop, modernise and professionally be more focussed and effective, progressive and dynamic—more they remain the same and thus being regressively backward-looking, rather than being visionary and forward-looking; and thereby always fail to dynamically remain in tune with the changing times and the advancing technologies.

Though historically it remains a different matter that in this 'Kargil-War' during the year 1999, militarily the Pakistani forces did suffer very embarrassing defeat at the Himalayan heights in the Kargil-region, but at the same time it is also equally true that onto the Indian multiply faceted Indian Intelligence-System's indeed highly complex, as well as internationally challenging fronts, the nation had to suffer all the more very embarrassing surprises and setbacks. In fact all these the then intelligence-related 'fiascos' turned out to be so disturbing and democratically 'chaotic' that immediately after the cessation of these hostilities, the Government of India under the then Prime Minister **Atal Bihari Vajpayee**, had to appoint a **'Kargil Review Committee', having following members: Mr. K, Subrahmanyam—Chairman; Lt. Gen. K.K. Hazari (Retd), Mr. B.G. Verghese, and Mr. Satish Chandra—all members of this committee**. And as per its terms of reference, the Committee did extensively go into all the details, covering the entire 'Kargil-episode' to do its massive investigative work; and finally submitted its voluminous Report on 15 December 1999, to the Prime Minister. In this report, the Committee did also give a number of the very valuable and well-considered suggestions-cum-recommendations, with an intention to indeed substantially augment and revitalize the almost entire working-systems of both defence and intelligence institutions of the Indian-nation. How implementation, as well as systemic improvements-related work has really been carried out and, precisely with what effectiveness and results, by the powerfully entrenched, naïve and indeed very frustratingly myopic—the 'politico-bureaucratic combine' in the Indian corridors of power in New Delhi—is anybody's guess because hardly any tangibly meaningful systemic-reforms and multiple improvements in the work-culture of India's various national security-related institutions, have really been seen in action on the ground.

Further it is also now well-known that the assertive rise of unilateralism during the post-Cold-War-era, as well as also, after the successful test of nuclear devices and the fairly 'long-range' missiles by both India and Pakistan, the entire geostrategic-cum-politico-military scenario had got altered in very far-reaching manners in the whole of South Asia, with its notable kinds of some global ramifications as well. As a result, this whole region has come onto some kinds of the continuous and also, very close-scrutiny through the 'geopolitical-radars' of the today's Great Powers. And then in these respects, Pakistan as a frontline State and being always sitting in the 'US lap' and thereby also being a fairly useful 'pawn' in the hands of today's sole Superpower the USA, and thus remains readily available, as a geopolitically valuable 'launch-pad-cum-base' for the US's numerous kinds of 'great-games' in the Middle-East region, as well as elsewhere. **Understandably as per the American interests, the USA very prudently remains inclined to ignore many misdeeds of the reckless and also the no less religiously fanatic Pakistani Establishment and thus consequently, even indirectly triggering some uncalled-for encouragements to the Pakistani military, towards very reckless-cum-unsettling kinds of the misadventures. 'Kargil-episode' nothing but in fact, has been a direct result of the one such misadventure on the part of Pakistani military against its much larger and more powerful Eastern neighbour, i.e. India.** In these respects, due to its numerous kinds of misunderstandings about itself and its capabilities; Pakistan even fails to realise that if it ever tries to 'bite many times more than what it is capable of chewing', then it is perforce bound to get rather irretrievably choked. Motivated by its irrational arrogance and also typically fixated misunderstandings, the Pakistan's military-dominated Establishment, after the crushing debacle during the 1971 'Bangladesh-War'; made audacious plans to launch proxy-war in the Indian State of J&K to 'teach suitable lessons' to the much larger and stronger India and in the process with indeed inevitably total futility, began squandering away its scarce economic and other resources, which could have been much more urgently utilised in its otherwise always paramountcy efforts to solve Pakistan's many burning socio-economic crises, in which its teeming millions are since already so haplessly left to suffer and thereby even being forced to gradually get 'devoured' by the so many highly dangerous fallouts of perhaps now being like a 'failed-State', with perhaps no end in sight. In precisely the above multiple-contexts concerning Pakistan lot many objective and well-meaning books and articles have already been published not only in India or in many other foreign nations, but also in Pakistan itself; but the so much arrogantly fixated, as well as the military-dominated Establishment in Pakistan, just doesn't care and even refuses to see the clear 'writings on the wall of its destiny'.

In nutshell we may even quite conclusively say now, that what the fanatically die-hard anti-India—since always being the military-dominated Pakistani-Establishment—wanted to do against the so vast and also the much stronger eastern neighbour i.e. India; has indeed now irreversibly boomeranged onto itself. Hence contextually with the desired objectivity in these regards, we may even say with heavy hearts at this juncture that in the process, the massive segments of the poor and hapless Pakistanis, are perforce made to suffer regardless; by the none else but the powerfully entrenched and also, extremely corrupt Pakistani civil and military elites. In any case here we may also say that, this is certainly no way of geopolitically indulging in any kind of the so-called 'great-game' by any Nation-State anywhere. But contextually in a little larger perspective, what we find that the matter in any case, has not at all ended here. Hence irrespective of the any 'crippling consequences on none other but onto itself, Pakistan remains in no mood to mend its ways and thereby, try to get rid of its irrationally fictitious thoughts and fixations to someday to be able to somehow by hook or crook, grab J&K from India. The Pakistani-Establishment stubbornly refuses to see the utterly foolish and thus in fact, the total impossibility of their misadventure succeeding ever. In spite of all these, **the question arises as to how Pakistan manages to still somehow survive; and here comes the handy and almost perpetually continuing factor of the 'US drips', which enable Pakistan to be still somehow going for the time being, and hence geopolitically in some long-term perspectives, it would remain interesting to see as to how long this continues**.

In precisely the above multiple-perspectives; General V.P. Malik has indeed very revealingly written for the interested people, "Of the three services, the Army takes the maximum time to complete its mobilization and is the most visible. Unless the top political leadership could declare a war, we had to achieve such mobilization without causing alarm, both in the country and in the rest of the world. In the existing circumstances, how soon could we launch an offensive, if permitted? Also, we needed to take stock of our important inventories and reassess urgently our capabilities for defensive and offensive operations. On 23 may (*1999*), I gave my own assessment of the situation and presented my thought processes to the VCOAS and the operational staff in the Operations Room. I then went to meet the Navy Chief, **Admiral Sushil Kumar**, whose office was in the same corridor. After bringing him up to date with the prevailing circumstances, **I discussed the need for enlarging the scope of the fighting and for carrying out joint services planning. My logic for such an integrated approach at the level of the COSC was simple. All three services are national security assets. A single-service approach to defence and operational planning at the level of the armed forces chiefs, though outdated, tends to continue in our country because of its peculiar higher defence control organization and because of the fact that there is no Chief of the Defence Staff or Chairman, Chiefs of Joint Staff. For any combat situation, we must employ all three services optimally, in an integrated manner**. The allocation of exact missions thereafter is a matter of detailed coordination, keeping in view factors such as the characteristics and capabilities of assets available with each service, the level of joint training and the degree of interaction among the services. **In our country, where the political leadership and its civil advisers, have virtually no knowledge or experience of warfare; differences in the opinions of the professionals are often played up. Such a tendency makes it extremely difficult for the political leadership to overrule any inter-service argument. Such differences must be resolved at the level of the Chiefs of Staff**.... I invited Sushil Kumar and **Anil Tipnis** (***being Air Chief Marshal: Chief of the Air Staff***) for a COSC meeting in my office at 4 p.m. on 23 may. As this was to be an important meeting and could become sensitive, I decided to keep it restricted to the three of us.... My picture of the ground analysis and the resultant discussion on the future course of action convinced my COSC colleagues about the validity of the suggestion. We took a unanimous decision to recommend the joint strategy and joint operational planning and action, including the use of naval and air power in the CCS meeting.... **My first impression of the CCS (*Cabinet Committee on Security*) meeting in the Operations Room the next day was an unusually large number of civilian officers from the Prime Minister's office, the Cabinet Secretariat, the Ministry of Defence, the R and AW, Intelligence Bureau and other ministries and agencies had walked in. (*Limit, look at this. How this was allowed and why the three Chiefs of Staff, led by the Chairman, Chiefs of Staff Committee didn't stop the 'briefing' and, strongly protested to the Prime Minister to order an immediate 'filtering of personnel, in the highest and also the then most urgently sensitive national-interests? After all a series the then most sensitive and vital national security details and data were involved in that briefing and so in any case, all these had to be made available on the 'need to know basis' only. Oh no, what a foolishly manifesting 'gloomy picture' of India: and still India aspires to be indeed a credible Power to reckon with. When shall the so much regressively-fixated—the Indian Establishment—would finally wake-up befittingly for the larger good of the Indian Nation-State? In any case, such a foolish-thing can only happen and be permitted in India.*)**. There was a great deal of inquisitiveness and anticipation on the part of the cabinet ministers and officials. However, many officials appeared to have turned up not for making any useful contribution but only to know 'what next'. The meeting went on for more than ninety minutes.... I briefed the CCS about various factors such as operational situation, the terrain, the gaps in our defences, the manner in which the intruders had managed to infiltrate into our territory, the heights they have been able to occupy till now, the overall intelligence picture and the conduct and the handling of the intrusion by the Pakistani Army personnel, although they continued to deny it officially.... My appraisal of the situation was that, lately, since Pakistan was losing ground in Jammu and Kashmir, it had launched this operation. The political objectives were to create a situation that would enable the Pakistanis to negotiate on Jammu and Kashmir from a position of strength and to internationalize the Kashmir issue once again. Pakistan's objectives, from a military viewpoint, (*In June 1999, we recovered a large cache of arms and ammunition from Turtuk village. Twenty-four suspects were arrested and handed over to the local police. They disclosed Pakistani plans to initiate insurgency in Turtuk, Chalungka, Thang, Tyakshi and Pharol.*), appeared to be as follows: (1) to cut-off the strategic Srinagar-Leh road (thereby creating a crisis situation for us); (2) to alter the status of the LOC permanently for strategic and territorial gains; (3) to divert attention from anti-terrorist operations in the Kashmir Valley; (4) to revive insurgency in Jammu and Kashmir (if the Dras and Mashkoh sectors came into Pakistan's possessions, it could drive a strategic wedge into our territory east of Zozi La and could facilitate infiltration towards Pahalgam and the Kashmir Valley; and (5) to capture Turtuk and part of the Central Glacier in the Siachen sector.... I then brought to the notice of the CCS the strategic discussions held in the COSC the previous day. It was necessary, I emphasized, to gain the strategic initiative to facilitate operations of 15 Corps and Northern Command. I also stated that we had to be prepared for war escalation, either by Pakistan or by us, and in such a situation, all the three services would have to be prepared and act cohesively.... Besides, a war effort at the national level required a great deal of preparation. A considerable amount of work had to be done on the diplomatic front, particularly because only the previous year (1998) India and Pakistan had blasted their way out of nuclear ambiguity and upset the United States of America and other powers, including the UK, China and Japan. The nuclear factor too must have been weighing on the mind of the Prime Minister and the CCS colleagues, though this aspect was never mentioned or discussed in the meeting.... In my summing up, I accepted that the operational situation was definitely more serious that made out by the assessment so far. I stated that we needed to investigate the intrusion further and to learn how that had happened. But I felt that this was not the time for such an investigation; it should be done later (*yes indeed.*).... By this time, I could sense that the long faces in the audience had become longer (*What else to expect in any case, from India's so much clueless, shortsighted and still the no less regressively, as well as powerfully entrenched—'politico-bureaucratic combine'—all across the nation's 'corridors of power'?*). The CCS had been conveyed an unambiguous picture of the ground realities and the implications militarily and politically. If there were doubts in anyone's mind these were clarified; if there were any expectations from Track-2 dialogues, these were dispelled. I concluded my briefing by informing the CCS that the Army had codenamed the operation in the Kargil sector as Operation Vijay (victory) and we would take all necessary action and ensure that ultimately *vijay* would be ours. There were no further discussions and the meeting ended on that note.... **That night, Prime Minister Vajpayee spoke to his Pakistani counterpart Nawaz Sharif. Vajpayee told him that 'we will not allow any intrusion to take place in our territory. We will clear our territory'.**" (Malik, 2006, pp. 120, 121, 122, 123, 124, 126 & 127)

As usual the above quoted paragraphs from General Malik's book, do also very clearly convey 'volumes' about the poor state of affairs in India's always paramountcy domain of **the 'multiple-aspects of essential strategic-culture and thoughts'**—whose crucial importance from the political, military, and the bureaucratic perspectives and also with respect to the entire gamut of the nation's leadership and governance, cannot be overemphasized. But still in India, not much importance is given to these indeed timelessly paramountcy aspects. At the same time, it remains always unquestionably established that in a truly vibrant and also healthy, as well as constitutionally functional democracy it remains settled—the matters of 'civil-supremacy over the military only and exclusively at the political-level—and thus lawfully, this unchallenged and well-accepted constitutional norms and practices in a democracy; cannot be disturbed or even remotely questioned by any extra-ambitious military chief or in any case by any kind of the perhaps, sinisterly manipulative-cum-subverting bureaucratic-class as being part of the government, while carrying out its official duties. And sadly this is not the reality since the malignantly persisting counter-productive and also highly damaging work-culture in the various ministries and departments of the Indian Government and at the same time thereby, all the more in the Ministry of Defence are still continuing regardless.

It remains to be seen when the concerned people sitting in the Indian Establishments will wake-up compulsively in the highest national-interests and thus, be driven to suitably change themselves and also get their thought-processes and work-culture proactively reformed? And that being clearly so far not happening in any tangible manners since the 15th August 1947 itself, periodically a series of very embarrassingly unpalatable, as well as highly damaging consequences with respect to defence and also the overall national well-being of the modern and sovereign democratic India, are getting recorded onto the continuingly multiplying pages of her living-history and also obviously, within the larger world in these regards. We may also recall that all these in fact, crucially far-reaching aspects have already been discussed earlier in their differently numerous, but indeed always being broadly inter-related-cum-relevant perspectives. And thereby we may contextually recollect that earlier in this sub-chapter itself, how strangely as well as all the more so naively and also immediately after India's independence in 1947 itself, the nation's first Prime Minister, Mr. Jawaharlal Nehru did not hesitate to rather irrationally 'blow his top' and even so shockingly advocated to 'disband' the Indian Armed Forces once for all with no remorse for such indeed unpardonably blind and highly self-injurious thoughts. In today's multiply overall contexts in these regards any fairly conscious and reasonably thinking Indian-citizen would even shudder to think about the inevitable tragic fate, that would have visited his/her nation, had Nehru's any such totally irreconcilable-cum-mad wishes would have somewhat got fulfilled. But now in retrospect one may understandably even say that, 'Thank God, India was not all that unlucky to undergo perhaps, any such self-destructive consequences due to some rather irrationally-fixated ideas of her first Prime Minister (*who otherwise happened to be remarkably scholarly and indeed visionary too.*)—courtesy: the audacious invasion of by the so called Pakistani Razakars into the then newly merged Indian State of J&K—in the month of October 1947 itself; and the rest is again so well-known history.

In precisely the above multiple-perspectives; India's **Field Marshal Sam Manekshaw**, who was Indian Army Chief during 1971 'Bangladesh War'; while addressing the student officers, faculty-members and their ladies of the Defence Services Staff college, Wellington (India), on 11th November, 1998 had very perceptively, this also to say, ".... A while ago I was invited to a seminar where the subject was youth; and people said that the youth of this country was not pulling its weight, that society generally was not satisfied with how the young were functioning. When I was asked what I thought about it, I said that the youngsters of this country are disappointed, disturbed and confused. They cannot understand why all these untoward things are happening in this country. They want to know who is to blame. Not them. If they want to study at night and there is no power, they want to know who is to blame. Not them. If they want to have a bath and there is no water, they want to know who is to blame. Not them. They want to go to college and university and they are told there are not any vacancies; they want to know who is to blame. Not them. They say—here is a country which was considered the brightest jewel in the British crown. What has happened to this bright jewel? No longer there are excuses with the old political masters saying that the reason why we are in this state is because we were under colonial rule for 250 years. They turn around and say that the British left us almost fifty years ago. What have you done? They point to Singapore, they point to Malaysia, they point to Indonesia, point to Hong Kong. They say, they were also under colonial rule and look at the progress those countries have made.... They point to Germany and to Japan, who fought a war for four and half years—whose youth was decimated and industry was destroyed. They were occupied, and they had to pay reparations; look at the progress those countries have made. The youngsters want an answer. So, ladies and gentlemen, I thought I should give you the answer.

(iv) The Problem with us is Lack of Leadership

Commandant, ladies and gentlemen, do not misunderstand me when I say a lack of leadership. I do not mean just political leadership. Of course, there is a lack of political leadership, but also a lack of leadership in every walk of life, whether it is political, administrative, in our education institutions, or whether it is in our sports organisations. Wherever you look, there is a lack of leadership. I do not know whether leaders are born or made. There is a school of thought that thinks that leaders are born. Ladies and Gentlemen, we have a population of 960 million (*Interesting, well it was so way back in the year 1998.*) and we procreate at the rate of seventeen million – equaling the total population of Australia—each year, and yet there is a dearth of leadership so, those of you who still contribute to the fact that leaders are born, may I suggest you throw away all your family planning, throw away every inhibiting factor and make it free for all. Then perhaps someday a leader may be born.... So, if leaders are not born, can leaders be made? My answer is yes. Give me a man or a woman with common sense and decency, and I can make a leader out of him or her.... We in India have tremendous pressures—pressures from the government, pressures from superior officers, pressures from families, pressures from wives, uncles, aunts, nieces, nephews and girlfriends, and we lack the courage to withstand those pressures. That takes me to the next attribute of leadership—moral and physical courage. Ladies and Gentlemen, I do not know which of these is more important. When I am talking to young officers and young soldiers, I should place emphasis on physical courage. But since I am talking to this gathering, I will lay emphasis on moral courage. What is moral courage? Moral courage is the ability to distinguish right from wrong and having done so, say so when asked, irrespective of what your superior might think or what your colleagues or what your subordinates might want. **A 'yes man' is a dangerous man. He may rise very high he might become the Managing Director of a company. He may do anything but he can never make a leader because he will be used by his superiors, disliked by his colleagues and despised by his subordinates. So shallow—the 'yes man'**.... I am going to illustrate from my own life an example of moral courage. In 1971, when Pakistan clamped on its province, East Pakistan, hundreds and thousands of refugees started pouring into India. The Prime Minister, Mrs. Gandhi had a cabinet meeting at ten o'clock in the morning. The following attended: the Foreign Minister, **Sardar Swaran Singh**, the Defence Minister, **Mr. Jagjivan Ram**, the Agriculture minister, **Fakruddin Ali Ahmed**, the Finance Minister, **Mr. Yashwant Rao Chavan**, and I was also order to be present.... Ladies and Gentleman, there is a very thin line between becoming a Field Marshal and being dismissed. A very angry Prime Minister read out messages from Chief Ministers of West Bengal, Assam and Tripura. All of them saying that hundreds and thousands of refugees had poured into their states, and they did not know what to do.... So the Prime Minister turned around to me and said, 'I want you to do something'. I said, 'What do you want me to do?' She said, 'I want you to enter East Pakistan.' I said, 'Do you know that means war?' She said, 'I do not mind if it is war.'.... I, in my usual stupid way said, 'Prime Minister, have you read the Bible?' and the Foreign Minister, Sardar Swaran Singh (a Punjabi Sikh) in his Punjabi accent said, 'What has Bible got to do with this?', and I said, 'The first book, the first chapter, the first paragraph, the first sentence, God said, 'let there be light' and there was light, You turn this round and say 'let there be war' and there will be war. What do you think? Are you ready for a war? Let me tell you—it's the 28th of April, the Himalayan Passes are opening now, and if the Chinese gave us an ultimatum, I will have to fight on two fronts'.... Again, Sardar Swaran Singh turned round and in his Punjabi English said, 'Will China give ultimatum?' I said, 'You are the Foreign minister. You tell me.'.... Then I turned to the Prime Minister and said, 'Prime minister, last year you wanted elections in West Bengal and you did not want the communists to win, so you asked me to deploy my soldiers in penny pockets in every village, in every little township in West Bengal. I have two divisions thus deployed in sections and platoons without their heavy weapons. It will take me at least a month to get them back in their units and to their formations. Further, I have a division in the Assam area, another division in Andhra Pradesh and the Armoured Division in the Jhansi-Babina area. It will take me at least a month to get them back and put them in their correct positions. I will require every road, every railway train, every truck, and every wagon to move them. We are harvesting in Haryana; we are also harvesting Uttar Pradesh. And you will not be able to move your harvest.'.... I turned to the Agriculture Minister, Mr. Fakruddin Ali Ahmed, 'I said if there is a famine in the country after wards, it will be you to blame, not me.' Then I said, 'My Armoured Division has only thirteen tanks which are functioning.' The Finance Minister, Mr. Chavan, a friend of mine, said, 'Sam, why only thirteen?' Because you are the Finance Minister, I have been asking for money for the last year and a half, and you keep saying there is no money. That is why.' Then I turned to the Prime

Minister and said, 'Prime Minister, it is the end of April. By the time I am ready to operate, the monsoon will have broken in that East Pakistan area. When it rains, it doesn't just rain, it pours. Rivers become like oceans. If you stand on one bank, you cannot see the other and the whole countryside is flooded. My movements will be restricted to roads. The Air Force will not be able to support me and, if you wish me to enter East Pakistan, I guarantee you a hundred percent defeat.' 'You are the government I said turning to the Prime Minister, 'now you give me your orders.'.... Ladies and Gentlemen, I have seldom seen a woman so angry, and I am including my wife in that. She was red in the face and I said, 'Let us see what happens.' She turned round and said, 'The cabinet will meet this evening at four o'clock'.... Everybody walked out. I, being the junior most man the last to leave. As I was leaving, she said, 'Chief, please will you stay behind?' I looked at her. I said, 'Prime Minister, before you open your mouth, would you like me to send my resignation on grounds of health, mental or physical?' 'No, sit down Sam. Was everything you told me the truth?' 'Yes, it is my job to tell you the truth. It is my job to fight and win, not to lose.' She smiled at me and said, 'All right, Sam. You know what I want. When you will be ready?'.... 'I cannot tell you now, Prime Minister', I said, 'but let me guarantee you this that, if you leave me alone, allow me to plan, make my arrangements, and fix a date, I guarantee you a hundred percent victory.'.... So, ladies and Gentlemen, as I told you, there is a very thin line between becoming a Field Marshal and being dismissed—Just an example of moral courage. Now, those of you who remember what happened in 1962, when the Chinese occupied the Thag-la ridge and **Mr. Nehru**, the Prime Minister, sent for the Army Chief (*the then **General P.N. Thapar***), in the month of December and said, 'I want you to throw the Chinese out.' That Army Chief did not have the moral courage to stand up to him and say, 'I am not ready, my troops are not acclimatized, I haven't the ammunition, or indeed anything.' But he accepted the Prime Minister's instructions, with the result that the Army was beaten and country humiliated.... **Mr. Krishna Menon**, the Defence minister, disliked me intensely, **General Kaul**, who was Chief of General Staff at the time, and the budding man for the next higher appointment, (***courtesy: the then relentlessly regardless, all kinds of the 'matters-military related goof-ups', by the Prime Minister Mr. Nehru.***) disliked me intensely. So, I was in disgrace, at the Staff College. There were charges against me – I will enumerate some of them—all engineered by Mr. Krishna Menon.... I do not know if you remember that in 1961 or 1960, **General Thimayya**, was the Army Chief (*Whom, Field Marshal Sam Manekshaw, had very emphatically called—the best Army Chief—India has ever had. And that was historically, a very befitting tribute having been paid by India's one military-legend to another one; and they both made highly commendable history indeed, in their very laudable soldierly manners; during their lives and times while in the Indian Army. But ironically they both had to suffer and nevertheless, very bravely endured some extremely suffocating periods, in their respective capacities then; courtesy: the indeed very nauseating—the **'Nehru-Krishna Menon-Kaul' nexus**—during the highly topsy-turvy pre-1962 era, then so naively prevailing through the apex-echelon in the Indian 'corridors of power'; and now the rest all so much unpalatable history.*). He had fallen out with Mr. Krishna Menon and had sent him his resignation. The Prime Minister, Mr. Nehru, persuaded General Thimayya to withdraw his resignation. The members of Parliament also disliked Mr. Krishna Menon, and they went hammer and tongs for the Prime Minister in Parliament. The Prime Minister made the following statement, 'I cannot understand why General Thimayya is saying that the Defence Ministry interferes with the working of the Army. Take the case of General Manekshaw. The Selection Board has approved his promotion to Lieutenant General over the heads of 23 other officers. The Government has accepted that.' (*Wah, look at Mr. Nehru's typical 'culture of fooling-around' in all sorts of manners, during the India's pre-1962 era; with no serious and committed concerns to the then rather befittingly strengthen defence build-up, to indeed effectively face the* *lurking dangers from across the borders.)....* I was the Commandant of the Staff College. I had been approved for promotion to Lieutenant General. Instead of making me the Lieutenant General, Mr. Krishna Menon levied charges against me. There were ten charges, I will enumerate only one or two of them – that I am more loyal to the Queen of England then to the President of India, that I am more British than Indian. That I have been alleged to have said that I will have no instructor in the Staff College whose wife looks like an ayah. These were the sort of charges against me.... For eighteen months my promotion was held back. An enquiry was made. Three Lieutenant Generals, including an Army Commander, sat at that enquiry. I was exonerated on every charge. The file went up to the Prime Minister who sent it up to the Cabinet Secretary, who wrote on the file, 'If anything happens to General Manekshaw, this case will go down as Dreyfus case.' So the file came back to the Prime Minister. He wrote on it, 'Orders may now issue' meaning I will now become a Lieutenant General. Instead of that, ladies and gentlemen, I received a letter from the Adjutant General saying the Defence Minister, Mr. Krishna Menon, has sent his severe displeasure to General Manekshaw, to be recorded. I had it in the office where the Commandant now sits. I sent that letter back to the Adjutant General saying what Mr. Krishna Menon could do with his displeasure, very vulgarly stated. It is still in my dossier.... Then the Chinese came to my help. Krishna Menon was sacked, Kaul was sacked and Nehru sent for me. He said, 'General, I have a vigorous enemy. I find out that you a vigorous General. Will you go and take over? I said, 'I have been waiting eighteen months for this opportunity,' and I went and took over." (Depinder Singh, 2002, pp. 237, 238, 241, 242, 243, 244, 245, 254 & 255)

One old saying in a rather general term is that, **'there can be no smoke without fire'**. In fact if there would have been even somewhat reasonably well, and that too in terms of both concerning the apex-level policy-matters inclusive of all their important dimensions and also, pertaining to the actual conduct of routine-cum-other strategically weighty activities, in the vast and extremely complex spheres of the democratic India's 'matters-military', then so many indeed intense debates and publications appearing in both electronic and the print media respectively, won't have been so vigorous, as well as so much devastatingly critical, and also as being the unending order of the day in intellectual-circles of the nation. **Right from the days of Nehru's Premiership to start with the politicians in power have been so superficially announcing from the top of a 'Qutub-Minar' that, all remain well with India's defence and thereby, the Government of the day always remains competent and also both geostrategic-cum-politico-militarily visionary enough to ensure that the nation's Armed Forces are quite unfailingly kept well-equipped and sufficiently well provided-for, so that no hostile power could make any breach into the nation's national-security systems in any manner whatsoever. Contextually could there be anything more distant from the truth in these regards?** Who in his/her senses, would even marginally believe in India's unscrupulous and corrupt politicians and their tall-claims which all in fact amount to nothing but indeed their routine 'culture of fooling around', to misguide and hoodwink the peoples of India, even in the presently ongoing 21st century. Right from the 1947-48 India-Pakistan Kashmir-War, almost all the conscious, patriotic, and critically thinking citizens as well as, similar members of Armed forces and the academia in the modern and democratic Indian nation-State and interestingly in addition also, a number of similarly well-wishing foreign nationals too have indeed been the no less honestly-cum-very analytically—writing, debating and pressing and sharply projecting in their highly focussed manners—India's numerously national defence related crucial issues; so that the requisite course-corrections are prioritized and also fruitfully activated, in India's long-term paramountcy national well-being and at the same time thereby to begin with, ensuring the overall regional peace and socio-cultural-cum-political tranquility in the whole of South Asia.

Hence precisely in their multiply larger geopolitical-perspectives and thus, while contemplatively deliberating upon these larger South-Asian regional issues, if the democratic India's apex-level political leadership-system gets more focussed, decisive and also becomes multiply confident and visionary in its always otherwise too, expectedly with essential strategic maturity and prudence then the above well-meaning contentions may not appear to be rather naïve or some kinds of the wishful 'tall-orders' only. In fact the multiply pivotal-points in these regards, are that for all these to fruitfully materialise and also to endure, India being the largest and also the key Nation-State, in the geopolitically volatile and also the no less strategic South-Asia in almost every conceivable manner—her government's top political leadership must first ensure for itself, a strong, fairly decisive, always being impeccably incorruptible, substantially far-sighted with the requisite geostrategic-cum-politico-military maturity and at the same time, also democratically having a commanding majority in the Indian Parliament—and the rest would follow, as the 'ball would obviously get set rolling' in these regards. Hence, it is rightly said that precisely in these contexts, the fault-lines run not so much externally but, all through the Indian corridors of power in New Delhi. We may quite unhesitatingly emphasise that ironically at times in the recent past itself, even if the apex-level Indian political-leadership then being in power, was commanding absolute majority in the Indian Parliament and thus ensuring the Government being stable and strong, its apex-level leadership either lacked strategic prudence and also geopolitically realistic-vision, or remained naively short-sighted and even being corrupt; and thereby very sadly, the above cited crucially long-term essentialities in these regards, could not be realised. In these specific perspectives, we do also know that at times during the tenure of any politically short-haul Government with all sorts of rather untenable groupings of the MPs, having hardly any strong base, both among the peoples and in the Parliament in New Delhi, it would have indeed been rather too much to expect any tangibly enduring achievements from them, in the above discussed larger South-Asian policy-framework, as well as India's broader geopolitical imperatives.

In fact in these multiply crucial perspectives, for India—the 'core-issues' are that to ensure stability and peace in the strategically important south-Asian region, India must be able to talk and conduct herself politico-diplomatically from an effectively persuasive positions of economic-cum-military strengths and geopolitical decisiveness—so that the balance of power in the region is not jeopardized and thus, triggering any irresponsible activities or misadventures in the region in any manners. Needless to say that the fundamentally paramountcy prerequisites in these regards remain that India must geopolitically remain in a position to visibly display and command her always proactive and also flexibly well-stationed: **'the multiple-aspects of her essential strategic-culture and thoughts'** for the needful. The pivotal-point is that if you aspire to be truly big; you must think big, become big in reality and also, act big with strategic-prudence as well as, multiply in positive manners and as far as possible, with politico-diplomatic reassurances to others. Since now even during the present 21st century, these vital aspects are not found to be vibrantly in place with their perpetually ongoing pro-activeness in the modern, fast developing, as well as the liberally democratic India's entire gamut of the national governance; her continuingly manifesting saga of thoughts, vision, policy-formulations, and thus decisions-making-cum-actions—in the highly complex areas of her 'matters-military', as well as indeed the multi-dimensional spheres of diplomacy—from the very beginning of entire post-1947 era so far; and hence these have sadly been far from reassuring and befitting for the Indian Nation-State. And thus we cannot wish away that ironically indeed, always in the very depressingly embarrassing manners; these overall and still broadly ongoing scenarios by themselves and also both regionally and globally, do speak volumes in many unpalatably glaring manners, for the Indian peoples, their leaders, and the decision-makers as well. And then on top of all these in India's now the presently further expanding national perspectives the still bigger irony persists that her so powerfully entrenched 'politico-bureaucratic combine' all along the nation's corridors of power, remains almost totally unconcerned of their such extremely worrisome 'work-culture', regardless of its multiply adverse consequences for the nation.

And as India entered into the present 21st century, interestingly a new Government, headed by the new Prime Minister, **Dr. Manmohan Singh**, came into power in 2004, with lot many kinds of the propagated expectations, as he happens to be a well-known economist with multiple kinds of experiences, and scholarship as well. **But in reality his/her actual performances, as well as the results-oriented deliveries in the arena of governance of that concerned Nation-State ironically, may not even marginally be matching and inspiring.** The unbroken 10 years of tenure of Manmohan Singh's so much unprecedentedly lack-luster Prime Ministership came to an end in 2014. Precisely with respect to the above, a noted Indian journalist and commentator, **Mr. T.N. Ninan**, in his analytically very perceptive article titled, 'The Adjustment PM'; since republished in *The Times of India*, Sunday, dated 04 May 2014 (*courtesy: 'Business Standard', India*); has written that, "Manmohan Singh has often said, especially as his prime ministership came under siege in UPA II, that he believes history will treat him more kindly than contemporary media has. If the 'ultimate measure of a man is not where he stands in moments of comfort and convenience, but where he stands at times of challenge and controversy', as **Martin Luther King** once said. How will history measure the man who was our PM for a full decade? As he prepares to move out of 7, Race Course Road (*i.e. 7, RCR, New Delhi: the Indian PM's official residence*).... Finally, if asked to choose between doing the right thing in a specific instance and pursuing the larger good (as in insisting on the auctioning of captive coal mines versus humouring an ally and keeping his coalition government going), he would and did choose the later. That also happened to be the expedient choice; yes survival was important too—especially since he believed that he could make a better job of being prime minister that the next man (*how naively ignorant and also, consistently wrong he had been.*). What he may not have reckoned was that too many expedient decisions would eventually take virtue away from survival.... So what about corruption and crony capitalism—which along with the economic slowdown has destroyed his prime ministership? Singh slipped up because he views issues from the perspective instrumentality rather than from the viewpoint of curbing wrongdoing by others—about which he is surprisingly non-judgmental for someone so fastidious about personal integrity.... He was on tap, not on top, as was rubbed in all too often, and most rudely by **Rahul Gandhi** (*limit. what more to say about such an extreme kind of a 'spineless man'?*) at the Press Club in New Delhi. More privately, at a dinner that he hosted in his official residence, guests crowded to get a place at Sonia Gandhi's table while Singh waited for someone to come to his own table. His philosophical detachment and escape into Urdu couplets must have helped him cope with the indignities.... These are the positions you would expect from a 'sarkari' (*i.e. a person on government pay-roll, like an employee*) economist who is immune to post-modern realities and oblivious of growing disenchantment. At the end of the day, Manmohan Singh remains the person he was 40 years ago: the reticent academic who moved from teaching to policy making and whom fortune pitch-forked into high office. Once there, his personality traits—evident decades back—combined with what Harold Macmillan called 'events' to create environment where most of his countrymen are eager to see the back of his government."

Hence it is rightly said that paradoxically sometimes even the real history is so bizarre, as well as broadly no less irreconcilable in its variously relevant contexts, that it indeed appears

to be stranger than any fiction itself. At this stage of our multiply continuing discussions, what more to say about the independent India's this 'ten-year period', being so much regressive, non-governing, non-decision-making, administratively static or rather being mostly in 'reverse-gear', and non-performing, unprecedentedly corruption-cum-nepotism ridden and thus also, rather mostly being unbelievably embarrassing and injurious for the entire nation onto its both internal and external fronts, so far as their various dimensions, within the whole gamut of the national governance, well-being and also the nation's future growth are concerned.

Now while critically reviewing with its always imperative objectivity, and also contextually in their multiple-perspectives—the so much indeed unprecedentedly depressing deeds or lack of deeds, and the numerous misdeeds of this ten years rather 'sick-life' of a Government headed by Dr. Manmohan Singh, even more sickening facts perforce do come in our minds as why and how a person of his learning and stature, could continue to be in the chair of the Prime Minister of India; and thus what precisely were the reasons or what all were perhaps the so overwhelmingly powerful factors or even some 'extra-constitutional power-centre/centres in action behind some 'curtains of perhaps even sinisterly conspiring duplicity'; which were so visibly capable to keep in 'chains' this Prime Minister and thus, were even rather nakedly continuing to make him 'dance to their tunes like a puppet', with no whisper from any other leader and the rank and files of theirs the then ruling Party in power. **Then as if all these were not enough, Prime Minister Dr. Manmohan Singh had an otherwise politically fairly experienced leader from Kerala as India's Defence Minister and interestingly he too, continued for the same ten long years and thus headed India's so critically important—the Ministry of Defence, located in New Delhi's South Block—where he perhaps didn't know as to what all were the always being nationally paramountcy job-profiles of India's Defence Minister, and also as how to successfully, as well as reassuringly provide the political-level synergy-creating leadership to the Chiefs of Staff of India's impeccably apolitical Armed Forces, in terms of the most vital national-interests and needless to say that now the rest is equally depressing-cum-sickening history in these regards.** In fact at this precise point and also in a somewhat lighter vein we may even say that, the duo of Prime Minister **Dr. Manmohan Singh and Mr. AK Antony**, as the Defence Minister; happened to be acting as ***'a pair of the political Laurel and Hardy'***, seen to be rather pathetically slipping around, while walking (*or perhaps dancing like puppets only?*), during their so-called 'work-assignments' in the Government of India's corridors of power in New Delhi.

No wonder as to why and how India happened to be both geostrategic and politico-militarily undergoing a series of humiliations, as well as the no less embarrassing setbacks for those ten long years; in the always highly dicey 'arena of the international power-politics', both in regional and global perspectives, and with not even a milligram of sincere concerns happened to be visible on the rather 'lifeless' faces of these two gentlemen who due to a bizarre 'twist of history' then, quite undeservedly happened to be indeed the no less damagingly, kept-on presiding over the destiny of the Indian Nation-State.

In precisely the above multiple-contexts and also concerning the state of India's defence preparedness, **Lt. Gen. (*Retd*) Ashok Choudhary, in his article, titled 'Navy Chief's Sacrifice Should Not be in Vain'** (*The Statesman*, Kolkata, dated 04 March 2014), and therein, he has written that, "Kudos to Admiral Joshi for being a professional and honourable Navy Chief and taking a principled stand in today's materialistic and rogue environment obtaining in the country. But this sacrifice by the Naval Chief (*since he resigned from his top naval job, after some battery-related explosion in an Indian Submarine, under the Indian Western naval Command*) should not be allowed to be used by the establishment to ward off discussion on the genesis and implications of such incidents by brushing the entire event under the carpet.... **Why has the frequency of accidents in the Armed Forces increased to alarming figures in the last few years? Obviously it is just not the standard of training. There are some major issues which basically come in the ambit of lack of strategic view, poor governance and pathetic functional ethos of the ministries at the national level. Take the examples of accidents while firing Artillery guns, Missiles, Tanks or flying aircrafts/ helicopters, or sailing ships/submarines.... The common threads are substandard equipment profile and their maintenance, unacceptably slow and unplanned modernisation and procurement of extremely substandard accessories as well as poor standard of ammunition produced in the country**.... The equipment profile of the Forces has not remained operationally optimal due to haphazard induction of arms and equipment, both from PSUs as well as other venders. The perspective plans projected by the Forces have never been seriously analysed or discussed at the level of ministry as they believe in processing the proposals on a case-to-case basis without taking a holistic view and in many cases the decisions are based on superfluous factors and in some cases on whims and fancies.... New equipment takes so much time to fructify that by the time it arrives in the user units, it is already vintage or does not match with other advanced accessories or communication systems.... **the so called functionaries of the government who neither understand nor have the time or inclination to initiate any changes in the existing procedures. They are also incapable of explaining the implications of operational requirements to the political hierarchy. They also do not allow the Cabinet Committee on Security (*i.e. CCS*) to be briefed directly by the representatives of the services as it would impinge on their projected importance and vice-like grip over the functioning of the Forces**.... In many case the Forces are left with no option as delays due to procedural hurdles could worsen their minimum acceptable holdings for training and day to day operational functioning.... Shortage of manpower in the services is yet another stark reality which directly impinges on training and maintenance of these vintage equipment.... **However neither has any constructive effort been made to mitigate these problems nor are they being highlighted and brought to the notice of the Nation on the pretext of national security which is really a façade to cover up their own incompetence and failures**."

While keeping in our minds the above quoted totally factual and hence, indeed irrefutable observations of Lt. Gen. Ashok Choudhary, we may recall that precisely in these being always nationally paramountcy perspectives we have since already contextually highlighted that since 1947 itself, the callously corrupt, arrogant and then also powerfully entrenched—the extremely clueless political and bureaucratic minions being only busy in warming-up their seats of power in the Indian Ministry of Defence—have so damagingly been indulging in their regressively fixated 'culture of fooling around'. More they falsely pretend to rather sincerely make the much needed changes in their 'work-culture and thought-processes', more things remain the same and at the most they only allow for some superficial 'window-dressing' only, and that too at the most only for public-consumption and to misguide the both national and international media. Thereby, with their total disregard of the numerously adverse and even some highly dangerous consequences to the nation. **But what all are being allowed, encouraged and even continuingly being indulged-in, by the 'politico-bureaucratic combine' in India's Ministry of Defence, especially with respect to the top policy-level management and decision-making systems in the vital domain of the nation's 'matters-military'; may at times be termed as hoodwinking the peoples of India and thus be taken as, even some severe 'acts of treason' against nation itself**. Hence it is clear that nobody in his/her right 'frame of mind' would ever disagree that the multiply far-reaching functional reforms as well as, systemic course-corrections in the entire decision-making processes in India's Ministry of Defence are urgently called-for, irrespective of the anymore and, any possible kinds of the man-made delays and obstructions. In these multiple-contexts, as we may critically study and review the today's overall geopolitical scenario prevailing all-round the Indian Nation-State and at the same time, while also keeping in minds the highly complex ongoing 'great-games' being played in the very slippery and dicey 'arena of the international power-politics', then it becomes crystal-clear that any more delays in these regards, if callously permitted in India's corridors of power would only be at the great risk to the overall security and wellbeing of the nation.

Precisely in the above multiple-perspectives, a Delhi-based doctor and a freelance writer, **Dr. Anjali Mehta**, in her noteworthy article has indeed very perceptively observed that, "From what we read and hear it becomes quite apparent that some old navy vessel are in a worn down condition and some outright unsafe. Some of the spare parts are of such an antiquated vintage that they are not manufactured anymore.... **Procurement and repair of naval vessels to my mind should be the sole prerogative of senior naval officers and should not be left to bureaucrats and politicians. Someone who has spent a lifetime knowing the ins and outs of vessels clearly has a better perspective. Officers who have reached senior positions have passed rigorous boards at several levels and have the capability and intelligence to handle big decisions themselves.... The current hierarchy which puts most of the decision making in the hands of the minister and bureaucrats holding the defence portfolio does not seem to be working too well**. The navy service professionals spend most of their time in work related tasks, but increasingly, their political masters are burdened with political turmoil leaving less time for understanding the finer intricacies of naval functioning.... Immediate and corrective steps should be taken in view of this incident. An urgent evaluation of all vessels by technically competent personnel and the ones declared unfit should be condemned. No personnel should be allowed to be on board an unfit vessel.... The services and their officers have always enjoyed a fine reputation and enviable traditions. Officers have a well-deserved special aura as they are brave and people of high integrity who give a lot and ask for very little in return. It is extremely demoralising for them when they are treated like cannon fodder and their safety disregarded.... **It is also a sad culmination, when an officer by dint of hard work and capability reaches the rank of chief, only to be made a scapegoat for the inefficiencies of others. I feel our officers, command and deserve far greater respect than that**." (Mehta, *The Statesman*, 11 March 2014)

It is needless to say and as emphasised earlier—that a comprehensively well thought-out, well-designed, and also futuristically flexible reforms are indeed urgently called-for, if India wishes to prevent any further fiascos, humiliations, and tragedies. It is also no less true that they persistently seem to be displaying no sincere concerns towards the geostrategic-cum-politico-militarily ensuring the nation's always paramountcy—the both internal and external security-aspects and wellbeing—in a rather unquestionably enduring manner. **Hence all these in fact boil down to a fundamental necessity for India that it is high time for her, to be bestowed with a democratically elected and powerful, truly honest, focussed, experienced and unwaveringly determined, and also the multiply visionary top-level political leader to take charge at her helm of affairs to constitutionally do the needful as per above and in addition, also in many other different aspects concerning her efficiently effective governance, multiply fruitful and enduring economic-development, national progress, sustained protection of the environment and ecology with their scientifically continuing regeneration as being always in harmony with the mother-Nature and thereby guaranteeing the overall wellbeing of the nation, indeed for the generations to come in**

future as well. Precisely in the above multiple perspectives, **Mr. Jay Bhattacharjee, in his hard-hitting but indeed the no less factual article has written that,** ***"In times of war and not before, God and the soldier we adore; But when the war is ended and all things righted, God is forgotten and the soldier slighted'***. The evocative lines have a disputed origin, some attribute them to an anonymous poet and others to that arch-imperialist **Kipling**, who despite his numerous other faults, wielded a fine pen a pen when it came to non-colonial themes. All Indians who have read these words will immediately relate to the way in which our country has treated its armed forces.... To be fair this deplorable conduct is not the fault of ordinary citizens. The average Indian has always admired our men and women in uniform and has readily come forward at every opportunity to honour and respect the nation's defenders. The only two groups that have formed a shabby alliance since independence to downgrade our armed forces, disrespect them and deny them the necessary wherewithal to defend our frontiers are the politicians and the bureaucrats. The behaviour of this coterie has been so venal and egregious that the phrase *neta-babu* cabal (*i.e. politician-bureaucrat nexus or sinister faction*) is now widely accepted and used in all studies and assessments of military issues.... Certainly the armed forces have shortcomings and deficiencies; after all our soldier shares the genes and culture of rest of the population and the armed forces do not live in isolation. We have had some disastrous military leaders, but they are invariably people who were favourites of the neta-babu combine (*i.e. as since repeatedly mentioned in numerously different contexts earlier: the 'politico-bureaucratic combine'*), foisted on the armed forces and given undeserved boosts in the services hierarchy.... **Nevertheless, the vast majority of our services personnel have been role-models as far as their commitment, dedication, honesty and valour are concerned. Whenever there has been a crisis or calamity, whether natural or defence-related, the military has come forward to serve the country without asking any questions or seeking any compensation or reward. It seems that this is what sticks in the gullets of our politicians and civil functionaries—that the nation has such a glorious icon, while they themselves, are so disliked and abhorred.... Since independence, the Nehruvian power-centre was not comfortable with the military**. Some analysts have attributed this to the perception that the Indian armed forces were much too colonial in their culture and traditions. Yet the same 'colonial' army saved India's bacon both in Kashmir and in Hyderabad. **It is now undisputed that Krishna Menon's influence on Nehru also contributed to the slow rift between the country's political leadership and the armed forces. Menon's abrasive and quixotic method of working led to the enormous resentment in the defence establishment, culminating in the Thimayya episode in 1959.... Taking a cue from their political basses, the civil services, led by the 1AS-IPS combine started their sniping campaign to deny services personnel their legitimate benefits and humiliate them in every possible manner.... Many of the services chiefs and senior brass were also not strong enough or adequately vigilant when the forces were being downgraded and humiliated. Too many Chiefs of Staff were glad to accept plum post-retirement postings and assignments that allowed them to retain their earlier life-styles (*irrefutably true and hence, facts are facts.*)**.... The lives of two valiant naval officers (*recently this year, i.e. in 2014 itself*) were snuffed out because their submarines were compelled to use batteries that had exhausted their service lives...all because a few *babus* (*i.e. faceless, callous, irresponsible, arrogant and corrupt bureaucrats in the Defence Ministry.*) had not released the requisite funds for replacing them. The Navy Chief **Admiral Joshi** resigned taking moral responsibility for the tragedy. There was not a whimper from the Defence Minister and the Defence Secretary (*and what about the Prime Minister, along with his same Defence Minister, i.e. this duo have since constituted in fact, nothing but the earlier mentioned 'pair of political Laurel and Hardy' only?*). This brings me to the oath that young officers take when they pass out of the Indian Military Academy. When they cross the *'Antim Pug'— (i.e. the well-known 'last-step', in that Academy.)*, they see the Chetwode motto bequeathed to us by a colonial military officer. The words proclaimed quite simply: 'The safety, honour and welfare of your country come first, always and every time. The honour, welfare and comfort of the men you command come next. Your own ease, comfort and safety come last, always and every time'. Independent India's armed forces have admirably honoured this splendid code of conduct. The same can hardly be said of our politicians and civil administrators." *(In fact anything like, the above mentioned 'oath/code of conduct' to perforce be made applicable for, with respect to India's politicians and civilian bureaucrats now; would be a great 'insult to these time-tested sacred 'words of honour', since exclusively related to the patriotic domain of the 'profession of arms'.).* (Bhattacharjee, *The Statesman*, 15 March 2014)

In the otherwise highly complex, as well as multiply vast areas of the 'matters-military' concerning most of the nations and also, perhaps few exceptions in today's ultra-modern world it is true that 'corruption in some form or other', indeed remains present and thereby interestingly from time to time, do also get manifested in their own sensational manners. But barring some kinds of the 'banana-republics', or some 'dictatorship-ridden nations' in today's world, India otherwise being even a well-noted fast developing, the world's largest and indeed a liberal parliamentary-democracy has also very unfortunately acquired an embarrassingly dubious and ill-famed position of rather a 'crowning-glory' in the some sort of 'international-hierarchy' of corruption and the malpractices in the multiple-areas concerning the democratic governance, and also the no less including that of the 'matters-military' as well. Moreover it would never be perhaps anyone's intention in his/her right frame of mind, to get the modern Indian Nation-State ever bracketed into the comity of still perceived as some 'failed-states, pseudo-democracies, or tin-pot dictatorships' in any manner whatsoever.

If that being so then it also can't be denied that some kinds of the indeed matters-military-related—stinking 'corruption and malpractices'—to the great detriment of the nation's both internal and external security-aspects, have undeniably been able to seep into the professional persona of some of the very senior and also the middle rank military-officials. Consequently, tarnishing the image and glory of the democratic India's 'profession of arms' and thereby, this malice is also very humiliatingly persisting as well as, even enlarging in the present 21st century too. We may recall that among the many currently being investigated and talked about political and bureaucratic corruption and malpractices in the huge and dicey domain of the 'matters-political-cum-civil', one fairly engaging issue has been that of the nation's telecommunication-related massive **'2G-scandal'**, as found to be getting considerable prominence in the nation's media-circuits. Interestingly in some indeed very different forms, a **'2G-disease'** is also very dangerously getting manifested as well as, affecting the very senior military-officers' fraternity, contaminating the domain of India's matters-military' in many different ways. The critically noteworthy points in these regards remain that, this now lately spreading phenomenon of '2G-disease' in the always so paramountcy domain of the nation's 'matters-military', just can't ever very wisely bracketed with the 'matters-political-cum-civil' related scandals and also the unending kinds of the acts of corruption and malpractices in any manner. As in addition, this now quite lately being observed '2G-disease' and its symptoms in fact, have the potential to considerably weaken the nation's defence related 'critical-armour and military-systems' and thereby, even ransack the soldierly qualities and also 'crack' the always being so magnificent as well as, indeed the time-tested war-fighting-spirits of its impeccably apolitical Armed Forces. **This now undeniably defence-services-related '2G-disease', in fact means the games of 'Greed and Grab',which unfortunately and are being played at the very**

senior-levels of the military command and staff positions and also at times, even at the very apex-level of military-hierarchy in the Indian Armed Forces with ulterior motives, and regardless of the consequences to the nation's highly complex, multi-dimensional, and also the super-high-tech oriented defence-aspects.

The larger problems are that the modern democratic India's political-classes, of the all possible kinds of shades and/or the so-called political ideologies or schools of thoughts, due to various reasons as well as the nation's past history are not at all suitably knowledgeable enough, conscious, and even experienced enough, so far as the multiply complex affairs of the political-level managements of the nation's matters-military aspects in a democracy are concerned. Above all, they have never shown any serious willingness or concerns towards rectifying these indeed quite serious lacunae or shortcomings among themselves in these regards. All these malignantly lingering shortcomings need to be urgently attended to and also suitably rectified by bringing-in the all concerned on-board, sincerely and firmly as well. **In fact at this precise point of our ongoing discussions, one gets duty-bound to rather even forcefully emphasise, while keeping in our minds the overall geopolitical scenario since so disturbingly prevailing around the Indian Nation-State, that at the highest national-level governance, the nation's multiply-complexmilitary related aspects and dimensionscan never be treated as any less important than the various crucial aspects of the agricultural, industrial and thereby overall economic, as well as many other kinds of developments of the nation. In fact any dedicated, visionary, honest, and patriotically committed—the apex-level political-leadership and its other politico-bureaucratic functionaries—need always to comprehensively understand that the modern democratic India's overall economic as well as, numerously different, the human value-systems-related developmental projects and activities in the 21st century's presently fast-moving-cum-super high-tech era world can only be fruitfully enduring, if the nation's geostrategic-cum-politico-military related security-aspects too, are indeed always effectively ensured**.

After 1947 and accordingly since the fairly long tenure during the 1950s and 60s, of India's first Prime Minister **Mr. Jawaharlal Nehru,** the matters-military-related saga of the arrogant disservices to the nation's always multi-dimensional strategic-thoughts and also the defence preparedness, as being rather unendingly meted out regardless by India's powerfully entrenched 'politico-bureaucratic combine', and then on top of it as per demands of the multiply paramountcy national-interests and thereby, in spite of the persuasions and coaxing, this regressively arrogant 'combine' is refusing to see reasons therein, and begin doing their duties in the service of their nation, for which they are regularly being paid for. They need to get their heads examined, if they think that they would ever be doing any favour to India's Armed Forces; if they mend their ways and earnestly begin to be doing their duties to facilitate the ongoing updating, as well as resort to making efforts towards the multiply reinforcing the national defence-systems without any hassles. **Irrespective of these, in fact the malignantly lingering 'mental-warps' all across the Indian corridors of power, it remains always critically important to keep insisting and rubbing contextually the above mentioned vital aspects so that modern and democratic India's 'politico-bureaucratic combine'may sooner or later be forced to wake-up and see reasons, and thus, fruitfully commence acting in their best possible manners.**

Hence at this precise point of our multiply continuing discussions, we may now contextually ask the mankind's 'living-history', and also the peoples of the Indian Nation-State, as well as their ruling-classes/decision-makers—as to what went wrong, or had been going wrong, or rather what were the principal reasons consequent to which India has traditionally been lacking the

'conceptual-wherewithal' or the 'strategic-tools', towards being able to essentially evolve the necessary ways and means to trigger the requisitely motivated and proactive innovations to cultivate, evolve, develop, groom, imbibe and flexibly place **'the multiple-aspects of essential strategic-culture and thoughts'** in India' highest national-interests? When the far-reaching and also indeed the unprecedentedly history-making **'Industrial-Revolution'** was momentously manifesting and historically unfolding itself, with its numerously valuable fallouts in the European continent, India was already a British colony and hence naturally with no blame onto the mankind's living-history in any case; but unfortunately for India, this momentous 'Industrial-Revolution' was made to almost totally by-pass the Indian subcontinent. This any 'accidental' or otherwise any 'chance-event' of history, did bring-in many adversely far-reaching consequences for India and thereby, almost debarring her, though indeed abnormally so late, to be able to fruitfully, as well as proactively get conscious of availing the so many unprecedented benefits out of the numerously multiple-fruits and dimensions of the 'Industrial-Revolution'; when its time had arrived to begin-with, and as per its timelines as well. And thereby further in addition, also be able to cultivate the above mentioned requisite **'conceptual-wherewithal'** or the **'strategic-tools'** to trigger the essential activities in a fast-paced manner and also, towards fruitfully-cum-proactively bringing-in **'the multiple-aspects of essential strategic-culture and thoughts'**, through the conceptual-processes as since already talked about a number of times in their indeed differently relevant contexts earlier in these regards. As we may critically reflect into the Indian subcontinent's history of its colonial-era, we do contextually notice that specifically in these geostrategic-cum-politico-military perspectives, the British imperialism naturally had left hardly anything in the hands of Indians and consequently, India had to perforce remain under the geopolitically overwhelming guardianship of her imperial-masters only. Consequently what to talk about acquiring any fruitful awareness regarding the always universally paramountcy **'the multiple-aspects of essential strategic-culture and thoughts'** even during the Second World-War (1939-45), the Indian leaders were hardly aware of the historic **'Atlantic-Charter'**, where the clearly agreed-upon post-war era's, the 'future of India' too were spelled out. The problem with India's otherwise highly competent professionals from various fields, and whether coming from civil or military domain, knowledgeable academia with great intellectual acumen and also, the no less patriotic and critically thinking vast citizenry since ages, has been that their perhaps bizarre mindsets have always been getting rigidly fixated with their tendencies to be mostly inward looking and thereby, not much caring for the external subject-matters and the outer world. These tendencies make them largely ignorant about the external aspects which in any case, are seldom any less important and hence consequently, any such 'topsy-turvy obsessions', mostly turn out to be rather counterproductive too. We may even say that these far-reaching negativities are in fact, nothing but quite undesirable fallouts of the almost all pervasively lingering lack the universally paramountcy: **'the multiple-aspects of essential strategic-culture and thoughts'**. In certain specific perspectives here, and the following quoted paragraphs too, may be found to be considerably substantiating the above noted perceptions in these regards.

In precisely the above multiple perspectives, **Captain Narendra Singh Sarila**, a World-War II era's young Indian Army officer and who happened to be an A.D.C. to Lord Mountbatten during his tenure as the Governor General in India has written that, "**On 12 August 1941, Roosevelt and Churchill had signed the Atlantic Charter**. This charter laid down certain common principles that the US and UK wanted to follow for a better future for the world and included the declaration that they [the US and the UK] respect the right of all peoples to choose the form of government under which they will live.... On 7 August 1942 the British Embassy in Washington reported to

London that on the first anniversary of the signing of the Atlantic Charter, the United States was planning to propose that telegrams be exchanged between the President and the Prime Minister and that the President's telegram would state the charter applied to Asia and Africa, as well as to Europe. When this message reached Churchill on 9 August 1942, he was in Cairo. But he immediately wired Roosevelt:

> 'Charter's proposed application to Asia and Africa requires much thought. Grave embarrassment would be caused to the defence of India at the present time by such a statement.... Here in the Middle East the Arabs might claim that, by majority they would expel the Jews from Palestine. I am strongly wedded to the Zionist policy of which I was one of the authors. This is only one of the unforeseen cases which will arise from new and further declarations.'

By raising the Jewish issue, Churchill obviously hoped to temper American enthusiasm to apply the Atlantic Charter to India and the rest of the British Empire.... Raising the question of freedom under the Atlantic Charter had been one way of applying indirect pressure on Britain. Another was to draw British attention to the growing public opinion in the US in support of India in the wake of Quit India movement.... **Halifax** had earlier warned London of the 'slipping back' of the gains made with the US public opinion after the Cripps mission failed. Around the same time, the London Times contended, in an editorial, that the increase in critical American opinion on the Indian question 'has been so great that it threatens to affect the 'Anglo-American relations'. The same anxiety was reflected in another report of this period sent by the British Embassy in the US to London: 'India has become the first test which the American, friendly or unfriendly, at once applies to British Imperial Policy. In the main it is critical, and it is frequently sought through with emotion'.... The American Government, at this stage, decided to twist the British lion's tail by yet another stratagem. By September 1942, the Quit India movement had turned violent, rupturing communications and transport to the eastern parts of India, which served as the base for supplies to China and to the Japanese front on the Indo-Burmese border. The British repression to break the movement was then at its peak. At that very moment, without giving any notice to the British, the American stopped the deliveries of supplies being made under the **Anglo-US lend-lease agreement** to India...although the concerns of the American Chiefs of Staff that these supplies may be looted or lost in the turmoil that had engulfed India were also taken into account. Halifax, knowing fully well that the agitation would shortly be smashed, took no action of the American move.... After the Quit India movement was crushed by November 1942 and the fear of a Japanese invasion of India receded, India went off the blip in the US Press. And the public interest in Indian developments faded once again. By the end of 1942, Churchill had, for all political purposes, won the 'press war on India' in the United States.... Puckle analysed the reasons for the Anglo-American misunderstanding as follows:

> 'In the whirl of crosscurrents, floating rubbish and sunken snags which make-up the stream of American opinion on the subject of India, two things steadily catch the attention, the interest and the ignorance. The interest is, because of the old psychology of the frontiers which still influences America, and makes them utterly disinclined to believe that, there is no quick and simple way out of a deadlock in India or anywhere else.... It is also there because of the presence of American troops in India and their view that China's deliverance will come from their base in India.... **America has suffered two major disasters in this war, Pearl Harbor and Pearl S. Buck, the second because of the sentimentalism she has spread in the public about China and India.' (*Quite interesting, isn't it?*).**

The British by then had crushed the Quit India movement and with the increase in American Naval and Air pressure on Japan in the Pacific, the possibility of that country invading India had become remote. **Churchill and Linlithgow (*i.e. the then British Viceroy in India*) were now fully confident of retaining India well after the end of war and were in no mood to encourage anyone, even if he be the American President's representative, to say or do anything that might instigate the Indian leaders or give ideas to the Americans to pressure Britain to start a dialogue with them (*what a the then historically 'powerful' example, on the part of British, of the colonial-era's so much rigidly fixated obsessions of imperialism.*). Phillips warned Washington that the US had to safeguard 'our own position in India as [a] military base against Japan as well as our future relations with the coloured races**. If the Viceroy can obstruct the representative of President to see Gandhi, Indians will lose confidence in his capacity to accomplish anything'. Roosevelt, upon reading this message, instructed Cordell Hull in February 1943 to bluntly tell Halifax to convey to London that 'Gandhi should not be allowed to die in prison', but Phillips was instructed not to make public in India this concern expressed by the President for the Mahatma.... Roosevelt's directive to Phillips, his successor, to help the cause of Indian freedom only through friendly pressure acceptable to British officials, gave Linlithgow the means to restrict contacts in India, which created misunderstandings in India about American policy. It was unknown in India, for example, that Phillips had made several efforts to meet the Congress Party leaders but was prevented from doing so by the Viceroy.... Most Indians, even the educated ones, had a most perfunctory knowledge of America and Americans at that time. American history taught in the universities ended with the battle of Yorktown and the American independence from Britain. People were more knowledgeable about Canada and Australia—which were part of the British Empire—than about the USA or indeed about neighbouring countries or territories, such as China, Central Asia, Afghanistan or Persia. Whatever curiosity about America that existed—came from acquaintance with American cars, and the Hollywood films. Fords and Chevrolets, with their higher clearances, were better suited for rough Indian roads than English cars and ruled the roads. Hollywood was the main source of information about the new world, even though these films never reached the villages, where 80 percent of the people lived. When newsreels or events connected with the war began to be shown in picture houses, people were aghast to know that America, not Britain, was the senior partner in the war (*quite interesting and revealing too*).

(v) ...Wavell Plays the Great Game

In the British Administrative System, the man on the spot traditionally enjoyed considerable authority. This was more so for a person appointed the Viceroy of India and who occupied the most important post in the British Empire, outside the British Isles at that time.... Field Marshal Wavell was not a member of the aristocracy, from whose ranks most of the Viceroys were drawn. He belonged to a class of society—the upper middle class—which, according to a British historian, 'was the mainstay of the British Raj and largely responsible for its character'. Wavell was the first soldier to hold the office of the Viceroy after the British Crown took over the governance of India from the East India Company in 1858. Before his appointment to the viceroyalty in September 1943, he was Commander-in-Chief in India and had supervised the defence of the British Empire in South-East Asia against Japan. Earlier, he had been commander of the British forces in the crucial front of the Middle East, where his victories over the Italians in Abyssinia and Cyrenica, in North Africa, at the start of the war had made him quite a popular hero in Britain. On coming to power Churchill had removed him from the Middle East Command and sent him to the relatively calm waters—before the Japanese attack—of India and Asia. And after the British defeat at Japanese hands in South-East Asia, the P.M. kicked him upstairs in an honourable way, to the viceroyalty. Churchill considered him overcautious and defeatist, 'eminently suited to run a provincial golf club', he once said. Churchill expected him to take no political initiative in India, during the course of the war. Wavell, on the other hand, remained in awe of his chief, 'the bigger man than either Roosevelt or Stalin', he noted in his diary (*strange, isn't it?*), but he used to complain that 'Churchill was always expecting rabbits to come out of empty hats'. The irony is that one, whom Churchill considered so mediocre, has come to be acknowledged by several historians as the most important Viceroy of India since Lord Curzon. His forte was his lack of illusions; and his achievement, the division of India.... Wavell profoundly disagreed with his predecessors view, expressed to him just before Linlithgow left Delhi that, 'Britain would have to continue responsibility for India for at least thirty years and that 'the country was in pretty good trim.... [with] with Gandhi and the Congress leaders out of the way in prison and the Muslims immensely strengthened during the last three or four years. Wavell's assessment was that the British position was deteriorating fast and that a plan was need for an organized and orderly retreat that would, nevertheless, protect Britain's most important asset in India, which he saw as the military base that it provided to control the Middle East and the vast Indian Ocean region and from where fighting manpower was recruited." (Sarila, 2006, pp. 152, 153, 154, 155, 156, 159, 162, 163, 167, 169 & 170)

So the numerous kinds of British great-games, as usual very strategically-cum-craftily planned and also being multi-dimensional in their contents, as well as futuristic in their substances too were very judiciously continuing regardless; all along their multiply imperialism-soaked colonial-era in the Indian subcontinent. Hence substantial amount of the indeed historically significant writings have already appeared both in India and the UK, thus throwing valuable lights in these regards; and which in turn, have also been able to remove 'lots of fog', hindering the clear understandings in their desirably objective perspectives. We may recall that in earlier chapters too and while analytically dealing with these subject-matters in their different but the no less relevant perspectives, we have since also discussed these significant historic aspects in some interesting details; and then at the same time, substantiating them by quoting some relevantly appropriate and informative paragraphs from the writings of both British and Indian authors as well. In these historically indeed very engaging perspectives, another interesting and also very relevant book coming into our minds, happens to be that of **Khan Abdul Wali Khan**, son of one of the Indian subcontinent's greatest freedom-fighters, **Khan Abdul Gaffar Khan and a great disciple of Mahatma Gandhi (and popularly known as the 'Frontier Gandhi' as well as, 'Badshah Khan' too.), and titled as 'Facts Are Facts'.**

Hence in his above mentioned book, **Khan Abdul Wali Khan** has indeed very informatively mentioned that, ".... I did, however, believe that there was an underlying element of truth in the views of the Congress and Badshah Khan. But never could I imagine that their allegations were only a pale reflection of the truth; the truth was uglier. Their mischief exceeded our wildest imagination. Badshah Khan's and Congress' allegations were far short of truth. **If there was the slightest doubt earlier, it was now totally removed because the documents preserved in the archives bore the official British seal. Housed in the world's most prestigious library, they were signed by the no less than the Viceroy and the Secretary of State for India. What more proof could one ask for? While reading certain portions of these documents, I found myself staggering with disbelief. I could not believe the facts that were staring me in the face. Often I was compelled to close the files and go out for a cup of coffee, before taking courage to look at that bitter truth.**

To give the devil its due, one cannot help admiring British loyalty which stopped at nothing to ensure maximum benefit for their country. Although it was difficult to accept the unscrupulousness with which these rational and educated people protected the empire, yet one could not help admire them for their straight forward rendering of the truth. There was neither bias towards their own kind, nor malice towards others. One by one they stripped every one, friends, companions, and fellow-conspirators, to reveal the truth. They did not mince words in describing all the underhand tasks they assigned to their native lackeys. (*Hence it is indeed appropriately said that who could ever deny that sometimes, even the facts can be more bizarre and stranger than any imaginable fiction too?*) They appeared unconcerned about what could happen if these documents became known to the Indian people and the masks of their so called leaders were stripped off. The Indian public would be aghast if they realized these pious leaders and patriots were trafficking in the integrity of their own country, and were the country's enemy number one, besides being traitors and agents of the British Government. The British sense of fair-play did not take any of these factors into consideration.... **My study of the twenty years of correspondence of [from 1922 to 1942] between the Viceroy and the Secretary of State for India revealed that the earlier study I had made of the books and diaries pertaining to the subject was futile. Every detail of the policy was clearly stated in these documents which I was reading. The anti-USSR policy of the British was staring me in the face; all doubts and misgivings were thus removed.... What caused me tremendous embarrassment was the account of the incidents which revealed the character of the leaders of the country. The worst offenders by far were the Muslim Leaguers. The allegations levelled against them by the Congress and/or Badshah Khan were a fraction of what I found on record in these British documents.... The fact that the British were now emphasising education, albeit for training clerk to look after their officers, had a positive influence upon the Indians. Education opened new vistas of understanding. Travel to other countries exposed their minds to contemporary realities. When they saw how democracy worked in other countries, they became aware of their rights and conscious of their deprivation. The majority of this new enlightened class consisted of Hindus. They constituted the majority, had the power and influence to take advantage of Western education, and produced successful scholars and prosperous businessmen. The new demands for self-determination arose from them.... Therefore, later that year, at a gathering of the Indian Association at Lahore, he said: 'I heartily wished to serve my country and my nation faithfully. In the word nation I include both Hindus and Mohammadans, because that is the only meaning I can attach to it.... These are the different grounds upon which I call both these races which inhabit India by one word, i.e. Hindu, meaning to say that they are the residents of Hindustan.'.... Sir Syed said that every inhabitant of India, no matter what his personal belief, can be called a Hindu by virtue of his belonging to Hindustan. The essence of his teaching was that Muslims had to unite with Hindus against the British** (*Hence the British too as a countermeasure, devised their infamous policy of 'divide and rule'.*)**. The Hindus being the favoured community were provided employment and education. The Muslims, on the other hand, were deliberately ignored. Therefore the only way out for them was to join forces with Hindus.... The British were becoming conscious of this emerging trend. They realized that their position would be considerably weakened if the two communities got together and presented united opposition. To counter this, the British decided to introduce reforms based on communalism. The first of these, Minto-Morley Reforms of 1909, was related to elections to local and municipal committees. According to these, the Muslim votes could be cast only to the Muslims and the Hindu votes for the Hindus. In this manner the British laid communalism as the foundation-stone of democracy. This was first blow they**

inflicted upon Sir Syed's ideals. By proposing a communal rather than a national basis for politics, they forced the Hindus and Muslims into a position whereby if they wanted to enter municipal or community politics, their electioneering was limited to wooing their religious brethren, and fighting on religious rather than political issues (*Thus the 'seeds for the birth of Pakistan' so sinisterly, as well as the no less clandestinely, were sown—with all the more their ruthlessly-obsessed imperialism—by the crafty British, while ruling over the Indian subcontinent; and the rest is now the so much highly chequered history in these regards.*). The British were delighted at the success of this policy.... Jinnah was against according any democratic right to the people who lived in the Princely States. **Of the 600 Princely States only ten had Muslim rulers**. The Muslims of Kashmir were striving for their rights against the Hindu Maharajah. In their anxiety to please the British, the Muslim League forgot the existence of their fellow Muslim brothers. They challenged the Congress for the rights of the Hindu rulers of Kashmir.... On 28 March 1939, the Viceroy wrote regarding another meeting with Jinnah, 'But he [Jinnah] was satisfied now, he thought that the present system would not work, and that a mistake had been committed in the going so far.'.... The above statement that Jinnah was annoyed about the rights given to the States. He maintained that India did not need any final transfer of power and those who made such demands were unaware of the existing conditions. Democracy could not prevail in a country like India. It was a mistake to have held elections in the States and to have handed over the rights to the elected representatives. This statement of Jinnah was a God-send for the British. The British had prayed long and hard for an indigenous political party which would become the sworn enemy of democracy, support British policy, and be prepared to fight its enemies. (*So in fact, that was that so far as the 'real inner self and character' of Jinnah were concerned. It was all motivated in the inner psyche of Jinnah to satisfy his highly inflated personal ego and thereby, leave no stone unturned to harm India in all possible manners. He finally stopped at nothing less than extremely ruthless and violent dissection of the Indian subcontinent and thus; ironically leading to a still-born, socio-cultural-cum-politically untenable—the 'artificial' Nation-State of Pakistan—which now remains in a rather perpetual state of the numerous kinds of turmoil and, even self-negation too. Where since 1947 itself, India is a vibrant liberal democracy and also a reasonably fast developing economic-cum-military regional-power as well.*).... By this time the British had spread into every corner of the world and they proudly claimed that the Sun never set on their Empire. Millions of men and women had become British subjects. The wealth of the colonies was siphoned off into the mother country. Its armies were comprised of the youths of these nations. Being an international power, the British policies had to conform to international expectations. They were concerned about the Congress victory in the assembly elections, but a far greater source of concern was the situation in Europe. Hitler was gradually establishing Nazi rule over the smaller countries of Europe. In Italy, Mussolini was holding sway. The British had exhausted all their material resources, and to quote a Pushtu proverb, when it comes to a battle for power, there is no scope for a battle of wits. In December 1939, they were forced to declare war against Germany. This was a formidable undertaking because the Germans were a dreaded enemy. The British were fortunate to have the English Channel between themselves and Germany. Else they would have fallen headlong into the Nazi pit.... The Mullahs were concerned neither with Islam, nor Quran, nor with the faith itself. These followers of the Holy Prophet were busy selling religion for ten to fifteen rupees a month. The British were their Muftis and Qazis. Their religious sanction came from the British. Having raised the sword in the name of Islam, they were prepared to pounce upon the 'enemy' at the slightest nod from the British. If the British called the Russians 'infidels', the mullahs echoed 'yes'. If the British declared the same for the Germans, the Mullahs nodded in agreement, although the Germans like the English were Ahl-e-Kitaab, The Mullahs did not concern themselves with principles: they waited upon the pleasure of the British. When it was a question of damning Italy, the mullahs echoed the British sentiment without recognizing that Rome was the seat of Christianity, and Vatican City, residence of the Pope.... When the British needed Islam to resolve some of the internal problems, they summoned their 'indentured slaves'. In the Frontier Province their problem was that their opponents were the Khudai Khidmatgars, who were all Muslims, and their leader Badshah Khan was a Hazi, schooled in the best traditions of Islam. He had founded Islamic schools, and served the Islamic cause in every possible manner. His lifelong mission was to throw out the British. The Mullahs, having sold their souls to the British, came out in open confrontation with the Khudai Khidmatgars. I have no personal objection if a religious scholar jumps into a political foray. But the traitor mullahs accosted the patriots and true Muslims in the name of Islam, all for gaining favours with the British." (Khan, 1987, pp. vii, viii, ix, 3, 4, 14, 15, 67 & 68)

So all that is history now but nonetheless, these aspects do also remain further history-making in very far-reaching manners; and hence even now largely politico-economically, as well as geopolitical-cum-diplomatically too, with their multiply positive fallouts, the finely tuned and enduring—the Indo-UK relationship aspects—which are moving from strengths to strengths, and that's something substantial in many ways for both the nations.

(vi) Politics and Politicians in Independent India

But as the consequent results above all: a proven, highly dedicated, politico-strategically always being so focussed in his multi-dimensional manners and thus, always being brilliantly visionary—the nation's indeed new, powerfully decisive and also the most popular apex-level political-leader in the persona of Mr. Narendra Modi—has indeed so inspiringly emerged onto the India's vast, diverse, and the always multi-coloured political-horizon, as her new Prime Minister, and thereby in the process, has been able to multiply draw the whole world's attention uncommonly in very many engaging ways. India is thus fast preparing herself from now on, towards writing her new 'living-history' in substantially very different and reassuring manners; not only for herself but also for the all concerned both being far and near, in the presently ongoing super high-tech modern-era's world. As a consequence naturally, needless to say that the BJP, India's another national political Party has decisively come to power at the centre through such hard-fought electoral-contests that, almost the BJP's entire opposition now electorally stands almost decimated across the board.

While keeping in minds precisely the above multiple-perspectives; we may recall that earlier in some differently relevant and also—in the 'great-games' oriented contexts—we have since also talked about that the great-games may be both international, as well as even intra-national too. We have since also discussed some aspects and examples, precisely concerning the intra-national 'great-games' as well. And it remains quite interesting to note that in case of both international, as well as the intra-national 'great-games' in any given situation, with its historic or otherwise timelines anywhere; the consequent fallouts in fact, can be both positive or negative or even mixed too, based on their many accompanying variables and also even differently emerging factors therein in these regards. Hence as the history's timelines keep-on perpetually moving further, these too are few of the numerously many kinds of ways and means, through which the mankind's living-history also remain continuingly in the making onto its expanding civilisational-canvas itself.

Now so far as the immediately passed-over full decade of the Prime Ministership of Dr. Manmohan Singh is concerned, it is true that many kinds of the governmental or otherwise largely unconstitutional, and/or even unlawful indeed—some intra-national great-games oriented—initiatives were triggered from time to time with all possible kinds of the mostly quite untenable motives—regardless of the so much consequential financial losses and/or other kinds of all the more highly damaging harms to the Indian peoples and their unfortunate Nation-State. But precisely in these indeed quite unpalatable contexts, it is also equally true that paradoxically the hapless Prime Minister, Dr. Manmohan Singh during his ten-years long tenure, never had his own any ulterior personal agenda and thereby in spite of his such high learning and scholarship and also largely clean intentions, he proved himself to be thoroughly incapable of proactively holding and controlling the apex-level 'the politico-governing reins of power', in spite of his otherwise the then being always unquestionably top constitutional position in India's official hierarchy. But tragically it turned out to be all in vain that during his tenure, there happened to be quite sinisterly, the two 'centres of power': one legitimate and constitutional and then the other one being equally illegitimate and unconstitutional. And then on top of it as if all these were not enough, the second one really so much unlawfully kept-on exercising the real power and calling the shots at the national-level domain of the multi-dimensional-cum-multi-layered governance and also in fact, with total disregard to the so many highly injurious consequences to the hapless Indian Nation-State.

In these precise contexts, Mr. Sanjaya Baru, who happened to be earlier the Prime minister Dr. Manmohan Singh's 'media Adviser' in his rather somewhat 'sensational' book, has quite revealingly written that, "In the PMO some officials shared my unfashionable scepticism about the efficacy of the RTI Act, but Sonia was so committed to this initiative that no one seriously resisted it. I was not convinced that transparency in terms of public access to internal government communication was a necessary condition to making the government more responsive to people's needs, and to good governance. By this token, few organizations, including most NGOs and the media as an institution, were 'transparent' even though they were more responsive. The glare of public scrutiny would not scare corrupt and inefficient officers, who would always find new means of playing old tricks, but it would certainly discourage honest officers from stating in writing views that might later be used to question their motives. I felt Dr Singh was sympathetic to my view though he never explicitly said so.... **Moreover, monitoring what others were doing was one thing, getting others to do what the PM wanted, quite another**. At the time I did not realize how the limits to the PM's political authority and the PMO's institutional weakness meant that there was very little control of the PM and his office over the misdemeanours of ministers. Whenever I heard a tale about ministerial corruption that was credible enough to bear repeating, I would relate this to Dr Singh. He would always listen with attention. Most of the tales belonging to ministers belonging to parties that were allies in the coalition, but a good many also related to Congress ministers. It was clear that Dr Singh wanted to know what was happening. I assumed that this information would help the PM to remain alert, especially when signing files, and that he would perhaps pull up the minister concerned (*Well, was this in fact, nothing else but just wishful thinking only?*).... **The politically fatal combination of responsibility without power, and governance without authority, meant that Dr Singh was unable, even when he was aware, of checking corruption in his ministry without disturbing the political arrangement over which he nominally presided**. Political power resided with the heads of parties of the coalition and, as PM, he could not dismiss ministers at will. He could, perhaps, have done more to discipline his ministers. One way in which this could have been done would have been to appoint upright and effective officers as secretaries under corrupt ministers. Here too the PM often failed to assert his authority, appointing as secretary a person that the minister concerned preferred.... The consequence of all this was to bring the PM into disrepute despite his own impressive record and reputation for personal probity and integrity. Still, I never imagined that charges of corruption of the kind that came to haunt him, in the manner they did, in years to come would so sully his reputation. For a long time the media was willing to give him the benefit of doubt on his role in questionable decisions, by accepting the view that his lack of political authority prevented him from disciplining his wayward ministers. But when the issue of corruption took centre stage in public discourse, the question that was relentlessly asked was why had the PN not prevented what was going on. That he 'yielded so much space' to other centres of power, so that he had little of his own to act, was not viewed as an adequate defence." (Baru, 2014, pp. 101, 102, 103 & 104)

It is indeed the no less shockingly amazing as to why and how a man of Dr. Manmohan Singh's personal integrity and academic standing continued to allow himself, to be so pathetically a helpless 'pawn' by the unscrupulously crafty UPA chairperson for ten long years? These questions are not going to easily fade-away as the history would rather chase him to demand clear answers and in any case thereby, the history as usual won't easily be kind to him and hence not going to spare him, as he while 'standing in the history's dock', he would find himself alone, as none from among the Congress Party's herds of the so many 'goons', would ever be willing to be present by his side there, with clear possibility of even likely to be led to the history's 'chopping-block' itself. Hence it is rightly said that sometimes, 'the facts are even stranger than fiction itself'. Nevertheless it is also true that earlier **Dr. Manmohan Singh as Finance Minister** in the erstwhile **Narsinhma Rao**'s government, did achieve marvellously visionary, as well as very far-reaching economic reforms and which enabled the nation to indeed decisively come out of the so-called very embarrassing and age-old *'Hindu-rate of growth'* forever. **Hence India's 'living-history' would also very laudably remember him as a great architect of its so path-breaking and enduring economic-reforms too.** But at the same time now, subsequent to the quite inglorious exit of the Manmohan Singh's government the 'silver-lining' in these regards for the country in fact is, that a new as well as shinning dawn has broken with great hopes and also indeed, with remarkable possibilities of the multiple accomplishments, for the Indian masses in years to come.

Now precisely with respect to the above multiple-perspectives, **Mr. Swaminathan S. Anklesaria Aiyar**, one of India's well-known politico-economic commentators; in his indeed greatly analytical article has written quite reflectively that, "Few people know or care about Lincoln's failings. Similarly Singh will be remembered as the FM (*i.e. finance Minister*) who launched reforms, and as a PM (*i.e. Prime minister*) under whom the economy grew at 8.5% for most of a decade.... **This is a party (*i.e. Congress Party*) where all members are supposed to sit-up and beg when any member of the Gandhi family whistles. Its members have no rationale, purpose or hope of power, save through the grace of the Gandhi's.... Manmohan Singh was fully aware of and agreeable to this when he accepted the Prime Ministership** (*Otherwise sometime back, he won't have publicly declared that he would be very happy and feel fortunate to even serve under the leadership of Mr. Rahul Gandhi—the so-called only dynasty-based leadership of an already defunct 'heir-apparent' of the same 'corrupt' dynasty—oh no, what a rather unhesitant display of one's spinelessness as well as that of, even his naked and stinking sycophancy too.*). **A seasoned technocrat, he was used to proposing ideas but retreating respectfully if opposed by his boss. He was ideal for Sonia, a respected technocrat with absolutely no political base or ambitions, who could serve but not threaten the**

dynasty.... This dynasty has generated more black money than all others combined. It found very useful Singh's unimpeachable reputation for integrity. This provided some cover to the dynasty when seams exploded. But it damaged Singh's own reputation. Remember the scene from the film *'Dabang'* where Malaika Arora dances in front of Salman Khan singing 'Munni badnam hui, darling tere liye?' This inspired a cartoon showing Manmohan Singh dancing in front of Sonia Gandhi, singing 'Munna badnam hua, darling tere liye'.... In dynastic terms, Singh should perhaps be judged not as a Prime Minister but as a regent, keeping the throne warm for the young prince, under the watchful gaze of the Dowager Empress. I wrote back in 2005, after his first year as Prime Minister, that his approach could be summed up in seven commandments: 1. Thou shalt not displease Sonia. 2. Thou shalt not displease the Left Front. 3. Thou shalt not test the limits of your powers. 4. Thou shalt focus on surviving a full term. 5. Thou shalt accommodate in the cabinet all criminals who can help this aim. 6. Thou shalt at the end of your regency hand over the reins to the true dynastic inheritors. 7. Thou shall, meanwhile be free to initiate policies that threaten neither the dynasty nor the coalition's survival (such as improving relations with Pakistan and China, and organizing buses to Muzaffarabad (*look at this, limit. Isn't it*?).... Alas, they failed on all three counts. GDP growth remained at just 4.5%, half the 9% achieved earlier. Consumer price inflation simply did not fall below the 8-10% range. As FDI in multi-brand retail were so loaded with cumbersome clauses that they have not yielded much investment. The subsidy on diesel was supposed to be phased out, but the crash in the rupee last year raised import price, so the diesel subsidy remain as high as ever. The Cabinet Committee on Investment cleared stuck projects worth Rs 6 lakh crores, yet this did not translate into any boom in orders for capital goods or construction contracts.... **Why? Because a new license-permit raj, had come up unnoticed. The old license raj was based on industrial licenses, import licenses and forex controls. The new license raj was based on the environment, forests, tribal areas and land acquisition. A veritable jungle of new controls in these areas was created at central and state levels. Initially, these new barriers were overcome by bribes. But once public anger exploded over corruption, clearances could not be bought, and the impenetrable nature of the new controls became evident. Judicial activism made bureaucrats wary of taking any decisions**.... Singh and **Chidambaram** claim that things would have been much worse but for their efforts (*limit, no shame in even saying or making this fictitious claim.*). May be, but that is hardly a winning election platform for most of his 10 years. **Singh was feted for his economic skills and integrity. He leaves office amidst economic travails and the smell of corruption."** (Aiyar, *The Times of India*, May 15, 2014)(*Hence all said and done, finally it is indeed a good riddance for the nation. Whatever, but the nation has to move-on regardless. Past damages have to be repaired, past mistakes and/or any willful misdeeds and blunders need to be effectively corrected, so that no scope or possibilities do remain any more for their recurrences in future. These too some of the very many tasks and challenges are awaiting now the newly installed and indeed so magnificently encouraging and dynamic Government of the India's* ***newly elected Prime Minister, Mr. Narendra Modi.*** *The Indian Nation-State is now indeed both definitively and also justifiably, looking to future with great hopes and encouraging expectations.*).

It would perhaps indeed be no less relevant to rather somewhat quite objectively-cum-critically visit some important aspects of the basic **'concepts of secularism'**, and that too particularly in the larger terms of the sovereign and democratic India's socio-cultural-cum-political and also the national governance related dimensions. It is noticed that nowadays it has become quite fashionable to look into these aspects of secularism, generally with all sorts of half-baked ideas. And naturally consequent to these, many kinds of the socio-political disturbances and turmoil do get generated and then at times these are even willfully created by certain vested interests, as well as also quite unscrupulously crafty politicians too, for their ulterior motives. What we have noticed in the 2014 general election in India that such corrupt, criminalized and also quite dirtily selfish politicians were busy in raising **now the already defunct bogey of 'communalism Vs secularism'** with no holds barred, shamelessly just for their immediate electoral gains. But they obviously could not succeed, as the general masses refused to be led like foolish herds of animals and thereby, they very befittingly threw such shadowy political parties and their leaders into 'garbage-bins' of the nation's democratic electoral polity. And the rest is so well-known, far-reaching recent history; now seems to be so reassuring for the nation's overall future well-being.

In fact all the more particularly in terms of the today's super high-tech and so fast moving modern world; if we quite scientifically as well as rationally, may wish to maintain that secularism in any case, just does never mean—to be totally non-religious or being even non-believer in God—in any manner whatsoever, while remaining busy in one's pursuits in life and also, while conducting any official-cum-diplomatic or political business of the State. We all know that in most of the democratic nations in the present modern-era's world the peoples' elected representatives, during their oath-taking ceremonies, do solemnly begin to administer their constitutionally mandatory oaths, in the name of Almighty (i.e. God) in their own different languages. In fact, being religious or a piously deep believer in God, in one's personal, social or even official life, does never mean that one is being communal as one can very comfortably be deeply religious and thus at the same time be quite impeccably secular, both in its letter and spirit too. **Being communal means being religiously fanatic, intolerant to other religions and thereby, having quite unbecoming prejudices against the peoples belonging to other religions.** But ironically, even these simple facts have deliberately been distorted, as well as made dangerously controversial too, by some people to quite shamelessly further their selfish and sinister agendas only. Hence nonetheless in India since the last few decades, especially after the most tragic, sudden and untimely demise of the second Prime Minister Lal Bahadur Shastri in the year 1966, many political parties and their selfish-cum-greedy political leaders, quite unscrupulously began to misuse these communalism-cum-secularism related sensitive issues in their very corrupt manners with no concern for any adverse consequences to the nation. It is a very sad commentary on the Indian societies and their democratic polity too.

In modern, sovereign and democratic India and then even all the more in the present 21st century as well, which Indian whether surviving or otherwise, can ever claim to be more dedicatedly secular than Father of the nation, Mahatma Gandhi? But even he was religious and then deeply spiritual too, in his thoughts and actions. In fact these earlier already well-settled issues in the multiply multi-cultural Indian societies, got unnecessarily somewhat 'disturbed' during the Prime Minister Nehru's times and that too, in largely quite uncalled for manners—so naively by the none else, but ironically by Nehru himself. It is true and also now historically well-settled that, nobody can raise any doubt on Nehru's secular credentials, his intellectual prowess, his scholarship, patriotism, his multiply great service to the then newly independent India, and his so much charismatic statesmanship and that too while laying now the historically so magnificent foundations of India's democratic polity, its modern scientific and technological temper and also, its zeal for the socio-economic progress and modernization. But it is also again equally true that in terms of the religious dedications and faiths of certain equally great and also intellectually brilliant Indian leaders and memorable personalities, he quite unnecessarily became extra-sensitive and also naively intolerant as well. It remains a great irony, so far as Nehru's otherwise magnificently great persona is concerned and therefore, it perhaps remains quite difficult to objectively rationalise in fact these few rather 'naively fluid' segments of his great personality.

Precisely in the multiple contexts of the above debated subject-matter, specifically concerning India's first Prime Minister; **Romit Bagchi**, Chief Sub-Editor of *The Statesman*, while quite justifiably praising the many great dimensions of Nehru's historic persona, in his article titled, 'Jawaharlal Nehru and Hindutva' since published in the same newspaper, dated 13 June 2014; he has written that, "**Nehru** was resentful, yet pathetically helpless, when the President of India, **Rajendra Prasad** decided to preside over the deity-installation ceremony at the reconstructed Somnath Temple in May 1951. Importantly, the initiative had been taken by **Sardar Vallabhbhai Patel**, Nehru's home minister until his death in 1950. Nehru asked Prasad not to attend the purely religious function which, he apprehended, would reopen the scars of the Partition in the national psyche to the detriment of peace, harmony and development. But Prasad remained unmoved.... 'Somnath temple has a historic significance...some people may not like the idea of a temple which was destroyed by the Muslim invaders of the time being rebuilt and revived. But I think you will agree that is not a reasonable attitude to take,' the Head of State wrote to the Head of the government.... At the installation ceremony, he said, 'The Somnath temple signifies that the power of reconstruction is always greater than the power of destruction'.... A senior minister in Nehru's Cabinet, **K.M. Munshi**, was even more aggressive when he wrote to Nehru—'I can assure you that the collective sub-conscious is happier with the scheme of reconstruction of Somnatha sponsored by the Government of India than with many other things that we have done and are doing.' A sullen Nehru wrote to G.B. Pant—'Communalism has invaded the minds and hearts of those who were the pillars of the Congress in the past' (*how naively wrong Nehru was here.*).... **We may recall what Nehru said while addressing the Madurai session of the All India Congress Committee in October 1961, where he spelled out rather unambiguously what he thought of cultural nationalism as the fulcrum holding the cultural unity of the country. 'India has for ages past, been a country of pilgrimages. All over the country, you find these ancient places, from Badrinath, Kedarnath and Amarnath, high up in the snowy Himalayas down to Kanyakumari in the south. What has drawn the people from the south to the north and from the north to the south in the great pilgrimages? It is the feeling of one country and this feeling has bound us together. Our ancient books have said that the land of Bharat is the land stretching from the Himalayas in the north to the southern seas. This conception of Bharat as the one great land, which the people considered a holy land, has come down the ages and has joined us together, even though we have had different political kingdoms and even though we may speak different languages and follow different faiths. This silken bond keeps us together in many ways'.** ***(Indeed greatly perceptive words from, a great man, i.e. Nehru.)***.... Left-inclined intellectuals say that the namby-pamby kind of secularism that is being practiced now, with temple-hopping, almost invariably followed by mosque-hopping, remaining the mainstay of the political domain is an outcome of Nehru's wishy-washy view of the State as equal to all religions. The Hindutva brigade, on the other hand, says that Nehru sought to obliterate Hinduism altogether by singing paeans for the composite culture of India, beginning with the 'Aryan invasion' and attaining consummation through the invasion of the British and the Islamic invasion remaining the most important part of her inclusive history."

And with respect to the above in fact interestingly, it is also the no less perceptively said further, that the real Nehru is seen to have been lost in the marshy bogs of the political/intellectual mumbo-jumbo, meaning that the crust blurring the kernel. Well, the point in any case is indeed well-taken but at the same time and perhaps the no less substantially, another point too arises historically so as to who, in fact is responsible for these kinds of some misunderstandings concerning Nehru's thoughts, perceptions, action and the persona, than principally Nehru himself? **Nehru had seemed many times to be unnecessarily overplaying the 'tune of secularism', etc. He should**

have been taking these aspects much more discreetly in his strides, like the other great leaders of his time and then all the more particularly as India's first Prime Minister, he was expected by the people (*who have always been traditionally very religious too*) to be more prudent and patiently tolerant, rather than being always harshly 'judgmental' in these regards. In other words, he should have been more willing to be giving more time to these sensitive aspects to gradually and discreetly settle-down into the peoples' larger psyche since otherwise too in fact, traditionally always thriving onto the India's multi-cultural, multi-religious and also, multi-lingual socio-cultural-cum-political canvas having great, as well as indeed highly chequered past, spanning over the thousands of years of the mankind's 'living-history'. Well in retrospect, we may visualise all these as now being already matters of the recent past; and hence obviously being part of the post-1947 era's Indian 'living-history' as pertaining to the very beginning years in these regards.

Nonetheless, it also remains equally true that in any modern and truly vibrant democracy, the vital criticality of the constitutionally proactive role, as well as official place and involvement of the nation's civil-services cannot be minimized and hence, the same now equally hold good for the sovereign, fast developing and democratic modern India as well. And ironically in these specific contexts, we do sadly notice **some serious chinks in the Indian-Nation's 'civil-services armour'** in fact, within of the overall gamut of its national governance. In these multiple-perspectives of indeed now of all the more great importance, we may recall that we have since already somewhat analytically discussed as to how the then British colonial-masters of the vast and also socio-historically so much complex Indian subcontinent, had so prudently been able to create an elite civil-service named as the ICS. The then imperially principal strategic-aim of the British was, that the officers belonging to this service as being the so-call 'steel-frame'; could be not only successful in forming the higher-level bureaucracy of the British administration, but would also mostly remain colonial-minded in their work-culture and thereby, do remain rather desirably-cum-perpetually effective in 'magnificently' facilitating and also strengthening the British colonial-rule in the subcontinent. And therefore historically, the British while finally agreeing to grant India her political freedom and thereby leaving the Indian shores finally; left behind this elite and bureaucratic civil-service as their legacy in India. **But whatever historically might have been so; it now remains to the great credit of the visionary leadership of Sardar Patel and also within the overall framework of the Prime Ministership of Jawaharlal Nehru, that the Indian members of the then ICS too, did their best to serve the country—then being a new and fledgling democracy—and thereby, also proactively began contributing in reorganizing the entire gamut of the civil-administration of the nation with totally changed systems and circumstances, under the sovereign India's democratic government, duly formed by the peoples' elected representatives. Soon and as per the great political statesmanship and vision of Sardar Patel, a new civil-service was born since known as the IAS, as a new replacement of the old and also the bygone colonial era's ICS in the newly independent democratic India.** But so sadly and ironically, the indigenously groomed members of this newly constituted IAS too in a very bizarre fashion did not lag behind to so regressively imbibe the bygone and socio-politically discarded 'colonial work-culture'; in fact to great detriment towards the constitutionally desired and the highly expected modern values-systems with democratic ethos of the dynamic and people-friendly civil-services. All these have been a very frustrating negation of the great vision of Sardar Patel, indeed in multiply damaging ways to the entire gamut of civil-administration in the country.

But in the befittingly larger perspectives of our ongoing critical discussions in these regards, so far as the malice very undesirably seeped into the 'work-culture and ethos' of now the democratic

India's modern civil-services, are concerned, it would indeed be very incorrect to lay the entire blame onto the doors of bureaucrats only, since the political-classes and that too all the more top-level political leaders of all shades and ideologies, in fact both historically as well as contemporarily, have always been much more callous and guilty. Such undesirable 'derailments' began with certain omissions and commissions on the part of India's first Prime Minister Pandit Jawaharlal Nehru only and thereby, a bizarre and also quite counterproductive legacy has now continued. And consequent to all these and to our great dismay, what we find today that though the British colonial rulers of the Indian subcontinent have since already finally left the Indian shores more than sixty years, their now already bygone and also the highly arrogant imperial-culture are still quite malignantly lingering among the thinking-cum-living of the democratic India's political leaders and also the government officials as well! In fact, in any other matured and truly vibrant democracy elsewhere, even any marginal continuation of any such legacy would never be thinkable at all. But even in today's ultra-modern and also so fast moving super high-tech era, the Indian democracy and the work-culture and ethos of its government machinery have its own bizarre peculiarities.

In the above multiple-contexts, the basic irony has rather very clearly been found that many a times even the very senior-level bureaucrats positioned in important places in the corridors of power in India's both national, as well as the various States' capitals across the country, have mostly been found to be very greedy, unscrupulously petty and thus, rather incapable of thinking big and acting courageously in the highest national interests. On top of their own so deeply entrenched selfishness and small-mindedness that they mostly remain professionally divided in fact, duly triggered by the most unbecoming and also petty considerations of caste, creed, 'foolish' provincialism, and also highly undesirable political affiliations too and thereby, they themselves so naively provide even open avenues to any corrupt and the extremely unscrupulous politicians in power, to exploit them and thus rather compel them to take wrong decisions with many damaging consequences. At this precise point, we may get compelled to again remember an old proverb that, 'one does commonly see the 'iron' getting rusted, but how shockingly frustrating it can be if anyone to his/her great dismay sees even the so called 'gold' getting rusted too'!

The point is that every year as the annual results of the 'civil-services' are announced by the nation's UPSC, one is able to see that most of the successful young boys and girls notably, have clearly had indeed very brilliant academic accomplishments/careers all along their respective schools and university years; and they do display lot of confidence, high patriotic sentiment and also praiseworthy zeal to always honestly serve the nation and its citizens to the best of their capabilities. **But the greatly frustrating paradox remains that within few years of their careers in the government service, all these high ideas and feelings promptly, as well as rather unstoppably get evaporated like high-octane gasoline. Here lies the 'rub' and needless to say that in spite of the stinking political-muck always found to be encircling them as the elite civil services, it is only they as nothing but patriotically united and professionally determined teams can decisively fight the variously sinister vested-interests confronting them and thus, effectively be able to powerfully push-away that 'muck' from their paths in the country's highest national interests. Are these 'tall-orders' for these selected bodies of men and women? Well certainly not—and as per their own constitutional oaths, they are capable to decisively rise above their numerously lingering petty-considerations and get patriotically united, as well as motivated teams—for the needful in these regards. But there are some other much larger and also seriously worrisome dimensions to it.**

Another point is as to how many of the aspiring youngsters today, while securing entries

into the civil-services, truly do so for the sake of serving the nation based on honestly considering themselves to be competitively having the truly enduring spirits and also correctly toughened aptitudes for the intended job-profiles, as well as the challenges therein? Why the presently ongoing mad-rush and rather some kinds of the perhaps even irrational scramble among so many finely educated youngsters, including good-products from the nation's IITs/IIMs, for this job? Hence there has to be some kinds of rather quite undesirable-cum-unscrupulous 'magnets' therein, and also should this so-called 'magnet' be allowed to exist any further, while appropriately keeping the larger long-term interests of the country, when as the nation is fast developing and, also economic-cum-industrially advancing in so many multiple ways, and thereby so many other more lucrative job-avenues too are opening-up for many of those youngsters. There have to be some unreasonably attractive reasons, but those also being not at all healthy and desirable in terms of larger national considerations, which would be pulling many of these indeed above-average youngsters towards these All India civil-services into their folds. And if these being so in the first place itself, and which rather clearly appear to be the facts; then these too also are not at all good for the larger interests and ethos of these civil-services themselves. These so loudly visible extra-ordinary reasons too in these regards, need to be looked into and critically studied in their requisite depths and details for devising necessary course-corrections, in the best long-terms interests of the country itself. In these broader perspectives concerning some fundamentally crucial aspects of the nation's civil-services, their modalities so far being followed by the UPSC to select suitable candidates every year from among indeed a very vast number of aspirants; also need to be critically examined and overhauled. In fact in the above multiple-perspectives, just a glimpse as to how deep are the maladies and rot in the nation's civil-services can be seen in the following quoted paragraphs from the articles written by few of the retired bureaucrats only.

Mr. MG Devasahayam, himself a retired IAS officer, in his very frank and precise article has very perceptively written that, "Civil servants have been dragged into the 'vicissitudes of convulsive politics, scams and scandals' against which they were supposed to be a bulwark. They have become accessories to the colossal corruption and its cover-up. **Due to the collective failure of civil-servants in living up to the covenant of the Constitution, India's democracy has diminished and democratic governance has decayed**.... As for acting on oral instructions of political bosses, it has been made mandatory for the 'officer superior' to confirm the same in writing. 'The civil-servant, has received such information, in turn, is required to seek confirmations of the directions in writing', the court has ruled. Most of these orders are already in place in civil-service rules and procedures, but have not been complied with because of compromise and cowardice on the part of senior civil servants at the level of Principal Secretaries, Chief Secretaries and Cabinet Secretary. The Supreme Court provides stilts to these civil servants, most of whom have broken their legs on their own volition. The court has clarified that by issuing these directions it has not overstepped its ambit but has acted because of the failure of these worthies and the political executive to frame a legislative framework to address these key concerns despite 'draft bills in circulation on such reforms.... But that was not to be. To understand why, we should go back to the genesis of post of post-independence All India Services (AIS), almost single-handedly brought about by Sardar Vallabhbhai Patel. In April 1948, he wrote to Prime Minister Nehru strongly advocating the formation of an independent civil service in the functioning of which 'political considerations, either in its recruitment or in its discipline and control, are reduced to the minimum, if not eliminated altogether.' This was strongly opposed by the Chief Ministers and many members of the Constituent Assembly (*interesting, isn't it?*). Retorting, the Sardar in his speech to this Assembly in October 1949, said. 'The Indian Union will go. You will not have a united India if you do not have a good All India Services which has independence to speak out its advice—if you do not adopt this course, then do not follow the present system, substitute something else.... Sardar had his way and AIS (IAS and IPS) were established to 'give a fair and just administration to the country and manage it on an even keel'. To ensure this and safeguard the civil servants from 'vicissitudes of political convulsions', these civil services were covenanted in the Constitution.... Set up under such challenges, All India Services were meant to implement the national agenda of delivering decentralized and honest governance to all citizens and uplift the poor and the downtrodden. Such governance would pursue an equitable, small is beautiful, need-based, humble-scale and balanced development while conserving nature and livelihoods. But the 'reform era' of the past two decades has turned this upside down. Most of the civil servants are touting the MNC agenda of turning the country into a 300 million rich/middle-class market through FDI-funded big-ticket projects by mortgaging the resources of the nation, leaving the 900 million 'laggards' in the lurch.... The fact is that despite constitutional protection, most civil servants have abdicated their independence and political neutrality and have become willing pawns of ruling politicians and resource-rich vested-interests. Serving corrupt carpetbaggers has become their *mantra*, let *aam aadmi* be damned. They have abdicated the civil service *mantra* and hence the broken legs. Supreme Court directives have not restored the legs, but have only provided stilts. While giving the directives the Court had hoped that these reforms would lead to 'preservation of integrity, fearlessness and independence of civil servants at the Centre and State levels in the country'. Using these stilts, conscientious and honest civil servants would assert themselves by reverting to the constitutional scheme of things from which they have drafted and reinvented themselves to become fearless, independent, honest and efficient entity bound by an esprit de corps which is appallingly absent now. Only then, there would be some hope. Mere walking on stilts will not do." (Devasahayam, *The Statesman*, 30 November 2013)

It is indeed equally true that many of the points in the above quoted paragraphs so powerfully elucidated in very precise manners, do need to be comprehensively as well as very intensely debated too, so that various inter-related issues can be analysed for putting them in honestly desired perspectives and thereby, facilitating the best possible decisions in these regards. Now that's that and hence needless to say that, if as per the always so much malignant callousness on the part of Indian political parties and their leaders, the intelligentsia, academia and also the nation's interestingly so much 'noisy' media too in fact, continue to remain regressively unconcerned and passive in these regards and thereby—the bureaucracy as an otherwise always indeed essentially important 'institution'—would only be allowed to continuously decay and drift; then rather most unfortunately, the Indian democracy would perforce sooner or later, fall into the deep pit of some very serious problems. Hence, some well-planned, visionary and also very comprehensive 'civil-services reforms' are urgently called-for to safeguard the health and larger interests of the Indian democracy, as well as also the nation's highest national-interests, and all the more so in the present era's super high-tech modern world. While keeping all these contextually very crucial aspects in our minds, and hence thereby analytically looking at them in a little different but indeed also the no less closely related perspectives, we may also very sadly say that if at the intake points India's these elite civil-services are able to attract such high young talents to start with every year—then why the civil-services, in the first place as a so much important institution since engaged in the overall gamut of ensuring the nation's civil administration effectively—have now been found to be the no less frustratingly as well as rather irreparably, in a 'defunct and rotten' state? This crucial question since many years but now all the more glaringly confronting the nation's highest level politico-bureaucratic leadership as a system, appears to be no more possible to be wished-away or brushed under the carpet in any manner whatsoever. Hence patriotically higher prudence in these regards now demands that these so much problematic lacunae plaguing the civil-services, may comprehensively be looked into and thereby, objective and long-term solutions are indeed proactively found so that India's highest national-interests are not allowed to suffer anymore.

Now thereby further in continuity as per the above; another retired bureaucrat, Mr. Debakinandan Mandal, a former Joint secretary, Government of West Bengal; in his article has written that, "Indian bureaucracy of late has come under increasing criticism. It has been rated as the worst in Asia with a 9.21 rating out of 10—ten being the lowest on a scale from 1 to 10—according to a report advanced by the Political and Economic Risk Consultancy Ltd, Hong Kong, and released in 2012. India fared worse than Vietnam (8.54), Indonesia (8.37), Philippines (7.57), and China (7.11). Singapore turned out to be the best with a rating of 2.25.... **The report makes it clear that India's inefficient bureaucracy is largely responsible for most of the serious complaints of the corporate class. And these include poor infrastructure and corruption. Officials are willing to accept under-the-table payments and companies are tempting to pay to overcome bureaucratic inertia and gain government favours**.... The Court asked Parliament to enact a Civil-Services Act under article 309 of the Constitution to provide for a Civil-Services Board, which can guide and advise the political executive on transfer, postings, disciplinary action etc. The ruling is a mix of positives and negatives. Since bureaucracy-bashing is a favourite national pastime, the court-order will hopefully improve governance and enable the bureaucracy to function better. Certain uncomfortable questions are bound to crop-up. Are we entering an era of 'government by judiciary?' Hasn't the Supreme Court stepped into the domain of the executive? Is it the business of the court to decide on the mechanism of the functioning of the government? Does it not strike at the basic feature of the Constitution—separation of powers? The only plausible explanation is that the weakest political executive under UPA-II has ceded its turf to the judiciary. Were the petitioners, all retired bureaucrats, true to their office and authority, when in service? Had Cabinet secretaries and Chief Secretaries been strictly neutral, there would have been no controversies as faced by **Ashok Khemka** and **Durga Shakti Nagpal**. Indeed politicians are calling the shots because senior bureaucrats have become wimpish. They tend to kowtow to their political masters.... Many civil servants do extraordinary work in equally extraordinary circumstances—during floods, drought, and elections, in confronting extremists, and of course maintaining law and order, and containing fissiparous tendencies.... In a representative democracy to quote **Prof. Harold Laski**, the Cabinet or the political executive is 'amateur' and 'temporary', while the civil service is 'permanent', consisting of 'experts and professionals'. UPA-II has turned out to be the weakest dispensation since independence. **India direly needs an unfettered and responsive bureaucracy. The problems afflicting the bureaucracy need to be identified and addressed suitably**. This would call for periodic review of performance.... It has been observed that while the government fares reasonably well on the policy-front, implementation and delivery are below par because of the absence of team-work. Lack of inter-departmental coordination hinders timely execution of projects. A lesson can be drawn from corporate governance instead of considering the problem from the narrow perspective of each department.... **After the 2G spectrum scam and coal-block allocation scandal, the bureaucracy is scared of such entities as the CAG, CVC, CBI and the judiciary. 'Level-jumping', which facilitates quick disposal of files, is never encouraged**. The example set by **E. Sridharan** should inspire the bureaucracy. He had deviated from procedures and never bothered about CAG reports which deal with violation of rules. This explains why Delhi got a world-class Metro and within the stipulated budget.... **The technocrat-bureaucrat controversy and inter-service rivalry have led to divisiveness. Politicians take full advantage of the situation and adopt a**

policy of divide and rule to pursue their objectives. The evolution of a consultative mechanism headed by the DM in the district, the Chief Secretary in the state headquarters, and the Cabinet secretary at the Centre will take carte of the respective roles of the technocrat as the provider of professional inputs and bureaucrat as the chief coordinator of governance. The protectors should not be poachers. The profile of a civil servant as penned by **P. Woodruff** is relevant even today—'Who gets through his files quickly and wastes no time on looking up the rules, who writes short, decisive judgments because he is clear in his own mind where the right lies and does not seek to justify himself and who expects his subordinates to do their own work and trusts them unless he has reasons not to.'.... The erosion of values is responsible for the maladies of the bureaucracy. The joint-family of the bureaucracy needs to be better organised both at the Centre and in the States." (Mandal, *The Statesman*, 18 February 2014)

Now just in the above few preceding pages, we have analysed and also critically discussed various aspects of now the modern and democratic India's civil-services in their variously relevant perspectives. It is true that immediately after the Indian independence, India's so much historically charismatic 'iron-man'—**Sardar Patel** as the nation's first home minister and the Deputy Prime Minister—was rather singularly responsible in reorganising the All India civil-services, as seen to be suited towards the nation's democratic governance in the post-colonial modern era and thereby, entirely within the constitutionally changed circumstances. From then on these civil servants were supposed to be dedicatedly serving the teeming Indian masses, rather than imperially ruling over them as during now the already gone-by British colonial times. These concepts by themselves were indeed brilliant, but what about the realities now so disappointingly prevailing all across the country? No doubt it was indeed very visionary and also then soon expected to be a very major overhaul of the system itself in these regards. But the genesis of these civil-services lay substantially in their colonial-past and hence, could not be completely delinked or wished away. IAS became the successor of the erstwhile ICS; and similarly but the no less ironically in a very counter-productive manner, the IPS turned out to be a new and rather 'comical' incarnation of that of the British colonial-era's IP (i.e. the 'Imperial Police') in sovereign democratic India, in a quite avoidably distorted manner; and the rest is nothing but a largely unpalatable saga of now the **substantially derailed police-services of the Indian Nation-State**.

In precisely the above multiple-perspectives first and foremost, we do need to understand that the constitution and design of the erstwhile colonial-era's civil and police services in the Indian subcontinent were one of the numerous segments of the their highly thoughtful as well as, the multi-dimensional British Imperialists' ongoing 'great-games' to geopolitically serve and safeguard their paramountcy colonial-interests as per the overall geostrategic-cum-politico-military dictates and compulsions of the then intercontinental British Empire. And thereby interestingly now in the historic perspectives of these aspects of the then British-Empire, if any temporary or futuristically permanent benefits did accrue substantially for the Indian-Nation and its teeming native populace, then it was largely nothing else but incidental only. But as we may recall—our somewhat extensive discussions in the present and earlier chapters too, as well as the various quotes suitably incorporated therein from the valuable writing of many learned authors—and accordingly then, we may also say that it would be grossly incorrect and naïve to blame the erstwhile British colonial rulers in these regards. In fact and also as discussed earlier, it is true that subsequent to the decline and decay of even the earlier Mughal Empire, the 'rudderless' and also so vast Indian subcontinent in any case, was just waiting to be colonised by any of the competing and seafaring European imperial-powers since having arrived onto the Indian shores earlier and hence, they had already begun building-up their different kinds of the citadels across many locations in the subcontinent to geostrategic-cum-politico-militarily firm-up their footholds and also enhance their very shrewd visibilities onto the Indian soils, for soon to bring-in many expected benefits to them in the near future. And keeping these historically important aspects in our minds and also in retrospect, we may finally say that the then Indian subcontinent was indeed very lucky to be finally colonised by the British imperialists only, rather than any other European imperial-power since already engaged in such struggles over here.

(vii) Police Reforms

In the earlier paragraph, we have mentioned that immediately after the Indian independence in 1947 and during the reorganisation and redesign of the civil-services, the present IPS (i.e. Indian Police Service) as an 'All India Police Service', turned out to be the new 'incarnation' of the erstwhile IP, i.e. the Imperial Police, which was functioning as a kind of the all-inclusive—the earlier colonial-era's Indian sub-continental-level higher Police-Service—for ensuring policing of this vast subcontinent under the British-rule and for safeguarding the British imperial interests and stakes, and thereby within the well-coordinated and also the overall much larger gamut of the then multi-continental British Empire. Hence in fact, the erstwhile British designed 'I.P.' constituting the then colonial Indian subcontinent's higher police-service under the British, was supposed to be imperially colonial in its training, ethos and professionalism and thereby principally engaged and dedicated to serve and nurture the British colonial interests in the subcontinent, and to be hardly any people-friendly at all. As per the then prevailing paramountcy British colonial compulsions in the subcontinent, quite understandably from the British points of views, all these were very shrewdly intended, planned and proactively continued to be implemented in the entire subcontinent as being a very vast colony within the empire. Hence this kind of the policing service happened to be nothing else but, exclusively imperial in nature and only suited to be and in tune with the then autocratic and also largely exploitative British colonial-rule in the Indian subcontinent. **Moreover, another very pertinent point to note happens to be that no such national-level higher 'Police-Service' has ever existed at any time within the British-Isles themselves which the then British imperialists conceived and enforced in the Indian subcontinent during their colonial-rule**.

In fact the crucially pivotal-point to be clearly understood in these respects; happens to be that this erstwhile higher Police-Service, both in its constitution, design, professional ethos and work-culture; never happened to be even remotely suitable and constitutionally desirable for any truly democratic-system anywhere and thus it can very perceptively, as well as the no less objectively said that—any such imperially designed and developed higher police-service since exclusively favoured for any colonial-rule to serve only its own imperially alien interests—as a system had to result in being nothing but largely counterproductive only, in any vibrant and liberal democracy and that too, in spite of having well-educated and talented persons to man the same; as now rather so 'devastatingly being seen in the modern, sovereign and democratic Indian Nation-State. **Hence objectively speaking, it would quite naively be wrong to blame the selected persons now regularly being part of this police-service as IPS officers since as nowhere else but, the crippling faults exclusively lie in very system and thus being as systemic-failures in this presently existing also so much malfunctioning Indian Police Service in the sovereign democratic India. And here in fact lies the centrally crippling 'rub' in these regards, and hence the basic 'disease' needs to be correctly identified, diagnosed and suitably treated for ensuring the much larger and also rather permanently enduring benefits to the nation as a whole; and hence needless to**

say that just by superficially attending to merely some symptoms here and there aimed at some sorts of 'window-dressing' only, would never lead us fruitfully anywhere and thus in the process—any intended essential 'police-reforms' as being so much talked about at present in the country—would only continue to remain nothing but distant 'pipe-dreams' only. Moreover, if without effectively attending to the already above mentioned 'systemic-diseases and failures' first and foremost, even if any sincerely well-intended and also well-conceived action-plans for achieving any comprehensive 'police-reforms' are launched across the country it would rather pathetically look like efforts of 'putting the cart before the horse' and hence, perforce result in being counterproductive only. The problems in these respects, are too complex and deep-rooted, to be successfully tackled just by securing some judicial directions from the Supreme Court, or even by getting some administrative-laws duly passed through the parliamentary processes. In fact so far as now the democratic India's whole gamut of the needs of desirable policing, the existing systems of her nation-wide police-services are concerned, first their inherent problems with all their depths and dimensions and also their existing malignancies need to be objectively understood without any biases or prejudices before proceeding further for the needful.

At the outset in fact, it remains basically wrong to club-in together the IAS and IPS, as now it is naively being done in India. It is not something like—Army, Navy and Air Force, the three segments of the nation's Armed Forces—as being closely interrelated as well as, both logistically-cum-professionally remaining interdependent; so as to getting always integrative and clubbed-in together at the higher strategic levels of planning, operations and so on, to successfully serve the vital national-interests. In fact, if we again look back into the history as earlier we do clearly notice that the British in India never did club-in together the erstwhile ICS and IP in any manner as has been done after the national independence, and they could maintain their colonial-rule in the subcontinent here, as their sole imperial aim was to administer this vast landmass of the Indian subcontinent as a 'Police-State' only, where there was then no question of any democratic norms applicable and hence thereby their constitutional compulsions as interestingly, back-home for the British themselves. Further to look-after their very shrewd imperial-interests in the subcontinent and also so far as their colonial civil-governance in these regards were concerned, both at the district and higher levels, they firmly maintained the supremacy of the ICS over the IP officers in their official functioning and so on. In order to officially firm-up these, they ensured that the selection-processes for both the ICS and IP too, were kept quite different and separate in their forms and contents during the British colonial-era for the subcontinent. After 1947, the Indian policy-makers couldn't grasp these aspects in requisite manners and hence, so much vagueness and confusion are being seen all across the country in the constitutional phenomena of its democratic governance; and even the further irony is that, the politico-bureaucratic combines stationed in the corridors of powers all across the country, are refusing to effectively see-through these serious problems and, to ensure enduring course-corrections for the desired better civil administration in the best long-term well-being of the masses. In fact how can all these essentials concerning the paramountcy of the 'first-class policing' within the overall gamut of the democratic civil-governance, be ever made possible, if there are always many kinds of the uncalled for and even at times highly irrational and also quite comically unbecoming conflicts and the so-called clashes of interests, as having been seen to be the matters of routine between the IAS and IPS fraternities in the country. Instead of dedicatedly attending to their official jobs both at individual as well as at team levels, they tend to be fixated with their futile tensions and ego-clashes between these so-called 'sister-services' regardless of the many adverse consequences for the country and its democratic-polity too.

Precisely in the above contexts, **Debakinandan Mandal**, a former civil servant with the West Bengal State Government in India again in his another article titled, 'Tensions in Civvy Street' (*The Statesman*, Kolkata, dated 26 December 2013) has written that, "**Sense of arrogance of the IAS and relentless campaign of the IPS to be at par with the former lie at the root of the declining faith in each other. Their petty in-fighting, the use of respective service associations to run each other down, and the squabbles in the social circuit are now almost a regular feature which contributes to divisiveness in administration. Even the National Police Commission was viewed with suspicion by some IAS officers.... Critics contend that inter-service rivalry has intensified**. Members of a particular all-India service grab an opportunity to outsmart and belittle member/s of the sister service, settle scores, and tarnish its image in the perception of the public. The impression is confirmed if one considers this year's election to the post of President of Delhi's 100-year old Gymkhana Club located behind the Prime Minister's residence on 7 Race Course Road.... It begs the question whether the successor services have inherited these traits from their predecessors? Is it the legacy of the colonial past? **KF Rustamji**, former member National Police commission (*and interestingly, himself a former IP officer*), has answered these questions incisively. I quote, 'All that I can recall is fellowship, *shikar* (*i.e. hunting expeditions*) and picnic, from four till darkness, club nights and much laughter. Many of us in the ICS and IP lived in chummeries, shared our horses, our towels and socks, and even adored the same woman sometimes (*Well...interesting isn't it?*). In our style of work too, the same feeling was continued. It was well understood that we had to work together. The informal decisions, the lunch meetings, telephone conversations—that was all. There was no exchange of office-drafted letter between the two district officers, and no encouragement from the top to quarrelling."

So that was that historically, but then the Indian subcontinent was a colony and being the 'Jewel in the Crown', was an important segment of the then intercontinental British-Empire and hence, was meant to be administered and maintained just as a 'Police-State' as per the larger geopolitical compulsions of the empire, and also at the same time very shrewdly serve exclusively the British imperial-interests in the subcontinent. **But after the Indian independence in 1947, and soon this former British colony getting transformed into a sovereign, democratic, and federal democratic entity; with a highly visionary-cum-futuristic Constitution duly imbibed and enforced and wherein also, the numerously former 'princely-states' too got socio-politically integrated into what constituted the modern democratic India; and accordingly now the newly prevailing times along with many other overall national compulsions also, got totally altered.** And hence in these overall broader perspectives these indeed importantly relevant aspects concerning the present sovereign democratic India, the presently ongoing high-tech and modern-era added many new dimensions and challenges to the entire gamut of civil-cum-police services in the nation. Ironically these newly emerging dimensions into the job-profiles of these services were perhaps willfully ignored, by both political-class and also the personnel manning them. **And consequently, as we have been critically discussing these indeed crucial aspects of the civil and police services, we are also perforce noticing many lingering lacunae, as well as even the strangely crippling functional hiccups within these services so frustratingly always acting as major stumbling-blocks in the paths of national democratic governance and its dynamics, and also creating many avoidable problems for its multi-dimensionally paramountcy socio-economic developments. Thus, resulting-in many sad negations of largely those very constitutional provisions, which are principally meant to bring-in the comprehensive welfare measures at the very doors of the teeming masses all across the country. So it needs to be objectively realised that these problems are too complex to just be considered in any simplistic and superficial**

manners whatsoever, and hence there are no rooms available for 'indulging in any culture of fooling around' in these regards.

Now at this precise point of our ongoing discussions and also in the above specific perspectives, we may need to bear in our minds that for any public or government service-organisation; first and foremost the following three aspects are to be essentially in order: (a) first, the basic organizational-structure or the organizational-framework of any particular service must be appropriately designed and brought to finally in functional-shape with no vagueness due to any possible vested-interests, any loose-ends, and also must never have room for any possible prejudices of any kind. Secondly, (b) the kinds of quality-parameters of the personnel needed to fruitfully man the concerned service-organization, must remain clearly defined and mandated to ensure that there remains no room for any manipulation, no scope for scuttling the laid-down rules and procedures, as well as no possibility of any foul-play or bypassing the strictly laid-down criteria of suitability and merit. And lastly, (c) there must be well-defined procedure for the selection of personnel at any intake points. Now keeping these three moot points in our minds, can we maintain that the systems as in practice for the nation's police-service are suitable at all? In fact, they are far from it. And moreover, this is not only the prevailing case at senior or junior-level officers, but all the more so at the recruitment-stages of the police constables too. And hence needless to say that in spite of the all kinds of so-called 'thoughtful' efforts from different quarters in the government from time to time, the entire police-service all across the country largely remains derailed and defunct, and all these in turn, is found to be a sure recipe for all-round corruption and foul-play within this service itself. So in the end sadly, the Indian citizens are forcibly made to suffer, as they are mostly denied a first-class and healthily effective 'policing' in their societies, for which they do not only remain constitutionally entitled to always in the first place, but also they are regularly contributing to the national exchequer as tax-paying citizens and also as per duly laid-down the nation's laws and policies.

Now so far as the recruitment-aspects of the civil-police constables at their intake point is concerned in a vibrant parliamentary democracy, it would always remain totally wrong and counterproductive to club-in these aspects along with those of the nation's Army, Para-military force;—and also then as India being a federal entity having many states within—those of any state's armed-police outfits. In no other truly developed, liberal and matured democracy; such a non-starter kind of system is in force for the civil-police recruitments. This is always a very crucial aspect for any democracy to succeed and be effective healthily at the grassroots level onward. But in India even now, a very sad scenario is indeed so much regressively lingering; as quite ironically and naively too in these specific contexts, **India is still refusing to successfully come out of the otherwise already obsolete and absurd yoke of the colonial-era; and thereby consequently the civil-police outfits at the constable-level itself to begin with, is still suffering from basically the so much hated and also ant-people colonial mindsets.** As in western and any other matured democracies, the each member of the civil-police constables as a body, is supposed to be educated, responsible, be of high professionalism and character, be always responsive to the people's feelings and legitimate expectations and then; also be in the know of law, and naturally thereby always, unfailingly be able to provide first-class policing-service to the peoples in any corner of the country. But sadly the image of the entire civil-police outfits all across India as a sovereign, liberal, and also a fast-developing democracy, remains far from anything desirable in these regards. Moreover, a civil-police constable should be considered and addressed as an 'officer' rather than be comically seen as—in fact what to talk of the Army-solders—mostly nothing but poor caricature of even the nation's Para-military jawans (i.e. soldiers), in his appearance and uniform. No wonder that unlike in Western democracies, civil-police constables are denied any befitting importance, respect and status in the society. And then they in turn, not only remain frustrated, having no sense of honour and self-respect, demoralized and thus, just not being fit and capable of providing any high quality response and policing-service to the citizens of India. Further the colour, pattern of uniform and the cap etc of the civil-police in a democracy, must never be khaki, which is always considered military in nature; but all these should be more soothing, well-matched, people-friendly, appropriate, and also impressive as well.

Now in order to meet the above mentioned qualitative-cum-fitness criteria of a civil-police constable to be deservedly addressed as 'officer' by the citizens, he/she must be properly educated and multi-dimensionally well-trained, including being skilled in handling of the small-arms and trained in unarmed-combat as well. In India, it is indeed possible to compulsorily lay-down the prescribed educational qualifications as not only a graduate but also being a law-graduate too. To ensure this, their pay and perks must remain appropriately befitting to begin with; so that quality materials in sizeable numbers get attracted to respond for enthusiastically going through the prescribed selection-processes. In India now, there is no shortage of such manpower for the purpose and hence, the need is to comprehensively change the policies and systems; so that the successful candidates after completion of training and as in the Western democracies, become fit to be called as 'officers' from the beginning of a respectful career of civil-police constables itself. In addition, their physical fitness as well as other allied attributes too, must also be as per the prescribed norms. Hence as their careers and experiences advance from among themselves, suitable candidates can be picked-up for promotions to the higher ranks in the civil-police departments and thus bringing-in the much sought-after overall police-reforms in the country. Then administration and also the command and control systems of such civil-police outfits also need to be suitably decentralized form the district-level onward itself; so that any undue political interference, corrupt manipulations, and their unlawful/unethical handling too, get substantially minimized.

The prevailing regressive culture of irrationally aping the Army, as so 'foolishly' being practiced in the Indian-police departments from top to bottom everywhere must forcible be done away with, before more and more disservices are being perforce inflicted onto the peoples all across the country. It is a peculiarly comical but the no less sad, and highly damaging scenario still prevailing in the Indian police departments; where 'a crow while trying to copy the style of walk of a swan, is developing the tendencies to forget even its own pattern of walk.' These are not any superficial or small issues but irrespective of the any kinds of biases or entrenched vested-interests, they do remain always crucial to the very fulcrum of the entire gamut of the democratically essential rendering of the very high quality, skillful as well as, professionally dynamic arts and crafts of the civil policing-services to the peoples residing in every nook and corner of the modern, democratic and also fast developing Indian Nation-State, as being seen as a model particularly in South Asia. Now every civil Police Station or even an Outpost as far as possible and also on the need-basis professionally, must have district armed-police contingents of suitable strengths at its beck and call to reinforce its muscle-power needed as an effective deterrent or even to be judiciously used in special circumstances under higher supervision and directions, for maintaining law and order in its area of responsibility. And at the same time without any kind of unwanted confusion, any such district armed-police contingent whenever having been called for deployment, must always remain under strict command and control of the senior most member of the civil-police officer irrespective of any ranks, etc. and also at the same time, under the overall higher-level professional-cum-operational command and control of the concerned district's Superintendent of Police. Any further district-level civil administration-related matters should come under the civil and field-level official authority of the concerned District magistrate. But it would always remain essential and administratively paramountcy for the District Magistrate to remain always in active consultation-cum-cooperation with his Superintendent of Police, in the crucial interests of the policing, as well as also in tune with the constitutionally highest interests of the State. In addition, it would always be part of the official responsibilities of the District Magistrate to befittingly cultivate and acquire the trusts of his Superintendent of Police. Nevertheless, it would remain equally mandatory on the part of concerned head of the district civil-police to be always actively being reciprocal in these regards. And thereby the multi-dimensionally essential civil administrative culture and traditions, in course of time would get healthily cultivated and hence also be quite desirably groomed.

The civil-police outfit organizationally should have to begin, the police constable as an 'officer' at the grass-root level, then corporal, sergeant, sub-inspector, inspector, Deputy Superintendent and finally, the District Superintendent at the district level and all these personnel must come-in from the suitable and experienced constables themselves in appropriately prescribed numbers, to man the organisation in the entire district. And the same system should prevail in all districts of the country. Moreover as per the present socio-cultural scenario in the country and also the people's needs for mobility everywhere, these recruitments-cum-placements may be made cosmopolitan in nature and content as far as possible. There is indeed no shortages in the availability of young, talented, and also educated manpower in the country today and hence; if the management of the entire systems in these regards is imaginatively carried out with desired professional vision, then any likely deficit in the processes of civil-police recruitments at the constable-level itself, could quite satisfactorily be ruled-out. At the district-level itself to start with, there should be an appropriately formed committee to control and supervise the performance of the District Superintendent of police. This district-level committee should have as members as suggested: the District Magistrate, the members of Parliament, whose constituencies are within the district, the Judicial Commissioner/District Judge concerned, leading citizens/personalities in district and, belonging to the different walks of life such as, medical, law and academia (i.e. well-known and highly respected doctors and lawyers, academic persons, like a senior professor or Vice-Chancellor of repute and standing, whether serving or retired), Mayor of the district-town and then the concerned Superintendent of Police too, who would officially be acting as member-secretary of this Committee, Board or the Governing-Body as since officially designated. The socially senior most and also the most revered among these members should be the Chairman of this committee. But the government officials and also the political personalities being part of this committee, should be debarred from being chairman, and thereby the chairmanship should only go to an independent, very senior and also socially well-respected member of this committee as per above.

Similarly above the district-level, committed and close-knit committees could be formed with still higher-level police officers of ranks of Senior Superintendents/DIGs, being both professional advisers as well as acting as member-secretaries of the concerned Committees. Now at the State level, the highest administrative authority should unambiguously rest with the Chief Minister, who would mostly be bound with the advice of the concerned State-level committee in this regard; except in any constitutionally extraordinary circumstances, which would in any case be only rare exceptions rather than rule. Further at the State-level there should no police-rank of IG and DG with being totally counterproductive military connotations-cum-displays here and thus, being always out of tune with the crucially important aspects of the civil policing of societies in a truly vibrant and liberal democracy, like the modern India. But again at the State-level, a still senior police-officer should be the Police-Adviser to the Chief Minister and at the same time, also officially act as member-secretary of the concerned State-level committee in this regard. The status, pay and perks of this Police-Adviser in the State won't in any case be, lower than the present DG of Police. Further, these senior police-officials would always gradually come as per the new police-hierarchy and also, from within the ranks of the civil-police, so that they are duly and proactively equipped with the rich professional knowledge, experience and the wisdom too, in the highest interests of the overall democratic civil-administration in the country. In addition, a sitting Justice of the concerned State's High Court and also as nominated by that High Court's Chief Justice himself, should also be an honourable member of this State-level Committee, being in protocol next only to the Chief Minister. Needless to say that, the Chief Secretary of the concerned State too would also officially be a member of this committee.

The above thoughts are sketchy but indeed quite precise outlines only, and are intended here to bring-in some specific suggestions to trigger more detailed and also, both critically and also comprehensively focused discussions with unwavering-cum-uncoloured objectivity in the modern, fast developing and democratic India's best national interests; so that still so much regressively lingering colonial hangovers in these regards, are effectively wiped-out once for all. Hence these in turn, would also multiply bring-in—the so far always totally elusive and also unprecedented fallouts as numerously far-reaching improvements—in the entire policing-patterns as well as, in the presently existing largely very corrupt, feudalistic, and the dirtily arrogant mindsets all across the police departments in the country. Such indeed so much quite justifiably hated mindsets since still thoughtlessly allowed to continue from top to bottom in the police departments, are not only instrumental in always nauseatingly breeding and promoting all sorts of corruption and attitudinal criminalities, but also promoting the highly damaging vested-interests among the policemen and their officers as well. In any such prevailing scenario, any talk of professional spirit-de-corps among the nation's policemen, would only be a wishful-thinking and thus also a naïve exercise in futility. **Hence the above discussed alternatives, if comprehensively formulated and implemented in detail would in turn trigger such all-round and also multiply magnificent police-reforms, which do remain unthinkable for the time being in the country. As quite appropriately said earlier that these are no simple subject-matters and hence any half-hearted approach based on some half-baked ideas and that too, without comprehensively looking into the many lacunae and diseases in the existing police-setup in the country, having their rotten and still existing roots in India's colonial-past would very superficially bring-in some 'window-dressing' only and thus, with no larger good to the nation's governance at all, and thereby the peoples in India would feel again nothing but cheated.** But at the same time, no individual mind anywhere can formulate any comprehensive blueprints for desired solutions and hence needless to say that, expert-level committees having members from among the peoples coming from different areas of human endeavour in the country, would need to go into all the suggestions and details and also need to bring-in more new ideas, for objectively working onto the much needed essential solutions and, reforms-cum-improvements in these regards.

It is obvious that not only very strong criticisms would naturally come-in, but also many kinds of the hue and cry would surface, especially from the variously entrenched vested-interests and also from other quarters too. The entrenched, as well as since already so much monstrously grown IPS-fraternity in the country would come out first to condemn these ideas, calling them nothing but absurd and bogus; and for the time being that remains always understandable. It is true that if now the sovereign and democratic India's civil-police is revolutionized and also reorganized in its entirely new reincarnations as suggested above then obviously the IPS as the presently existing 'All India Service' would totally lose its relevance and hence, would become redundant and thus financially wasteful. **But the highest long-term national-interests should unfailingly always override all other considerations, as per the changing timelines of the nation's living-history—where always nothing but irrefutably the nation and its highest interests—must reign supreme. Hence, no kind of the any 'vested-interests', whether incubating and coming from among the nation's political-classes, the military, the bureaucracy, the police, or constitutionally any other similar socio-political and official segments whatsoever can ever be permitted to ride over the always unquestionable supremacy of the nation's vital-interests, as well as the peoples' well-being in a democracy.** We do need to historically be aware of the facts that, during the now already bygone colonial-era; the British imperialists while ruling over the Indian subcontinent as their colony, had very shrewdly and imaginatively created the IP and that too in a very restricted manner, as the colony's higher police-service for ensuring command and control over the quite sizable native-constabulary; for policing and administering the subcontinent as a 'police-state', and obviously within the larger framework of the British Empire's own vested-interests. Moreover interestingly, the same British back home never felt any need for any such the so called *'elite police-service'*; for democratically governing their home-lands. It seems to be indeed ridiculously strange that after more that so many decades of India's national independence and that too even after the dawn of 21st century onto the entire humanity itself, the politico-bureaucratic combine in India, now ironically no more enjoying any tangible confidence amongst the peoples, but nevertheless also constitutionally occupying the decision-making seats in nation's corridors of power are still not caring to see through these crippling lacunae in the police-departments and thus, begin to make comprehensive efforts to effectively solve the problems once and for all. It is a great paradox that on one side the democratic Indian Nation-State is otherwise very befittingly aspiring to make multiply great-leaps forward, but on the other hand, its civil-police departments still continue to be in a kind of the very pathetic 'time-warp', and thus so much regressively continue to remain a 'non-starter'. And needless to say that consequently they also again pathetically remain incapable of proudly identifying and cultivating their indeed unquestionably exclusive professionalism, always paramountcy investigative-skills, praiseworthy honesty, and also their special kinds of leadership qualities as well. Moreover there are no alternatives to these inescapable imperatives at all.

The colonial-era is all gone into the dust-bin of history and presently since the 1947 itself, India now being a sovereign and thriving democracy, but instead of winding-up these outdated legacies; the new power-lobbies of the politico-bureaucratic combine for their own highly selfish vested-interests, so callously disregarding the long term national-interests as discussed above, and thus very craftily have not only allowed the prop-up of the IPS, but also permitted it to keep expanding in hugely wasteful as well as, so much counterproductively bizarre and monstrous proportions. This particular and indeed very much intentional lapse or even sin on the part of the independent India's new power-centres have consistently gone a long way in severely damaging the multiple-aspects of her policing-outfits and thus consequently, the peoples of India have been not only been suffering but also so much avoidably being made to bear the wasteful financial burdens too. All said and done, but at the same time, all these are no deriding the multiple-qualities of IPS officers at their individual-levels. It is true that as in case of others in the country, the existing IPS-fraternity too has its own quite sizable share of 'black-sheeps', and thus tragically putting them into the all kinds of disreputes too. Further it is also equally true that, many IPS officers as individuals have been known to be fine specimens of patriotism, high intellect and education; having wonderful creative qualities and also, known to possess quite high administrative abilities. No wonder that, many of them after retirement from service have been known to also excel in many other fields of human endeavour. Hence a bulk of them can be fruitfully engaged in the IAS, public-sector organisations, Diplomatic-Corps, appropriately in academic areas, and also in the corporate world as well.

Hence, what precisely is intended here in the country's best national-interests, is to establish absurdities of the 'IPS as an All India Service' as in any other matured, vibrant, and developing/ developed democracy, no such futile national-level police service exists; which is nothing but now an absurd legacy of the British colonial-era, which has long outlived its utility; even if it ever was so in any manner for the then British imperialists in the Indian subcontinent. And all the more paradoxically in the modern and democratic India; the already 'redundant' IPS, as since so much mindlessly permitted to quite irrationally expand rather without any essentially governance-oriented checks and balance or necessities; soon the IPS officers themselves, in fact the no less thoughtlessly began to rather 'comically' copy—the military-trappings, badges of ranks, ceremonials and many other 'displays' (such as flags, star-plates, ceremonial 'pilot-cars' and so on) etc—of the senior Army-Officers, without any meaning or suitability, rationale, and also any kind of the professional relevance specific to the senior Police-Officers. In addition many of the IPS officers rather all the more ironically, themselves do remain knowledgably aware that in the Armed Forces, these aspects are both professionally-cum-ceremonially, are always differently defined and maintained applicable to the senior military-officers' command and staff appointments appropriately since centuries; and thus all these do have their different traditions-cum-situations' based soldierly significance and also relevance. Further the colour, designs of badges and insignias, etc., of these military symbols are totally different for the Army, Navy and the Air Force, and each one of them is also varying for different units, regiments and other similar military-segments within them too having their own different history, legacies and traditions, since going back to the centuries of their historic pasts, and at the same time into all these aspects always exclusive to them, only the military personnel belonging to the above mentioned three wings of the Armed Forces, do always rightfully have their soldierly-pride, sources of motivations and so on. Hence these always exclusively military-aspects, just quite befittingly do not have any meaning or rationale for the civil-police and its officers, and therefore obviously no relevance to their otherwise the no less laudable profession and identities.

Precisely in the above multiple-contexts, in an article written by **Mr. R.K. Raghavan** he has mentioned that, "**When Sir Robert Peel, the then Home Secretary who went on to become the British Prime Minister, created the London 'Bobby' through the Metropolitan Police Act of 1829, he was sowing the seeds of a system that was to be the model for builders of police forces in several countries. Peel's hypothesis was that if you succeeded in bringing in the best of youth into law enforcement, the community had little to worry about in terms of order and crime. The Bobby (named after 'Bob', the Home Secretary) was soon to become a household name for dignity, alacrity and personal integrity**. This was, however, the creation of time when England was homogenous society, with little or no immigration from the rest of the world.... Things were to change substantially in the later part of the 20th century, when burgeoning opportunities attracted people from a variety of cultures and different parts of the globe. Aided by a liberal and lax immigration policy, the country was soon to be a melting pot of religions and faiths that made policing a complex task. The expanded entity of the United Kingdom, which brought England, Wales and Scotland together, is now a political arrangement that calls for high skills and sagacity, not only in general administration but policing as well. It is a test of the U.K. Police how well it will keep different sections of the community—the British white and black, South Asians, Africans and a host of others—separate, yet together, so that they can retain their identities while, at the same time, understand one another without getting into an adversarial relationship. This is a gargantuan task that has weighed down the U.K. Police, especially in London, during the past few decades. The problem has been exacerbated by the recent happenings, which have virtually polarized society between one small group that relies on terrorism as a means to bring about change and the much larger one, which places a premium on order and tolerance.... The policing system conceived more than 175 years ago by Peel has generally coped with the pressures, especially of the kind seen immediately after the July explosions this year (*i.e. in 2005*). **There are, however, few signs of wear and tear that cause some anxiety to the Home Office, which administers the police. This department is convinced that certain fundamental reforms in the style of policing are warranted if society is to be held together. The assessment especially, is that, while tough measures against terrorism are required, the emphasis should be on taking care of the sensitivities of individual groups within the community.... What is remarkable is the open debate that governs policing matters in the country, and a willingness to adapt to technology. My critical appraisal of the U.K. Police is that, underneath all current exercises at raising standards, there is a popular faith in the quality of current police leadership which will help to support the much needed reforms, that is, once a consensus emerges about what such reforms should be**. (*Well, all that is so far as the U.K. Police is concerned; but can anything like these be even remotely ever said—so far as the presently existing Indian civil-police affairs—are concerned? And the rest is all history as well as, the currently indeed so well-known overall scenario in these regards, as so disgustingly being witnessed by the Indian citizens all across the country.*).... **Armed with a degree from Oxford, in the elitist subject of Modern History, one would hardly expect a bright youth to opt for a career in the police force, especially in the U.K. where, unlike in India, one has to start from the lowest rung of Constable (*Well, the system prevailing in the U.K., is always the befittingly correct system in any healthy, matured and dynamically vibrant democracy; and certainly not the presently existing defunct system in the Indian-Police departments, which remain so counterproductive as well as grossly unsuited, for various reasons; in the so much 'heat and dust infested' and thereby the still maturing Indian democracy.*)**. Peter (*i.e.* ***Peter Neyroud****, who the then headed the Thames Valley Police in the UK.*) attributes his decision to the influence of a classmate at Oxford who was a Thames Valley Inspector. His four-day vacation course with the force convinced him that his only place was in the police. He has never looked back since joining the Hampshire Police as a Constable, way back in 1980. And it was rightly so, because his progress within the police has been phenomenal. He rose to the position of Sergeant (equivalent to our Sub-Inspector) in 1986 and to Inspector in 1988. He went on to become a Superintendent in 1993, just 13 years after he was enlisted; a record that can hardly be beaten. In 1998, he moved to the West Mercia Constabulary, to be first as Assistant Chief Constable and later a Deputy.... Peter knows about the Indian Police Service (IPS), and its strengths and weaknesses. He feels that beginning at the foot of the ladder, as U.K. policemen do, lends, greater credibility to a leader vis-à-vis the other ranks. In fact, a major criticism against IPS officers is that they are hardly knowledgeable about field difficulties, because they start at the middle-level, and their exposure to the grassroots of policing is only for a brief spell during training in the Districts. Peter enjoys patrolling as a constable, even now as the head of the force. How many would do this in India is a matter of debate." (Raghavan, 2005). (***Well perhaps none whatsoever. As informatively***

revealed above, these are some of the numerously very distinguished as well as, the indeed highly laudable reasons: as to why and how the U.K. Police, professionally so befittingly never look at always a so different professional world of the British Armed Forces in any manner; and never try to absurdly ape them for rather comically enhance their socio-professional image and status, or even remotely try to achieve any superficial distinction whatsoever. That has been and still is the British-Police, to always remain dedicated, highly competent, technically suave and also unwaveringly being truly proud of their esteemed profession, having its great historic pasts and the legacies as well. Do we now need to say anything more here? Now it is up to the Indian civil-police and its politico-bureaucratic masters to do their own sincere soul-searching for the needful, in the modern and fast developing democratic India's highest national-interests.).

In a number of the above preceding pages, we have here somewhat managed to critically discuss the numerously crucial aspects of the modern and democratic India's apex-level civil-services (i.e. especially the IAS and IPS) as they are presently existing and also as far as possible; without jeopardizing any objectivity in these regards. It is true that all these since always nationally so much important aspects are indeed very complex subject-matters and hence, they need to be both extensively, as well as the no less constructively, debated openly without any prejudice, all across the nation, so that a much desired process of 'thinking out of the box' gets patriotically triggered, and thereby throw-up a comprehensively new 'blue-prints of reforms', which have so far always very dauntingly eluding our Nation-State. It is now quite irrefutably true that these civil-services are now in urgent needs of the many comprehensively drastic-reforms in the highest national-interests as indeed, all these are becoming rather abnormally late since the Indian peoples have already been made to suffer unnecessarily too long. As the time keeps moving-on regardless, multiple-kinds of reforms become crucially essential and hence these can never be wished away. Moreover why only the civil-services, different kinds of the almost equally important reforms and professional improvements have also become equally pressing in the nation's military-services too, as we may recall that precisely these have also been discussed and analysed in some considerable details earlier. And in the process, we have also been able to first identify and critically discuss some newly emerging kinds of unbecoming shortcomings in the indeed somewhat 'exclusive' professional domain of now the modern and democratic India's Armed Forces as well. In these overall perspectives in fact, the moot point that always rather supremely prevails that, in any presently ongoing kind of the 'strategically critical journey in any broadly specified perspectives', nothing more important but truth and only the truth, and regardless of pertaining to any area of governance-related activities/initiatives of any kinds, must be allowed to prevail and thus indeed therein, facilitating any desired reforms/course-corrections.

(viii) Other Multiple Reforms

In the similar manners, what is so clearly seen in today's India that along with the numerous kinds of reforms/improvements/course-corrections in the Government's many areas, the equally pressing judicial-reforms too are also called-for, and in addition at the same time these are also very frankly been critically discussed in not only the media and academia, but also all the more loudly in the nation's Parliament as well. Of late interestingly indeed, the very strong intent to press for many kinds of the nation's governance-related reforms and delivery-systems in the best possible interests of peoples, and then also some kinds of the institutions-related improvements/course-corrections nationwide, relentless movements have now been seen to be going-on, under the remarkably proactive leadership of a well-known, as well as highly respected

social-activist and also dedicated reformer: Shri Anna Hazare. These in turn have also succeeded in rather forcing the Parliament to pass some new laws and resolutions as well. How fruitful with all these developments be in the Indian citizens' best interests, only time will tell; but so far all these have indeed been able to trigger some phenomenal socio-cultural-cum-political debates and churning for the time being.

Now while coming back to the various aspects of any far-reaching reforms in fact quite interestingly we now come to a very befitting remark of an American author, **David Brin**, and that being, *"**When it comes to privacy and accountability, people always demand the former for themselves and the later for everyone else.**"* Well this is perhaps inherently typical to the human nature and, which remains always universally true as well. In fact it is so powerfully true that in India now, when anyone tries to bring-in the subjects pertaining to reforms, as since already having been discussed above in some specific contexts; David Brin's above quoted 'statement of fact', does always come into play rather unstoppably in all sorts of manners and at times quite interestingly, some 'affected' as well as also vested-interests-oriented persons, do not shy away from even quite untenably taking 'shelter' behind the same 'statement of fact'. **Well that's the way it is but regardless of all these, reforms in various places—becoming now more and more inescapably paramountcy and urgent—have to be very prudently, comprehensively, and proactively thought-over and also indeed, relentlessly pressed-for and then fruitfully implanted at the earliest, in the country's highest national interests**. Nothing is above paramountcy national-interests and hence, nobody or anybody can ever be allowed to superficially and also somewhat subjectively gloss-over his/her vested-interests and thus hoodwink such reforms. As emphasised earlier that no amount of mainly the superficial 'window-dressings' with nothing but half-baked ideas are going to work and therefore the nation's powers that be, the opinion-makers, as well as the decision-makers, would not only be indulging in many absurdities or a 'culture of fooling-around', if they don't listen to these imperatives, but also unpardonably doing great disservices to the largely innocent but now so dreamily aspiring: the peoples of modern and democratic India. More than 65 years have passed since the nation achieved its independence from colonial subjugation and slavery and that too, through such memorable non-violent means; but sadly the same nation is now getting submerged in all sorts of internal socio-cultural-cum-political violence. But regardless of all these, the people occupying responsible seats in all across the corridors of power are still remaining confused and thus, democratically incapable of effectively securing the requisite hold over the nitty-gritty of governance. It is indeed a very depressing saga and that too now in the presently ongoing 21st century and thus, giving not only a very unfortunate but also a highly unbecoming display of the functioning of Indian democracy before the modern and very discerning international community. It is high time as since already being very late, that people in the nation's corridors of power; effectively learn the essentials of modern-era's geopolitics and also, the 'rope-tricks' concerning the nation's democratic governance.

Now concerning modern India's essential aspects of the judicial-reforms in the present 21st century: the editorial-column of *The Times of India* (Bangalore Edition, 3rd July 2014 issue) gives very perceptive insight into the precise imperatives by stating that, "By focusing on external threats, the higher judiciary has distracted itself from dangerous threats to judicial independence that result also from its opaque collegiums system of appointment of Judges. On part of the government, its job is to build not destroy institutions. Given that collegiums system continues to be in force, the NDA government should not interfere in the process of selection of Judges. Instead, it should strive towards long-term reform of the judiciary.... The government could consider **Arun Jaitley's** suggestion during the previous NDA's regime that a National Judicial Commission (NJC) be appointed to select Judges. It is in the context of frequent criticism that the collegiums system is a clubby arrangement that lacks transparency and accountability, making the higher judiciary the sole arbiter of appointments, that Jaitley's suggestion deserves serious consideration. The **Narendra Modi government** could amend the 2013 Judicial Appointments Commission Bill by expanding the proposed seven member authority to include representatives from judiciary, executive, legislature (including opposition leaders) and eminent legal academics drawn from civil society. This will ensure that the judiciary is not exempt from the requirement of accountability to the people it serves.... Since the judiciary is a co-equal branch of government, the proposed NJC give due weightage to a public consultation and foolproof vetting process whose records should be publicly available. The judiciary should not have any reservations on vetting if it believes that judicial independence does not excuse it from compliance with appropriate standards of accountability. But it is equally critical that some of the Supreme Court's senior most Judges themselves define and communicate in an open manner the standards by which they must be selected. The public has the right to expect that Judges will be willing and able to behave in accordance with the highest professional and ethical standards."

In precisely the above multiple-contexts, one is reminded of an indeed so thoughtful saying of **Thomas Jefferson** that, ***'To consider judges as ultimate arbiters of all constitutional questions...would place us under the despotism of an oligarchy.'*** Further in these multiply important perspectives, and that too specifically concerning India's judicial reforms *The Times of India* again in its editorial column, has the no less emphatically stated that, "Allegations levelled by former Supreme Court Judge **Justice Markandey Katju** (*which have also been clearly supported by a couple of more retired Supreme Court Judges too.*)—against three past Chief Justices of India and the PMO under UPA—have reverberated the Parliament, prompting the Narendra Modi government to declare that improving the system of appointment of judges is imperative.... The disclosures strike a hammer blow at the strongest argument made in favour of continuing todays clubby and opaque 'collegium' system for appointment of judges—that preserves judicial independence. They should spur the government to swiftly institute long-term reforms that will not only cleanse the taint of corruption eating away into the vitals of an institution that survives on public trust, but also take a slew of interlinked measures to make delivery of justice fast, effective and equitable.... NJC (*i.e. National Judicial Commission*) should include representatives from the judiciary, executive, legislature, eminent jurists and legal academics from the civil society, suitably weighted to preserve judicial independence. Deliberations of the commission should be recorded and available for public review. NJC should also be empowered to examine and investigate instances of misdemeanours, corruption and malfeasance and order appropriate punishment. (*Well, won't that be rather stretching it too far and hence; won't that be constitutionally more 'not legal and tenable', and also quite befittingly—the administration of justice-oriented and thus apolitical—the 'punishment' part, if any, is ensured through a proper and fast-track proceedings in a separate, credible, and also independently constituted Court for the specific purpose; and in addition also, with prudently built-in safeguards against any possible legal hassles and confusion? And at the same time, any concerned 'accused/tainted' judge for the time being, should be debarred from holding the court.*) This will ensure that the judiciary is not exempt from the requirement of accountability.... Alongside institutional reforms, efforts must be made to modernise and streamline the judiciary—bottom up—so that lower and higher courts run efficiently and professionally.... Legal records should be digitized to make them easily accessible. Long seasonal vacations should be abolished: (*since still continuing—the already an absurd and thus wasteful legacy of now the long bygone colonial-era—which was indeed feudal-cum-imperialistic and hence, totally* *undemocratic as well.*), and courts should function round the year. It should not be lost on either the government or judiciary that the real test of judicial performance is when courts serve to the general satisfaction of the public." (24 July 2014, Bangalore edition)

Based on our fairly critical and also quite prolonged discussions as since recorded above, we may precisely conclude that in the democratic and indeed fast-developing India's highest interests, and also for constitutionally effective and no less reassuringly result-oriented governance, now multiply sectorial, quite prudently comprehensive and enduring reforms are urgently called-for. Today at the national-level—relentless and extensive debates must be carried out to fruitfully bring-up the multiply-new ideas—towards formulating the multi-dimensional action-plans—for triggering the multi-sectored reforms, as since already critically deliberated-upon in so many the above preceding pages. We have talked about the effective and also multi-sectored reforms, indeed with constitutionally multi-dimensional action-plans; always keeping in view the nation's long-term and vital interests, in the so much high-tech-oriented and also accelerating-cum-fast-moving modern-era where ruthless competition remains the order of the day. And moreover particularly today in the Indian contexts, the multi-sectored reforms are not only urgent but also, they are inseparably interconnected, as well as mutually influencing too. These are hard realities and thus, cannot be wished away. **Now in a recent development, Parliament has already passed the much needed 'Judicial Appointments commission' and it is hoped that this would subsequently bring-in many reforms in higher judiciary, through selection of better candidates as suitable Judges**.

For example, if the politico-bureaucratic gamut of civil-administration and also the entire civil policing-system in the country continue to be so corrupt and defunct, as being the case today, then how can even the ideally reformed-judiciary be able to serve the people, as being so forcefully desired and debated now? Similarly, how can the overall gamut of indeed the multi-dimensional democratic-governance of the country be ensured for the peoples if both the internal, as well as external security aspects of the nation, with its always multiply dicey and also highly complex dimensions are not effectively, professionally, and also the no less apolitically—catered for comprehensively, along with also eliminating all kinds of the deeply entrenched vested-interests therein—with always keeping the vital national-interests politico-strategically in sharp-focus, at the country's highest political-level leadership, having the desired vision and the geopolitical tenacity? In addition to all these, the **well-planned and comprehensive reforms in the all-round Education-Sectors** too, are no less importantly crucial. But precisely nevertheless, the peoples too, since belonging to the all strata of Indian societies, do also need to shed their age-old and also so much regressively lingering lethargic inertia both in their thoughts and actions, to create the vital synergy that the great and highly effective beginnings are made without any callousness of any kinds anywhere. **Leadership, discipline, and the impeccable integrity,** are needed at all levels and everywhere. **Media too needs lot many reforms** and also, required to cultivate seriously constructive attitudes towards its nationally paramountcy profession. **There is indeed colossal corruption in media-circles too** and, which needs to be sincerely addressed to by themselves because the issues involved are so sensitive that, any kind of well-intentioned and even slight interference from any outside areas, may amount to the uncalled-for encroachment onto the freedom of press as guaranteed by the Constitution. Hence these aspects all the more bring-in heavy responsibilities onto the media's shoulders only and obviously therefore, along with its nationally vital profession, media itself needs to reform and live up to the peoples' expectations. Hence concerning all spheres of the modern democratic India, mantra of the multiple-aspects of patriotism, leadership, unwavering discipline, and also impeccable integrity, everywhere and at all levels cannot be overemphasized. Indians as citizens of modern democratic India, are quite befittingly capable of doing all these and so they must ask themselves, as to how long they would remain stuck with all kinds of 'inferiority-complex and the so much damaging inertia-syndrome'?

In precisely the above multiple-perspectives, Mrs. Kiran Bedi, India's first woman IPS officer and also an internationally well-known thinker and social-activist has so thoughtfully mentioned that, "The headlines in the newspapers and visuals on television are becoming more predictable and depressing lately. Violence and bloodshed grab the headlines. Gang lords, terrorists and their groups today appear to have reserved space in the media to their advantage. For the media, it means a huge circulation, with sufficient financial returns.... And this is exactly my anxiety. Why are we not giving choices to the viewers and readers? Is it because we have none left? Who are the directors of the focus lights who dictate the priorities? Some of these questions got partially answered in a recent experience with the electronic and print media. The occasion was meeting of Magsaysay awardees and the city was Jaipur. Over 20 laureates were present there. This time the Magsaysay Foundation trustees decided to hold their Annual Conference in India. It obviously was an important event for the city media, if not for the country. A press briefing was done by the trustees.... Later, as soon as the electronic media left, the scribes also came in large numbers, wanting to ask questions. And the queries which they put to us gave us some interesting insights into the priorities of some of the members of the media.... And the final question (now the trustees intervened to keep the briefing short) was, 'Why do you select only controversial individuals for the award?' To which the trustees retaliated by saying, 'because in our view, controversial individuals are doers'.... When does leadership, begin? Can we not demystify it? These questions kept coming back to me as I heard **Lee Kuan Yew, Rudy Giuliani, Francis Ford Coppola, and Warren Bennis** speak at the Global Brand Forum in Singapore in 2004 on 'Defining Leadership', where I, too, was a speaker and a panelist. I felt that we said the right things about leadership but took it to a conceptual plane from where it needed to be demystified. In addressing the question about when does leadership begin, we may succeed in placing it in the hands of all. What made me eager to find an answer to this question was when the learned speakers told the audience—consisting of over a thousand bright and keen people—that leadership is about mission statements, inspiring others, skills, communication, optimism and vision, strategies, hard work and discipline, team spirit, direction, risk taking and courage, observation and listening, values and integrity. I pinched myself to ask when and how do these qualities come to us. Do they fall from heaven? Are they learnt or acquired, or is one born with them, or does one develop them as one grows in life? And the one quick answer—in the form of a question—that flashed into my mind was that do we not begin first with 'learning to be led' before we grow up to lead others? Also, do we not 'learn to lead our own selves, before we get into 'leading others'? I found the answer that demystified leadership for me. It was in the 'Power of I'. In other words, it was in self-growth—growth through one's parents and teachers and one's environment. All of which teach one to love and be loved. Give and receive. Share and learn. Inspire and be inspired. Accept and accommodate. Listen and express. Cry and laugh. Patience and anger, experience life and death. Win and lose. Sweat and rest and then: fear and courage, respect and humiliation, honesty and deceit, enjoyment and responsibility, task and performance, friends and adversaries. Mine and thine.... My thoughts went on a rewind mode and I started to ask myself the following questions: Did leadership mean following one's inner voice? Is it about sowing the right seeds of leadership by parents, teachers, and the environment, or is it a question of self-seeding? The additional answer was: consistency and authenticity along with the spirit of giving to those who did their own gardening persistently. **They planted for others from day one. That is what made Gandhi the Mahatma, Gautama the Buddha, Mother Teresa into Saint Teresa, and Nelson Mandela, Ramon Magsaysay, Vivekananda and Florence Nightingale for all that they stood for**. They were what they were within. They were followed without seeking followers. They were visible and living examples for all times. And they were authentic. They were their own gardeners. But from decade one onwards and they stayed on course. Leadership is simple. It's daily. It's persistent. It's selective and ethics based. It's vigilant and nurturing. It's just careful gardening." (Bedi, 2003, pp. 7, 8, 341-344)

How true and rightly so, as Mrs. Kiran Bedi is linking the leadership aspects to be the powers of being selective, as well as ethics-based and that too at the same time, these need to be vigilant and nurturing as all these aspects amount to be just careful gardening in the best interests of one's nation and at the same time, always being conscious of the well-being of larger humanity as well. In these highly complex and multi-layered challenges, in any authentic leader's capacity to be able to work out an innovative reconciliation, lies the keys to success. Now while keeping these in minds and then coming back to the earlier discussed various aspects as particularly being India-specific, it is historically true that during India's freedom-struggle era and then also continuing into approximately the first two decades during post-1947 period too, the nation was indeed gifted with a galaxy of top-grade political leaders—commanding also a very high intellectual prowess, learning, and vision too—who were charged by the history to very responsibly preside over the nation's 'tryst with destiny', during those critically formative years. To begin with then, what went wrong at the nation's top political-levels and that too, in the hands of these erstwhile history-making statesmen, resulting in now so well-known numerously colossal 'derailment': regarding systems of the newly independent and now the democratic India's politico-bureaucratic governance all across its sovereign landmass in today's changed world? Moreover soon after those great political leaders and statesmen departed from the scene, rot began to further set-in especially with Mrs. Indira Gandhi coming to power and consequently, all round systemic decay began to gather accelerated momentum. It is historically indeed a greatly tragic irony, which appears to be not easy to decipher and get to some enlightening answers. But the nation has to proactively struggle to remove the engulfing thick-fog, find requisite answers and thus be able to come to light by reaching fast at the end of this 'messy-tunnel', before it is too late.

With indeed well-intentioned aim to throw some directly sharp lights on the above discussed, multiple-aspects—concerning the very sad derailments of the constitutional aspects of India's politico-bureaucratic governance—**Mr. P.C. Parakh, a retired IAS officer and a former Secretary, Ministry of Coal. Government of India; has recently written a very revealing book titled *Crusader or Conspirator: Coalgate and Other Truths*,** which has caused many ripples in India's current political circles. **Not that the truths contained in this book were not known, but bringing-in some very unbecoming and direct blame onto the PMO of the then Prime Minister Dr Manmohan Singh and to his personal leadership aspects as well; has really been not only shocking but nationally very unsettling too.** This book containing first-hand account of a very senior IAS officer, again irrefutably proves as to the rots since already set-in the entire systems of the decision-making processes, concerning the nation's politico-bureaucratic governance, are indeed of very alarming proportions. **It is thus true that all kinds of fault-lines do very damagingly run all over the places not only in the government but also, in the Indian societies as well. Many memoirs have already since appeared, written by the modern. Democratic and sovereign India's both civil and military services' personalities and many more would keep-on appearing in the future too;**

but would the nation wake-up and enforce comprehensive systemic-reforms and thus, ensure the truly vibrant and healthy enforcement of the nation's Constitution?

In Mr. Parakh's above referred book; a well-known and very senior bureaucrat, **Mr. TSR Subramanian, IAS (Retd),** has very frankly mentioned that, "The bureaucracy man's posts at all levels, from the village to the block, to the district and all the way to the State and Central Secretariats; and is primarily responsible for the implementation of tasks handed to it by the political executive and the senior bureaucracy. At its higher levels, the civil service acts as an adviser to the political executive and assists in policy formulation. While wider national policies are, ideally, debated and directed by the party in power, administrative policies are established by the executive. By very nature of things, the work of the political executive and the permanent executive have a very large and complex overlap; their smooth and harmonious interaction is an essential prerequisite for good governance.... In short, entering politics is a business venture; much as a young entrepreneur would invest in a business, with the ultimate aim of making a fortune. With honourable exception, this has sadly become the rule of the day. Fortunately for the politicians, the courts generally are benign on the grounds of 'privilege of the legislator', and there is no equivalent of a SEBI to watch over malpractices, in short, there is no regulator for the politicians—we still await the Lokpal.... Parakh's book amply demonstrates, chapter and verse, the sharp deterioration in the standards of the Central Government. **The naked venality and corruption that is seen in the districts and in many state secretariats, clearly has reached the Centre. The rot is now fully set in, and is corroding the administrative process. Parakh has correctly noted 'most ills in governance arise due to corruption'. Herein lies the explanation as to why seven decades after independence, 70 percent of the population is in poverty, education standards and health condition of the populace in abysmal condition, and India is an 'also ran' among the comity of nations, rather be seen in the same league, rivaling if not excelling China and USA in world leadership. Parakh has shown a mirror to our system of governance**.... Suffice to say that the Prime Minister (*i.e. Dr Manmohan Singh*) comes out as a mute, ineffective spectator. Note that the same Prime Minister was able to take a firm stand on the nuclear deal and in FDI in retail issue; however, he was wimp in implementing a decision taken by him, in the face of the corrupt forces surrounding him. Clearly the Prime Minister was 'prisoner of circumstances', mostly created by himself. The book also clearly debunks the theory that the CAG's calculation of loss was a spectacular over-assessment. **Most insiders, familiar with operations in the Government of India and the State Governments are aware of how CMDs and Directors of public sector corporations are appointed, and how in most such cases the minister concerned and other politicians have a direct personal financial interest. What the book throws light on is only the tip of the iceberg; but the reality is harsh**. Equally we have a sample of how the Members of Parliament, who are expected to be policy-makers, but in real life interfere with every element of implementation, almost exclusively for their personal benefit—public interest is nowhere in the picture. The overall fall in administrative standards have been brought out if anything, with some understatement.... While people like **Parakh and Khemka and Durga Nagpal** have been in the public eye for having stood up for their principles, the reality is that the system has sharply deteriorated; the political executive will not brook any resistance or 'logic' from the official; the demand that their needs be met instantly is strident." (Parakh, 2014, pp. 13, 14, 16-18)

It is indeed a portrait of extreme despair and thus, it appears that the entire nation has been forced into a 'dark and suffocating tunnel' with perhaps no end in sight. But to all these national-ills, there is one additional as well as, a very menacing and expanding dimension to it, and ironically no one is caring to talk about and forcefully warn the nation about this impending catastrophe. **And that dangerously looming dimension is India's unchecked population-growth, and the seeds of this menace were sown, due to a very arrogantly callous attitude of Late Jawaharlal Nehru, during the very early years of his Premiership itself. We may like to go through an article by this author's , which he has not got published so far and in this regard since reproduced as 'Appendix-A' at the end of this book. It remains irrefutable that this menace of unchecked population-growth has already made this country nauseatingly over-populated and thus, has perhaps transformed it into indeed an ungovernable nation. We may recall that earlier we also discussed the devastating ransacking of the nation's very fragile ecology as well as also, very dangerous-level of its environmental degradation, since being principally caused by this over-population, and also its unchecked growth**. All these issues are indeed interlinked and mutually influencing too in either manner, and hence, nothing can effectively be done for any long-term national good ever in isolation.

Now coming back to the decay and rot in the nation's civil-governance system, **Mr. P.C. Parakh** has very frankly written that, "Commenting on the quality of civil services in post-independent India, **Paul Appleby** said, 'India shared with Britain the distinction of having the best body of generalist civil servants in the world.' In contrast, Hong Kong-based *Political and Economic Risk Consultancy*, in a report released in 2012, ranked and rated the Indian bureaucracy as the worst in Asia, with a score of 9.21 out of 10. Thus on the eve of independence, India was fortunate to have an efficient, honest and fearless civil service and a highly respected and competent political leadership, with an exemplary record of great sacrifices of winning freedom for the country. But we have failed to build on these strengths (*How true*). (Parakh, 2014, pp. 165, 166 & 167)

Degeneration in the Political System: Degeneration in the political system started rather early after independence. Reading from a letter from a Congressman, at his prayer meeting, **Mahatma Gandhi** said:

> 'This rot in the Congress is that of people's representatives in the legislative bodies of provinces. They take money from people to get licenses, indulge in black-marketing, trade in ignorance of masses, corrupt the source of justice, and force administrative machinery to get transfers of administrative personnel.'

Weeks before his assassination, the Mahatma read from another letter from a friend, **Konda Venkatappaiah**:

> 'Several MLAs and MLCs are following the policy of making hay while the sun shines, making money by the use of influence, even to the extent of obstructing administration of justice, in criminal courts. Even the District Collector and other revenue officials are not free to discharge their duties on account of frequent interference by MLAs on behalf of their partisans. A strict and honest officer cannot hold his position, for false reports are carried against him to the ministers, who easily lend their ears to those unprincipled self-seekers.'

Partisan interference in day-to-day administration and harassment of honest and straightforward officers, which has now become endemic, thus started soon after independence. *(And ironically, the so 'tall' and indeed very charismatic Late Jawaharlal Nehru, as India's first Prime Minister then, had no time to decisively look into this menace, as he remained off and on very busy in improving his stature internationally by lecturing to big and industrial nations 'on principles of ethics,* *justice, fair-play and so on; and on whose doll India was then feeding itself, 'remember PL – 480?', and thus surviving.).*

Degeneration in the Civil Service: The civil service stood its ground for some time, but could not do so for long as even the celebrated steel frame of the ICS started to give way. In an article, **CS Chandra, ICS,** then Secretary, Rehabilitation, Government of India, wrote:

> 'A marked decline in the morale of the service became apparent in the early years of independence. **Sardar Patel**, who was in-charge of the home portfolio, was the one man who defended the service, whenever it was attacked inside or outside Parliament. His exit from the political stage was the death knell of the ICS whose decline now became quite evident. **Interference in the day-to-day administration became frequent; ICS officers who yielded to political pressure were promoted over the heads of their seniors, whereas those who resisted such pressures fell out of favour. A new class amongst the ICS sprang up. Instead of devoting time and energy to work allocated to them, their sole aim was to please their ministers. The weaker member of the service began to realise that it was not hard and conscientious work that was the passport to promotion, but sycophancy. Senior members of the service, in order to secure extension, after superannuating or to get other lucrative appointments after retirement, were only too ready to carry out the behest of the ministers even when they found such requests to be contrary to the dictates of their conscience. Far more serious was another factor, which I call intellectual dishonesty. Some members of the service were clever enough to find out which way the minds of their ministers were working. They would record notes contrary to the dictates of their conscience to ingratiate themselves to the ministers.**' (*Facts are fact. So soon after independence, that became the condition of the erstwhile so famous ICS; then what to talk of its later period 'poor' cousin the IAS, when the rot became deeper and deeper? Hence perhaps less said the better; but we may recall that in a number of earlier paragraphs in this sub-chapter itself in any case, we have since already discussed all kinds of the rot and decay in the civil services from top to bottom now, in somewhat substantial details*).

A number of Commissions appointed during the **Nehru era** spoke about corruption and favouritism in high places and consequent demoralization of the civil services." (*But the then first Prime Minister perhaps had no time to seriously look into the reports of these Commissions and thus, 'catch the bull by its horns for twisting its neck' to effectively crush this highly dangerous menace in the beginning itself. Well now in any case, the reports of these Commissions; along with also the reports of so many later-period Commissions/Committees since appointed from time to time on so many important subjects/issues, are already gathering dusts in the cobweb-infested archives of the various ministries/departments all over in the nation's governmental corridors of power.*).

After the no less greatly relieving demise of the 'Manmohan regime' or perhaps that of this now bygone 'extra-constitutional authority', interestingly many revealing books are getting published for the rightful information of the peoples. It is also true that more such publications are bound to come as time passes-by and hence, indeed it is in the larger interests of the Indian democracy, that the real truths must always see the light of the day at earliest; and thus leading to since the always so much desired more and more awareness among the peoples, so that any repetition of such shameful blunders are possibly prevented in future. (*Now it is very interestingly understood that, Mr. Shyam Saran retired IFS Officer and former Foreign Secretary, has also written and published a book wherein—he has been very precise and revealed that—how the government of former PM Dr. Manmohan Singh was so callously willing to compromise India's vital national security aspects and* *even barter away the 'Siachen Sector' in the Himalayas to Pakistan in a most irresponsible manner.*) One of the former Indian foreign ministers, **Mr. Natwar Singh**, and also a former CAG, **Mr. Vinod Rai** too have now brought out their books, which have succeeded in their aims of even 'dare all and bare all' and thus, revealing all sorts of deep-rooted political-rots since playing havoc with the entire gamut of India's top-level decision-making processes, and the overall national governance as well. *At this precise juncture of our continuing so critical discussions, we may also like to refer to* **Mr. M.K. Dhar's book** (a Retd. Joint Director of I.B.) too; such as (i) 'Inside India's Intelligence Services', (ii) 'Secret Agents in Pakistan' and also, the latest one, i.e. (iii) 'The Open Secret', so befittingly in these regards. In the previous pages, we have discussed the so far unstoppable decline and the colossal systemic-decay at both the apex political-level, as well as at the bureaucratic-levels; and further the politico-bureaucratic combine have together been very arrogantly busy in harming the legitimate stature, rights and dues of the nation's Armed Forces and thereby in the process, have even very adversely affected, the nation's both internal and external: vital security imperatives. All these in turn has not only caused some serious chinks in the Indian military's armour, but has very callously even forced their so far 'exclusive—anti-national corruption-culture'—into not only the military's top-level command system, but also into its ranks and files below. These need to be effectively taken care of, in highest national-interests.

In the overall above larger perspectives, we may relevantly prefer to end this sub-chapter by quoting a couple of paragraphs from an article, titled 'Evolution of a Nation: Government has expanded massively, but so must its power to deliver public goods to Indians', written by **Professor Devesh Kapur**, Director, Centre for the Advanced Study of India, University of Pennsylvania. And therein the Professor has very reflectively mentioned that, "A singular weakness of the debate on India's public institutions is that attention is almost exclusively focussed on entry as attested by the recent uproar on the modalities of the civil services exam conducted by UPSC and the enormous energy spent on expanding reservations. But what should be the starting point of debates on public institutions has unfortunately also become the end point. What happens post-entry gets short-shrift. In particular there is little debate that employment and representation is only one goal of public institutions. Serving the public is surely more important. This requires that organisations build internal cohesiveness, a shared sense of purpose and an esprit de corps.... Astrictive identities can be the basis of recruitment, but post-recruitment the organization's identity must be paramountcy. Else, the result is fragmentation and the absence of shared goals, and critically the absence of any accountability. The result is that India's public institutions end up serving their employees more than the public.... **One public institution that at least today recruits from a reasonably broad social base but then drills and trains the recruits so that they share a common organizational ethos is the army. It offers a lesson on how debates on public institutions needs to shift from who gets in, to what happens to them once they get in. When that happens, India's public institutions will perform better and state power will better serve the public good. And India's political evolution will become more mature**." (Kapur, *The Times of India*, 13 August 2014)

In the 21st century now, the somewhat—so called the 'vote-bank politics', as rather routinely indulged-in by the various political Parties in today's Indian polity—are only understandable and also acceptable in very limited manners and that too, if it doesn't lead to even any marginal harms to India's democratic polity and also, to her multiply-vital national-interests. The crucial points precisely in these regards; do irrefutably remain that only if the modern and democratic India's both the political-leaders and political Parties—do sincerely and with always unwavering consistency, adhere to these moral-cum-patriotic imperatives on their parts—then only in turn, they would be able to comprehensively reform, the nation's both public-sector Corporations as well as, all the strata of its civil-services, along with their presently existing malfunctioning policies-cum-rules and regulations, systems and sub-systems and so on, for enabling them to be able to truly serve her long-term national-interests. There are no short-cuts to all these paramountcy imperatives and hence, the roads ahead in these regards are indeed very bumpy and challenging; and thus need to be accepted with true steely determination, to make the nation succeed, as wished for.

*

CHAPTER 9

Conclusion

We may recall as since we began our this 'Strategically Critical Journey: through humanity's past, present and possibly future, specifically in global geostrategic and politico-military perspectives'; we have indeed experienced a very long and demanding journey all through the past five to seven thousand years of the humanity's highly checkered, as well as also the no less multiply developing, as well as also the continuously progressive advancements of the various human-civilisations onto the mankind's continuously expanding civilisational-canvas. **In fact all these have indeed been journeys all along conceptually through—very dauntingly bumpy and challenging paths—which in turn, have been so much soaked with human sweats, tears and blood. Any such consistently somewhat prolonged 'mental journey', would not only remain memorable but also the no less leaving that person mentally somewhat 'shaken' as well. At many junctures all through this very challengingly 'scaring' journey, it appeared to this author that the tasks involved are indeed beyond his very average and insignificant mental and learning calibre and hence during those moments at times he felt like abandoning this project and then even at times began to curse himself as to why he embarked upon this daunting 'journey', for which he was never capable of in the first place.** Such situations at times forced him to even close his 'laptop' for a couple of days to muster some courage to reflect afresh on the main subject-matters and begin revising his earlier studies. To his some pleasant surprise, this exercise from time to time turned-out to be indeed rejuvenating to even his very average mental-calibre, as well as his very ordinary knowledge-base and thereby, he was able to muster some extra will-power to further continue with this 'journey'.

Moreover, there are certain people whose encouragement and support I must acknowledge. I was very kindly given some time by a the then living-legend soldier, scholar, author, diplomat, and outstanding thinker: **Lt. Gen. S.K. Sinha** (*who in fact, doesn't need any introduction anywhere.*). He very patiently went through the main subject-matter and contents and indeed very encouragingly appreciated the 'wide-canvas' of the main subject and the effort I was able to make till that point. I appreciate his kind gestures of appreciation and his faith in me that encouraged me to complete my book, of finally whatever worth it is.

My uncle now deceased, since settled in Patna (*i.e. husband of my late father's only younger sister)*, **Shri Basant Prasad, IAS (Retd)**, went through the 'Introduction' chapter and commended me and wholeheartedly encouraged me to remain consistent in this effort and complete the book. His encouragement and blessings too, gave me all the more confidence in finally completing this project.

Last but not the least, **Dr. S. Prasad**, a prominent academician-cum-research scientist from the engineering-education fields, who is unfortunately no more, and who so affectionately treated me as his younger brother, remained always very encouraging and blessed me to complete this book. In fact, I have no words to explain the extent of affection that I was fortunate to have received from **Dr. Prasad**, all through the days he was alive. Principally, what this book has wanted to convey that the mankind's military-history, since its birth along with the advent of very ancient human-civilisations themselves on this planet-Earth, since being both conceptually as well as intellectually—a living-entity—and also simultaneously being an inseparable segment of the much larger and always living human-history itself, which in turn remains an important part of the all-encompassing, very vast, highly complex, multi-dimensional, and also continuously expanding store-house of the human-knowledge in this world. And thereby simultaneously as per the above the mankind's larger living-history too, remains continuously in its multiple-processes of being made afresh all over the world onto the man's expanding civilisational-canvas.

Now in addition the principal causative-factor, as well as the 'fulcrum' for indeed the continuously multi-dimensional-cum-multi-layered advancements of the human-civilisations since thousands of years, and as also being getting recorded onto the already multiplying pages of the mankind's living-history itself, has been **mankind's 'continuously and analytically thinking and creative mind'**, having been bestowed onto him by the mother-Nature in the very beginning only, in fact as a singularly unique and invaluable 'gift' in these regards. We may then recall and since already discussed in the initial chapters themselves that this aspect in particular, has made and also have all along been making all the differences for the man with respect to all other mammals and living-beings on this planet-Earth from the very beginning of human-evolution itself. So we may also say that this aspect in fact, has put the mankind rather permanently in a kind of the indeed an unassailably advantageous position in the overall and continuingly evolutionary scheme of things of the supreme mother-Nature on this planet-Earth. Further, we may recall that as the advancing human-civilisations in course of time, gradually began acquiring more and more new kinds of the civilisational confidence, progress, and wherewithal, conceptually one particularly far-reaching factor that began fast getting evolved and thereby became more and more decisive, was **'the multiple-aspects of essential strategic-culture and thoughts'** in variously complex manners. In fact this particularly unique factor in these regards, has been a much later development onto the mankind's advancing civilisational-canvas, as being understood today, but nevertheless since its advent and gradually fast processes of getting gelled into the man's dynamically proactive civilisational-psyche itself, has been making all the differences in his otherwise unstoppably relentless and multi-dimensional endeavours towards enhancing his acquisitions of the very many kinds of assets, riches, as well as numerously different wherewithal, over his any rival(s) in the world. **What precisely do we mean when we talk of and emphasise upon the 'multiple-aspects of essential strategic-culture and thoughts'**? We may recall and since we may have noticed that we have indeed tried to very briefly define their precise meaning and details there-in in the earlier chapters too. The important points are that the above mentioned aspects concerning the essential strategic-culture and thoughts in fact do always remain most decisive and all-encompassing with respect to all possible kinds of human-endeavours in this world, broadly since the very dawn of human-civilisations themselves. And hence as human-civilisations got more and more multi-dimensionally advanced and complex, as well as multiply flowered, and also as the mankind's living-history too began moving forward in the state of its continuous making as per its advancing timelines; various kinds of tribal/feudal fiefdoms, kingdoms, empires and other kinds of socio-cultural-cum-political entities began getting manifested in the world; and thus further making of man's living-history began getting more and more decisively steered in fact, by the dynamic and also the gradually fast intensifying interplay of the 'multiple-aspects of essential strategic-culture and thoughts' in any given situation and also anywhere in this world. **Hence thereby in these multiply-engaging contexts, we may indeed say that how and in what all manners, any particular socio-cultural-cum-political entity since also claiming to be sovereign and thus, at any given point of its historic-timelines could be able to conduct its overall governance, ensure its overall security and well-being while progressing in this otherwise mostly the no less dicey and 'hostile' world, would have decisively depended upon its own 'multiple-aspects of essential strategic-culture and thoughts.** In fact in nutshell, this specific perception is able to perhaps convey the principal points in these regards.

While proactively keeping the above discussed and even if some kinds of a fairly abstract 'conceptual-framework' in our minds, we may perhaps be able to strive to somewhat explain and highlight the main—'thoughts and contents' thus contained in the time-proven ideas and concepts and hence the principal meanings therein, concerning the 'multiple-aspects of essential strategic-culture and thoughts'—by listing the following points:

(a) Any particular sovereign political-entity/nation's level, intensity and rationale of its inherent feelings with respect to its aspects of nationalism, and thus no less also simultaneously being coupled with its multiple-aspects of the wise and fruitful management of since the inherent feelings of its 'aspects of nationalism' themselves, are in fact decisively governed by the overall state of health and dynamics of that nation's 'multiple-aspects of essential strategic-culture and thoughts'.

(b) Any political-entity/nation's level and also proactive effectiveness of the multi-dimensional perceptions, as well as the essential wherewithal being in action: concerning its overall successful governance, both in terms of its internal and external compulsions are again results of the same.

(c) Degrees of successes being achieved in the governance-oriented and thus indeed continuing managements of always perpetually challenging, as well as the multiply-complex tasks concerning both the internal and external security-aspects by the ruling-elite-cum-decision-makers of that concerned nation, at any given point of its historic timelines too, are dependent on the degree of effective vibrancy of the same.

(d) Degrees of well-regulated pro-activeness and successes, as well as level of the overall health of the concerned nation's well-placed and duly activated systems/mechanisms as per the socio-political culture, since concerning the aspects of visionary-cum-realistic policies-formulations and their meaningful implementations within the overall gamut of being the always proactive wherewithal of its governance also, get flowered by the same attributes. These important aspects in fact, always include not only the aspects of the policy-formulations concerning not only the essentials of that specific nation's internal governance, but also all the more concerning the making of foreign-policies, successfully fruitful conduct of its foreign-affairs as well. **In fact, these paramountcy imperatives**

concerning any nation's qualities of the aspects of its overall governance do decisively depend upon again the overall effectiveness of the above mentioned attributes only.

In other words, nothing else but only the well-cultivated, well-groomed, dynamically imbibed, being always multiply-proactive and flexibly well-placed—**the multiple-aspects of essential strategic-culture and thoughts**—within the comprehensively overall gamut of the governance of any specific sovereign Nation-State, would be able to desirably ensure the following:

(i) Continuing visionary updating-cum-fine-tuning of its 'multiple-aspects of the essential strategic-culture and thoughts.

(ii) The imaginative strategic-development, well-rounded and consistently dedicated nursing of the nation's matters-military, as well as the related military-wherewithal too.

(iii) Continuous visionary development-cum-grooming of the always paramountcy strategic-skills concerning the comprehensive statecrafts with respect to that nation's multiple-aspects of internal-governance, successful conducts of its foreign-policies-cum-security aspects in its best long-term national-interests.

(iv) Proactively successful cultivation of one's arts and crafts concerning the ongoing fruitful conduct of one's inescapable—perhaps the so-called 'great-games'—in the highly challenging, extremely dicey and slippery, and since being mostly the no less turmoil-ridden 'arena of the international power-politics'.

Hence in short, we may thus say that almost anything and everything concerning one's nation's overall total gamut of governance, the time-tested and enduringly multi-dimensional progress and well-being in fact, decisively depend upon the overall state of health, and multiple-dynamics of its 'multiple-aspects of essential strategic-culture and thoughts'. Hence we may recall that repeatedly, as well as in the indeed differently relevant contexts, **we have since been emphasising the historically timeless-paramountcy of these 'multiple-aspects of the essential strategic-culture and thoughts' with respect to the particular nation's very survival, multi-dimensional economic development, ensuring its overall effective-security and well-being, continuing scientific and technological advancements, intellectual-cum-spiritual enrichments, and thereby ensuring at least always a reasonably praiseworthy overall 'quality of life' and so on**. Hence while discussing with analytical criticality—the 'multiple-aspects of essential strategic-culture and thoughts'—in the end rather quite appropriately, we may wish to conclude these important discussions by citing two non-western but the no-less far-reaching, as well as history-making examples in these regards:

(a) First prime-example, highlighting the positively very enduring and also, the internationally history-making outcome and thus at the same time, establishing the fulfillments of a nation's perseverance and principally focussed aim; **if that particular nation tenaciously, as well as with all possible sincerity, sticks to the universal importance-cum-decisiveness of its 'multiple-aspects of essential strategic-culture and thoughts' then that in fact pertains to the mid-20th century rebirth of a very ancient Jewish Nation-State, as a highly confident and strong, geostrategically very pragmatic, democratic, multiply very fast developing, and modern and sovereign Israel onto the eastern-coast of the Mediterranean.**

(b) In somewhat identical perspectives as per above but in fact, in not only very differently history-making manners, but also having almost entirely dissimilar pasts the second example pertains to the modern, democratic **India** since belonging to the post-1947 era, which thus does display in quite depressing history-making manners for herself and her hapless peoples—**courtesy: the very bizarre and arrogant geopolitical-naivety on part of the first Indian Prime Minister, along with his the no-less and consistent disregard of this the then newly independent Indian-Nation's paramountcy of the 'multiple-aspects of essential strategic-culture and thoughts'—which resulted in her very pathetic humiliation in 1962, in the hands of Chinese invaders.** Here again we may recall that we have since also critically discussed India's recent historic-fiascos in their some substantial details, in earlier chapters. But it appears to be contemporarily the no less relevant to also quote some passages from a well-known book titled, *Tibet: The Lost Frontiers*, since very scholarly authored by **Mr. Claude Arpi**, a very famous columnist and author, and also a well-known expert on Tibetan and Chinese affairs.

Certain irrefutable important facts

In his above cited book **Claude Arpi** has indeed in very revealing, as well as the no-less devastating manners; has written that, "In ancient Greek tragedies, the tradition was to start by a prologue, in which the main performers were introduced to the audience. The *dramatis personae* was the list of characters in order of hierarchy, with Gods first, followed by semi-Gods, ending with ordinary humans. The motivations of the different players and the main 'knot' of the drama were also explained to the public. In some cases, the choir would intervene between each act to explain further the intricacies of the play. We will dispense with a choir, but before plunging into the history of the Roof of the World, it is interesting to thus get introduced to the main characters and the forces moving them, consciously or unconsciously.... The main 'knot' of our 'tragedy' took place in 1950. During this 'fateful' year the dice of fate were thrown. There are turning points in history when it is possible for events to go one way or another,—when the tide of time seem poised between the flood and the ebb, when fate awaits our choice to strike its glorious or somber note, when destiny appears to leave us the choice of our fate. The year 1950 was certainly one such crucial year in the destinies of India, Tibet and China. The three nations had the choice of going towards peace and collaboration, or tension and confrontation. Each one chose its fate with all the consequences which were to follow.... **Decisions can be made with all good intensions—as in the case of Nehru who believed in an 'eternal friendship' with China—or with less good motives as for Mao. Decisions can be made out of weakness, greed, pragmatism, ignorance or fear, but once a choice is made, consequences unfold for years and decades thereafter.** (Arpi, 2008, pp. 11, 12, 13, 15-17)

The personae: The end of the forties saw the entry of a new player in the Great Game of free Asia: Red China. A new Emperor, **Mao Zedong** ascended the throne of the Middle Kingdom; he is the lead character of our story. Whether he was a man, a God or a titan, is left to the readers to decide.

His counterpart was a British-educated Indian, **Jawaharlal Nehru**.

On one side of the chessboard was the Great Helmsman, the supreme pragmatist for whom means were irrelevant; he believed in real life only. 'The philosophers have so far only interpreted the world the point is to change it.' On the other side was the Indian Prime Minister, an idealist, not to say a dreamer.

A well-known anecdote is revelatory. Shortly after independence the Indian Army Chief of Staff had drafted the first paper on the threats to India's security. The paper contained recommendations for dealing with the newly independent nation's security and asked for a government directive on defence policy. When General Sir Robert Lockhart took this paper to the Indian Prime Minister, the later remarked: *'rubbish, total rubbish. We don't need a defence plan. Our policy is non-violence. We foresee no military threats. Scrap the Army. The police are good enough to meet our security needs.'* Thus spoke Nehru, India's first Prime Minister in 1947. (*We may now recall that in some other identical contexts, we have since also discussed the same episode indicating the so much 'derailed strategic thinking' of Nehru on India's security needs and the matters-military as well.*).

Mao knew his Indian interlocutor perfectly: no real danger could come from a leader who had decided to be the champion of non-violence, who wanted to prove to the world that problems and crises could be solved without recourse to force. The first casualty would be Kashmir. Nehru lost Kashmir trying to stick to 'principles'. On the Indian side, there was another great persona, **Sardar Vallabhbhai Patel, the Indian Deputy Prime Minister, also a pragmatist, who foresaw all the consequences of Tibet's 'liberation'. The opinion of Patel was diametrically opposite to Nehru's, but unfortunately fate removed him from the stage hardly 2 months after the People's Liberation Army walked into Tibet**. Another pragmatist was a 15-year old kid, the **Dalai Lama**. On November 17, the God spoke through the Nechung State Oracle in Lhasa: 'Make him King'. Thus the young boy was enthroned as Tensin Gyatso, the Fourteenth Dalai Lama. His task would be the most arduous in view of his young age and inexperience. By religion and temperament, he shared many convictions dear to Nehru, but increasingly agreed with Mao that country needed 'socialist' reforms. However he would never be allowed to implement them on the roof of the world.

It is necessary to briefly introduce two smaller players—**K.M. Panikkar** and **V.K. Krishna Menon**, who greatly influenced Nehru, thereby playing a role in the tragedy to unfold. Both believed in the greatness of the Chinese revolution. Panikkar believed that Mao was the 'chosen leader of a resurgent people, driving out those who had sold out the Chinese Revolution and pushing back to the sea, from whence they came, the western nations who had enslaved the nations of Asia' (*What a misguided naivety in one's so called perhaps any understanding, if any in these regards.*). His views would have far-reaching and tragic consequences for India's security.... **For Mao and his comrades, only one thing mattered; the final goal. Further, while in course of her history, India has never had an expansionist propensity; China—whatever the colour of the regime in Beijing—always had strong imperialist tendencies**. The historian, **Dr. R.C. Majumdar** sheds light on the traditional Chinese way of thinking and acting:

> 'There is, however, one aspect of Chinese culture that is little known outside the circle of professional historians. It is the aggressive imperialism that characterized the politics of China throughout the course of her history, at least during the part of which is well known to us.... It is characteristic of China that a region once acknowledged her nominal suzerainty even for a short period, she should regard it as a part of her empire forever and would automatically revive her claim over it even after a thousand years whenever there was a chance of enforcing it.'

Unfortunately, Nehru was unable to see clearly. Further, besides being the inheritor of the traditional Chinese expansionism, the new regime in Beijing was also the dynastic inheritor of the imperial mantle. Mao was the new 'son of Heaven'. This tradition added to the Marxists' revolutionary zeal and the Han Chinese strong superiority complex, gave a most formidable mix.... Traditionally, the Son of Heaven's mandate to rule came directly from Heaven; he had to rule 'all under Heaven.' The administration of the Emperor was not made it had a cosmic dimension (*sic.*).

As the Chairman of the People's Republic and head of the People's Liberation Army, Mao had de facto become the new Emperor.

Another inheritance from traditional China was the contempt of the mainland Chinese (the Hans) for their 'vassals' of 'Outer China'. Let us not forget that the Great Wall of China was built for the purpose of separating Chinese from 'non-Chinese'.

In the minds of the Chinese leaders, 'revolution' was never limited to the mainland but had to be spread to 'Outer China' and then to India, the whole of Asia and finally the rest of the world. Mao's earlier-quoted letter to the Indian Communists was clear on this point: 'India had to be liberated'. (*Look at this.*).... While Nehru or the Dalai Lama, both adepts in the philosophy of non-violence, was ready to accept many compromises to avoid struggle or conflict, the Chinese did not find anything wrong in war and upheaval.

Mao noted in *Problems of War and Strategy*: 'Some people have ridiculed us as the advocates of omnipotence of war. Yes, we are: we are the advocates of omnipotence of the revolutionary war, which is not bad at all, but good and is Marxist.' (*Such bizarre thoughts anywhere, can only perhaps be crushed by force; or sometimes more effectively, by some more superiorly appealing socio-political ideas, coupled with much superior conducts of the business of geopolitical statecrafts; as the USA's erstwhile 'Ping-Pong' diplomacy and so on.*).

It should not be forgotten that the Communist Party in China had 'grown up as an army and not a civilian organization like any other Communist party.' Mao's strength was the People's Liberation Army, 'military virtues and military men have been elevated to a position of new prestige in the Chinese Society.'

Until recently, the army has been the main pillar of the Communist regime and it is only because of the conservative elements of the Liberation Army that, in Tiananmen in 1989, the hard-liners were able to crush the student movement and keep alive the so-called dictatorship of the proletariat.

For the Chinese, it was therefore a 'sacred duty' to liberate Tibet and 'to make the East wind prevail' (*limit of what, arrogant madness? Well...*).

General Li Chi-Min went to the other extreme when he wrote, 'Modern revisionists have exaggerated the consequences of nuclear war, and the results will not be the annihilation of mankind. Over the debris of a dead imperialism, the victorious people would create very swiftly a civilization thousand times higher than the capitalist system and a truly beautiful future for themselves.' (*Incurable minds indeed, but what other option than to effectively counter such rubbish?*).

The Chinese people have exemplified two characteristics during their five-thousand year history: one is their attachment to 'their land' (or what they perceive as 'their land'), and the second is their obsession with power—'power comes out of the barrel of the gun'. A third can be added: the Chinese did not like to lose their face. These characteristics explain most of the actions of the Chinese leadership, from 1949 until today. (*How precise and true too.*).

The Indian leaders only fooled themselves in believing in the Hindi-Chini Bhai-Bhai doctrine: the Chinese aims were always clear, loud and publicly announced; nowhere did these plans even hint towards a friendship with India." (*Can the author of the above passages, i.e. Claude Arpi himself, ever be faulted on these, irrefutable facts here?* ***Will then, the mankind's living-history in fact, ever be able to forgive Nehru's damaging matters-military related naivety****—with respect to the* *cool arts and sciences of the statecrafts—in mostly the turmoil-ridden, very dicey, extremely slippery, and also highly challenging 'arena of the international power-politics; during his innings as India's first Prime Minister, since immediately after her independence?).*

Consequent to some phenomenally quick and far-reaching major policy-decisions, as having been already taken by **the new Indian Prime Minister, Mr. Narendra Modi,** the overall geopolitical-cum-economic attitudes of the USA, Japan, Russia, and also many of the leading European nations too have suddenly undergone major alterations in favour of India's long-term national-interests. **World at large have suddenly begun to realise that, India as a sovereign democratic nation, now no longer is stuck with many kinds of the extremely ironical *'policy-paralyses regime'* of her previous leadership and its so much corruption-ridden government**. As a result now, even China's attitudes towards India, seems to be fast changing for the better. But even now, India has to be all the more not only geopolitically careful, but also politico-militarily very cautious too, so far her dealings with China are concerned. It must never be politico-diplomatically lost sight of that even now China is not refraining from arrogantly continuing with its all possible kinds of 'coercive-diplomacy' against India; by deliberately creating futile tensions between the Chinese and Indian militaries, at India's northern borders. 'Doklam' is a indeed, a recent 'case in point'. China also has to realise that India is no longer any *'push-over'* nation as its southern-neighbour Moreover in the present super high-tech era, any border-dispute between two neighbouring nations, and that too between two leading nuclear-powers of Asia just cannot be settled by any threat or use of force in any manner. **It appears that now India's new and always very prudently strategic, as well as also geopolitically charismatic Prime Minister, Mr. Narendra Modi, with his unique politico-diplomatic acumen would soon be able to somewhat geopolitically—*'rein-in'* China, as per the best interests of the both nations—to obviously ensure a long-term and vibrantly enduring peace and tranquility in the whole of Asia and thereby globally as well.**

But in precisely the above multiple-contexts, the gradual shaping-up of since the always so much desired—the ultimate success-stories—would also mostly the no-less critically depend upon any concerned nation's requisite 'military-wherewithal' as indeed—the always paramountcy geostrategic and politico-militarily credible back-up—in any situation-specific, its compulsive 'great-game(s). **Hence in any case, the any contextual importance of the requisite 'military-wherewithal' may be only wished away, at the great risk of possibly any consequential peril for that concerned/ affected nation**. The persistently very 'callous-cum-arrogant' neglect of the indeed, these essentially crucial geopolitical 'nitty-gritty' in these regards; in fact during his **Prime Ministership, Mr. Jawaharlal Nehru**, and that too in spite of his indeed great learning and authorship; with so much charismatic leadership qualities, and also being duly equipped with—India's the then prevailing situation-based his great politico-economic vision—Nehru so naively caused such historic humiliation to the nation; which indeed perhaps defies all possible logic here. And the Indian peoples took almost two decades (*i.e. only after their historic victory in 1971-war against Pakistan.*) in any case, to only somewhat come out it. Hence now we may see that in these multiple-contexts, the earlier above cited examples too, considerably substantiate these concepts here. It is true that the presently ongoing 21st century's and beyond is no 1962, and also neither the China of 1962 nor the India of 1962 are the same in any manner.

As usual new kinds of challenges and also some new 'upheavals/turmoil too, do keep-on surfacing and begin to threateningly stare sometimes in the eyes of even almost the entire global humanity at large. These are essentially needed to be faced and also confronted with imperatively visionary prudence, by the frontline leading leadership of the world before it is too late. **Hence in**

any case, in any such highly consequential, but indeed the necessarily paramountcy 'great-game(s), inevitably being triggered or likely to be triggered, the 'multiple-aspects of essential strategic-culture and thoughts' of the leading nation(s), will always turn-out to be decisively crucial towards subsequent shaping-up of the final 'success-story' in the best possible long-term interests of the global humanity too. And here-in lies the 'rub' and ironically therefore many times—the present world's sole Super-Power, the USA—has been found wanting; and that too at times, to the great dismay and costs to the larger global humanity itself. At times some failures or at the most some partial successes only, in any concerned 'great-games', with respect to the particularly any emerging challenges/situations in any specific region of the world—and that too all the more consequently on the part of the Super-Power itself—may always have some very far-reaching and also seriously worry-some fallouts as well. Further these adverse developments may also lead to the fast emergence of even some new threatening situations, where geostrategic-cum-politico-militarily for the same leading Powers it may ironically become an inescapable necessity to launch a new set of 'great-games', to neutralise now the somewhat different and also altogether the newly emerging threats and challenges.

And hence to the great dismay, as well as a shock to largely the hapless global-humanity; that's what has precisely happened now ironically for the mightiest USA, in dicey and unpredictable Middle-East today. **Further, any unimaginative or faltering indulgence in any particular situation-based crucial 'great-game(s)' anywhere, and that too by any globally important Super-Power, could soon lead to the cropping-up of a new set of all the more menacing and unexpected problems/threats, which would rather inescapably call for launching of even more risky and costly set of 'great-games' to neutralise those new and the much more complicated 'problems/threats', to any existing lawful world-order and peace. And ironically and that too in quite bizarre manners, that precisely has happened in 2014 for the USA, in indeed the mind-boggling 'geopolitical-cobweb' of the oil-rich desert-landmass of the Middle-East.**

In precisely the above multiple-contexts, but also historically as well as contemporarily in much larger perspectives; a well-known author and thinker, **Mr. George Soros** has very vividly brought out an indeed critically analytical perceptions in these regards in his book, "My thinking has evolved out of many influences and experiences. The most pervasive influence, other than my father's, came from **Karl Popper**, who was my tutor during my last year at the London School of Economics. Having lived through both Nazi and Soviet occupation in my native Hungary, Popper's book *The Open Society and Its Enemies* struck me with the force of revelation and prompted me to explore the author's philosophy. He argued that the Nazi and Communist ideologies have something in common—they both claim to be in possession of the ultimate truth. Since ultimate truth is beyond human reach, both ideologies had to be based on a biased and distorted interpretation of reality (*How true indeed.*); consequently they could be imposed on society only by the use of repressive methods. He juxtaposed a different principle of social organization that is based on the recognition that claims to the ultimate truth cannot be validated. Popper called this principle 'open society', and he held it out as preferable to definitive design. He never defined exactly what an open society means; in view of our imperfect understanding, he did not like quibbling about definitions. In my case, an open society needs to be constantly redefined by the people who live in it, otherwise it might become a definitive design.... It is clear that the concept of an open society is closely associated with the concept of democracy, but it is an epistemological concept, not a political one. Historically, the concept of democracy has emerged out of a consideration of the role of power in society. People are likely to use their power to further their own interests, not the common interest. One way to protect the common interest is to insist on the division of powers. This became the cornerstone of the Constitution of the United States of America. Our Founding Fathers, particularly **James Madison, John Adams and Thomas Jefferson**, were acutely aware of our imperfect understanding, but that was not the foundation on which the Constitution was built. That is an important point to remember.... **American democracy predates the concept of open society. The Constitution is a product of Reason; open society belongs to the Age of Fallibility**. The difference has led to a curious result: America is an open society that does not fully understand the concept of open society.... **The United States has thoroughly destabilized the Middle East by invading Iraq. The task of the occupying forces is no longer confined fighting a Sunni insurrection; they have to contain an incipient civil war** (*i.e. the 2006 scenario.*). The country has divided along sectarian lines and each faction has established a fighting capacity. The Ministry of Interior is in the hands of a radical Shiite Islamist Party whose death squads maintain prisons and engage in extra-judicial executions. Some of the most reliable fighting units in the army consist of Kurdish Peshmerga. The level of fighting is contained by the presence of the United States but the primary objective of the occupying forces is to protect themselves, not the population. The civilian population can see the civil war coming and seeks safety by lining up with one side or the other. People relocate within the country or, if they can, abroad. Real estate prices in Amman, Jordan have gone through the roof. Those in mixed areas such as Baghdad, Kirkuk and Mosul are the most exposed and already suffer the most. **At the time of writing (April 2006), the political process is in a stalemate and it is difficult to see a resolution in the absence of an overarching agreement with Iran. Even then it is doubtful whether Iran can exercise sufficient influence over the Shiite factions. On the Sunni side, al Qaeda and other Salafist factions are hell bent on provoking sectarian violence and the population is more afraid of the United States and Iraqi armies than the terrorists. The situation slowly but inexorably deteriorating and could widen into a regional Sunni-Shia conflict, especially if the United States reduces or withdraws its troops.... Iran has a vested interest in keeping Iraq friendly, weak and somewhat unsettled, but it does not want a regional Shia-Sunni conflict or to have Iraq fall apart. (*How prophetic it has now turned out to be in 2014; with indeed the menacingly consequential 'birth of the ISIS', i.e. 'Islamic States of Iraq and Syria' or so, now the 'IS', i.e. "Islamic States'.*)**— After all, it (*i.e. Iran*) was bogged down in a brutal war with Iraq for eight years and suffered immense casualties.... Iran is the major beneficiary of the invasion, which removed its enemy Saddam Husain from power, tied up American forces in a task that they are ill-prepared to perform, and tightened the supply of oil. What a change from the situation that prevailed prior to 9/11.... The world order is based on the sovereignty of states. Sovereignty is an anachronistic concept born out of an era when society consisted of rulers and subjects, not citizens. Sovereignty became the cornerstone of international relations with the Treaty of Westphalia in 1648. After thirty years of religious wars, it was agreed that the ruler had the right to determine the religion of his subjects. *Cuinus regio eius religio*. The French Revolution overthrew King Louis XVI and the people seized sovereignty. In principle, sovereignty has belonged to the people ever since.... There has always been rules governing the relationship between states, but these could always be broken by the use of superior force. There has never been a world order capable of preventing war, although some arrangements have been more satisfactory than the others. Nevertheless, the idea that there is no world order other than the use of force is a fallacy—a companion piece to the misinterpretation of the nature of power. The idea appealed to the advocates of American supremacy because it would have allowed the United States to impose its will on the world. But it did not work.... The world cannot be ruled by military force. Military power is only one of the many ingredients that a country needs to exercise its influence over others. (*True, but it is one of the mostly paramountcy ingredients for the purpose; and hence precisely concerning the post-1947 India here, needless to say that Nehru persistently lived with his lingering 'geostrategic-cum-politico-military naivety'—in his 'own rather make-believe world of the geopolitical-darkness'—and consequently with great harm and humiliations to the Indian Nation-State during his times.*). Imperial Powers did not succeed by the force of arms alone. Even the Ottoman Empire which was built by conquest, had an elaborate system of maintaining peace and justice and the empire disintegrated when the system broke down.... The United States did not become the dominant power by military means. Victory in the Second World War was followed by the establishment of the United Nations and the Bretton Woods Institutions (the International Monetary Fund and the World Bank) and the Marshall Plan. The behavior of the United States was not always exemplary—-the CIA was busy hatching plots, planning assassinations, and arranging coups d'etat; but most of these activities were clandestine, and when they were revealed, the CIA was reined in. The Vietnam War ended in defeat and broke the can-do-spirit that had characterized America until then. (Lyndon Johnson has put forward his Great Society project before he announced his candidacy for reelection on account of the Vietnam War). The United States continued to fight proxy wars and to sustain authoritarian regimes. Remember the Iran-Contra affair? But these were aberrations. On the whole, the United States fulfilled its role as leader of the free world reasonably well. The rest of the free world willingly accepted American leadership against the Communist threat, and the United States was supportive of its allies. For example, it played a constructive role in the evolution of the European Union and fostered the economic development of Japan and the other East Asian Tigers. The United States could be a superpower and the leader of the free world at the same time. The Communist threat served as a force of cohesion in a society characterized by the pursuit of individual self-interest and self-satisfaction.... This comfortable identity was disrupted by the collapse of the Soviet empire. Being the sole superpower and the leader of the free world was no longer the same thing, but the United States failed to recognize it. America used its dominant position to promote its national self-interest in every sphere, economic as well as military. That is not the leader of the world is supposed to do. In a world order consisting of sovereign states, the United States, as the dominant power, has to concern itself with the well-being of humanity in addition to pursuing its self-interest. This unique responsibility derives from the privileged position that the United States occupies in the world. It is in charge of the agenda. It cannot unilaterally impose its will on the world, but no collective, cooperative action is possible without its leadership or active cooperation. The United States can block every international institution. It has veto power in the United Nations Security Council and it is the only country that has a blocking minority in the Bretton Woods institutions. Washington determines the direction in which the world will move, but the rest of the world does not have a vote in Congress; therefore it falls to the leadership in Washington to give due weight to the common interests of humanity in addition to the national interests of the United States." (*Hence internationally perhaps, the most weighty and far-reaching point that emerges here that whenever the USA has played this role with due pragmatism and geopolitical-vision, the mankind's consequent global 'living-history' has turned out to be rather magnificent for the entire humanity; and when not, the multiply-adverse fallouts for almost the whole world, have rather been catastrophic too. And that's precisely the burden, the world's sole Şuperpower has to perforce bear, and also always bear well.*). (Soros, 2006, pp. 16, 17, 111, 130,131 & 132)

Now precisely in the above multiple-contexts, we may obviously see that the 'keys' to success or otherwise in these regards rest in the crucial 'twin-domains' of the **'multiple-aspects of essential strategic-culture and thoughts'**, as well as the degree-cum-range of the acumen to participate in any specific situation-based impending or manifesting **'great-games'**, geopolitically therein. In these crucial contexts, as the USA has rather so naively faltered in its earlier geostrategic-cum-politico-military engagements in Iraq during the present 21st century's first decade only, that the present world's this sole Superpower is again being dragged into the most complex and dicey geopolitical 'whirlpool' in the oil-rich and also the no less—Islamic religious-fanaticism, as well as mostly senseless jealousy-cum-conflict infested—vast desert-mass of the oil-rich Middle-East. In fact, the Middle-East got thoroughly destabilized not just only by the US military invasion of Iraq—to liberate Kuwait and thus, also to get rid of **Saddam Hussain** too in the first place—but most paradoxically while doing all these, the USA due to its typical—geopolitical shortsightedness and also its military-commands naivety—was concentrating on winning the battles only, and not caring to win the war; and hence as the much more adversely consequential fallouts, the almost whole of Middle-East itself, got thoroughly destabilized. Hence in the battlefields of Iraq, when the US forces were engaging the Saddam's 'Republican-Guards' it just very shortsightedly won the battles only, thus losing the whole War itself. Now to the great dismay of USA, those victories in the Iraqi battlefields have now turned out to be totally futile resulting in senseless waste of valuable human and other resources.

In fact, had the USA cared to win the whole War in Iraq during the **'Op-Desert-Storm'** then, with its all possible dimensions, from the planning-stage itself, and hence thereby, as they militarily advanced in Iraq while conquering territories, it would have also meticulously cared to keep-on restoring all the essential gamut of civil-administration, civil-supplies, and utility-services, and finally on surrender of the Saddam's forces, if the US military-command would also have thoroughly disarmed them in custody, and thus not allowed them to melt-away into the Iraqi civil-population spread over the Iraqi interiors and also in addition, if as the then occupying Power in Iraq, the USA would have been geopolitically far-sighted enough to restore the new Iraqi civil-government firmly in its seats of power and thus make it desirably functional too; the subsequent scenario in the same largely 'fanatically regressive' Middle-East, would have been entirely different today. All these would have in fact, magnificently resulted in a 'win-win' kind of feather in the Uncle Sam's hat, for a long time to come and thereby the long-term and very positive fallouts not only for the USA but also for the whole world, can obviously be well-imagined. But alas in these specific challenges, the then Washington's top politico-military leadership was not able to measure up to the tasks in Saddam's Iraq, and the rest is indeed not any 'palatable-history' over there.

Historically in our larger perspectives, we may therefore conclude that the USA's role in the then Saddam's Iraq though in quite limited manners only, appears to be in total contrast to its earlier such magnificent performances for nearly five years or so, under **General Douglas MacArthur** in much more challenging tasks in Japan during 1945 onward. These historically momentous and also very far-reaching episodes and that too concerning the USA, as World's Superpower, by themselves on closely discerning reviews, do indeed speak volumes about the **timeless paramountcycy of the 'multiple-aspects of essential strategic-culture and thoughts'**, since coupled with always the crucially decisive multiple-acumen with respect to any challengingly staring 'great-games' anywhere and in any given situation; as the ultimate 'keys' to any history-making success or fiasco—either way with great consequences—for the entire humanity for a long time to come. And hence in their broader perspectives, and thus in any given situational challenges anywhere and as per their historically advancing timelines, the highly complex depths and dimensions of any intended 'trysts with destiny' therein, need to be understood and fully grasped from the beginning itself, with no room for any half-baked ideas and plans to ensure success and thus to avoid even any 'partial-fiasco' as well.

Now in somewhat globally larger perspectives but also excluding the Superpower and other developed countries in the world and then, with particular reference to the Middle-East again, **Mr. George Soros** in his already above referred book has very relevantly said that, "Developing countries that are rich in natural resources tend to be just as poor as the countries that are less well-endowed; what distinguishes them is that they usually have more corrupt governments and they are often wrecked by armed conflicts. This has come to be known as the **resource curse**. The tale of the anti-resource movement is instructive because it is illustrative of what I call a **fertile fallacy**." (Soros, 2006, p. 178). (*How true and also, both historically as well as contemporarily. Moreover contextually even in our futuristic perspectives too, if we may critically look into now all the more turmoil-cum-violence-ridden developments that are presently taking place almost in the entire since already extremely unsettled Middle-East region; then indeed very depressingly it appears that this 'resource-cursed' region is rather sooner or later, quite unstoppably slipping towards some kind of an 'abyss'. These are in no way any good 'omen' for the mankind itself and thereby, for successfully facing this highly complex menace, globally joint and also truly visionary efforts, with their requisite consistency are called for.* ***Hence all the above dimensions, since pertaining to the Islamic-States in the Middle-East—socio-psychologically-cum-culturally, as well as also socio-politically too—in fact, do form an indeed very 'lethal-combination' not only for themselves, but for the whole world as well in many ways.***).

A number of countries in the Middle-East, having got so well-endowed by an extremely valuable natural-resource—i.e. 'oil' in this case over—and then since they being also not at all endowed with any tradition of democratic governance, and thereby in addition, also not having any historic legacy of the 'multiple-aspects of essential strategic-culture and thoughts', their fabulous richness of a particularly valuable natural-resource, being also of great strategic values, and hence also especially sought after, all over the world—i.e. the 'oil' as being their greatly substantial natural-resource—does very paradoxically turn out to be finally nothing much, but indeed a very daunting kind of the 'resource-curse'. But for almost the whole world and that too in the no less bizarre manners, this present phenomenon in the deserts of Middle East appears to be the paradox of paradoxes. In somewhat lighter vein here, we may historically call the highly chequered 'history' with respect to the ***'Helen of Troy'*** as more paradoxically bizarre; since it is said that during the bygone ancient times, the 'beautiful face of *Helen* then was able to launch thousand ships'. But 'oil' in any case is no 'beauty-queen', but for obvious techno-economic reasons the world over; it probably commands so far, perhaps an indeed the unsurpassed importance and values in multiply-financial terms across the continents. World needs always more and more oil, but it hardly cares for enduring peace, as well as the always paramountcy socio-political tranquility with any tangible sincerity in that 'oil-rich' Middle-Eastern region.

As immediately after settling-down in his now such very far-reaching assignments as India's Prime Minister; **Mr. Narendra Modi** first embarked on his visits to nations in the Indian neighbourhood, and then also to **Japan** and the sole Superpower **USA** as well; new and very encouraging chapters in India's domain of foreign-relations, had already got begun to be written. Then in-between the events of Modi's visit first to Japan and then to USA respectively, **China**'s President came to India on his State-visit and this too, resulted in remarkable understandings, frank talks, and signing of some important agreements as well. In additions to all these during the month of August 2014 only, the Prime Minister of **Australia** too, came to India on his State-visit and thereby, triggering the new beginnings of some very far-reaching and also enduring relationships between the two countries. All these indeed very remarkable and also the no less history-making developments during the initial months of Modi's Premiership of India, are some unique 'feathers in his cap', and these appear to be the beginnings only; and hence it looks like that the 'future of the Indian Nation-State' itself, is rather looking at this new Indian Prime Minister, with great expectations and aspirations in spite of many great challenges already confronting him.

During Modi's recent and the very first visit to the USA as India's newly installed Prime Minister, even the world-media gave a wide coverage to it; and needless to say that the Indian media in fact, went truly 'hyper-active', with almost all the Indian TV news-channels resorting to very detailed and highly informative live-coverage from both New York and Washington. In New York, Modi got the opportunity to address UN General Assembly as well. Pakistani Prime Minister too addressed the same world body, as since its session was going-on over there; but he could politico-diplomatically in an indeed already untenable manner, was able to only harp on the 'fictitious' Kashmir-issue and nothing else. Media has covered all these in considerable details. **Mr. Pranab Goswami** of the Indian TV news-channel 'Times Now', along with India's one of the leading and highly competent strategic-analysts, **Major Maroof Raza** were able to hold some analytically very interesting debates on their channel, on these issues and Pakistan's rather somewhat illogical fixations therein, by so imaginatively inviting few of both Indian and Pakistani knowledgeable personalities in these regards.

On the other hand, the oil-rich and also few other countries of the Middle-East; are rather unstoppably 'drowning into their vast sea'—of a typical kind of a 'heady-cocktail' of very naively their own making—and also thereby, this 'sea' being consisting of their mindless Islamic fundamentalism and the irrational culture of Jihadi-fanaticism, their rather irreconcilable rivalry and jealousy; their immature arrogance as being propped-up by their 'oil-wealth', and then also their the no less all-round knowledge-deficit; since also coupled with their relative darkness concerning now indeed the invaluable disciplines of the modern sciences and technologies. Future well-being of this so much 'cursed' region ironically, is no more in hands of the Middle-East countries alone, but indeed substantially rests with the so called 'wisdom' of the larger world-community. And here the present world's sole Superpower the USA—with all its multiple-wisdom, geostrategic and politico-military vision, as well as its far-sighted geopolitical-prudence—is perforce required to be sincerely accommodative of all the important players and stakeholders, and also in addition coupled with the UN's support and proactive cooperation; and thereby take a comprehensively patient-cum-decisive lead to finally write the most desired 'success-story' in these regards, in the best possible long-term interests of the entire humanity on this Earth. And hence only future will tell later, as to what it contains in its wombs.

*

Post-Script

It is true that after the end of World War-II way back during the year 1945, so far the global humanity has indeed quite mercifully been spared the any colossal agony of the deaths, destructions and devastations, and also the unimaginably extreme sufferings, due to any further global-war. But still it just cannot be anybody's take anywhere that otherwise always largely—the hapless humanity in now the today's super high-tech-oriented and so fast-changing world—has also been mercifully experiencing any truly all-encompassing, and enduring peace. In fact even before the Second World War ended the then highly menacing seeds of an altogether new kind of conflicts-cum-clashes of interests, in the form of 'Cold-War' and that too again unstoppably on a global-scale, had already sprouted and also, begun to flower rather very fast. Now historically the root cause for this was the sudden emergence of—ideologically-cum-diametrically exactly opposite and also at the same time, entirely irreconcilable—two Superpowers in the then post-war world as the USA and the erstwhile USSR. And needless to say that the proverbial 'icing on the cake of this Cold-War'; was the possession of abnormally high stock-piles of the all sorts of nuclear weapons and their multiply different kinds of the delivery-systems, by both these Superpowers in fact always, mutually threatening each other's 'camps' and their geostrategic-cum-politico-military alignments. Hence along with all these colossal military-wherewithal in both the rival 'Superpower-camps', it was also true that during now the since already bygone Cold-War era—the animosity between the two erstwhile Superpowers, had reached such monstrous dimensions and proportions—that this the then prevailing and also indeed frightening 'balance of terror'—had truly reached a lingering state of the mutually assured total destructions-cum-devastations and that too in fact, many times over for each other. And thereby the whole world in many geopolitical mode, remained in a state of frightening uncertainty, even for somewhat reasonably assured continuation of its survival. In fact it was commonly talked about all across the whole world then—that if at all the then since prevailing Cold-War, ever gets transformed into an all-out hot-war—then the planet Earth would deeply get sunk in a state of nothing but unimaginably prolonged 'nuclear-winter'.

But alas, the contemporary point still remains that has the world even in the presently ongoing 21st century, been able to reach a state of geopolitically, geo-economically and also geo-culturally—in a true state of truly reassuring and enduring peace and tranquility—so that the multi-dimensionally fragrant advancements of both the human-civilisations and human-values henceforth are allowed to continue without any further extractions in fact, of the so much painful prices in terms of the human tears, blood, and also colossal sufferings? This is the moot-point and hence—it remains a rather frightening 'question' in fact that as of now, has already begun to stare 'loudly'—

onto the face of today's global humanity as well as, into the eyes of the so-called 'masters of its destiny' since occupying various kinds of the powerful 'seats of powers' all across the world. Hence in the presently ongoing super high-tech modern-era, the so-called 'masters of destiny' of the global-humanity all across the world, since manifesting in various kinds of the ideological, and also socio-political-cum-economic shades and fixations, are not only duly mandated to take the necessary calls in these regards, but they are also in fact, rather duty-bound to ensure that—the global-peace and tranquility, all-round human progress, further continuance of the civilisational-advancements; and also not the least, the multiply-effective preservation of the Earth's otherwise indeed fragile ecology and environment—are always and unwaveringly ensured everywhere in the world. But it is now an irrefutably sad commentary that, all these above mentioned paramountcy aspects concerning rather the entire global-humanity in the highly challenging present times of the 21st century, are found to be rather ruthlessly trampled-upon, even for some very minor, short-sighted, and even also for any short-term gains; by the same above mentioned so called 'masters of its destiny'.

Now in the highly daunting present-times, while objectively coming to the overall global geopolitical perspectives, we find that the so much religiously fanatic, highly criminalized, and also the very demonic birth as well as, monstrously sudden rise of the **ISIS** (***i.e. 'Islamic States of Iraq and Syria'; now since also known as 'IS', i.e. 'Islamic States'.***); has taken almost the whole world by indeed, the hugely unnerving surprises. Well it is indeed too current a problem and hence, both the print and electronic media almost all across the world are since already very intensely occupied in giving this so much demonic phenomenon—very critical, closely informative, and also a very largely daily coverage—and that in spite being quite bizarre, may also be somewhat perhaps understandable too. But in today's I.T. and also now since a kind of instant media-age quite paradoxically in the process, the ISIS goons and their indeed extremely demonic chief, **Al Baghdadi** too, are getting much of their undeserved publicity; and which in turn rather helping them to maintain the morale of their fanatic-cadres and also ironically, giving them considerable mileage in their efforts to clandestinely rope-in some new, young and misguided recruits as ISIS militants; from even different continents as well. In fact in the last 'Iraq-war' during the US Presidency of **George Bush (Junior)**; if the US Command in Iraq then—would have had a multi-dimensionally well-crafted 'Plan-B'—with requisite wherewithal and also ready to be put in action, immediately after the fall of **Saddam Hussain** and surrender of his military forces and also along with, totally effective disarming of the Iraqi Republican Guards Divisions, other military units and Iraqi Militias—then the entire subsequent scenario in the post-Saddam era Iraq would have been considerably positive and favourable too, for further neutralization of the various kinds of 'root-causes' for many types of 'Islamic-Jihadi terrorism' since otherwise then sprouting in various regions of the world. And thereby at the same time, there would have been no ISIS as well. Now at last, its good for the whole humanity that, 'ISIS' is finally getting crushed, by the world's leading powers.

It is not only the **'Islamic-Jihadi terrorism'**, that the world is so much menacingly confronting today but there are other kinds of geopolitical problems as well—such as **'the multiple-rise of China as a Power in Asia'**, its otherwise indeed very arrogant expansionist-designs in almost in all directions across the peripheries of its national-borders—and at the same time, China's since already strengthening **'hegemonic-nexus'** with almost *'failing'* Islamic-State of Pakistan; which the truly democratic India has many reasons to get worried about. Needless to say that in Asia, China sees democratic India as a fast rising rival; and which China is just not willing to be friendly with on equal terms, for sincerely cultivating any good and enduring relations as neighbours. But the rivals too, can prudently and meaningfully cooperate with each other. China is giving lot many kinds of assistance to Pakistan and also, propping it up all the more as a military-threat to

India. It is good that now democratically with so much awaited regime-change in India and thereby, **Mr. Narendra Modi** becoming new Prime Minister of the country; USA has very farsightedly, as well as multiply begun to be diplomatically briefed and also engaged by the new Indian Prime Minister and also by his highly talented team not only for the good of the Indian nation-State, but also for the much long-term good of the world-peace as well. Hence thereby in the present much changed scenario, very encouragingly the USA too has now very seriously begun to see and substantially understand the somewhat 'turmoil-ridden' ground-realities in South Asia—largely courtesy: Pakistan, and is also able to see through the Pakistan's 'double-games' as well as, its 'very crafty-lies'.

In fact not only in South-Asia but also with respect to the much larger world-arena, since in these otherwise very 'fluid and dicey' geostrategic-cum-politico-militarily multiple-perspectives, it remains both very important as well as overall good for almost the whole world—that the sole Superpower USA remains better informed and also geopolitically persuaded, to clearly and objectively see the real nitty-gritty of the various kinds of 'problems and issues' as since prevailing in various parts of the today's world—so that America is able to more prudently deploy her colossal resources and military-wherewithal, for solving these problems/issues as the needs be—since not only her own good, but also for the overall long-term good of the world-peace as well. Howsoever one since sitting in any corner of the world may criticise or even 'dislike' USA, the undeniably hard geopolitical fact remains that, no tricky international-problem can truly and reassuringly be solved without the USA's involvement as well as, her timely and also any substantially positive contributions. At this precise juncture of our ongoing discussions, and also keeping in our minds, the present-era's now so visibly prevailing—the multiply-complex international geopolitical-realities—we may recall certain paragraphs from **George Soros**'s book: *The Age of Fallibility: The Consequences of the War on Terror*, as since already very relevantly quoted in the 9th chapter on 'Conclusion' earlier, which do spell out that why and how, the USA's involvements today, has become so much necessary and also desirable in these regards. Hence as since her birth as a sovereign Nation-State in the North American Continent, the USA has quite healthily continued to remain a truly liberal democracy; and at the same time India too, since her so much indeed the history-making independence in 1947; also quite unexpectedly and also uninterruptedly too, is continuing to be a truly liberal democracy in South Asia; and hence there should never be any major difficulty for the USA and India—to almost whole-heartedly as well as sincerely click with each other in numerously positive and also far-reaching manners—for not only their own indeed enduringly mutual-benefits, but also for the larger world-peace and overall global tranquility as well.

A very dynamic former President of the USA, John F. Kennedy; has indeed known to have very perceptively said that the, "Purpose of foreign policy is not to provide an outlet for our own sentiments, it's (*in fact*) to shape real events in a real world". What could be more precise and visionary too, in these regards? But sadly so far as the then 'fledgling and yet to diplomatically shape-up', the domain of the nitty-gritty of newly independent but nevertheless; the most praiseworthily as well as, so much courageously fast firming-up—of the magnificent tryst with her democracy and, in spite of its all kinds of the accompanying *'heat and dust'*—the India under her otherwise so charismatic, scholarly, and also the no less multiply-visionary first Prime Minister **Jawaharlal Nehru**; seldom seriously and consistently cared for the timeless 'ideas concerning the diplomatic wisdom', as since contained in the above quoted words of John Kennedy; and thus for India, the rest is nothing but in many ways, a quite 'unpalatable-history' only. We may also recall that all these aspects, have since already been analytically discussed in the numerously different contexts, in the earlier chapters here. But as the saying goes that the mankind's 'living-history' with all its any conceivable depths and dimensions regardless, and since always continuingly in the making onto its expanding civilisational-canvas; has never remained stagnant and inert in any manner whatsoever. Hence India too has also come a long way while entering into now the present 21st century and thus already geared to move-ahead as duly equipped with multiple strategic-strengths and that too, after her 2014 general elections for her altogether new and also substantially more self-assuring 'tryst with new destiny' in the years ahead. India's new Prime Minister **Narendra Modi**, within just few months of his office as the nation's political 'Chief Executive' has already given clear signals both within and also outside the country, that he is no 'day-dreamer' or any sentimentally 'poetic'; but a cool realist, multiply-strategist, and also a hard task-master, but quite humane leader who means business and tenaciously wishes and acts for India's bright future as well as, all round betterment. Hence consequently no wonder that, now the India's entire gamut of the foreign-policy architecture, and its visionary steering too under this new and also irrefutably phenomenal Prime Minister Narendra Modi, has already undergone and also further undergoing a 'sea-change'.

The Times of India (Bangalore, dated 12th November 2014) came out with an 'Editorial-column' titled 'Strategic Swivel'; that very reflectively mentioned, "As Prime Minister **Narendra Modi** embarks on a high-profile three-nation tour, New Delhi should capitalize on this opportunity and adopt a more assertive approach to foreign policy. Modi's first port of call is Myanmar where he is slated to attend the ASEAN-India Summit. Both meetings come at a time when geopolitical focus is increasingly moving to Asia with the rise of China, an assertive **Shinzo Abe** regime in Japan and the US pivot towards the region. It's unfortunate that India has hitherto only made half-hearted attempts at integration with Asian multilateral mechanism, a fact made apparent by its absence at the APEC meetings in Beijing over the past week.... **In order to transform its Look East Policy into an Act East Policy, the Modi government must increase its footprint in East Asia. This would include pushing for India's membership at APEC while assiduously seeking a stake in free trade blocs such as the Regional Comprehensive Economic Partnership, the proposed Free Trade Area of the Asia-Pacific and even the US-led Trans-Pacific Partnership**. Doing so would not only provide India with strategic leverage in the region, but also help revive economic growth at home. **In fact, Modi will do well to keep in mind on the Australia leg of his tour that would see him attend the G20 Summit. While there Modi is also scheduled to address a joint session of the Australian Parliament—a first for an Indian Premier—highlighting the increasing strategic complementarities between New Delhi and Canberra. Australia also presents—huge economic opportunities for India—with significant room for improving ties in areas such as energy, mining, services and education. Plus along with Japan and the US, the two countries can become the fulcrum of a new strategic platform to ensure security and freedom of navigation in the Asia-Pacific.... Which is why Modi's final destination of Fiji serves as an important signal that India is ready to assert its interest in the Far-East. With a substantial Indian Diaspora, the South Pacific island nation (*i.e. Fiji*) can serve as an important base for New Delhi and enhance its reach in the region**. It's time India's foreign policy sheds its inertia and imparts new momentum to its Asia-Pacific strategic swivel."

It is all very true and hence the no less encouraging that now, the *'new'* India's so much momentous piloting of the quite dynamically altering foreign-policy architecture, under the highly proactive leadership of Prime Minister **Narendra Modi** has already fast begun shedding its earlier rather proverbial inertia and thereby along with in other regions of the world, has also begun imparting the much needed new momentum to its Asia-Pacific strategic swivel too. In fact the so much almost instantly fructifying—both the I.T., as well as the Media age, along with their numerous kinds of so much sweeping manifestations almost all across the world—these developments concerning today's any modern Nation-State, cannot ever be kept under any wrap and moreover in any case, the new Indian Premier in the best interests of his nation, understandably loves publicity and thus never appears shy of the same. In the multiple perspectives of her national-security concerns India has to now first and foremost in fact—both geostrategic and politico-militarily—worry about of the always lingering *'nuisance'*, i.e. rather the 'incurably failing State of Pakistan' in South-Asia itself, and howsoever one may desire, this problem as some kind of a 'geopolitical pain' in India's 'neck', just cannot be wished away for the at least some considerable years to come. **Hence needless to say that whatever any diplomatic or socio-cultural posturing from other side of the highly 'charged-borders' India just cannot afford to trust Pakistan and thus, take this 'turmoil-cum-terrorism fixated and failing' neighbour on its any face-value for granted, for building-up any long-term relationship.**

China, both geopolitically and geo-economically, is altogether a different 'ball-game' as since the modern China being powerful and also multiply-fast rising Power, visualizes India as her Asian rival and it is true that in any larger geopolitical-framework, the two rising Powers and moreover as giant rivals too can also very prudently cooperate with each other in many ways, for their long-term mutual interests. And nonetheless, China also feels being somewhat geopolitically at 'ease', to talk and do business with another strong and also fast rising democratic neighbour. Hence, India must consistently put-in her best to make herself stronger—both economic-cum-industrially as well as militarily too—to be multiply-confident; to fruitfully engage China for long-term peace at their borders, and also for numerously the many kinds of mutual-benefits to each other. **Moreover with a long-term geopolitical vision Prime Minister Narendra Modi has indeed very prudently appointed one of India's most well-known national-security experts, Mr. Ajit Doval, a former IPS officer, and a well-acclaimed man of deep understanding, knowledge, experience, and credibility too as India's new National Security Adviser, and since expected from the day one of assuming office, he is already proving to be a great asset to the Prime Minister, as remaining very silently behind the scene both in New Delhi as well as, during his always so much phenomenal foreign visits as well.**

Professor Kishore Mahbubani, Dean of the Lee Kuan Yew School of Public Policy at the National University of Singapore, in his article has very perceptively as well as with greatly futuristic-vision; has written that, "When the history of the 21st century is written and a list is made of the history's greatest missed opportunities, the visit of **President Xi** would probably be on it. No, the visit was not a failure. **But it failed to seize the great opportunity that beckons China and India in this century. Why was that? The simple answer is that the minds of Chinese and Indian policy-makers are polluted with a European concept invented 366 years ago, in the Treaty of Westphalia. That concept is 'sovereignty'**. Why is this concept destructive? Let me explain.... Just imagine that China and India were companies, not countries. As companies they would study each other's strengths and weaknesses objectively and see whether economic synergies could be exploited to make both companies profitable. Any such objective study would show that enhanced economic cooperation between China and India would be akin to a marriage made in heaven. India badly needs world-class infrastructure. As an infrastructure superpower, China has demonstrated that it can deliver super-highways, fast trains, and cheap power stations to India. And it can even fund them. At the same time China is running short of labour, India has labour in abundance. Chinese manufacturers could become globally competitive by manufacturing in India. India would then become manufacturing power.... If I were a Mckinsey or Bain consultant looking around for companies with synergistic opportunities I could not possibly find a better economic partnership. **In his heart of hearts, Prime Minister Modi understands this, because he visited China four times as Chief Minister of Gujarat to seize these opportunities. This is why Gujarat has about 31% of Chinese investment in India. President Xi also understood these opportunities. This is why he took the bold step to be among the first foreign leaders to call on PM Modi**. Before the visit, the Chinese Consul General of China in Mumbai, Liu Youfa, said that $100 billion of deals could be signed. So what went wrong? Sovereignty reared its ugly head. While President Xi was in India, there was a Chinese incursion into disputed territory. (*And the rest is nothing new, but some minor, but avoidably unpleasant episode of history.*).... **Once again, just imagine that China and India were companies. The Company of China has a land territory 9.6 million sq. km. The company of India has a land territory of 3.3 million sq. km. If these companies were located side by side and discovered that they had a dispute over 7,770 sq. km (or 0.08% of China's territory and 0.2% of India's territory), they would consider it absolutely irrational to allow a trivial piece of territory to interfere with the massive growth opportunities they could have**. I am not saying that money is the most important value. If China and India cooperate and significantly enhance their economic growth, hundreds of millions of Chinese and Indians will be rescued from poverty. What is morally more important? Rescuing millions from poverty or quarreling over bits of territory? If P.M. Modi and President Xi were company CEOs, they would both focus on the larger economic gains and push aside territorial disputes. Unfortunately if they do so as heads of governments, they would be immediately accused of being 'traitors' who gave up 'sovereign' territory. Sadly, the people making these accusations from their moral high horses would not be aware that their minds are trapped by **a 17th century European concept which no longer makes sense in the small, dense, and the inter-dependent world we have created.... From the year 1 to 1820, the two largest economies of the world, as documented by economist Angus Maddison, were China and India. The last 200 years of Western of world history, have been a major historical aberration (*well, courtesy: the unprecedented advent and advancement of now the so much new 'history-making'—the Industrial-Revolution.*).... it would be perfectly natural for China and India to once again have the largest economies in the world.... even though the Asian countries have been politically decolonized their minds remain mentally colonized by European concepts. When two major Asian countries ignore a great historical opportunity and instead focus on 'sovereignty', the only question the rest of the world will ask as they watch China and India bicker: is 'Can Asians think?"** (*Well, it is now a prime duty for both these Asian Giants; to very prudently prove and show to the whole world with a great strategic-vision that: 'Yes, the Asians too can think with indeed, the no less geopolitical-cum-geo-economic vision.'*). (Madhubani, The Times of India, 16 November 2014)

In these very multiply-precise perspectives here; as we pause a little and begin to reflect in depth and also with a critical sense of history—onto the above quoted paragraphs, since both historically as well as contemporarily; reflecting upon the two giant and fast emerging Asian economies, i.e. China and India—we do indeed find that, there are a number of very paths-breaking and also thought-provoking points in above referred article of **Professor Mahbubani**. Even though in fact, the two Chinese and Indian civilisations, happen to be now the two oldest surviving civilisations but ironically they seem to still not be able to purposefully come out of some of the both medieval-era as well as, even the last 20th century grooves of their lingering pasts. ***As we find that, since already very persuasively deliberated upon, in the above referred article—the certain momentously unique economic opportunities now, seems to be knocking at the doors of today's India and China—and thereby if these two Asian Giants still refuses to get engaged with their an altogether 'new trysts with their destinies' and thus, ignoring the bold writings on the walls of***

mankind's, specifically futuristic 'living-history', then that won't be sensible at all. **The two largest Asian neighbours should sincerely come together and muster the requisite courage with a great sense of history, to jointly take the 'calls of the future'.** *The 'million-dollar question' remains that will these phenomenally unique economic opportunities soon be able to decisively override some regressive geopolitical limitations, since working as shackles for these two Asian Giants in these regards or would they so naively be betraying the history now, even in the presently so much fast moving and the super high-tech era of this 21st century*? As clearly indicated, the whole world has already begun watching them with great expectations. **Now while contextually coming back to an indeed highly eventful visit of the Chinese President to India, and hence during his absence from China, a question also arises diplomatically as to that—who had begun calling the shots in Beijing—that in turn induced some obviously uncalled-for and even rather perhaps an 'unbecoming', but minor misadventure at India's northern borders with China? Thereby if the Chinese leadership was really being so geo-economically visionary—as since some advance statements already having been made by the Chinese Consul-General in Mumbai, and also the substantially very positive signals as so clearly given by the Chinese President himself while in India—then suddenly during his visit itself, what went geopolitically 'wrong' or perhaps 'disjointed' in Beijing, resulting in some rather 'senseless' but minor incursions by the Chinese troops into the northern border territories of India?***It is true that Chinese military since beginning from the times of Chinese communist revolution itself, has always been part of the ruling Chinese Communist Party's political establishment; but it is also true that Mao Ze Dong during his life-time itself as China's supreme leader and the unquestionable power-center, has finally established that it is the Party which would continue to irrefutably rule the gun, and certainly not vice-versa.* And hence it was all the more bizarre and even no less meaningless display of some rather 'naïve jingoism', by the so large and powerful Chinese military; and which was just not befitting for indeed the biggest Asian Power vis-à-vis its southern biggest and powerful neighbour India, with whom China now also admittedly desires to so much phenomenally engage itself; through investments of some billions of dollars in India's very large and long-term infrastructural development projects. Was it just an unintended minor aberration due to some communication-gap within the Chinese establishment or was it in fact, geostrategic-cum-politico-militarily and also 'sinisterly-planned', some 'games of one-upmanship' for some secretive purposes, or even perhaps the Chinese military strangely itself rather 'out of turn', even not wanting—now the already so much publicized huge Chinese long-term and also multiplying investments into India for their mutual benefits? Whatever may be the truth, but nobody in his or her right frame of mind, would ever imagine that China was even rather remotely inclined to behave like the 'failing State Pakistan'.

If these definitely uncalled-for Chinese some minor off and on misadventures at India's northern borders, are causing some headache only, for the Indian establishment but nevertheless, the same too are also resulting into some considerably awkward embarrassments as well as, all the more loss of face for the Chinese establishment, before the world as well. Any worthwhile modern Nation-State in today's times, must clearly understand that the present-era world has already considerably as well as, also irreversibly changed for the better in many very far-reaching manners, where any naked military-force now, just can never succeed in 'pushing-around' another big nation's also considerably powerful military, for any intended 'gains or even for securing any diplomatic mileage'. Hence the only credible avenue available, for any two powerful and also successful nations is to diplomatically engage in sincere and consistent talks, for solving any problems or misunderstandings. Needless to say, that the whole world is quite expectantly now, hoping that these two Asian Giants would soon very prudently-cum-multiply—take requisite politico-diplomatic steps to sincerely display their befitting statesmanship—to solve these otherwise indeed, very minor irritants and move ahead for the all-round good of the world and also, well-being of entire global humanity.

Interestingly at this precise juncture of our ongoing discussions a thought may also most probably come in our minds, as to why India's both the 'foreign-policy mandarins', and also the apex-level politico-diplomatic-cum-matters-military related, as well as professionally-focused the 'national security-establishment' be so much forewarned and thus, be doubly cautious vis-à-vis Pakistan in these regards? The point is that we have since for very many years, been sincerely talking to Pakistan, but with no avail whatsoever. Pakistan has been displaying a rather chronic habit of going back on its own words and thus, without just caring for its any diplomatic standing and reputation. In fact as a typically-fixated and also the highly regressive exercise in futility, Pakistan has been rather incurably harping on Kashmir always quite knowingly pretending to forget that the state of Jammu and Kashmir is already an inseparably integral part of the Indian Nation-State and hence under no circumstances, India would ever permit any change of its geographical position in these regards. So far Pakistan has tried all kinds of unacceptably foul means to somehow grab Kashmir including by both military misadventure as well as now, through the so many kinds of clandestine cross-border Islamic Jihadi-terrorism too, but to no avail to itself whatsoever. Due to these extremely foul acts on its part, Pakistan has gained nothing but in fact, suffered a lot in many ways, including its own dismemberment leading to the loss of once its own Eastern-Wing. But the both Pakistani military and civil-leadership establishments are even now, refusing to learn any important lessons for own good. **Just for argument-sake for a moment: suppose the state of J&K is 'handed-over' to Pakistan, will it change its ways and make it peaceful? If anyone in both India and Pakistan, or even in any other corner of the world is ever inclined to believe so then, he or she should only get his or her 'head examined at the earliest opportunity', to get any myopic-vision corrected for good**.

In fact, even if Pakistan is 'given J&K' the Pakistani establishment and mindset won't change even slightly but on the contrary, would even start laying absurdly fictitious claims on some other parts of the otherwise since always irrefutably sovereign Indian territories. Hence no wonder that today, Pakistan finds itself totally exposed before the world and thus commands very little credibility anywhere. This is the real problem or 'disease' with Pakistan since its birth itself and hence, ironically and even quite regressively, Pakistan finds this as the only way to be able to somehow maintain some relevance for its own existence before its own misguided and frustrated masses. And then in turn even quite 'humorously', Pakistan finds itself almost forced by its own consistent misdeeds, to keep-on indulging in all kinds of the absurd ***'Takia-Kalam'***, as since clearly told to some Pakistani representatives earlier, by India's one well-known defence-analyst, **Major Maroof Raza**; but Pakistan quite stubbornly, refuses to see reasons and to get 'normal' for her own good. In fact Pakistan's extremely regressive as well as, this typically-fixated 'culture of fooling around', had begun almost with the birth of this present-day largely stagnant and 'failing' State. **In any case, the irrefutably basic fact remains that in order to practice any 'culture of fooling around' at any level anywhere; the only workable vehicle remains, is telling all around lies only**. Hence due to its own misguided and also quite foolishly cultivated—Pakistan's some untenable and the 'so-called necessities'—Pakistan was quick to master this 'art of her culture of fooling around'. And thereby Pakistan in the process, developed its own rather comical-art of practicing its absurd skill for ***'Takia-Kalam'***; and thus Pakistan kept on doing the same and is still quite ironically doing the same, both internally and also externally irrespective of its so much loss of face and also any credibility it once had, in any global diplomatic-circuit.

In certain contemporary perceptions and also due to contextually some larger reasons on USA's part herself; it is also now true that in the process, Pakistan was able to take even the Superpower USA *'through a garden-path'* and thereby, has also been successful in continuously reaping substantial bonanza in terms of somewhat free flow of dollars as well as, the military hardware (*which in any case, it never deserved at any given timeline of its limping existence.*). And now even needless to say that during all these years, Pakistan has only been rather criminally misusing and squandering-away most of these undeserved assets. **In these specific contexts, ironically USA had her own geopolitical compulsions; in fact in the 'arena of her some importantly competitive great-games', in the Middle-East, in both South and Central Asia, and also in China—vis-à-vis the then another Superpower USSR, during now the already bygone Cold-War era—and for all these, the USA had no other option but to use Pakistan has her 'client frontline State' for obvious reasons also, to her greatly advantageous effects. And the rest is now indeed all the more well-known history, concerning the final demise of the then infamous Cold-War and in addition, historic disintegration of the former USSR and also the positively far-reaching transformations of even the then erstwhile fanatically communism-fixated China too.** Sometimes it remains true, that some historic-facts are indeed stranger than fiction. Moreover as since the saying goes that internally in any truly liberal democratic State, if the competitive electoral-politics remains a game of possibilities then it is all the more phenomenally true that in the highly dicey-cum-slippery 'international arena of the geopolitical power-politics', any conceivable and all the more, especially for any Superpower its international great-games' too, remain the games of some much more variables, chances, and the possibilities.

In other words, we may also say that when two big Powers 'clash', whether in a kind of 'Hot even Cold War'; the chances remain that any much smaller and poor nation too can get trapped and 'crushed', but to its good-luck it could also be possible that the same otherwise inconsequential and small geographical entity, can also very craftily be able to reap some great and at times, even largely quite undeserved benefits. And that's what precisely has so far happened to Pakistan (*as since already discussed above*) while acting as a readily available, since the USA's 'client frontline-State' in these regards. But now as a great disadvantage to in fact—the same artificially created, and so a typically still-born and thus unviable-cum-untenable, the so-called 'Nation-State' of Pakistan—where ironically for all these years, the USA had to waste billions of dollars in terms of both hard-cash as well as, also the military hardware; to finally see clearly through the Pakistan's all kinds lies and *'Takia-Kalam'*, and also its since very dirtily-crafty 'culture of fooling-around'; and hence Pakistan's 'games seem to be up' now and naturally thus, for the overall much larger good of the Indian subcontinent. That's why the USA has now begun listening to the 'new' India, under her current Prime Minister **Narendra Modi. But lo and behold, Pakistan with its typically very crafty and old habit of playing as 'innocently helpless victim'—is now putting across a new 'card of getting very menacingly cornered and thus about to be overwhelmed by the new and also much more dangerous monster, i.e. ISIS'—to seek even further US economic and military aid by indulging in the drama of all kinds of 'chest-beating and also shedding crocodile's tears', and so trying to again take the mighty USA, *'through the same old garden-path'*. It is hoped that now this time at least, the USA doesn't again very naively falls in the trap of this great 'bluffer', i.e. Pakistan and thus it is essential that the today's sole Superpower very prudently takes India too on-board, while more effectively dealing with Pakistan**. And further in any case now, the USA doesn't need Pakistan so compulsively anymore, as like during the erstwhile Cold-War era times.

Now internally in India, so far as the multiple-aspects of governance are concerned here

too Prime Minister Narendra Modi is proving to be the no less dynamic, focussed and decisive, and also considerably results-oriented too, as the nation's new Chief Political Executive. It is true that almost immediately after so phenomenally taking-over the reins of the modern and democratic Indian Nation-State, PM Modi has very ambitiously but at the same time, very systematically planned the no less visionary, and multi-dimensional reforms-measures and programmes in rather unprecedentedly action-oriented manners.And needless to say that in these crucial spheres of policy-formulations and governance, and that too in a large, complex, and also the so much diversified country like India much patience, perseverance, as well as the equally patriotic, motivating, impeccably honest, and tenaciously focused—top-level political leadership with a strong-cum-decisive mandate—is always an unquestionable prerequisite. But some past experiences in the post-independence India shows that here too remains a very dicey catch in the sense that, in spite of the above factors being largely in place—the personal integrity, experience, and the well-gelled knowledge and political wisdom too at the very top-level of otherwise very popular leadership, and simultaneously even if this leadership commanding some unchallenged authority over his or her own Party, if ever comes under some lingering shadow of doubts and thus becomes directly or indirectly questionable—then the entire plans-cum-programmes can get derailed and thus very tragically, force the multiply great costs, setbacks, and disadvantages, onto the hapless Indian masses across the nation.

In India, there are many reforms yet to be formalized at apex policy-levels, such as—crucially essential amendments and reforms in the nation's labour-laws, critical course-corrections in the laws of sales-taxes and so on, by bringing-in the much desired 'GST' in some well-formulated and also accepted manners across the board; reforms in the existing and thus highly problematic systems-cum-laws concerning the acquisitions of lands for industries, and also for other highly valuable developmental-projects and so on; persuasively bringing-in the various state-governments on-board, as since India constitutionally having an otherwise much advantageous federal-structure too; and last but not the least—**the prime need to multiply modernise, reform, and revamp the bureaucracy at various-levels**—since being the essential segments of democratic governance-systems across the country as well.

In these multiple contexts, **Dr. A.P.J. Abdul Kalam**, former President of India, and **Dr. A. Sivathanu Pillai** Chief Controller (R&S of DRDO) and also CEO-cum-MD of the Brahmos Aerospace (An India Russia Joint-Venture)—**in their Book titled, 'Thoughts for Change—We Can Do it'**,—and Published by Pentagon Press, New Delhi, in 2013 and then represented in 2017; and thereby in the Preface Chapter itself (in the pp. xiv, xv , & xvi), has very befittingly said that—"We need to set our targets high and work hard to achieve them unitedly. We may encounter genuine problems on the way but we have to defeat them so that the problems do not absorb us and become our masters..., India has seen significant progress in certain mission mode projects. In this pursuit, the human resources of India became the key to its growth, particularly in many niche technology areas. Advanced management practices are also important to keep pace with-global growth in high technology areas. Consortium of institutions and matrix organisations provided excellence in high technology projects through synergy and cross flow of knowledge. Apart from indigenous development of products, collaborations and joint ventures helped in the fast development of certain critical technologies and the production of systems. This multipronged approach not only helped in acquiring key technologies in the shortest possible time but also opened up new frontiers in the developments of new products leading to economic growth. There are the few examples the newer technologies that are going to revolutionize the future and empower mankind. India should utilize this opportunity in developing these technologies and to become a Developed Nation on par with-industrially developed nations. But many question whether India will become a developed nation? This is the result of an inferiority complex that was built over the years, when we were ruled by others. It is time that the youth of India must come out of this defeatist spirit and herald the message to the world that we are knowledge power. We must shed the mindset of the defeatism and energize ourselves with the spirit of confidence and victory with indomitable spirit to achieve our cherished dream of a developed nation for India."

These are no easy jobs, but these are all the more inescapably essential too, which would always remain beyond the capabilities and political-visions of any average and thus, easily dispensable politicians since belonging to any Political Party, and also thus masquerading as 'leaders' almost everywhere in the country, including some of those in the Prime Minister Narendra Modi's Party itself. He as Prime Minister himself, fully understands the nitty-gritty of this ground-reality and so always remains aware of it; but it is also the no less important that peoples of India too, as citizens of the world's largest, as well as a fast developing democracy equally understand these and thus, always very discerningly remain alert in these regards. Now so far as very precisely here, the multiply crucial aspects of the multi-dimensional reforms, modernisation, and also revamping of the bureaucracy and thereby, those of the nation's civil-services are concerned—**the Modi-Government's dynamically proactive-cum-visionary policies and programmes—have very revealingly been brought-out in the editorial-columns of *The Times of India* (Bangalore edition), dated 27th October 2014.**

The above cited very important editorial-column, titled as 'Open Babudom's Gate: Later entry will improve governance and waft a fresh breeze through musty bureaucratic corridors'; has very aptly stated that, "The NDA government is working to institutionalize lateral entry from academia and the private-sector into some senior government positions. This is long overdue reform with far-reaching potential. Changing the way bureaucracy works and moving from a closed to an open system for recruiting civil-servants are prerequisites to enhancing the quality of governance. Suggestions to institutionalise lateral entry have come in the past from the government's own groups of experts, but have not been followed-up. Hopefully on this occasion, NDA will see this important reform through to its logical end (*Needless to say that this is a must—in democratic and modern India's multiply highest national interests—and therefore in spite of any kinds of resistance emanating from any quarters, the* ***Prime Minister Narendra Modi*** *must decisively push-ahead such indeed very far-reaching reform measures concerning the bureaucracy as usual; true to his words, thoughts, and also so dynamically governance-oriented initiatives systematically to their logical fruition.*).... Lateral entry has always existed in independent India's civil services. **Nandan Nilekani**, who set-up the Aadhaar scheme which can transform India's social welfare system, is an illustrious recent example. The process, however, has been ad hoc in nature. **Given the strong link between governance and propriety and growing complexity in the way societies function, countries such as the UK, US, Australia, Holland and Belgium throw open specific government positions to qualified personnel from all walks. It's the best way to net the right person for the job.... Domain competence and a combination of relevant knowledge and skills are an essential requirement in governance. These attributes are not always present in a cadre of generalists; moreover one that is recruited increasingly through reserved quotas which limit competition. The second administrative reforms commission, which presented an elaborate blueprint a few years ago, envisaged a shift from a career-based approach to a post-based approach in the top-tier of government jobs.... Without independent authority, there is a danger that lateral entry which enhances the quality of governance will be thwarted by the entrenched IAS lobby. Indeed this process of recruitment from outside could be used to pries open the stranglehold the IAS lobby has on key**

appointments, thus enhancing the quality of governance. For sure a change of this nature will not be easy as there is bound to be resistance from within babudom. NDA, however, should push ahead with this reform as India's interest is greater than the IAS interest."

Indeed now at this unique and also highly expectant juncture of the Indian polity the above noted and particularly far-reaching, as well as the no less long awaited 'Reform-item', which in turn has the potential to soon begin bringing-in the so much valuably multi-dimensional benefits, at the top policy-advisory levels' of the bureaucratic processes in decision-making and national governance, would also be a 'test-case' for the multiple-aspects of the Prime Minister Narendra Modi's visionary leadership and decisiveness. **We may recall that earlier in the last sub-chapter of the 8th Chapter, we have since also critically discussed similar aspects concerning the 'Police-Reforms', as well as the IPS, i.e. 'Indian Police Service; and hence these too need to be similarly attended to, with equal vigour.** But the present Indian Prime Minister almost regularly and that too at any earliest opportunity, has been proactively presenting his personal, as well as very inspiring examples of quite unhesitatingly **'thinking out of the box'** to push-ahead the nation's multi-dimensionally fine-tuned developmental-agenda—for ensuring economic progress and the peoples' welfare, meeting the nation's overall security needs and also enriching its military wherewithal, meeting the concerns for internal and regional-peace and thereby, trying to fruitfully contribute towards strengthening the essentials of world-peace by striving to establish both diplomatically-dynamic and also personal friendship-based relationships with the almost all leaders of powerful nations of the present times. **In India's multiply important contexts today and also as the recently some depressing experiences go to show; these are no small matters and hence, these do point towards very reassuringly encouraging future now, for the modern Indian Nation-State.**

So far as the multiply vital 'foreign policy aspects' are concerned—India's **oneretired foreign secretary Mr. M.K. Rasgotra, in his well-received book titled, "A Life in Diplomacy"** (published in 2016 by Penguin-Viking Group, New Delhi), and therein in the p. 368 has very precisely stated that, "The foreign policy of a state is a framework of principles and practical considerations in the light of which it seeks friendship and cooperation of other countries to pursue and strengthen world peace and to safeguard its security and other national interests, on a reciprocal basis. Foreign policy making is a dynamic process which must remain open to an adaptation, adjustment and change in accord with changing conditions of the world.

The foreign policy of a state begins with its neighbours, but in must not remain trapped in the region to the neglect of the state's interests and role in the larger world. India will grow to greatness only if its vision and policies transcend the subcontinent's truncated geography." And hence needless to say that, precisely as per the well-gelled nitty-gritty of a Nation-state's foreign policy aspects; **India's most popular and dynamic Prime Minister, Mr. Narendra Modi** is, indeed, very effectively and also very fruitfully triggering his multiple-directional initiatives-in both India's neighbourhood as well as in the larger world.

And hence precisely in the above specific contexts, with all their multi-faceted connotations and their magnificent historicity too let's quote here, certain excerpts from **Henry Kissinger's book. Some very relevant excerpts from the above mentioned book have recently been published as an 'article' by this well-known author and an erstwhile American diplomat, *The Times of India* (Bangalore Edition, dated 21 November 2014). And therein, Henry Kissinger has very reflectively said that,** "World Order in Hindu cosmology was governed by immutable cycles of an almost inconceivably vast scale—millions of years long. Kingdoms would fall, and the universe would be destroyed, but it would be recreated, and new kingdoms would rise again. The true nature of human experience was known only to those who endured and transcended these temporal upheavals.... The Hindu classic **Bhagavad Gita** framed these spirited tests in terms of the relationship between morality and power. Arjuna, 'overwhelmed by sorrow' on the eve of battle at the horrors he is about to unleash, wonders what can justify the terrible consequences of war. This is the wrong question, Krishna rejoins; because life is eternal and cyclical and, the essence of the universe is indestructible. Redemption will come through the fulfillment of a pre-assigned duty, paired with recognition that its outward manifestations are illusory because 'the impermanent has no reality, reality lies in the eternal. Arjuna, a warrior, has been presented with a war he did not seek. He should accept the circumstances with equanimity and fulfill his role with honor, and strive to kill and prevail and 'should not grieve'.... While Lord Krishna's appeal to duty prevails and Arjuna professes himself freed from doubt, the cataclysms of the war—described in detail in the rest of the epic—add resonances to his earlier qualms. This central work of Hindu thought embodied both an exhortation to war and the importance not so much of avoiding but of transcending it. Morality was not rejected, but in any given situation the immediate considerations were dominant, while eternity provided a curative perspective. What some readers lauded as a call to fearlessness in battle, **Gandhi** would praise as his spiritual dictionary.... The pioneering exemplar of this school was the 4th century BC minister **Kautilya (*i.e. Chanakya*)**, credited with engineering the rise of India's Maurya dynasty, which expelled Alexander the Great's successors from northern India and unified the subcontinent for the first time under a single rule.... Kautilya wrote about an India comparable in structure to Europe before the peace of Westphalia. He describes a collection of states potentially in permanent conflict with each other. Like **Machiavelli**'s, his is an analysis of the world as he found it; it offers a practical, not a normative guide to action.... The Arthashastra sets out, with dispassionate clarity, a vision of how to establish and guard a state while neutralizing, subverting, and (when opportune conditions have been established) conquering its neighbours. **The Arthashastra encompasses a world of practical statecraft, not philosophical disputation. For Kautilya, power was dominant reality. It was multidimensional, and its factors were interdependent. All elements in a given situation were relevant, calculable, and amenable to manipulation toward a leader's strategic aims. Geography, finance, military strength, diplomacy, espionage, law, agriculture, cultural traditions, morale and popular opinion, rumors and legends, and men's vices and weaknesses needed to be shaped as a unit by a wise king to strengthen and expand his realm—much as a modern orchestra conductor shapes the instruments in his charge into a coherent tune. It was a combination of Machiavelli and Clausewitz**.... What our time has labeled covert intelligence operations were described in the Arthashastra as an important tool.... The Arthashastra's exhaustive and matter-of-fact catalogue of the imperatives of success led the distinguished 20th century political theorist **Max Weber** to conclude that the Arthashastra exemplified 'truly radical Machiavellianism'...compared to it, Machiavelli's 'The Prince is harmless.' Unlike Machiavelli, Kautilya exhibits no nostalgia for the virtues of a better age.... Whether following the Arthashastra's prescriptions or not, India reached its high-water mark of territorial extent in the third century BC, when its revered **Emperor Ashoka** governed a territory comprising all of today's India, Bangladesh, Pakistan, and part of Afghanistan and Iran." (Kissinger)

In these multiply vast and always timelessly relevant contexts we may recall that in the earlier chapters, we have fairly analytically discussed the principal contents as enshrined in the **Kautilya**'s Arthashastra, and at the same time, along with while discussing **Sun-Tzu**'s equally far-reaching treatises on 'War and the Art-cum-Principles of War', we have also quite critically compared and correlated both these immortal Masters' 'ever-green' creations; as since passed-on to the mankind, as unique and also ever fresh storehouses of knowledge and wisdom. But a distinct feature of Chanakya's Treatise, i.e. Kautilya's Arthashastra is that it not only deals with the multiply-crucial aspects of war, but also very enlighteningly delves into the entire gamut of governance and that too along with, very comprehensively encompassing onto the **'multiple-aspects of essential strategic-culture and thoughts'**; and thereby establishing their inseparable-cum-integrative relationships with the **'multiple-aspects of governance'** as well, for the first time in the annals of mankind's 'living-history'; since continuingly in the making onto its expanding civilisational-canvas in this world. These precisely otherwise too well-known aspects here, have again been very aptly revisited by **Henry Kissinger** as well, in the above quoted paragraphs from his writings.

So far as **Bhagavad Gita** is concerned, it is always believed since the very ancient of times that, this Immortal Epic's Canvas is not only colossally expansive but in fact, it is truly nothing else but Infinite. Gita indeed through the timelessly unique and also otherwise extensively tormenting saga of the numerously different kinds of 'trials and tribulations'; has been teaching and conveying the essence of 'True Knowledge', and which alone in this illusory world can lead one to the immortal bliss and enlightenment. No wonder that since the very beginning of times and the advents of human-civilisations, i.e. in whatever manners we may prefer to delve into it, we find that many Peers, Rishis, and much later even **Mahatma Gandhi** too, brought Gita into their lives of both meditation and action not only as their spiritual compass or dictionary, but also as their friend, philosopher and guide as well. **As an Epic, Gita is not very voluminous or vast—but its true essence is not only timeless, but truly unfathomable in all its infinite depths and dimensions—by means of the worldly human intelligence faculties and instruments.** In other words, we may also say that the Gita's truly complete and comprehensive interpretations are in fact, beyond any human mental or even spiritual capabilities. But through close and also truly involved reading or delving into Gita, in one's undisturbed solitary states can somewhat experience the true essence of Gita in one's life, not only unimaginably blissful, but also mesmerizingly enriching too. **In fact, the both scientific, as well as spiritual fulfillment is not in succeeding to 'measure Gita' both quantitatively and also qualitatively—as an 'Infinite source of the true realisation and bliss in all its manifestations'—as since it remains just impossible to do so but in so captivatingly entering into this 'infinite ocean of unique-nectar', and then equally be most willing to get 'drowned into it' as well.** *And one may say in any context and also in a layman's words that, that's what Gita is all about.*

Interestingly in the above quoted excerpts from the above cited Henry Kissinger's book, we find that he has very perceptively mentioned about the great historicity of the then establishment of an unprecedentedly massive empire during the very ancient era, by India's one so much glorious **Emperor Ashoka** and; and that's what is indeed true and also, the no less well-known all across the world. **But it is also all the more historically mesmerizing that the same glorious Emperor, after the extremely bloody 'Kalinga War', when came in contact with the radiating Divinity of Lord Buddha and as he most sincerely surrendered his mighty sword in front of the Lord's lotus feet, he achieved true enlightenment and then, denounced violence and war forever. And thereby the same mighty Emperor Ashoka, indeed from his unchallenged position of might and sway, sent his historic peace-missions and learned emissaries, including his daughter into almost all the kingdoms of the then South-East and also Far-East Asian regions; and the rest is—all the more socio-culturally as well as in human-civilisational terms—the very glorious-cum-multiply-enriching chapters of the mankind's 'living-history', since pertaining to the Emperor Ashoka's times and their relics even today, remain the no less captivating for the whole world.**

In humanity's these magnificent civilisational processes of socio-cultural-cum-spiritual interactions and historic exchanges—between the ancient era's Indian societies, as well as those of the then regions of South-East and also the Far-East Asia—and wherein subsequently, a couple of the gloriously resourceful kingdoms of South-India also very proactively participated; and thereby onto the mankind's expanding civilisational-canvas, the gloriously immortal epics of Ramayana, Mahabharata and Gita too, phenomenally travelled into those Asiatic regions. And needless to say that even today, the very attractively captivating imprints of these unique Indian Epics are found to be alive. **Therefore the mankind's 'living-history' since continuingly in the making onto its expanding civilisational-canvas and thus, in any larger civilisational perspectives, that's the power of socio-cultural interactions and exchanges and hence thereby, altogether even quite newer kinds of their subsequent fruitions are seen in various human societies across the world.** In any case, these aspects in all their socio-cultural-cum-heritage connotations and dimensions too, do always form living-segments of the multi-dimensional comprehensiveness of, since the already substantially discussed in numerously different contexts in a number of earlier chapters here, and that's what being nothing else but precisely, **'the multiple-aspects of essential strategic-culture and thoughts'** only.

It is true that the rules, substances, aims, and modalities of the various kinds of 'great-games', which the nations do keep-on playing onto the very dicey and slippery "arena of the international power-politics"; are in fact governed-by, or remain subjected to the numerous kinds of indeed always the very unpredictable and also the highly complex-cum-age-specific variables, since pertaining precisely to specifics of contemporary geopolitics and geo-economic compulsions as well. But the precise and always the larger governing contents-cum-values of the—**'multiple-aspects of essential strategic-culture and thoughts'**—do remain largely constant. Keeping this 'conceptual-framework' in our minds as we do embark on our critically-discerning journey through various kinds of the highly engaging passages of the mankind's 'living-history', as per their advancing timelines in historically chronological-orders; we do begin to feel and experience the meanings having been conveyed in the above recorded statement, since specifically pertaining to **'the multiple-aspects of essential strategic-culture and thoughts'**. And thereby in these ongoing processes, as we do come now to the very fast-moving super high-tech and also the ultra-modern era of the present 21st century; we find that democracy in all its connotations and socio-political perspectives—is the highest kind of and also the most revered concept-cum-tool for the man's civilisational-existence and the entire gamut of governance for any worthwhile, truly successful, and also the really valued nation in the global comity of nations in today's world. Finally and that too once and for all now in fact, gone are days of any kind of imperialism or 'empire-building' and similarly as per the same unalterable ground-realities; gone are also the days by expanding one's territorial geographies, or even altering any international boundaries between any two neighbouring nations; by sheer and naked military means. Hence any nation, if due to any reason or any kind of regressive fixations or geopolitical-hangover, still chooses to believe otherwise, then it is living in its own self-created 'time-warp' and thus, remain trapped in its own futile 'culture of fooling around'.

Having said that and also keeping in our minds—the paramountcycy of 'the multiple-aspects of essential strategic-culture and thoughts'—it can so far never be any sensible proposition that, possession of—the adequate, well-crafted and, always the dynamic-cum-perpetually-modernising as well as, militarily-strengthening Armed Forces and, their geostrategic-cum-politico-military wherewithal; would ever be any wastage of national resources and efforts. In fact due to any otherwise foggy reasons or thoughts, anybody who harbours any such ideas or beliefs, should get his or her 'head examined' at the earliest; and in addition also persuade anyone of the similar kind to do so as well, to effectively get cured of any such absurd fallacies. **Both**

geopolitical-cum-geo-economically adequate and vibrantly multi-dimensional military-wherewithal even today, remains essentially inescapable to not only be able to maintain peace, but also to effectively believe-in so and talk-peace; always from one's strategically tangible as well as, duly well-recognised positions of strengths in any given or emerging external situation(s), in the very dicey and highly slippery arena of international power-politics. Hence in the today's fast-track-cum-ultra-modern era that also in fact, precisely means 'geopolitics' in many important ways and thus, sure 'keys to the always realistic and cool business of the statecraft', in any truly vibrant and successful democracy. And all the more at the same time, it never means any naïve 'saber-rattling' or allowing any 'militarism or war-lordship' under any circumstances, to even slightly be allowed to creep-in into one's thoughts, policies or actions, whatsoever.

In precisely these multiple-perspectives now, it is heartening to learn and watch the new Indian Prime Minister **Narendra Modi**, to remain totally free from any such duality or conceptual confusion at his apex level of geopolitical-cum-diplomatic engagements with any foreign leaders, during his official-visits abroad. As the timelessly factual saying goes that in the always ongoing geopolitical-cum-geo-economic 'great-games' in the comity of nations, there are no permanent friends or enemies, but there are only one's permanently compulsive national-interests. **In a democracy, there is always some 'lurking danger' for even otherwise a very vibrant Nation-State to suffer setbacks; if the elected leaders happen to be corrupt, are not sincere and incapable of understanding the crucial nitty-gritty of the above discussed geopolitical-cum-geo-economic imperatives vis-à-vis their conduct of the concerned nation's foreign-policies and thereby, successfully serving its highest national-interests and also, the future well-being of their nation's peoples. So the peoples in a democracy are always needed to be all the more alert and vigilant in these regards; as it is true that in any given situation, there is always very slander threads separating the fruits of success, from the turmoil of any failure or pain even with some temporary setbacks.**

To substantiate the aspects concerning the nation's macroeconomic fronts, we may like to refer to a very precisely relevant article by **Mr. Sajjid Z. Chinoy, Chief India Economist, J.P. Morgan,** and wherein, he being indeed a very distinguished expert has quite perceptively mentioned that, "INDIA HAS UNSURISINGLY brushed-off any post-Brexit impact. Some of this is because emerging markets, themselves, have come under little stress so far. But a lot of this is because, in a matter of three years, India has gone from being the poster-child of emerging market vulnerability to becoming the safe haven in that universe. Inflation has halved, the central government fiscal deficit has narrowed by a third, the current account deficit (CAD) has gone from 5 percent of the GDP in 2013 to 1 percent of GDP in 2016, and foreign exchange reserves are a record high. There are several proximate causes for the turnaround: lower oil prices have helped across the board, the government's bold FDI reforms have improved the quantum and quality of capital and weak investment has narrowed the CAD. All are true at some level. **But the bedrock of macroeconomic stability has been provided by sound fiscal and monetary policies post the taper tantrum, evidenced by the fact that the macros began to improve well before the oil-prices fell.** The question is how institutionalised are these policy frameworks. Can they survive regime changes? Reforms are also about ensuring that they are institutionalised such that future governments find it hard to reverse them for politically expedient reasons. Viewed from this lens, there has been dramatic progress on the new monetary policy framework. Just 30 months ago, monetary policy was still being run via a multiple indicator approach, creating uncertainty about the focus of policy. Inflation? Growth? Financial stability? Exchange rate? What combination thereof? Within inflation, was it CPI or WPI? Headline or core? Little wonder it was hard to anchor expectations. But things moved rapidly from January 2014. The RBI implemented Urjit Patel Committee Report—Which recommended a decisive pivot to headline CPI targeting and a glide path to 8 percent, 6 percent and then 4 percent with a band of +/-2 percent in the steady state. The new government was quickly on board with FM arguing for a 'modern monetary policy framework' in his maiden budget speech in July 2014. By February 2015, the ministry of finance and the RBI had entered into a monetary policy agreement which gave an unambiguous primacy to price stability and spoke about targeting inflation at 4 percent with 2 percent band on either side for 'all subsequent years'. These may have been small changes in language. But they represented a huge leap in thinking.... In just two years, India has seen a remarkable transformation of its monetary policy framework, for which the government and the RBI deserve credit. The sooner the government notifies the inflation target (with the hope that the goalpost is not changed) the faster the speculation will end, and we can appreciate just how much has been accomplished." (Chinoy, The Indian Express, 19 July 2016) (*All these aspects are indeed true and also both policy as well as overall national governance-wise, they are very far-reaching and enduring too. No wonder that this government has now been able to get the much talked-about GST Bill along with the related necessary constitutional-amendments Bills as well; duly approved and also passed by both Houses of the Parliament. Now needless to say that these so important Bills were kept hanging since last 25 years or so, due to all kinds of political mud-slinging and also many kinds of equally unworthy political-games of one-upmanship only.*).

Similarly, we also find that the present Modi Government has quite consistently been highly innovative and focused, be it nation's both internal and external security compulsions, internal political stability, all-round efforts to bring-in multi-dimensional improvements in political governance, bureaucratic functioning, educational-systems in the country, matters concerning Centre-States relationships and so on. And also above all, this Government's sincerity to consistently maintain maximum possible transparency in its all segments and practices of democratic polity as well as laid-down constitutional-framework—so that the peoples of India remain participative and feel empowered too. While doing so decisively and numerous positive things, nothing is ever easy and then big achievements do take also their own mandatory timelines; and hence due patience too becomes mandatory. Many big-tickets beginnings have since already been made and thereby, relevant activities on the ground too, are showing their encouraging visibilities; but it is understandable that consequent fruitful results would in any case take their own time, and we have still miles to go, before we even pause for a rest. Of paramountcy importance is that, numerously proactive and multi-dimensional beginnings have since already been made and simultaneously newer beginnings in these regards too, are being very innovatively triggered, and thus making the peoples encouragingly optimistic as well. Just to cite few examples: we may notice, the Prime Minister's programmes and decisive initiatives such as—Peoples' 'Dhan-Jan Yojna', Ujjuala Yojna, Swachchhata Abhiyan, linkages based on 'Adhaar Cards', voluntary giving-up of LPG-cylinder subsidies, giving LPG connections to so far deprived sections of our societies, farmers' 'soil-cards' system, farmers' crop-insurance schemes, programmes concerning 'keeping the girl-child safe, loving and also to encourage their due education too; Start-up India, Skill-India, Make in India, one rank-one pension (OROP) for the Defence Services and so on.

Hence we may also say that, these too are principally the pivotal lessons from the mankind's 'living-history'. In fact in today's indeed ultra high-tech and also super fast-track-oriented world; the dimensions of the many mind-bogglingly ongoing advancements in the numerous areas of sciences and technologies, including those in the space-age technologies always multiply assume very far-reaching-cum-futuristic implications, including those in the highly complex areas of the 'geopolitics' as well. **Hence, these aspects too are needed to be kept in the minds of top political-leadership of the any concerned truly democratic Nation-State.**

Now precisely with respect to the above and thereby specifically concerning to the extraordinarily latest advancements in some classified and also highly 'guarded' areas of the 'Space-Sciences'; **Tara Macisaac** therein, says that, "Google Moon shows what appears to be a massive alien spacecraft on the moon. It is L-shaped and what appears to be seven lights dotting its edge. The Google Moon imaging is taken from NASA's Lunar Reconnaissance Orbital Camera (LROC).... Turner Radio Network reports that the US government has been secretly tracking two huge objects approaching Earth for two years, citing an outside consultant from NASA's Meteoroid Environment Office. In January 2013, the objects disappeared all of a sudden about 200,000 miles past Mars, reports the network, but it seems that one of them has landed on the moon.... On Jan. 15, NASA launched three Terrier-Orion suborbital rockets from the Wallops Flight Facility in Virginia for the Department of Defense (DoD). The NASA press release states, 'At the request of DoD project managers, no real-time status updates will be available. The launch will not be shown live on the Internet, not will launch status updates be provided on social media once the countdown begins.'.... As the Google Moon image gains attention, some have speculated that the launch could be related.... Turner Radio Network named its source as 'Dr. Eric Norton', a long time consultant for the National Security Agency (NSA) and NASA. Norton described the objects he was called in by NASA to help monitor at the MacDonald Observatory in Texas.... 'What I saw was an array of massive, three-dimensional, black structures in space, in straight-line formation, advancing in direction of planet Earth.' he said. He was shown the objects over a three-month period and saw they 'had moved millions upon millions of miles closer within just months.'.... **Analyzing the objects, he and his team hypothesized that they must be made of a material several thousand times harder than the hardest materials known to humans on Earth.... The objects have been a cause of great concern, Norton said, with many questions arising and few answers.... Turner Radio Network bills itself as a media source that doesn't make it to mainstream media.... Tech and Gadget News speculates, 'The gigantic shape actually looks like the leading edge of an immense, triangular space ship, similar to so far, super-secret stealth aircraft technology, but a much larger than any airplane ever built on Earth**.... The video showing the Google Moon sighting has had more than 500,000 views on You Tube." (*Indeed quite unimaginably mindboggling, isn't it? But that's what means precisely, in the truly multi-dimensional ongoing advancements and such astonishing breakthroughs in the latest frontiers of the extraordinarily Outer-Space related technologies; since continuingly multiplying and also expanding as highly specialised domains of the human-knowledge.* ***But what about the so-called, and yet all the more little-known Aliens; who appear to be much more advanced than we humans on this planet Earth? Well as time passes-by, only future will tell and unfold many more secrets in these regards.Hence so far that's all as it is and in any case, what more to say?***). (Macisaac, *Epoch Times*, 21 January 2014)

*

Appendix–A

INDIA AND THE MULTIPLE-ASPECTS OF 'STRATEGIC-CULTURE AND THOUGHTS': THE DEMONIC MENACE OF HER IRREFUTABLE OVERPOPULATION

Major C.B. Verma (Retd.)

(N.B.: *An unpublished article written before this book, around 2008*).

It is a proven historic fact that, since thousands of years India has been lacking the **"multiple-aspects of essential strategic culture and thoughts"** and at the same time, since throughout the continuously ongoing making of India's 'living-history' onto her expanding civilisational-canvas, we as a nation, have never even cared to give any meaningful thoughts to these vital aspects and therefore, never any fruitfully considered to gradually cultivate and imbibe them with requisite and proactive synergy. This specifically bizarre aspect in fact, may objectively be considered to be even a great 'negation' of since, otherwise so much multiply glorious: the Indian-civilisation. These crippling and continuous failings of ourselves, as a 'Nation' have been of disastrous consequences to the very 'idea of India' as well as also been bringing a repeated 'shame' onto ourselves. It also resulted in very sad-commentaries' on our history. Even bigger tragedy is that today ironically, already with the dawn of twenty first-century, we are still refusing to recognize and understand these continuous and shameful failings of ours in a truly historic as well as, in the fast changing modern-era perspectives. It is indeed a great failing and negation of we Indians as a fast developing and rising Nation-State. Another important aspect in this regard is that, this dangerous and highly regressive age-old fixation has been making us naively slip towards an 'abyss'. Are we so shamefully 'foolish' that even today, we continue to refuse to see "writings on the wall" and not bother to start urgent and comprehensive 'course-corrections'?

Shri Abhijit Bhattacharyya (an aluminous of the National Defence College of India and a Member of the International Institute of Strategic Studies, London), in his article on "The Victor of 1971: Glorious Triumph of a Historically Inglorious Military", (published in the *Statesman*, Kolkata, dated 10 July 2008), in an indisputably scholarly manner, has very precisely written that, "An honest historical analysis will inevitably raise unpleasant questions pertaining to the State and the war machine. This will not be easy as the reality is bitter. For all the wonder that was India, its military history recounts a rather insipid track record of the garrison, stories of sporadic bravery and victory notwithstanding. The reality is that India has never been able to extricate herself from the clutches of foreigners—**from Alexander's invasion, in 327-326 B.C. to the Kargil incursion in 1999**. It is a country that has almost always been aggrieved as the victim of aggression. ...**To cut a long story short, one would attribute the defeats to the failure of command, not necessarily of**

the military. It was a failure of statecraft, the lack of vision and unity to anticipate and act. To this can also be added the frequent acts of treachery; again not necessarily from amongst the military, but of the divided establishment of the day. Indeed, the recurring invasion of India stopped for the first time in 1000 years as the plethora of 'Armies in India' transformed into the 'Army of the British India'. Since independence however, there has been a spurt of foreign invasions. ...**One is inclined to quote HG Wells' assessment of the 3rd century B.C. king, Ashoka's place in Indian history—'Almost the tens of thousands of names of monarchs that crowd the columns of history, the name of Ashoka shines and shines almost like a star.' In the context of Sam Manekshaw, Wells' assessment of Ashoka could be modified thus: Almost tens of thousands of names of defeated marshals and treacherous monarchs that crowd the columns of history, the name of Manekshaw shines and shines indisputably as the brightest star**."

The above reproduced paragraphs from a highly thought-provoking and intellectually stimulating article of a noted scholar, thinker and writer, perforce make us sit-up and think, as to what has mostly been wrong with us and also, what all we, as an ancient civilization as well as 'magnificent people' of our this great Indian-Nation, are continuously been lacking and so refusing to even gradually evolve, cultivate and imbibe in our psyche, thought-processes, intellectual and techno-economic pursuits, politico-military planning and also, while facing many other diverse challenges confronting us from time to time, since more than last three thousand years? **Now, straight forward answer to the question of this '*What*', as since raised above in simple terms is that of our almost total lack of 'the multiple-aspects of essential strategic-culture and thoughts"**. This most tragic and crippling 'reality' of ours, has been unambiguously confirmed from time to time, by many scholars and thinkers. At the same time, another very ironic and shameful failure of ours has been that, at no point of our earlier as well as post-independence history, we have even remotely cared to meaningfully and proactively study ancient India's one of the most prominently glorious 'Treatises', i.e. Chanakya's comprehensive and well-recorded and intellectually reflective visionary-thoughts and writings onto the politico-military subjects and also, on almost all aspects of the statecrafts since also known as **Kautilya's 'Arthashastra'**."

In this suggested 'study', it is absolutely essential that the whole exercise is always undertaken with desired and altering perspectives and also, with necessary 'in-built auto-correction mechanism'; and similarly, it was equally imperative as to gather the crucially strategic 'inputs' into our gradually 'evolving and critically-analytical thought-processes'; bringing-in requisite changes, as 'dictated' by the major 'strategic factors' during different periods, over a long 'canvas of time' of more than two thousand years of our turbulent history. Here, we should note that surprisingly many other nations (including Pakistan.) have been studying this indisputably one of the most valuable 'Treatises' of the whole world, with a sincere strategic aim to learn and benefit in many ways. In these multiple-aspects in fact, **Chanakya** indeed ranks parallel with **Sun Tzu** of ancient China.

Now, before further deliberating on the details of many-sided harms and humiliations to our this ancient nation, caused by our perpetual lack of this **"multiple-aspect of essential strategic-culture and thoughts"**; in the fitness of things, we are required to reflect upon, as to how and what this singular shortcoming on our part, especially since the year 1947; has resulted in an uncontrolled and monstrous rise of a most worrisome **'mother-problem'**, which in turn has been giving births to many kinds of the multi-layered and multi-dimensional new problems, as a 'cobweb'; to trap our nation with a fledgling democracy and also, with disastrous consequences too. This 'mother problem', with its multiple fallouts, has been discussed in the subsequent paragraphs.

Certain important points regarding India's ailments: We must understand that, in any context, for any number of the mutually inter-related multiple-problems, there is always a **'Mother-Problem'**, which gives 'birth' to such multiple 'problems'. This basic truth is also true, with respect to India's ailments. Now, the question arises as to what is that 'Mother-Problem', in the context of India's unending 'ailments'? **Here, that mother-problem is India's continuously increasing "Population" India's 'power-centers/power-brokers' are in fact, constantly refusing to honestly acknowledge/ accept even the existence of this all pervasive 'Mother-Problem'.** This is, indeed, a great ironical tragedy for our ancient Indian Society as well as our modern, democratic and fast-developing Indian-Nation. At the same time, especially in today's multiple-contexts the much bigger ironical truth is that this highly crippling tragedy, to start with, was absolutely preventable.

Now, first of all, we are required to understand that what are these so-called 'Power-centers'. **In the present day democratic India, the "Politico-bureaucratic Combine" is the power-brokers/ power-centers.** After our national independence and first general election immediately, with Late **Jawaharlal Nehru** as the first Prime Minister of India, the various political leaders (mainly from the then Congress Party) got themselves 'entrenched' into the seats of power, as new 'ruling-elites', with almost 'unquestioned' power and authority, as new 'rulers' of our newly independent and democratic Nation-State, with hardly any genuine concern and love for extremely poor, half-fed, and half-clad toiling masses residing in the vast expanse of rural India and of course nevertheless; with few highly gifted and illustrious leaders as honourable exceptions, belonging to almost all shades of political ideologies. And at the same time, first P.M., Nehru appeared to be even *'highly satisfied'* to preside over this political ruling-class. The tragic irony is that he with his extremely busy schedule hardly cared to spare any time and thought to take effective and sustained efforts to change this highly regressive and feudalistic mind-set. Further, arrogant and racial 'white-sahibs' of the erstwhile colonial-era, got replaced by the new and so-called 'Brown-sahibs'; who turned out be even more arrogant and irrationally feudalistic in their thinking and action. So, nothing substantially really changed.

Howsoever, we may try to ignore them or wish them away, these irrefutable facts, just cannot cease to exist by our sweet-will or day-dreaming. In fact, it is already abnormally very late and, if we sincerely desire, both individually and collectively as a Nation-State then, we have to be ready to make many painful sacrifices and make a proactive, all inclusive, multi-dimensional, well-formulated and comprehensive 'plans of action' and immediately launch the same, on war-footing along with a most determined and unshakable aim to bring the same to a successful conclusion. We must be able to clearly see the ' writings on the wall' and start acting without any further loss of time, both socio-culturally and officially, with a steely determination and unshakable 'ruthlessness', before it is too late and most tragically, INDIA starts 'sinking' under the unbearable weight of its own exploding and so far, unstoppable growth of its population.

Now, having put these problems in a perspective, we are required to give a comprehensive thought to the sets of possibilities and then, try to formulate as to what precisely the best possible 'solutions' could be and then, **make an all-inclusive "SWOT" analysis**, before proceeding with the designing of a multi-layered, multi-dimensional and pro-active "Plans of action". As already brought-out earlier in some detail, as to what exactly is the "Mother-problem" in this context, giving birth to all sorts of problems for our society and the Nation-State. At the risk of repeating, we need to again emphasize that, **this mother-problem is the continuing population-growth in our already over-populated country, which in any case, is highly 'derailing' for our 'unfortunate' country in many ways.**

Hence, what exactly is the "SWOT" analysis? SWOT means:

S = strength, W = weakness, O = opportunity, T = threat.

Now, this so-called 'Mother-problem' has resulted in many disastrous consequences for our nation. It would be obvious as to why this mother-problem has not been discussed since the year 1947, while not taking into consideration the earlier period. One interesting point to note here is that, at the time of our independence and also after partition of our country, India's population happened to be just around 34 or 35 crores only. Therefore, the moot-point emerging here is that, after our national independence, as we were then starting with a 'clean-slate' and also, with our own sovereign democratic government, under our duly elected first Prime Minister; was this 'monstrous-growth' of our already over-grown population, really preventable with time-bound and consistently effective official policies and hence, with grass-root level onward proactive programme? **The simple and precise answer is that, of course, it was preventable**. In this context, it is worth quoting certain most significant lines from a historic address delivered by Late **JRD Tata** at a Special Convocation of the University of Bombay (now Mumbai) held to confer on him, the Honorary Degree of Doctor of Laws (LL.D.), on April 10, 1981, gratefully copied here (pages 3, 4, 5 and 6 of the book " KEYNOTE" and published by Tata Press Limited, Mumbai, under **Jamshetji Tata** Trust in 1986), as follows: " The economic plight of our people and what was to be done to relieve it, soon became an obsession with me which has lasted to this day. In my then innocence and impatience, I had dreams of what people in business and industry like myself, could do once political freedom was won, to help build a new India in which our people could live a decent life, secure from want and fear.... In my youthful eyes Jawaharlal was the knight in shining armour, who would bring us freedom and lead us to a bright new world. Although we were good friends, it was a matter of great regret to me that I never managed to pierce his aloofness, perhaps he sensed that, despite my love and admiration for him, we were not on the same wavelength in economic matters and he evaded my attempts at discussing the subject with him by the simple expedient of looking out of the window in what I came to recognize as his habitual form of courteous dismissal. I found myself more easily in tune with Vallabhbhai Patel, that great and wonderful man whose forbidding features concealed a kindly heart and a ready wit...the magnitude and complexity of the problems and tasks that faced a new and inexperienced Government, grappling with such major problems as those of the aftermath of the War, the tragedy of partition, the integration into India of the Princely States, the framing and adoption of a monumental constitution, not speak of droughts, floods and other calamities. These initial tasks and difficulties were tackled with great competence, authority and dedication by our first Government, but history cannot absolve them and their successors of the responsibility for the failure to achieve in subsequent decades a significant improvement in living standards of our people, surely the principal object of any development plan. They first failed to perceive paramountcy need to curb the rate of our population growth, about which I was perhaps the first to sound the alarm in a public speech I made in 1951 as way backs thirty years ago (that was in 1981, and now it is around 60 years.). Not only was what I said was ignored, but I was smartly rapped over the knuckles in the highest quarters, on the ground that the larger the population the greater the strength of a nation. Even when realisation of the magnitude of the threat dawned in Delhi, family planning measures and the resources allocated to them were pitifully inadequate and still are. As a result, the monstrous growth of our population in the past thirty years (now, a devastating reality of the present day i.e. after around sixty years or so, and still these crippling and enlarging problems not being attended to in any manner whatsoever.) has deprived our people of much of the benefits of such economic growth as we have achieved."

It clearly appears to be a fact that, the doyen of Indian industry, Late J.R.D. Tata was indeed the first person to sound 'loud' and repeated alarm in the context of this "Mother-Problem", which is continuously enlarging like an ever growing and highly venomous 'dragon/monster', which appears to gradually 'crush and swallow' our country, if not checked and effectively tackled' while there is still time. Therefore, we all, including the 'Establishment'; the various social, academic, religious (inclusive of all religions) and charitable/philanthropic, and also many NGOs', Institutions/organizations must get awakened, gear-up themselves and get involved with their dedicated leadership at all levels, with singular and totally focused 'aim' to effectively solve our nation's this most damaging "Mother-problem" with maximum possible synergy and also, in a time-bound programme, once and for all; well before it is too late and then, making us naively reach a 'point of no return'.

Now, the question arises as to how to go about it and also to start with, how to get organized and thus, activate the essential 'unity of highest-level political leadership'? Once this is achieved then, the leadership at subsequent levels and in various areas of people's endeavour in multiple directions will naturally get activated and focused. These in turn, would lead to simultaneous and well-planned action-oriented programme; with essential co-ordination and synergy. These appear to be 'tall-orders' and extremely complicated and difficult tasks. This is understandable but at the same time, absolutely inescapable since earlier, we did not even care to listen to the quite early 'alarm', so sincerely and patriotically sounded by Late JRD Tata, way back in the year 1951 itself. History just does not spare anyone and therefore, it will lay the blame on the doors of the then highest level-political leadership and all others 'silently' associated with it. Now is there a question of any 'easy-option' at this abnormally late and also a crisis-ridden stage?

It is essential that, the dynamics of this "Mother-problem" is put comprehensively in a clear perspective and also, in its 'problem creation and multiplication mode' and at the same time also, in its 'methodology of some effective solution-mode'. When JRD Tata first sounded the alarm in this regard way back in 1951, India's population was around 34 to 35 crores only. If to our good luck, the then our 'Power-centers' would have cared sincerely and also with essential sense of visionary and strategic patriotism, then to start with, it would have been much simpler to initiate a comprehensive action-oriented plan and programme on 'war-footing'. It is obvious that there was never any 'magic-wand' to solve this 'mother problem' and hence, after initiation of multi-layered, multi-dimensional action-plan and programme it would always logically be a gradual and time consuming affair. Therefore, if to our good luck, this the then would have started way back in 1951 as mentioned above, then gradually to the desired and cumulative effect; India's population would have reached nearly around 60 crores by the years 1965 to 1970, with almost zero growth-rate, as duly achieved. One can easily visualize, with our then also a developing economy, what would have been the overall socio-psychological, political and socio-economic conditions in our country and at the same time; what would been the average quality of life of our people at large. Further in continuity today, and also with a great contrast, India would have already become a fully developed nation, and consequently this monstrous "Mother-problem" also would have vanished long ago. But most tragically for our nation, we have already missed the quite early 'bus' and also the initial opportunity and time, which have been so naively wasted and these are just not going to come back. Again it is obvious that if, this 'mother-problem' would have been permanently tackled, as explained above then, so many different kinds of 'fall-out problems', getting cropped-up in our midst today and menacing our country so devastatingly, won't have taken birth in the first place. Ironically, to some people with 'lingering scepticism', all these analytical submissions may appear to be matters of conjectures only, but these well-established and irrefutable points are based on hard logic and hence, totally factual.

It is certainly nobody's stand that after 1947 at no stage, the Government or more appropriately, the so called (India specific.) 'Power-centres' have not thought of or taken any initiative or action, to even remotely attend to the nation's this crippling problem of population-growth. It has been done in the past and being done even now, and in this respect, a separate 'Ministry of Family-Welfare' has also been set-up since long. But all these have always been attempted in an extremely callous, inconsistent, isolated, ill-planned and without any time-bound, all inclusive, strategically action-oriented, multi-dimensional, pre-set targets-based, multi-layered programme with essential continuity; and also nevertheless having desired flexibility and synergy. The great 'rub' concerning our disastrous national failure are lying here. We must remember with 'great shame' that this is not just a failure but a great 'blunder' since, its 'fallouts' are many and hence, are perforce leading to almost total undoing of a series of Five-year Plans and also, negation of our so many other hard-work based, governmental and non-governmental development-activities, since the advent of India's independence. Another ironical development is that since India's so-called economic liberalization of late, we have made ourselves 'prisoners' of "India-shining" and our fast-track economic upswing, just because F.D.Is (Foreign Direct Investments) are coming in, our stock-markets are jumping-up and large-sized Indian middle-classes are further expanding with multiplying purchasing-powers. These in turn, have been 'impressively' increasing our foreign-exchange reserves leading to multiple import liberalizations. Our shops everywhere in the country are 'over-flowing' with all sorts of Indian and foreign make consumer-goods and consumer-durables. Here, the most pertinent questions arise as to that, are these completely true and all inclusive, with respect to our national progress and 'socio-economic well-being'? These aspects just cannot be confirmed in the affirmative in any manner whatsoever; irrespective of **the 'vote-bank fixated' political leaders belonging to any political party/alignment**, as may like to indulge in any 'fabricated' argument either way since, facts remain staring in our face so loud and clear that even 'deaf, dumb or blind' too, can't miss them at all. Logically at this stage, it is needless to say that this extremely depressing state of affairs are disastrously there in our country for all to see so clearly, and entirely due to the earlier explained crippling **"Mother-problem"**, with all its manifestations and also 'fallouts'; as so many continuingly enlarging problems in our politico-psychological, socio-political, and also socio-economic thoughts-processes, as well as ways of life, as an independent nation and also, as a multi-religious, multi-lingual and multi-cultural ancient society as well.

Earlier, it has also been mentioned that in the larger context of this crippling "Mother-problem", we as a nation, have become a 'comical' example of 'putting cart before the horse' and by this; it is being conveyed that instead of at the outset, first to attempt solving this expanding "Mother-problem" then, at its initial stage itself, and truly on 'war-footing' along with the beginning of our multiple developmental efforts, as since suggested by Late J.R.D. Tata himself; in fact, we totally ignored the then clear 'birth' of this "Mother-problem" so callously. The focus was only to develop India's scientific, industrial, technological, infra-structural and socio-economic aspects in all sorts of manners for achieving accelerated progress; irrespective of their many inevitable consequences. Therefore, multiple 'negative results', as mentioned have not only cropped-up in 'isolation' but these have also resulted in multiply-complex, as well as nationally disastrous consequences, that we are so shockingly forced to see that perhaps, our this 'unfortunate' nation now appears to even slowly *'drift towards an abyss'*. Most tragically, this has also resulted in **a typical 'catch-22 situation'**, as to *'while Rome was burning, its Emperor Nero then, was fiddling with his musical instrument'*. Similarly, our today's so-called "power-centers" are almost incurably suffering from this most regressive 'Nero-syndrome'. What else could be a more disastrous tragedy at this stage, with the beginning of the present 21st century?

Now, a word about our multi-dimensional developmental activities: This aspect too, due to our lingering many-sided naivety, has also resulted in another connected *'catch-22 situation'* in the sense that—along with our multiple developmental-activities—we are perforce also made to relentlessly keep chasing many enlarging 'fallouts' of this ever present and continuously expanding "Mother-problem". Since our population-growth itself is almost unstoppably increasing; so we are 'trapped' by the perpetually more and more, many types of the mind-boggling 'needs' of our ever-growing population. Along with many other unavoidable consequences whatsoever, now in fact, howsoever relentless simultaneous multiple-efforts we do keep on making, we find ourselves totally helpless and tragically powerless to such ever increasing demands.**The 'point' is that how many houses, hospitals, schools, colleges, universities, roads, night-shelters, utility services (such as power-supply, water-supply and so on) consumer items, consumer durables, buses, trains and many other kinds of transport-services and also, many other allied infra-structural facilities, we would continue to build? In any case, there is no end in sight of the ever expanding demands for all these endless multiple-needs. Consequently, we are so pathetically made to 'senselessly' run in ever-increasing 'circles' and are unable to 'conclusively' arrive at our planned 'targets' in any area/areas of our multiply diverse developmental-activities. Needless to say that, the 'root-cause' for all these perpetually problematic state of affairs since as compelling us to remain and keep living in an never ending, extremely disturbing, and also hopelessly depressing scenario all around us; is the same consistently enlarging "Mother-problem" along with its so many menacingly problematic 'fallouts' too.** Another very disturbing new 'dimension' to all these, has also taken 'birth' of late, such as rampant production of spurious medicines and even life-saving drugs, fake consumer goods, fake milk along with various popular milk-products; resulting in making of spurious sweetmeats and even tea, and also many such allied items of daily human consumption. Added to them, the old habits of routinely indulging in the adulteration/ dilution of these items, on the part of many dishonest manufacturers, suppliers and traders at various levels have become common. All these are extremely important points to nationally deliberate upon, with strategic open-mindedness and objectivity.

While elaborating upon the so many menacingly problematic 'fallouts' of this "Mother-problem", in the contexts of 'issues' under discussion we find that most of these 'problems' since continuing in our country, and even a number of somewhat older problems confronting us today; are the direct off-shoots of this disastrously crippling **"Mother-problem"**. Be it shortages of housing, public-transport, health-care, other basic public amenities, employment avenues/opportunities, alarming 'derailment' of the 'law and order' situation and other allied matters, multiplication in the number of criminals among all age-groups and in almost all strata of our society and thereby, increase in all sorts of unimaginable criminal activities and violence resulting in extremely worrisome 'smashing' of the essential social peace and tranquility and also, deterioration of both internal as well as external security scenario and so on; are just a few prominent ones of the vast number of the menacingly problematic 'fallouts', in this regard. This point cannot be over-emphasized since, it is not possible to refute the glaring facts, and deny their existence in any manner whatsoever. Further, we must always bear in mind that our nation's geographical area and for that matter any piece of land is not like a highly flexible 'rubber' and hence, it just can't be 'stretched' in anyway. **Now since, this "Mother problem", i.e. the unchecked 'population growth'**

in our country has not been sincerely attended to at any stage so far, since as earlier discussed in no uncertain terms; we are thus compelled to even hastily initiate all sorts of the 'half-baked' developmental activities in many areas without any well-meaning co-ordination, co-operation and essential synergy and also, without much 'checks and balances'. And the results are there for all to see. The entire exercises in these regard, can logically be termed as a modern classic example of *'putting cart before the horse'* and thereby, naturally *'landing in a 'Catch-22 situation'*.

Other extremely damaging results of all these 'reckless' activities, are most tragic and continuous decay of our fragile ecology and unchecked environmental pollution. No wonder that, we are now even frequently witnessing the many natural, as well as man-made calamities; with their numerously tragic and disastrous consequences, for our peoples all over the country. It is indeed a fact that these aspects at 'macro-level', are truly global problems and everything is not under India's control. But coming to our own contexts, if we would have made our beginning after 1947 with requisite pragmatism and futuristic vision, as then so repeatedly and timely suggested by Late JRD Tata; our entire progressive scenario would have been a brilliant success, and a 'shinning' example to other under-developed and developing countries and also at the same time; our overall ecology and environment too, would have been in a much better state. Making mistake is one thing since, people at all levels do commit mistakes and they can be controlled and corrected too. But if a top-level politician-cum-statesman, as 'ruler' of any nation; willfully 'refuses' to listen to the highly visionary and timely advice and as a result, somewhat 'arrogantly' commits a serious "blunders" and then, even refuses to make valuable course-corrections, while there is still adequate time, and also in spite of the repeated 'alarm-bells' then having been sounded, and then as a result, extremely tragic consequences with crippling damages perforce befall on that unfortunate concerned nation. This is perhaps the singularly biggest tragedy of the modern, independent, and democratic India after 1947; bringing in a series of disastrously multiplying damages as great misfortunes to her. What more is needed to say in this regard? It is nobody's intention to sound unnecessarily pessimistic since, as citizens of our fast-developing democratic nation, more so in the 21st century, we are always expected to be dynamically positive and optimistic; but certainly with essentially strategic pragmatism, as that would naturally be desirable for today's India and the Indians.

Naturally, the obvious question arises as to 'where' precisely this depressingly unpleasant 'scenario', has been able to 'drag' our nation? We find ourselves in a singularly 'twin-tragedy', i.e. landing in a typical 'catch-22 situation' and at the same time, getting trapped in the most regressive fixation of putting 'cart before the horse' syndrome. What could be more disastrously tragic narration of our Nation's present on-going internal situation? **Logically therefore, the immediate need is for comprehensive and strategically well-planned all-inclusive "COURSE CORRECTIONS", with appropriately futuristic perceptions of all the inter-connected issues at all levels and in all areas. There is no alternative and also, no 'shortcut' available to this imperative**.

There can be and, naturally there will be many views/arguments against this analysis in the perspective of India's this "Mother-problem" and its numerous fall-outs; as all sorts of serious 'road-blocks'/problems are getting created so menacingly, in the 'bumpy' multiple-paths for our relentless multi-dimensional developmental efforts at all levels. But, arguments/counter-arguments, in this regard are one thing and, critically constructive as well as truly comprehensive appraisal of the studies, in some depth and detail, as put across in this article, are indeed quite another. It can logically be emphasized that, since last few years, particularly the U.S.A. and other leading western-world countries have emphatically acknowledged that India is a fast developing economy and even an emerging 'Super-power' and therefore, what is the point in taking so much trouble in expressing such 'depressing' views? Well, it is nobody's 'brief' or any intention to even remotely indulge in any 'culture of fooling around'. But, it would definitely not be in our short as well as long-terms national-interests, to altogether ignore these glaring and continuously enlarging **"Mother-problem"** and its many menacing and multiplying 'fallouts' or try to take a willfully 'callous views' on them. Problems of such multiple-nature and disastrous dimensions must be faced directly with resolute determination as well as with duly thought-out, broadly time-bound, action-oriented and programme-based 'Plans' leading to desired solutions.

On desired 'solutions': It won't be even remotely fair in this regard, to expect to 'design' or suggest any "Plan of action" by any one individual, with any kind of expertise and at any level. The entire detailed, multi-disciplinary and desired solution-oriented 'Exercise' and to start with, the already suggested 'SWOT analysis', will have to be undertaken by a comprehensive 'Team'/a number of Institutions with multiple discipline and expertise, and with unreserved full support from all possible quarters and at all levels. If the need be then, such Institutions may even include suitable foreign Institution(s) and reputed 'Consultancy-Organizations', without any 'unhealthy' or dogmatic reservations. Therefore, all necessary steps will be required to be undertaken with totally positive attitude, open mindedness and consistently focused attention.

But before all these, all the political Parties, especially the two largest and most powerful Parties, i.e. the Congress and the BJP, will have to perforce take the lead and in this context, rise above their short-term electoral-interests and 'narrow' politics, in the best national interests for all of us. These may appear to be 'tall-orders' here but, all these are inescapable necessities. Moreover, the 'moot-point' at this juncture, particularly pertaining to these two leading Political Parties is that, who will be able to come to power to form the Government and rule if, INDIA herself, 'perishes' under the crushing 'weight' of her this menacingly expanding "Mother-problem", and also its multiple and destructively problematic 'fallouts'? So in this perspective, the prominent and senior national-level leaders, especially those belonging to these two leading Political Parties, will be expected to keep in mind these points of paramountcy importance before 'showing' any lack of essential interest or due consideration, in any manner whatsoever. The greatest tragedy for our 'Bharat-Mata', down the pages of 'turmoil-ridden' centuries of our history has been our highly destructive and callous attitude towards our vital national interests leading to the most damagingly regressive habit of 'letting down' our nation in some way or the other, and thus causing repeatedly great misfortunes to ourselves. Now, let all of us 'gear-up' and start paying serious and consistent attention to the 'writings on the wall', which are so loud and clear.

Hope this article is able to generate some fruitful and analytical debate, on certain extremely important "issues", as having been discussed here; and thus leading to some desirably positive-action, in a well-planned, all-inclusive and strategically comprehensive manners as since already put-across earlier, in the above paragraphs. It is also a fact that this so-called "Mother-problem" not only in India but also remains a demonic menace in every South-Asian nation, and in some South-East, as well as, in many African countries as well. So India, as big and fast developing countries in these regions, can most appropriately take the suitable 'leadership-based' initiative for effectively solving these most disastrous problems, as suggested above, as an 'example', for others to follow. All these definitely would be a singularly one of the most glorious 'chapters' in the modern 'international history' since already in the making; during the present 21st century and that too, in some wonderfully memorable manners. 'Ball' is in India's 'Court' and therefore, let India not even now 'internationally' let-down this 'history in the making'. Now, one last 'point' but not the least important, and that pertains to the availability as well as mobilization of required financial resources on yearly and also, on long-terms basis. Well, today's India will always be able to arrange substantial funds through her 'internal' resources. In this context to start with, India will certainly be able to easily arrange requisite funds for getting the systematic, comprehensive and detailed "PROJECTS'-REPORTS", as suggested earlier in this article, at the earliest possible and then, the same with all the allied and other necessary documents, can be submitted to the 'WORLD BANK' or even to similar other financial Institutions for the further needful in these regards.

Earlier pages of this article might have appeared to be showing this author as an unnecessarily "pessimistic" person, but the above 'lines' in the concluding paragraphs, would definitely be able to clear such 'misapprehension, if any, to a large extent. It is naturally expected from every citizen of our country to always remain optimistic and positively hopeful. In the fitness of things, it is 'inescapably' essential to put a desired "rider" in this regard, in the sense that, it is no use to remain 'blindly' just optimistic alone but, our nation always expects us to remain optimistic with due flexibility, realism and vision, and at the same time, as also remain dictated by the 'dynamically strategic pragmatism' as well.

*

Appendix–B

MODI IS THE LAST HOPE TO REVITALISE THE ECONOMY

Annat Jain

(Annat Jain is the Founder of Acropolis capital Group, a private equity firm that invests in India. He can be reached at anatjain@gmail.com. The views expressed are personal.

Recent weeks have witnessed an enormous amount of hand—wringing in the press about India's economy. These articles share a common thematic trajectory: a recitation of statistics, their use to settle personal scores, and, in the case of the political opposition, using them to prepare oneself for 2019 with nervous glee. None of this hand-wringing is accompanied with any solutions beyond the obvious.

The fox knows many things. In a now famous essay written when he was an Oxford Don in the 1930's, Isaiah Berlin classified people into foxes and hedge-hogs. Foxes, according to Berlin, may know many things, but a coherent worldview is beyond their comprehension. The hedgehog, however knows one great truth, and steadfast in its pursuit, remains unreconciled until he/she reaches it.

Narendra Modi is the unusual hedgehog who may not only know what India needs to do to become prosperous, but importantly, is also willing to act upon it. Indeed, he may be the only politician in contemporary Indian history who has undertaken Structural reforms out of choice rather then, like Narasimha Rao in the 1990's, being compelled to do so.

Structural reform is thankless: costs are borne upfront, and rewards come later. (The currency for this transaction is political capital, and Arun Jaitley has paid the most for his role as the able knight who is the face of such change.) For years, it was the very absence of these structural reforms that armchair foxes have bemoaned. Now that the reforms are occurring, the foxes are unwilling to pay the price.

Through his decade-long executive leadership of one of India's richest states, Modi knows what the foxes don't: First, that there's never a right time to make a hard choice, and second, the slowdown is the result of a fundamental fragility of the Indian economy baked into India's economic foundation at its creation: the government's overwhelming role in the economy.

Modi is seeking to eliminate this fragility by recalibrating the entire engine of India's economic growth methodically: Jaitley's increasing of the states' share in the divisible pool of union taxes in 2015, digitization /demonetization of the economy in 2016, the GST in 2017, Piyush Goyal's work in the power sector over the last three years, Nitin Gadkari's ongoing work in transportation, all are transformative contributions to the structure of the economy, and designed to create a robust equilibrium from which the economy can propel itself higher.

That these reforms are happening without any taint on any senior politician of the government is in-itself a first in India's modern history. The republic was lucky to survive through years of pillage under the Congress, and if it were nothing else but just that the BJP is taint-free, we would still have much to be thankful for.

But Modi's toughest test is yet to come. He must now dismantle the socialist economic superstructure which is India's lasting curse. To do so involves treating the economic organization of society not merely issue inextricably linked to individual rights and dignity, and moving whole heartedly towards the only economic system that provides for such: a free-market system adapted to help those on India's economic margins.

This then, is the call to arms. In addition to a dogged focus on anti-corruption (a necessary issue for 2014 but insufficiently ambitious for 2019), Modi needs to speak to the country in civilisational terms about the manner of its economic organisation, and seek the mandate to put the nation on the path to double-digit growth for decades to come.

In practical terms, and as a first step, it requires the sale of the government-owned banks and PSU's in their entirety. Not re-capitalization, not mare NPA—resolution, not partial—disinvestment of loss-making PSU's. A complete sale. These banks and PSU's are giant black holes that destroy capital (and our soul), or at best, use it sub-optimally. Modi cannot rebuild a nation without removing the termites from its foundation. Only on its conclusion will Modi have lived up to his own revolutionary declaration of 2014: Government has no role in business.

Modi may represent the last best hope for the Indian economy to achieve its true potential. If Modi doesn't do it now, India will be condemned to this snakes-and-ladders growth for another several generations. If not Modi, then who? If not now, then when?

The hedgehog knows.

(Reproduced from a leading Indian Daily *Hindustan Times*, New Delhi, dated 30 September 2017—and thereby also, most gratefully acknowledged)

*

Appendix–C

KEEP CHANGING

Meghnad Desai

Memories are short. When 26 years ago Narsimha Rao and Dr Manmohan Singh launched their revolution in economic policy with a double devaluation, there was near unanimous denunciation by the leading economists of what they had done. Ruination was predicted. The World Bank was thought to be dictating India's policy. Even after all these years, the CPM/CPI keep denouncing neoliberalism. They prefer socialist stagnation.

For a government to be criticized for undertaking deep reforms is almost invariably a compliment. The BJP should be proud that it has undertaken several policy initiatives which have begun to clear the decades of debris left from earlier economic policies. Lenin said you could not make omelets without breaking eggs, nor a revolution without breaking heads. There is still unfinished business which will bring forth cries of anguish; ignore them.

At the anniversary of Demo (demonetization), it is worth repeating that old currency was not devalued. It was converted into bank deposits. Total money supply was unchanged. This is shown by what people cite as the failure of the policy, i.e., 96 percent of the outstanding cash came into the banks! The difficulty was Remo (remonetisation) not Demo; new currency did not arrive in time or in sufficient quantity during the first two months after Demo was launched. The new Rs 2,000 notes being unsuitable for ATMs compounded the difficulty. This was an implementation failure by the bureaucracy, not a policy error. The problem of cash shortage was sorted by end of the first quarter of 2017. The loss in growth rate of GDP was not more than 0.5 percent, for the two quarters.

The government has been untiring in trying to bring money stashed abroad, as much as it will be in chasing the black money which came to the surface during Demo. It has tackled the insolvency legislation which has harmed the economy all these many decades. That then opens up the possibility of dealing NPAs. The State Bank of India has been consolidated to give India, for the first time, a place in the 50 largest banks in the world.

The pace of reform needs to be kept up. PSU banks have had some recapitalization, but the real solution is to consolidate them by merging and, when the climate is right, privatizing them. Indira Gandhi's move to nationalise banks was the biggest step in corrupting Indian politics. If Modi is serious about tackling corruption, PSU banks have to be divested. Otherwise another mountain of NPAs will rise up.

There is more unfinished business. The 2013 Land Acquisition Act is the most serious obstacle to growth and, if anything, the most anti-farmer, despite what its proponents claim. To increase infrastructure and manufacturing investment, the government needs to make the release of lands from agriculture to the larger economy easier. It will help many subsistence farmers sell land, get out of their hopeless situation and take up productive occupations.

The final challenge is labour market reform. All trade unions will fight these reforms. But if Indian manufacturing has stagnated for 50 years, it is due to these laws. Labour laws defend a minority of workers, leaving the vast majority insecure in the small and medium enterprises or informal sector. Clear this obstacle to see Indians grow rich.

[Reproduced from a leading Indian Daily, *The Sunday, Indian Express*, Kolkata, Dated 12th November 2017 and thereby also, most gratefully acknowledged].

Appendix–D

THE MANY MEANINGS OF MAGNA CARTA: AN IDEAL BATTLEGROUND

Sunanda K. Datta-Ray

The little grey woman saw me looking at the picture. "It's Gandhi," she said excitedly in a broad Lancashire accent. "My friend saw him when he came to our mill town!" We were in the Magna Carta: Law: Liberty, Legacy exhibition in London's British Library and I was't surprised to find Mohandas Karamchand Gandhi there. The exhibition's purpose was to stress that the rights of man received formal legal recognition only when England's King John conceded the Charter of Runnymede to his rebellious barons 800 year ago. But the point to emerge was that even if Magna Carta competes with the English language as Britain's greatest export, the fight for freedom has been carried out by individuals of courage in every age.

The massive, meticulously organized display of books, pictures, coins, seals, statues and other artifacts testifies to that struggle. A couple up from Portsmouth warned us to wear sturdy shoes: they had spent two hours tramping from showcase to showcase. We spent three and must confess to being physically too tired towards the end to do full justice to the nine rooms. But it was clear there was little merit in Dean Acheson's jibe that Britain had lost an empire without finding a role. The role of inspiring not merely Magna Carta but also the fundamental rights enshrined in India's Constitution, the American Bill of Rights, the Universal Declaration of Human Rights and the European Convention on Human Rights had never been lost.

Much of this is a puestion of faith. As a British chief justice, Lord Bingham, put it, "the significance of Magna Carta lay not only in what it actually said, but in what later generations claimed and believed it has said". That could make it a dangerous proposition, explaining why in 1947 the British colonial Office rejected a plea for annual Magna Carta Day celebrations throughout the empire. It feared "ill-disposed colonial politicians" with great enthusiasm for something they hadn't read "but which they presume to contain guarantees of every so-called 'right' they might be interested that moment in claiming" would exploit the myth. The American War of Independence demonstrated such misguided colonials didn't have to be black, brown or yellow. Europeans in India could also be recalcitrant. *On display is The Humble Petition of the British Subjects residing in the Province of Bengal, Behar and Orissa and their Several Dependencies* submitted on February 26, 1779 demanding the "inherent, unalienable and indefeasible" right under Magna Carta to trial by jury.

The petition concerned a public works department superintendent called James Creassy accused of assault, battery and illegal detention against two Indian carpenters. Creassy and his servant had apparently locked up the men and beaten them with rattan canes. The Supreme court in Fort William had limited jury trial in 1773. Six years later, the notorious Elijah Impey, the chief justice, ruled that Magna Carta was "merely a local influence" in Britain without wider application in the colonies. He warned that Creassy "was running his Head against a Wall, and would dash his Brains out" if he persisted. Unfortunately, the exhibition doesn't say whether Creassy saved his brains.

A certain deliberate attempt to play down Magna Carta's importance has exactly the opposite effect. Visitors are reminded the original document was only a peace treaty between the king and his warring lords. John was notorious tyrant, and a chronicler's words when he died- "Foul as it is, Hell itself is made fouler by the presence of King John"—are prominently inscribed high on one wall. Pope Innocent III, to whom John swore fealty in 1213, declared the pact "base and shameful, null and void" and annulled it within weeks, The papal annulment provoked civil war and French invasion followed by John's death when "Magna Carta seemed to be a failure without a future"

Almost all its clauses have been repealed; those relating to forest law being transferred to a separate Forest Charter as long ago as 1217. Even in 1215, only "free men" were entitled to the document's assurance of justice and protection of the law. "Free men" were an elite minority which is why the author of 1066 and All That mischievously commented that Magna Carta was "a Good Thing for everyone (except the Common People)".

No doubt, that was one reason for refusing universal celebrations Magna Carta became all things to all men. The royalists cited it against Parliament and Parliament invoked it against King Charles I. Radical politicians demanding parliamentary reform as well as Chartists seeking to extend the franchise quoted the document. It was used to challenge press censorship, imprisonment without trial and even King George IV's divorce of his hapless queen. Magna Carta became even more popular after being printed in 1508, inspiring individual resistance and empowering popular protest. But like the petitioners from "Bengal, Behar and Orissa and their Several Dependencies", expatriates found to their dismay that British justice and the right to trial by jury did not always apply overseas.

The second Indian invocation seems to be the East India Company's dispatch of 1854 which was "called the Magna Carta of English education in India". Then—and somewhat more justifiably —came "The Queen's Proclamation, the Magna Carta of the people of India" promising equality irrespective of race. Victoria held herself bound to Indians "by the same obligations of duty" which bound her to her other subjects and which she vowed she would "faithfully and conscientiously fulfill". Satyendranath Tagore, Behari Lal Gupta, Romesh Chunger Dutt and Surendranath Banerjea took advantage of the more pertinent commitment that her "subjects, of whatever race or creed, be freely and impartially admitted to offices" solely on the basis of "education, ability, and integrity".

The Proclamation prompted *a Congratulatory Address from the Inhabitants of Calcutta* with 11 pages of signatures that the sheriff forwarded on November 22, 1858. Rajidra Mullick heads the list of Mullicks, Ghoses, Dutts and other loyal Bengalis. But it was the fierce resistance to the Ilbert Bill to give Indian magistrates the same jurisdiction over Europeans that Europeans in the Indian Civil Service enjoyed that brought powerful tribal passions to the surface. The European and Anglo-Indian Defense Association thought the measure trampled on its ancient right to trial by jury. For Kristo Das Pal, editor of the Hindoo Patriot and member of the legislative council, the Bill was Magna Carta, "the battle ground on which the whole principle of legislation for India was being fought".

Sadly, the exhibition doesn't quote Gandhi on any of this. As indicated by the photograph of his South African attorney's office, his involvement is restricted to South Africa's 1914 Indian Relief Act which Gandhi hailed as "the Magna Carta of our liberty in this land". It abolished the 3 pound tax on former indentured labourers, validated Hindu and Muslim marriages, and enabled children in India to join their parents in South Africa. Thereby, it also confirmed the British constitutional theory "that there would be no legal, racial in-equality between different subjects of the Crown" and "vindicated Passive Resistance as a lawful clean weapon".

The little grey woman knew nothing of either arena of strife. But she was from Preston, and her friend in nearby Darwin had never forgotten how as a 12-year-old girl she had watched her mother and other mill workers cheer the dark little man in a lion cloth in adjacent Nelson. There are other memories of Gandhi's visit to Lancashire: a Bolton journalist remembers asking for a message as his train passed through the town and being rewarded with perplexed silence. But 84 years after the encounter, the Darwin woman's excitement seems as infectious and enduring as Magna Carta's appeal.

[Reproduced from a leading indian daily, '*The Telegraph*', Kolkata, 11 April 2015 and thereby also, most gratefully acknowledged].

Appendix–E

PROSPERITY FOR PEACE

Khaled Ahmed, consulting editor, Newsweek Pakistan

Free trade could end Pakistan's siege mentality, ensure peace with India

A British think tank has noted that Pakistan's army chief, General Qamar Javed Bajwa, "invited the Indian military attaché and his team in Pakistan to the Pakistan Day military parade in Islamabad; and then General Bajwa said two weeks later that the Pakistan military wanted peace and dialogue with India".

There is no doubt about the wisdom which is another name for strategic flexibility of seeking peace with neighbours. An economically pressured Pakistan simply can't afford to have its fronts heating up east and west while there is instability within. In the eyes of the world, however, Pakistan must take the initiative because the world rightly or wrongly thinks that Pakistan, as a "revisionist" state, has bothered India with covert cross-border attacks. There are no takers for Pakistan's explication of the disastrous Kargil Operation of 1999.

General Bajwa must retrace other missteps taken at home with those who sought to "normalise" relations with India, starting with the Vajpayee visit in 1999 that ended with a joint Lahore Declaration signed with then Prime Minister Nawaz Sharif, which the then army chief Pervez Musharraf didn't like too much. Earlier, it was the PPP which sought a policy change with India and suffered for it. On the eve of the 2013 election, Sharif and his party, Pakistan Muslim League-Nawaz, declared their intent to seek peace with India and won the election, which meant a popular mandate for the pledged India policy. The army didn't like that and Sharif as prime minister was soon in trouble.

It is only logical that General Bajwa undoes some of the policies pursued inland before approaching India, if only for the sake of credibility. So far, India has rebuffed his gesture and the world too is unimpressed.

Given this scenario, the Indian High Commissioner to Pakistan, Ajay Bisaria, is doing his duty of focusing on optimism. He was at the Lahore Chamber of Commerce and Industry that had responded positively to Sharif's call for open trade with India. He knew his government was harassing Pakistani diplomats in New Delhi to highlight its unhappiness over Kashmir. So Bisaria zeroed in on "the traders on both sides of the Indo-Pak border who face visa issues, which is among the reasons why the two sides have failed to actualise the potential of $30 billion trade mark estimated by the World Bank". He was unhappy that given the potential, the two countries were officially trading only $2.2 billion. Bisaria pointed to 70 years of bilateral conflict that had brought no good to the region.

There is no doubt that Pakistan is under pressure to normalize relations with its neighbours. Going against its martial grain of "Castle Pakistan", it has allowed the China Pakistan Economic Corridor (CPEC) to puncture its territory and has not yet weighed the changes this has the potential to bring about. While it is true that New Delhi has rebuffed China's offer to join the CPEC, India's new-found anti-China "adjustment" with a mercurial President Trump may not last.

In Pakistan, however, some analysts have ventured to break from the idea of Castle Pakistan—more accurately "Islam ka Qila"—and are thinking of a new identity. The most recent statement of this new identity was made by a senior journalist M Ziauddin: "Pakistan's strategic location makes it an ideal country to adopt the warehouse/transshipment economic model. So, in order to make the most of the opportunity offered by CPEC, we need to invite Russsia as well to set up its own trade corridor through Pakistan and also let India trade with Afghanistan via Pakistani land, facilitating New Delhi to go beyond to Central Asia and also reach western China while China could access western India through this corridor which would also hopefully open up a trade window to the landlocked India-occupied Kashmir. *(whole of J&K is Indian territory)*

Trade puts an end to war because of the vested interest in mutual prosperity. Nothing destroys borders more effectively than free trade, which Pakistan and India should move to as nuclear powers no longer able to wage war. May be the process of "becoming like the enemy you hate" in India will come to an end and Prime Minister Narendra Modi's original instinct of economic advantage will prevail.

(Reproduced from a leading Indian Daily: *The Indian Express*, Kolkata, dated 12th May' 2018; and thereby also, most gratefully acknowledged)

* * *

N.B.

With respect to the above wonderful ideas for peace – 'the ball is squarely in the court of the Pakistan Army' and hence, it all depends upon the consistent sincerity, decisions and also positive actions on the part of Pakistan Army.

Author's Note

At the outset, I wish to share that even after completion of this book, I, as a person of an average mind, am unable to clearly remember as to how and also through what all thought-processes, if any worthwhile, such a topic came into my mind. Moreover, today as I present this book to you, it seems like a dream that how I began the work, finalised the content in each chapter, gave apt titles, structured and sequenced the chapters. I remember jotting-down some ideas and also few disjointed thoughts as they began coming to my mind and thus, the outcome is now placed before the readers. In the process of reading the book one will see that the topic and its related content does indeed command such a multi-dimensional approach, on a vast canvas in order to do justice to them. It will undoubtedly have some limitations that may kindly be overlooked.

Now very briefly coming to the personal background etc of myself to begin with; I would say that I was born on 07 August 1942, in a very small and old town of Jamalpur in the district of Munger on the southern bank of the river Ganges in Bihar; and this makes me now 72 years old. Though my official date of birth as since recorded, happens to be 14 December 1944, and this too also has an interesting story. As the fairly common practice prevailing in the society then, especially during pre-1947 era; whatever was generally said by an elder of any concerned family, was taken as final and thus recorded. My family happens to be a joint-family and where during those times; any uncle, elder brother or even sister could generally take the child to school for admission, and verbally mention the child's birthday and year for the needful. I now myself do not know as to who from my family took me to the then very magnificent and spacious kindergarten-school located very near to my father's official residence in the Jamalpur's railway-officers' colony, sometime during 1946-47 or so; and thus very informally my official birthday was recorded. Anyway very interestingly then, people used to commonly provide some 'age-related cushion' to the child as a precaution for the future for some obvious reasons. And that's the way interestingly in my case too, this matter finally rests. Anyway during those days in that erstwhile now already bygone-era, and thus in Railway's that wonderful kindergarten-school then at Jamalpur; I was indeed fortunate to be taken care-of and also taught by a few British ladies as teachers; who were actually wives of some British officers, since residing then in the same residential-colony. I still very fondly cherish the memories of their strictness towards discipline, punctuality, manners and ethics and also not the least, their impeccable sense of honesty and the purpose therein as well. Sadly now in our independent country gradually after 1947; instead of trying to further augment, expand, and thereby enrich these invaluable attributes in life; we rather display no hesitation in even discarding or not caring at all for these precious value-systems in life. And what more to say—the rest is now all extremely sad, as well as also so well-known and rotten our modern socio-cultural values in our country perhaps all over—in these regards.

Further this otherwise quite insignificant town of Jamalpur, which indeed also happens to be very unique and well-known destination for the Indian Railways; as the Asia's oldest and thus more than one and half centuries old but still indeed very vibrantly functional railway-Workshop, and thus based on the same—the Indian Railways' historically unique engineering-training Institutions—since located here. These Institutions since also continue to cater for the training-cum-education of the Railway's 'Special-Class Apprentices' as well. My father, **Late Surendra Narayan Verma** (then commonly known as Mr. S.N. Verma, and also rather intimately as Verma Saheb in his social network over there) was in railways; in which at a very young-age only, he was selected to join his service as a 'First-Class Railway Apprentice' sometime around 1924 or so; when India was under the then British colonial-rule. Hence I still feel fairly fortunate to be rather somewhat relatively privileged to spend my childhood in the then substantially exclusive environment of the railway-officers' colony located in Jamalpur; as my father was working in the same Workshop as mentioned above. Now the same residential colony of the railway-officers; so unfortunately happens to be not even a pale-shadow of its olden glory, upkeep, pleasant-cum-soothing environment; and thus the so-called exclusiveness of the bygone era. Whatever may presently be the Indian Railways' progress, as well as indeed quite laudable and thus obviously the no less essential techno-economic modernisation of that very old Jamalpur Workshop today; but while somewhat discerningly revisiting the same official residential-colony over there—at its socio-cultural and also its overall efficiency levels—one finds typical 'decay and deterioration' only; and hence nothing to feel proud or happy as one finds nothing to be nostalgic about but only a 'sad-commentary' and thereby, it is all for the present-era's railway-authorities to take further call in these regards.

After my KG education, I had my further schooling up to the then matriculation-standard in Jamalpur only in the railways' own the then fairly old but indeed well-established and reputed Middle and High Schools; and thus completing my matriculation in the year 1958. But for me midway during my childhood only; great family tragedies visited my life; as in 1956 itself when I happened to be a student of only 9th standard in my school, I lost my father very prematurely, and this indeed very sadly brought a great upheaval in my highly sheltered and also reasonably comfortable life. As a child I felt absolutely shattered then, as I had already lost my mother even much earlier, when I was yet to even begin my schooling too. Needless to say here that I have had a relatively quite heavy share of the tragedies in my life, since my childhood days only. But as the much larger saying goes that when the most crucial doors in one's life are shut so shatteringly; the Supreme Divine Grace too comes—visiting so benevolently into one's life and thus, providing great and very timely rescues as well. Hence who can ever even marginally understand or evaluate that Divine-Grace in any manner in this mortal world.

My father in addition to his own children then, also had so loving my grand-parents (*i.e. the parents of my father*), and also his two younger brothers and one youngest sister; who as my aunt (*i.e. Fua/Fufi or Bua, or Khala, since as also known in India*), was very simple and saintly and hence, very loving to we brothers and sisters. Though she has very sadly passed away recently, but her uniquely sweet and loving memories would always remain cherished in my heart. We had been four brothers and four sisters, and my sole brother **Late Schchidanand Verma** was the eldest and thus senior-most among us and who too being very loving to all of us and on top of it, he happened to be also exceptionally large-hearted as well. Then I have three elder sisters and also, one youngest sister who being the youngest in fact, happens to be younger to even my two younger brothers. My all sisters too have always been very kind and affectionate to us. Unfortunately few years back, I also lost my eldest sister. My two uncles (i.e. younger brothers of my father), **Late Ram Chandra Verma** and **Late Krishna Chandra Verma** (*i.e. Mr. K.C. Verma, Advocate; who began his professional career in the Munger Dist. Bar*) respectively; were also exceptionally attached to all of us and thus very loving and caring to us. And also their wives too (*i.e. my aunties*), along with the wife of elder brother (*my bhabhi, i.e. my elder sister-in-law*),—thus all of these ladies in the family throughout their lives—were the no less caring and loving to us. But my grandmother in spite of the so many tragedies in the family; she always remained so loving, affectionate and caring towards all of us, till she breathed her last in around 1969; that she was both like an umbrella, as well as an anchor for all the family members. My youngest uncle, (*i.e. Shri K.C. Verma, Advocate*), from his student days in fact, ; happened to be and also remained all through his life, always very active, visionary and efficient guardian of our entire joint-family and thereby, also remained the most trusted '*lieutenant*' of my father so long he remained alive; and thus sadly after my father's sudden and so premature demise, the entire 'family-burden' fell onto his shoulders, when he himself was fairly young and also in only the beginning years of his legal-career. Those were very trying times for the entire family; but during those days due to the then extremely close and affectionate kinship among all family members, as well as some of the very close relations as well, things could be taken care of and managed. In today's fast-moving, so strenuous, principally self-oriented, and also rather very fast-changing and trying times; these aspects can't even be marginally dreamt-off in any manner whatsoever. One can now imagine as to how and how far the societies in our country have got changed and thus up to what extent, and most of the individuals coming from any walks of life today, have gone so heavily and also indeed perforce now self-obsessed; that hardly anybody has any spare time or perhaps even any consideration left for any other member of his/her family, other than own 'miniscule-unit'. But I was indeed very fortunate not to experience any such rather 'suffocating' environment during my childhood, as well as my student-days. In fact then luckily, these kinds of times since so commonly prevailing today; was yet to so much overwhelmingly dawn onto the Indian societies.

My youngest uncle who happened to be an Advocate (*and since also popularly known as 'Vakil Saheb' in the localities of the twin-towns of Jamalpur and Munger*); and along with my sole elder brother, as since mentioned above; were both very actively supportive to the entire family and thus; were in fact, decisively instrumental in letting me complete my engineering-degree education in 1964, as among the very first batch from the State Government's College of Engineering at Bhagalpur; so comfortably as one of the boarders over there and also since, located at hardly 55-60 kms from Munger. My uncle as so effective a guardian for the whole family and also in spite of his such responsibilities and busy schedule; was always keeping very close eyes on me and thus in fact, was regularly visiting my college and the hostel, to ensure that a very close watch was maintained on me. All these are highly cherished memories of mine and hence, the least I can do is to always remain indebted to all them in my heart. In addition to them, so far my mental grooming, as well as development during my formative years was concerned; my two maternal uncles (i.e. elder and younger brothers of my mother respectively), along with my sole paternal uncle too (*i.e. husband of the sole younger sister of my father, as since already mentioned above*); have served as a sort of 'beacon light' in my life and thus, always very unhesitatingly acted like an ***"icing on the 'cake' of my every effort in life"***, finally being of whatever worth and that too wholly due to my personal limitations; have been great sources of inspirations to me in whatever pursuits I have been keeping myself engaged-in from time to time in my life. In fact even now I feel myself quite closely attached to them, as being a member of such a large and extended family of mine; even during the present almost totally altered, 'topsy-turvy', and also the so much recklessly fast-moving world. Well, that's the way I have almost all along been looking and also since reflecting on my life so far. It is true that one can always look and reflect on his/her life in indeed the numerously different ways and thereby, be at liberty to paint typically differing kinds of pictures onto his/her always a fairly 'tattered', and also being of quite a limited size 'canvas of life' in this unsparing world. Especially for a person of any very average kind like myself, this aspect remains fairly common almost anywhere; though for any outstanding and also the history-making personality coming from any walk of life anywhere, it indeed remains altogether a different 'story' and 'ball-game' in these regards.

In the above paragraph, I have just mentioned about two brothers of my mother and the husband of my father's sole younger sister and in addition here; I would also wish to mention about another very elderly uncle of mine (*i.e. husband of the sole elder sister of my mother too.*), who all have been of great sources of inspirations and knowledge for me throughout their lives. Out of these four now, I am still so fortunate to have been continuing to receive a fatherly affection and inspirations from only one (*i.e. the husband of the sole younger sister of my father*), **Shri Basant Prasad, IAS (Retd).** Along with his and then also that of also Late **Dr. S. Prasad's**, as about them; I have already very briefly mentioned in the last chapter on 'Conclusion' in the end. The names of my two maternal uncles (i.e. one elder and one younger brothers of my mother), and also that of the husband of the sole elder sister of mu mother; have been **Shri Harnandan Prasad, Advocate**, (*then a very leading, famous, and also the indeed a highly flourishing lawyer of his times, in the Dumka District Court; as Dumka being an important district headquarters since the bygone British-era; and now belonging to the present Jharkhand State, as also one of the divisional headquarters of civil administration network of that State in Eastern India*); **Dr. Lala Surajnandan Prasad**, then settled in Patna, (*a very well-known pediatrician of his times in the country and also, earlier the Founder-President of the Indian Academy of Pediatrics, having its headquarters in Bombay*); and **Shri Krishna Prasad Sinha, Advocate**, the then a very senior lawyer in Biharsharieff, headquarters of the Nalanda District in the Indian State of Bihar, respectively; and needless to say that so unfortunately for me, none of the above three of them is obviously no more alive today.

So far as my cherished memories regarding my association, along with also very informative-cum-educative interaction with my younger maternal uncle, **Dr. Lala S.N. Prasad**—the senior most pediatrician of Bihar state, as well as founder of the Department of Pediatrics plus the Children's Hospital in Patna Medical College—is concerned; I wish to very briefly tell just three stories here:

(a) He in fact narrated to me in his uniquely educative and lucid style that, since he always felt very fortunate to have met and talked to **Mahatma Gandhi** sometime in 1930s, and that too when he was a college student as he was then doing his MBBS course in the Patna medical College. During that time, Gandhiji had come to Patna as per his some programme; and by chance my uncle came to know that the next day at about four o'clock in the morning when it would still be dark, he would be crossing the holy-river Ganga to go across in north Bihar, as per his further programme. He also came to know about the selected/secluded spot on the river-bank, where Gandhiji would be boarding a county-boat and that happened to be very near my uncle's medical college hostel. So my uncle along with his couple of close friends came to the spot much earlier than Gandhiji's arrival and waited with lot of thrill and expectations. As scheduled a car very quietly came and Gandhiji along with **Mira Behn** came out of the car; and was little surprised to see a couple of very excited young men and he obviously asked as to what the few young boys were doing there at such an hour. My uncle replied that they were medical college students and came to so exclusively see their revered Bapu, and didn't want to miss this lifetime's opportunity. Gandhiji very affectionately blessed them and told them to serve especially the poor after becoming a doctor. By that time a couple of press photographers too had arrived and closely clicked their cameras as Gandhiji walked and boarded the boat to sit on a blanket, already spread for him by Mira Ben onto a wooden-plank in the boat. My uncle the then himself a boy only, was also standing closest to the bank with folded-hands; and the Mahatma with his so loving and captivatingly innocent smile, quite amusingly asked Mira Ben as why he starts his morning always with a photograph clicked; and she replied that, 'Bapu, your sitting on a country-boat makes a history.' When my uncle narrated this story to me, I too was amused and talked at length about such a great-man, like whom, rarely ever walks on this Earth.

(b) My uncle recollecting from his memories further told me that, when he was sailing back from UK after completing his higher studies and also specialisation in the medical-subject of Pediatrics, in a ship of the then a very famous P&O Liner in 1947; the 15th August dawned when he was still in the mid-sea only. So he along with his other Indian co-passengers requested the British Captain of the ship to permit them to stitch an Indian tri-colour for hoisting on the top-deck of the ship and hence having been permitted; they did so and thus celebrated the historic Indian independence, with a lot of fanfare in 1947 in the mid-sea only.

(c) Further very interestingly my uncle explained to me as to why he puts on always a very brightly coloured and printed ties, when he is in his clinic for his busy professional work? And he very informatively explained to me that when he leans over his 'examination-table' to examine a sick baby or child, his/her attention is diverted to my tie and while he is playful with it, he is more effectively able to examine his patient in very close-details for deciding the further line of treatment.

Well, all these may be some 'passing-stories' only, but they do convey a lot if one is inclined to care; and these are few of the indeed numerous ways to rather somewhat fruitfully interact with one's elders. These elders in the family so long they remain alive, do become great sources of the different kinds of knowledge, and also provide many valuable and inspiring information about one's family's past and thus, also needed to cherish them as future sources of energy for fighting one's further 'battles' in life more effectively. Hence, while talking about my elders and mentors; I cannot miss very nostalgically recalling the no less another very affectionate person, **Shri Surendra Prasad Sinha** (*i.e. the only elder brother of the husband of my second elder sister; and affectionately known as 'Barka Bhaiyya' among us, thereby meaning: 'a very elderly brother'.*), who was a scholar of English and thus, remained always very kindly available to me to correct and polish my any writings in English, during my collage days. He remained unmarried and was working in Railways. Even after retirement he remained dedicated to his passion for reading and research in English and did his PhD in English too, way back in the 1960s and thereby he has also written a couple of books on English literature, and about some noted British poets of the empire-days as well. Unfortunately, we have lost him just about 7 to 8 years back; but before he breathed his last, he very kindly presented the copies of his books from his residence at Bhagalpur, through my second elder sister.

My this second elder sister, **Mrs. Archana Sinha, D.Lit.** and who earlier retired as Principal of one of the oldest and most well-known girls' college in Patna; has been a professor of Sociology, during her academic profession. During her active life as teacher of Sociology, she has also written a well-researched book on *'Sarvodaya philosophy'* and thus years back after its publication, she too has very affectionately presented a copy of her book to me. Naturally all these books do also now form part of my very valuable collections, in my small personal library. It would always be advantageous for anyone to pause and indulge in such pastimes while on leave or vacation; to not only avail very valuable relaxations, but also to gain some very interesting/enriching information; instead of always getting stuck in all sorts of routine drag and also indeed some largely avoidable activities.

After my graduation in electrical engineering in 1964; on my own initiative and even before the university-results of mine were published, I joined as an unpaid 'engineer' trainee in the then Patna Electric Supply Co. Ltd, which happened to be a private ltd. Company then; so that I could gain indeed some worthwhile real on-the-job training, instead of just wiling-away my spare time of few months; and also nevertheless, this Company too was quite willing to get a 'technically-educated' hand to get some job done without incurring any expenses. So in whatever any small way, it did happen to be a quite 'win-win situation' for both of us and thereby; though all these now pertaining to some very distant-past and that too of the indeed half a century back in my life; I still very fondly continue to remember my interesting and also quite hard-working and fairly tight-schedule days with lot of nostalgia, during my association with that company in Patna. Those were months in fact, when I was waiting in my younger maternal-uncle's house in Patna; to receive my instructions from the Army HQ, to join the Indian Military Academy (IMA), Dehradun, in 1965 for training; as I was already selected for the purpose through some rigorous competition and tests as per the well laid-down system in these regards. So when the scheduled time came; I travelled to Dehradun and joined my since already programmed training-course and then did pass-out as a very young engineering graduate Lieutenant, as commissioned in 1966, in the Corps of Engineers of the indeed being always top-class Indian Army; though needless to say without any award or distinction which I indeed never deserved, at the time of the highly cherished 'Passing-Out Parade of mine in the IMA. Well, what the Indian Army means to any conscious and discerning Indian citizen—being so professional, patriotic, duty-bound to always defend the country from kind of threats, as well as coming to the peoples' rescues whenever required; and then always remaining impeccably and unquestionably apolitical too; as an Institution—would clearly see to be discussed in some vivid details in the relevant chapters, for anyone who may care to spare his time for reading this book since as intended.

Though it's a different matter that I in spite of being a permanent commissioned officer; I could serve only for twelve and half years, and had to take my premature retirement on my own request as a young major due to some fresh family tragedies. My request for premature retirement was finally approved with initial rejection, and that too after lot of questioning by the authorities concerned; but that has been my personal bad fate-lines only. And whatever, but I do also most sincerely wish to say that even now, I still cherish my uniquely valuable times in the Army, along with so many invaluable training-cum-experience-inputs in my life; which would always remain its exclusively invaluable gifts to me, and also continue to being multiply fruitful in my life till I breathe my last in this mortal world. Those twelve years of my life in the Army, did not only open many new windows of my life as a young officer then; but in addition for me they were also very educative and enriching as well, in many unique ways. And hence, I would always feel very fortunate to have received them as then a fairly young man while in the Army; and those unique years are obviously never to return to my life again in this world. While in the Army, in addition to many educative experiences; in course of my few periodical transfers, I was able to stay and work in a number of different places and far-off places in the country, such as—Pune, Udhampur in J&K, Tezpur, Vishakhapatnam, Port Blair, and Bombay etc—and thus was also able to gain valuable learning and experiences too; and this aspect is indeed quite an unique feature of our Indian Armed Forces.

During my so much engaging and also the multiply-fruitful service in the Army; in March 1972 I got married and was fortunate to get such an affectionate and extremely tolerant young girl as my wife and interestingly, whose name also happens to be rather very befittingly Priti. And who has always been standing by my side in any kind of the 'thick and thin' in life, in this indeed unsparing world. On this score especially when I am now fast moving towards the evening of my life; I find myself rather quite uncommonly fortunate to get her as not just the better, but indeed the best half of my life. And then in June 1973 our first son, Rahul was born; who after his graduation joined the Army in its infantry-arm, and now happens to be at the moment a fully dedicated and hard-working full-Colonel. Well by the Divine-Grace of the All Mighty; we were again blessed after exactly two years with another son, Rohit, and who after his graduation and also completing his MBA from IIM, Bangalore; is now working in USA, in a successful managerial-position career with a leading multi-national American I.T. Company and now, he has started his own IT Company in USA, and doing well. He is now settled in Florida and remains busy with his professional activities. And thereby life moves-on as usual and hence, for any ordinary person like me; all these developments are always bound to bring-in adequate satisfaction and thus keeping me busy in my pursuits with any requisite energy, health and motivation too; and obviously commensurate to my age etc. Moreover, my wife's younger brother, **Ashok Sinha** too being a brilliant 'IITian' himself; along with his very affectionate wife: ***'Vikki'*** originally from the USA only, have also been always very loving and caring to us. And thus in spite of his very busy schedule in USA; Ashok has always been ever ready to send me any books that I have been requesting from time to time as great inputs and help to me in my rather 'crazy' endeavour. I would always cherish Ashok's such feelings, love and the indeed genuine gestures towards we both and our children. Further always the no less, Ashok's mother too, i.e. **my mother-in-law, who sadly is no more amongst us now**; had been very affectionate and in addition, was always having some special and loving considerations towards me, in spite of my occasional and the indeed very naive indiscretions as well. Now in particular, I would especially remain grateful to my very distinguished four maternal Uncle-in-law all through my life, for their very valuable affections and guidance, so long they remained in this world; namely, **Late Baleshwar Prasad**, old-time Retd. Chief Engineer (PWD), Govt. of Bihar; **Late N.P. Sinha**, Retd E-in-C, BSEB, Patna; **Late Professor Dr. Bimala Prasad** of JNU and also earlier our Ambassador in Nepal; and then, **Late Kamala Prasad**, IAS (Retd) as well. Now by the very kind Divine Grace, only one very distinguished Uncle-in-law of mine, namely, **Professor Dr. Kamta Prasad** (a very learned man in the discipline of Economics) is very much around us and hence, always available to me for any valuable guidance.

Further, I am also grateful to **Mr. R.K. Sinha, IAS (Retd),** from erstwhile Bihar, for his so much affectionate considerations for me always.

Now, with indeed very special affection, I must mention about one my nephew, **Mr. S.N. Verma** (Suman), elder son of my late eldest sister who has always been of unique help and support to me). He is also an electrical engineer like me and at present working as Managing Director of

Uttarakhand State Hydel Power Corporation, Dehradun. Further I have always considered myself very fortunate that in spite of myself being a person of just average IQ, but even then by the Divine-Grace—from my school-days itself, I have been very fond of reading books, journals and also articles on diversified-subjects as much as possible—along with my many varied commitments; as the life too has naturally been moving-on regardless. And then to my always great fortune that along with my both sons, the no less one nephew of mine, **Prabir** (*being Sanju in the family, and the youngest son of my third elder sister*) too; have been regularly and always very lovingly, sending many books that I have been needing from time to time to my great advantage. And needless to say that all these have greatly helped me to gradually build-up my personal library.

Well while also reflecting on all these, I do also sometimes feel that how selfish the man, including myself, has been; but again I also the no less feel like telling myself that, that's the way perhaps the All Mighty has wished and thus has gone ahead to proceed with His multiple-creations within His infinite Universe; and thus all these phenomena have indeed been beyond any kind of the 'total' comprehension on the part of any mortal in this world. But for we humans and accordingly within our any understandable contexts, life must go on and so it does always go on regardless. I and my wife do also feel ourselves 11 to be very fortunate to have been blessed with two very affectionate, as well as all the more understanding daughters-in-law, **Mridula and Jaya**; and thereby we now have four lovely grand-children—**Anjali, Anandi, Shaan and Nikita**, (i.e. two daughters from my elder son's side and then, one son and a daughter from my younger son respectively); and these four kids have always been so lovingly keeping us on our toes. I am always indebted to **Mr. Kalyan Roy**, a very leading Senior Advocate at Ranchi High Court for his numerous assistance and also his friends' feelings towards me. And at the same time, the same goes to **Late Gopi Babu**, a leading and well-known Advocate at Hazaribagh District Court, in many special ways. In addition to all these but certainly not the least, we have one more 'nephew', **Shri Sanjay Kumar, Advocate**, in Ranchi; who conventionally or as per the any social norms or so, may or may not be any family relation of mine; but he has been a unique source of love, strengths and inspiration in my life since our chance acquaintance some years back—as a very pleasant and perhaps indeed also an uniquely destined surprise in my life—and I don't have any words to express as to what he means to me in many ways in this, as usual, so much unsparing world. Sanjay's wife **Manisha**, their two lovely daughters, **Varsha and Megha**, along with their younger brother, **Krishna** too (though at present a fast-growing child only); also very affectionately bestowing great additional sweetness in our life, and by treating us as their grand-parents. And further-on interestingly, I have also been having a long 'associate' of mine and that too since my arrival in Ranchi, and named **Shri Parasmal Jain**, a 'cranky' businessman of Ranchi (or is he really any 'cranky' in anyway); and we hardly have anything in common in tastes, way of life, but still we 'click' in very special manners and though otherwise too, he hardly 'clicks' with anyone and thus all the more, which defies all logic for me. It is true that sometimes, such strange things do happen in one's life, which always seems to be stranger than even any bizarre fiction itself. Another very close associate of mine, **Mr. P.S. Rohilla** of Bhuvneshwar, has always been a very trusted friend of mine, indeed in many ways. As being obvious in fact, we do remain in constant touch, as he is a great source of strength for me all the more, in the evening of my life. It is true that according to my basic constitution and the way of life; fortunately I have never been inclined, as well as perhaps, also not being capable of cultivating any large number of friends and acquaintances in my life, and that in any case has also been more manageable and comfortable.

I have been indeed very fortunate to get acquainted with a quite young, dynamic and also professionally competent young Army Captain from Corps of Engineers, named **Gautam Bannerji** in Ranchi way back in 1980-82 or so; when finally I shifted over there after taking my early retirement from Army. Gautam has been very considerate to me like a very affectionate younger brother. As years passed and also as expected during his profession of soldiering, he rose very befittingly to the rank of Lt. General be fire his retirement. He too has been very helpful to me in so nicely facilitating me in the so difficult task of getting this book published. I would always remain grateful to him for his help and encouragements.

It is also my most sincere duty to pay my most sincere and heart-felt regards to **Swami Madhusudanji**, Professor of Yoga, I.A.S. Administrative National Academy, Mussoorie, and Dehradun; for his most kind and affectionate blessings to me and all my family members. I do feel indeed extremely fortunate to remain in touch with him and thereby, keep receiving his blessings.

I would also my affectionately mention here, one my close friend, **Mr. Prem Verma** (an old-timer I.I.Tian, Kharagpur!) for his versatile personality and many achievements and also for the least his very special feelings towards me. I do also cherish the fond memories of my Engg. College days very close friend Air Commodore (Retd) Ajit Sahai of Bhagalpur as well.

And my same feelings go to **Late Md. Yusuf** (a village man from Ranchi), being very popular person of many qualities. I had the liberty to call him at any time, for my any work to be done and he was always available to me, till he breathed his last.

As right from my school days, I was extremely fortunate to also having reading-habits regularly for hours together, and whenever I could get whatever time to do the same. But it was only in the Army that I was able further diversify and so fruitfully enhance the scope of reading-habits and their profiles too. And thus to my great advantage even after leaving the Army; my same habit has further continued. I find it as a great source of many benefits in my life like—cultivation of some knowledge, relaxation, quietly soothing of nerves, spending time with much better sense of solace and contentment, and so on—but nonetheless, slowly the enriching of my personal library too, in whatever limited manner have also been going-on even today; which later on may be of some use for my children and grand-children etc, even after my departure from this world. Whatever, but at least that's the way I think and look at these in my life. It is obviously needless to say that without all these my personal thought-processes and activities in rather somewhat consistently in the first place; perhaps in any case, this book won't have been able to see light of the day. And hence for all these and above all, I am ever grateful to the Supreme Benevolent Almighty; besides my many elders, seniors and even few celebrities of the Army, and other associates in my life for all the encouragements and kindness throughout in my life. Interestingly after somehow securing premature retirement and then having got stuck in Ranchi with my extended family; in the beginning itself and with great but rather quite misconceived enthusiasm or so, I even tried my hands in some business in small-industry too; and initially I did have some success and luckily, could make some money too; but that was indeed just a passing affairs only. And soon my naivety too came into play, as in any case fundamentally I was not at all suited to wade-through with any success, in always a 'dicey-cum-murky pool' of any business activities; and the rest is now just memories for me at the back of my mind, of my some bygone years. **But it is also the no less true that—adversity teaches the most in many ways in one's life—and hence it is rightly said that, one must learn to 'bear one's own cross' in this world.**

*

Bibliography

Abhijit Bhattacharyya, Author: "The Victor of 1971: Glorious Triumph in a History of Inglorious Military", *The Statesman*, Kolkata, dated 10 July 2008.

Abhimanyu Ghosh, Brigadier (Retd): A technical paper on "Cyber Threats and Military Challenges", USI Journal, October–December 2011, New Delhi.

Adam Smith, Economist: *Wealth of Nations*, 1776.

Adimurthy, V, Satish Dhawan Professor and Senior Adviser (Interplanetary Missions), ISRO HQ, and Dean (R&D), Indian Institute of Space Science and Technology: A technical paper on "Orbital Safety Challenges in Future Space Missions", Technical Journal of the Institution of Engineers (India), 8, Gokhale Road, Kolkata, December 2012,

Adolf Hitler, former Dictator of Nazi Germany, and also war monger: *Mein Kampf*, Indian edition, New Delhi: A.B.C. Publishing House, 1968.

Afsir Karim, Maj. Gen. (Retd): A book on "Terrorism...." as 'Lancer Papers – 5', New Delhi: Lancer International, 1993.

Air Vice Marshal Arjun Subramaniam: *India's Wars: A Military History (1947-71)*, India: Harper Collins Publishers, 2016.

Akbar, M.J., Author and journalist: *The Shade of Swords: Jihad and the Conflict between Islam and Christianity*, New Delhi: Roli Books (Lotus Collection), 2002.

Akbar, M.J., Journalist and author: "The Politics of Not Having to Apologise", *The Times of India* (Ranchi Edition), dated 24 February 2013.

Al Gore, a former US Vice President: *An Inconvenient Truth*, Produced by 'Melcher Media' and then, published by the 'Rodale' USA, 2006.

Aleksandr I, Solzhenitsyn, Author from erstwhile USSR: "The Gulag of Archipelago: 1918–1956" (Volumes III and IV), Published originally in Russian; and then translated in English by Thomas P. Whitney, and then published in 1974-75, by Harper and Row, Publishers Inc., New York.

Allen Dulles (A former CIA Director, USA): *The Craft of Intelligence*, Signet Book, by the New American Library and, under the arrangements with Harper and Row Publishers, New York, 1964-65.

Alvin Toffler, author and thinker: *The Third Wave*, India: Macmillan India Limited (second reprint), 1981.

Alvin Toffler: *Future Shock*, London and Sydney: Pan Books, 1975.

Amartya Sen, Professor and author, and also a Nobel Laureate: *Identity of Violence: The Illusion of Destiny*, London and New York: W.W. Norton, 2007.

Amit Pathak: a medical doctor: *1857...Living History*, India, Government Museum, Jhansi, 2008.

An interesting **'News-item'** on 'Cuban Missile Crisis of the 1960s', *The Times of India*, Ranchi; dated 02 August 2013.

Andre Beteille, Professor, social scientist, and author: *Chronicles of Our Time*, India: Penguin Books, 2000.

Andrew Roberts, military historian: *The Storm of War: A New History of the Second World War*, London: Allen Lane, an imprint of Penguin Books, 2009.

Anjali Mehta (Dr): "An Officer and a Scapegoat", *The Statesman*, Kolkata, dated 11 March 2014.

Anklesaria Aiyar S. Swaminathan, Economist, Journalist and Columnist: "History Will Be Kind To Manmohan", *The Times of India*, Bangalore Edition, dated 15 May 2014.

Ann Cramer: *Mandela: The Rebel Who Led His Nation to Freedom* London: New Burlington (Marshall Edition), 2005.

Anuj Dhar: "India's Biggest Cover-Up" (On Netaji's Death), New Delhi: Vitasta Publishing Pvt. Ltd., 2012.

APJ Abdul Kalam and A. Sivathanu Pillai, *Thoughts for Change—We Can do It*, New Delhi: Pentagon Press, 2013 (reprint 2017).

APJ Abdul Kalam, Scientist, author, and a former President of India: *Ignited Minds*, India: Viking by Penguin Books, 2002.

Apratim Mukarji, Journalist: *War in Sri Lanka: Unending Conflict*, New Delhi: Har-Anand Publishers Pvt. Ltd., 2000.

Arthur Birnie, military historian: *The Art of War*, London: Nelson and Sons Limited, 1942.

Ashish Sonal, Vr C, (Captain, Indian Army): *Terrorism and Insurgency in India: A Study of the Human Element*, New Delhi: Lancer Publishers Pvt. Ltd. and Spantech and Lancer, 1994.

Ashok Choudhary, Lt. Gen. (Retd): "Navy chief's Sacrifice Should Not Be in Vain", *The Statesman*, Kolkata, dated 04 March 2014.

Ashok V. Desai: "Sanskrit has relatives all over Europe—Indo-Europeans' Origins", *The Telegraph*, Kolkata, 2nd Feb. 2016.

Ayub Khan, Field Marshal (Pakistan): A political autobiography, *Friends Not Masters*, Pakistan: Oxford University Press, 1967; and Dacca: Press Section of Pakistan Cooperative Society Ltd.

Bagchi, Romit: "Jawaharlal Nehru and Hindutva", *The Statesman*, Kolkata, dated 13 June 2014.

Bandyopadhya, N.R., Professor (Dr) and Editor: *Technorama*, a technical journal of the Institution of engineers (India), 8, Gokhale Road, Kolkata – 700020: "Editorial Column" in the December 2012 Issue of the above mentioned Journal.

Banerjee, Gautham, Lt. Gen. (Retd): *India's National Defence Reforms and Military Modernisation*, New Delhi: Pentagon Press, 2017.

Banerjee, Gautham, Lt. Gen. (Retd): *The Kashmir Narrative: Time For Vigorous Adaptations*, New Delhi: Vivekananda International Foundation and Pentagon Press, 2017.

Barnett, Thomas, *Pentagon's New Map: War and Peace in 21st Century*, New York: G.P. Putnam's Sons, (a member of the Penguin Group Inc., USA), 2004.

Barnett, Thomas, Thinker, strategist analyst and author, USA: *Great Powers: America and the World after Bush*, New York: G.P. Putnam's Sons, 2009.

Baru Sanjaya: *The Accidental Prime Minister: The Making and Unmaking of Manmohan Singh*, Viking: Penguin Books India; 2014.

Batra, N.D., Professor in Norwich University, USA: Two articles on "When Our Daily Lives Becomes Hyper-Personal"; and on "Global Civic Management through Technology", *The Statesman*, Kolkata, dated 10th and 31st October 2007, respectively.

Bergamini, David: Editor of the *Universe*, A 'Time-Life' Book, Time Inc., USA and Canada, 1962–69.

Bertrand Russell: *Philosophical Essays*, London: George Allen and Unwin Ltd.; Digital Printing in 2003, by Routledge, London.

Bertrand Russell: *Power: A New Social Analysis* (6th impression), George Allen and Unwin Ltd, London, 1948

Bhagat, P.S., V.C., Lt. Gen. (Retd): *Defence of India and south-East Asia (Forging the Shield)*, Dehra Dun (India): BY E.B.D. Publishing and Distributing Co., 1967.

Bhaskar Chakravorti: "The return of boundaries – Nativist sentiments and a growing tendency towards looking inwards imperil globalisation", *The Indian Express*, Kolkata, 27th July 2016.

Bhattacharjee, Jay: "Babu-neta and fauji", *The Statesman*, Kolkata, dated 15 March 2014.

Bhattacharya, Subhendu:"Vivekananda and Vedanta", *The Statesman*, Kolkata, dated 07 November 2013.

Binns, Stewart, a TV Series Producer and Writer: *The British Empire in colour*, London: Carlton Books Ltd., 2002.

Bucaille Maurice: *The Bible, The Qur'an and Science* Paris: Seghars, 1976; 8th edition, India: Crescent Publishing Co., 1982.

Carr, E.H., Professor of History: *What Is History*, Australia: Penguin Group, 2008.

Carr, EH, Professor, "What Is History"—Compiled as Collection of Professor Carr's lectures, as the George Macaulay television lectures; and since delivered in the University of Cambridge, during January-March 1961.

Castleden, Rodney: *Encounters That Changed the world*, UK: Futura, 2009; and printed in India.

Chaudhuri, JN, General and a former Chief of the Indian Army: *Arms, Aims and Aspects*, Mumbai: Manaktalas, September 1966.

Christer, Jorgensen, General Editor: *Great Battles: Decisive Conflicts That Have Shaped History*, UK: Paragon, UK, 2011.

Clarke, Peter, Professor, UK: *Last Thousand Days of British Empire* London: Allen Lane, an imprint of the Penguin Book, 2007.

Claude Arpi, *Tibet: The Lost Frontiers*, First published in the USA; New Delhi: Lancer Publishers & Distributors, 2008.

Cornelius Ryan, "The Last Battle", New York: Pocket Books, 1967.

Creigton, Louise: "Heroes of European History", London: Longmans Green & Co., 1917.

Dalrymple, William: *White Mughals*, India: Penguin Group, 2004.

Davis, Paul K.: *100 Decisive Battles: From Ancient Times to the Present*, New York: Oxford University Press, 2001.

Dejevsky, Mary: Interviewer of the erstwhile Soviet Leader Mikhail Gorbachev—and later also published in *The Statesman*, Kolkata, dated 08 June 2010.

Depinder Singh, Lt. Gen. (Retd): *Field marshal Sam Manekshaw, MC: soldiering with Dignity*, India: Natraj Publishers, Dehra Dun, 2002.

Desai, Lord Meghnad: *Rediscovery of India*, India: Allen Lane, an imprint of Penguin Books, 2009.

Devasahayam, MG, (IAS, Retd): "Walking the Stilts" (*i.e. concerning Indian civil bureaucracy*) *The Statesman*, Kolkata, dated 30 November 2013.

Dictionary: *New Webster's Thesaurus and Medical Dictionary*, New York: Book Essentials Inc., 1991 Edition.

Dixit, JN (IFS retired),: "Future Conflict, Clash of Civilisations—Validity of Huntington's Thesis and the Security Challenges of this Paradigm to India", National Security Paper, USI (India), New Delhi, 1997.

Dorril, Stephen: *MI - 6: Fifty Years of Special Operations*, London: Fourth Estate, 2000.

Dr APJ Abdul Kalam: A former President of India: *Wings of Fire*, Hyderabad: University Press India Ltd, 1999.

Dyer, Gwyyne, a London-based Journalist: "The Urge for Democracy is Universal", *The Pioneer*, dated 02 March 2011.

Ebon, Martin: *Svetlana: The Story of Stalin's Daughter*, New York, USA: Signet book by the New American Library Inc., 1967; Bombay, India: Pearl Publications Pvt. Ltd., 1974-75, and distributed by the India Book House.

'Editorial Column': *The Statesman*, Kolkata, 01 September 2013.

'Editorial column': *The Statesman*, Kolkata, 19 March 2014.

'Editorial Column': *The Times of India*, Bangalore, 03 July 2014.

'Editorial Column': *The Times of India*, Bangalore, 12 November 2014.

'Editorial Column': *The Times of India*, Bangalore, 24 July 2014.

'Editorial Column': *The Times of India*, Bangalore, 27 October 2014.

'Editorial Column': *The Times of India*, New Delhi/Ranchi, dated 13 January 2005.

Edward Said, Professor: *Culture and Imperialism*, New York: Vintage Books Edition, 1994.

Edward Said, Professor: *Orientalism*, New York: Vintage Books Edition, New York, 1979.

Eisenhower, Dwight D., General of the Army (USA): *Crusade in Europe: A Personal Account of World War II*, New York: Doubleday and Company Inc., 1948.

Eisenhower, Dwight D., General of the Army (USA): *Dear General': Eisenhower's Wartime Letters to Marshal*, Compiled and edited by Joseph P. Hobbs; Baltimore, Maryland: John Hopkins Press.

Farago, Ladislas: *Patton's Ordeal and Triumph*, Mayflower Books.

Ferguson, Niall—Academician and author: *Colossus: Rise and Fall of the American Empire*, India: Penguin, 2005.

Fisk, Robert: "Guarding Secrets of the Syrian War", *The Statesman*, Kolkata, dated 22October 2012 (courtesy: 'The Independent', London).

Friedman, Thomas L., "The US and Its Double Game", *The Sunday Indian*, dated 15 August 2010.

Friedman, Thomas L., "The World is Flat: The Globalised World in Twenty-first Century" USA: Farrar Strauss and Giroux, 2005; London: Penguin Group, 2006.

Fuller, Maj. Gen. (Retd), UK: "American Civil War: 1861-65"—and later published in India, under the arrangement with Messrs Eyre and Spottiswoode (Publishers) Ltd, London; by the Current Events, 17, Rajpur Road, Dehra Dun, in the year 1959.

Galbraith, John Kenneth, Professor, economist, author and diplomat: *Economics and the Public Purpose*, USA: Signet Books, New American Library, 1973; and then the Paperback Reprint in 1975.

Galbraith, John Kenneth, *The Affluent Society*, London: Hamish Hamilton, 1958; and then Paperback edition and subsequent reprint in 1974, by Penguin Books, in association with Hamish Hamilton.

Garratt, G.T. (Editor): *The Legacy of India* London, 1937; and later reprinted in India in the year 2005, by 'Black and White', an imprint of Sundeep Publishers, New Delhi.

George Gaylord Simpson, Professor: *The Meaning of Evolution: A Study of the History of Life and of Its Significance for Man*, Copyright: Yale University Press – and first Indian Edition brought by Oxford and IBH Publishing Co., 1965.

George Soros: *The Age of Fallibility: The Consequence of the War on Terror*, New York: Public Affairs, 2006.

Ghosh, SK, A Research Officer with the IDSA, New Delhi: An article on "Insurgency around the World: 1968-69", IDSA Journal, New Delhi, October, 1969.

Gorbachev, Mikhail, Former President of the erstwhile Superpower USSR: *Perestroika - New Thinking for Our Country and the World*, New York: Harper and Row Publishers, 1987.

Grechko, A.A., Marshal of the Soviet Union: Rendered 'Preface' and also edited—"Liberation Mission" (A Soviet era publication, depicting the then Soviet version of the contributions of Soviet Armed Forces in World War –II—and published originally in Russian, by Progress Publishers, Moscow; and later translated in English by **David Eidlon** in 1975, and published by the same Publishers in Moscow.

Greider, William, *Who Will Tell the People: The Betrayal of American Democracy*, New York: Simon and Schuster, 1993.

Gupta, Swapan Das, Columnist and author: "The Politics of Not Having to Apologise", *The Times of India*, Ranchi edition, dated 24 February 2013.

Gurcharan Singh Sandhu, Maj. Gen. (Retd): *A Military History of Medieval India*, New Delh: Vision Books, 2003.

Gurcharan Singh Sandhu, Maj. Gen. (Retd): *India: Military History of Ancient India*, New Delhi: Vision Books, 2000.

H.R. Trevor-Roper, Editor: *Hitler's War Directives*, London: Pan Books Ltd, 1966.

Harbakhsh Singh, Lt. Gen. (Retd): *India: In Line of Duty*, New Delhi: Lancer Publishers & Distributors, 2000.

Harbhajan Singh, Lt. Gen. (Retd): "Cyber Warfare: Dangerous Trends", April-June 2011, Issue of the USI (India) Journal, New Delhi.

Hasluck, E.L., Historian: *The Second World War*, London and Glasgow: Blackie and Son Ltd.

Hawking, Stephens, Professor, U.K.: *Black Holes And Baby Universe, and other Essays*, London: Bantam Books, 1994.

Hellberg, Thomas and Magnus Jansson: *Alfred Nobel*, A highly illustrative and also historically informative 'Coffee table' book.

Hirst, John, Professor, Australia: *Sense and Nonsense of Australian History*, Melbourne, Black Inc. Agenda, in 2005.

Hobbs, Joseph P.: *Dear General: Eisenhower's Wartime Letters to Marchal*, Maryland, USA: John Hopkins Press, Baltimore.

Holmes, Richard: *Sahib: The British soldier in India (1750-1914)*, London: Harper Collins Publishers, 2005; and then by Harper Perennial edition, 2006.

100 ***Decisive Battles*** *from Ancient Times to the Present (The World's main battles and How They Shaped History*, Santa Barbara, CA: ABC – CLIO, Inc., 1999;—Paperback edition brought by the Oxford University Press, New York, 2001.

Huntington, Samuel P, Professor, USA: *The Clash of Civilisations And the Remaking of World Order*, India: Penguin Books India Ltd, 1977 (Indian edition).

India Today, dated 29 September 2008; giving some details from 'The 2007 Report of Terrorism', by the National counter-terrorism Centre, USA.

India Today, New Delhi, dated 12 September 2005.

Indian Newsmagazine: *Outlook* (Weekly), dated 27 July 2009.

Israel: *Secure and Recognized Boundaries: Israel's Right to Live in Peace within Defensible Frontiers*, Jerusalem, Israel: Carta Publication, 1971..

Israeli Government's official publication: "NAHAL", Israel's Ministry of Defence, Tel Aviv; 1970; and prepared by Mordechai Noar.

Jacob, J.F.R., Lt. Gen. (Retd): *India: Surrender at Dacca: Birth of a Nation*, New Delhi: Manohar Publishers & Distributors, 1997.

James, Truslow Adam: *The Epic of America*, USA: Little Brown, Atlantic Monthly Press, USA, 1931; and then as Pocket Books, USA, with Reprint in 1956.

Jaswant Singh, an erstwhile Army Officer, veteran Parliamentarian, thinker and author: *Defending India*, UK: Macmillan Press Ltd; and then in USA and India in 1999.

Jean Dreze and Amartya Sen, Professors: *An Uncertain Glory: India and its Contradictions*, New Delhi: Allen Lane, imprint of Penguin Books, 2013.

Jhunjhunwala, L.N.: "My Experience of Spiritual and Mystic Phenomenon", *South Asia Politics*, New Delhi, December 2010.

Jitatmananda, Swami, Secretary, Ramkrishna Mission, Institute of Culture, Kolkata:"Hinduism and Modern Times"—as the Fourth Chapter, in the book on "Understanding India: Relevance of Hinduism"; New Delhi: Vitasta Publishing Pvt. Ltd, 2007.

John Keay, Historian and author: *China: A History*, Great Britain, 2008.

John Keay: *India: A History*, London and Delhi: Harper Collins Publishers, 2000.

John Simon, Sir, Erstwhile member of the British Parliament; and also of the 'Simon Commission' fame in India, during the British colonial-era: *Comments and Criticism*, edited by D. Rowland Evans, London: Hodder and Stoughton Ltd., 1930.

Jonathan Steele: *Ghosts of Afghanistan: The Haunted Battle ground*, London: Portobellow Books, 2011.

Joshi, Manoj, strategic analyst and author: *The Lost Rebellion: Kashmir in the Nineties*, New Delhi: Penguin Books India, 1999.

Kaldor, Mary: *New and Old Wars: Organized Violence in a Global Era*, Polity Press, in association with Blackwell Publishing Ltd, 1999; and then first Indian, brought out by Natraj Publishers, Dehra Dun, 2005.

Kapur, Devesh, Professor, Director, Centre for the Advanced Study of India, University of Pennsylvania, USA: An article on "Evolution of a Nation: Government Has Expanded massively, But Must its Power to Deliver public goods to Indians", *The Times of India*, dated 13 August 2014.

Katoch, P.C., Lt Gen (Retd), India:"Impact of Technology on Warfare", Journal of USI (India), New Delhi, July-September, 2011.

Kautilya: The Arthashastra,—Edited, rearranged, translated, and introduced by **L.N. Rangarajan,** India: Penguin Book, 1992.

Khan Abdul Wali Khan: "Facts Are Facts"—originally written in Urdu and since published (*but banned in Pakistan*), and subsequently translated into English by **Dr. Syeda Saiyidain Hameed,** Vikas Publishing House, New Delhi, 1987.

Khanduri, B.C., Brig (Retd), India: *Thimayya: An Amazing Life*, (under authorization of the USI, New Delhi) of a legendary Indian General, and former Indian Army Chief, 'Knowledge World' Publishers, New Delhi, 2006; and with 2nd and 3rd Reprint in the year 2009 itself.

Khushwant Singh, Veteran Journalist, columnist, editor, and author: *Truth, Love and a Little Malice*, Delhi: Viking, in association with Ravi Dayal Publisher, and Penguin Books, India, 2002.

Kiran Bedi, IPS (Retd): India's First Woman IPS Officer: "As I See....", New Delhi: Sterling Publishers, 2003.

Kissinger, Henry; Former US Diplomat, US Secretary of State, and author: Excerpts from his book titled, *World Order: Reflections on the Character of Nations and the Course of History*, India: Penguin India, 2014.

Krishna, B.: *Biography of Sardar Patel*, New Delhi: Rupa & Co., 2005.

Krishnan, Simatini: "Recasting the Emergency", *The Sunday Statesman*, Kolkata, dated 12 January 2014.

Kuldip Nayar, Veteran columnist, commentator, and author: "Indira, Thatcher and the "Blue Star' Controversy", *The Statesman*, Kolkata, dated 30 January, 2014.

Leon Uris: *Exodus (on the birth of Israel as a sovereign nation and also, the consistently tenacious struggles of the Jews, since ages.)*, USA: Bantam Books, 1959.

Lind, S. William of USA: *Maneuver Warfare Handbook*, India: Natraj Publishers (as the first Indian Edition), 2005.

Lionel Giles: *The Art of War (on Sun Tzu's oldest 'Military Treatise' in the world)*; and edited by **Ngawang Phuntsok**—and since first Indian Edition, 2008.

Lynn Montross: *Washington and the Revolution*, New York: Ladder Edition, by the Washington Square Press Inc., January 1962.

M.K. Rasgotra, IFS (Retd): *A Life In Diplomacy*, New Delhi: Penguin Group (Viking), 2016.

MacArthur, Douglas, General of the Army (USA): *Reminiscences*, Connecticut, USA: Fawcett Publications Inc., and then first Crest Printing, as paperback edition, in the year 1965.

Macmillan, Sir Harold (a former Prime Minister of UK): *The Blast of War 1939-45* a political history of the Second World War; New York and Evanston: Harper and Row Publishers, 1967-68.

Mahbubani, Kishore: Dean of the Lee Kuan Yew School of Public Policy of the National University of Singapore: An article on "Why Can't Countries Think Like Companies", *The Times of India*, Bangalore, dated Sunday, 16 November 2014.

Malik, V.P., General: *Kargil: From Surprise to Victory*, New Delhi: Harper Collins Publishers India, 2006.

Mandal, Debakinandan, a retired bureaucrat from West Bengal: "Bureaucratic Maladies", *The Statesman*, Kolkata, dated 18 February 2014.

Mandal, Debakinandan: "Tensions in Civvy Streets", *The Statesman*, Kolkata, dated 26 December 2013.

Marquis of Zetland: *Steps towards Indian Home Rule*, Hutchinson and Co., London, 1934-35.

Masani, R.P.: "Caste and the Structure of Society" in *Legacy of India*, ***G.T. Garratt*** (ed.), First published in London, 1937 and then, reprinted and published in India in 2005, by 'Black and White', and imprint of Sundeep Publishers, New Delhi.

Maulana Wahiduddin Khan: *Jihad, Peace, and Inter-Community Relations in Islam*", originally written and Published in Urdu; and then translated in English by Yoginder Sikand and published by Rupa & Co., New Delhi, in the year 2010.

Montgomery, Field Marshal: *Memoirs of Field Marshal Montgomery*—and then later brought out as a paperback edition in 1960; by the Collins Fontana Books, London.

Montgomery, Viscount of Alamein, Field Marshal: *A Concise History of Warfare*, William Morrow & Co., 1983.

Mudur, G.S.: "A Scientific Report; as In the Fast Lane of Evolution", Indian Daily, *The Telegraph*, Calcutta, 14 December 2007.

Mukherjee, Subrata, Professor, Faculty of Political Science, University of Delhi: "Coup that Changed Iran", *The Statesman* Kolkata, dated 23 September 2013.

Mukherjee, Subrata, Professor: "Cultural Churning", *The Statesman*, Kolkata, dated 04 December 2013.

Mukherjee, Subrata, Professor:"Unforgettable Soldier" (i.e. legendary Vietnamese General, VO Nguyen Giap: 1911-2013) *The Statesman*, Kolkata, dated 24 October 2013.

"Mysterious Tune Played by An Unknown Piper", *The Times of India*, Patna/Ranchi dated 23 July 2005.

Narlikar, J.V., a well-known Astrophysicist, Professor and author: "Science of the Times: From Abstract Concepts to Daily Life", *The Times of India*, dated 01 March 2000.

Natwar Singh: *One Life Is Not Enough* (An autobiography), New Delhi: Rupa Publications India Pvt. Ltd, 2014.

Neamatollah Nojumi: *The Rise of the Taliban in Afghanistan: Mass Mobilization, Civil War and the Future of the Region*, USA: Palgrave, the new global publishing report of St. Martin Press, USA, 2002.

Nehru, Jawaharlal: *Glimpses of World History*, New Delhi: 'Nehru Memorial Trust', 1934.

Nehru, Jawaharlal: *The Discovery of India*, New Delhi: 'Nehru Memorial Trust': Teen Murti House, 2002.

Neville Maxwell: The 'controversial' author of *the book, "India's China War – 1962"*: An article on "The **Henderson Brooke Report** Sees Light of the Day", *The Statesman*, Kolkata, dated 19 March 2014.

'News Item', on "You bl.... di Neanderthal.... what a beauty", *The Statesman*, Kolkata, dated 12 January 2010.

'News-Item' (*Technical*): "On LiFi or Light Fidelity technology", *Technorama* (*December, 2012 Issue*) of the Institution of Engineers (India), 8, Gokhale Road, Kolkata – 700020.

'Newspaper Report' on the "Appointment of the Chief of 'National Security Agency' (NSA) of USA", *The times of India*, Ranchi edition, dated 16 June 2013.

'Newspaper Report', *The Times of India*, Ranchi edition, dated 19 June 2013.

'Newspaper Report': *The Times of India*, Patna/Ranchi, Dated 12 November, 2003; since referring to an earlier 'news-report', citing an interview given in Singapore by **Mr. Robert L. Lindstorm**—Founder-President of the 'C-60 Communications', USA.

Niazi, AAK, Lt Gen (Retd), Pakistan: "The Betrayal of East Pakistan", India: Manohar Publishers and Distributors, 1998.

Ninan, T.N., columnist and author: An article on "The Adjustment P.M." (*Interestingly on India's former Prime Minister, Dr. Manmohan Singh.*), *The Times of India*, Sunday, dated 04 May 2014 (courtesy: *Business Standard*, India).

Noam Chomsky: *Failed States: The abuse of power and the assault on democracy*, USA: Allen and Unwin, USA; and later as the first South-Asian Edition in 2007, and printed in India.

Norman, Copeland, UK: *Psychology and the Soldier*, India: English Book Depot, Dehra Dun.

Overy, Richard: *The Times History of the World*, London: Times Books, 2008.

Palit, D.K., Maj. Gen. (Retd), Vr.C.: *Major General AA Rudra: His Service in Three Armies and Two world Wars*, New Delhi: Reliance Publishing House, 1997.

Palkhiwala, Nani, A., an eminent Jurist and a former Diplomat: *We the Nation*, New Delhi: UBS Publishers and distributors Ltd., with 12th Reprint in 1994.

Parakh, P.C., IAS (Retd): *Crusader or Conspirator: Coalgate and Other Truths*, New Delhi: Manas Publications, 2014.

Petley, Vijay; R. Anbazhagan, M. Tamilselvi, and U. Chandrasekhar: all scientists; since associated with the Gas Turbine Research Establishment, Bangalore; along with **Thulsiram and G. Mohan Rao** of the Indian Institute of Science, Bangalore: A technical article on "Health Monitoring of Aero Gas Turbine Engine through Conformal Thin Film Sensor"—and since published in the already earlier referred Technical Journal 'Technorama'—as in the item No. 152, listed above.

Plutarch: *The Lives of Noble Grecians and Romans*, since translated in English by **John Dryden**, and the then revised by **Hugh Clough**—and published for the Modern Library, by Random House, New York, sometime in the 1930s.

Powell, Collin, General (Retd), former Chairman of the Joint Chiefs of Staff, USA: *My American Journey*, an autobiography—and first published as Ballantine Books Edition, New York, 1996.

Prasad, Dr. Rajendra; The First President of India: his first address to the Nation; as the President: on his swearing-in, on 13 May 1952 (Ref: "Official Publication, as 'Speeches of President Rajendra Prasad delivered from 1952 to 1956')—and published by the Publications Division, Government of India, New Delhi, in January 1958.

Prasad, Dr. S., Professor: 'Proceedings of the National Conference' on the "Emerging Technologies: Nanotechnology and Cryogenics"—since held on 30-31 October 2009, at the Bajaj Institute of Technology and Management in Greater NOIDA (India)—and therein, Dr. Prasad being also the Presiding Member of the Committee of editors, responsible for publication of the Proceedings of this Seminar.

Qadir, Abdul: 'The Cultural Influences of Islam', in The legacy of India, G. T. Garratt (ed.), Oxford: The Clarendon Press, 1937.

R. Rama Rao, Colonel: "A Review-article", with respect to few expertly knowledgeable books; concerning the multiple-aspects of NBC Warfare—and since published in the January 1970 Issue of the Institute of Defense Studies and Analyses (IDSA) Journal, New Delhi.

Radhakrishnan, Dr. S., *The Present Crisis of Faith,* Delhi: Hind Pocket Books (P) Ltd., 1970.

Raghavan, R.K., IPS (Retd):"An Emerging British Model" (on the multiple-aspects of policing societies), *Frontline,* dated 07 October 2005.

Rajapaksa Mahinda: President of Sri Lanka: "The Sri Lankan President's Interview", conducted by Mr. Shekhar Gupta; and published in *The Indian Express,* Chandigarh Edition; dated 1 June 2009.

Ramachandran Shastri: "Nehru's Stubbornness Led to 1962-War with China", Published in the Indian Daily, 1'The Times of India', Patna Edition; dated 20 December 2010.

Ramakrishnan, S., Professor; Director of the Vikram Sarabhai Space Centre, Thiruananthapuram (India): "Earth to Orbit: The Crucial first Step in Accessing Space", *Technorama* as per the Item – 152, listed above.

Ravindra Singh, Lt. Col., Of the corps of engineers, Indian Army: on "NBC Warfare....", Published in October 2012 Issue of the Journal of the Institution of Military Engineers (India); CME, Pune.

Reader's Digest: "Illustrated Story of World War II"—Printed and Published in the year 1969, in USA.

Ricks, Thomas, E.: *Fiasco: The American Military Adventure in Iraq,* England: Penguin Books, 2006.

'Rommel Papers'—Compiled by B.H. Liddell Hart and Rommel's son, Manfred Rommel—and since also published in India, by the Natraj Publishers, Dehra Dun; in the year 1977.

Roopinder Singh: *Arjan Singh, DFC: Marshal of the Indian Air Force,* New Delhi: Rupa & Co., 2002.

S.R. Sharma, Professor of History (India): *A Brief Survey of Human History,* Bombay: Karnataka Publishing House, 1948.

Sajid Z. Chinoy: "A Strong Foundation: India has institutionalised its monetary policy framework in two years; this has to be taken to its logical conclusion", *The Indian Express* Kolkata, 19 July 2016.

Sarila, Narendra Singh, a former Indian Army Captain; and later India's Foreign Service Officer, and a Diplomat: *The Shadow of the Great Game: The Untold Story of India's Partition,* New Delhi: Published by Harper Collins Publishers; and subsequently the third edition of this book, having come out in the year 2006.

Schaffer, Ronald; Professor: *Wings of Judgement: American Bombing in World War – II*, New York: Oxford University Press, 1985.

Singh, R.K., Group Captain:"Indian Anti-Satellite Weapon: Necessity, Urgency and the Way Ahead" and since published in the January-March, 2013, Issue of the USI (India) Journal, New Delhi.

Singh, V.K., General (Retd): a former Indian Army Chief of Staff: "Courage and Conviction" – An autobiography, published in 2013, by Aleph Book Company, New Delhi.

Sinha, S.K., Lt Gen (Retd): a former Indian Army Vice Chief of Staff: "National Security Lecture – 1990"—since Delivered by General Sinha in the USI (India), New Delhi; during two sittings in two days—and then published subsequently, in a 'book-form'; by the USI (India), New Delhi.

Sinha, S.K., Lt Gen (Retd): India's a former Vice Chief of Army Staff, a well-known and impeccably upright soldier, Statesman, Diplomat, learned thinker and author: *A Soldier Recall*s, New Delhi: Lancer International, in 1992.

***Special Forces**: Doctrine, Structure and Employment across the Spectrum of Conflict in the Indian Context*, New Delhi: Knowledge World, 2006.

Sri Aurobindo, Thinker, scholar, author, and spiritual leader: *Dayananda, Tilak, Bankim: Pioneers of Indian Renaissance*, New Delhi: D.A.V. Publications, Chitragupta Road, 2001.

Subrahmanyam, K., IAS (Retd): a well-known strategic analyst Committee: K. Subrahmanyam; Members: Lt Gen (Retd) K.K. Hazari; B.G. Varghese, a well-known Journalist; and Satish Chandra, IAS (Retd)—and since the 'Report' submitted to the then Prime Minister of India, Mr. A. B. Vajpayee, on 15 December, 1999; and subsequently published as well.

Sugata Srinivasaraju: an article "Years of the Guru", Indian News Magazine, "Outlook", 27 July 2009.

Sundarji, K., General (Retd): a former Indian Army Chief of Staff: "Of Some Consequence: A Soldier Remembers" (*tragically an unfinished book*)—and since published in the year 2000, by Harper Collins Publishers, New Delhi.

Swaminathan, M.S., Dr.: *From Rio De Janeiro to Johannesburg*, Chennai: East-West Books (Madras) Pvt. Ltd., 2002.

Tagore, Rabindranath: *Nationalism*, London: Macmillan and Co., 1937.

Tara, Macisaac: An article on "Huge Alien Object Spotted on Moon via NASA Imaging (+Photos), dated 21 January 2014; and last updated January 21 2014, 2.42 pm; and since published in the 'Epoch Times' USA.

Tata, JRD, a legendary Indian Industrialist, upright thinker, great patriot, and a visionary: "KEYNOTE" – a compilation of the selected speeches of JRD Tata—and since published by the Tata Press Ltd., Mumbai, in 1986; under the Jamshetji Tata Trust, Mumbai.

Thorat, SSP, Lt Gen (Retd): *India: "Reveille to Retreat"*, New Delhi: Allied Publishers Pvt. Ltd., 1986.

Tom Clancy, with Gen. Carl Stiner (Retd), and Tony Koltz of the USA: *Shadow Warriors*.

Udaya Chandar, Major: *The Art of Military Leadership*, Bombay: Jaico Publishing House.

Vancouver Bush, considered as the father of modern computer: "Science Is Not Enough: Reflections for the Present and Future" (The Builders' copyright 1945, by Alumni Association of Massachusetts Institute of Technology)—since first published in 1965, by William Morrow and Co. Inc., New York—and subsequently also published in India in 1969, by Vakils, Feffer and Simons Pvt. Ltd., Bombay.

Varma, Pavan K, a former Indian Diplomat: *Being Indian*, India: Penguin Books, 2004/2005.

Vidyalankar, Pandit Satyakam: "The Holy Vedas: A Golden Treasure"—since blessed by **Swami**

Satya Prakash Sarasvati—since first published in 1983, by Clarion Books (associated with Hind Pocket Books), J-40 Jorbagh Lane, New Delhi-110003; with 7th Reprint having come out in 2005.

Vijay Shankar, Vice Admiral (Retd): Indian Navy: An article on "Civil-Military Relations in India: The Soldier and the Mantri (*i.e. Minister*)", 'November 2012 Issue' of *Defence and the Security Alert*, New Delhi.

Vivekananda, Swami: "Vedanta: Voice of Freedom", Published by Advaita Ashrama, Kolkata; with 6th Impression having come out in 2002.

Wells, H.G., An erstwhile and even now, a famous historian: "Outline of History".

Yousaf, Nasser: "Choosing not to see the Real Enemy", *The Statesman*, Kolkata, dated 23 September, 2013.

Zhong, Jingwen, Professor (China); Head of the Chinese Department, University of Beijing (1920s-1930s): "Favourite Folklores of China", since originally written and published in Chinese; and subsequently translated into English by **Professor John Minford** (with 'Introduction' by **Professor Jingwen** himself) and then published with first English-Edition in 1983; by New World Press, 24, Baiwanzhuang Road, Beijing, China.

*

Index

B

J

N

O

S

*